Fundamental Chemistry

Fundamental

Chemistry

by Donald H. Andrews
B. N. Baker Professor of Chemistry
The Johns Hopkins University

and Richard J. Kokes
Associate Professor of Chemistry
The Johns Hopkins University

John Wiley & Sons, Inc., New York • London

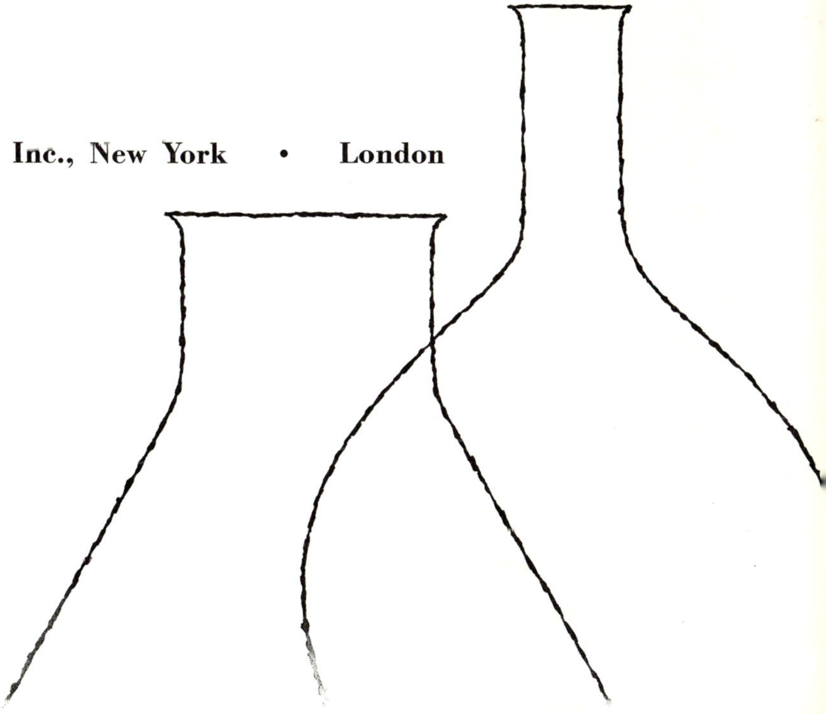

Second Printing, April, 1963

Library of Congress Catalog Card Number: 62-15329
Printed in the United States of America

to Elizabeth H. Andrews and Marian T. Kokes

Preface

THIS BOOK IS WRITTEN WITH THE CONVICTION THAT AN INTRODUCTORY CHEMISTRY COURSE must include discussions of quantum mechanics, thermodynamics, and chemical kinetics in order to give students a true picture of modern chemistry; and this conviction is coupled with the belief that such discussions can be made intelligible and interesting to the beginner in chemistry. For example, this approach can reveal the way in which the periodic relations of the chemical elements stem directly from the quantum structure of atoms. It makes the octet rule reasonable and also shows the rule's limitations. It encourages students to rationalize exceptional behavior but utters ample *caveats* on the dangers of oversimplification; in short, we believe that such an approach makes chemistry more understandable.

For the concepts of quantum mechanics, thermodynamics, and chemical kinetics to be useful, the principles must be developed in a logical manner; little is gained if principles are set forth as dogma, for this merely substitutes one mystery for another. The ideal program provides the conceptual basis of these three areas as a framework to support the subsequent discussion of the chemistry of the elements; this, of course, must be done without exceeding the mathematical comprehension of the students.

Ambitious as this approach is, it has been used with success for several years at the Johns Hopkins University. The advanced topics are developed in a way that requires only simple algebra. No knowledge of calculus is assumed, and sufficient algebra of finite differences is introduced to handle the few situations that involve concepts of rate, integration, and the like. Beginning students are particularly sensitive to gaps in reasoning, so we have tried hard to avoid them. Whenever principles have been stated rather than developed, we have tried to show that the principles are reasonable. Finally, we have found that algebraic equations are not nearly so vivid to beginners as physical pictures; so we have complemented most equations with a verbal description of the meaning and have based most mathematical discussions on a concrete model. Throughout we have kept in mind the ultimate objective: to provide a robust framework on which the student can hang his chemical facts.

At the Johns Hopkins University only one introductory chemistry course is offered; the same course is taken by chemistry majors, engineers, pre-medical students, social science majors, and humanities majors. This book is designed for such a course. Although some students have had excellent science courses in high school, the variability in training makes it necessary to plan the work on the assumption that the student has had no previous background in chemistry or physics. Accordingly, the first chapter surveys the domain of chemistry and the questions it seeks to answer in a way to meet the needs of students who are studying chemistry for the first time. The second chapter discusses the basic concepts of physics and mathematics that are needed later in order to understand chemical principles. For many students these chapters are largely a review; for others, this material is an essential foundation on which to build a sound understanding of chemistry. In the Johns Hopkins course, these introductory chapters are covered largely in the recitations where the students with poorer preparation are grouped together in special sections where they get extra help. The students with better preparation are placed in other sections where more advanced material is discussed.

In the first ten chapters a broad fundamental base is provided for the understanding of chemical principles. There are discussions of correlation diagrams for diatomic molecules, Boltzmann's equation, elementary nuclear structure, atomic weights and stoichiometry, and gas-law problems. Throughout the book chapters on more complex theory alternate with chapters dealing with simpler theories or facts in order to reduce the impact of less familiar material.

Quantum theory is introduced in historical perspective by a brief discussion of black body radiation, the Bohr theory, and the Heisenberg Uncertainty Principle. The concept of probability density is developed in an elementary way; quantum numbers are related to momentum and spatial orientation and extension; the periodic structure of the elements is rationalized on the basis of coulombic attraction between charge clouds and the nucleus. Discussion of molecular structure includes hybridization, resonance, dipolar bonds, electronegativities, and bond energies. An attempt is made to arrive at these concepts in a reasonable manner, to point out their shortcomings, and to illustrate their applications. The chapter on molecular structure is followed immediately by one on the covalent bond in organic compounds in order to show the applicability of these principles to chemistry.

Chapters on thermodynamics and equilibrium develop the concept of the driving force of a chemical reaction. The student now has a physical model of atoms, molecules, gases, liquids, solids, and solutions that is based on a kinetic picture. It is then shown how the thermodynamic laws lead to conclusions about these systems without the necessity of assuming a model. For example, the law of mass action is derived from the concept of free energy without recourse to the dynamic nature of equilibrium.

Kinetics is discussed from the point of view of both the collision theory and the pseudo-thermodynamic theory. Formation of hydrogen halides from the elements is used as an example of the varieties of chemical behavior. The common meeting ground of thermodynamics and kinetics is stressed.

In the description of the chemistry of the elements these principles are used liberally. Thus thermodynamics appears again not only in the chapter on electrochemistry but also in the chapter on metals and alloys; throughout the discussion of descriptive chemistry, thermodynamic data or principles are employed whenever feasible. Principles of molecular structure are used throughout the discussion of descriptive chemistry; kinetic aspects of chemical reactions are stressed repeatedly, especially in the chapter on redox reactions and the chemistry of complex ions. Practical limits have to be set, however. It requires more space to describe a specific reaction in terms of structure, thermodynamics, and kinetics than it does to set down the facts. We cannot elaborate on all chemical data presented, but we can discuss some to show how they fit into chemical theory and hope the student will do likewise on his own. In this respect, the problems at the ends of these chapters are an important aspect of this text. These problems are designed not only to test the student on what he has read, but also to extend the principles to topics not covered in the text. More advanced problems are starred.

Several other topics have been covered in detail in recognition of their growing importance. Chapter 9 is a systematic survey of the structure of metals and various oxides designed to show the relation of crystal structure to chemical type. Problems dealing with equilibrium in solution include not only acid and base ionization and solubility, but also stepwise dissociation of complex ions and competing equilibria such as the hydrolysis of ammonium cyanide. Metals and alloys have been discussed at length from the standpoint of structure and thermodynamics. The treatment of complex ions provides an introduction to the magnetism of bond types, inner and outer orbital complexes, and ligand field theory.

An introductory course cannot compete in content with an advanced course, but not all students go on to physical and organic chemistry. The budding biologist, physicist, or even historian forms his impression of chemistry from the introductory course. For such a student, we believe that this course has value as part of his liberal education. Perhaps it may convince him that chemistry is worth further study. Perhaps it will modify the image of the chemist as an eccentric puttering with fuming retorts. But chemistry is sufficiently interesting that a representative picture needs no justification; it speaks for itself. Our aim is to present such a picture—one useful to the terminal student as an intellectual discipline—one useful to the beginning student as an introduction of things to come.

The text reflects our likes and our prejudices. In our course we try to cover all the material in the text, although portions of some chapters have been omitted occasionally. Since time is limited, not all chapters are discussed in the lectures. Stoichiometry is covered in recitation. The two chapters on solution equilibrium are also covered in recitation in the second semester along with the topics on qualitative analysis. Most of the remaining material is taken up both in lecture and in recitation. Personal taste may dictate other omissions. Parts of the chapters on the covalent bond in organic chemistry, kinetics, chemistry of non-metals, metals, and alloys, complex equilibria, and complexes are examples. They may be skipped without seriously detracting from the continuity of presentation. However, an attempt has been made to give an integrated approach to chemistry and many chapters are interdependent. Elimination beyond that suggested above should be done with care.

In conclusion, we should note that we have found the course demanding and also rewarding, for instructor and student alike. Students recognize quantum mechanics, thermodynamics, and kinetics as sophisticated material; they are surprised to find that they can comprehend so much of it; they enjoy it, work hard at it, and learn a good deal of it. The instructor, who has to rephrase in simple terms material he knows in complex terms, finds that the introductory course puts stimulating demands on all his previous training.

No book is written in a vacuum. We are grateful to a number of our colleagues at Johns Hopkins for helping us to avoid incorrect statements of principles. Many have helped us correct ambiguities and errors. We also express our gratitude to the students who watched the birth of this text, who were enthusiastic from the start and were patient with our mistakes. We thank especially Professor Everett Johnson of the Stevens Institute of Technology who used this book in preliminary edition and offered many useful criticisms that were incorporated into the final manuscript. Other reviewers who have been helpful in the preparation of the book are: Professor John Gryder of the Johns Hopkins University who offered many suggestions on the chapters on atomic and molecular structure; Professor Walter S. Koski, who reviewed the chapters on nuclear chemistry; Professors Alex Nickon and Emil White who reviewed the chapters on organic chemistry; Professor J. D. H. Donnay and Dr. G. H. Donnay who reviewed the chapter on crystal structure; and Professor Charles Thomas who reviewed the section on biochemistry. Thanks are also due to the many graduate instructors whose enthusiastic response to the teaching of the course relieved many of our anxieties about its difficulty. Finally a special note of thanks is due to our wives who endured the vexations of a book-writing spouse with admirable good humor and contributed so much help in so many ways, when help was critically needed.

It is not feasible to mention all the source material in the limited acknowledgment possible in an elementary text. We wish to mention especially the JANAF (Joint Army Navy Air Force) Tables (Ref. 1), edited by Daniel R. Stull and his collaborators, as a unique and invaluable source of thermodynamic data; Circular No. 500 of the National Bureau of Standards for other thermodynamic data; *Elementary Theory of Nuclear Shell Structure*, by Maria Goeppert Mayer and J. Hans D. Jensen, for nuclear data and concepts; and *Inorganic Chemistry*, by Therald Moeller, for material in advanced inorganic chemistry. These and other helpful sources are listed in the section on references in the Appendix.

We are grateful to the following publishers for permission to quote from their publications: Prentice-Hall, Inc., to quote from *Oxidation Potentials* by Wendell M. Latimer; McGraw-Hill Book Co., Inc., to quote from *The Constitution of Binary Alloys* by Max Hansen; the Cornell University Press to quote from *The Nature of the Chemical Bond* by Linus Pauling.

We express our thanks to Professor J. A. Bearden and his collaborators and to Mr. A. G. McNish of the National Bureau of Standards for recent values of the fundamental physical constants based on C^{12}. The shift in the basis of the atomic weights from atmospheric oxygen to the isotope carbon-12 not only shifts the values of atomic weights but also has a significant effect on many other numbers, such as Avogadro's number, the Faraday equivalent, and the gas constant (R). If readers are startled to find unfamiliar values of constants in this book, we hope that they will be aware of the reasons for these changes to this new basis that will, in all probability, be the numerical cornerstone of chemistry for many years to come.

Baltimore, Maryland
April 15, 1962

DONALD H. ANDREWS
RICHARD J. KOKES

Contents

1

The architecture of the atom

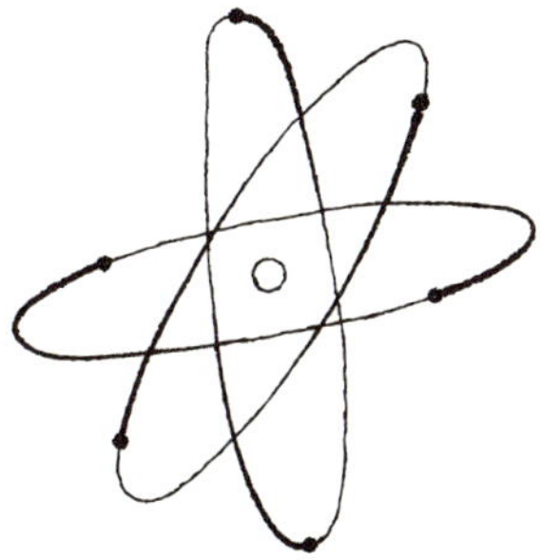

Suppose that you could build a supermicroscope that would magnify without limit, far beyond the power of any real microscope existing today; and with this supermicroscope you take a look at the skin of your hand. You turn up the magnification a million, a billion, or even a trillion times. What will you see? What are the shapes of atoms, electrons, and protons? How big are they? How are they moving? How are they tied together? These are some of the fundamental questions that chemistry and physics are seeking to answer. And, to get a first bird's-eye view of the architecture of atoms, we shall begin our exploration of chemistry by describing what we think we might "see" with such a magic supermicroscope.

A LOOK AT THE ATOM

It is known from actual observation what a magnification of a hundred to ten thousand times will show. For example, if you take a microscope with glass lenses, you can look at your finger and see the coarse structure of your skin, Fig. 1. You will see level places like flat land; and also here and there you will observe little wells running down beneath the surface, the *pores* through which you perspire. You turn up the magnification some more and look at the side wall of one of these pores. Now you find that the skin is really made up of little sacs or little bags, joined to one another at their edges. These are the biological *cells* that form the fabric of your skin, very much as the little pockets filled with honey in a honeycomb form the structure of the comb.

We now point our microscope at one of these cells and turn the magnification up still higher. We take a look at the membrane on the outside of the cell and

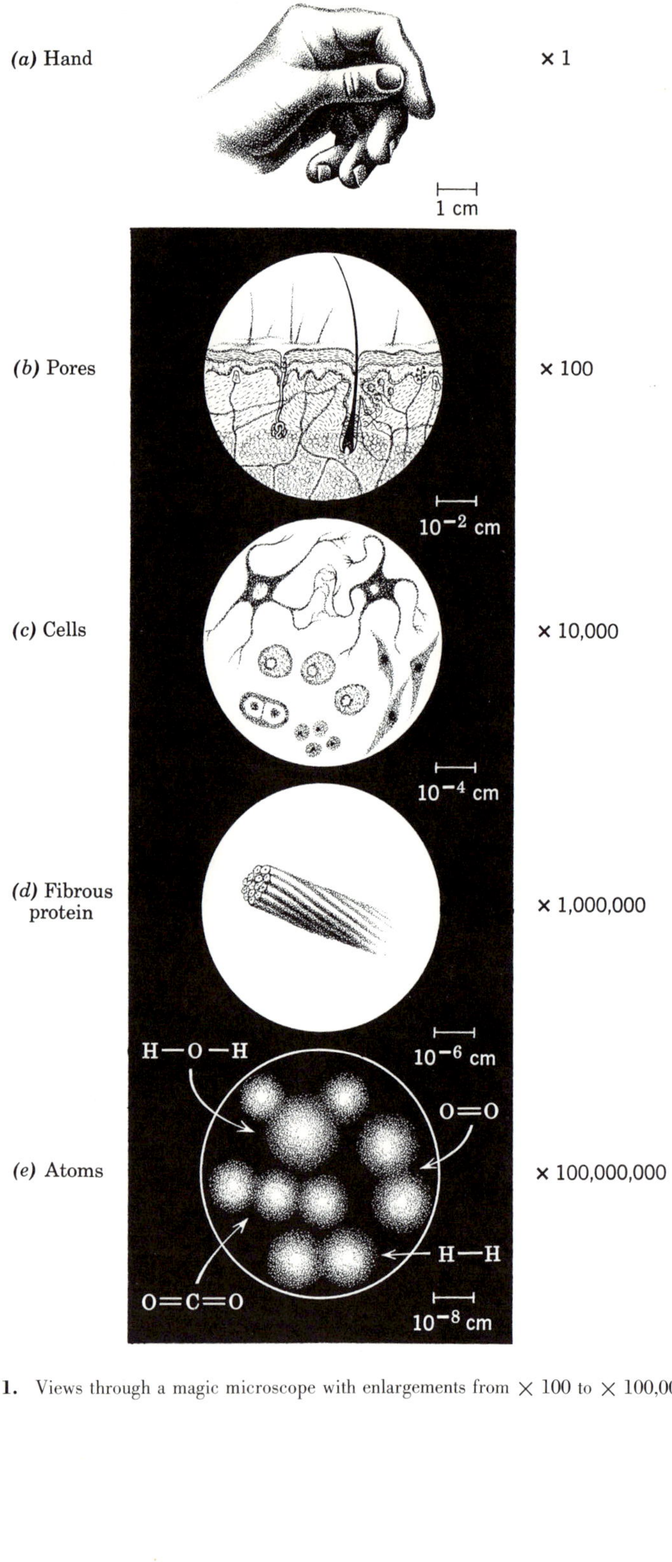

Fig. 1. Views through a magic microscope with enlargements from $\times$ 100 to $\times$ 100,000,000.

see that it is woven together with threads. At this point, our supermicroscope has passed beyond what we see with glass lenses and is showing us the kinds of things that we see with the *electron microscope,* one of the newest instruments that helps us explore the realm of the very small.

Next we take a look at one of these membrane threads, turning the dial on our supermicroscope so that it magnifies at about the limit of the power of any known instrument; and we see that the thread consists of strands wound together like the strands of a rope. Finally, we look more closely at a single strand and see that it is made up of rows and bunches of little balls. If someone asks you what these little balls are, you may guess that they are *atoms* and you will be right. If you look at the control dial of your supermicroscope, you will see that the needle points at ten million; you are making these atoms look about ten million times larger than they are; and each of the atoms appears to be about the size of the head of a pin. Some may look a little bigger and some may look a little smaller, but this is the average size.

The nature of the atom. In order to see more clearly what an atom is like we next turn up the dial of the supermicroscope to a hundred million and focus on one of the larger atoms. Under this magnification it looks about the size of a marble, and, if you observe closely, you notice several things. (We shall show shortly that undistorted observation of the atom is more than we can hope for in the world as we know it today; hence our microscope is truly a *magic* supermicroscope.) First of all, the atom is not standing still; it is bouncing back and forth, hitting against some of the neighboring atoms that are clustered like a bunch of grapes. You may observe also that the atom does not have a solid surface but is more like a ball of smoke or fog. And, if you look closely, you see that this fog is shimmering with waves, much like the ripples on the surface of a pond when a stone is dropped into it. This fog is the cloud of negative electricity that makes up the outer part of every atom, the cloud of the *electrons* that are the units of negative charge. Down at the center of this cloud you also see a tiny speck, *the nucleus* of the atom, containing the positive electricity as well as practically all the *mass* of the atom.

What we have just described is the "wave picture" of the atom which, we believe, comes closest to portraying correctly its true nature. Forty years ago the picture was quite different. Scientists believed then that these atomic electrons were not clouds but particles of electricity circulating around the nucleus like planets around the sun. Such a planetary model has elements of truth and correlates some aspects of atomic behavior. We shall discuss this picture of the atom at some length later in this chapter; and in the following chapters we shall introduce the modern and more accurate wave description that is so necessary for an understanding of chemistry. Examples of the particle picture and the wave picture are shown in Figs. 2*a* and 2*b*.

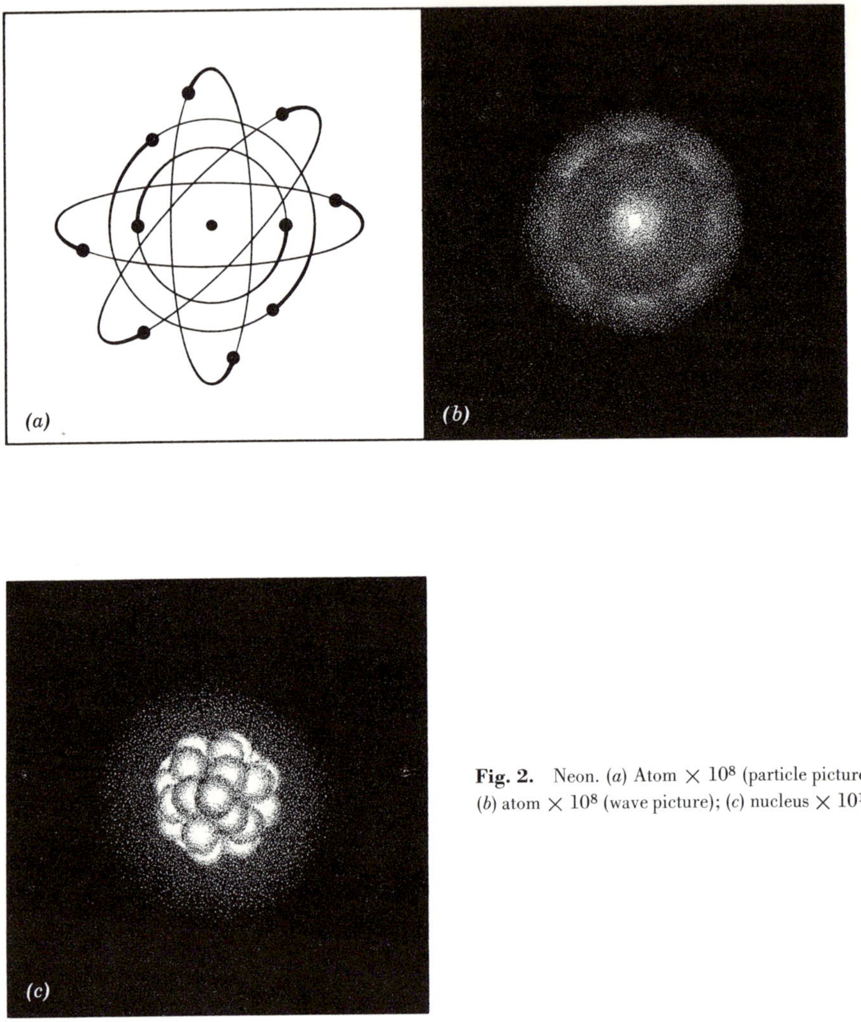

Fig. 2. Neon. (*a*) Atom $\times$ 10^8 (particle picture); (*b*) atom $\times$ 10^8 (wave picture); (*c*) nucleus $\times$ 10^{12}.

We now increase our microscope magnification still further to get a better look at the atomic nucleus. Actually, we have to increase the magnification *ten thousand times more* to make this nucleus readily visible. On this scale we have magnified the outer part of the atom to the point where it would be a hundred yards across, and big enough to hold a football field. Thus, if we could stand inside such a magnified atom, we would see the electron cloud over our heads, down at the sides, and under our feet like a great ball a hundred yards in diameter; but the nucleus at the center, even in an atom large enough to hold a football field, would still be smaller than a marble.

The nature of the nucleus. Undoubtedly, you would like to have a closer look at this tiny nucleus that holds the secret of atomic power. So you turn up the dial on your supermicroscope again, increasing the magnification ten times more, until the nucleus itself appears to be about as big as a baseball; and now what will you see?

Often you will find an illustration (Fig. 2*c*) showing the nucleus looking like a baseball made of peas, all glued together. This is not a true picture, but it does convey the idea that the nucleus is made up of some kind of combination of particles crowded closely together, quite the opposite of the outer part of the atom in which the electrons (regarded as particles) are widely separated in relatively empty space. This close crowding in the nucleus gives it an unbelievably great density. If a real baseball had the same number of grams of mass per cubic centimeter as are found in the actual nucleus, it would weigh 10,000,000,000 tons. In our galaxy there are some stars that have extraordinarily high densities, as great as a billion tons per cubic inch; it is believed that these stars may consist of nuclei much more closely packed than in matter found on earth.

The idea of the atom. Many times during the course of history, men believed that the solid matter, out of which the different things in the world are made, was really *continuous* matter. They thought that if you could look at the surface of a stone with magnification unlimited you would always see a *continuous* surface, no matter how much you turned up the magnification. They believed that you would *not* find individual particles like atoms and electrons. They argued that, if you had a magic knife by which you could cut the stone into smaller and smaller pieces, you could continue cutting it up indefinitely and could make the pieces as small as you like—a trillion, quadrillion, quintillion times small or even infinitely small. But, some twenty-five hundred years ago, there lived in Greece a group of men with inquiring minds who perceived that this argument for continuous matter was not sound. Facing the fact that the world contains so many different kinds of matter stones, metals, vegetable matter, animal matter, solids and liquids of so many forms —they argued that it was more logical to believe that all these varieties of matter are brought about by coupling together a relatively few kinds of particles, which could not be cut up any further. Because this indivisibility was the fundamental property of these particles, they gave them the name *atom*, or *a-tom*, which in Greek means *un-cut-able*. It is astounding that this purely philosophical guess should be confirmed so completely two thousand five hundred years later. Of course, today, we recognize that nearly all atoms *can* be cut up or broken up into smaller particles, such as the electrons in the outer shells and the fundamental particles in the inside of the nucleus; but we do believe that the truly fundamental particles, of which *atoms* are made, are really un-cut-able; and they are the true building blocks of all matter. So these Greeks were right in the ultimate sense of the argument.

The nature of a model. We have just taken a look at the over-all picture of matter and the fundamental particles of which it is made. We have glimpsed the way in which these fundamental particles are put together in order to produce something that you can see and feel like the skin on your finger. On first presentation, this picture is necessarily sketchy; there are many points that we will have to examine in more detail; but, most important of all, we have to realize from the very beginning that any picture of matter always must be only a partial truth. Just as a road map reveals the pattern of the roads in a given section of the country, a picture or map of the atom gives an idea of its pattern. But, no matter how large the scale on which we draw the road map, it never shows all the details of the road, it never *is* the road; and no matter how detailed we make the picture of the atom it never shows all the structure of the atom, it never *is* the atom.

This picture of the atom that we have presented is frequently referred to as a *model.* This word, *model,* is used to convey the idea that the picture is not a *complete* picture of what is present in the atom; but, although it is incomplete and although it would be quite wrong if we took it too literally, it is extremely helpful in guiding our thinking as we try to find out more and more about atoms and the nature of the world around us. In learning to speak a language we have to learn the alphabet and then the combinations of letters which make up words and finally how to put these words together into sentences, paragraphs, and essays, before we can express adequately our thoughts and communicate them to other people; so in the same way we have to learn bit by bit the alphabet and the vocabulary of chemistry and its grammar and syntax, before we can grasp the broader aspects of its ideas. This is illustrated in Fig. 3.

Consider the analogy between letters and atoms in a little more detail. For example, take three letters: d, g, and o. Separately, they are just three letters with no special meaning; but put them together as *dog* and they convey a distinct idea, the four-footed canine animal. If we wish to be more precise, we can draw a picture of a dog and show that it is large or small, long-legged like a wolfhound or short-legged like a dachshund; and we can make the picture in minute detail representing almost every hair of the animal's coat if we have that much patience. But, no matter how much detail we draw into the picture, we never think that the picture *is* the dog. There is always the unbelievably complex reality of the animal itself, of which the word *dog* is only the symbol, and the picture of the dog only a very inadequate representation.

In the same way we use letters as symbols for different kinds of atoms. H stands for *hydrogen* and O stands for *oxygen.* HOH is the symbol for the water molecule, sometimes abbreviated to H_2O. This molecule has an oxygen atom in the center with a hydrogen atom on each side, tied to it by a chemical bond. We can represent it by the symbol H—O—H or we can draw pictures of it showing the atoms as round balls. Such a picture is useful in thinking of the ways in which

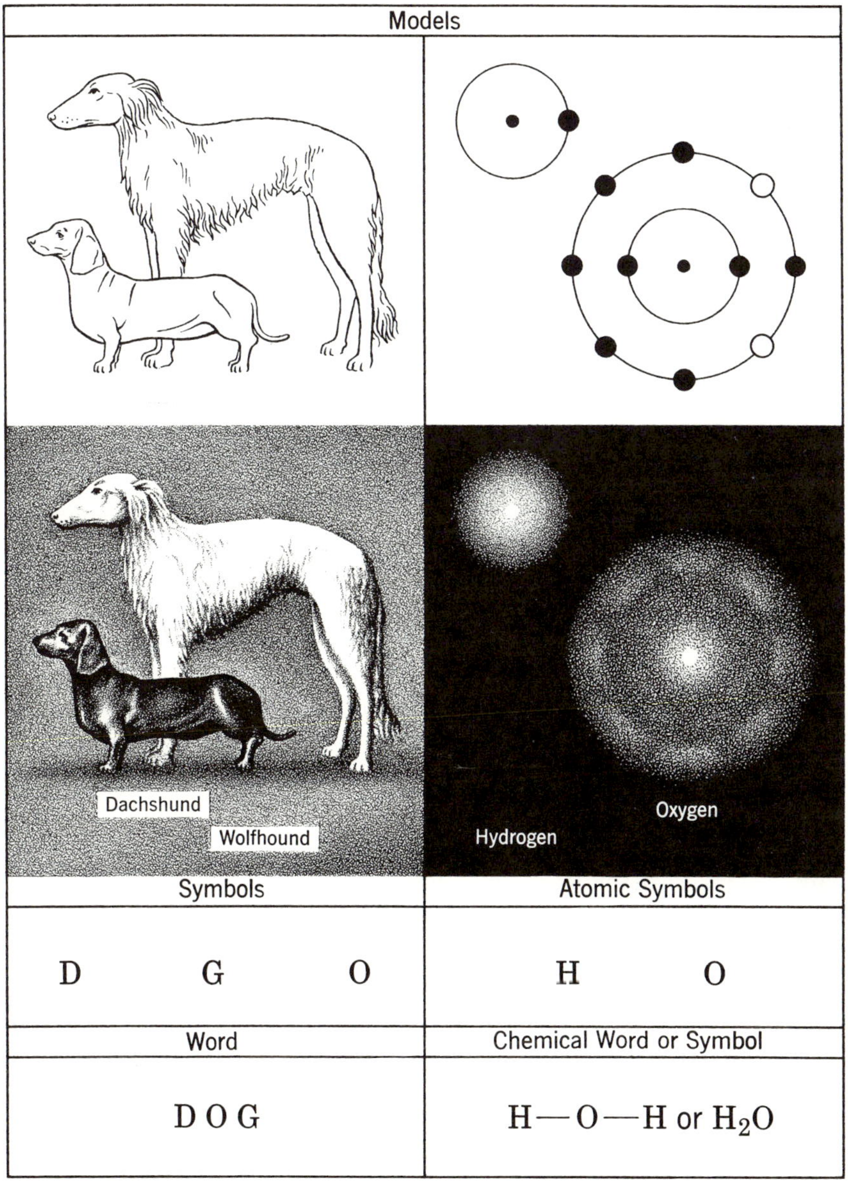

Fig. 3. Models.

atoms in the gaseous state behave. We can also draw a picture of the atoms showing the nucleus at the center and the electrons moving in orbits. This helps us to understand how the atom acts when a high-speed particle hits it and knocks an electron into a larger orbit or even knocks an electron completely out of the atom.

But this is still only a picture, a model, useful as an aid to our thinking but far from reality. We know today that electrons behave more as if they were *waves* than *particles* and that the *wave-cloud picture* enables us to understand even more clearly how the atom acts. But it is still a picture, a model, and far from the ultimate reality. So, as we proceed in our study of chemistry, we shall use these model-pictures constantly, and they will enable us to think about atoms in a much more fruitful and coherent way than we could without their help; but we must always keep in mind their limitations.

Whenever there is discussion of a physical model on which theories are based, you should keep in mind also that such a model is merely part of a *portrait* of nature. When an artist paints the portrait of a man, he strives not for a photographic reproduction but rather to produce on the canvas his own special impression of the subject. We can picture scientists as artists creating such a portrait of *nature.* One will be most interested in the atom; another in the electron; still another in the nucleus. Each draws his own part of the portrait with some inaccuracies. Then, after a few years, another artist looks at the portrait, realizes that a few lines need changing, and makes the alteration. Rarely is it necessary to erase a large part of the picture and to start over again from scratch. Sometimes, however, these artists, although they do not have any better way of presenting some aspect of nature, know that something is wrong with a certain part of the portrait. Then they can only wait until more about this point is known, and, for the time being, acknowledge that this detail is not wholly correct. These models, therefore, are in a constant state of change; and the current theories reflect in whole or in part the work of many artists. A picture of nature fifty years old still roughly resembles nature, but the present picture is more accurately true to nature. Thus, what you learn today still will be largely the basis of the picture of nature which scientists will have fifty years hence. Therefore, keep in mind that models constitute a *portrait* of nature and not a *duplication.* They help us to recognize nature but they are not, nor can they ever be, perfect reproductions of nature.

LARGE NUMBERS

To express the difference in size between ordinary objects and very small objects such as atoms and their nuclei, we have already used numbers like million and billion. The number of atoms in the human body is about 1,000,000,000,000,000,000,000,000,000; the name for this number is *octillion* and the usual symbol for it is 10^{27}. Since this number 10 raised to the power of 27 (or 10 with the exponent 27) is the same as the number 1 followed by 27 zeros, this *exponential* notation is more convenient than writing out 27 zeros or using a name like "octillion." We shall use the exponential notation almost exclusively, and it is important for you to learn how to use it. The simple rules for combining exponents in equations are given in the Appendix.

You should also try to get the "feel" of these large numbers. Even though you write the number 1 followed by 27 zeros, can you really appreciate the magnitude of this number? One way to grasp such size is to visualize a collection of a large number of familiar objects.

We start with a few smaller numbers. For example, suppose that we have a million (10^6) dollars in one-dollar bills. Stack these on top of one another and they make a pile about six hundred feet high, roughly fifty feet higher than the Washington Monument. Lay this stack horizontally and it is twice as long as a football field. Now we ask what a billion (10^9) dollars would look like. If we had a billion one-dollar bills, and we placed these in the same way horizontally, this stack would reach a hundred miles, from New York to Philadelphia. So we see that, whereas a million reaches twice the length of a football field, a billion reaches a hundred miles; and this shows us how greatly we increase the number of things when we jump from a million to a billion by just adding three ciphers, going from 10^6 to 10^9.

Next, suppose that we have a trillion (10^{12}) dollar bills. This stack of dollar bills would be one hundred thousand miles long, enough to go around the world four times. So we see that, as we go beyond the trillions, we deal with distances just as difficult to visualize as the numbers themselves. Therefore, let us try another approach.

We take a collection of peas as the next example. If you count out a hundred (10^2) average-sized peas, you will find that they occupy roughly the volume of a cubic inch. A million (10^6) peas are just about enough to fill an ordinary household refrigerator, and a billion (10^9) peas will fill a whole house from cellar to attic; thus a trillion (10^{12}) peas will fill a thousand houses, the number you might find in a medium-sized town; and a quadrillion (10^{15}) peas will fill all the buildings in one of our larger cities like Philadelphia or Baltimore.

You can see that you will run out of buildings very soon at this rate, so let us try a larger measure, for instance, the state of Pennsylvania. Suppose that there is a blizzard over Pennsylvania, but, instead of snowing snow, it snows peas. We get Pennsylvania covered with a blanket of peas about four feet deep all the way from New Jersey out to Ohio and all the way from Maryland to New York State. This blanket of peas drifts over the roads and banks up against the sides of the houses, and covers all the fields and forests. Think of flying across the state with the blanket of peas extending out as far as you can see. This gives you an idea of our next number, for there will be in this blanket about a quintillion (10^{18}) peas.

Imagine that this blizzard of peas falls over the entire land area of the globe, North America, South America, Europe, Australia, Asia, and Africa, so that all the continents are covered with peas four feet deep. This global blanket will contain sextillion (10^{21}) peas. Then imagine that the oceans are frozen over and the blanket of peas covers the entire area of the earth. Go out among the neighboring stars and collect two hundred and fifty planets each the size of the earth and cover each of these with a blanket of peas four feet deep. Then you have septillion (10^{24}) peas. Finally, go out into the farthest reaches of the Milky Way, and collect two hundred and fifty thousand planets, each the size of the earth; cover each of these with a blanket of peas four feet deep; and then, at last, you have octillion (10^{27}) peas, the number corresponding to the number of atoms in your body. So you see what an extraordinarily small object an atom is and how complicated *you* are. It is this small size of the atom that causes so much difficulty in the effort to uncover its structure. In our initial discussion of certain large numbers we shall use this pea-picture to emphasize their enormity. Often the description will seem ludicrous, but it will aid in grasping the idea.

CHEMICAL ELEMENTS

Now that we have had a bird's-eye view of atoms, let us examine in more detail how they are put together. Atoms are the elementary building blocks out of which everything in the universe is made. If you could see the surface of your *hand* through a supermicroscope that magnified a hundred million (10^8) times, your skin would look as if it were made of a layer of little balls. These balls would not be all exactly the same size. If you looked carefully, you would see at least a dozen kinds, differing in size and in appearance. By way of contrast, if you look at the surface of a *diamond,* this would also appear to be made up of little balls arranged in regular rows; but, in this case, the balls are all the same size, the same shape, the same kind, because they are all atoms of *carbon.* No matter what object we pick—a stone, a piece of iron, a piece of wood, or a drop of water—you will always find, if you look at it with sufficient magnification, that it is made up of these small round objects: these atoms. Because *everything* is made of a limited variety of atoms, the kinds of atoms are called the *chemical elements.*

If you use your supermicroscope to examine thousands of different objects, and then make a list of all the different kinds of atoms you recognize, you will find that there are about a hundred different kinds. Each kind of atom has been given a name and symbol. Thus the smallest of all the atoms is *hydrogen,* and it is also the lightest; the symbol for it is H. The heaviest of all naturally occurring atoms is *uranium,* which weighs about 238 times as much as hydrogen; this atom provides the explosive power in the atomic bomb; and the symbol for it is U. Another kind of atom, which is found in your body, in water, in trees, and in the air, is *oxygen,* which has the symbol O. This atom is about sixteen times as heavy as hydrogen. As far as we know at this time, the newest atom to be recognized is nobelium (symbol No), which has been produced artificially by bombarding the nuclei of a lighter element with helium ions. This atom is the heaviest known, about 253 times heavier than hydrogen.

These one hundred-odd varieties of atoms constitute the chemical elements. The reason why salt is something different from water lies in the fact that salt (NaCl) is made of the two elements sodium (Na) and chlorine (Cl), while water (H_2O) is made of the two elements hydrogen (H) and oxygen (O). Again, common table salt, *sodium chloride*, is different from the "salt," *calcium chloride* ($CaCl_2$), which is sprinkled on the roads in winter to melt the ice and snow. Both of these substances are loosely referred to as *salt;* but calcium chloride, sprinkled on the meat that you eat, would not enhance the flavor. Particle diagrams of NaCl and $CaCl_2$ are shown in Fig. 4.

There are also substances made of the *same number and kind of elements* that differ only because their atoms are *linked together* in a *different way*. Thus *ethyl alcohol* (familiar *grain* alcohol) is composed of two atoms of carbon and five

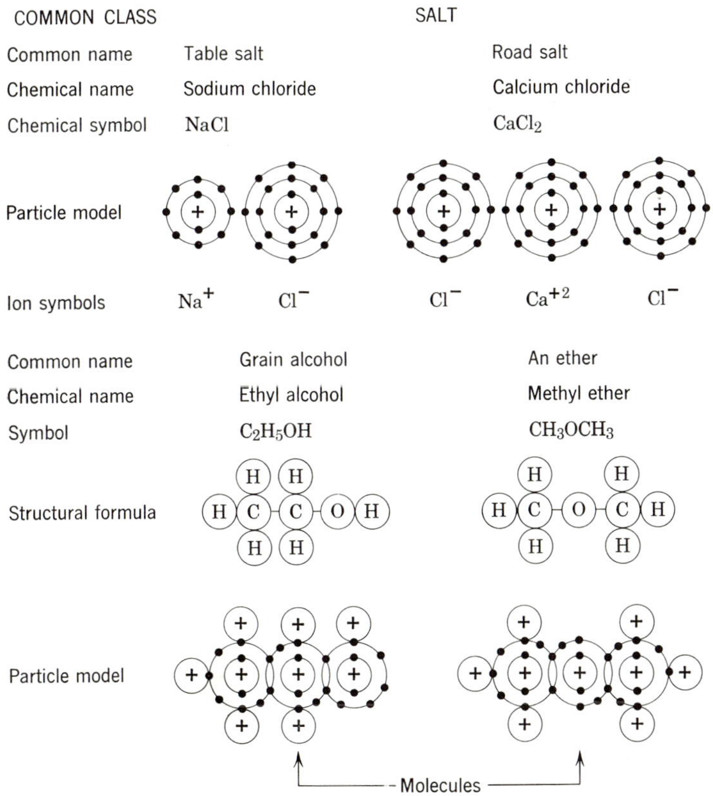

Fig. 4.

atoms of hydrogen linked to an oxygen-hydrogen group at one end (C_2H_5OH); *methyl ether* has exactly the same number and the same kinds of atoms, but in this compound the oxygen atom is in the middle (H_3COCH_3). At room temperature, ethyl alcohol is a liquid, potable and digestible in small amounts; methyl ether is a gas that must be cooled to $-24°C$ before it liquefies, and it has quite different chemical and physiological properties.

We now consider the first ten chemical elements in terms of *the particle picture.* This *particle picture* provided the best understanding that we had of the atom forty years ago; particle pictures of the first ten elements are shown in Fig. 5. Today, the *wave* picture gives a more accurate and detailed understanding of the atom, but it is harder to grasp without a thorough discussion. Since there is still con-

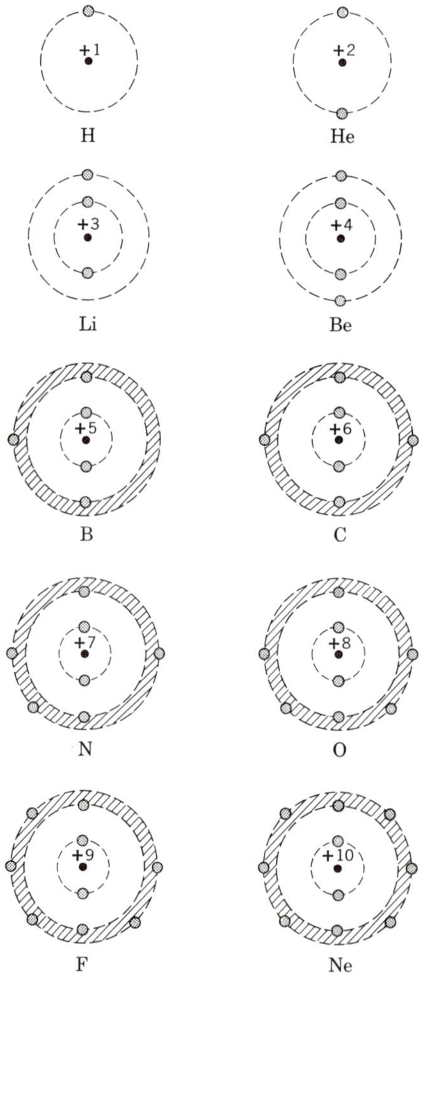

Fig. 5. Schematic diagrams of the first ten elements.

siderable truth in the simpler particle picture, it is a good starting place for our initial exploration of these lighter elements.

Hydrogen. In the simplest of all the atoms, *hydrogen* (H), there is only one electron whirling around a very small nucleus at the center. The diameter of the orbit of the electron is about one-hundred-millionth (10^{-8}) centimeter,† and this is the figure appropriate to take for the diameter of the atom itself; it is one of the important *small* numbers that we shall be using frequently. Since this distance is used so much in chemistry and physics, it is regarded as a special atomic unit, called the *Ångstrom* unit. Just as the millimeter is a "little" meter (one-thousandth of a meter), the Ångstrom unit is a still smaller unit (one-hundred-millionth of a centimeter or one-ten-billionth of a meter). The diameter of the atomic *nucleus* of hydrogen is believed to be about 10^{-13} cm. This is a much smaller length, and for dealing with distances involving the nucleus it has been proposed that we use 10^{-13} cm as a still smaller unit of length. Let us now examine these two parts of the hydrogen atom, the electron and the nucleus, in more detail. We want to find out how heavy they are, how big they are, and what it is that holds them together in this particular form.

The electron. First of all, note that the electron is found not only in hydrogen but in every kind of atom. The atoms of each chemical element have electrons in their outer parts and all these electrons are identical. The electron that can be taken from the hydrogen atom has the same mass and charge as those from the uranium atom. Moreover, it is electrons that make up electricity. When electricity flows through a copper wire to light a lamp or to put power into an electric motor, the electric current consists of a stream of electrons flowing through the wire. In fact, one of the chief characteristics of substances that we call metals is their ability to permit electrons to flow through them, much like water flowing through a pipe.

As we have seen, the electron is very small. It is a little difficult to say exactly how big a free electron is because it does not have a definite surface but is more like a ball-shaped cloud of smoke thinning out indefinitely from the center. The effective size depends on what the electron is doing, but at present we can think of a free electron in space as the elementary *particle* of electricity having a diameter between 10^{-11} and 10^{-12} cm. Magnified a trillion (10^{12}) times, it would be roughly the size of a golf ball. When an electron is no longer free, but is bound in the outer part of an atom, it usually behaves as if it had expanded enormously; on this same magnified scale it might be the size of a balloon one hun-

† As explained in more detail in the Appendix, the negative exponent (10^{-8}) means that this number is the equivalent of 1 followed by 8 zeros in the *denominator* of a fraction (1/100,000,000), so that 10^{-8} cm is the equivalent of one-hundred-millionth centimeter.

dred yards in diameter. As part of an electric current in a copper wire, it expands still more into something like a balloon hundreds of miles across, when viewed with 10^{12} magnification. So we have to be careful in thinking about the size of electrons since size depends so much on where they are and what they are doing.

The most important property of the electron is not size but electric charge. What electricity or electric charge *is* we cannot say; it is just one of the fundamental things in the universe. However, we can describe what this charge *does* and thus get some idea of its nature. Two kinds of charge have been observed. One is called *positive* and is the kind found in the nucleus of the hydrogen atom; the other is called *negative* and is the kind present in the electron. Both these charges possess a *field* of influence that extends out from them and produces a *force* when it interacts with the fields from other charges. *Unlike charges* tend to pull each other together closer; that is, positive charges *attract* negative charges. *Like charges* tend to push away from each other; that is, a positive charge *repels* other positive charges and a negative charge repels other negative charges.

Just as the standard unit of distance is the centimeter and the standard unit of mass is the gram, the standard unit of electrostatic charge is called one *absolute electrostatic unit.* But, just as we frequently use the *meter* for measuring length, and the kilogram for mass, we shall frequently use the *coulomb* as the unit of charge. It is 6.242×10^{18} times as large as the charge on an individual electron.

Frequently we are interested in the flow of electrons, a current of electricity. When one coulomb per second passes through a wire, we say that the strength of the current is one *ampere.* When a one-hundred-watt electric light bulb is lighted with a potential of 100 volts, we find one coulomb per second of electric charge flowing through it, or a current of one ampere. And we know exactly how many electrons must flow through the wire of the lamp per second to make up the one ampere current, namely, 6.24×10^{18} electrons.

If you recall our original discussion of atoms, you will remember that it took 10^{18} peas to cover the state of Pennsylvania four feet deep. Thus, if you had the number of peas evenly spread over Pennsylvania corresponding to the number of electrons flowing through a 100-watt lamp each second, you would have Pennsylvania covered to a depth of about twenty-five feet. In our common unit phraseology, this would be twenty-five "foot-pennsylvanias" of peas. So you see how small this elementary unit of electricity is. Expressing this in another way, we say that the charge on the electron is 1.6×10^{-19} coulomb. This tiny particle of electricity naturally has a very small mass. It is just under 9.109×10^{-28} gm.

The proton. Let us now turn our attention from the electron in the outer part of the atom to the nucleus at the center. As we have pointed out, the diameter of this nucleus is very small; in the case of hydrogen it is probably about 10^{-13} cm. But this minute speck contains nearly all the mass of the atom and the charge of *positive* electricity that results in the force of attraction that holds the atom together. This positive charge has exactly the same magnitude as the negative charge on the electron so that they balance each other; thus the ordinary individual atom is in the condition that we call electrically *neutral.*

The nucleus of the hydrogen atom is called the *proton.* Just as the electron is the unit of negative electricity and is regarded as one of the fundamental particles of which the universe is made, so the proton is regarded as one of the fundamental units of mass; it weighs 1.672×10^{-24} gm. We believe that the force of electrical attraction between the positively charged proton and the negatively charged electron is balanced by an equal and opposite force caused by the motion of the electron. If we think of the electron as a particle whirling in the orbit around the nucleus, there will be *centrifugal force* which will tend to pull the electron away from the nucleus. A common example of a centrifugal force is the pull you feel if you attach a string to a stone and then whirl the stone around your head. As long as you hold on to the string, the stone continues to whirl around in a circle, but you can feel its pull against your hand as it whirls. This is centrifugal force. If you let go of the string, the stone then moves away from you in a straight line. In the case of the electron whirling around the nucleus, the outward centrifugal force produced by the circular motion of the electron exactly balances the inward electrical pull between the positive charge on the proton and the negative charge on the electron.

There are several other facts about hydrogen which should be discussed at this time. In its normal state at room temperature, hydrogen is a gas. Moreover, it is not in the form of separate individual atoms; instead the atoms are linked in pairs. If you could "see" one of these pairs of hydrogen atoms, it would look like a gymnasium dumbbell. In order to condense this gas to a liquid, it must be cooled to 253° *below zero* centigrade (−253°C). It sounds even more impressive to say that hydrogen boils at 423° *below zero* on the *Fahrenheit* temperature scale. It crystallizes or freezes at −259°C. There is so little force of attraction between one hydrogen "dumbbell" and another H_2, that they have to be cooled to this very low temperature to be condensed to a liquid or a solid.

Helium. Let us now take a preliminary look at some of the other chemical elements that we will soon be studying. In order of ascending weight, the next element is *helium.* An atom of helium has two positive charges of electricity on the nucleus and, consequently, two electrons circling around it. As we encounter it in nature, helium is a gas and traces of it are found in our earth's atmosphere. Helium boils at −269°C, a temperature even lower than the boiling point of hydrogen. It can be made to freeze only when a large pressure, about twenty-five times that of our atmosphere, is applied to force the atoms together.

It is believed that there is a great deal of helium in the sun; in fact, that is the origin of the name of this element, coming from the Greek word meaning *sun.*

You may ask at this point whether there could be a chemical element with one and one-half charges of positive electricity on the nucleus; the answer is *no.* We never find any bits of electricity floating around with one-half or one-quarter the charge on the electron; in other words, there are never any "baby" electrons; all electrons have the same charge. By the same token, we never find any split positive charges on the nucleus. They are always *integral multiples* of the electron charge, that is, *once, twice, three* times, and so on, the charge on the electron. It is a situation something like our coinage; we never find coins that are one-half cent or one-quarter cent (at least today in the United States); the one-cent coin is the unit of coinage just as the single charge on the electron is the unit of charge; and we have coins which are integral multiples of this unit, like the five-cent, ten-cent, twenty-five-cent, and fifty-cent coins. Nature is more prolific with multiple charges; the helium nucleus has 2, the *lithium* nucleus has 3, the *beryllium* nucleus has 4, and so on; every integer is represented up to 102, *nobelium* with 102 units of positive charge on its nucleus; but there are never any nuclei with fractional charges like $2\frac{3}{4}$.

You might expect to find the nucleus of the helium atom weighing twice as much as the nucleus of the hydrogen atom under these circumstances; but, curiously enough, with mass, we run into a somewhat more complicated situation than with charge. We find that the nucleus of helium weighs not twice but about four times the nucleus of the hydrogen atom. The reason for this lies in the fact that this helium nucleus is made up, not of two protons stuck together, but of two protons plus two particles of another kind that we have not encountered before, *neutrons.* This neutron has about the same mass as a proton, 1.675×10^{-24} gm, as compared with the mass of the proton, 1.672×10^{-24} gm; but it has no charge at all. No one even suspected its existence until about thirty years ago. Until that time, scientists thought that a helium nucleus was made up of four protons, somehow stuck together with two electrons so that the four positive charges on the protons were partly balanced by the two negative charges on the electrons, leaving the total helium nucleus with a net positive charge of 2. Then, in 1932, Sir James Chadwick discovered this strange particle, the neutron; and we now believe that all nuclei heavier than hydrogen are made up of neutrons and protons and that the electrons cannot exist independently within the nucleus.

The point can well be raised whether it is proper to call the neutron a really independent fundamental particle, because when it is left alone it breaks up spontaneously and flies apart, producing an electron and a proton. This question "when is a fundamental particle really a fundamental particle" is being discussed at great length by nuclear scientists today. We know that a nucleus like that of the helium atom can be broken up into two protons and two neutrons, but we do not know whether it is correct to say that these particles actually are present *as such* in the nucleus. In fact, you could well have raised the question: "How can the two protons stick together in the helium nucleus?" For the two positive charges have a really enormous force of repulsion on each other when they are brought so close together that they are only 10^{-13} cm apart. We are just beginning to get an idea of what some of these nuclear forces are, but the concepts are so complicated that it is necessary to become familiar with a great deal more of chemistry and physics before they can be discussed readily. At the moment, we shall take it for granted that there is a special kind of nuclear force that we will regard as an elementary force along with the force of gravity and the force of electrical attraction or repulsion. It is this nuclear force that ties the two protons and the two neutrons together into the helium nucleus.

The helium nucleus is very stable. We find this combination of two protons and two neutrons suddenly being ejected from the nuclei of larger atoms by the kind of nuclear explosion which takes place in radium and uranium, two of the heavier elements that we shall be discussing shortly. This "stripped" nucleus of the helium atom is called the *alpha*-particle. In other words, it is a helium atom which has lost its two electrons and is going around as a bare nucleus with its two positive charges unneutralized.

The other light elements. Finally, we want to discuss the chemical elements that follow helium in the series where the charge on the nucleus, the *atomic number,* increases stepwise from 3 to 10. The atom with three charges on the nucleus and three electrons circulating around it is called lithium (Li). Its atomic number is 3, though it weighs about seven times as much as hydrogen. Lithium is a metal and will conduct electricity; electrons will flow in one end and out the other end of a wire of lithium if a battery is connected across it. Lithium melts at 181°C and boils at 1347°C. Thus, as we pass from helium with two planetary electrons to lithium with three, we are passing from a gaseous element boiling at −269° *below* zero to a metallic element boiling at a temperature over 1000° above zero. Why should the addition of just one electron cause this great change from a gas where the atoms naturally fly apart to a metal where the atoms are so tightly bound to each other that they have to be heated red-hot to get shaken apart? That is one of the fundamental questions that chemistry undertakes to answer, and it will be at the center of our discussions throughout the whole course of our study.

When we increase the number of charges on the nucleus to four, we find the element *beryllium* (Be). This nucleus weighs about nine times as much as hydrogen and is metallic in nature like lithium. It melts at 1283°C and boils at 2477°C.

Increasing the charge on the nucleus to five, we form the element *boron* (B), which weighs about eleven times as much as hydrogen. Here we have five electrons circulating around in the outer shell, and the element begins to lose its metallic quality. One might think, that the more electrons that circulate around the outside, the better the substance would conduct electricity, but the situation is more complicated than that. Boron melts at 2030°C and boils above 3600°C. It is interesting that, in its elementary form, boron is the hardest of the elements next to carbon in the form of diamond.

When we increase the number of charges on the nucleus to six, we arrive at the familiar element so important to us all, the element *carbon* (C) that weighs twelve times as much as hydrogen. Carbon is the most vital constituent of our bodies and our food. Pure carbon is found in two forms which differ from each other greatly in their physical properties. One is graphite, a black solid that will conduct electricity, though not nearly so easily as the metals lithium and beryllium. This form of carbon vaporizes at 4347°C. The other form of carbon is the non-conducting diamond, the clear, hard, bright crystal that is so prized as a jewel. We shall see later that it is the way in which the outer electrons in this six-charge atom combine that decide whether it has the form of black opaque graphite or the clear sparkling diamond. These carbon atoms have the ability to form long chains, and this enables them to produce the variety of substances which make up many of the different chemicals in our bodies.

With seven charges on the nucleus, we find another familiar and important chemical element, nitrogen (N), fourteen times as heavy as hydrogen. As contrasted with carbon atoms that go together in large solid masses to form graphite or diamond, nitrogen atoms have a great preference for going around in pairs, in the gaseous dumbbell form that we observed with hydrogen. It is in this gaseous form that nitrogen (N_2) is found in the atmosphere. Thus the difference in the physical state and the structure of carbon and nitrogen is spectacular. Elementary carbon is characterized by a very high melting point and boiling point; nitrogen, with a nuclear charge greater than carbon by only one, melts at −210°C and boils at −196°C. In the solid form it is a non-conductor.

When we have eight charges in the nucleus, we get the element also important for life, and common in our atmosphere, oxygen (O). Oxygen atoms are found in pairs as a gas at ordinary temperatures; hence it is in this form (O_2) that oxygen is found in the atmosphere. An atom of oxygen is about sixteen times as heavy as an atom of hydrogen. In its physical properties oxygen is similar to nitrogen; it melts at −219°C, boils at −183°C, and is a non-conductor.

Finally, we examine two more elements. With nine charges on the nucleus, we have the element *fluorine* (symbol F), about nineteen times as heavy as hydrogen. Fluorine is one of the most active of the chemical elements; that is, it readily combines with other elements. It is normally found as a diatomic molecule boiling at −188°C and freezing at −218°C.

Finally, with ten charges on the nucleus, we reach the element *neon* (symbol Ne), which is also a gas but surprisingly one of the most inactive of chemical elements. It is a little over twenty times as heavy as hydrogen. It boils at −246°C and freezes at $-249°C$. Thus, both in its chemical inactivity and in its low boiling point, neon resembles helium; it is a "lone-wolf," a hermit not wishing to associate either with its own kind or any other kind of atom.

In this survey of the first ten of the chemical elements in the ascending order from the lightest, hydrogen, moving toward the heavier elements, we have seen that, as the positive charge in the nucleus is increased stepwise, the number of electrons circulating around the nucleus in the electrically neutral atom also increases stepwise. We also note that the physical and chemical behavior changes remarkably. Hydrogen and helium are gases; hydrogen atoms go around in pairs (H_2); helium atoms remain single and apart, not combining at all either with themselves or any other atoms. Lithium and beryllium are metallic solids conducting electricity. Nitrogen and oxygen are elements that normally go around in pairs as gases but which can be coupled very firmly to other chemical elements. Then there is fluorine, the very active chemical element in the form of a gas that has a strong tendency to couple to other elements, and next to it there is the very inactive element, neon. Thus, in going from the second to the tenth element, we have passed from a gas (He) through metals (Li, Be) and semimetals (B, C), through gaseous elements (N, O, F) that combine readily with many other elements to the almost completely inert and unreactive element (Ne) that ends the first decade of atoms. We shall find this pattern of change—gas to solid to gas—repeated over and over again as we proceed to elements with larger numbers of electrons. It is such repetition that gives the series of chemical elements its *periodic* character.

This variety of action was one of the most puzzling of problems in chemistry for well over a century. Actually, it is only within the last thirty years that we have begun to understand the underlying reasons for this strange pattern. The discussion and explanation of such chemical behavior will occupy us for a great part of this book.

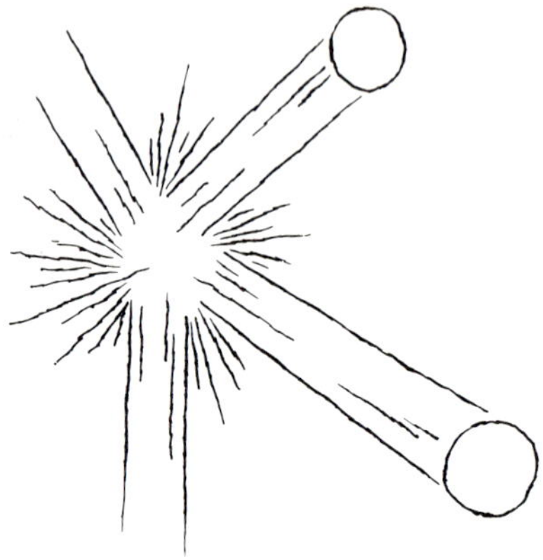

2 The energy of atoms

Chemistry is the study of atoms in action, not of atoms in specimen bottles on the shelves of a museum. In chemical research we try to find out how atoms act in processes like the explosive combination of carbon, hydrogen, and oxygen atoms in the cylinder of a gasoline engine; we study how atoms combine in a life-process like breathing where oxygen atoms join to a complicated molecule in the blood stream containing carbon, hydrogen, nitrogen, and iron; we design manufacturing processes where atoms are linked together to form new drugs, dyes, plastics, and thousands of other products. And, in such chemical processes, we find that the action depends primarily on the *energy* of these atomic groupings; different kinds of atoms react differently because of the different energies of their atomic combinations. *Energy* is the key to action.

THE NATURE OF ENERGY

What is energy? We are all familiar with some of the forms of energy. Suppose you have a pingpong ball and a baseball. You throw both balls at a window each with a speed of 100 cm/sec. The pingpong ball bounces off harmlessly, but the baseball shatters the glass. Why? Because the baseball has more energy. It is the energy in the motion of the baseball that at the moment of impact surges into the window pane and tears apart the chemical bonds that held the glass together.

The form of energy associated with motion of this sort is called *kinetic* energy, and it will be denoted by the symbol: E_k. It is related to the mass (m) and velocity (v) of a moving object by the equation

$$E_k = \frac{1}{2}mv^2$$

In words, this states that the kinetic energy is equal to one-half the product of the mass and the square of the velocity.

Considering our example of the two balls that are thrown with the same velocity, we find that, if the baseball weighs one hundred times as much as the ping-pong ball, it will have one hundred times more energy. It is not the ball, but the *energy* of the ball, that breaks the window; it is not gasoline, but the energy released *by the combination* of the atoms of carbon and hydrogen with oxygen in the engine-cylinder, that makes your car go; it is not the food, but the energy of the food, that keeps you alive. This is why the study of chemistry is essentially the study of energy.

Since the things that happen when atoms combine depend largely on how *much* energy is released, we need to have a measure for energy in order to discuss chemical reactions, just as we need the yard or the meter to discuss distances. The formula for the kinetic form of energy involves mass and velocity, so, in order to arrive at a measure for energy, we take a look first at these two quantities on which energy depends.

Mass. Mass is the measure of the quantity of substance. In the English-speaking countries, mass is frequently measured in pounds and ounces; but, in order to have units that are simply related by powers of ten, the scientific world has adopted the cgs (centimeter-gram-second) system of units. In this system the ultimate unit of mass is the mass of a standard block of metal, made of a hard inert platinum-iridium alloy and kept in a vault at Sèvres, France. This is the standard *kilogram*, *kilo* meaning a *thousand*. The *gram* is defined as one-thousandth of this reference standard mass.

Length. The precise unit of length is the *meter*, defined as 1,650,763.73 times the wavelength of orange-red light emitted by the chemical element krypton (Kr^{86}) in passing from the $2p_{10}$ to the $5d_5$ level of energy (Paschen notation). A *centimeter* is one-hundredth of this meter length. These units are chosen so that a cubic centimeter of pure water at a temperature of 4°C has a mass of one gram.

It is sometimes convenient to use large units like kilograms or pounds of mass and kilometers or miles of length; sometimes it is more convenient to use smaller units like grams or ounces of mass and centimeters or inches of length. In the Appendix are listed conversion factors to be used in changing from one kind of unit to another.

Time. The unit of time is the *second* defined in terms of astronomical measurements involving the average length of time from noon until noon the following day; thus a mean solar day is 24 hours, or 1440 minutes, or 86,400 seconds.

Weight. In defining a unit for matter, it is necessary to distinguish between *mass* and *weight*. *Mass* is the *amount* of substance; *weight* involves the downward pull on that substance exerted by the force of gravity, the attraction between the substance and the mass of the earth. One gram of mass is the same amount of substance anywhere on earth; but the *weight* of this one gram of substance depends on where it is placed. At sea level in New York City it will have one value; at Paris it will have another; on top of Mt. Everest it will have a smaller value because it is farther away from the center of mass of the earth and the pull of gravity on it is less.

We can make a secondary reference block of metal with exactly one kilogram of mass by comparing its weight with that of the primary standard kilogram. If the same force of gravity acts on the primary standard and on the secondary block, we know that they are equal in mass. The secondary block may be taken to some other location where it weighs more or less, but any third block that balances it on a scale will also have the same *mass* as the primary block.

In practice, we determine the mass of an amount of substance by the use of a device called the balance. This device is essentially a finely machined see-saw. On the one side we place the substance of unknown mass and then we add pieces of metals of known mass to the other side. When the instrument is in balance, the *weight* of the substance on the one side is equal to that of the pieces of metal on the other. Since at a given place on the earth the weight is the same for all substances of the same mass, the *mass* of the substance is equal to the mass of the pieces of metal. Such reference masses are called *weights*.

Speed and velocity. We say that the pingpong ball has a *speed* of one hundred centimeters per second as it is thrown toward the window, for speed is the time-rate of motion, the distance covered in unit time. If, in one second, the ball travels one hundred centimeters, the speed or rate of motion is 100 cm per sec or 100 cm/sec.

As the pingpong ball bounces back off the window and flies back at us, it still has approximately the same speed.† On the other hand, if we say that the *velocity* was one hundred centimeters per second (100 cm/sec) before hitting the window, we must take the value of the *returning* velocity as *minus* one hundred centimeters per second (−100 cm/sec), because velocity is defined both in terms of rate of distance covered in unit time *and* direction of motion. After the collision with the window, the ball is traveling in the *opposite* direction, so the velocity must have the *opposite* sign.

Quantities like speed and mass that bear no relation to any direction are known as *scalar* quantities, while quantities like velocity are known as *vector* quantities

† In all discussions of moving bodies we shall ignore the effects of friction.

because they refer to a property with a definite direction; the algebraic sign of a vector quantity is plus or minus depending on the direction.

Acceleration. Frequently we shall be concerned with changing velocity. For example, an electron moving toward the bare nucleus of a hydrogen atom (a proton) approaches it with an increasing speed. The charge of positive electricity on the proton attracts the negatively charged electron and causes it to move faster as the two particles get closer together. This is called accelerated motion. In a similar way a baseball, tossed from your hand as you stand on the top balcony of the Empire State Building, will fall downward faster and faster. If it hit someone on the head only a yard below your hand, it would be painful; but, if it hit the head of a pedestrian on the sidewalk below, it would be fatal because of the increased velocity.

The rate of the increase of velocity with time is called *acceleration*. With the baseball, the increase in velocity is due to the downward pull or force of gravity acting on the ball. If you could measure the rate at which the ball falls you would find that it was falling at the rate of 980 cm/sec (centimeters per second) at the end of the first second of fall and at 1960 cm/sec after two seconds. The acceleration or rate of change of velocity per second is therefore 980 cm/sec. Thus $v = at$, where v is velocity, a is constant acceleration, and t is the time during which the acceleration is effective. In this instance the acceleration is constant as the ball falls toward the ground. For a meteor approaching the earth from outer space, the acceleration increases as the meteor comes closer because the attraction of gravity increases as the distance between the two objects gets less. For an electron approaching the proton, the acceleration also increases as the distance of separation decreases. Acceleration, like velocity, is a vector quantity because it refers to increase of velocity in a definite *direction*.

Momentum. Consider another experiment with moving objects on the icy surface of a frozen pond. The ice is slippery, so slippery that a stone given a push along the surface will slide for many yards before stopping. So we assume that in studying the sliding motion over a few feet we can neglect any slowing down due to friction. In other words, we assume that the surface of the ice is an *ideal* surface with no significant friction. This is the assumption of ideal conditions that we shall make many times in analyzing the principle illustrated in an experiment.

We now take a cubical block of wood with a mass of 1000 gm or 1.000 kilogram (kg), and a block of stone with a mass of 5000 gm or 5.000 kg. We slide them toward each other so that each is traveling at the same rate of 100.0 cm/sec, though in opposite directions; and we observe what happens when the two blocks meet in a head-on collision. As you might expect, the block of stone keeps right on moving after the collision, but more slowly; the block of wood bounces and starts sliding back toward the place it originally came from.

The quantitative change in motion can be calculated from the law of the conservation of momentum combined with the law of the conservation of energy. We shall discuss first the nature of *momentum,* customarily denoted by the symbol p. This quantity is defined as the product of velocity and mass, mv. Thus, before the collision, the momentum p_s of the block of stone is $m_s v_s$ or 5000 gm × 100.0 cm/sec = 5.000 × 10^5 gm cm/sec, and the momentum p_w of the block of wood is $m_w v_w$ or 1000 gm × (−100.0 cm/sec) = −1.000 × 10^5 gm cm/sec, where we take the direction of travel of the wood as negative. The subscripts s and w refer respectively to the stone and to the wood. Thus it appears that momentum is a vector quantity, since it is the product of a scalar (mass) and a vector (velocity); and the product of a scalar and a vector is always a vector.

As we shall show in detail when we conclude the discussion of energy, the complete relation of the momenta before and after the collision is given by the following equations where the quantities denoting the values after the collision are marked with primes (′).

Before Collision				After Collision		
$m_s v_s$	$+$	$m_w v_w$	$=$	$m_s v_s'$	$+$	$m_w v_w'$
$5000\ \text{gm}\ 100.0\ \frac{\text{cm}}{\text{sec}}$	$+$	$1000\ \text{gm}\left(-100.0\ \frac{\text{cm}}{\text{sec}}\right)$	$=$	$5000\ \text{gm}\ 33.40\ \frac{\text{cm}}{\text{sec}}$	$+$	$1000\ \text{gm}\left(+233.0\ \frac{\text{cm}}{\text{sec}}\right)$
p_s	$+$	p_w	$=$	p_s'	$+$	p_w'
$5.000 \times 10^5\ \frac{\text{gm cm}}{\text{sec}}$	$+$	$\left(-1.000 \times 10^5\ \frac{\text{gm cm}}{\text{sec}}\right)$	$=$	$1.670 \times 10^5\ \frac{\text{gm cm}}{\text{sec}}$	$+$	$2.330 \times 10^5\ \frac{\text{gm cm}}{\text{sec}}$
$4.000 \times 10^5\ \frac{\text{gm cm}}{\text{sec}}$			$=$	$4.000 \times 10^5\ \frac{\text{gm cm}}{\text{sec}}$		

As the last equation shows, the value of the momentum before the collision is the same as the value after the collision, i.e., the momentum remains constant and is *conserved.*

Kinetic energy. A brass ball having the mass of one gram will be about the size of a pea. If you gently toss this small brass ball at a window with the velocity of one centimeter per second the energy will be

$$E_k = \tfrac{1}{2}mv^2 = \tfrac{1}{2}(1\ \text{gm})\ (1\ \text{cm/sec})^2$$
$$E_k = \tfrac{1}{2}\ \text{gm cm}^2/\text{sec}^2 = \tfrac{1}{2}(\text{cgs unit of energy}) = \tfrac{1}{2}\ \text{erg}$$

This unit of energy with the dimensions of gm cm^2/sec^2 is given the name *erg.* It is a very small unit. The energy given out in the combustion of one small drop of gasoline is about 10^8 ergs. If you wished to produce this same amount of kinetic energy by tossing a brass ball at one cm/sec, the ball would have to be 50 feet in diameter and would weigh 220 tons. This shows the surprising magnitude of the chemical energy available in a combustible fluid like gasoline.

Because the erg is such a small unit, it is replaced in many calculations by the unit called a *joule,* equal to 10^7 ergs. Another common unit of energy is the *calorie,* defined as 4.184 joules, and equivalent to the amount of heat needed to raise the temperature of one gram of water by 1°C.

Potential energy. How can all this energy be hidden in a drop of gasoline? We do not see any particles in the gasoline exhibiting motion like that of the ball. We conclude that the energy is *stored* somehow in the atoms. The nature of this *stored* or *potential* energy is illustrated by a baseball tossed vertically in the air.

Throw a ball weighing 500.0 gm upward with a velocity of 9.800 meters/sec. It rises in the air, moving more and more slowly; just one second after it was thrown, it stops its upward motion at a height of 4.900 meters and then falls faster and faster and returns to your hand again. As it reaches your hand it is traveling downward with a velocity of 9.800 meters/sec, just the same speed with which you initially impelled it upward. This flight is shown in Fig. 1.

Calculating the kinetic energy E_k of the ball, we find that as the ball leaves your hand it has 2.401×10^8 ergs.

$$\begin{aligned} E_k &= \tfrac{1}{2}mv^2 = \tfrac{1}{2} \times 500.0\ \text{gm}(980.0\ \text{cm/sec})^2 \\ &= 2.401 \times 10^8\ \text{ergs} \end{aligned}$$

At the top of its flight, the motion ceases, the velocity is zero, and the ball has no kinetic energy. When it returns to your hand the velocity is -980.0 cm/sec, so that the value of E_k is again

$$\begin{aligned} E_k &= \tfrac{1}{2}mv^2 = \tfrac{1}{2} \times 500.0(-980.0\ \text{cm/sec})^2 \\ &= \tfrac{1}{2} \times 500.0 \times 960.4 \times 10^3 = 2.401 \times 10^8\ \text{ergs} \end{aligned}$$

We know that the velocity gets less in the upward flight because the ball is rising against the force of gravity. We say that the ball does *work* against this force and thereby the energy is transformed from kinetic E_k to potential energy E_p.

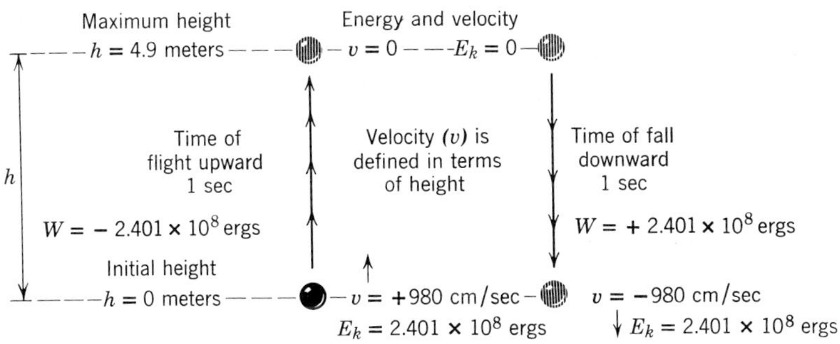

Fig. 1. A ball weighing 500 gm tossed into air and falling back to ground.

Thus the visible kinetic energy of motion is changed into the *invisible* potential energy stored in the gravitational field. When motion ceases, the force of the downward pull is still operating, and its action, as the ball falls, increases the speed so that the ball regains all its kinetic energy when it returns to your hand. The invisible potential energy in the gravitational field is transformed again into the visible kinetic energy of motion.

Work, force, and pressure. When a mass like the ball moves to a higher position against the force of gravity, work is done against this force, and potential energy is gained. The force of gravity under standard conditions at sea level is such that a body will attain a velocity of -980.0 cm/sec in one second of free *fall;* and in each subsequent second of free *fall* the body gains an additional velocity of -980.0 cm/sec; therefore the acceleration of gravity is -980.0 cm/sec per sec. The force F acting on the body is equal to its mass m multiplied by the acceleration a; that is, $F = ma$. Therefore the force on the ball is $-500.0 \text{ gm} \times 980.0 \text{ cm/sec}^2$ or $-4.900 \times 10^5 \text{ gm cm/sec}^2$. The unit of force, 1 gm cm sec^{-2}, is called the *dyne;* a force of one dyne will accelerate a mass of one gram by *1 cm per sec* per sec. As the force of gravity acts on the ball through its climb of a distance d of 490.0 cm, the work W done is the negative of the increase in E_p;

$$W = F \times d$$
$$W = -4.900 \times 10^5 \text{ dynes} \times 490.0 \text{ cm}$$
$$W = -2.401 \times 10^8 \text{ dyne cm} = -2.401 \times 10^8 \text{ ergs};\ E_p = +2.401 \times 10^8 \text{ ergs}$$

As a second example of potential energy, imagine a baseball attached to the end of a long spiral spring as shown in Fig. 2; this is a more vivid example of stored energy where the mechanism for storing the energy can be seen in action. The outward-moving ball stretches the spring, and the force of the spring slows the ball down in the same way that the force of gravity slowed it down when it was thrown vertically. Its motion ceases momentarily, and the force of the spring then pulls it back to your hand; and, when you catch it, it is traveling toward you with the same speed that it had when you threw it away from you. Thus the pattern of motion is much the same as when the ball was thrown vertically and returned. However, there is a difference because the force of gravity remains essentially constant throughout the entire sequence of motion while the force of the spring increases as a result of the stretching as the ball goes out, and decreases as the stretching is relaxed when the ball returns. In this experiment, kinetic energy visible as motion is transformed into potential energy visible as the stretched spring and then transformed again into kinetic energy visible as the return motion.

This type of action is especially interesting to us in our study of chemistry because chemical bonds act much like springs, and the electric fields around electrons bear a close resemblance to gravitational fields. For example, the heat

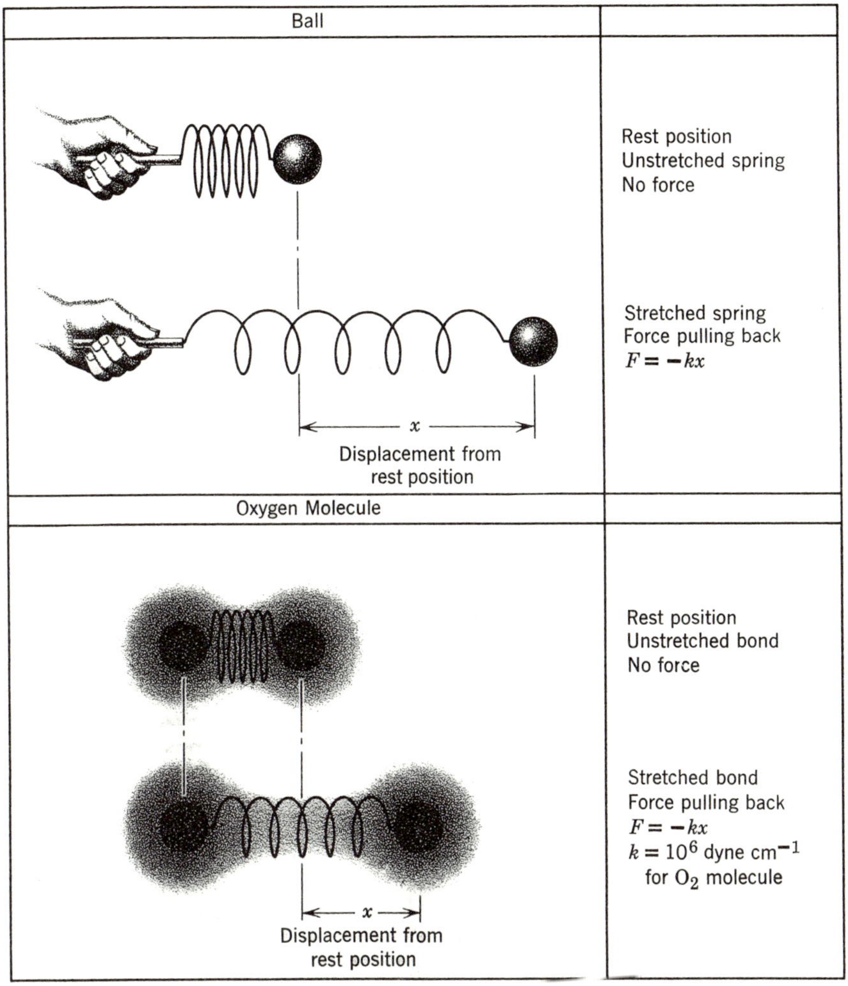

Fig. 2.

present in an oxygen molecule makes the two atoms move out away from each other and back toward each other, stretching and compressing the spring-like chemical bond. This action is shown also in Fig. 2.

The formulas for the force in a gravitational field and for the stretched-"spring"-chemical-bond express these relations precisely. For the gravitational field:

$$F = \text{constant}$$
$$W = F \times d$$

Because the force of gravity remains essentially constant, we can multiply it by the distance to calculate the work. In the case of the spring or bond, the force is not constant but increases as the spring is stretched.

Imagine an oxygen molecule lying still on a flat surface. The two atoms of oxygen lie touching one another so that the spring-bond between them is not stretched and therefore exerts no force. We now slowly pull the oxygen atoms apart. As the atoms move apart and the bond between them stretches like a spiral spring, the force F between the atoms increases from zero and is proportional to the distance x by which the atoms have been moved from their original position. This is expressed by the equation

$$F = -kx$$

where k is the proportionality constant that is a measure of the elasticity of the bond. The stronger the bond, the larger is the value of k.

If we move the atoms from their original position, where the force is zero, to a distance of separation x, the average force F_{av} during the change is given by taking the sum of the initial force and the final force and dividing by 2:

$$F_{av} = \tfrac{1}{2}(F_{initial} + F_{final})$$
$$F_{av} = \tfrac{1}{2}[0 + (-kx)]$$
$$F_{av} = -\tfrac{1}{2}kx$$

Thus the work done as the chemical bond is stretched can be expressed as

$$W = F_{av}x$$
$$W = (-\tfrac{1}{2}kx)x$$
$$W = -\tfrac{1}{2}kx^2$$

As another example of potential energy, consider what happens when air is compressed in a bicycle pump as shown in Fig. 3. Suppose that the tube outlet, normally attached to the tire, is stopped up so that no air can escape. Push the handle down and you feel the piston resist your push more and more as you compress the air in the cylinder of the pump. In this instance it is the impact of the individual gas molecules against the bottom of the piston that resists your push. The molecules contain heat; this heat is really energy of motion. When the gas is compressed, the molecules are crowded more closely together, more of them hit the piston and their push up is therefore greater as the piston is lowered.

It is customary to express this force exerted by the impact of the gas molecules as *pressure.* Pressure is the force *per unit area* exerted by these molecules. When air is at the normal pressure of the atmosphere that surrounds us, it exerts a pressure of one standard *atmosphere* (one atm); in cgs units this is a pressure of 1,013,250 dynes per square centimeter. An ordinary bicycle pump has a piston with a cross-sectional area of about 10.00 square centimeters (10.00 cm^2). The

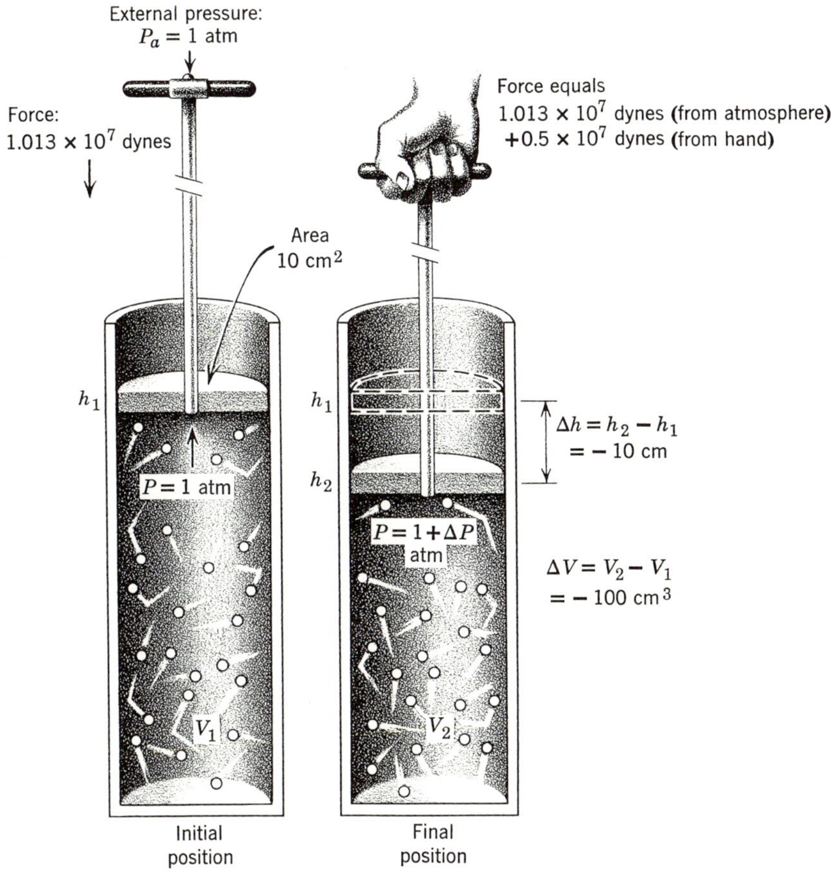

Fig. 3. Forced compression of gas.

force F, pushing down on the piston due to the pressure (P_a) of the atmosphere above it, is thus the pressure P_a multiplied by the area A:

$$F = P_a \times A$$
$$1.013 \times 10^7 \text{ dynes} = 1.013 \times 10^6 \text{ dynes/cm}^2 \times 10.00 \text{ cm}^2$$

Before you push it down, the piston has pressure exerted on it by the air above it but this is balanced by the air inside the cylinder (neglecting the weight of the piston).

We now introduce a notation that will be useful in studying the changes, like the change in volume when a piston is pushed down as shown in Fig. 3. We use the symbol Δ (delta, the Greek capital letter corresponding to our English D) in designating this change in the height of the piston.

This change is always expressed as the difference between the final value at the end of the process (h_f) and the initial value at the beginning of the process (h_i) written as $h_f - h_i$. If we push the piston down, h_f is less than h_i and Δh will be a negative quantity. If the piston is rising, Δh will be positive. For the same reason, in compression the change in volume, ΔV, is negative and in expansion ΔV is positive. Expressing such a relation as an equation,

$$\Delta h = h_f - h_i$$

where h_i is the initial height on the piston at the beginning of the experiment and h_f is the final lower height at the end of the experiment. Note that Δ is not something that is multiplied by the symbol h; Δh should be read "the change in h," and ΔV should be read "the change in V."

In experiments with gases we are more interested in pressure and volume changes than in force and distance changes. We therefore convert our formulas to these new factors. We can express the volume V of the cylinder as equal to its height h multiplied by the cross-sectional area A of the piston:

$$V = A \times h$$

We express the conversion from force and distance-change to pressure and volume-change by writing first

$$-F_{av} \times \Delta h = W$$

Then, dividing F_{av} by A and multiplying h by A,

$$-\frac{(F_{av})}{A} A(\Delta h) = W$$

This gives us

$$-P_{av}\,\Delta V = W$$

Heat and temperature. If you push the piston of the pump down rapidly and hold your hand on the cylinder, you will notice that momentarily the cylinder gets hotter; if you let the piston move back up, the cylinder gets colder. You may reason correctly that in pushing the piston down you have transformed the work done in the downward push into heat energy; but you might also conclude that heat is just another form of potential energy and this is not strictly true; although heat is a kind of stored energy, the transformation into storage and out again is subject to certain restrictions.

In order to see the nature of heat more clearly we must first take a look at temperature. The most common instrument for measuring temperature is a mercury thermometer. As temperature goes up, the thread in the thermometer goes up, because, as temperature increases, the liquid mercury in the bulb expands faster than the glass and so part of the mercury is forced into the capillary in the upper part of the thermometer and the mercury level rises. To establish the centigrade temperature scale, the one most commonly used in scientific work, we mark the point on the thermometer where the top of the mercury thread rests when the bulb is in a mixture of ice and water, and we call this the zero point of 0° C. We next place the bulb in water boiling at 1 atm pressure and mark this height 100° C. The distance between the two marks is divided into one hundred parts each of which represents a degree of temperature.†

Temperature difference indicates the direction of heat flow; heat always flows from a place at a higher temperature to a place at a lower temperature. This is the most significant meaning of temperature. It is the index or parameter that shows which way heat will flow; but it is also a measure of the amount of internal heat in a body, as we shall see later.

A cylinder with a piston, like that shown in Fig. 4, also behaves much like a thermometer. Let us discuss such a cylinder made of glass equipped with a gas-tight piston and having an area of 10.00 cm^2; on the piston there rests a weight so that the total downward force on the piston is the equivalent of 10.33 kg. Above the cylinder the space is evacuated so that the force on the cylinder is due entirely to the weight. In the space below the piston we have helium gas. Under these conditions this gas will be under a pressure of 1.000 atm at all times. This constitutes a constant-pressure gas thermometer.

We place the cylinder first in an ice and water mixture and put a 0° mark at the place where the piston comes to rest. We then place the cylinder in water boiling under 1.000 atm pressure and label the point where the piston comes to rest as 100.0° C. We then divide the distance between these points into a hundred intervals, each equivalent to a degree of centigrade temperature. If we compare the readings of our mercury thermometer with this gas thermometer we find that they closely coincide. Actually the gas thermometer is the standard for establishing the scientific centigrade temperature scale, but the mercury thermometer is more convenient for most purposes.

We now cool the gas thermometer farther and farther below 0° C and the piston drops down, finally approaching the bottom of the cylinder. What would be the meaning if the piston finally reached the bottom? Of course the volume of the gas would then be zero. This would never happen with any real gas like helium because it would always condense to a liquid with a small though appreciable volume that would never disappear no matter how low we made the temperature. But we can imagine an *ideal* gas that has no force of attraction acting between molecules that are themselves of negligible size. If such a gas were cooled sufficiently, the

† On the Fahrenheit scale the freezing point of ice is 32° and the boiling point of water is 212° C. Thus a 100° C change corresponds to a 180° F change and 1° C is $\frac{9}{5}$° F. Conversion from one scale to another is simple on this basis. Thus +20° C is 20 centigrade units above the freezing point (f. p.) of ice. It is $20 \times \frac{9}{5} = 36$° F above the freezing point (+32° F) or +68° F.

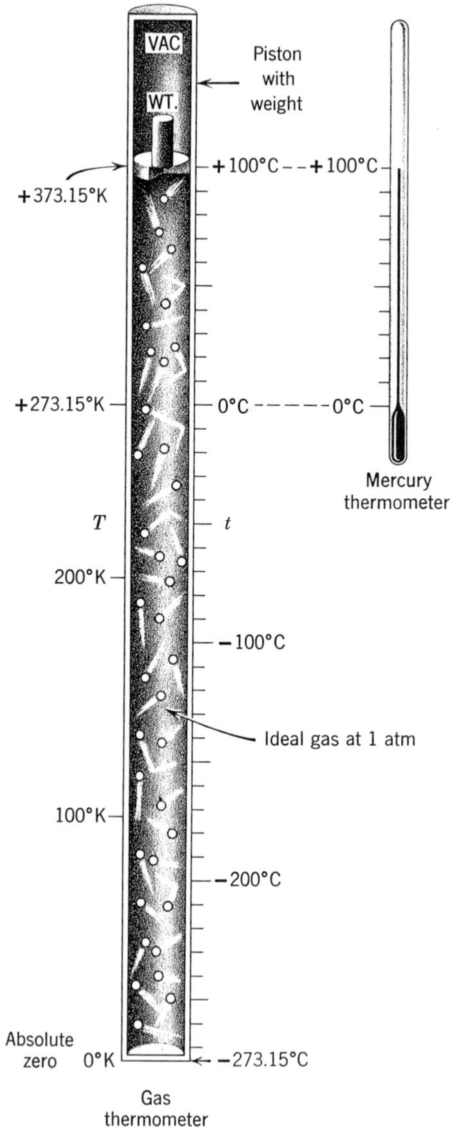

Fig. 4. Temperature scales.

piston would finally come to rest on the bottom. This means that the volume of the gas is zero, and that the pressure is zero. We have abstracted all the heat which gave the molecules motion; they are therefore motionless and still; and, since nothing can be stiller than being stopped, this represents the bottom of the temperature scale. Lord Kelvin suggested that this is a natural zero point for the temperature scale and set up an absolute scale of temperature with this point as zero. If on the cylinder of the gas thermometer we measure the distance from the ice point to the bottom of the cylinder, we find that the bottom is 273.15°C below the freezing point of water. In other words, the absolute zero is at −273.15°C. If we designate Kelvin's scale by using the letter K, then on this absolute scale the freezing point of water is at +273.15°K. The relation between temperature on the centigrade scale (t) and on the Kelvin scale (T) is thus

$$t + 273.15° = T$$

As will be shown later, the absolute scale of temperature derives its primary significance and usefulness from the laws of thermodynamics that deal with the interconvertibility of heat and work; but this absolute scale is also important because of its relation to the behavior of gases. It is found that the pressure, volume, and temperature of a *mole* of atoms (a standard "package" containing 6.0229×10^{23} atoms) are related by the equation

$$PV = RT$$

If pressure P is in atmospheres, volume V is in cubic centimeters, and temperature T is in absolute degrees (°K), then R is a constant called the *gas constant* with a numerical value of 82.056 cm^3 atm deg^{-1}. This equation will be used in studying the relation of energy to chemistry and will be discussed in detail in Chapter 8.

Heat and mechanical energy. As we have just seen in the experiments with the bicycle pump, it is possible to convert energy derived from a force doing work into energy in the form of heat. Another way to do this is to have a force that rotates paddle wheels in water so that the mechanical energy of the force is transformed into heat in the water. This heat energy produces many effects in the atoms into which it enters; they move faster, they may stretch their chemical bonds; ice can be made to melt, and water can be changed into steam. For the moment we are primarily interested in the amount of heat it takes to heat one gram of water from 14.5 to 15.5°C. This amount of heat is called a *calorie* (15°). For our purposes we can take this calorie as equal to 4.184×10^7 ergs. In order to avoid writing the power of 10 when dealing with quantities of heat of this large size, a unit of heat equal to 10^7 ergs is called a *joule*. Thus 1 calorie = 4.184 joules.

CONSERVATION LAWS

Conservation of mass. If a ball of paper is sealed inside a glass bulb filled with oxygen and an electric spark is induced inside the bulb, the paper will burn to ashes and produce the gas carbon dioxide (CO_2) and water vapor (H_2O) together with small amounts of other gases. As long as the bulb is gas-tight, however, one can weigh the bulb with paper and gas before combustion, and the bulb with ash and new gas content after combustion, and the weight before and after is always found to be precisely the same. No matter how many variations with other substances and other chemical reactions are tried, the transformation of atoms from one combination to another combination never results in any chemically significant gain or loss of mass. Therefore we deduce the important law that mass never can be created or destroyed by any chemical or physical process. Some nuclear processes seem to deviate from this law, but at the moment this does not concern us.†

This law is an example of a conservation principle, and it is called the principle of *conservation of mass.* Because *mass* is generally visible, tangible, or directly measurable in one way or another, this principle has been readily accepted by all people except those who still believe in magic.

Conservation of energy. Exactly the same principle applies to *energy.* Like mass, energy can never be created or destroyed. It took scientists longer to discover the energy conservation principle, because there is no way of determining energy directly and unequivocally in the way in which mass is determined by weighing. But thousands of experiments have demonstrated indubitably that no matter how energy is transformed from kinetic to potential to chemical or any other form, back and forth, round and round, energy always behaves like an indestructible fluid; you always end up having exactly the same amount with which you started, provided that you have kept it in a closed system like the paper in the glass bulb.

In the example of the ball tossed into the air we were dealing with a *closed system;* i.e., there was no way for energy to be added to the ball or taken away from the ball after you tossed it and it left your hand. (We specified that air friction was to be neglected.) Therefore, when the kinetic energy of motion disappeared, it must have been changed into some other kind of energy still present though invisible. We found this to be the potential or stored energy of the ball due to its increased height in the gravitational field. When the ball reached the top of its flight and started to fall again, this invisible potential energy changed back into visible kinetic energy of motion. But throughout the entire flight of the ball the sum of the kinetic and potential energy remained constant, even though part of the first was transformed into the second and vice versa. No energy was created or destroyed.

It is this energy conservation law that makes the study of chemical energy so important. The branch of chemistry primarily concerned with energy is gener-

† Einstein's relation, $E = mc^2$, tells us that in such a reaction a small amount of mass appears to be transformed into energy. In ordinary chemical reactions the amount of mass that apparently disappears is so small that it cannot be detected.

ally called *chemical thermodynamics,* and it establishes the logical quantitative connections between chemical changes and energy changes.

As a numerical illustration of the principle of conservation of energy, we take the example of the block of stone and block of wood sliding toward each other on the ice, colliding and exchanging momentum and energy. Just as the conservation of momentum principle requires that the sum of the momenta of the two blocks before and after collision must be the same, so the conservation of energy principle requires that the sum of the energies of the two blocks before and after collision must be the same. These relations are expressed in the following equations.

A. Conservation of Momentum

Before collision: After collision:

$$m_s v_s + m_w v_w = p_{\text{total}} = m_s v_s' + m_w v_w'$$

B. Conservation of Energy

Before collision: After collision:

$$\tfrac{1}{2} m_s v_s^2 + \tfrac{1}{2} m_w v_w^2 = E_{\text{total}} = \tfrac{1}{2} m_s (v_s')^2 + \tfrac{1}{2} m_w (v_w')^2$$

As before, we select the following values for these quantities:

m_s = mass of stone = 5000 gm
m_w = mass of wood = 1000 gm
v_s = velocity of stone before collision = 100.0 cm sec^{-1}
v_w = velocity of wood before collision = −100.0 cm sec^{-1}

Then it is simple to calculate the values

p_{total} = total momentum = 4.000×10^5 gm cm sec^{-1}
E_{total} = total energy = 3.000×10^7 ergs = 3.000 joules

This leaves two unknowns in these equations, the values of the velocities v_s' and v_w' after the collision takes place. Since there are two equations and two unknowns, the equations can be solved algebraically, and we find that the velocities after the collision are $v_s' = 33.40$ cm sec^{-1} and $v_w' = 233.0$ cm sec^{-1}.

Pictorially we can represent the result of this collision as shown in the diagram on p. 36. Thus, before the collision the two blocks are sliding toward each other with the same speed but in opposite directions; after the collision the block of wood, having bounced off the *moving* block of stone, picks up both energy and momentum and speeds back in the opposite direction, moving over twice as fast as it did originally, while the block of stone continues in the same direction but moving only about one-third as fast as it did originally, having lost both momentum and energy to the wood.

Collisions of this sort between atoms and molecules are taking place constantly all around us in the atmosphere and even inside our bodies. Except in a few

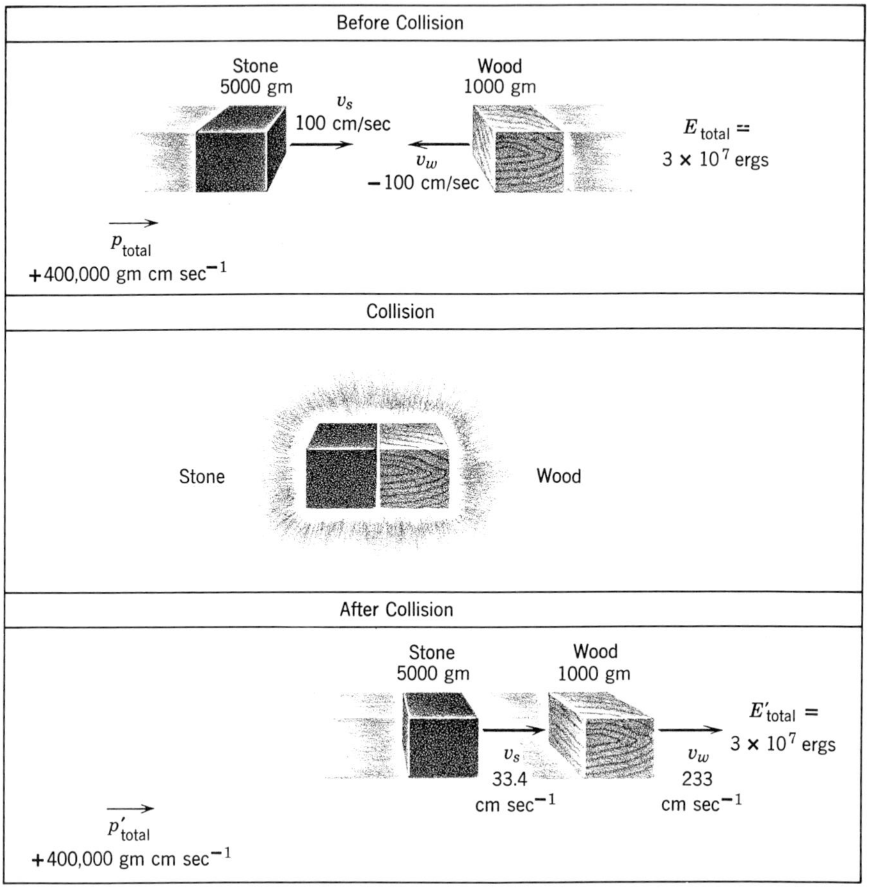

cases of theoretical states of very special systems at 0°K, there is a constant exchange of energy and momentum taking place every moment of time between all adjacent atoms, whether these happen to be existing as gases, liquids, or solids. This is the reason why it is so important to attain a clear understanding of the nature of momentum and energy and to acquire facility in making calculations involving these quantities.

In a collision between a real block of wood and a real block of stone some of the energy is always transformed into heat because of the imperfect elasticity of these substances. On the other hand, there are many collisions between individual atoms and molecules where no kinetic energy is transformed into heat because these particles are actually perfectly elastic. Such collisions are called *elastic collisions.*

GRAPHS AND EQUATIONS

In order to make clear many quantitative relations like those we have just discussed in connection with energy, we shall frequently use mathematical pictures called *graphs,* since pictures often enable us to see at a glance the nature of these mathematical relations or *equations.* These graphs and equations are an important part of the *language* of chemistry. You may feel that a discussion of these topics is a detour from the direct road to an understanding of chemistry. But, just as you have to spend many days learning the vocabulary of French words and the rules of French grammar before you can read, write, and speak the French language, it is essential to have a good working knowledge of graphs and equations before you attempt to get a real understanding of chemistry.

Linear equations. As a first example, consider the ideal gas thermometer made from the glass bicycle pump. We fill this pump with the molecular-standard number of atoms; then the pressure, volume, and temperature are related by the equation

$$PV = RT$$

where R is a constant. At the moment we are only interested in the mathematical fact that

$$V = (R/P)T$$

In our thermometer, the pressure is kept constant by the weight on the piston; so the ratio R/P is also a constant quantity and we shall designate it by the single letter B. Thus

$$R/P = (82.06\ \text{cm}^3\ \text{atm/deg})/(1.0000\ \text{atm}) = B$$

and

$$V = BT = (82.06\ \text{cm}^3/\text{deg})\,T$$

Here we round off the value of R to four significant figures.

We can thus say that V is a *function* of T. This means that for every value of T there will be a distinct value of V. Suppose that we want to know the values of the volume at precisely $T = 100°, 200°, 300°, 400°$, and $500°$. A table of these values prepared from our formula is given in Table 1. We construct a graph of this function by making a horizontal scale to show temperature and a vertical scale to show volume. Our table tells us that at 100.00°K the volume is 8206 cm^3, at 200.00°K the volume is 16,412 cm^3, and so on. Above the point on the temperature scale at 100.00° we place a point at the height corresponding to 8206 cm^3; above the point at 200.00° we place a point at the height corresponding to 16,412 cm^3, and so on. This is shown in Fig. 5*a*.

Table 1

T (°K)	V (cm^3)
0.000	0.000
100.00	8.206×10^3
200.00	16.412×10^3
300.00	24.618×10^3
400.00	32.82×10^3
500.00	41.03×10^3

If we place a ruler on the diagram we find that we can draw a straight line that passes through all these points. Thus the *picture* or *graph* of the mathematical relation between volume and temperature in our ideal gas thermometer is a straight line which passes through the zero point for the two scales where the horizontal and vertical lines intersect.

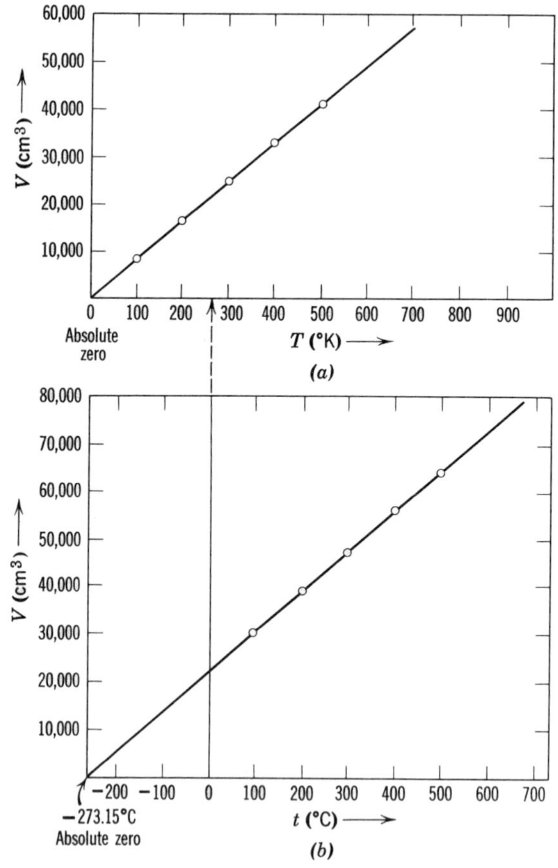

Fig. 5. Graphs of temperature scales. (*a*) Absolute (Kelvin); (*b*) centigrade.

The equation just quoted is an example of a general class of equations of the type

$$y = bx$$

x is called the independent variable because we select a value of x to try out our function; y is called the dependent variable because its value depends on the value of x selected. If in Fig. 5*a* we select 100° for x, then we find that y must have the value of 8206 cm^3. In general, we plot x horizontally and y vertically and have the intersection represent zero on both scales. This common zero point is called the *origin*.

Let us examine next the relation between V and temperature expressed not in degrees absolute T, but in degrees centigrade t. Since we have the relation

$$T = 273.15 + t$$

we can write

$$V = B(273.15 + t)$$

or

$$V = V_0 + Bt$$

where $V_0 = 273.15 \times B = 22{,}414$ cm^3. In other words, V_0 is the volume of our standard amount of an ideal gas at a temperature of 273.15°K or 0.00°C and a pressure of 1.0000 atm.

We can use our equation to calculate values of V for t and we get those shown in Table 2. We make a plot of these values and get a straight line again, but one which does not go through the origin and instead crosses the axis of zero temperature at the value of 22,414 cm^3 for the volume. This equation is a more general linear equation of the type

$$y = a + bx$$

This reduces to the simpler linear equation when $a = 0$. Whenever a mathematical relation is expressible by these equations, its graph will be a straight line; whenever a relation shows a straight-line graph, it is expressible by these equations.

Table 2

V (cm^3)	t (°C)	T (°K)
22,414	0.00	(273.15)
30,619	100.00	(373.15)
38,825	200.00	(473.15)
47,030	300.00	(573.15)
55,236	400.00	(673.15)
63,442	500.00	(773.15)

Logarithmic graphs. In order to make an accurate plot of a curve it is frequently necessary to calculate many points. By way of contrast, the graph of a linear equation is a straight line, and it is only necessary to know two points in order to plot a straight line. Thus there is a great advantage to be gained in transforming a curve into a straight line. Frequently this can be done with the help of logarithms.

If we take the logarithm of both sides of the gas equation

$$PV = RT$$

we get

$$\log P + \log V = \log R + \log T$$

or

$$\log P = -\log V + \log RT$$

Although the plot of P against V for a constant value of T gives a line called an *isothermal* (constant temperature), which is a hyperbola and requires the calculation of a number of points to obtain an accurate graph, the plot of log P against log V is a straight line. By plotting the two points for $V = 10.00$ and $V = 100.0$ ($\log V = 1.000$ and $\log V = 2.000$), one can readily find the corresponding values of P and draw the straight line. Figure 5*c* shows the graph of P versus V for several values of temperature. Figure 5*d* shows the same relations on a log-log plot. This kind of plot is useful in making many kinds of calculations. For example, it might take many hours to calculate directly and plot twenty isothermal *curves*

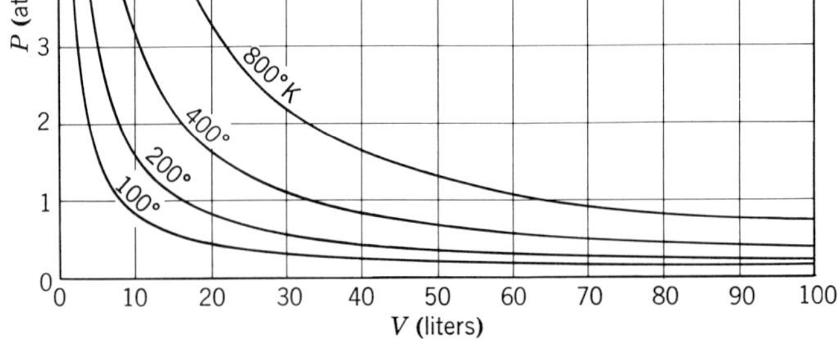

(*c*)

Fig. 5c. Plot of ideal gas equation.

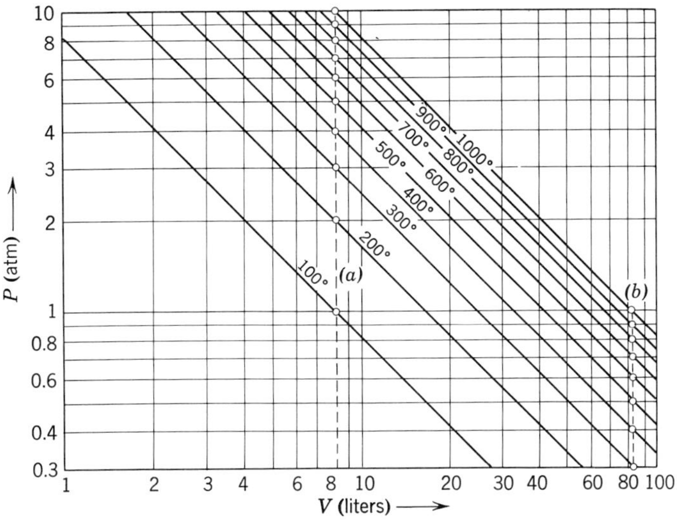

(d)

Fig. 5d. Plot of ideal gas equation on log-log scale. Where the vertical dotted line (*a*) crosses the horizontal pressure lines, we find $T = 100P$; similarly along dotted line (*b*) we find $T = 1000P$. By drawing straight lines through these intersections the temperature lines are directly established.

on a graph of P against V. By using the plot of log P against log V, the corresponding *straight lines* can be located and drawn in less than five minutes. We shall encounter many relationships where the proper use of logarithmic coordinates greatly facilitates an understanding of the mathematical structure and the computation of numerical values.

Rate. The constant quantity, B, appearing before the variable T or t as a multiplier, tells us how much the volume increases for each *unit* increase in temperature. For each increase of a degree (when $P = 1$ atm) the volume goes up by 82.06 cm^3. Figure 6 shows what a small portion of the graph would look like if we viewed it through a magnifying glass. We have an enlarged scale of temperature showing each degree and an enlarged scale of volume showing each hundred cubic centimeters. We see that, when the temperature goes up by *one* degree, the volume increases by 82.06 cm^3. This is the value of the constant B. It is called a rate because it is what we get in volume *for each* unit of temperature. If you sell candy for $2.00 a pound, you get two dollars *for each* unit (pound) of candy that you sell. $2.00 is the price or *rate* of return on the candy. Thus our constant B tells us the *rate* of increase of volume with temperature increase.

Slope. This quantity is also called the *slope* of the line on the graph. Suppose we have a road climbing a hill such that the cross section looks like the line shown

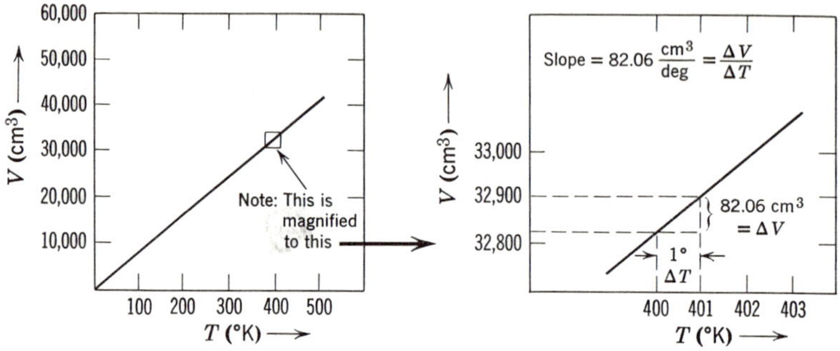

Fig. 6.

in Fig. 7. Here the slope of the road is 4 inches per foot. In the same way we say that the slope of the graph in Fig. 6 is 82.06 cm^3 per degree.

Because the slope represents a *change* in the value of the volume for a unit *change* in the temperature, we shall use frequently the delta notation, expressing the slope as $\Delta V/\Delta T$. Here $\Delta V = 82.06$ cm^3 and $\Delta t = 1.0000$ degree. It is important to note that for the plot in Fig. 6 we would get the same value for the slope if we took any increase in temperature, say 200.00°, and the corresponding increase in volume, 16,412 cm^3, and formed the ratio $\Delta V/\Delta T = 16{,}412$ cm^3/200.00°, which is again equal to 82.06 cm^3 per degree.

Another important example of a linear relation is found in the equation relating the increase in potential energy (ΔE_p) to the product of force (F) and the distance through which it acts (Δh):

$$\Delta E_p = -F(\Delta h) = -F(h_f - h_i) = -(-4.900 \times 10^5 \text{ dyne})(490.0 \text{ cm} - 0 \text{ cm})$$

In our example of the 500.0-gm ball rising against the force of gravity the increase in potential energy is given by the formula above.

Suppose that the ball starts at zero height ($h = 0$) with an upward velocity of 980 cm/sec. It will rise for one second with decreasing velocity and stop momentarily at a height of 490.0 cm or just under 5 meters. The graph of its potential energy is shown in Fig. 8. We see that the magnitude of the force is equal to $-\Delta E_p/\Delta h$, the slope or the *rate* at which work is done, and has the value -4.900×10^5 dynes.

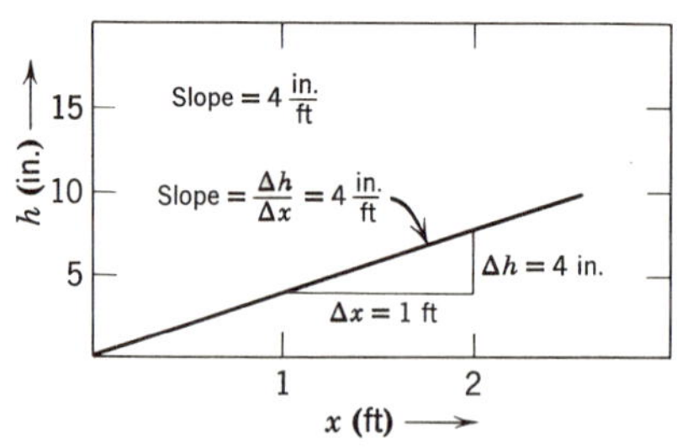

Fig. 7.

In the example of the 500.0-gm ball falling under the influence of gravity we found that the force exerted on the ball was also -4.900×10^5 dyne. The ball

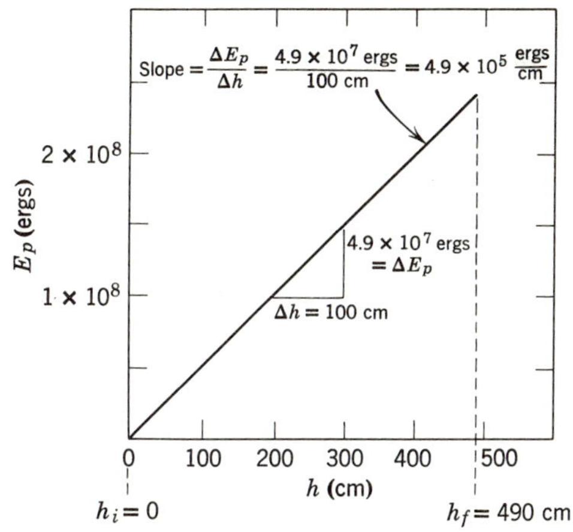

Fig. 8. Rising ball.

falls from a height of 4.900 meters, having initially a velocity of zero, and no kinetic energy. The increase in kinetic energy (ΔE_k) due to the work performed by the force of gravity (F) is given by

$$\Delta E_k = F \cdot \Delta h = F(h_f - h_i) = -4.900 \times 10^5 \text{ dyne}(0 - 490.0 \text{ cm}) = 2.401 \times 10^8 \text{ ergs}$$

where h_i is the initial height and h_f is the final height at which the ball has the energy E_k. We can plot the relation between E_k and h as shown in Fig. 9. Here the energy increases as the ball falls. F is the rate of increase of energy as the height decreases. The sign of the slope is negative because kinetic energy increases as height decreases.

Non-linear equations. While linear relations play an important role in chemistry and physics, non-linear relations are even more important. In the falling

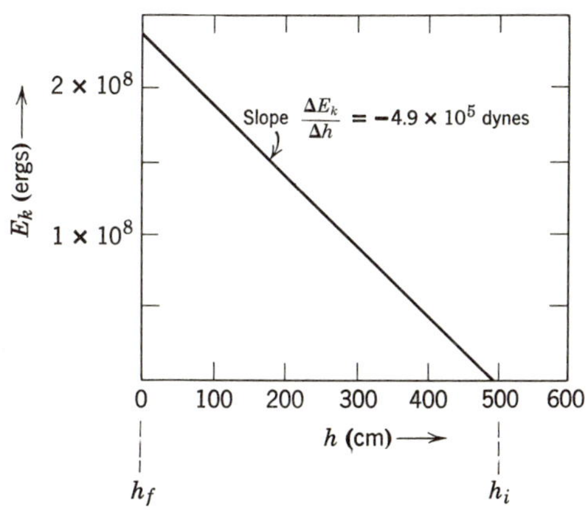

Fig. 9. Falling ball.

ball, velocity increases linearly with the time, but energy increases non-linearly with the square of time. We now calculate both quantities at 10.000 seconds after the ball starts to fall:

$$v = at = -980.0 \text{ cm/sec}^2 \times 10.000 \text{ sec} = -9800 \text{ cm/sec}$$
$$E_k = \tfrac{1}{2}mv^2 = \tfrac{1}{2}ma^2t^2$$
$$E_k = \tfrac{1}{2}(500.0 \text{ gm}) (-980.0 \text{ cm/sec}^2)^2 (10.000 \text{ sec})^2$$
$$E_k = 2.401 \times 10^{10} \text{ ergs}$$
$$E_k = 2401 \text{ joules}$$

Table 3 shows both the velocity and the energy for five values of the time, calculated as in the example above.

Plots of these quantities against time are shown in Fig. 10. The graph of velocity is a straight line while the graph of energy is a curve; i.e., it is non-linear.

We now observe the slope of each of these graphs at the point of time 10.000 seconds after the fall starts. We have seen how to get the slope of the straight line in Fig. 10*a*. It is the velocity gained in unit time. We can find it from the graphs by actually measuring it. The value is -980.0 cm/sec of velocity gained each second. Or we can get it immediately from the equation; since this is a linear equation, we know that the slope is the value of the constant a that multiplies the independent variable t, and this is -980.0 cm/sec as the equation shows on the right. This slope is the same everywhere on the straight line.

Non-linear slope. But what is the slope of the curve in Fig. 10*b?* If we observe the increase of energy in the unit time from 2 to 3 seconds, we find the energy is 96 joules at 2 sec, and 216 joules at 3 sec, an increase of 120 joules; if we observe the increase of energy from 6 sec to 7 sec, we find an increase of 312 joules; if we observe the increase in energy from 10 sec to 11 sec, we find an increase of 504 joules. Clearly the slope or steepness of our graph increases as time goes on; this is obvious just from looking at the graph. How, then, can we arrive at a clear and unambiguous meaning for slope or steepness of the curve?

To proceed, let us examine exactly what $\Delta E/\Delta t$ represents on our graph at, say, the point corresponding to 10 sec. We will make an enlarged drawing of this part

Table 3

Time	Velocity	(Time)^2	Energy
0 sec	0 cm/sec	0 sec^2	0 joules
2.000	1960	4.000	96.00
4.000	3920	16.000	380.00
6.000	5880	36.000	864.00
8.000	7840	64.00	1537.10
10.000	9800	100.00	2401.0

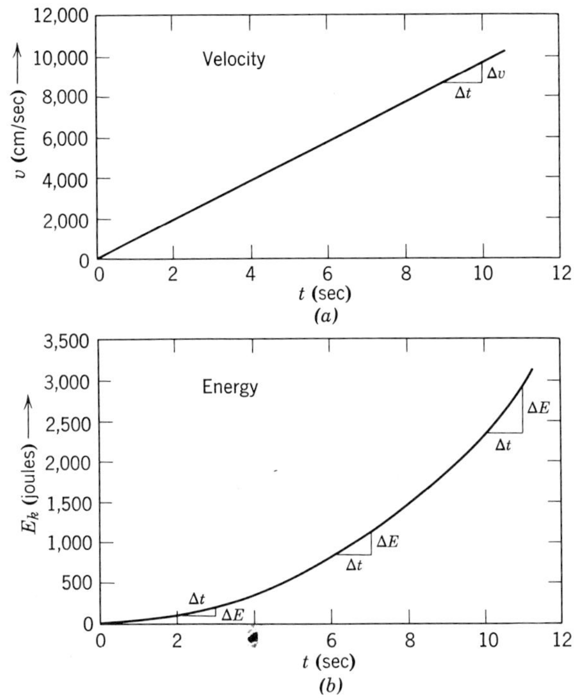

Fig. 10.

of the curve as shown in Fig. 11. If we draw a straight line through the point representing the value of the energy at 10 sec and the point representing the value of the energy at 11 sec, we find that $\Delta E/\Delta t$ represents the slope of this line (a, b) (i.e., 504 joules/sec). Now, a line (c, d) parallel to this line will be tangent to the real curve at a point roughly midway between 10 sec and 11 sec as shown.

Now let us draw a graph (Fig. 12) to show the increase in energy from 10 sec after the fall begins to 10.1 sec after initial time, an interval of only one-tenth of a second instead of the unit interval of one second which we used originally. Now Δt is only 0.1 sec and, if we calculate ΔE, we find that it has an approximate value of 48 joules:

E_k at 10.1 sec	2449 joules
E_k at 10 sec	2401 joules
ΔE	48 joules

Thus $\Delta E/\Delta t \cong 48/0.1 \cong 480$ joules/sec.†

† The symbol $\sim$ or $\cong$ means "approximately equal to."

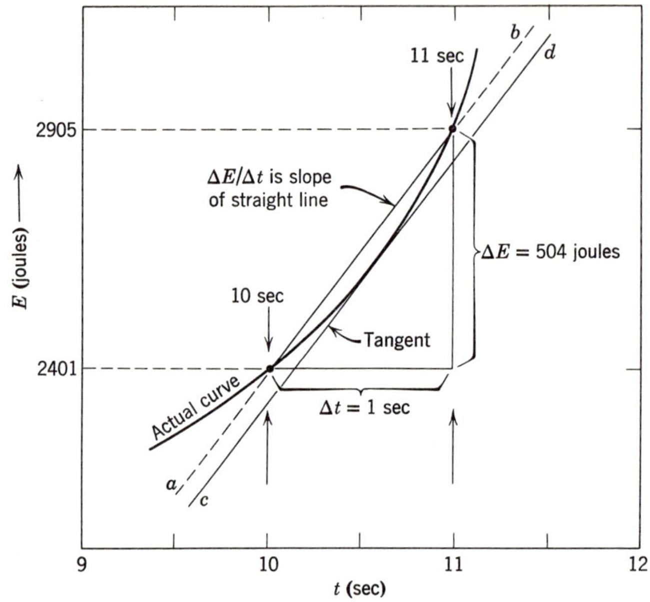

Fig. 11.

We note that on the drawing the line (a', b') passing through the points corresponding to 10 and 10.1 sec is so close to the line (c', d'), tangent to our actual curve, that we cannot distinguish between them. Mathematically we let ΔE and Δt become so small (infinitesimal) that their ratio is effectively a constant, and this ratio is the slope of our curve at the time 10 sec after fall. In other words,

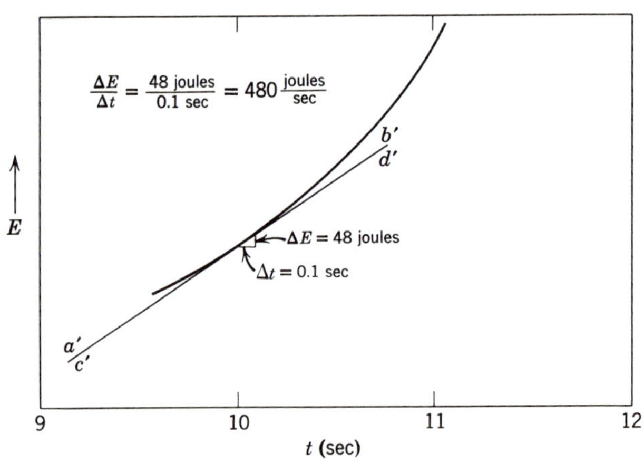

Fig. 12.

the slope of the straight line, tangent to the curve at 10 sec, is the slope of the curve itself at 10 sec.

Derivative. To show this more precisely, the values of t, Δt, E, ΔE, and $\Delta E/\Delta t$ are listed in Table 4 for decreasing values of Δt. The length to which this calculation should be carried depends on the number of digits in our answers that are significant for our purpose. This question will be discussed more in detail later. Since we are specifying the acceleration of gravity by the number *980* with three digits, then the value of $\Delta E/\Delta t$ for $\Delta t = 0.0001000$ sec has not changed appreciably from the value at $\Delta t = 0.001000$ sec, since both round off to 480.2 joules/sec. Therefore we can take the value of 480.2 joules/sec as the true value of the slope at 10 sec. For the precise value of the slope at a given point it is customary to use the symbol dE/dt, employing the small d instead of the capital delta (Δ) in order to emphasize that the increments are infinitesimally small. This is called the *derivative* of E with respect to t.

As in the use of the delta symbol (Δ), the two letters dE do *not* mean d multiplied by E; on the contrary, d means an *infinitesimal increment* of E. Thus the joint symbol dE/dt means the *ratio* of an infinitesimal increment of E to the corresponding infinitesimal increment of t; and this ratio is the slope of the graph of E against t at the point in question.

The value of dE/dt can also be arrived at analytically. Let us consider the change in E when the time changes by an amount Δt. In general,

$$E_k \text{ at } t = \tfrac{1}{2}ma^2t^2$$

At $t = t + \Delta t$ we will have

$$E_k \text{ at } (t + \Delta t) = \tfrac{1}{2}ma^2(t + \Delta t)^2 = \tfrac{1}{2}ma^2(t^2 + 2t\,\Delta t + \Delta t^2)$$

Thus we find

$$E_k \text{ at } (t + \Delta t) - E_k \text{ at } t = \Delta E_1 = \tfrac{1}{2}ma^2(2t\,\Delta t + \Delta t^2)$$

or

$$\Delta E_k/\Delta t = ma^2t + \tfrac{1}{2}ma^2\,\Delta t$$

Table 4

t (sec)	Δt (sec)	E (joules)	ΔE (joules)	$\Delta E/\Delta t$ (joules/sec)
11	1.00000	2905.00	504.00	504.000
10.1	0.100000	2449.2601	48.2601	482.601
10.01	0.0100000	2405.8044	4.8044	480.440
10.001	0.00100000	2401.480224	0.480224	480.224
10.0001	0.000100000	2401.0480202	0.0480202	480.202
10	0	2401	0	480.200 = dE/dt

When we make the interval Δt smaller and smaller, the value of $\Delta E_k/\Delta t$ approaches the value ma^2t and $\frac{1}{2}ma^2\,\Delta t$ approaches zero since Δt approaches zero. Thus

$$dE_k/dt = ma^2t$$

With the help of this formula we can determine dE/dt by substitution. Thus we find

$$\begin{aligned}(dE_k/dt)_{\text{at 10 sec}} &= ma^2t\\(dE_k/dt)_{\text{at 10 sec}} &= (500.0 \text{ gm})(980.0 \text{ cm/sec}^2)^2\ (10.00 \text{ sec})\\(dE_k/dt)_{\text{at 10 sec}} &= 480.2 \times 10^7 \text{ ergs/sec}\\(dE_k/dt)_{\text{at 10 sec}} &= 480.2 \text{ joules/sec}\end{aligned}$$

This is the same value we obtained by direct measurements on the graph.

Thus we have effectively defined the *slope* of a *curve*. If you have studied the calculus you are aware of the profound mathematical logic behind this concept of slope. Since we are at present making a study of chemistry and not of mathematics, we cannot take the time at the moment to discuss these ideas more deeply. It is to be hoped that, if you have studied the calculus, the chemical examples which follow will give you a better grasp of the meaning of these relations; if you have not studied the calculus, you will get at least a working knowledge of these relations and will acquire a sense of their importance in chemistry and physics; a deeper mathematical knowledge can be added later.

To summarize this example, we list the equations and the slopes.

$$\begin{cases} v = at \\ dv/dt = a \end{cases}$$

$$\begin{cases} E = (\frac{1}{2}ma^2)t^2 \\ dE/dt = (\frac{1}{2}ma^2)2t = ma^2t \end{cases}$$

In general algebraic terms, these equations are examples of the type-equations:

Linear:

$$\begin{aligned} y &= bx \\ dy/dx &= b \end{aligned}$$

Quadratic:

$$\begin{aligned} y &= bx^2 \\ dy/dx &= 2bx \end{aligned}$$

Constant force. One of the most important relations involving derivatives is that between force (F) and potential energy (E_p). In the example of the ball thrown directly upward, we have, as previously shown, the relations

$$\begin{aligned} \Delta E_p &= -F\,\Delta h \\ dE_p/dh &= -F \end{aligned}$$

In words, the potential energy generated equals the negative of the force times the distance through which it acts; thus the *rate* at which energy is gained as the height increases is equal to the negative of the force. Here, since the height *up* is measured as a positive quantity and the force is a downward pull, the force is given a negative value. To make this specific, if the ball weighs 500.0 gm, the force acting on it will be $-(500.0 \text{ gm} \times 980.0 \text{ dynes/gm}) = -4.900 \times 10^5$ dyne. This downward force is constant throughout the upward motion of the ball.

Variable force. If an atom is tied by a chemical bond to another atom and motion takes place that moves the atoms apart thus stretching the bond, work is done against the elastic force of the bond; at first this force increases as the bond is stretched; it resembles the force encountered when a spiral spring is stretched. This type of force, called a linear elastic force, is zero when the spring is unstretched but increases linearly as the spring is extended. Again the sign of the force is taken as negative because the force pulls back as the atom moves out.

For a stretching of less than 0.1 Å, an approximate equation for the force is

$$F = -kx$$

where x is the distance in centimeters out from the initial position where the atoms rest when the bond is not stretched; and k is the elastic constant of the force in dynes per centimeter. The bond between two hydrogen atoms in the hydrogen molecule H_2 acts like a spiral spring when the atoms are pulled apart from a rest separation of about 0.74 Å (Ångstrom unit $\equiv 10^{-8}$ cm) to 0.84 Å; here the amount of stretch (Δx) is 0.10 Å $= 1.0 \times 10^{-9}$ and the value of k is roughly 5×10^5 dynes/cm. Thus stretching the bond by 10 per cent produces the force as shown by the following equation with numerical values inserted:

$$-kx = -(5 \times 10^5 \text{ dyne/cm})\,(1.0 \times 10^{-9} \text{ cm}) = -5 \times 10^{-4} \text{ dyne}$$

To calculate the potential energy increase due to this stretching we have to take into account the fact that the force is changing during the movement. One graphical way to do this is shown in Fig. 13. When the constant force of gravity (F_g) acts as the ball moves upward through a distance Δh, the work done is the product of F_g and Δh; this is equal to the area shaded on Fig. 13*b*. If the force is not constant when the atoms are pulled apart, the work done is still equal to the shaded area. In this case the force at the beginning is zero, and at the end it is -5×10^{-4} dyne; since it changes linearly, the average force is one-half of the maximum force at the end, or -2.5×10^{-4} dyne. When this acts over a distance of 1.0×10^{-9} cm, the work done is -2.5×10^{-13} erg. Thus $-W$ or the increase in potential energy is the area equal to minus the product of the average force and the distance (x) through which it has acted. But this is exactly equal to the area of the triangle in Fig. 13*d*.

The fundamental reasoning behind these relations is again part of the calculus, but the principles are so simple in the cases we meet in chemistry that one can understand them with the aid of simple diagrams like these. To see the deeper underlying principle let us re-examine the problem of the ball thrown into the air. Imagine that we photograph the upward flight of the ball with a moving picture camera. The speed of the photography is timed so that the position of the

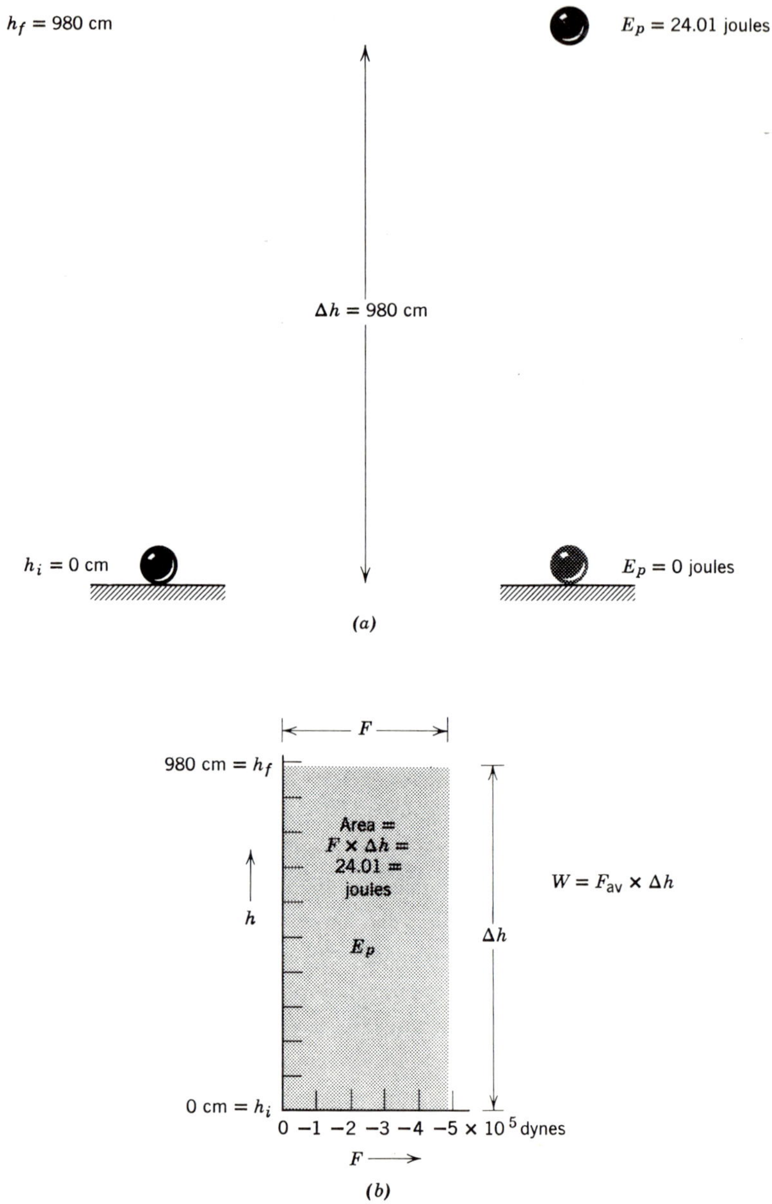

Fig. 13. Potential energy of ball tossed in air.

ball as it moves upward is photographed at equal increments of height and is shown in each of eleven frames of the cinema film strip as represented in Fig. 14. We call the frame showing the initial position of the ball frame No. 0. The frame showing the final position of the ball is frame No. 10. The total work

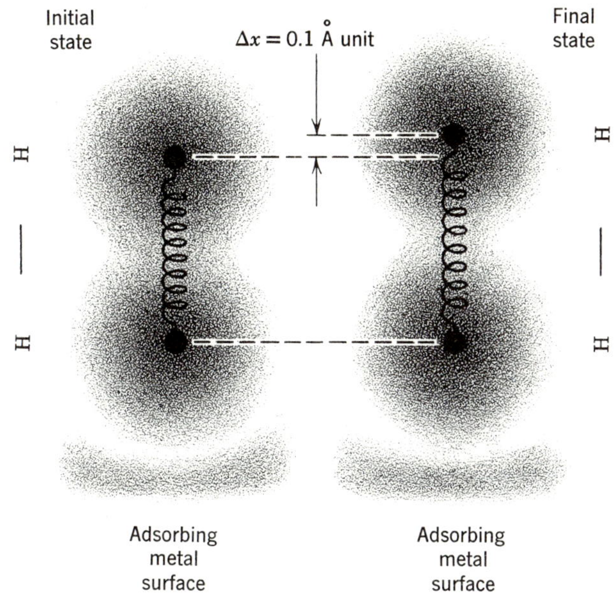

(c)

$$W = \int_i^f F \, \Delta x$$

$$W = F_{av} \, \Delta x$$

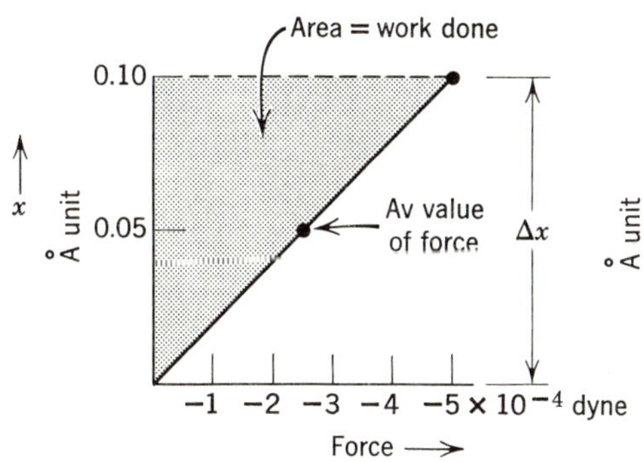

(d)

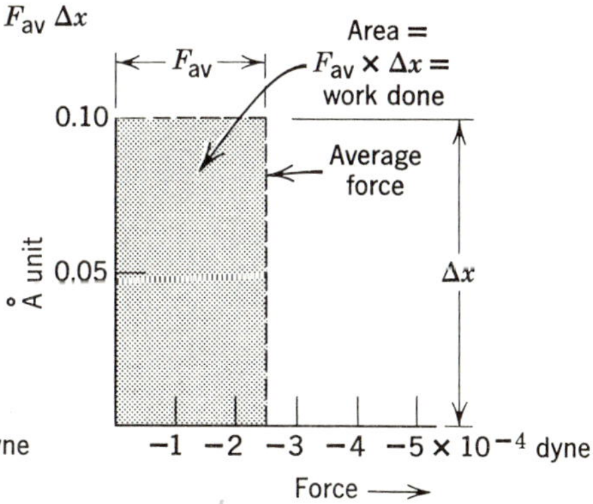

(e)

Fig. 13 (Continued). Work done in stretching H—H molecule (lower end adsorbed on surface).

done in passing from the position in frame No. 0 to the position in frame No. 1 is

$$\Delta W_{0\rightarrow 1} = F \, \Delta h_{0\rightarrow 1}$$

On the diagram, the value of this work is equal to the size of the area that is shown shaded at the left of the drawing. The area is rectangular, so its value

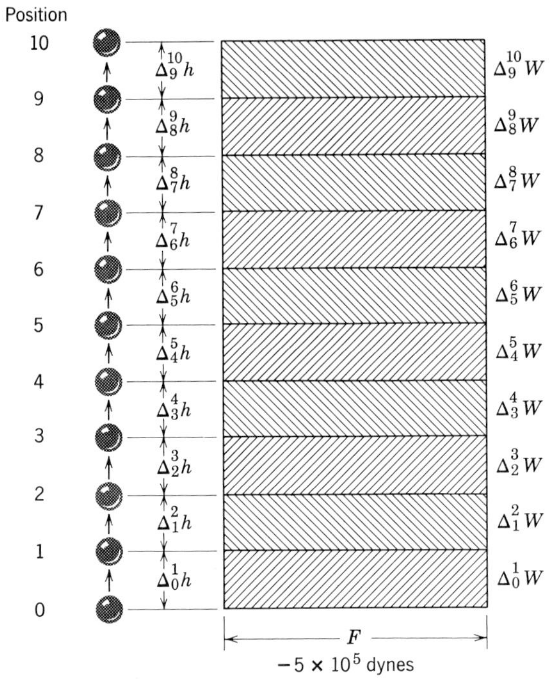

Fig. 14. Work done, ball rising against constant force.

is given by its height (F) multiplied by its breadth (Δh), and work done in going from position 0 to position 1 is given by this small area. Similarly, in going all the way from the position shown in frame No. 0 to the position shown in frame No. 10, the total work ($\Delta W_{0\to10}$) will be

$$\Delta W_{0\to10} = \Delta W_{0\to1} + \Delta W_{1\to2} + \cdots + \Delta W_{9\to10}$$

$$= F\,\Delta h_{0\to1} + F\,\Delta h_{1\to2} + \cdots + F\,\Delta h_{9\to10}$$

This is also represented pictorially in Fig. 13*a*. Just as the work done in the first frame is the area indicated by the shading in the graph of force versus height; the total work done is the area under the line in the plot of F versus h since this rectangular area is equal to the product of the lengths of its two sides, F and h.

To get the total work done (the negative of the potential energy) due to stretching of a chemical bond, we must take into account the fact that the force changes during the movement. Imagine that we have a hydrogen atom adsorbed on a vertical surface. The force of the chemical bond holding it on the surface closely resembles the force of a spiral spring; so we will show it in this form in the picture in Fig. 15. We also imagine that we can use a moving picture camera to photograph the different positions of the atom as it moves out from the surface, just as we photographed the tossed ball. As its kinetic energy is changed to potential energy, the atom will move more and more slowly but the speed of our camera is slowed down correspondingly so that we get pictures of the atom at equal increments of distance. As in the previous case, the work done in each increment will correspond to an area under the curve which represents F plotted against Δx. But, because F is not a constant as in the case of the ball moving against gravity, we do not have perfectly rectangular areas. We can calculate each area easily, however, by taking the value of the force midway across the top of the area, the value that corresponds to the average value over the interval. This average force (F_{av}) multiplied by the width of the rectangle (Δx) will give us the area.

Specifically, let us calculate the work done in passing from position 0 to position 1. We designate this by the symbol $\Delta_0^1 W$, which is read as the change (Δ)

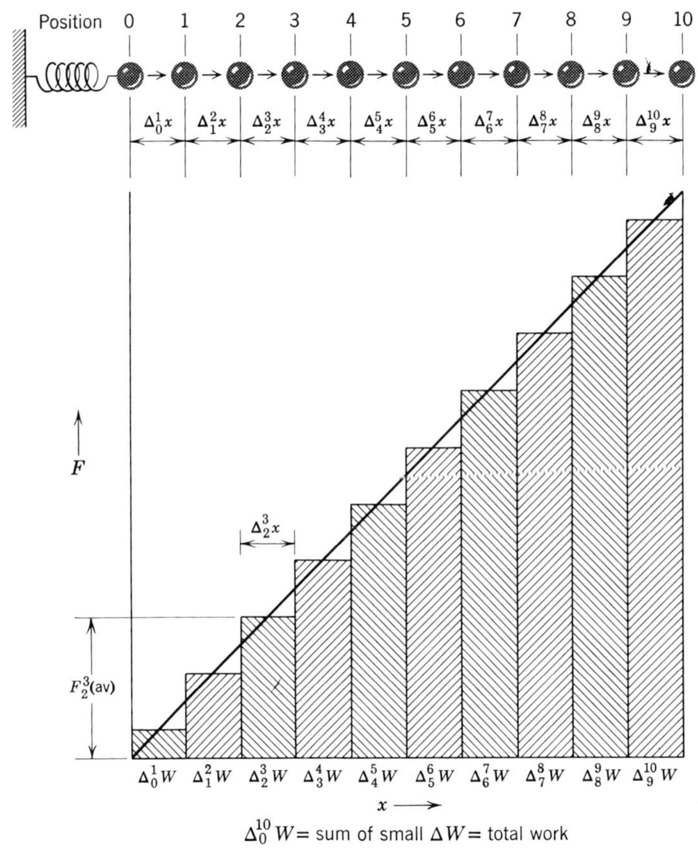

Fig. 15. Work done by the ball moving against linearly increasing force (hydrogen atom moving out from surface).

in passing from 0 position (lower index figure) to the 1 position (upper index figure) that we find in the work (W). We see that

$$\Delta_0^1 W = F_{av}\,\Delta_0^1 x = F_{av}(x_1 - x_0)$$

The midpoint of this interval is the average of the value of x at the beginning of the interval (x_0) and the value at the end of the interval (x_1). This average is

$$x_{av} = \frac{x_1 + x_0}{2}$$

Since the value of the force is given by the equation

$$F = -kx$$

the value of the average force will be

$$F_{av} = -kx_{av} = -k\,\frac{(x_1 + x_0)}{2}$$

If we put this value in the expression for the work we find

$$\Delta_0^1 W = -k\,\frac{(x_1 + x_0)}{2}\,(x_1 - x_0)$$

From algebra we know that the product of the two parentheses yields

$$\Delta_0^1 W = -k\,\frac{(x_1^2 - x_0^2)}{2}$$

Now if we examine the total work done in going from position 0 to position 10 we find

$$\Delta_0^{10} W = \Delta_0^1 W + \Delta_1^2 W + \cdots + \Delta_9^{10} W$$

Substituting the values of each of the terms on the right expressed as functions of x:

$$\Delta_0^{10} W = -k\,\tfrac{1}{2}(x_1^2 - x_0^2) - k\,\tfrac{1}{2}(x_2^2 - x_1^2) - \cdots - k\,\tfrac{1}{2}(x_{10}^2 - x_9^2)$$

If we add all these together we see that all the x's cancel each other in pairs with the exception of the first x_0^2 and the last x_{10}^2. So we find

$$\Delta_0^{10} W = -\frac{k}{2}\,(x_{10}^2 - x_0^2) = -\Delta_0^{10} E_p$$

This gives us the exact value of the area that determines both the work done as the bond is stretched and the increase in the potential energy of the atom during the stretching process.

This method can be applied to any shape of curve representing a change of force with stretching. For example, if we continue pulling the hydrogen atom away from the surface, the force starts off increasing linearly with the distance, but then increases more slowly. By the time we get out to 1 Å distance we may find that further stretching of the chemical bond actually decreases the force. Finally when we get out to 10 Å there is no longer any appreciable force tending to pull the atom back to the surface. These changes in the force are shown in Fig. 16. At 10 Å the atom is completely separated chemically from the surface. If we know the way in which the force varies with the distance we can make a plot of force against distance as shown in the figure, and by drawing little rectangles we can make an accurate estimate of the work done in separating the atom from the surface and thus find the increase in potential energy possessed by the atom when it escapes and is free in space. This method of calculating the value of a product like $F\,\Delta x$ is called "graphical integration."

Finite differences. There are many problems in physics and chemistry where it is necessary to calculate the effect of a variable factor operating over a range, like a variable force doing work over a distance so large that the variation is significant. As a first approach, one can break this range up into small ranges each of which is so small that the force may be regarded as constant. As we have seen, this problem can be handled by *finite difference* equations. The great utility of finite difference equations lies in the fact that, if we can write an equation for a small change, we can tell what will happen in a large change. If we have a formula of the type

$$\Delta Z = Q\,\Delta A$$

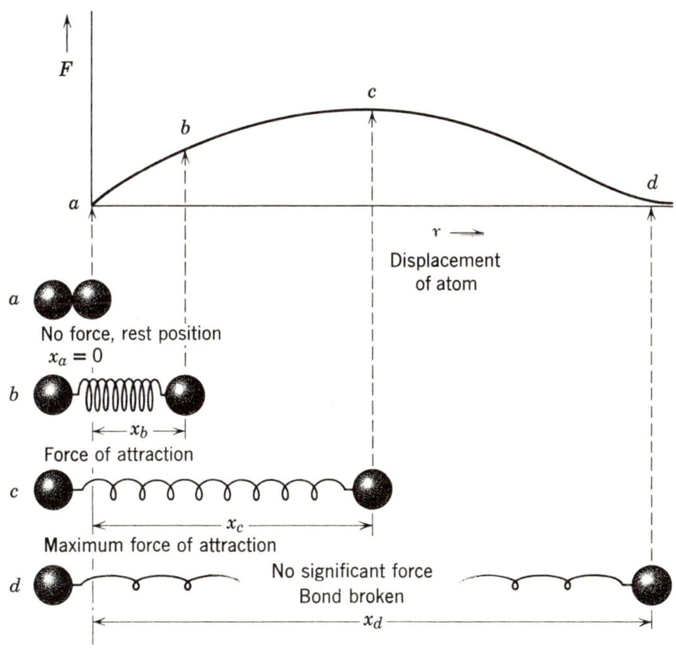

Fig. 16. Variation of chemical bond force with distance of separation of atoms.

wherein Q may change with A, the change in Z when A changes a large amount from A_0 to A_n is the area under a plot of Q versus A in the interval between A_0 and A_n. If, by use of the principle of finite differences, we can arrange any equation for small change in the form

$$\Delta Z = \Delta(RA)$$

wherein R may depend on A in any fashion, the result for a large change is

$$Z_{\text{at } A_n} - Z_{\text{at } A_0} = (RA)_{\text{at } A_n} - (RA)_{\text{at } A_0}$$

In order to apply the principles set forth in the foregoing section it is helpful to have some practice in the algebra of finite difference. Let us therefore derive several relations that will be useful in our later discussions. The reasoning involved in deriving these formulas may be foreign to you. Our immediate concern is to be able to write a formula

$$\Delta Z = Q\,\Delta A$$

in the alternate form

$$\Delta Z = \Delta(RA)$$

so that we can determine the change in Z for a large change in A without resorting to graphical procedures. To this end we have summarized the results in Table 5, wherein m represents any constant.

We have already seen that logarithms are very useful for solving various arithmetic problems. In the use of finite differences, which is really an elementary form of calculus, a different kind of logarithm is quite useful. The logarithms we discussed earlier are logarithms to the base 10; that is, these logarithms represent the power to which 10 must be raised to give you the number in question. There is also another base, namely, e, which has the value 2.718. Clearly, these two logarithms will be related. The relationship is

$$\log_e x = 2.303 \log_{10} x = \ln x$$

This equation should be read log to the base e of x is equal to 2.303 times the log to the base 10 of x. We will represent the $\log_e$, called the natural logarithm, as ln rather than

Table 5

Small Difference	Total Difference
$Q\,\Delta A$	$\Delta(RA)$
$m\,\Delta x$	$\Delta(mx)$
$mx\,\Delta x$	$\Delta\left(\frac{m}{2}x^2\right)$
$Z\,\Delta x + x\,\Delta Z$	$\Delta(xZ)$
$\frac{1}{x}\Delta x$	$\Delta(\ln x)$
$\frac{1}{x^2}\Delta x$	$-\Delta\left(\frac{1}{x}\right)$

log. The importance of this natural logarithm is that, provided a is much much less than 1, the following relationship holds true:

$$\ln(1 - a) = -a \tag{a}$$

Now suppose we have the equation

$$y = \ln x$$

and we wanted to find the change in y per given change in x. Clearly, by the same method which we used before, this would be

$$\Delta y = \Delta \ln x = \ln(x + \Delta x) - \ln x$$

Because of the properties of a log, we can rewrite this as

$$\Delta y = \Delta \ln x = \ln\left(\frac{x + \Delta x}{x}\right)$$

If we divide through the quantity in parentheses, this will become

$$\Delta y = \Delta \ln x = \ln\left[1 - \left(\frac{-\Delta x}{x}\right)\right]$$

Provided $\Delta x/x$ is much less than 1, Eq. (a) applies and it is clear that

$$\Delta y = \Delta \ln x = \frac{\Delta x}{x}$$

Thus far, we have been considering cases where y was the function of only one variable. Let us now consider a case where y is a function of two variables. For example,

$$y = zx$$

If we wanted to get the change in y for a given change in zx, it is apparent that we could say that, at first, the value of y would be given by the equation

$$y = zx$$

If both z and x were changed slightly, this would in general bring about a corresponding change in y, Δy. The new values for this equation would become

$$(y + \Delta y) = (z + \Delta z)(x + \Delta x)$$

If we subtract the former from the latter, we obtain

$$\Delta y = \Delta(zx) = (z + \Delta z)(x + \Delta x) - xz$$

This will simplify to

$$\Delta y = x\,\Delta z + z\,\Delta x + \Delta z\,\Delta x$$

And, if we make the approximation that Δz and Δx are so small that their product is negligible we can write

$$\Delta y = x\,\Delta z + z\,\Delta x$$

Thus we see that, when y involves a product of two variables, the change in y is given by the change in the product which would be obtained if we first consider one of the variables fixed and then consider the other variable fixed.

Integral calculus. If we know the algebraic formula for the variation of one quantity with another ($F = -kx$), then we can find the value of a related quantity like $\Delta_0^{10} W$. In this case we know that the increment in W over a small interval is given by the formula

$$\Delta_0^1 W = F \Delta_0^1 x$$

We express the relation for the whole distance from x_0 to x_{10} by writing

$$\Delta_i^f W = \int_{x_i}^{x_f} F\,dx = \int_0^{10} F\,dx$$

The symbol $\int$ is a distorted letter S and is used to signify that we are in effect taking the sum of a very large number of small areas, each with the width dx and with varying heights F as we go from the initial value of x (in this case $x_i = 0$) to the final value of x (say $x_f = 10$ Å). By means of the mathematical analysis developed in integral calculus, we find that we can derive a formula directly for ΔW as a function of x. Actually we have done this by the method of finite differences as we have just seen and have obtained

$$\Delta_i^f W = \int_{x_i}^{x_f} F\,dx = -\tfrac{1}{2}k(x_f^2 - x_i^2) = \int_0^{10} F\,dx = -\tfrac{1}{2}k(x_{10}^2 - x_0^2)$$

Knowing such a formula, we do not need to count little areas under the curve of F against x since we can calculate the value of ΔW directly from the formula. Many times, however, we do *not* know the formula for F as a function of x and we have to use the method of graphical integration.

In trying to understand the nature of chemistry through the study of the energy changes in chemical reactions, it is essential to see how energy depends on many factors such as the distances by which atoms are separated. Sometimes we try to study the energy of the atoms by finding the actual stretch of the chemical bonds; at other times, as in the study of gases, we shall be concerned with the relation between energy and the volume that the gas occupies, as this volume is in turn a measure of atomic separation. The bigger the volume, the farther apart the atoms are.

Constant force in expansion of gas. As an example consider the work done when a gas is expanded against a constant downward force, as shown in Fig. 17. We have seen that the downward force on the piston is related to the pressure as follows:

$$\underset{\text{Force}}{F_d} = \underset{\substack{\text{Area of}\\ \text{piston}}}{-A} \times \underset{\text{Pressure}}{P}$$

Also the distance through which the force acts is given by

$$(h_2 - h_1) = x$$

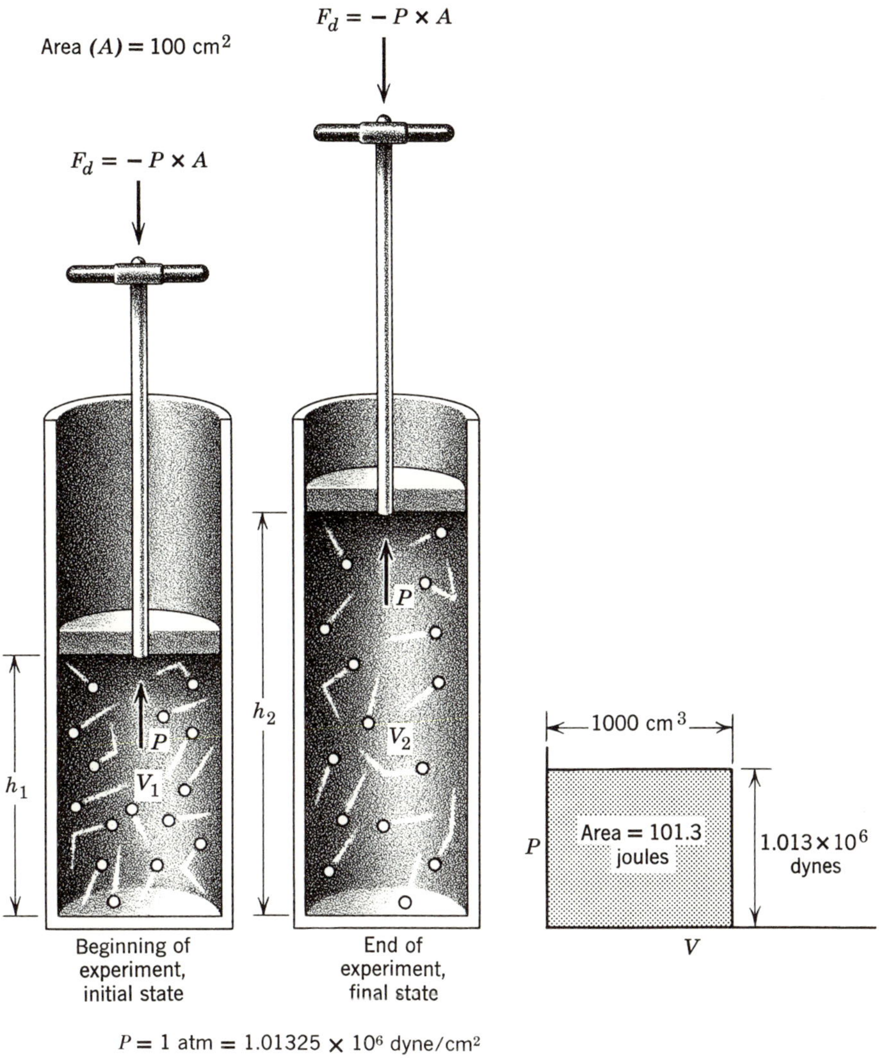

Fig. 17. Work of expansion at constant pressure.

Volume and height are related by

$$hA = V$$

The work done is therefore

$$\text{Work} = F_d \cdot x = F_d(h_2 - h_1) = \frac{F_d}{A}(h_2A - h_1A)$$

$$\text{Work} = -P(V_2 - V_1)$$

In the constant-pressure gas thermometer operating at 1.000 atm, with a piston cross section of 100.0 cm^2, we can calculate the work done when the volume increases by 1000 cm^3.

$$
\begin{aligned}
F = -P \cdot A &= -1.013 \times 10^6 \text{ dynes/cm}^2 \times 100.0 \text{ cm}^2 \\
&= -1.013 \times 10^8 \text{ dynes} \\
\text{Work} = F(h_2 - h_1) &= -1.013 \times 10^8 \text{ dynes} \times 10.00 \text{ cm} \\
&= -1.013 \times 10^9 \text{ ergs} = -101.3 \text{ joules} \\
\text{Work} = -P(V_2 - V_1) &= -1.013 \times 10^6 \text{ dynes/cm}^2 \times 1000 \text{ cm}^3 \\
&= -101.3 \text{ joules} \\
&= -24.22 \text{ calories}
\end{aligned}
$$

If we plot P against V we see that the work done is equal to the negative of the graph of P in Fig. 17.

Variable force in isothermal expansion of gas. If, instead of maintaining constant pressure (P), we maintain constant temperature (T) and allow the expansion to take place by very slowly decreasing the pressure, the situation is quite different. There is a general equation that applies to a standard quantity of any gas; that is,

$$PV = RT$$

or

$$P = \frac{RT}{V}$$

Since R and T are both constants, the product RT is also a constant. The graph of P against V is of the form shown in Fig. 18. But, since P is variable, we must consider first the work done when the gas expanded by a small amount dV, so small that P may be regarded as essentially constant *over this small range.* Thus $DW = -P\,dV$, and, substituting the value of P given above, we get $DW = -RT(dV/V)$. Using the result in Table 5, when the form is $\Delta x/x$ the total change is $\Delta \ln x$ or $\ln x_2 - \ln x_1$. The difference of two logarithms is equal to the logarithm of the ratio, $\ln (x_2/x_1)$. Therefore, if a small amount of work is $-RT(dV/V)$, the total work is $-RT \ln (V_2/V_1) = -2.303\ RT \log (V_2/V_1)$. If we have one mole of gas expanding at $T = 300.0°\text{K}$ from a volume of 1.000 liter to a volume of 2.000 liters, then the work is given by

$$
\begin{aligned}
\text{Work} &= -2.303 \times 1.9872 \times 300.0 \times \log 2.000 \\
&= -413.2 \text{ calories} = -1.729 \times 10^4 \text{ joules}
\end{aligned}
$$

As we shall see in our more detailed study of gases later on, a gas like this kept at constant temperature has a constant amount of energy maintained within it. If, by the expansion, energy flows out of the top of the cylinder as work is done against the pressure maintained there, where does this energy come from?

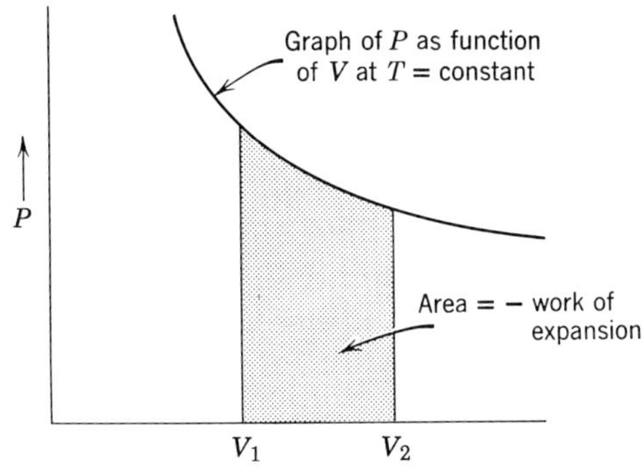

Fig. 18. Work of expansion at variable pressure.

The conservation of energy principle tells us that it must have a source. Actually if we carry out the experiment carefully, we note that heat must flow in as the gas expands in order to keep the temperature constant. This heat flowing in is exactly equal to the work done or energy flowing out. This is an example of the energy balance principle that will guide us throughout our study of chemistry; it is called the First Law of Thermodynamics.

PRECISION AND SIGNIFICANT FIGURES

Any measured quantity is subject to some error. This means that, if we want a complete scientific statement when we express a measured quantity, we must give not only the number and the units but also some indication of how precise this number is, i.e., to what sort of error this value is subject. A concise system for expressing this uncertainty has already been worked out. Basically, it requires that you do not write down any numbers which are uncertain. Thus, if you measure the weight of a large crystal of salt, you may find that it takes 11.6594 gm of brass weights to balance exactly the weight of the crystal on a chemical balance. But, if it is a moist day, the salt may absorb from zero to 0.006 gm of moisture from the air around it. Therefore, if you do not know how much moisture it has absorbed, you should not write down the last two digits in your result; but you should add half of the uncertainty, 0.003 gm, to your result (11.6594) so that you get 11.6624 gm. This number should then be rounded off so that only the last digit is uncertain and then written: 11.662 gm.

The problem of ascertaining the uncertainty in an experimental measurement is not always easy. For example, suppose a student measures the distance across a desk with a 15-cm ruler that is graduated in tenths of a centimeter. Since he can estimate the distance between the markings, we expect the measurements to be precise to 0.01 cm. When these measurements are made three times, he obtains the values 149.56 cm, 150.12 cm, and 151.23 cm. The straight arithmetic average of these figures is 150.303 cm. But this is misleading. This number gives the impression that he knows the length of the desk to within 0.001 cm. The actual measurements, however, suggest that he can only be sure of the length of the desk to within 1 cm. Thus it is best to write this number as 150 cm. When we express a measured quantity in science, we always write it so that only the last digit is uncertain.

Many of the experimentally determined quantities in science involve several actual measurements. Let us see if there is any way in which to determine the precision of such a quantity from the precision of the individual measurements. For example, let us put our student back to work and this time have him measure the volume of a large piece of sheet metal by measuring the length, width, and height. We will assume, this time, that he has learned to measure sufficiently well that he can actually obtain these measurements to within 0.01 cm. The values he obtains for the length, width, and thickness are 100.00 cm, 50.00 cm, and 0.20 cm. The volume is the product of these three quantities, and we might be tempted to write it as 1000.00 cm^3. This is *not* correct. The thickness could just as easily be 0.21 or 0.19 cm as 0.20 cm. If it were the higher value, the calculated volume would be 1050.00 cm^3. If it were the lower value, the calculated volume would be 950.00 cm^3. According to the convention of significant figures, therefore, we must express this value as 1.0×10^3 cm^3, which would indicate the experimental error inherent in this measured volume. We can approach this problem more directly. The calculation involved the product of three figures; the first was known accurately to 0.01 cm in 100.00 cm or 1 part in 10,000. The second was known to 1 part in 5000, and the third to 1 part in 20. As the result of the calculations above, we found that we could express the answer to a precision of roughly 1 part in 20. Thus it appears that *in any product involving experimentally determined quantities the precision of the product can be no greater than the precision of the least accurate factor that makes up the product.*

In operations involving addition and subtraction, special care must be taken to keep track of the significant figures. For example, 8397 − 8396.43 involves subtracting a figure with 6 significant figures from one with 4 significant figures. The correct answer is 1. Let's take another example. Consider the addition of 0.06 to 8396.37. The correct answer is 8396.43. In other words, *the extension to the right of the decimal point of a quantity derived by addition or subtraction of experimental numbers can never exceed that of the numbers which are used in the calculation.*

Before leaving this subject of significant figures, let us discuss a very common difficulty. The words precision and accuracy, according to the dictionary, mean just about the same thing. This is not so when we are referring to data. The precision of experimental data refers to the reproducibility or the repeatability of the data. Accuracy refers to the correctness of the data. In other words, it is perfectly possible (but not very likely) to have data that have great precision but poor accuracy.

The examples which we have covered in our study of graphs and functions illustrate these principles. In our examination of the slope of a curve, we saw that, if the section of the curve is small enough, it may be regarded as a straight line; and the slope of the curve is the slope of this straight line to within the precision which is significant for the problem under consideration. The principle of significant precision must be kept in mind in studying all the quantitative relations that will be discussed in the chapters that follow.

PROBLEMS

1. 6.02×10^{23} molecules of oxygen weigh 32 gm. Calculate the weight of a single molecule and of a single atom of oxygen.

2. A typical molecule of oxygen in the atmosphere at 0°C moves with a velocity of 4.61×10^4 cm/sec. Calculate the velocity of this molecule in miles per hour.

3. Calculate the kinetic energy of the molecule in the preceding problem in ergs.

4. Calculate the kinetic energy of 32 gm of oxygen in joules and calories, assuming each molecule to be moving at the velocity obtained in Problem 2.

Ans. 3.4×10^3 joules; 8.1×10^2 cal.

5. A man who is 5 ft 10 in. high runs at a speed such that he covers twice his height in each second. Calculate this speed in centimeters per second and in miles per hour.

6. If this man weighs 150 lb, calculate his kinetic energy in joules and in calories.

7. While running at this speed, this man dives into a swimming pool and his kinetic energy is changed into heat. How much is the temperature of the pool raised? The pool is 75 ft long, 25 ft wide, and 8 ft deep. Assume that 1.0 calorie is needed to raise the temperature of 1 gm of water 1°C and that each cubic centimeter of water weighs 1.0 gm.

8. How fast would such a man have to be moving to raise the temperature of the pool 1.0°C?

Ans. 1.6×10^4 mi/hr.

9. If a 10-ton truck falls off a cliff that is 1000 ft high, what is the kinetic energy of the truck on striking the ground, if the acceleration of gravity is 980.3 cm $\sec^{-2}$?

10. The acceleration of gravity at New York is 980.3 cm $\sec^{-2}$; at the Lick Observatory on Mt. Hamilton, California, it is 979.7 cm $\sec^{-2}$; and at St. Michael, Alaska, it is 982 cm $\sec^{-2}$. If a man weighs 150 lb in New York, calculate his weight at the other two locations.

11. Calculate the momentum of the man running at the speed given in Problem 5, and of the truck at the end of its fall in Problem 9.

12. A meteor of iron enters the earth's atmosphere and becomes white hot in a matter of a few seconds. If the meteor weighs 100 kg and its temperature increases from $-250°C$ to $1500°C$, what speed would give it the kinetic energy equivalent to this amount of heat? Assume that it takes 0.13 cal gm^{-1} to raise the temperature of the meteor 1°C. *Ans.* 3.1×10^3 mi/hr.

13. The elastic force constant k for the chemical bond in the H_2 molecule has a value of 5×10^5 dyne cm^{-1}. How much work is done in increasing the distance of separation of the two H atoms from the rest position of 0.74 Å to 1.3 Å, assuming that the elastic force constant has the above value over this range.

14. Assuming that the elastic force constant keeps the value in the previous problem over the range of separation of the two H atoms from 0.5 Å to 1.3 Å, make a plot of force against distance of separation over this range; make a plot of the logarithm of potential energy against logarithm of distance of separation over the same range, using log-log paper; from the graph obtained, make a plot of the direct values of potential energy against distance of separation over this range.

15. Using the same assumptions and method of plotting, make graphs for potential energy against distance of separation of Cl_2 and O_2, using the values $k_{Cl_2} = 3.3 \times 10^5$ dyne cm^{-1} and $k_{O_2} = 11.8 \times 10^5$ dyne cm^{-1}.

16. If the total surface area of a human body is 1.5 $meter^2$, what is the total force exerted on it by 1 atm pressure. Express the answer in kilograms and in tons.

17. The piston in a hydraulic jack has a radius of 10 cm. When the piston rises by 30 cm, what is ΔV? If it raises a weight of 2000 kg, what is the work performed? *Ans.* 5.8×10^3 joules.

18. An atom of hydrogen has an elastic collision with an atom of oxygen. Using the values $m_H = 1.673 \times 10^{-24}$ and $m_O = 2.656 \times 10^{-23}$ gm, $v_H = 10^3$ cm sec^{-1}, and $v_O = -10^3$ cm sec^{-1}, calculate the velocity of each atom after the collision.

19. Make a plot of the momentum of the hydrogen atom as a function of its velocity over the range 0 to 10,000 cm sec^{-1}. To what quantity is the slope of this line equal?

20. If the total momentum of the colliding H and O atoms remains constant as found in Problem 18, make a plot of the velocity after the collision of the H atom as a function of the velocity of the O atom after the collision.

21. Using pressure in atmospheres for the ordinate and temperature in degrees Kelvin for the abscissa, make a graph showing the linear relation between P and T when V has the values 2, 5, 10, 20, 50, and 100 liters in the equation $PV = RT$. Let T vary between 0° and 1000°K and P between 0 and 10 atm on the coordinate axes.

22. Using volume in liters as the ordinate and absolute temperature as the abscissa, make a graph showing the linear relation between V and T when P has the values 0.2,

0.5, 1, 2, 5, and 10 atm in the equation $PV = RT$. Let T vary from 0° to 1000°K and V between 0 and 100 liters on the coordinate axes.

23. Make a graph of log P in atmospheres against log V in liters showing the linear relation between these two quantities when T has the values 200°, 400°, 600°, 800°, and 1000°K in the equation $PV = RT$.

24. Make a plot of V against P for 1 mole of gas at 200°, 400°, 600°, 800°, and 1000°K by reading the values from the linear plot in the preceding problem.

25. Using the values given in Chapter 1, calculate the boiling points of hydrogen, nitrogen, oxygen, fluorine, and neon in degrees Fahrenheit. ($t_{\text{Fahrenheit}} = \frac{9}{5}t_{\text{centigrade}} + 32°$.)

26. What is the value of the slope of each line plotted in Problem 21? *Ans.* R/V.

27. What is the rate of change of volume with temperature on each of the lines plotted in Problem 22?

28. By drawing the tangent to the curve for V against P at $T = 400°$K in Problem 24, estimate the slope of the curve at $V = 10$, 20, and 40 liters.

29. Calculate the rate at which a hydrogen molecule ($k = 5 \times 10^5$ dyne/cm) is gaining energy as it is stretched and the distance between the two nuclei is (*a*) 0.75 Å, (*b*) 0.85 Å, (*c*) 0.95 Å.

30. Calculate the rate at which a hydrogen molecule ($k = 5 \times 10^5$ dyne/cm) is gaining energy as it is compressed and the distance between the two nuclei is (*a*) 0.73 Å, (*b*) 0.70 Å, (*c*) 0.65 Å.

31. From the graph prepared in Problem 24, estimate graphically the work done in compressing 1 mole of gas from a volume of 30.0 liters to a volume of 20.0 liters with the temperature held constant at 800°K.

32. From the formula for work done in isothermal expansion or compression calculate the work value in the previous problem and compare with the value estimated graphically.

33. Calculate the work done in compressing the same gas from 30.0 liters to 20.0 liters if the pressure is held constant instead of the temperature. *Ans.* 21.8 liter atm.

34. With what accuracy must the temperature be measured and held constant in the experiment described in Problem 31, in order to have the value for the work done at the same accuracy as the values given for the volume?

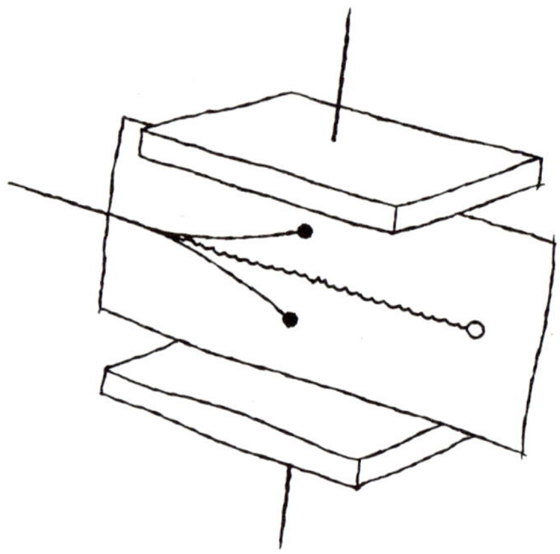

3
The nucleus

In using the word "atom" for the fundamental particles out of which the world is made, the Greeks expressed their belief that these particles never could be broken down into anything smaller. From the time of Dalton (1805) until the end of the nineteenth century, this immutability of the atoms was the cornerstone of chemical theory. In view of this, one can understand the shock when it was discovered that certain atoms spontaneously break up. Surprisingly enough, these discoveries did not signal the demise of the atomic theory; instead, they made possible experiments that gave birth to a newer, more vigorous theory of the atom.

FUNDAMENTAL PARTICLES

In 1896, Becquerel found that uranium minerals gave off invisible radiation that would pass through opaque objects. Shortly afterward, Pierre and Marie Curie isolated from these minerals two new chemical elements (polonium and radium) that produced even more intense radiation than uranium itself. Rutherford and Soddy later established that the radiation was the result of an atomic *disintegration* in which atoms *spontaneously* exploded. This phenomenon was called *radioactivity*.

Detailed investigations next revealed that there are three kinds of radiation, called alpha (α), beta (β), and gamma (γ) radiation, emitted by exploding atoms. These types of radiation can be distinguished by studies of the type depicted in Fig. 1. In this apparatus, a narrow beam of the radiation is obtained by placing the radioactive substance in a lead "gun barrel"; only radiation in the direction of the muzzle escapes. This radiation is made visible by a fluorescent screen,

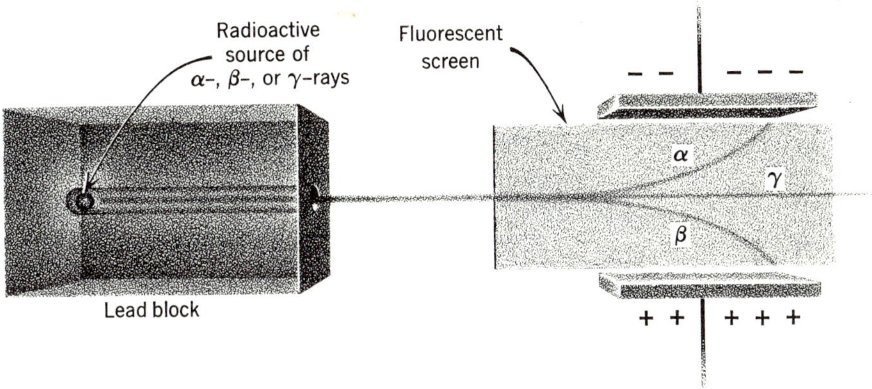

Fig. 1. Characteristics of α-, β-, and γ-radiation.

which glows when struck by α-, β-, or γ-rays. If the beam is directed between two oppositely charged plates, an α beam is bent toward the negative plate, a β beam is bent toward the positive plate, and a γ beam is not deflected. In view of this, it is clear that α-radiation is a beam of positively charged particles, β-radiation is a beam of negatively charged particles, and γ-radiation is a beam of invisible light or neutral particles.

The α-particle was shown to be a helium atom that has been stripped of all its planetary electrons. These particles emerge from the nucleus and travel at speeds as high as 10,000 miles per second. Since they carry with them two positive charges, they leave behind a nucleus having two *less* positive charges, that is, a *transmuted* atom.†

The *β-particle* was identified as an electron. Like the α-particles, these electrons also come out of the nucleus at very high speeds. When an electron leaves the *nucleus* of the atom, this nucleus loses a negative charge (or equivalently gains a positive charge), and transmutation occurs.

The γ-radiation was identified as a type of electromagnetic radiation similar to visible light. It is not visible to the eye because, like X-rays, it has a short wavelength. Since the emission of a γ-ray results in no change in nuclear charge, the parent atom is *not* transmuted.

Since 1910, we have become increasingly aware of the existence of other nuclear fragments that can be produced by the artificial and natural disintegration of atoms. Three of the more important ones are listed in Table 1, together with

† The number of *positive* charges on the nucleus of an atom determines how many *negatively* charged planetary electrons must be present to produce electrical neutrality. The number of planetary electrons determines the chemical properties; so, when the nucleus loses two positive charges by emitting an α-particle, the atom is transmuted to a different chemical element.

Table 1

Name	Mass a.m.u.[a]	Charge[b]	Origin
Positron	0.00055	+1	Produced by nuclear reactions
Proton	1.0073	+1	Hydrogen atom nucleus
Neutron	1.0087	0	Produced by nuclear reaction—unstable when uncombined; $t_{1/2} = 13$ min
α-Particle	4.0015	+2	Helium atom nucleus
β-Particle	0.00055	−1	Electron
γ-Particle	0.0000	0	Hard X-rays from radioactive decay

[a] An a.m.u. is a small mass unit often used for comparing masses of atoms and subatomic particles. 1 a.m.u. $= 1.66 \times 10^{-24}$ gm.

[b] Charge is expressed in units such that the charge on the electron is −1.

their mass, charge, and properties. For the sake of comparison, we have also indicated the properties of α-, β-, and γ-particles.

Two of these newer particles, the neutron and the positron, were discovered in 1932. The positron is a positively charged counterpart of the electron. The neutron is a neutral particle with a mass comparable to that of the proton. It is *not* an electron bound intimately to a proton. A free neutron is unstable; it has a fifty-fifty chance of disintegrating in 13 minutes. When it does, it forms an electron and a proton with a large amount of energy.

RUTHERFORD'S EXPERIMENT AND THE NATURE OF THE ATOM

It was soon recognized that the α-particles emitted during radioactive decay could be employed as a sensitive probe in experiments dealing with the nature of matter. By 1910, such experiments were under way in Rutherford's laboratory. Rutherford and Geiger used the apparatus shown in Fig. 2 for the study of the deflections of directed α-rays by sheets of gold foil about 1000 atoms in thickness. The entire apparatus was enclosed in an evacuated chamber to prevent extraneous deflections caused by air molecules. A cylindrical fluorescent screen was placed around the gold foil and a hole was cut in this to permit the α-ray beam to impinge directly on the gold foil. The course of the α-particle could be deduced from the position of the flash of light whenever an α-particle collided with the screen. By such studies, it was established that the bulk of the beam passed through the gold foil without deflection. Very few α-particles were deflected at all, and, even when this occurred, the deflection was slight.

At the time when Geiger was performing these experiments with the gold foil, a young scientist named Marsden arrived at Rutherford's laboratory to study for a few months and learn some of the experimental techniques. To this end, Rutherford and Geiger suggested that Marsden carry out a series of experiments in order to see if any of the α-particles bounced back from the gold foil. (Since then, Rutherford, himself, remarked that he considered this as unlikely as a 3-inch shell bouncing back from a sheet of tissue paper.) When Marsden performed these experiments, he found that a few of the α-particles actually did bounce back in the direction from which they came. The number that did were directly proportional to the thickness of the gold foil, and he was able to estimate that, if the gold foil were one atom thick about one out of 10^8 α-particles directed at the foil would bounce back. These observations formed the first good experimental evidence for the atom as we know it today.

A complete interpretation of the results obtained by Rutherford and co-workers requires an elaborate mathematical treatment, but an elementary interpretation will reveal the qualitative consequences of this experiment. Let us assume for simplicity that the atoms are little cubes, and that a sheet of foil one atom thick could be made by arranging these atoms in a plane. In Fig. 3*a* we show a stream of α-particles passing through such a cubical atom. The part of the atom that they hit must be determined by chance, so that we can say that, if only one α-particle out of 10^8 is deflected, this must mean that most of the atom is effectively empty space; most of its mass must be concentrated in a very small fraction of the total space occupied by the atom. To obtain an estimate of the size of this region of concentrated mass, this *nucleus,* let us consider the front view of the atom, shown in Fig. 3*b*. If an atom is such that half of the area presented is effectively empty space and half is massive and impenetrable, then half of the α-particles that strike this atom strike empty space and go through unimpeded, whereas the other half

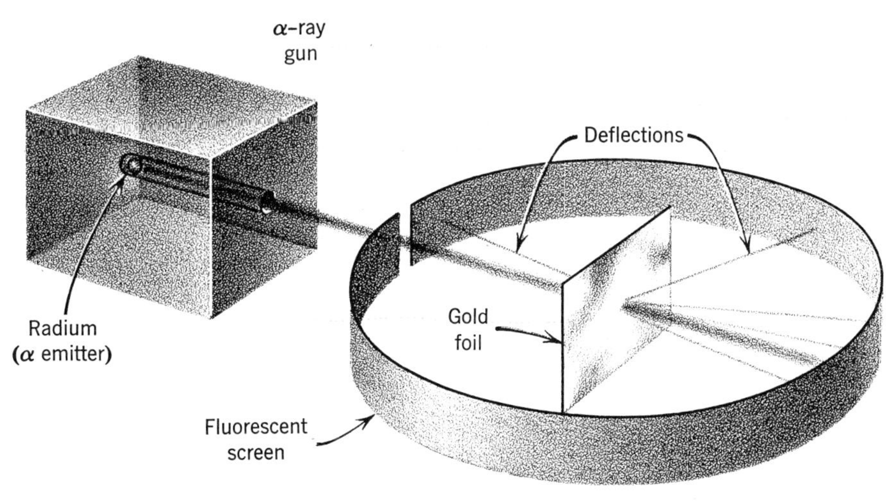

Fig. 2. Rutherford's apparatus.

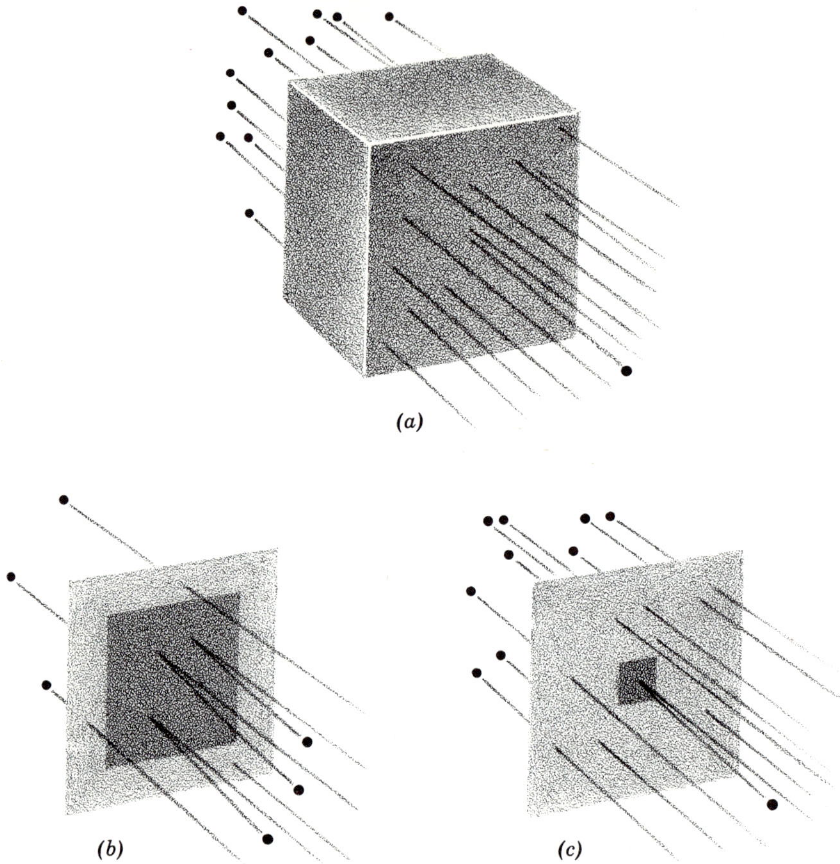

Fig. 3. α-Ray bombardment of atoms (schematic).

hit the heavy nucleus and bounce back. If we make the nucleus of this hypothetical atom still smaller, as in Fig. 3*c*, where the ratio of the cross-sectional areas is 1 to 10, only 1 out of 10 of the incident α-particles will bounce back. In general, we can say

$$\frac{\text{Number deflected}}{\text{Number striking}} = \frac{A_n}{A_a}$$

where A_n represents the area presented by the nucleus and A_a represents the area presented by the whole atom. On this basis, the observations of Rutherford, Geiger, and Marsden suggest that the area presented by the atom is 10^{+8} times larger than the area presented by the nucleus. The diameter of atoms is roughly 10^{-8} cm; hence the area presented by an atom is $(10^{-8})^2$ cm^2; consequently, the effective area of the nucleus is 10^{-24} cm^2 and the diameter of this nucleus is roughly 10^{-12} cm.

From these experimental observations and conclusions, it is only a step to a planetary model for atoms. It was known at this time that, compared to the proton, the electron has a small mass. Rutherford showed that the results of his experiments were consistent with the planetary model in which the heavy nucleus, only 10^{-12} cm across, takes the role of a miniature sun with the light planetary electrons moving in orbits about 10^{-8} cm across. The size of these orbits determines the size of the over-all atom, even though 99.9% of the mass is concentrated in 1/1,000,000,000,000 (10^{-12}) of the over-all atomic volume. By way of comparison, the atom has a larger fraction of empty space than our galaxy.

THE NUCLEUS AND ATOMIC MASS

The nuclei of all the elements can be viewed as if they were made up exclusively of protons and neutrons. Although an electron (β-particle) is sometimes emitted in radioactive decay, there is evidence that this comes from a break-up of one of the neutrons in the nucleus into a positively charged proton, which stays in the nucleus, and a negatively charged electron, which is ejected. This picture of the nucleus is far from complete; but it is sufficiently accurate for our present purposes.

In order to specify the nuclear structure of atoms, we write two numbers with the symbol of the element. In this notation, the lower number, preceding the symbol, indicates the number of protons in the nucleus; the upper number, after the symbol, indicates the number of protons plus neutrons. Examples of this notation are given in Table 2.

Such symbols not only tell us the structure of the nucleus, they also tell us the mass of the atom. (For such purposes, we can neglect the mass of the electron, which is only about 1/1835 that of the proton.) In general, each proton in the nucleus adds 1 a.m.u. to the mass of an atom; similarly, each neutron contributes 1 a.m.u.; therefore the total mass of an atom in atomic mass units is approximately equal to the number of protons plus neutrons. Thus, the mass of

Table 2 Nuclear Symbols

	Number of Protons	Number of Neutrons	Total Number of Particles
$_1H^1$	1	0	1
$_1H^2$	1	1	2
$_2He^4$	2	2	4
$_3Li^7$	3	4	7
$_4Be^9$	4	5	9
$_5B^{11}$	5	6	11
$_6C^{12}$	6	6	12

$_1H^1$ is 1 a.m.u. On the other hand, the $_2He^4$ nucleus has two protons and two neutrons; therefore a helium atom weighs 4 a.m.u., and the nucleus has a +2 charge. Similar remarks apply to the other nuclei listed in Table 2.

Table 2 lists two forms of hydrogen nuclei, $_1H^1$ and $_1H^2$. The symbol $_1H^2$ stands for *deuterium*, the "heavy" hydrogen discovered about thirty years ago. This substance has chemical properties virtually identical with hydrogen, but an atom of deuterium weighs almost twice as much as an atom of hydrogen. When two different species of atoms have the same nuclear charge but different masses, they are called *isotopes.* The chemical properties of these different atoms are virtually the same because the arrangement of electrons around a neutral atom determines the chemical properties and this arrangement depends solely on the nuclear charge.

Many of the elements occur in several stable isotopic forms. For example, there are three stable isotopes of oxygen, $_8O^{16}$, $_8O^{17}$, and $_8O^{18}$, and two stable isotopes of lithium, $_3Li^6$ and $_3Li^7$. (The $_6C^{12}$ isotope has a mass of exactly 12 a.m.u., by definition; this is the basic standard of the a.m.u. scale for mass). With the more complex heavier elements, the variety of isotopes is even more impressive. Krypton occurs principally in nature as $_{36}Kr^{84}$, but more than twenty other isotopes are known. Such variety leads to complications. In talking about the masses of atoms, we must always specify whether we mean the mass of a particular isotope or the *average* mass of a large number of atoms as they occur in nature. Thus the weight of $_3Li^7$ is 7.00 a.m.u., but lithium, as it occurs in nature, consists of 92.6% $_3Li^7$ and 7.4% $_3Li^6$; hence the average atomic mass of naturally occurring lithium is 6.94 a.m.u.

In an imaginary experiment, we could prepare all the elements from protons, neutrons, and electrons. Such a process actually takes place at temperatures of millions of degrees, but, at this time, we are not concerned with details. We could form helium from two protons, two neutrons, and two electrons as indicated below:

$$2\ _1H^1 + 2\ _0n^1 + 2\ _{-1}e^0 \rightarrow\ _2He^4$$

where the symbol, $_0n^1$, stands for a neutron. In this equation, we have indicated only the approximate mass of each species involved; with the data in Table 1, we can make a more accurate calculation of the masses of the reactants and products. The exact mass of two neutrons plus two protons and two electrons is 4.03300 a.m.u.; the exact mass of an α-particle plus two electrons is 4.0026 a.m.u. Thus, in this process, 0.0304 a.m.u. of mass apparently disappears. Actually, it does not really disappear but it is now seen as an amount of energy consistent with Einstein's equation, $E = mc^2$. According to this equation, the amount of energy that appears after m grams of mass disappear is equal to this mass multiplied by the speed of light squared. The amount of mass that is thus "annihilated" when an element is synthesized from the proper number of neutrons, protons, and electrons is called the *mass defect.* Because of it, the mass of an isotope will not be exactly equal to the mass of the protons, neutrons, and electrons from which it could be synthesized.

The existence of a mass defect tells us that the neutrons and protons bound within the nucleus must be fundamentally different from the neutrons and protons existing independently. This conclusion is further supported by the fact that unattached neutrons have a mean life expectancy of only 13 minutes, whereas "neutrons" in the nucleus are as long-lived as the atoms.

NUCLEAR CHANGES AND EQUATIONS

Nuclear changes such as the synthesis of a helium atom from the required number of protons, neutrons, and electrons are accompanied by changes in mass. For the most part, however, such changes in mass are small compared to the over-all mass of the atoms or fundamental particles involved; therefore we can assume that the principle of the conservation of mass holds approximately even for nuclear reactions.

Nuclear reactions can be succinctly described by equations. The radioactive decay of U^{238} can be represented as follows;

$$_{92}U^{238} \rightarrow {}_{2}He^{4} + {}_{90}Th^{234}$$

In this process energy is released and there is a measurable decrease in mass. This decrease, however, is so small that it is only detectable when instruments of great precision are used. In other words, the uranium atom has a mass of 238 a.m.u.; by disintegration, a helium atom with a mass of 4 a.m.u. and a thorium atom with a mass of 234 a.m.u. is formed. To a fair approximation, therefore, there is conservation of mass in this reaction.

The principle of conservation of charge always applies.

The two conservation rules provide the key to the balancing of equations describing nuclear change. If we know the particle emitted in a radioactive decay process, we can always deduce the product of the disintegration. For example, $_{6}C^{14}$ emits a β-particle when it decays. The over-all reaction must be

$$_{6}C^{14} \rightarrow {}_{-1}e^{0} + {}_{7}N^{14}$$

since the product must have the same mass as the $_{6}C^{14}$ nucleus and also a nuclear charge of $+7$. (The nuclear charge of $+7$ identifies the product as nitrogen.) In all such equations, the sum of the superscripts on the left equals the sum of the superscripts on the right; the sum of the subscripts on the left equals the sum of the subscripts on the right. As another example, consider the radioactive decay of $_{14}Si^{27}$. This decay yields a positron, a particle with the same mass as the electron but a positive charge. The equation describing this process must be

$$_{14}Si^{27} \rightarrow {}_{1}e^{0} + {}_{13}Al^{27}$$

In addition to the naturally occurring radioactive processes, many nuclear reactions are now produced in the laboratory. Today, there are a number of atom-smashing machines that will accelerate elementary particles to such high velocities that they can penetrate into the nucleus of an atom and thereby cause a reaction. Long before this, in 1919, Rutherford observed the first artificial nuclear transformation. He found that the α-particles emitted by radium could react with a nitrogen nucleus as shown in the equation

$$_{7}N^{14} + {}_{2}He^{4} \rightarrow {}_{8}O^{17} + {}_{1}H^{1}$$

Today, a good many such nuclear reactions are known. Samples of them are

$$_{4}Be^{9} + {}_{2}He^{4} \rightarrow {}_{6}C^{12} + {}_{0}n^{1}$$
$$_{0}n^{1} + {}_{7}N^{14} \rightarrow {}_{6}C^{14} + {}_{1}H^{1}$$
$$_{4}Be^{9} + {}_{0}\gamma^{0} \rightarrow {}_{4}Be^{8} + {}_{0}n^{1}$$

You can verify for yourself that in each of these cases the equations are consistent with the conservation of mass and charge.

RADIOACTIVITY

The surprising thing is not that a uranium nucleus spontaneously emits an α-particle but that it hangs together at all. According to our discussion thus far, the uranium nucleus has 92 protons squeezed together with neutrons to form a nucleus 10^{-12} cm across. If only coulombic forces were acting, the mutual repulsion between protons separated by such small distances would be fantastic. Thus, in order for the nucleus to exist, we must invoke some special type of force, nuclear force. This force must be sufficiently great to hold the nucleus together in spite of the very high repulsion between protons. (We are, of course, assuming that like charges still repel at distances shorter than 10^{-12} cm.) At this time, let us accept the existence of such forces as a necessary postulate.

The radioactive decay process is a consequence of the nuclear structure of atoms. Apart from this, however, it provides an interesting illustration of the way in which we can arrive at a mathematical description of a physical phenomenon on the basis of a very few observations. Let us consider the radioactive process

$$_{92}U^{238} \rightarrow {}_{2}He^{4} + {}_{90}Th^{234}$$

If we examine a large number of different samples, we will find that the number of such disintegrations per minute is directly proportional to the number of

$_{92}U^{238}$ atoms present. We could, if we like, rephrase this observation and say that the fraction of the $_{92}U^{238}$ atoms that decay in a minute is a constant; in each of these samples a fixed fraction of the uranium atoms decay each minute. This statement is very much like a statistician's prediction that in a year one-ninety-eighth of the eighteen-year-old males in the United States will marry. The interpretation is also similar. Unless a statistician has more information, he will not tell *which ones* of these males will marry; to him, his prediction means that, *by chance,* a certain fraction of these males will marry by the end of the year. Similarly, we might suppose that *by chance* a certain fraction of these uranium atoms will break up each minute.

This observation is universal for all types of radioactive decay. Provided the number of atoms disintegrating is small compared to the total number present,† a certain fraction of the atoms present decompose in a given instant. The longer the instant, the more atoms will decompose. We can formulate this mathematically as follows:

$$\Delta N/N = -k\,\Delta t \tag{1}$$

ΔN stands for the change in the number of atoms present in the instant Δt, and $\Delta N/N$ represents the fraction of the atoms present that decompose in an instant Δt. The negative sign indicates that N decreases. Obviously, if we double Δt, we double the fraction that decompose. The actual value of the constant k is determined by the nuclear structure of the radioactive isotope we are examining. In other words, k will be different for $_{92}U^{238}$ and $_{6}C^{14}$.

The equation above is a valid approximation only if ΔN is small compared with N. Unless this is true, the fraction $\Delta N/N$ becomes imprecise. The fraction $\Delta N/N$ is unitless; hence the product $k\,\Delta t$ must also be unitless; and, if we measure Δt in seconds, the units of k must be $\sec^{-1}$.

In problems involving radioactive decay, it often becomes necessary to make calculations for cases in which $\Delta N/N$ is large, i.e., more than about 0.05. We can derive the appropriate exact formula by the application of the algebra of finite differences discussed in Chapter 2. On the basis of that discussion, Eq. 1 can be rephrased as follows:

$$\Delta \ln N = -k\,\Delta t \tag{2}$$

or

$$\ln N - \ln N_0 = -k(t - t_0) = -kt$$

† The situation for radioactive decay is different from that for eighteen-year-olds. The supply of eighteen-year-olds is maintained almost constant by aging seventeen-year-olds; when a radioactive atom decays, the number of radioactive atoms is decreased. To define concretely the fraction decomposing, this fraction must remain small. For example, if initially there are 200 radioactive atoms and 100 decompose in 10 minutes, should we say 10/200 decompose per minute or 10/150 decompose per minute? If the fraction is small, this difficulty does not arise. If initially there are 1,000,000 radioactive atoms and 100 decompose in 10 minutes, then 1/100,000 decompose each minute.

In this form, N stands for the number of atoms remaining at time, t, after the start of our observation; N_0 stands for the number of atoms present at the start of our observation at $t_0 = 0$. On the basis of the properties of logarithms, Eq. 2 can be rearranged to yield

$$\ln \frac{N}{N_0} = -kt \tag{3a}$$

or

$$N/N_0 = e^{-kt} \tag{3b}$$

These are the general equations for the rate of any radioactive decay. Differences in the rate for different processes stem from differences in the value of k. The constant k characterizes the decay rate of a given radioactive isotope; hence k is termed the *rate constant.*

The implications of these equations can be rendered more tangible if we consider an actual case. Suppose that we start an experiment with N_0 radioactive particles at $t = 0$ and that at various later times we determine how many radioactive atoms are left. We can plot these numbers as a function of time; to make the plot correspond closely to the experimental data, let us join each of the determined points corresponding to an observation by straight lines as indicated in Fig. 4.

According to the plot in Fig. 4, which is merely another way of presenting Eq. 3, the number of radioactive atoms left at any time becomes steadily less as we

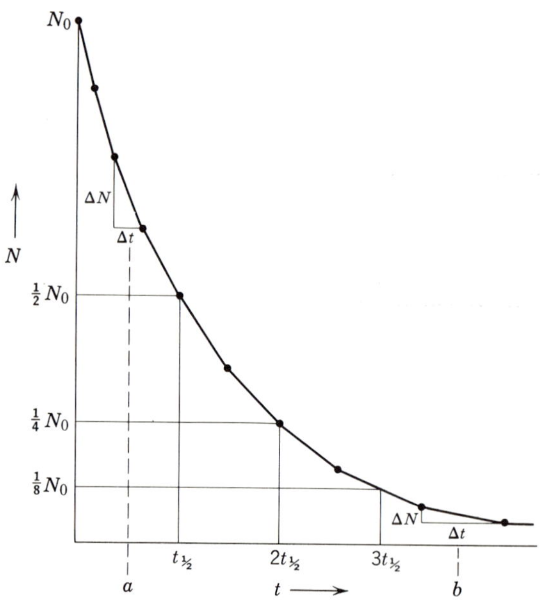

Fig. 4. Radioactive decay law.

wait longer and longer. If we were to extend the time axis of this plot, we would find that N would become zero only after an infinite time. This feature of the decay process can be expressed in still another manner. Let us define the *half-life* of a radioactive substance as the time required for half of the radioactive atoms to decay. In the plot in Fig. 4, this half-life, $t_{1/2}$, is the time required for the number of remaining atoms to fall to $\frac{1}{2}N_0$, half the initial value. If we wait for two half-lives, do all the radioactive atoms disintegrate? *No.* In the first half-life, half of them will disintegrate. In the second half-life, half *of those remaining* will disintegrate, leaving one-quarter of the atoms undisintegrated. Thus, during a period of time $t_{1/2}$, each of the radioactive atoms has a fifty-fifty chance of decomposing; after a period of time $t_{1/2}$, half of those initially present decompose. It should be clear that, after the elapse of five half-lives, the number of radioactive atoms halve themselves five times and $\frac{1}{2} \times \frac{1}{2} \times \frac{1}{2} \times \frac{1}{2} \times \frac{1}{2} = \frac{1}{32}$ of the initial number of radioactive atoms are left. Mathematically no matter how many half-lives we wait, we will never reach that point where there are no radioactive atoms.

As one might expect, there is a simple relation between the half-life, $t_{1/2}$, and the radioactive decay constant, k. This relation is

$$k = \frac{0.693}{t_{1/2}} \tag{4}$$

Since $t_{1/2}$ is that time when N/N_0 is equal to one-half, it is a simple matter to verify this relation by substitution into Eq. 3a. This is left as a problem for the student.

The relation expressed in Fig. 4 can also be discussed in terms of Eq. 1. Consider the observation over the interval indicated by a. During Δt, we observe a decrease, ΔN, in the number of radioactive atoms present. Specifically, this change may be 100 atoms in 0.101 second, so that we say that, at time a, these radioactive atoms are disintegrating at the rate of 991 per second. Suppose, however, that we chose a smaller time interval, say, 0.0100 second, and that, during this Δt, ΔN is equal to 10; then we say that the rate of decay is 1000 atoms per second. As Δt, the time interval for observation, becomes smaller and smaller, the rate at the time a will be given by the ratio $\Delta N/\Delta t$. This ratio will not be small even though both ΔN and Δt are small. If Δt is so small that the observation is virtually instantaneous, the ratio $\Delta N/\Delta t$ is called the *instantaneous rate*. Under such conditions, Eq. 1 is exact.

Equation 1 can be rewritten

$$\Delta N/\Delta t = -kN$$

Although this is an exact relation only for a truly instantaneous rate, in practice it is a good approximation whenever ΔN is much smaller than N. We can see that this is true from Fig. 4. The rate for interval a is much greater than the rate for interval *b*. If we calculate these two rates from the graph, we will find that they are directly proportional to the average number of atoms present during the observation in accord with the equation above.

Example Problem 1: A chemist prepares 1.00 gm of pure $_6C^{11}$. This atom is radioactive and decays as follows:

$$_6C^{11} \rightarrow {_1e^0} + {_5B^{11}}$$

The half-life is 21 minutes. What are the number of disintegrations per second immediately after its preparation? How many grams of carbon are left exactly one day after its preparation?

In working this problem, we must make use of Eqs. 1, 3, and 4. Since in these equations both N and ΔN refer to the number of atoms, the first task will be to compute the number of atoms in 1.00 gm of $_6C^{11}$. To do so, we make use of the facts that 1 a.m.u. is equivalent to 1.66×10^{-24} gm (Table 1) and that a single atom of $_6C^{11}$ weighs 11.0 a.m.u. Thus we can write

$$\frac{1 \text{ at. } {_6C^{11}}}{11.0 \text{ \sout{a.m.u.}}} \times \frac{1. \text{ \sout{a.m.u.}}}{1.66 \times 10^{-24} \text{ gm } {_6C^{11}}} = \frac{1 \text{ at. } {_6C^{11}}}{18.3 \times 10^{-24} \text{ gm } {_6C^{11}}}$$

From this factor, we can compute the number of atoms of C^{11} in 1 gm as follows:

$$\frac{1 \text{ at. } {_6C^{11}}}{18.3 \times 10^{-24} \text{ gm } {_6C^{11}}} \times 1.00 \text{ gm } {_6C^{11}} = 5.5 \times 10^{22} \text{ at. } {_6C^{11}}$$

Now, it is not the half-life, $t_{1/2}$, but k that appears in Eq. 3. Before we can apply these equations, we must compute k with the help of Eq. 4. We obtain

$$k = \frac{0.693}{21 \text{ min}} = \frac{33 \times 10^{-3}}{\text{min}}$$

The exact relation describing radioactive decay is Eq. 3. It is inconvenient, however, to work with negative logarithms, so we will rearrange Eq. 3 as follows:

$$-kt = \ln \frac{N}{N_0} = -\ln \frac{N_0}{N}$$

This follows from the properties of logarithms.

If we know N_0, the number of $_6C^{11}$ atoms at the start of the experiment, we can compute N, the number remaining after 1 second, from Eq. 3. From this we can determine the number that must have decomposed. Therefore let us write

$$2.303 \log \frac{N_0}{N} = \ln \frac{N_0}{N} = kt = \frac{33 \times 10^{-3}}{\text{\sout{min}}} \times \frac{1}{60} \text{ \sout{min}}$$

(Note that we express the time in appropriate units, so that the product kt is unitless!) This reduces to

$$\log \frac{N_0}{N} = 0.00024$$

The value of this logarithm is so small that it is difficult to find the antilog accurately with the ordinary four- or five-place logarithms; hence N_0 is nearly equal to N, and the number of atoms that disintegrated, ΔN, must be small compared with the number of radioactive atoms present. In view of this, Eq. 1 becomes almost exact, and we can write

$$\frac{\Delta N}{N} = -kt = 0.55 \times 10^{-3}$$

where N is the number of carbon atoms at the start, and ΔN the number that disintegrates in 1 second. Substituting for N, we find

$$\Delta N = 0.55 \times 10^{-3} \times 5.5 \times 10^{22}\ {}_6C^{11}\ \text{at.}$$

or

$$\Delta N = -3.0 \times 10^{19}\ {}_6C^{11}\ \text{at.}$$

The negative sign denotes a decrease in number.

The second part of the problem is such that Eq. 1 is no longer a valid approximation. Since the number of ${}_6C^{11}$ atoms halves every 21 minutes, ΔN in 24 hours is no longer small compared with N; consequently, Eq. 3 must be used. Once again, we must express t and k so that kt is unitless. We can do so if we express 1 day as minutes. This is done as follows:

$$\frac{24\ \text{hr}}{1\ \text{d}} \times \frac{60\ \text{min}}{1\ \text{hr}} = \frac{1440\ \text{min}}{1\ \text{d}}$$

The remainder of the problem is straight forward, viz:

$$\ln \frac{N_0}{N} = kt = \frac{33 \times 10^{-3}}{\text{min}} \times 1440\ \text{min}$$

$$2.303 \log \frac{N_0}{N} = \ln \frac{N_0}{N} = 47.5$$

and

$$\log \frac{N_0}{N} = 20.6 = 0.6 + 20$$

According to the properties of logarithms, we find

$$\frac{N_0}{N} = (\text{antilog } .6) \times (\text{antilog } 20)$$

$$\frac{N_0}{N} = 4.0 \times 10^{20}$$

This leads to

$$\frac{N_0\ \text{at.}\ {}_6C^{11}}{4.0 \times 10^{20}} = N = \frac{5.5 \times 10^{22}}{4.0 \times 10^{20}}\ \text{at.}\ {}_6C^{11} = 1.4 \times 10^2\ \text{at.}\ {}_6C^{11}$$

In other words, after 1 day the number C^{11} atoms remaining is approximately 140. We are asked how many grams of C^{11} remain; from a simple application of units, we find

$$1.4 \times 10^2\ \text{at.}\ C^{11} \times \frac{18.3 \times 10^{-24}\ \text{gm}}{\text{at.}\ C^{11}} = 25 \times 10^{-22}\ \text{gm}\ C^{11}$$

PROBLEMS

1. Compute the number of atoms in: 16.0 gm of ${}_8O^{16}$; 1.0 gm of ${}_1H^1$; 4.0 gm of ${}_2He^4$.

2. Chlorine as it occurs in nature is 24% ${}_{17}Cl^{37}$ and 76% ${}_{17}Cl^{35}$. What is the average weight per atom of naturally occurring chlorine?

3. If the following disintegration were to occur completely,

$$ {}_{92}U^{238} \rightarrow {}_2He^4 + {}_{90}Th^{234} $$

how many grams of ${}_2He^4$ would be produced from 1.0 gm of ${}_{92}U^{238}$? *Ans.* 0.017 gm.

4. Complete the following equations.

(*a*) ${}_3Li^7 + ? \rightarrow 2\,{}_2He^4$. *Ans.* ${}_1H^1$.
(*b*) ${}_3Li^7 + {}_1H^1 \rightarrow ? + {}_0n^1$.
(*c*) ${}_{15}P^{30} \rightarrow {}_{14}Si^{30} + ?$.
(*d*) ${}_1H^2 + {}_1H^3 \rightarrow {}_2He^4 + ?$ *Ans.* ${}_0n^1$.
(*e*) ${}_7N^{14} + {}_0n^1 \rightarrow ? + {}_1H^1$.
(*f*) ${}_{15}P^{31} + {}_1H^2 \rightarrow ? + {}_1H^1$.
(*g*) ${}_6C^{14} \rightarrow {}_7N^{14} + ?$. *Ans.* ${}_{-1}e^0$.
(*h*) ${}_1H^3 \rightarrow {}_{-1}e^0 + ?$.
(*i*) ${}_{15}P^{30} \rightarrow ? + {}_{+1}e^0$. *Ans.* ${}_{14}Si^{30}$.
(*j*) ${}_5B^{10} + ? \rightarrow {}_7N^{13} + {}_0n^1$.

For the nuclear charge (atomic number) of the elements see the table on the inside of the front cover.

5. If 1 gm of ${}_1H^2$ reacted completely with an excess of ${}_{92}U^{238}$ in accord with the equation

$$ {}_{92}U^{238} + {}_1H^2 \rightarrow {}_{93}Np^{238} + 2\,{}_0n^1 $$

what would be the weight of ${}_{93}Np^{238}$ produced?

6. In Problem 5, compute the weight of the neutrons produced.

7. In a specific reaction of the type

$$ {}_{95}Am^{241} + {}_2He^4 \rightarrow {}_{97}Bk^{243} + 2\,{}_0n^1 $$

12.0×10^{23} neutrons are produced. How many grams of ${}_2He^4$ and ${}_{95}Am^{241}$ reacted?
Ans. 4 gm of ${}_2He^4$ and 241 gm of ${}_{95}Am^{241}$.

8. Verify by substitution in Eq. 3 the relation

$$ k = \frac{0.693}{t_{1/2}} $$

9. Show from Eq. 3 that an alternative form of this law would be that the fraction N/N_0 halves every $t_{1/2}$; that is,

$$ \frac{N}{N_0} = \left\{\frac{1}{2}\right\}^{t/t_{1/2}} $$

10. If we were to observe the changes that occur in uranium during its lifetime, we would find that 238.105 gm of uranium produced 32.0016 gm of α-particles, 0.0033 gm of β-particles, and 206.052 gm of lead. Within the indicated precision of these results, is this consistent with the principle of the conservation of mass? Can you explain this in detail?

11. Indicate by symbol the following:

Helium with 2 neutrons and 2 protons in the nucleus.
Uranium with 142 neutrons and 92 protons in the nucleus.
Carbon with 5 neutrons and 6 protons in the nucleus.
Oxygen with 7 neutrons and 8 protons in the nucleus.
Nitrogen with 8 neutrons and 7 protons in the nucleus.
Krypton with 50 neutrons and 36 protons in the nucleus.

√**12.** Radium E is a β-emitter with a half-life of 5.0 days. If a sample contains 1.0×10^{19} atoms, what is the activity (in disintegrations per minute) of the sample?

Ans. 1.0×10^{15} per minute.

13. The most accurate determination of the half-life of ${}_6C^{14}$ suggests that this value is 5.58×10^3 years. ${}_6C^{14}$ is constantly produced in the upper atmosphere by interaction of cosmic rays with air. This radioactive carbon mixes with the non-radioactive carbon present in the air as carbon dioxide. The rate of production of ${}_6C^{14}$ from cosmic rays is equal to the rate of disintegration; consequently, a certain concentration of ${}_6C^{14}$ is maintained in the air. The carbon in the air (as carbon dioxide) dissolves in the waters of the earth and is also utilized by plant life and, subsequently, by animal life. For this reason, all *living* things have the same ${}_6C^{14}$ content as the air in which they live. This concentration is such that each gram of carbon undergoes 15.3 disintegrations per minute. What fraction of these carbon atoms are ${}_6C^{14}$?

√**14.** Carbon extracted from the door frame of a house built in the time of Hammurabi of Babylon has an activity of 9.3 disintegrations per minute per gram of carbon. If the house was built right after the tree which supplied the wood for the door frame was cut down (killed), can you estimate when Hammurabi lived?

Ans. 2100 B.C. ($\pm$100 yr).

15. Carbon extracted from the linen wrappings of one of the Dead Sea Scrolls was found to have an activity of 12.0 disintegrations per minute per gram of carbon. Can you estimate the age of the Dead Sea Scrolls? Since 12.0 disintegrations/min/gm of carbon probably means between about 11.9 and 12.1, what sort of error can be present in this estimate of age?

16. ${}_{14}Si^{31}$ is radioactive with a half-life of 150 minutes. Calculate the number of disintegrations per minute in 1.4 gm of ${}_{14}Si^{31}$.

√**17.** A chemist carries out a reaction in which he starts with 1.4 gm of ${}_{14}Si^{31}$. After 16.00 hr, the reaction is complete. How many grams of ${}_{14}Si^{31}$ are left?

Ans. 0.0165 gm.

18. If, in the problem above, the chemist is lacking in skill and only 1% of the ${}_{14}Si^{31}$ remaining after 16.00 hr has reacted, what would be the number of disintegrations per minute in the product?

4

Atomic structure

At the end of the nineteenth century, there were many complacent scientists. The theories of light, mechanics, and electricity were accepted by virtually all, and the aim of many investigations in physics was merely to obtain established results more accurately. The shock was still to come. In the next few decades, the wave theory of light—one of the triumphs of nineteenth century science—was to be modified by a corpuscular theory; Newtonian mechanics, the cornerstone of physics, was to be done over completely; concepts of mass, distance, and time were to be altered; and it would be found that behavior of particles is described better by a wave pattern than by a corpuscular one. These unnatural-sounding descriptions are actually more natural than the old ones, inasmuch as they yield a more correct description of the facts. Clearly, this better state of affairs was possible only because a few courageous, clear-sighted scientists were willing to assault the foundations of physics, if physics, as it existed, could not yield theories consistent with the experimental results.

Although these newer concepts revolutionized physics, they did not require the complete destruction of the basic foundations of physics. Any true advance of physics created by new fundamental theories can only amplify and generalize such established areas as Newtonian mechanics. Thus the modern quantum mechanics and the classical Newtonian mechanics are both valid pictures of nature; quantum mechanics simply provides a more complete, *but still incomplete*, picture of nature.

As a consequence of this revolution, the modern theory of the atom was conceived, and, to understand it completely, we have to be familiar with the newer concepts of physics. A complete understanding requires a lengthy discussion with sophisticated mathematics. This we cannot do here. On the other hand, to ignore these developments completely, and to give you the end results only, would be unfair and misleading. So we shall present a brief sketch of these developments; and, to do so coherently, we shall first discuss the classical wave theory of light.

WAVE THEORY

Waves. Most of us are familiar with waves. We have seen the small ripples on a pond, or the large breakers that roll in from the ocean at a bathing beach. If we were to fly over the ocean in an airplane in order to examine the waves far below, we would find that the crests of these waves extend in regular patterns as far as the eye can see. These waves are periodic in space; in other words, the distance from the crest of one wave to the crest of its neighbor is always the same. This distance from crest to crest is known as the wavelength, λ.

Now, suppose that we are not flying in an airplane but are sitting on the beach and watching the waves as they roll in toward us. Then we will notice that the crests appear regularly, perhaps one every 20 seconds. Thus waves are also periodic in time. The number of crests per unit of time is known as the frequency, ν. In this case, the frequency is 3 per minute.

The third property of waves is the speed at which they travel. From our observations in the airplane and on the beach, we know that a wave arrives at the beach every 20 seconds and we know that the distance between these waves is 10 feet. If they are to remain evenly spaced, the crests must be moving at 30 ft/min. The symbol v is often used for the speed, but in the case of light it is customary to use the symbol c. There is an obvious relationship between the wavelength, the frequency, and the velocity. It is

$$\lambda \times \nu = v$$

and it holds for all types of waves.

Now, consider a wave that does *not* travel. Such waves are found in a vibrating violin string that is tied at both ends to the wood of the violin but is free to move up and down. Let us examine the up-and-down motion when we pluck the middle of such a string. If we take a motion picture of the string, we will find that it vibrates up and down in the middle while the ends remain fixed. A single frame of such a motion picture would resemble the sketch in Fig. 1*a*. This picture could also be looked upon as a graph of the instantaneous displacement (y) of the string from its equilibrium position as a function of the distance (x) from one end. The frequency of this vibration is characteristic of the violin string. If we pluck it harder, we may increase the displacement of the string from its equilibrium position, that is, its *amplitude*, but we will not alter its frequency.

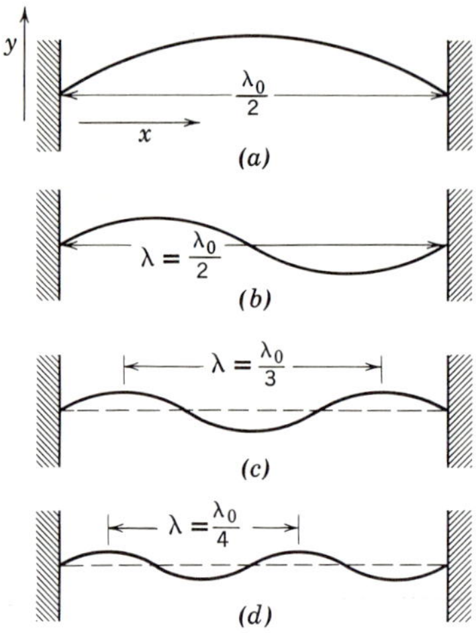

Fig. 1. Standing waves.

This wave in the violin string is known as a *standing wave* because its crest remains always in the same position and does not travel. The wave pictured in Fig. 1*a* represents only half a wave or one *loop*. A full wave will have two loops, one convex and the other concave, as we have indicated in Fig. 1*b*.

If we pluck the violin string not in the middle but a quarter of the way from one end, we will hear a tone considerably higher than the original one. A musician would recognize that this new note is one octave higher. An instantaneous picture of the string will reveal that we now have the whole wave pictured in Fig. 1*b*. The maximum occurs one-quarter of the way down the string, and the minimum three-quarters of the way down the string. We can produce other tones from this violin string. For example, we can pluck the violin string one-sixth of its length from one end and get the behavior indicated in Fig. 1*c*. Or, by plucking the violin string at one-eighth of its length from one end, we get the behavior indicated in Fig. 1*d*. Underneath each of these figures, we have noted the wavelengths in terms of λ_0, the apparent wavelength of the fundamental frequency.

Only certain wavelengths are possible in a violin string because the two ends are fixed; these waves contain an integral number of loops. In this description, λ_0 is referred to as the *fundamental* wavelength. The waves corresponding to Figs. 1*b*, 1*c*, and 1*d* are called the first overtone, the second overtone, and the third overtone; theoretically, an infinite number of these are allowed. In general, the wavelengths of the fundamental plus all the overtones are given by the expression

$$\lambda = \lambda_0/n$$

where n is any positive integer. Thus, in this simple one-dimensional case, any standing wave is specified by the value of the integer, n.

What do the vibrations of a square drumhead look like? In any section across this drumhead (Fig. 2), we find behavior of the type found with the violin string, namely, an even number of loops between the two fixed ends (indicated a-a' and b-b'). To specify the waves in such a drumhead, therefore, we need two sets of numbers (N and L), one for the a-a' axis, and one for the b-b' axis. Each of these has a series of possible values corresponding to positive integers.

With three-dimensional vibrations, we need *three* numbers in order to specify a particular type of vibration. Let us call these numbers N, L, and M. In some cases, these three numbers are independent; the fact that N assumed the value 1 makes no restriction on the number assumed by L. Occasionally, however, the coordinates used to describe the system are such that these three numbers are not independent and, if N assumes a certain value, this automatically restricts the values of L and M.

Lastly, we should note that, when several waves occur at the same place, the observed results are a net result. Thus, if we have two waves, a and b, traveling together as indicated in Figs. 3A and 3B, the result we observe is a wave formed by the sum of these two separate waves. In part (A), the two separate waves reinforce each other, so that the result is a wave of greater amplitude than either of the contributing waves. In part (B), wave a is going up while wave b is going down; hence the result is a wave smaller in amplitude than the component waves.

As a third example of the behavior of waves, consider the situation shown in Fig. 3C. In this plot, we have portrayed a group of waves of the same amplitude and frequency, differing only in phase, that is, the point at which they become zero. If we add these waves together, we find that a cancels a', b cancels b', and c cancels c'; hence the resultant is zero. It is a general principle that waves differing only in phase by random amounts always cancel each other.

Electromagnetic radiation. Visible light has a wave character. The speed of light is 3×10^{10} cm/sec; the wavelength varies from roughly 8000 Å to 4000 Å. In addition to this visible type of *electromagnetic radiation*, there are also invisible types that move at the same speed and have similar properties. Radar, for example, has a wavelength of about 1 cm. Radio waves are also a type of electromagnetic radiation, and the wavelengths go up to 1000 meters. Heat waves fit this classification; their wavelengths range from roughly 0.8 micron

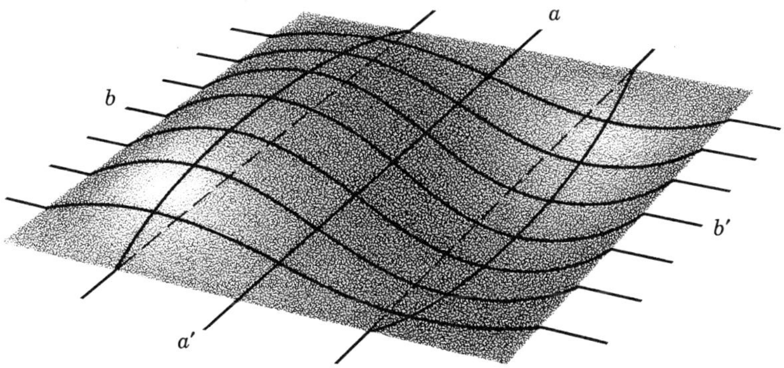

Fig. 2. Waves in square drumhead.

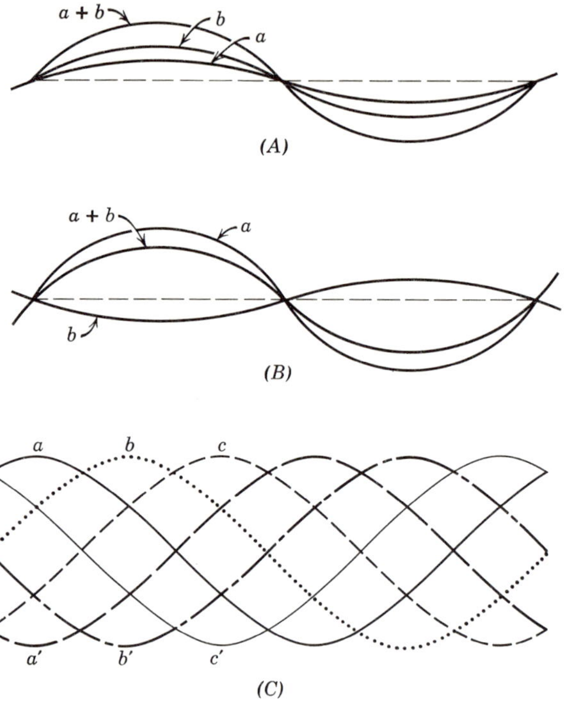

Fig. 3. Interference.

(μ) to 1000 microns. All of these types of radiation have a longer wavelength than visible light. On the shorter side of the scale, there is ultraviolet light, which has a wavelength between about 4000 and 100 Å, and X-rays, which have wavelengths between 10^{-6} and 10^{-9} cm. At still shorter wavelengths are the γ-rays given off in radioactive transformations and cosmic rays from outer space. All of these are examples of electromagnetic radiation.

The wave amplitude of a vibrating violin string relates to the displacement. With what, however, are we to identify the amplitude of a light wave? The detailed theory of light reveals that the *square* of the amplitude of a light wave is proportional to the energy or *intensity* of the light.

Consider the origin of light. Since radio waves are a part of the electromagnetic spectrum, let us look first at these. We can generate radio waves by passing a high-frequency alternating current into an antenna. In other words, a current of electrons flows into an antenna—first up, then down, and then up again. This wave generation occurs as the result of the acceleration of an electrical charge. If we could connect a proton and an electron by a spring, and cause these to vibrate, this would be a miniature antenna and electromagnetic radiation would be radiated from it. The frequency of the radio wave is the same as the frequency of the oscillating current in the antenna; also, the light generated by this spring-coupled electron and proton would have the same frequency as the mechanical vibrations. This is the *classical* picture of the origin of electromagnetic radiation.

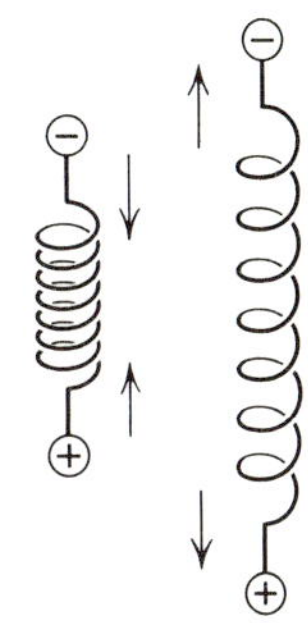

Fig. 4. Oscillators.

QUANTUM THEORY, 1900–1910

Appropriately enough, quantum theory was born in the first year of the twentieth century. At that time, physicists were puzzled by the details of the radiation of heat waves from incandescent bodies. These heat waves, a form of electromagnetic radiation, were described *as if* their source were a collection of oscillators of the type pictured in Fig. 4. Such oscillators would vibrate with increasing amplitude when heated and emit light just as if they were miniature antennae. On the basis of this model, it was possible to explain some, but not all, of the experimental observations.

In 1900, Planck showed that agreement of the theory with facts could be made essentially complete if the following postulates were made:

1. For each oscillator there is an *atom* or *quantum* of energy. Therefore the energy of an oscillator can only be a multiple of this quantum of energy. The value of this quantum is related to the frequency of the oscillator; one quantum is $h \times \nu$, where h is a universal constant equal to 6.625×10^{-27} erg sec.
2. An oscillator with the energy $E = nh\nu$ does not emit radiation continuously but in bursts containing one or more quantums of energy.

These postulates of Planck are summarized by the diagram in Fig. 5. This is called an energy diagram. If we supply energy to one of Planck's oscillators, the amplitude will not increase continuously. It will suddenly start vibrating with an amplitude corresponding to $E = h\nu$. If we supply still more energy, the amplitude will not increase unless we supply enough energy to increase the amplitude to a value corresponding to $E = 2h\nu$. Thus energy gain can occur only in increments of $h\nu$. If we could watch such an imaginary oscillator while it emitted an amount of light with an energy $h\nu$, we would find that the amplitude did not slowly die down as it emits the light; instead, the amplitude would instantaneously change from that corresponding to $E = 2h\nu$ to that corresponding to $E = h\nu$. At the same time a burst of radiation would be emitted.

As formulated by Planck, this theory dealt only with the oscillators themselves. In other words, although the energy changes of the oscillator were discontinuous, the light emitted was believed to behave in the traditional wave fashion. It

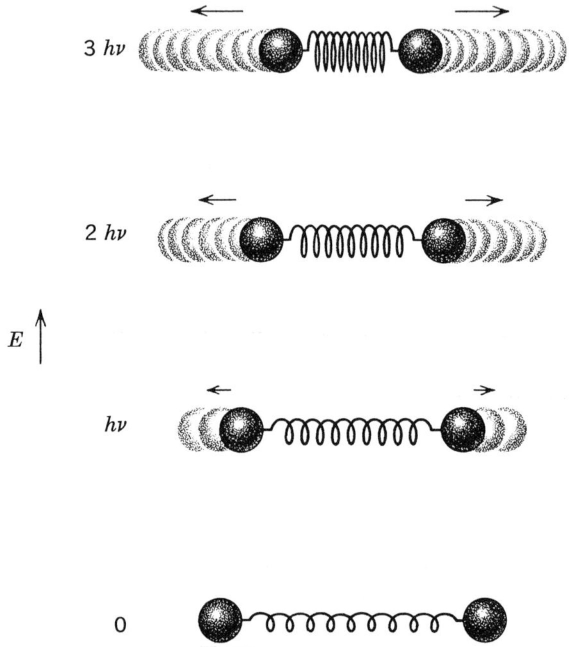

Fig. 5. Energy states in oscillators.

remained for Einstein, in the year 1905, to demonstrate that the photoelectric effect (that is, the ejection of electrons from a metal when exposed to light) could be best described by the assumption that light itself was corpuscular and that each corpuscle of light, each *photon,* had an energy equal to $h\nu$.

QUANTUM THEORY, 1910–1920

By this time, it was suspected that a hydrogen atom consisted of a proton and an electron. But how were they held together? Should an electron establish a stable orbit around the proton like a satellite around the earth? *No!* If an electron were to rotate around a proton, it would be constantly accelerating (see below); hence, according to the classical picture of light (p. 85), it would give off energy as electromagnetic radiation. This loss of energy would slow down the electron and, in less than a microsecond, it would plunge into the nucleus.

When a satellite revolves around the earth in a stable circular orbit, its speed is constant, but, since its direction of motion changes, the velocity is not constant. By definition, change in velocity is acceleration; hence a satellite (or a planetary electron) in a stable orbit must be subjected to a constant force ($f = ma$). With a satellite, the force is supplied by gravitational attraction. (With a planetary electron, the force is supplied by coulombic attraction.) This balance is given by

$$\text{Attraction} = m \times a$$

where m is essentially the mass of the satellite and a is the rate of change of velocity in a circular orbit. This acceleration can be shown to be related to the speed of the satellite, and its distance from the earth by

$$v^2/r = a$$

The $m \times a$ term, mv^2/r, is an effective force that tends to keep the velocity constant in direction, that is, a force that acts to make the circulating satellite (or electron) move away from the center; hence mv^2/r is often called the centrifugal force.

Bohr suggested that there might be some orbits in which this radiation does not occur. For such orbits,

$$\text{Centrifugal force} = mv^2/r = e^2/r^2 = \text{coulombic attraction}$$

where e is the electronic charge.* It is convenient to restate this condition in terms of angular momentum (the angular momentum is defined by the formula, $p_a \equiv mvr$) and rewrite the preceding equation as

$$p_a^2/mr^3 = e^2/r^2 \tag{i}$$

The total energy of such an orbit is given by

$$E = -e^2/r + p_a^2/2mr^2 \tag{ii}$$

The first term on the right represents the potential energy of the electron in the hydrogen atom *relative* to the energy when the electron and proton are separated by an "infinite" distance. The second term, as you can verify by substitution of the definition for angular momentum, is simply the kinetic energy, $\frac{1}{2}mv^2$.

Bohr postulated that such orbits were stable only if the angular momentum assumes certain definite values. These values are given by the equation

$$p_a = nh/2\pi \qquad n = 1, 2, 3, \ldots \tag{iii}$$

where n is any integer and h is Planck's constant. If we substitute Eq. iii in Eq. i we find that the radii of the stable orbits are

$$r = n^2h^2/4\pi^2me^2 \tag{iv}$$

With this, and Eq. ii, we can derive an expression for the energy of each of these stable orbits; that is,

$$E = -2\pi^2me^4/n^2h^2$$

Thus there is a series of possible orbits for the hydrogen atom, similar to those depicted in Fig. 6. An electron can move from a larger orbit to one closer to the

* See Appendix A.

nucleus, by a discontinuous jump, but all orbits between those indicated by Eq. iv are impossible.

When a gas composed of atoms is heated strongly or subjected to an electric discharge, it often gives off energy in the form of electromagnetic radiation. The "light" thus emitted is characteristic of the atom and is called its *spectrum.* To explain the spectrum of the hydrogen atom, Bohr made a second postulate. He assumed that, when a hydrogen atom was excited so as to emit light, it did not do so by continuous emission. Bohr said light was emitted by quantum jumps from one energy level (or one orbit) to another. The frequency of such light was assumed to be related to the difference in energy of the two orbits, ΔE, by the equation

$$\Delta E = h\nu$$

Accordingly, when a hydrogen atom is excited, the electron moves out to a higher energy orbit farther from the nucleus. When it emits light, the electron drops from the higher to a lower energy orbit, as depicted in Fig. 6.

We can summarize the energy levels of a hydrogen atom by the diagram in Fig. 7. In this diagram, the lowest line corresponds to the most stable position of the electron, that is, the smallest orbit. The next to the smallest orbit would be represented by the second line from the bottom, and so forth. If we excite a gas composed of hydrogen atoms, we might add enough energy that some of the electrons would go into the $n = 5$ level. These electrons can then drop down to any of the four energy levels below them. Each one of these quantum jumps results in the emission of a quantum of light with a frequency defined by the equation $\Delta E = h\nu$. On this basis, Bohr could predict all but the finer details of the spectrum of a hydrogen atom.

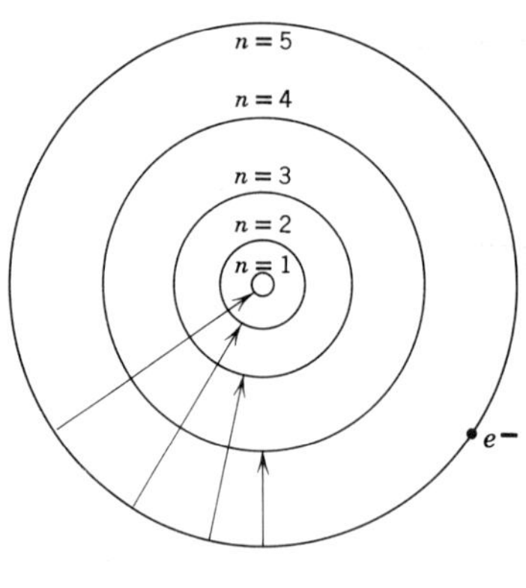

Fig. 6. Bohr orbits.

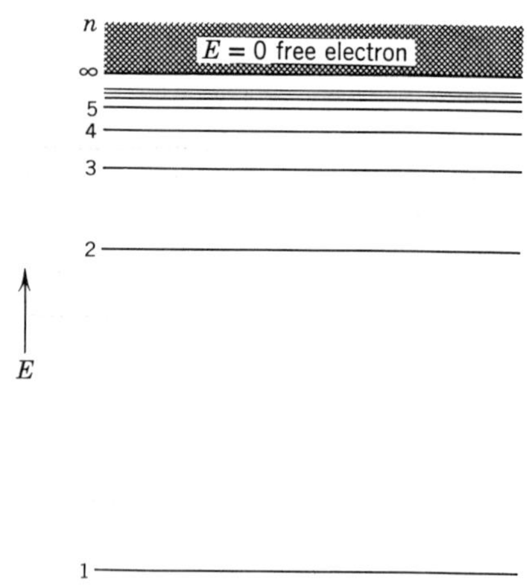

Fig. 7. Energy levels of a hydrogen atom.

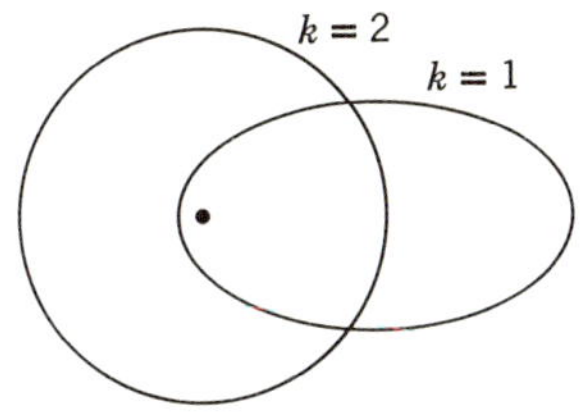

Fig. 8. Orbits of Bohr-Sommerfeld atom for $n = 2$.

Let us calculate some of the quantities that characterize a hydrogen atom in its lowest energy state, with $n = 1$. The radius of the orbit is 0.524 Å, and the speed of the electron is 2×10^8 cm/sec, roughly 1% the speed of light. This speed becomes quite impressive when we consider how small the atom is; the electron circles the proton *6.63 × 10¹⁶ times per second.*

This simplified Bohr treatment of the hydrogen atom ignored the fact that the rotation of the electron is accompanied by a corresponding rotation of the proton around the center of mass. If we take this mutual rotation into account, the mass in the equation above will not be the mass of the electron but will differ from it by 0.06 per cent.

The motion ascribed to the electron in the Bohr atom is, of course, quite similar to that encountered in the solar system. In the solar system, however, orbits are not only circular but are also often elliptical. Sommerfeld made the suggestion that we should consider the possibility of such elliptical orbits wherein there is not only motion normal to a radial line but also motion in the radial direction. To be consistent, both types of motion should be quantized. (The radial momentum varies from place to place in an elliptical orbit; therefore the postulated quantum condition for radial motion is somewhat complicated.) The net effect of this modification was that the *quantum* number n becomes a sum of two quantum numbers; that is,

$$n = (k + n_r)$$

The quantum number k is associated with the angular momentum in the same simple fashion as n was in the original picture. The quantum number n_r is associated with the radial momentum. When n_r is equal to 0, we again have a circular orbit with no radial motion. In view of the defining equation for angular momentum, the larger the value of k for an elliptical orbit, the further the electron will be, on the average, from the nucleus. A sketch showing two possible orbits is given in Fig. 8.

QUANTUM THEORY, 1924–?

The Bohr-Sommerfeld theory of the hydrogen atom was successful in explaining many of the details of the atomic spectra of hydrogen. It proved, however, to be limited and was incapable of explaining the details of the atomic structure of the larger atoms. More important than this, in certain cases where it was believed to be applicable, the predictions of the theory were inconsistent with experimental results. Thus the Bohr theory offered only a partial explanation of the atomic structure of matter.

While the Bohr theory was being tested and found wanting, experiments in other areas were performed which ultimately led to the modern picture of the atom. Once again, one of the most important of these was one dealing with light.

Corpuscle-like waves. In 1905, Einstein suggested that not only was the process of radiation from black bodies discontinuous but also the radiation itself was discontinuous. This suggests that light waves have a corpuscular character. If we are to describe light waves as corpuscles, what will be the property of these corpuscles? How will they behave? The answer was given in 1922 by A. H. Compton. In studies on X-rays, it was found that light of sufficiently short wavelength could collide with an electron and in so doing transmit kinetic energy. This is exactly what we would expect in a collision between two material particles. In addition to this, he was able to say that the momentum p of a photon was given by the formula

$$p = h/\lambda$$

where h is Planck's constant and λ is the wavelength of the light. This equation shows us that with every photon we must associate momentum.

γ-Ray microscope. Heisenberg seized upon this new mode of description in order to demonstrate that there are rigid limits to the accuracy of observations. To do so, he devised an imaginary experiment in which a beam of particles of known momentum was examined by the imaginary microscope shown in Fig. 9. In this experiment, a beam of light is used to determine as precisely as possible the position of a particle. The principles of classical optics tell us that with a given wavelength of light there is a definite limit on the accuracy of measurement with a microscope. The minimum uncertainty, Δx, in a position microscopically determined is called the resolving power, and for *all* microscopes this limit is specified by the formula

$$\Delta x = \lambda/(2 \sin \theta)$$

where θ is the angle indicated in Fig. 9. Thus, by using an appropriately short wavelength, we can decrease this uncertainty in position, Δx, to an arbitrarily small figure. In so doing, however, we increase the momentum of the light photon. In a collision between two particles (or a particle and a corpuscular photon) a change in the individual momentum of each particle occurs even though the sum of the two momenta remains the same before and after the collision. The

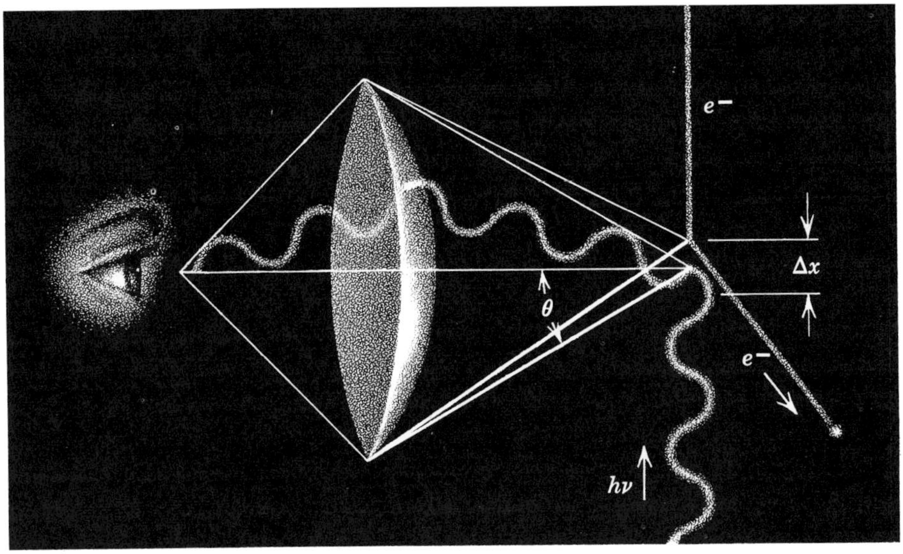

Fig. 9. Heisenberg's γ-ray microscope.

amount of momentum gained or lost by the particle during the collision increases with the momentum of the light photon. Therefore, if we attempt to make Δx small by making λ small, we increase the momentum of the photon *and* we increase Δp, the uncertainty in the particle momentum introduced by the observation. Thus the uncertainty in particle position, Δx, and the uncertainty in momentum, Δp, are inversely related. This is called the *Heisenberg Uncertainty Principle* and can be stated mathematically as follows:

$$\Delta p \, \Delta x = h$$

Thus, according to this relation, the more precisely we measure the position, the more uncertain we make the momentum. To state it less precisely, if we knew where the particle was, we would not know how fast it was going; and, if we knew how fast it was going, we would not know where it was.

It is natural to wonder if this uncertainty relation of Heisenberg's is due to the faultiness of our measuring instruments or is a fundamental property of nature. The answer is not particularly important. This uncertainty is a fault in all instruments as they exist today, and, in addition, it is the fault of any instrument we can conceive. Since in science we are concerned only with things we can measure, let us accept this uncertainty as a fundamental rule of nature. It implies that we cannot measure anything without disturbing it to a small degree. This is certainly reasonable. To observe something, our senses must be coupled with it. The coupling may occur only through a beam of light, but none the less it exists.

Hence a small and not always measurable disturbance is a part of all measurements. Many eminent scientists of our day have regarded Heisenberg's uncertainty principle suspiciously and some have declared it to be wrong. To date, however, no one has proposed even an imaginary experiment that could be carried out without such an uncertainty.

This uncertainty in measurement cannot be significant for large bodies; if it were, it would invalidate the laws of mechanics. Thus, for this principle to be correct, the uncertainty predicted for large objects should be trivial. Let us see if this is the case. Suppose we wanted to locate the position of a pea weighing 1 gm to within 10^{-8} cm (1 Å). If we definitely located the position to within 1 Å, we would be uncertain of the momentum and could *not* say that it was standing still. According to the equation, $\Delta p \, \Delta x = h$, it could be moving to the right or to the left with speeds as high as 2 Å per millenium. The situation is not so ridiculous when we are considering a small particle like the electron. According to the uncertainty principle, if we located this electron with an uncertainty of 1 Å, the momentum again would be uncertain. In this case, however, the electron could be moving to the right or to the left with a speed of roughly 1 per cent the speed of light. Because in discussing the structure of the atom we are interested in these small particles, we must take this uncertainty into account. For larger particles, visible to the eye, it is no longer an important factor.

If we accept the uncertainty principle, we must reject the Bohr picture of the hydrogen atom. The assignment of a *definite* orbit to an electron is unacceptable; planetary models of the atom are wrong.

Derivation of the uncertainty relation. The uncertainty principle can be derived in a general form applicable to any type of observation. For simplicity in this derivation, let us restrict ourselves to the experimental procedure involving a γ-ray microscope. The lens of the microscope will not differentiate between light quantum 1, deflected through a small angle, $\pi/2 - \theta$, and light quantum 2, deflected through a large angle, $\pi/2 + \theta$. These and all photons with intermediate deflection will be indistinguishable when we look through the microscope. If we take the particle viewpoint, we see that photon 1 lost a small amount of momentum in the x direction and photon 2 lost an even larger amount. The change in momentum *in the x direction* in these extreme cases will be given by the formulas

Case 1: $$p_i - p_f = p_i - p_i \sin\theta = h/\lambda \cdot (1 - \sin\theta)$$

Case 2: $$p_i - p_f = p_i + p_i \sin\theta = h/\lambda \cdot (1 + \sin\theta)$$

According to the law of the conservation of momentum, this change could occur if and only if the particle we are observing gained or lost x-momentum.† From our observations, this gain or loss of momentum could be anywhere from $p_i - p_i \sin\theta$ to

† Strictly speaking, this is an approximate treatment since relativistic corrections are ignored.

$p_i + p_i \sin\theta$. Thus, in determining the position of this particle, we have made the x-momentum uncertain to the following degree:

$$\Delta p = 2\frac{h}{\lambda}\sin\theta$$

There will also be an uncertainty in the position of the particle due to the limited resolving power, Δx, of the microscope. The product of the uncertainty in the momentum and the uncertainty in the position turns out to be

$$(\Delta p\,\Delta x) = \left(\frac{2h}{\lambda}\sin\theta\,\frac{\lambda}{2\sin\theta}\right)$$

which reduces to

$$\Delta p\,\Delta x = h$$

Thus the product of uncertainty is completely independent of the wavelength of light used; it will be the same for all such experiments.

Wave-like corpuscles. Shortly after Compton's experiments, de Broglie showed that it was reasonable (on the basis of relativity) to expect particles to show wave-like properties. The wavelength of a moving particle was regarded as determined by an equation similar to that used to describe the particle character of light waves. That is,

$$p = mv = \frac{h}{\lambda}$$

At that time, the interpretation of this "wave" was not clear. It was supposed that, where an electron "was" according to a classical description, a wave "was."

On this basis, Bohr orbits would correspond to waves like those pictured in Fig. 10. These waves are what we would get if we bent the violin string, discussed earlier, and connected the two ends. For such a picture to be reasonable, each orbit should contain a whole number of wavelengths. If it did not, waves that had completed a different number of cycles would be out of phase and, as we

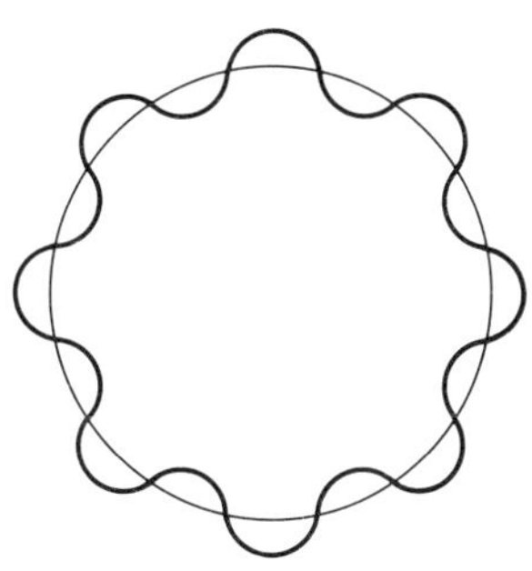

Fig. 10. Simple electron wave in H atom.

stated earlier (Fig. 3*C*), the superposition of such waves results in cancellation of the net wave. If this occurs, there could be no electron in such an "orbit." In this description, the quantum conditions become particularly simple. For an orbit of radius a, the condition for an integral number of waves is

$$\lambda n = 2\pi a$$

where n is an integer. If we express λ in terms of particle momentum, we obtain

$$nh = 2\pi amv$$

This, by rearrangement, becomes

$$\frac{nh}{2\pi} = p_a$$

which is the same as Bohr's quantum condition.

How should we interpret this wave? With light waves, the square of the amplitude, ψ^2, is assumed to be proportional to the intensity of the light. This means that ψ^2 is proportional to the number of photons at that point. In Fig. 11, we have sketched a wave function, ψ, and the square of the amplitude, ψ^2, at a given instant. For a light wave, the height, A, is proportional to the number of photons present in the region, R, at that instant. This interpretation is straightforward provided we are working with an intense light ray. But suppose we are working with weak light. Then the height, A, might correspond, mathematically, to 0.1 of a photon or even 0.01 of a photon. In such a case, the only reasonable interpretation is that A represents the *probability* of finding a photon at this point. This probability must depend on the size of the region R, as well as on the value of ψ^2. In other words, if the region R were made half as great, the probability of finding a photon would be half as great. Thus the probability of finding a photon in a region of size R is ψ^2 at that point *times* R. In three dimensions, R represents a volume and $\psi^2 R$ represents the probability of finding a photon in that volume. Thus ψ^2 is a probability per unit volume; it is often referred to as the *probability density*.

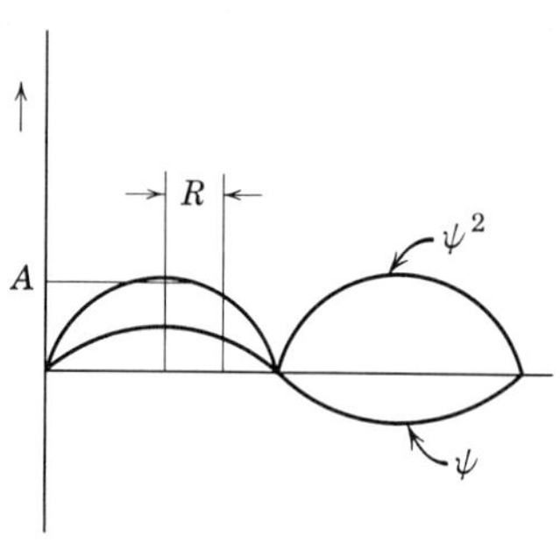

Fig. 11. Wave function.

If the wave function in Fig. 11 represents the wave associated with an electron, the square of the amplitude should be related to the "intensity" of the electron. This means that the value, A, is proportional to the probability of finding an electron in the region, R. Thus the position of the electron can no longer be pinpointed. We can discuss the probability of its being in a given region, but we can no longer say that the electron is definitely at a fixed point.

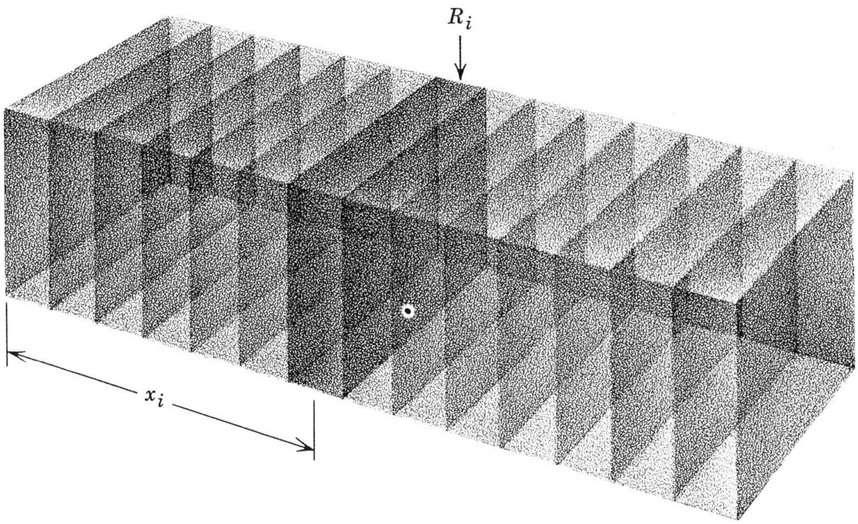

Fig. 12. Interpretation of wave functions.

In order to obtain a clearer picture of the meaning of an electron wave function, consider an imaginary experiment in which an electron is confined to a box like that in Fig. 12. Let us take the approach that an electron is indeed a particle and attempt to measure its position. When we measure its position, we find that repeat measurements do not agree; sometimes the electron is in one region of the box; other times it is in another region of the box. To describe these results, let us imagine that we divide up the box into small regions like that indicated in Fig. 12. We will number these regions starting from the left-hand side of the box and indicate the corresponding volumes R_1, R_2, etc. Mathematically, we would say

$$\sum_i R_i = \text{volume of box}$$

where the sign $\sum_i$ reads "sum over all possible i." In other words, the sum of these volumes is equal to the volume of the box. Now, we find that there is a wave function ψ that will describe our results. In this description, let ψ_i equal the value of ψ in the ith volume; then $R_i\psi_i^2$ represents the probability that we will find the electron in the ith volume. We can assume that the variation in the position of the electron is brought about by constant motion of the electron and that $R_i\psi_i^2$ represents the fraction of the time it spends in the region, R_i. We can also assume that $R_i\psi_i^2$ represents the fraction of times we will find an electron at R_i in a large number of observations on identical boxes. In either case, this experiment will be described best by the wave function ψ_i.

Let us extend this statistical interpretation and measure the average distance, $\bar{x}$, of the electron from the left side of the box. Let us call x_i the distance of R_i from the left side of the box. Then we can write

$$\bar{x} = \sum_i \psi_i^2 R_i x_i$$

The correctness of this equation is best seen by a similar example. Suppose that of 100 nickels 30 weigh 5 ± 1 gm, 40 weigh 7 ± 1 gm, and 30 weigh 9 ± 1 gm. The average weight will be

$$\frac{30}{100}\,5\text{ gm} + \frac{40}{100}\,7\text{ gm} + \frac{30}{100}\,9\text{ gm} = \text{average weight}$$

In order to get the average weight, we took the probability that a nickel weighs 5 gm and multiplied it by 5 gm and added it to like terms for the other possibilities. This is completely analogous to what we did to find the average position of the electron; in this calculation $\psi_i^2 R_i$ represents the probability that the position is x_i.

A similar approach can be used to calculate the average of any quantity Q that depends on position, that is,

$$\overline{Q} = \sum_i \psi_i^2 R_i Q_i$$

This would apply, for example, if there were a force that varied with distance acting on the electron. The average force would be given by the equation above.

What have we gained by the use of this description of an electron? If we accept this indefiniteness of the position of an electron, we are able to calculate quite precisely the momentum and energy of the electron. This holds regardless of whether we are talking about an electron confined to a box or attracted by a proton to form a hydrogen atom. To be a valid treatment, however, the results must also be consistent with the uncertainty principle, and the real triumph of the wave description is that it assures this consistency. Thus the electron appears (statistically) not as a particle but as an unsplittable charged cloud that, *in toto*, has the spin, mass, and charge associated with the electron.

MODERN THEORY OF THE HYDROGEN ATOM, 1924–?

Schroedinger devised an equation to describe the properties of these waves as a function of the kinetic and potential energy of the particles. The mathematics is involved but yields wave functions *formally* analogous to those found for elastic waves like those in a vibrating violin string. In the case of a hydrogen atom, however, the wave has three dimensions; therefore, to specify the allowed waves, we need three numbers, n, l, and m. The most accurate description is obtained when the consequences of Einstein's theory of relativity are considered. Such considerations (which might be said to recognize the necessity of a four-dimensional description involving spacetime coordinates) suggest that for a complete description of matter waves we need four numbers, the three mentioned above plus an additional one which is *formally* associated with the spin of an elementary particle. These four numbers are not all independent. The allowed values are

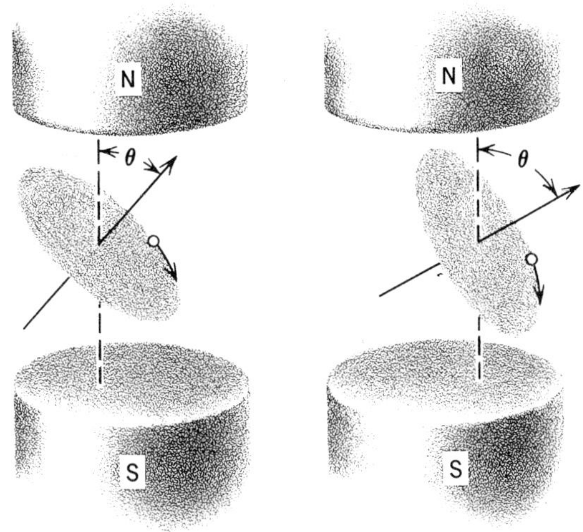

Fig. 13. Orientations in magnetic field.

$$
\begin{aligned}
n &= 1, 2, 3, \ldots, \text{any integer} \\
l &= 0, 1, \ldots, n - 1 \\
m &= +l, l - 1, \ldots, -(l - 1), -l \\
s &= +\tfrac{1}{2}, -\tfrac{1}{2}
\end{aligned}
$$

The number n has a composite character and is similar to the principal quantum number in the Bohr-Sommerfeld theory. The number l corresponds to the quantum number k in the Bohr theory and specifies the expected value of the angular momentum.† The lower the value of l, the lower is the angular momentum, and the more time the electron spends in the neighborhood of the nucleus; the higher the value of l, the higher is the angular momentum, and the more time the electron will spend at distances far away from the nucleus.

The quantum number m is associated with the orientation of the angular momentum or, classically, the axis of the orbit of the electron. An electron, in the classical Bohr orbit, is essentially a current in a loop. The theory of electricity and magnetism tells us that this would be a small electromagnet. If we put this between the poles of a magnet, there would be a tendency for it to orient so that the north pole of the atom would be as close as possible to the south pole face of the magnet. Details of the theory of quantum mechanics tell us that the orientations in a magnetic field are limited to a certain fixed number as indicated in Fig. 13. If the angles of these allowed orientations are measured in terms of θ, then $\cos \theta = m/[l(l + 1)]^{1/2}$. In other words, there is an allowed orientation in a magnetic field for each value of m. In the absence of a magnetic field, which

† In the Bohr theory, the angular momentum is given by $p = (hk/2\pi)$. In the modern theory, the angular momentum is given by $p = (h/2\pi)\ \sqrt{l(l + 1)}$.

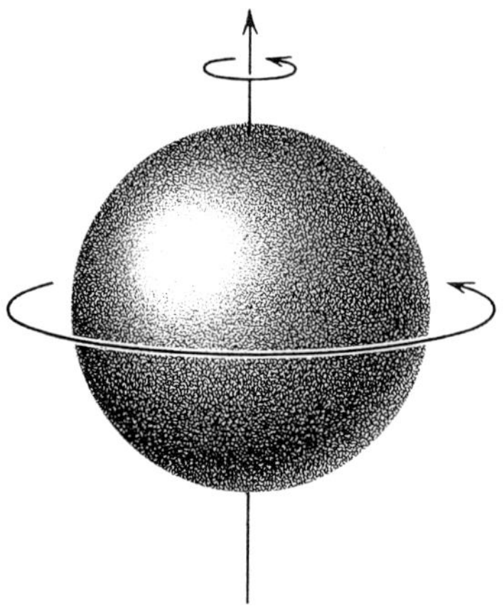

Fig. 14. Electromagnet spinning electron.

essentially establishes a special direction in space, atoms differing only in m value are indistinguishable.

The quantum number s, although it originates from a relativistic treatment, can be associated with a spin of the electron. Such a spin, indicated in Fig. 14, would have a quantized angular momentum. This momentum is given by the formula $(h/2\pi)\sqrt{S(S+1)}$. Since there is a circulating electrical charge, a spinning electron behaves like an elementary magnet and between the pole faces of an external magnet it can only take up certain allowed orientations. Two orientations can occur, one with the north pole of the spinning electron pointing toward the south pole face, and the other with the north pole of the electron pointing toward the north pole face. The first orientation is referred to as the parallel or + orientation; the second is termed the anti-parallel or − orientation.†

In effect, these four quantum numbers label the electrons so that the specific description that we lost because of the uncertainty principle is replaced by quantum numbers. Accordingly, there are many different ways in which an electron can be bound to a hydrogen atom. A few of these are summarized in Table 1, which lists all possible quantum numbers for an electron in a hydrogen atom with $n = 1, 2$, and 3. There are two different labels for the electrons corresponding to $n = 1$; eight for $n = 2$; and 18 for $n = 3$. We get more and more labels as n increases, and it would be rather cumbersome to include all of them in a single table; instead, let us merely note that, for each specification of n and l, there will be $(2l + 1) \times 2$ labels. The $2l + 1$ indicates the number of the m specifications consistent with n and l, and the factor 2 indicates that for each one of these specifications we can have the electron with a + or − spin.

According to the modern theory, the energy of the hydrogen atom in empty space is the same as that given by the older description of Bohr:

$$E_n = \frac{-2\pi^2 e^4 m}{h^2 n^2}$$

Thus the energy depends on the principal quantum number alone. In the presence of a magnetic field, however, the energy is different for states having dif-

† The orientation is not really parallel to the field. According to quantum mechanics, perfect alignment with the field is impossible both for the axis of spin and for the orbital axis. Perfect alignment would require a violation of the uncertainty principle.

Table 1

n	l	m	s
1	0	0	$+\frac{1}{2}$
			$-\frac{1}{2}$
2	0	0	$+\frac{1}{2}$
			$-\frac{1}{2}$
	1	+1	$+\frac{1}{2}$
			$-\frac{1}{2}$
		0	$+\frac{1}{2}$
			$-\frac{1}{2}$
		−1	$+\frac{1}{2}$
			$-\frac{1}{2}$
3	0	0	$+\frac{1}{2}$
			$-\frac{1}{2}$
	1	+1	$+\frac{1}{2}$
			$-\frac{1}{2}$
		0	$+\frac{1}{2}$
			$-\frac{1}{2}$
		−1	$+\frac{1}{2}$
			$-\frac{1}{2}$
	2	+2	$+\frac{1}{2}$
			$-\frac{1}{2}$
		+1	$+\frac{1}{2}$
			$-\frac{1}{2}$
		0	$+\frac{1}{2}$
			$-\frac{1}{2}$
		−1	$+\frac{1}{2}$
			$-\frac{1}{2}$
		−2	$+\frac{1}{2}$
			$-\frac{1}{2}$

ferent values of m and s. Thus, in a magnetic field, a state with $n = 2$ has one of several values for the energy. If we ignore the effect of spin, the energy state $n = 2$, $l = 1$ splits up into three separate energy levels, one for each value of m as shown in Fig. 15. If we consider the spin, even further splitting occurs.

A nomenclature has been developed to describe the states of an atom. States with different quantum numbers but the same energy are referred to as *degenerate*. Thus the eight states of equal energy corresponding to $n = 2$ (Table 1) are degenerate states. In specifying the various states, it is quite confusing to use

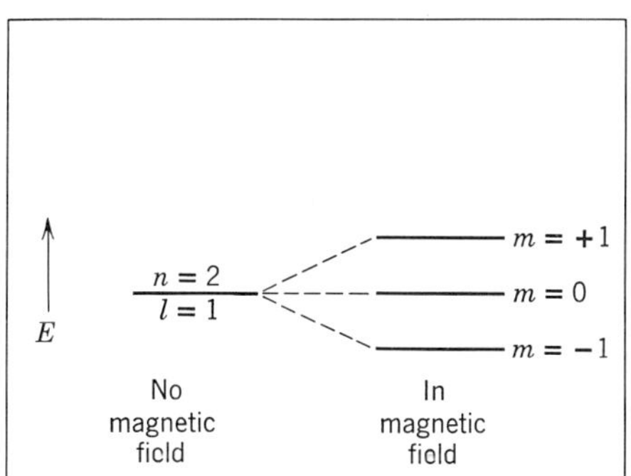

Fig. 15. Splitting of levels in magnetic field.

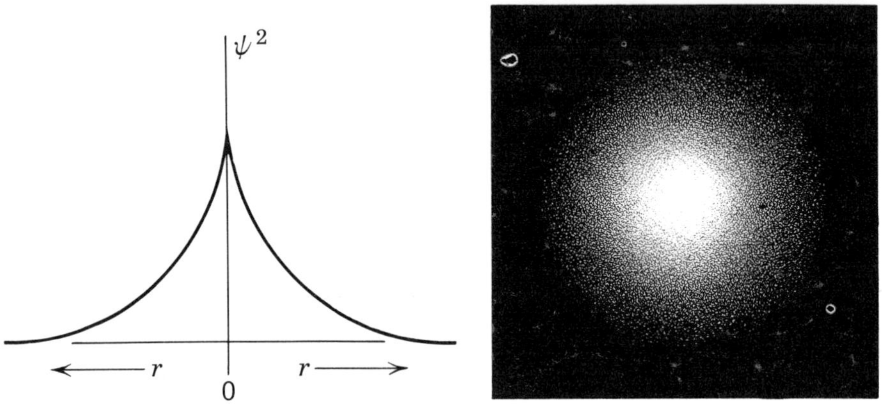

Fig. 16. Wave function for H atom: "picture" of H atom.

four sets of numbers. To avoid this, letters are often substituted for the quantum number l. According to this convention, $l = 0, 1, 2, 3, \ldots$ is also written $l = s, p, d, f, \ldots$.

The concept of degenerate states was contained in the Bohr-Sommerfeld picture. Thus it was possible to have the state $n = 2$ correspond to two different orbits that accidentally have the same energy, that is, a circular orbit and an elliptical orbit. Although these two states would have the same energy, the position of the electron in these two cases would be different. A similar situation holds with the wave theory, only now we have to couch our description in terms of probabilities.

Figure 16 is a graph of the probability density for an electron in a hydrogen atom in the $n = 1$ state, as a function of distance from the nucleus. Along with this, we have included a photograph of the hydrogen atom, imaginary, of course, that represents what we would see if the electron functions as a point source of light. In this picture, the lightness is proportional to the probability density at any given point.

The plot in Fig. 16 represents the probability density along a radial line from the center of the hydrogen atom. It is also of interest to consider the probability of finding an electron in a given spherical shell of fixed thickness. If we do so, we find that the most probable shell is at a distance of 0.524 Å from the origin;† that is, the electron is most likely to be found at a distance corresponding to the Bohr orbit.

† The volume of a thin spherical shell is $4\pi r^2\tau$ where r is the distance and τ is the thickness. The probability of finding an electron in such a shell is $4\pi r^2\tau \times \psi^2$. The first factor increases as r increases; the second factor decreases. It is this product that has a maximum at $r = 0.524$ Å.

Should we then look upon the wave description for the hydrogen atom as a "fuzzed-out" Bohr description? No. The Bohr description leads to the incorrect conclusion that the hydrogen atom in its lowest state possesses angular momentum; the wave description correctly predicts that the atom in its lowest state has no angular momentum.

In Fig. 17 we have shown some hypothetical pictures for a hydrogen atom in a magnetic field with $n = 2$. For s states, i.e., $l = 0$, the charge cloud is spherically symmetric. By this we mean that, as we go along *any* radial line, the probability density will vary in the same manner. On the other hand, for p states, i.e., $l = 1$, the charge cloud is not spherically symmetric.

Although the wave description is the only correct description, it is often useful to discuss various aspects of atomic and molecular structure in terms of the charge cloud we associate with the electrons. These charge clouds really represent the probability density of electrons in a dynamic atom; but many properties of atoms and molecules can be explained with the assumption that the electron is an unsplittable but deformable static charge cloud.

WAVE THEORY OF THE ELEMENTS

The major triumph of the wave theory was that it could be used not only to describe the hydrogen atom but also other atoms. Description of these other atoms is complicated by the fact that there are several electrons around the nucleus; hence, in describing the properties of these atoms, we must consider not only the attractive forces between the electrons and the nuclei, but also the repulsive forces

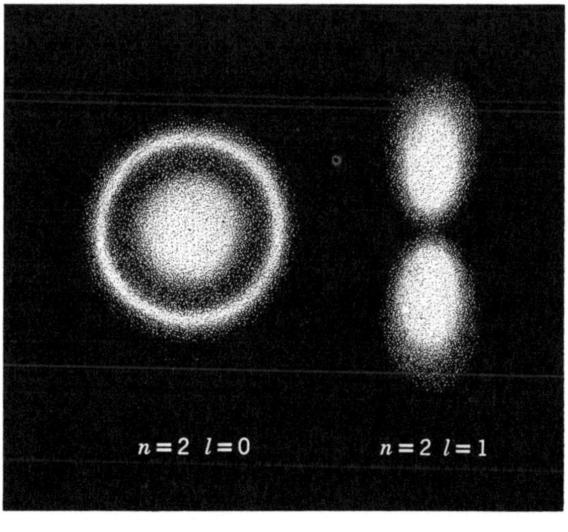

Fig. 17. "Pictures" of H atom.

between the electrons themselves. We will not consider these repulsive forces in detail but will only attempt to give some rationalized working rules. Let us now build the atoms that make up the periodic chart. We will do so by starting with the hydrogen atom and adding appropriate numbers of protons and neutrons to the nuclei and electrons to the outer shell.

Pauli exclusion principle. The guiding rule in this building-up process is the Pauli Exclusion Principle. This principle states that each electron in an atom must have a distinctive label; in other words, *no two electrons in an atom can have all four quantum numbers the same.*

When we form a helium atom from a hydrogen atom by increasing the nuclear charge from 1 to 2 and adding an additional electron, we find that for the electrons to be in their lowest energy states ($n = 1$) the spins must be opposed or Pauli's principle will be violated, cf. Table 1. In the lithium atom, however, only two electrons can fit into the $n = 1$ shell; the lowest possible energy level for the third electron is one of the eight corresponding to the $n = 2$ level.

Penetration effect. With the hydrogen atom, all eight of the $n = 2$ levels were equal in energy; for the lithium atom, they are not. In the lithium atom, the nuclear charge of $+3$ is surrounded by the spherical charge cloud of the two 1*s* electrons. If the last added electron has appreciable angular momentum, e.g., $l = 1$, it will spend most of its time outside this core charge cloud, and the effective nuclear charge will be $+1$. On the other hand, if this electron has zero angular momentum ($l = 0$), it will spend a good deal of its time close to the nucleus inside the core charge clouds. Thus *s* electrons will often "see" the nucleus at almost full strength and will be more firmly held than the *p* electrons. By similar reasoning, we can argue that the *p* electrons will be more firmly held than the *d* electrons, and so forth. This effect has been referred to as the *penetration* effect. Thus, in general, for atoms with more than one electron, the most stable energy state for a given value of n will be that with the lowest value for l.

Table 2 indicates the order of energy states for neutral atoms†. For the proper use of this table, we must keep in mind that each m level can accommodate two electrons. The *p* subshell has three values of m, each of which can accommodate electrons with + or − spin. Hence the *p* subshell can accommodate six electrons. Similarly, the *d* subshell can accommodate ten, and the *f* subshell can accommodate fourteen electrons.

Hund's rule. Before we discuss the structure of all the atoms of the periodic chart, we need one more rule. Consider the levels of the six electrons in a carbon atom. The lowest energy state, or *ground state*, according to Table 2, would have two electrons in the 1*s* subshell, two in the 2*s* subshell, and two in the 2*p*

† These levels shift with atomic number. This is the apparent order as judged by unexcited states of neutral atoms.

Table 2

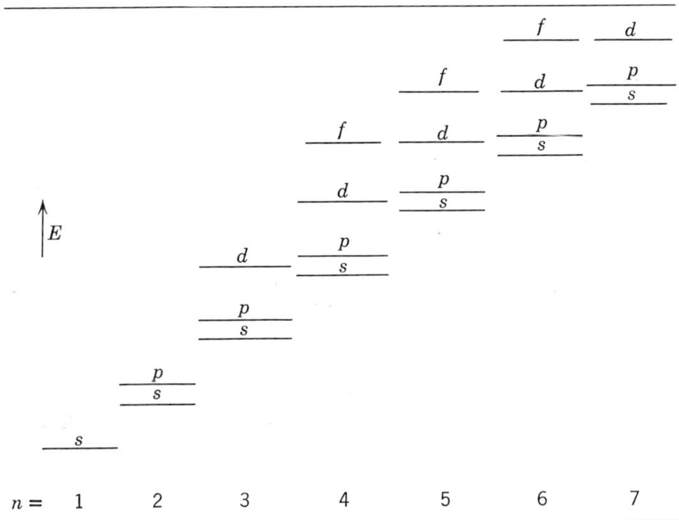

subshell. There are several ways of indicating this. The most common way is that shown in Fig. 18*a*. It involves a numeral to specify the value of n followed by a letter to specify the value of l. A superscript is used with a letter to indicate how many electrons are in this subshell. An alternative method, sometimes more revealing, is shown in Fig. 18*b*. Here we have a table in which the numerals indicate the number of electrons. Those in the same column correspond to the same principal quantum number; those in the same row correspond to the same orbital quantum number. For simplicity, we often omit the labeling. Both designations in Fig. 18 are incomplete. Two electrons in a $2p$ state can either have the same value of m, with antiparallel spins, or they can have different values of m with parallel or antiparallel spins. In the latter instance, the electrons will be further separated; different values of m correspond to orbitals located in different regions of space. Since there are repulsive forces between electrons, this arrangement would be the more stable. We can guarantee this minimum in repulsion if both the electrons have parallel spins; then it would be a violation of the Pauli exclusion principle to put them in the same state. This, in essence, is the interpretation given to the principle (Hund's Rule) that requires electrons in the same orbital state to have parallel spins, if possible.

$1s^2 \; 2s^2 \; 2p^2$

(*a*)

	1	2	3	4
f				
d				
p		2		
s	2	2		

(*b*)

Fig. 18. Electronic structure of carbon.

Chemical activity. We shall see in a later chapter that chemical activity of atoms occurs only if there is an unfilled subshell or, equivalently, unpaired electrons. Often, however, the atom in the ground state may have a filled subshell and still be active chemically. For example, beryllium with four electrons has the 1*s* and the 2*s* subshells completely filled. However, it requires only a relatively small amount of energy to move one of the electrons from the 2*s* shell to the 2*p* shell (Table 2), and once a chemical reaction occurs enough energy is released so that the *over-all* process is accompained by a decrease in energy. With some atoms having closed shells, the energy needed to create partially empty subshells is prohibitive. For example, an atom of helium with two electrons in 1*s* state is an unreactive molecule. True, it is possible to create unfilled subshells by supplying enough energy for one of the electrons in the 1*s* shell to be excited into the 2*s* shell. Under such conditions, helium can react to form a molecule, but the initial creation of the partially filled shell requires a good deal of energy (Table 2) and the over-all process of molecule formation will involve an increase in energy; hence the molecule is unstable. In summary, therefore, it would appear that *atoms with completely filled subshells can be reactive provided the higher energy level is fairly close.* If it is not, an atom with filled subshells will be unreactive.

Structure of the elements. Let us now resume our study of the building-up process of the elements. We have already discussed lithium with atomic number 3. With the addition of one more electron, we fill up the 2*s* subshell, and the next several added electrons must go into the 2*p* subshell. Thus boron has one electron in the 2*p* subshell. Carbon and nitrogen, atomic number 6 and 7, have two and three electrons, respectively, in the 2*p* subshell; these electrons have parallel spins. When we add three more electrons and form neon, we have filled the $n = 2$ levels and are in a level far below the next energy level. Thus we expect neon to be unreactive chemically, as it is.

The addition of the next eight electrons in the $n = 3$ shell follows essentially the same pattern. The element argon (atomic number 18) has the 3*s* and 3*p* subshells filled. In contrast to neon, however, the $n = 3$ shell is not filled at this point; the 3*d* subshell is still empty. Since electrons in the *d* subshell do not penetrate the core electrons, their energy is considerably higher than that of the other $n = 3$ subshells. Therefore argon is chemically unreactive just like neon.

At this stage, there is irregular behavior. The penetration effect is such that the 4*s* level is actually lower than the 3*d* level. Thus the next two electrons are added to the 4*s* subshell even though the $n = 3$ shell is not yet filled. Upon filling the 4*s* subshell, the next energy levels are the 3*d* levels. Ten electrons fit into these 3*d* levels and, when this 3*d* shell is filled, we have the element zinc, atomic number 30. The next six electrons go into the 4*p* subshell and, when this is completely filled, we have formed the inert gas krypton.

The procedure is again repeated for the next eighteen elements. First we fill up the 5*s* subshell; then the 4*d* subshell and the 5*p* subshell. When this filling is complete, we form the chemically inert rare gas xenon.

In the next period the situation again becomes irregular. After we add two more electrons, we fill up the 6*s* subshell. With the third electron, the complication arises. Between the 6*s* and 6*p* levels, there is the 5*d* level, which is still unfilled, and the 4*f* level, which is also unfilled. These have nearly the same energy, but the next added electron goes into the 5*d* subshell. The second electron, however, starts to fill up the 4*f* subshell. This continues until we have added the complement of fourteen electrons to the 4*f* subshell, that is, up to hafnium, atomic number 72. With the 4*f* subshell filled, we proceed to fill the 5*d* subshell. This is completed by the time we get to mercury, atomic number 80. Once again, in the usual fashion, we fill up the 6*p* subshell to form another inert rare gas, radon.

We have slightly oversimplified the description of the mode in which the levels are filled up. For example, vanadium has three electrons in the 3*d* level and two electrons in the 4*s* level. At the adjacent atom, chromium, two things happen. We add one electron to the *d* subshell, and one electron comes from the 4*s* subshell into the 3*d* subshell to yield five 3*d* and one 4*s* electrons. Instances of such irregularity are fairly common when we are considering the filling up of inner shells. Rather than discuss all these exceptions in detail, we have tabulated the electronic structures of the elements on the back cover pages.

Many of the elements are classified by their electronic structure. For example, elements which do not have partially filled *d* or *f* subshells are designated by Roman numerals. Elements which have partially filled *d* subshells are called transition elements; elements which have partially filled *f* subshells are called rare earth elements, or *lanthanons* (4*f* electrons) and *actinons* (5*f* electrons).

From the electronic structure of the elements, we can readily understand the periodic character of the chart on the inside of the back cover. Elements with atomic numbers 3, 11, 19, 37, 55, and 87 have a structure with one electron more than an inert gas. This electron will be in an *s* subshell. Elements with atomic numbers 4, 12, 20, 38, 56, and 88 have a rare gas core with two electrons in an *s* subshell. Elements with atomic numbers 9, 17, 35, 53, and 85 have the electronic structure of a rare gas with one missing electron, and for these elements we will always have a *p* subshell containing only five electrons. Such groups as these three, with similar electronic structure for their outer electrons, are said to be members of the same *group* or *family* in the periodic chart. Since it is largely the nature of these outer electrons that determines the details of chemical reactivity, we would expect "family resemblances" among elements of the same group.

Among the rare earths and the transition elements, there is often as much similarity between neighboring elements in the same row as there is between elements of the same group. This is not difficult to understand. Titanium and vanadium both have two *s* electrons in their outer shell. They are different inasmuch as the number of electrons in their inner *d* shell is two and three respectively. We expect the chemical properties to be determined mainly by the electrons in the outermost shell; consequently, we might expect titanium and vanadium to be similar in behavior. This similarity of adjacent groups is more pronounced for the lanthanons and actinons than for transition elements because the partially filled subshell (*f*) is even farther from the outer electrons that characterize the element chemically.

PERIODIC RELATIONS

We shall spend some time in this book discussing the various elements and showing how we can use the periodic chart as a tool for systematizing and understanding the reactions of the elements. At this stage, let us discuss the relation to atomic structure of three very important quantities: the radius of an atom, the ionization energy, and the electron affinity.

In view of the indefinite size of the electron, the *radius* of the atom is a somewhat ambiguous quantity. For our purposes, we shall take it to be one-half of the distance of the closest approach of two atoms in a molecule or in a crystal.

The *ionization energy* is the energy required to remove an electron completely from an atom to form a positively charged species called a positive *ion*. If we remove the first electron from an atom, we designate the required energy by I_z^{I}. For removal of the second electron from an atom we use I_z^{II}, and for removal of the third electron we use I_z^{III}.

Many neutral atoms still have a residual attraction for electrons. This arises because the electronic shell of the neutral atom does not completely shield the nucleus; that is, the electron cloud around the neutral atom does not completely "cover" the positive charge. Hence an electron in the neighborhood of the atom can still feel the effect of the positively charged nucleus even though the atom itself is neutral. In such cases, energy will be evolved on adding an extra electron to the atom. Such negatively charged atoms are called negative *ions*. The amount of this energy change is called the *electron affinity* and is designated E_a.

Table 3 lists the radius of the atom, the ionization potential, and the electron affinity for a number of the elements. Atomic radii are given in Ångstroms. Electron affinity and ionization potential are given in electron-volts. The electron-volt is a unit of energy equal to the amount of energy an electron gains when accelerated by a potential difference of 1 volt.

Shielding. Before we consider these properties, we must examine an important aspect of atomic structure. In Fig. 19, we have a sketch of singly charged lithium and beryllium ions. Suppose now that we bring an electron to the lithium

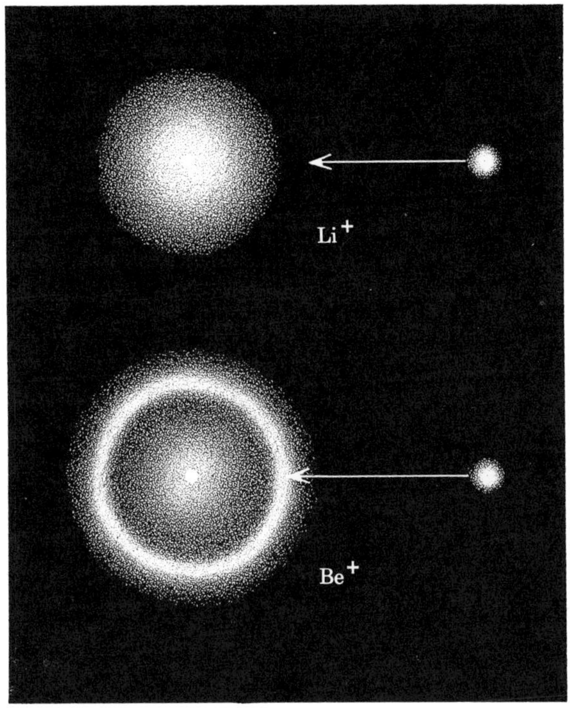

Fig. 19. Attraction of an electron by an Li and a Be ion.

ion. What would be the energy change involved? The electron would be attracted to the ion by the positive net charge of a +3 nucleus which has superimposed on it a spherical electron cloud with a −2 charge. From the beginning to the end of the process of bringing the electron up to the lithium ion, the electron would "see" an effective charge of +1. The case with beryllium is somewhat different. Here, again, at distances far from the nucleus, the electron would "see" an effective charge of +1. As it approached the nucleus, however, the electron in the same subshell, i.e., the 2*s* subshell, would become less effective in canceling part of the nuclear charge and, shortly before the added electron assumed its final position in the atom, it would be seeing an effective positive charge of nearly 2. Thus, at small distances, the inner electrons of an atom are effective in shielding the nuclear charge, but electrons in the same shell are not effective.

In general, when we consider elements having the same unfilled outer shell, the *effective* charge of the nucleus should increase as the atomic number increases. Estimates of this effective charge have been calculated, and for $n = 2$ elements the effective nuclear charge for a lithium atom is 1.3; for beryllium, it is 1.95; for carbon, it is 3.25; and for neon it is 5.85. When we start filling in the $n = 3$ shell, the eight $n = 2$ electrons become inner shell electrons and are again effective in shielding the nucleus; hence we get a repetition of properties.

Table 3 lists the atomic radii and ionization potentials of elements in the third and fourth rows of the periodic table in order to illustrate how the properties we have just discussed vary with position in the periodic chart. We have also given similar data for the elements in the first group of the periodic chart. Together with these, we have listed values for the electron affinity for those elements where this value is well established.†

Horizontal relations. Let us now focus our attention on the third row elements in order to show the horizontal relations in the periodic chart. The atomic radius decreases with increasing atomic number. This decrease is related to the increase in the effective nuclear charge as the atomic number increases. Since all these electrons are in the same shell, this effect draws the shell closer to the nucleus. Naturally, this effect is most pronounced the fewer the electrons we have in the outer shell. As this number gets larger, the tendency toward contraction of the charge cloud will be counterbalanced, at least in part, by a tendency for expansion due to interelectronic repulsions.

After completion of this third shell, the addition of one more electron forms a potassium atom. This last electron is in the fourth shell; hence we expect the potassium atom to be somewhat bigger than the chlorine atom. In addition to this, however, the electrons in the third shell are more effective in "shielding" the nucleus for the $4s$ electrons than they are for $3p$ electrons. Hence two effects make the potassium atom larger; first, the effective nuclear charge has been reduced and, in addition, the electron is now in a shell farther from the nucleus.

Values of the ionization potential for atoms of the same size should increase when the effective nuclear charge increases. This alone would lead us to suspect that the ionization potential should increase with increasing atomic number for elements in the same period of the periodic chart. The decrease in size of the atom with increasing atomic number accentuates this trend.

The *gross* features of the variation of the first ionization potential for the third shell element are explicable in terms of classical coulombic forces. But we must never forget that this type of description can only be approximate. Accurate description requires a more sophisticated approach. From our simple picture the first ionization potential of aluminum should be greater than that of magnesium because the effective nuclear charge is greater. We have, however, superimposed on this the predictions of quantum mechanics that the s electron should be more stable, i.e., harder to remove, than the p electron. In the competition of these two effects, the latter wins out and I_z^I is lower for aluminum than for magnesium.

† Values for E_a have been estimated for atoms such as Na, Be. These values are often subject to very large errors. For example, estimates of E_a for Na range from -1.2 ev to $+0.1$ ev. The safest conclusion is that, since the effective nuclear change is small and the radii large for the metal atoms, values of E_a are small but not zero.

Table 3 Periodic Relations

Third Row Elements

Element	Na	Mg	Al	Si	P	S	Cl	Ar	(K)
Atomic radius, Å	1.86	1.60	1.48	1.17	1.0	1.06	0.97	—[b]	2.27
I_z^{I}, ev[a]	5.14	7.65	5.99	8.15	10.98	10.35	12.96	15.76	4.34
I_z^{II}, ev[a]	47.29	15.03	18.82	16.34	19.65	23.41	23.41	27.6	31.81
I_z^{III}, ev[a]		80.12	28.44	33.49	30.16	35.05	39.91		45.7
I_z^{IV}, ev			120	45.13	51.35				
Effective nuclear charge	2.2	2.9	3.5	4.2	4.8	5.5	6.1	6.8	2.2
Electronic structure *p*	6	6	6 1	6 2	6 3	6 4	6 5	6 6	
s	2 2 1	2 2 2	2 2 2	2 2 2	2 2 2	2 2 2	2 2 2	2 2 2	

Group I Elements

Element	Li	Na	K	Rb	Cs
Atomic radius, Å	1.50	1.86	2.27	2.43	2.62
I_z, ev[a]	5.4	5.1	4.4	4.2	3.9
Electronic structure *d*				10	10 10
p		6	6 6	6 6 6	6 6 6 6
s	2 1	2 2 1	2 2 2 1	2 2 2 2 1	2 2 2 2 2 1

Fourth Row Elements

Element	K	Ca
Atomic radius, Å	2.27	1.97
I_z^{I}, ev[a]	4.34	6.1

Element	Sc	Ti	V	Cr	Mn	Fe	Co	Ni	Cu	Zn
Atomic radius, Å	1.51	1.45	1.31	1.27	1.25	1.25	1.25	1.24	1.28	1.33
I_z^{I}, ev[a]	6.7	6.8	6.7	6.7	7.4	7.8	7.8	7.6	7.7	9.4

Element	Ga	Ge	As	Se	Br
Atomic radius, Å	1.33	1.22	1.25	1.16	1.13
I_z^{I}, ev[a]	6.0	8.1	10.5	9.7	11.8

Electron Affinity

Element	H	F	Cl	Br	I
E_a ev[a]	−0.72	−3.7	−4.0	−3.8	−3.4

[a] By convention, positive values for energies mean that energy must be supplied; negative values for energies mean that energy is given off.

[b] Since Ar forms no compounds, its atomic radius cannot be expressed on a basis consistent with that for other elements.

A similar but less pronounced irregularity is encountered as we go from phosphorus to sulfur. Can you explain this?

An interesting result appears when we examine the ionization potential for removal of the second, third, and fourth electrons. For example, look at sodium. The energy required to remove one electron is relatively small; to remove the second electron requires nearly ten times as much energy. This enormous change is a result of two factors; first, the net charge remaining is doubled; second, the $2p$ electron we are now removing is in a shell closer to the nucleus and the other seven electrons in this shell are no longer effectively shielding the nucleus. If we compare I_z^{I} and I_z^{II} for sodium with I_z^{I} and I_z^{II} for magnesium, we are forced to conclude that the shielding effect is by far the most important. Effectively, this means that the core, rare gas arrangement of electrons is very stable. The reasons, however, are not mysterious; they are explicable in coulombic terms. This effect is apparent not only for sodium but also for magnesium and aluminum. For these atoms, the energy required to remove the third and fourth electrons respectively is far greater than that required to remove the $n = 3$ electrons.

Let us examine the first three ionization potentials of aluminum. On the basis of the penetration effect, we would expect that it would be harder to remove the second electron from an aluminum atom than the second electron from the magnesium atom. We predict this because in the first case, the aluminum atom, the p electrons should be somewhat less stable than the s electrons; hence for aluminum it should be relatively easy to remove the first electron ($3p$) and quite a bit more difficult to remove the second electron ($3s$). In the case of magnesium, both the first and the second electron are removed from the s subshell. Therefore the increase in ionization potential should be largely ascribable to simple coulombic interactions. Thus, *very approximately*, the first ionization potential for magnesium should indicate the work required to pull apart an Mg^+ ion and an electron. The second ionization potential would be at least double the first because now, to the departing electron, the magnesium appears to have a $+2$ charge. This tentative conclusion is borne out by the fact that I_z^{II} is twice I_z^{I}. In the case of aluminum, if our crude arguments are correct, the ratio $I_z^{\mathrm{II}}/I_z^{\mathrm{I}}$ should be a good deal larger than 2 and, in fact, it is 3. If we extend this approach, we say that $I_z^{\mathrm{I}}/I_z^{\mathrm{II}}$ for silicon, which involves only $3p$ electrons, should be roughly $\frac{1}{2}$, and, indeed it is. Once again, there is the expected increase for I^{III} which involves removal of an electron from the $3s$ subshell.

Transition elements. As we start across the fourth row (cf. Table 3), we find that potassium and calcium behave in the predicted manner. With the addition of the third electron, however, we begin to fill up the $3d$ subshell and form *transition elements.* These elements show the same trends as we observed for the elements in the third shell, but the change in the effective nuclear charge is not so much because the $3d$ electrons are effective in shielding the outer $4s$ electrons that determine the size and I_z^I. Thus the variation in the radius of the atom and the ionization potential is less pronounced than for the elements in the third shell. Once the $3d$ shell has been filled, the remaining elements formed by filling up the $4p$ shell behave in a manner similar to the corresponding elements in the third shell.

Vertical relations. Vertical relations in the periodic chart are straightforward. Within a given family the size of the atoms should increase with increasing atomic number, because the outer shell electrons are in shells further removed from the nucleus. In addition, the ionization potential should decrease since, if the atom gets bigger, the electron must be initially farther from the nucleus and easier to remove. Both of these predictions are confirmed by the data in Table 3 for Group I elements.

Electron affinity. There is a residual attraction for electrons by neutral atoms having nearly filled shells. The origins of this attraction are the result of incomplete shielding of the nucleus by unfilled shells. The attraction should be greater on the basis of coulombic considerations for the smallest atoms in a given family. The data in Table 3 indicate a trend in this direction.

Since the ineffectiveness of shielding by outermost electrons is most pronounced with the atoms having an almost complete shell, fluorine, chlorine, bromine, and iodine should have the greatest electron affinity in a given period. They do.

The rare gases have a completed octet and "look" very well shielded to an approaching electron that could only take up residence in the next shell out; consequently, we would expect these atoms to have no electron affinity. On the other hand, in the case of metal atoms such as lithium, boron, etc., we would *not* expect the shielding to be perfect, and it is not at all unreasonable to assume that these atoms have some tendency to pick up an electron. Since the tendency is relatively small, we will assume it is nearly zero in all our semiquantitative discussions. In other words, we will assume that only the atoms with unfilled shells *on the right-hand side* of the periodic chart have a significant tendency to pick up electrons and form negative ions. This results in the common generalization that the rare gas structure is so stable that atoms which can achieve it by gaining one or two electrons will do so.

PROBLEMS

1. Calculate the energy required to remove an electron completely from a hydrogen atom in the ground state. *Ans.* 21.7×10^{-12} ergs.

2. A helium atom that has lost one electron can be described by an equation similar to that used by Bohr for hydrogen. Calculate the value of I_z^{II} for helium.

3. What will be the wavelength of the light emitted when an excited hydrogen atom in the $n = 2$ orbit undergoes a transition to the ground state? What is the radius of the Bohr orbit for $n = 2$?

★ **4.** Calculate the frequency of electron rotation in the $n = 10$ state. What is the frequency of light emitted when an electron drops from the $n = 10$ state to the $n = 9$ state?

Ans. 6.7×10^{12}/sec; 7.7×10^{12}/sec.

5. Assume that Bohr's formula holds for a sodium atom. (It doesn't!) Using the effective charge given in Table 3, calculate the atomic radius and the I_z. (Use $n = 3$.) Calculate the wavelength of the light emitted when this electron drops from $n = 4$ to $n = 3$.

Ans. 2.1Å; 7.2 ev; 3960Å.

6. What is the de Broglie wavelength of an electron in a hydrogen atom in the ground state?

7. Calculate the uncertainty in the momentum of a proton introduced when its position is measured to within $\pm$1Å. Compare this to the uncertainty introduced for the same measurement on an electron. Repeat this calculation for an iron nucleus.

8. If an excited hydrogen atom ($n = 2$, $l = p$) is placed in a magnetic field, certain orientations are possible for the orbital axis and the electron axis. How many different orientations are possible?

Ans. Six.

9. Indicate a probable electronic structure of the element with atomic number 102.

★ **10.** In an excited lithium atom, the energy of a $3d$ level is below that of the $4s$ level. The opposite is true for the first transition series. Can you explain this?

11. What would be the reasonable lowest excited state for a chromium atom?

Ans. Ar core + $4s^2 3d^4$.

12. The atomic number of the mythical element centaurium (Ct) is 117. (*a*) Indicate the likely atomic structure. (*b*) In which family will it be found?

13. Compare the atomic radius, ionization energy, and electron affinity of Ra and Ct.

14. Compare the atomic radius, ionization energy, and electron affinity of At and Ct.

Ans. Radius Ct > At; I_z^I At > I_z^I Ct; E_A At > E_a Ct.

15. If in the seventh period electron affinities of the elements were twice that of the elements in the second period (it is not!), what would occur when Fr is mixed with Ct?

★ **16.** Show from the properties of the quantum numbers l and m that the elementary magnet formed by a hydrogen atom in the $n = 2$, $l = p$ state can never be lined up parallel to poles of a magnet. What is the nearest it can come to being lined up?

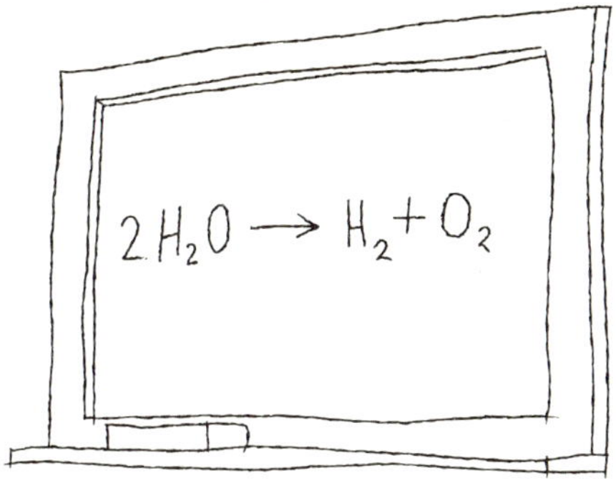

5

Chemical equations and stoichiometry

CHEMICAL SYMBOLS

All pure substances are made up of the 102 chemical elements either alone or in combination. As they were discovered, each of these elements was given a name and a symbol. Thus the silvery reactive metal obtained by electrolysis of fused table salt was given the name *sodium*, and the symbol Na; as is often the case, this symbol is an abbreviation of the Latin name for the element, *natrium*. Through usage, the name and the symbol for the element have acquired a slightly different connotation. The word sodium refers to the metal; the chunk of metal referred to may be as small as a pinhead or as big as a marble; *sodium* is the name of the substance. On the other hand, the symbol Na usually denotes an atom of sodium, or a specific number of atoms of sodium. In other words, the symbol Na often refers not only to the substance *sodium*, but also to a specific quantity of sodium. In this chapter, we shall adhere rigidly to this usage.

The symbols for the 102 elements make up a chemical alphabet with which we can describe any of the known substances. For example, when we write Na, we refer to an atom of sodium; when we write Cl, we refer to an atom of chlorine. Both of these, especially in atomic form, are violently reactive and poisonous. However, atoms of sodium and chlorine combine in a 1 : 1 ratio to form table salt. This combination, indicated by the formula NaCl, is essential to, rather than harmful to, life. Whenever we join the symbols together in this manner, we refer to a new substance totally different from the component atoms.

The elements occur in many structural forms; for example, the element hydrogen can occur in a gas as individual atoms, or combined into diatomic hydrogen molecules. If we were to adopt 2H as the symbol for a diatomic hydrogen molecule, we should not be able to tell whether we were referring to a diatomic hydrogen molecule or two single hydrogen atoms. Therefore, when referring to molecular elementary species, we indicate the number of atoms by a subscript. On this basis, H_2 refers to a molecule of diatomic hydrogen, whereas 2H refers to two hydrogen atoms. Many other elements are found in combined form. Thus we find oxygen as a diatomic molecule, O_2, and a triatomic molecule, O_3. We find sulfur occurring as monatomic, diatomic, tetratomic, hexatomic, and octatomic molecules: S, S_2, S_4, S_6, and S_8. Whenever we have an atom occurring more than once in a given molecule, we make use of the subscript notation to avoid confusion.

More often than not, molecules contain two or more different elements in combination. Water is a molecule consisting of the two atoms of hydrogen and one of oxygen. According to what we have said, we could write the formula H_2O, or OH_2. By convention, we always indicate the symbol for hydrogen first in water; hence H_2O is the correct formula. (The actual convention is to write the more metallic element first. We shall see shortly that the elements in the lower left corner of the periodic chart are more metallic than those in the upper right-hand corner. Therefore, in chemical formulas, the elements lying nearer the lower left-hand corner of the periodic chart are indicated first.) An important compound encountered early in introductory chemistry is H_2SO_4, sulfuric acid. The formula indicates that a single molecule contains two atoms of hydrogen, one of sulfur, and four of oxygen. A more interesting example is the formula for benzene, C_6H_6. The formula indicates that a single molecule contains six atoms of carbon and six atoms of hydrogen. It is totally different from the atoms that make it up and also totally different from $3C_2H_2$, which stands for three molecules of the compound *acetylene*.

The main purpose of the chemical formula is to convey as concisely as possible the make-up of individual molecules. For this reason, we often indicate the grouping of the atoms in the chemical formula. For example, the formula for methyl alcohol, written according to the above convention, would be H_4CO. It is known, however, that methyl alcohol consists of one carbon atom joined to three hydrogen atoms, and this group is linked to a group consisting of an oxygen and hydrogen atom. Therefore it is much more informative to write methyl alcohol as CH_3OH. This type of notation conveys additional structural information, but it sometimes does more than this. Often within molecules there are subgroups of atoms that are relatively stable, and during many chemical reactions they maintain their integrity. Groupings of this type are the hydroxide group, OH^-, the nitrate group, NO_3^-, and the phosphate group, PO_4^{-3}. Thus we recognize the stability of these groups by a special name, and in writing chemical formulas we generally indicate their presence as a group. Below we have listed examples of compounds containing these groups. On the left, we have indicated the formula according to rules developed in the preceding paragraph. On the right, we have written more common formulas that indicate the presence of the subgroups.

MgH_2O_2	$Mg(OH)_2$
MgN_2O_6	$Mg(NO_3)_2$
$Mg_3P_2O_8$	$Mg_3(PO_4)_2$

CHEMICAL EQUATIONS

In principle, a molecule of benzene, C_6H_6, can be decomposed into its constituent atoms, and then these atoms can be recombined to form a variety of new products. As long as we are talking about ordinary chemical reactions, the law of the conservation of mass will apply, and we can be sure that if we start with 78 pounds of benzene we will end up with 78 pounds of carbon and hydrogen atoms, or 78 pounds will be the weight of the reactants and the weight of the products. We can represent the simple decomposition by the equation

$$C_6H_6 = 6C + 6H$$

This equation is very much like an algebraic equation in which we have x and y on one side and x and y on the other. The number of carbon atoms on the left side must equal the number of carbon atoms on the right side; the number of hydrogen atoms on the left side must equal the number of hydrogen atoms on the right side. This balance guarantees the conservation of mass, but it is not the only way in which we can achieve a conservation of mass. For example, we can imagine the reaction

$$C_6H_6 = 6{}_7N^{13}$$

Here we have a molecule which weighs 78 a.m.u. on the left-hand side reacting to form six nitrogen atoms which have a total weight of 78 a.m.u. Although this equation is consistent with the conservation of mass, it does not represent an ordinary chemical equation. *In ordinary chemical equations, there are no transmutations of atoms.* In brief, equations describing chemical reactions must be consistent with the conservation of mass *and* the conservation of atoms. In a chemical equation, we must have the same number and variety of atoms on the left side of the equation as we have on the right side of the equation.

A very few of the possible reactions of benzene molecules are listed below. Each of these equations is consistent with both the conservation of charge and the conservation of atoms.

$$\begin{aligned}
C_6H_6 &= 6C + 6H \\
C_6H_6 &= 3C_2H_2 \\
C_6H_6 &= C_2H_6 + 4C \\
C_6H_6 &= C_2H_4 + C_2H_2 + 2C \\
2C_6H_6 &= C_6H_{12} + 6C
\end{aligned}$$

Each is a valid chemical equation and is similar to algebraic equations in that it has the same number of C's and the same number of H's on each side of the equality sign. In many respects, chemical equations are treated like algebraic equations. For example, consider the equation

$$2C_6H_6 + 5H_2 = C_6H_{12} + C_6H_6 + 2H_2$$

In this equation, we have some identical terms on both sides. We can simplify the equation if we subtract $2H_2$ and $1C_6H_6$ from both sides and write

$$C_6H_6 + 3H_2 = C_6H_{12}$$

By convention, chemical equations are always written in the simplest form with the lowest possible whole number coefficients.

At 150° in the presence of certain metals, benzene, C_6H_6, combines with hydrogen to form C_6H_{12}. The equation for this can be written

$$C_6H_6 + 3H_2 \rightarrow C_6H_{12}$$

In this particular case, we use an arrow instead of an equal sign; the arrow indicates the direction in which the reaction takes place. This equation tells us that to form C_6H_{12} we must add three molecules of hydrogen to one molecule of benzene. Like any algebraic equation, this equation not only tells us that three molecules of hydrogen react with one molecule of benzene; it also tells us that three hundred molecules of hydrogen can react completely with one hundred molecules of benzene and form one hundred molecules of cyclohexane; or that three dozen molecules of hydrogen can react with one dozen molecules of benzene. In other words, the equation above represents the simplest type of reaction. We can multiply each side of the equation by a dozen, a gross, a million, a billion, a trillion, and the equation will still be true. In all such equations we will have three molecules of hydrogen reacting with *each* benzene molecule regardless of the total number of molecules we are considering.

We must always be conscious of the fact that different atoms have different masses, and, because of this, different molecules will have different masses. Chemical equations tell us the relative *numbers* of molecules that will react with each other; they say nothing about the weight. The equations above tell us that three molecules of hydrogen react with one molecule of benzene. It does *not* tell us that 3 pounds of hydrogen react with one pound of benzene. *Chemical equations say nothing about the relative masses of the reactants.*

The simplest type of chemical equations are balanced by trial and error, but the result must be consistent with the two conservation rules. The procedure can be illustrated best by a series of applications. For example, suppose that we know that benzene reacts with oxygen to form carbon dioxide, CO_2, and water. We could write the unbalanced equation in the following way:

$$C_6H_6 + O_2 \rightarrow CO_2 + H_2O$$

If we are going to have the atoms conserved, we must produce six molecules of CO_2, each containing one atom of carbon, to account for the carbon atoms, and three molecules of water, each containing two atoms of hydrogen, to account for the hydrogen atoms. This leads us to the equation

$$C_6H_6 + O_2 \rightarrow 6CO_2 + 3H_2O$$

Now this equation is balanced as far as the carbon and hydrogen go, but we have fifteen atoms of oxygen on the right and only two on the left. To balance this equation, the number of oxygen atoms must balance; hence we write

$$C_6H_6 + 7\tfrac{1}{2}O_2 \rightarrow 6CO_2 + 3H_2O$$

This equation is completely correct; by convention, however, we write the coefficient before each of the formulas as an integer. Therefore we multiply both sides of the equation by 2 to obtain

$$2C_6H_6 + 15O_2 \rightarrow 12CO_2 + 6H_2O$$

Consider another example. Potassium chlorate, the compound which has the formula $KClO_3$, reacts when heated to form potassium chloride, KCl, and oxygen. The unbalanced equation for this is

$$KClO_3 \rightarrow KCl + O_2$$

Examination of the equation shows us that we have the right number of potassium and chlorine atoms on both sides, but the oxygen atoms do not balance. An oxygen balance can be effected as follows:

$$KClO_3 \rightarrow KCl + \tfrac{3}{2}O_2$$

However, since all coefficients should be whole numbers, we multiply by 2 to get the correct equation,

$$2KClO_3 \rightarrow 2KCl + 3O_2$$

Now consider the more complicated case in which phosphorus and iodine combine with water to form phosphorous acid and hydrogen iodide. Phosphorus occurs as a molecule containing four phosphorus atoms in a group, and the iodine occurs as a diatomic molecule. At the start, ignore this and write the unbalanced equation in the following way:

$$P + I + H_2O \rightarrow H_3PO_3 + HI$$

First note that we have three oxygens on the right side and must have three oxygens on the left side. If we multiply H_2O by 3, we will have six hydrogen atoms on the left side and only four on the right side. So we must put a coefficient of 3 before the HI on the right side. Thus, from this one attempt to balance the equation, we get

$$P + 3I + 3H_2O \rightarrow H_3PO_3 + 3HI$$

This equation is completely balanced, but we have not taken into consideration the fact that iodine exists as diatomic molecules and that phosphorus exists as tetratomic molecules. We can do this by rewriting the preceding equation as

$$\tfrac{1}{4}P_4 + \tfrac{3}{2}I_2 + 3H_2O \rightarrow H_3PO_3 + 3HI$$

This, however, does not conform to our convention that the coefficients of the various formulas can only be integers. Therefore we multiply both sides of the equation by 4 in order to obtain the final equation,

$$P_4 + 6I_2 + 12H_2O \rightarrow 4H_3PO_3 + 12HI$$

This equation is balanced and consistent with convention.

ATOMIC WEIGHT

In Chapter 3, we expressed the weights of fundamental particles in a.m.u. This is a convenient *a*tomic *m*ass *u*nit selected so that the $_6C^{12}$ carbon isotope has a mass of precisely 12, and also so that all isotopes have masses very close to integers. Until 1961, the physicists' atomic weight scale was based on $_8O^{16}$ = 16 a.m.u., whereas the chemists' atomic weight scale was based on a definition that set the atomic weight of oxygen of isotopic content found in the earth's atmosphere as exactly 16.0000. Today both chemists and physicists agree on the defined base that sets $_6C^{12}$ = 12.0000.

In some cases, the naturally occurring element is almost isotopically pure. For example, hydrogen is about 99.98% $_1H^1$, and 0.015% $_1H^2$. The precise mass of the lighter isotope is 1.0078 a.m.u., and that of the heavier isotope is 2.0141 a.m.u. The average atomic mass for naturally occurring hydrogen is 1.0080 a.m.u., very close to the atomic mass of $_1H^1$. This is not always the case. Boron† occurs in two varieties. B^{10} (atomic mass = 10.01 a.m.u.) constitutes 18.8% of the naturally occurring boron; the remainder is B^{11} (atomic mass = 11.01 a.m.u.). Therefore the atomic weight of the naturally occurring elements is 10.82, a value significantly different from the weight of either of the component isotopes.

Atomic masses such as those shown on the inside of the front cover are used by chemists to describe the behavior of a large number of atoms. Although we realize that these represent the *average* masses of the atoms *as they occur in nature* and that these average atoms are made up of one or more isotopes, we carry out chemical calculations just as if the group of atoms we are considering consisted of only one type, each having the average atomic mass. In other words, we are dealing with large numbers of atoms and are interested only in their average behavior; hence we describe this average behavior as if it were the behavior of an

† It is unnecessary to indicate the atomic number when the symbol for an atom is given. Thus the isotope of boron, $_5B^{11}$, can be identified by simply B^{11}. We shall follow this abbreviated usage except for nuclear equations.

average atom having average atomic mass. The average atomic masses are referred to as atomic weights. Although they are masses, not weights, we shall conform to common usage and refer to them hereafter as atomic weights.†

The table of atomic weights is a chemical dictionary that allows us to translate from the relative numbers of atoms, given in chemical formulas and chemical equations, into the relative weights of these atoms. To demonstrate this, let us consider a few example problems.

Example Problem 1: What is the per cent hydrogen by weight in pure hydrogen fluoride (formula: HF)?

This formula tells us that a sample of hydrogen fluoride contains a 1:1 ratio of hydrogen to fluorine atoms. It tells us nothing about the relative weights of hydrogen and fluorine in this compound. The table of atomic weights tells us that a fluorine atom is roughly 19 times as heavy as a hydrogen atom. Now the question is: What must be the relative mass of hydrogen and fluorine so that the ratios of the numbers of atoms is 1:1? The situation is completely parallel to a problem in which we want to pair up one grapefruit with one cherry, but we want to do this by determining the required weight of grapefruits and cherries. If the grapefruit weighs 100 times as much as a cherry, we will have to take 100 lb of grapefruit for every pound of cherries in order to have one cherry for each grapefruit. Similarly, if a fluorine atom weighs 19 times as much as a hydrogen atom, we must take 19 weights of fluorine for every weight of hydrogen in order to obtain a compound with a 1:1 atomic ratio. If the weights we decided upon were grams, we would have 1 gm of hydrogen combining with 19 gm of fluorine to form 20 gm of hydrogen fluoride. Hence the per cent hydrogen would be

$$\frac{1 \text{ gm hydrogen}}{20 \text{ gm hydrogen fluoride}} \times 100 = 5\%$$

Example Problem 2: A compound of carbon and chlorine contains about 90% chlorine. What is the simplest formula of this compound?

For any reasonable compound of carbon and chlorine, the formula must be in the form of C_pCl_q where the ratio $p:q$ is the ratio of carbon to chlorine atoms. Generally, p and q will be small whole numbers. Thus the information we have can be summarized as follows:

$$\frac{\text{No. of atoms of C}}{\text{No. of atoms of Cl}} = \frac{p}{q}$$

The mass of a sample of this compound can be expressed in *a*tomic *w*eight *u*nits. For generality, let us assume that our sample compound has a total mass of 100 ONVA a.w.u. (The ONVA is a convenient number like a dozen or a gross.) The formula indicates only the relative numbers of atoms in a compound, and it does not depend on how much sample we have. Of this 100 ONVA a.w.u. of C_pCl_q, 10 ONVA a.w.u. will be carbon, 90 ONVA a.w.u. will be chlorine. We can obtain the numbers of each type of atom in the sample as follows:

$$\frac{10 \text{ ONVA a.w.u. C}}{\text{sample}} \times \frac{1 \text{ C at.}}{12 \text{ a.w.u C}} = \frac{0.83 \text{ ONVA C at.}}{\text{sample}}$$

† To be consistent we will express the weight of an average atom in a.w.u. units where 1 a.w.u. ≡ 1 a.m.u.

and

$$\frac{90 \text{ ONVA a.w.u. Cl}}{\text{sample}} \times \frac{1 \text{ Cl at.}}{36 \text{ a.w.u. Cl}} = \frac{2.50 \text{ ONVA Cl at.}}{\text{sample}}$$

(We have rounded off the atomic weights for simplicity.) The ratio of these numbers of atoms must be the same as the ratio $p:q$. Hence

$$\frac{p}{q} = \frac{0.83 \text{ ONVA C at.}}{2.50 \text{ ONVA Cl at.}} = \frac{0.83 \text{ C at.}}{2.50 \text{ Cl at.}} = \frac{1 \text{ C at.}}{3 \text{ Cl at.}}$$

Thus, possible formulas for this compound are CCl_3 *or* C_2Cl_6, C_3Cl_9, and so forth. These calculations give us only the ratio of p to q; hence the formula could be any of the type C_nCl_{3n}, where n is an integer. It is important to note that the value of the ONVA is unimportant; the formula is determined by the relative weights of the atoms.

We could have done this calculation more briefly. The ratio of the weight of carbon to chlorine in this compound is 10:90. Since carbon weighs roughly one-third as much as chlorine, we must have three times as many atoms in the 10 weights that are carbon as we would have if the 10 weights were chlorine. Hence the ratio of the atoms would be 30:90. This reasoning leads to the same formula as does the longer calculation.

THE MOLE AND THE GRAM ATOM

The last two example problems were done in terms of the atomic weight units defined earlier (1 a.w.u. = 1.66×10^{-24} gm). This unit is inconveniently small for the chemist. The table of atomic weights establishes the relative masses of the atom regardless of the unit assigned to the numbers. If we wish, we can say that the numbers refer to grams. Obviously, we are no longer referring to masses of individual atoms, but just as 1 a.w.u. of hydrogen contains the same number of atoms (one) as 16 a.w.u. of oxygen, 1 gm of hydrogen contains the same number of atoms (*not* one) as 16 gm of oxygen.

This point can be made a good deal clearer by an application. Consider the reaction discussed earlier:

$$C_6H_6 + 3H_2 \rightarrow C_6H_{12}$$

This equation tells us that three molecules of hydrogen react with each molecule of benzene to form cyclohexane. The coefficient 1 (understood) and the coefficient 3 mean that one molecule of benzene reacts with three molecules of hydrogen; they also mean one ONVA molecule of benzene reacts with three ONVA molecules of hydrogen. The equation tells us the *ratio* of the numbers of reacting molecules. With the help of atomic weight data we can, if we restrict our interpretation of the chemical equation to one involving one benzene molecule, write

$$78 \text{ a.w.u. of benzene} + 6 \text{ a.w.u. of } H_2 \rightarrow 84 \text{ a.w.u. cyclohexane}$$

Now this tells us that 78 a.w.u. of benzene react with 6 a.w.u. of hydrogen, but it also tells us that 78 tons of benzene will react with 6 tons of hydrogen. Since this equation indicates, *in any units,* the relative mass of reactants and products, we might as well choose a convenient one, like grams. Thus we could write the equation

$$78 \text{ gm of benzene} + 6 \text{ gm of hydrogen} \rightarrow 84 \text{ gm of cyclohexane}$$

To be consistent with this, the coefficients in the *chemical* equation would no longer indicate numbers of molecules but would indicate numbers of *moles* of molecules, and the *mole* would be the number of molecules in 78 gm of benzene, or the number of molecules in 2 gm of hydrogen, or the number of molecules in 84 gm of cyclohexane. Thus we find that the chemical equation indicates not only relative numbers of molecules but also relative numbers of moles; *the weight of a single molecule in atomic weight units is the same, numerically, as the weight of a mole in grams.* The mole is a convenient package, like a dozen or a gross; but much larger.†

In chemical reactions of the type

$$2H + O \rightarrow H_2O$$

the elements are not in their usual molecular form but are in atomic form. Under such conditions, we interpret these equations to read two atoms of hydrogen plus an atom of oxygen form one molecule of water, or two moles of hydrogen atoms plus one mole of oxygen atoms form one mole of water. By convention, the number corresponding to a mole *when applied to atoms* is called a *gram atom.* Thus we should read the preceding equation not in terms of moles but as follows: two gram atoms of hydrogen and one gram atom of oxygen combine to form one mole of water. *The number corresponding to a mole is called a mole when it refers to a molecule but is called a gram atom when it refers to atoms.* The weight of one gram atom of any element is the atomic weight *in grams.* The weight of one mole of any molecule is the sum of the weight of the gram atoms of elements required to form one mole of that molecule. This follows from the conservation of mass applied to equations like that above. The weight of one gram atom of an element is referred to as the *gram-atomic weight;* the weight of one mole of molecules is referred to as the *gram-molecular weight.* In practice, these labels are often simplified to *atomic weight* and *molecular weight.*

STOICHIOMETRY

Stoichiometry is that branch of chemical mathematics that deals with weight relations determined by chemical equations and formulas. We have already seen

† One mole is a package containing 6.0229×10^{23} molecules. The number is large, interesting, *but not important in any of the problems in this chapter.* In other words, we would get the same answer to any of the problems we do using the concept of a mole if 1 mole were 6.0229×10^{3}.

that chemical equations coupled with atomic weights provide the key to these relations. Now we want to express these weight relations on the larger scale involving gram moles and gram atoms, rather than the smaller scale involving individual atoms and molecules. A good deal of anxiety will be avoided if, in doing these problems, it is remembered that the mole is a number, like a dozen or a gross. It was selected primarily because it represented a number of molecules that could be conveniently weighed on an analytical balance, and this number is such that one mole of any single type of atom (strictly speaking, one gram atom) has a mass equal to the atomic weight in grams. This concept provides an equivalent alternative for handling stoichiometric problems. We can interpret an equation as applying to individual molecules with the mass expressed in atomic weight units, or we can interpret an equation to apply to moles of molecules with the mass expressed in grams.

Example Problem 3: An oxide is known to have the formula M_2O_5. In its pure form, this compound contains 75.27% M by weight. What is the atomic weight of M? M is one of the known elements listed in the table of atomic weights; which one is it?

In any sample of M_2O_5, there will be two gram atoms of M for every five gram atoms of oxygen. Moreover, a 100 gm sample of M_2O_5 will contain 75.27 gm of M and 24.73 gm of oxygen. If we know the number of gram atoms of oxygen in this sample, we can calculate the number of gram atoms of M in 75.27 gm. This will yield the atomic weight.

First let us compute the number of gram-atoms in 24.73 gm of oxygen, with the help of our unit notation.

$$24.73 \text{ gm O} \times \frac{1 \text{ gm at. O}}{16.00 \text{ gm O}} = 1.546 \text{ gm at. O}$$

We can convert gram atoms of oxygen to gram atoms of M as follows:

$$1.546 \text{ gm at. O} \times \frac{2 \text{ gm at. M}}{5 \text{ gm at. O}} = 0.6181 \text{ gm at. M}$$

Once we know the number of gram atoms of M in 75.27 gm of M, we can write

$$\frac{75.27 \text{ gm M}}{0.6181 \text{ gm at. M}} = \frac{121.8 \text{ gm M}}{1 \text{ gm at. M}}$$

Thus the atomic weight is 121.8, and M corresponds to the element antimony.

Example Problem 4: What is the per cent aluminum in Al_2O_3?

We compute the weight of one mole of Al_2O_3 as follows:

$$2 \times \text{gm at. wt. Al} + 3 \times \text{gm at. wt. O} = \text{mole wt. } Al_2O_3$$

or

$$2 \times 26.98 + 3 \times 16.00 = \frac{101.96 \text{ gm } Al_2O_3}{\text{mole } Al_2O_3}$$

We see that in every 101.96 gm of Al_2O_3 we have 53.96 gm of aluminum. On this basis, the per cent aluminum is simply

$$\frac{53.96 \text{ gm Al}}{101.96 \text{ gm } Al_2O_3} \times 100 = 52.92\% \text{ Al}$$

Example Problem 5: A compound of phosphorus and sulfur contains 56.37% sulfur. What is the simplest formula?

The general formula for a compound of phosphorus and sulfur is P_pS_q, where p and q represent whole numbers. In a 100-gm sample of this compound, we have 56.37 gm of sulfur and 43.63 gm of phosphorus. If we use the definition,

$$\text{No. gm at. P in 100 gm } P_pS_q \equiv X$$

$$\text{No. gm at. S in 100 gm } P_pS_q \equiv Y$$

Then, since the ratios of the numbers of gram atoms of phosphorus and sulfur must be the same in a sample of the compound of any size we can write:

$$\frac{X}{Y} = \frac{p}{q}$$

X and Y can be calculated as follows:

$$X = 43.63 \text{ gm P} \times \frac{1 \text{ gm at. P}}{30.97 \text{ gm P}} = 1.409 \text{ gm at. P}$$

$$Y = 56.37 \text{ gm S} \times \frac{1 \text{ gm at. S}}{32.06 \text{ gm S}} = 1.758 \text{ gm at. S}$$

From this,

$$\frac{X}{Y} = \frac{p}{q} = \frac{1.409}{1.758} = 0.801 = \frac{4}{5}$$

Hence the formula for this compound of sulfur must be of the type

$$P_{4n}S_{5n}$$

wherein n represents an integer. The simplest formula for this compound would correspond to $n = 1$, or to P_4S_5.

Example Problem 6: How many grams of phosphorus and iodine are needed according to the equation

$$P_4 + 6I_2 + 12H_2O \rightarrow 4H_3PO_3 + 12HI$$

to produce 14.0 gm of HI?

Let us first calculate the number of grams of phosphorus required. The equation tells us that in this reaction we need one mole of tetratomic phosphorus to produce twelve moles of hydrogen iodide. In our unit notation, this becomes

$$\frac{1 \text{ mole } P_4}{12 \text{ moles HI}}$$

To convert this relation to a weight relation, we must know the molecular weight of these two molecules. This is simply the sum of the gram-atomic weight of each atom in the molecule, or

$$\text{Molecular weight } P_4\text{: } \frac{123.90 \text{ gm } P_4}{\text{mole } P_4}$$

$$\text{Molecular weight HI: } \frac{127.92 \text{ gm HI}}{\text{mole HI}}$$

With the help of our unit notation, we can write

$$\frac{1 \text{ mole } P_4}{12 \text{ moles HI}} \times \frac{123.90 \text{ gm } P_4}{1 \text{ mole } P_4} \times \frac{1 \text{ mole HI}}{127.92 \text{ gm HI}} \times \frac{14.0 \text{ gm HI}}{\text{Reaction}} = \frac{1.13 \text{ gm } P_4}{\text{Reaction}}$$

The first factor of this equation is derived from the chemical equation. The second and third factors convert from moles of phosphorus and hydrogen iodide to grams of phosphorus and hydrogen iodide. Combination of these first three yields the grams of phosphorus needed per gram of hydrogen iodide. In this particular reaction, however, we wish to form 14 gm HI; hence, to obtain the final answer, we multiply by the last factor.

The weight of iodine required to produce 14.0 gm HI can be obtained in a different fashion. According to the equation, all the iodine in the hydrogen iodide comes from the reactant, iodine. Each mole of hydrogen iodide (weighing 127.92 gm) will contain one gram atom of iodine (weighing 126.90 gm). Thus, in the unit notation, we can write

$$\frac{126.90 \text{ gm iodine}}{127.92 \text{ gm HI}} \times \frac{14.0 \text{ gm HI}}{\text{Reaction}} = \frac{13.9 \text{ gm } I_2}{\text{Reaction}}$$

Example Problem 7: A sample of the compound of carbon and hydrogen that weighs 0.280 gm is burned completely in air and yields 0.360 gm of water and 0.880 gm of carbon dioxide. What is the simplest formula?

As usual, let us assume that the formula of the compound is C_pH_q, where p and q are integers. If this compound reacts with excess oxygen, we will get one mole of carbon dioxide, CO_2, for each gram atom of carbon present. Since water, H_2O, contains two gram atoms of hydrogen, we will get half a mole of water for each gram atom of hydrogen in the compound C_pH_q. We could represent this by the equation

$$C_pH_q + \text{excess } O_2 \rightarrow pCO_2 + \frac{q}{2} H_2O$$

Now, according to the equation, the ratio of p to q is defined by the relative amount of carbon dioxide and water formed; that is,

$$\frac{\text{No. moles } CO_2}{\text{No. moles } H_2O} = 2\frac{p}{q}$$

Thus, all we have to do is compute the number of moles of carbon dioxide and water produced. This can be done as follows:

$$0.360 \text{ gm } H_2O \times \frac{1 \text{ mole } H_2O}{18.0 \text{ gm } H_2O} = 0.0200 \text{ mole } H_2O$$

$$0.880 \text{ gm} \times \frac{1 \text{ mole } CO_2}{44.0 \text{ gm } CO_2} = 0.0200 \text{ mole } CO_2$$

And from these results we can write

$$2\frac{p}{q} = \frac{0.0200 \text{ mole } CO_2}{0.0200 \text{ mole } H_2O} = 1$$

$$\frac{p}{q} = \frac{1}{2}$$

Thus the formula for the compound can be represented as

$$C_nH_{2n}$$

where n represents any integer. The simplest possible formula is CH_2.

In stoichiometric calculations, we make use of the formula or the equation to derive the conversion factor from numbers of one type of molecule or atom to numbers of other types of molecules or atoms. The table of atomic weights provides us with another conversion factor from numbers of molecules or moles to weights. From this point on, we need only follow the standard forms set forth in our discussion of the use of units in calculations. In other words, the chemistry of the problem appears at the very beginning when the problem is set up. From there on, the calculations involve only simple unit arithmetic or algebra.

PROBLEMS

1. Balance the following equations.

(*a*) $NaHCO_3 \rightarrow Na_2CO_3 + H_2O + CO_2$
(*b*) $H_2 + Fe_3O_4 \rightarrow H_2O + Fe$
(*c*) $Fe_2O_3 + CO \rightarrow CO_2 + Fe$
(*d*) $Zn + HCl \rightarrow ZnCl_2 + H_2$
(*e*) $Ca(OH)_2 + HCl \rightarrow CaCl_2 + H_2O$
(*f*) $NH_3 + CuO \rightarrow H_2O + N_2 + Cu$
(*g*) $LiH + H_2O \rightarrow Li(OH) + H_2$
(*h*) $Fe + H_2O \rightarrow Fe_3O_4 + H_2$
(*i*) $H_2 + CO \rightarrow C_2H_6O + H_2O$

2. 0.48 gm of oxygen will combine completely with 0.54 gm of aluminum to form a compound. What is the ratio of the number of oxygen atoms to the number of aluminum atoms? On the basis of these ratios, what would you expect the simplest formula for this compound to be? *Ans.* 3:2; Al_2O_3.

3. 0.355 gm of chlorine combines with 0.4875 gm of platinum. What is the ratio of platinum atoms to chlorine atoms in such a compound? What is the simplest formula for this compound?

4. 0.0700 gm of element M combine with 0.2000 gm of oxygen to form the compound M_2O_5. Determine the atomic weight of M and, from the table of atomic weights, determine what element M is.

5. 0.128 gm of oxygen combines with 0.190 gm of X to give the compound XO_8. What is the atomic weight of X? What is X? *Ans.* 190; osmium.

6. Calculate the per cent of oxygen in Fe_2O_3.

7. The ore carnotite has the formula KU_2VO_6. This ore is an important source of uranium. Calculate the per cent uranium in such an ore. *Ans.* 72%.

8. Calculate the yield of silver per ton from an ore consisting of pure silver sulfide, Ag_2S.

9. Principal minerals from which vanadium is obtained are V_2S_5 (patronite) and $Pb_5V_3O_{12}Cl$ (vanadinite). Compute the per cent vanadium in both of these ores.

10. Boron has two isotopes, $_5B^{10}$ and $_5B^{11}$. From the known atomic weight, estimate the relative amounts of these two isotopes.

11. Assume that lithium consists of only two isotopes:

$$_3Li^6 \quad \text{and} \quad _3Li^7$$

In nature, the former constitutes 7.9% of the lithium and the latter constitutes 92.1% of the lithium. Calculate roughly the value of the atomic weight of lithium.

12. A compound of carbon and hydrogen when burned in air completely yields 0.270 gm of water and 0.440 gm of carbon dioxide. Calculate the ratio of carbon atoms to hydrogen atoms in this compound. Give the simplest formula you can write for this compound. *Ans.* 1:3; CH_3.

13. A compound of carbon and hydrogen burned completely in air yields 0.090 gm of water and 0.440 gm of carbon dioxide. Calculate the ratio of carbon atoms to hydrogen atoms in this compound. Write the simplest formula for this compound. Does it correspond to the formula of any of the simple carbon hydrogen compounds we have discussed?

14. Using the equations given in Problem 1, calculate how many grams of hydrogen are required to convert 100 gm of magnetite, Fe_3O_4, to metallic iron. *Ans.* 3.46 gm H_2.

15. How many pounds of zinc are needed to obtain 2 lb of hydrogen if the reaction indicated in Problem 1*d* is used?

16. According to Problem 1*f*, how many grams of ammonia are needed to convert 16 gm of copper oxide to metallic copper?

17. How many grams of iron are needed to produce enough hydrogen to yield 7 gm of C_2H_6O? Assume that the reaction in Problem 1*h* is used as the source of hydrogen and that this hydrogen is used in reaction li.

★ **18.** Nitric acid can be prepared from ammonia by the following steps.

$$4NH_3 + 5O_2 \rightarrow 6H_2O + 4NO$$
$$2NO + O_2 \rightarrow 2NO_2$$
$$2NO_2 + H_2O \rightarrow HNO_3 + HNO_2$$
$$3HNO_2 \rightarrow HNO_3 + 2NO + H_2O$$

If all steps take place consecutively, how much NH_3 is needed to prepare 10 gm of HNO_3?

19. A 1.000 gm sample of iron ore consists of pure Fe_3O_4 and pure Fe_2O_3. By controlled treatment of the Fe_2O_3 with hydrogen the following reaction occurs:

$$3Fe_2O_{3s} + H_{2g} \rightarrow 2Fe_3O_{4s} + H_2O_g$$

When the sample is treated in this manner, the weight of the resulting solid is 0.986 gm. How many grams of Fe_2O_3 were present in the original sample?

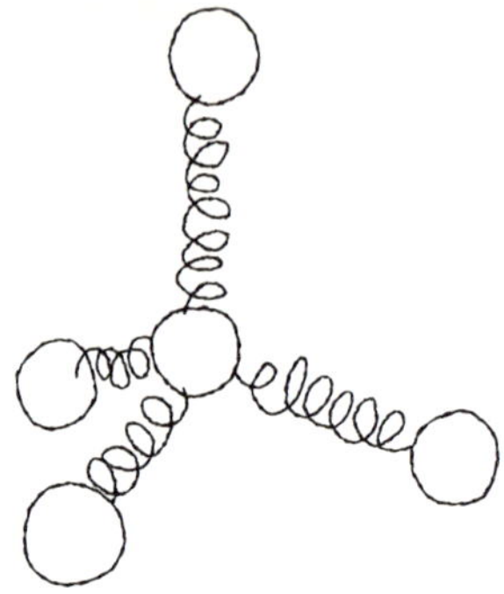

6 The chemical bond

Atoms, as such, rarely occur in nature. The simplest of all elements, hydrogen, occurs as a gas, and the fundamental unit is two atoms bound together to form a molecule. For the more complicated atoms, the elementary form often has a more complicated structure. For example, phosphorus occurs as a tetratomic molecule; that is, it consists of four phosphorus atoms bound together. In sulfur, eight sulfur atoms are bound together to form a molecule; a crystal of lithium is a single giant molecule and can contain an astronomically large number of atoms. What holds atoms together in a molecule? We can only give a partial answer even for the simpler molecules, but even this incomplete answer contributes greatly to our understanding of chemical reactions.

In the sections that follow, we shall discuss *ionic* and *covalent* bonds. These are two of the *extremes* of the chemical bond. Both types of bonds are the result of coulombic interaction as described by wave mechanics. The difference results from the fact that an ionic bond can be described rather well in terms of the attractive force between two oppositely charged *particles*, whereas the covalent bond requires a more detailed recognition of the wave nature of the electron.

THE IONIC BOND

The ionic bond joins two different elements having opposite electrical charges. This type of bonding is generally encountered in a crystal containing large numbers of atoms. Before we treat this more complicated case, however, let us first look at such a bond between two atoms; the principles involved in the formation of an ionic diatomic molecule are not different from those involved in the formation of an ionic crystal.

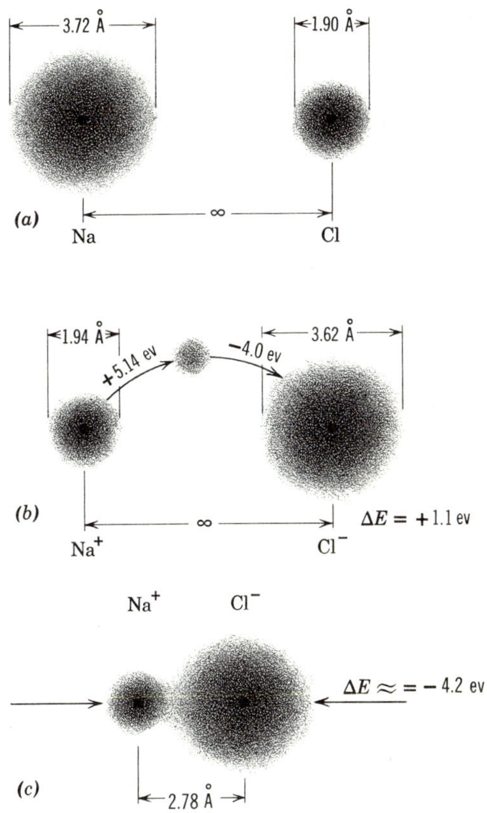

Fig. 1. The NaCl molecule.

The sodium chloride molecule. In Fig. 1*a*, we have depicted a sodium atom and a chlorine atom separated by an infinite distance. Both the idea of infinity and the concept of atomic diameter are imprecise. For our purposes, let infinity be a distance great enough that the sodium atom is not "aware" of the chlorine atoms, and vice versa. The indicated diameter of 3.72 Å for the sodium atom means that most of the charge cloud will be found in such a sphere; if we could press two sodium atoms together we would find that there was little resistance to this pressure until the distance between the centers was 3.72 Å.

The energy required to remove an electron from an atom is termed the ionization potential; the energy gained when a neutral atom acquires an electron is termed the electron affinity. What happens when we remove an electron from the sodium atom, take it very far away, and insert it into the chlorine atom? When we remove an electron from a sodium atom, we must expend the ionization energy, 5.14 ev, and, when the electron, now infinitely removed from the sodium atom, is put into a chlorine atom, we gain 4.0 ev. The *net* cost of this process is 1.1 ev. In doing this, we change the effective size of the atoms. Upon the removal of one electron, the sodium atom contracts because the interaction of the nucleus with the remaining electrons is more effective. When we put the electron into a chlorine atom, it expands for similar reasons. We can note, by the following equations, the processes taking place in Fig. 1*b*.

$$\begin{array}{ll} Na \rightarrow Na^+ + e^- & \Delta E + 5.14 \text{ ev} \\ e^- + Cl \rightarrow Cl^- & \Delta E - 4.0 \text{ ev} \\ \hline Na + Cl \rightarrow Na^+ + Cl^- \text{ (infinitely separated)} & \\ \Delta E = +1.1 \text{ ev} & \end{array}$$

A positive value for ΔE, the change in energy, indicates that we must supply this energy; a negative value of ΔE indicates that energy is evolved.

To form a molecule, we must bring the sodium ion up to the chloride ion. In so doing, we have the help of the coulombic attraction between the two ions. As these ions are brought together from a separation of $r + \Delta r$ to a separation of r, there will be a force acting through the distance Δr and this will result in the change in energy indicated in the following equation.

$$-\frac{e^2}{r^2} \cdot \Delta r = E_{\text{at } r+\Delta r} - E_{\text{at } r} = \Delta E$$

With the aid of calculus (cf. p. 56) we can add up the changes in energy that occur as we bring the two separated ions together. This turns out to be

$$E_{\text{at } r_0} - E_{\text{at } \infty} = -\frac{e^2}{r_0}$$

The quantity e is the charge on the electron, -4.803×10^{-10} esu, and r_0 is the separation of the two atoms when they are just touching. This joining of ions results in a decrease of energy of about 5.3 ev. If we add this to the ion formation energy requirement we find that 4.2 ev of energy is released when a sodium chloride molecule is formed in which the sodium atom has a positive and the chlorine atom has a negative charge. Since energy is released in this process the molecule is stable.

Would the Cl^+Na^- molecule be stable? We can answer this quite definitely from a consideration of the energy requirements. For the ion formation steps we write

$$\begin{array}{ll} Cl \rightarrow Cl^+ + e^- & \Delta E = +13.0 \text{ ev} \\ Na + e^- \rightarrow Na^- & \Delta E \sim 0 \end{array}$$

or

$$Na + Cl \rightarrow Na^- + Cl^+ \text{ (infinitely separated)} \qquad \Delta E = +13 \text{ ev}$$

The energy of ion-joining would lower the energy by about 5 ev. Therefore the formation of Cl^+Na^- from the atoms would *require* energy (10 ev) rather than give off energy and the isolated atoms would be more stable than the molecule.

To form a stable ionic bond, the ionization energy of the cation must be small and the electron affinity of the anion must be large. These requirements are usually met if the cation is found on the left side of the periodic chart and the anion is found on the right side of the periodic chart. If these conditions are fulfilled, an ionic bond will be formed, and, *when* they are fulfilled, the larger the charge on the ions and/or the smaller the size of the resulting ions, the stronger will be the ionic attraction and the stronger will be the ionic bond.

For an exact description of the ionic bond we must use a wave description. In most cases, the wave description merely modifies the distance of closest approach. Thus, at small interionic distances, there is a large repulsive force, so that, if we plot the energy versus interionic distance for the NaCl molecule, we obtain a plot like the curve $a + r$ in Fig. 2. The total energy $(a + r)$ is the net result of two opposing forces. There is an attractive force, stemming largely from a classical coulombic attraction between the cation and anion; this results in the energy term labeled a. There is also a repulsive term, stemming from coulombic repulsions of the charge clouds; this results in the energy term labeled r. The most stable position of the two ions occurs when the separation is r_0 and these opposing forces are balanced. According to this picture, the equilibrium interionic distance is not rigidly fixed but can be changed by an external force like pressure; that is, the ion does not behave like a "billiard ball" but more like a sponge rubber ball.

Ionic crystals. Let us consider an ionic bond with more than just two atoms. Below, we have depicted a one-dimensional crystal where the + sign indicates a cation, and the − charge sign indicates an anion.

$$- + - + - \oplus - + - + -$$

Consider the change in potential energy of *the encircled cation* when we bring it from infinity to a crystal in which the positions of the other ions are already fixed. If we

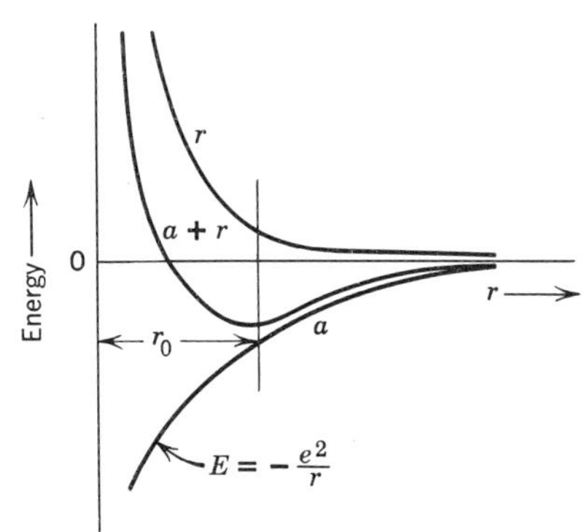

Fig. 2. Energy of NaCl molecule as a function of distance.

consider just the two nearest anions, the energy decrease is $-2(e^2/r_0)$, twice that found for the sodium chloride molecule. If we also consider the two nearest cations, however, the energy is increased $2(e^2/2r_0)$; this increase due to the nearest cations is less than the decrease due to the nearest anions because the cations are twice as far away as the anions. Consideration of all the charges in the crystal results in the energy change given below.

$$\Delta E = -\frac{2e^2}{r_0} + \frac{2e^2}{2r_0} - \frac{2e^2}{3r_0} + \frac{2e^2}{4r_0} + \cdots$$

We can rewrite this as

$$\Delta E = -\frac{2e^2}{r_0}\left[+1 - \frac{1}{2} + \frac{1}{3} - \frac{1}{4} + \frac{1}{5} - \frac{1}{6} + \cdots\right] = -\frac{e^2 A}{r_0}$$

where A is a constant defined by the structure. If we group the terms in the parentheses, we can see what is going to happen. Take terms one and two together, plus terms three and four together, plus terms five and six together, etc. The first set of terms makes the energy equal to that of the sodium chloride molecule. The second set of terms makes it a little bit more stable since $\frac{1}{3}$ is greater than $\frac{1}{4}$. The third set of terms does likewise, and so forth. Thus, in forming a crystal, the decrease in energy is somewhat more than that in forming an isolated molecule; and, therefore, the crystal is more stable than the molecule.

A drawing of part of a three-dimensional sodium chloride crystal is shown in Fig. 3. Although the situation is fairly complicated, the steps indicated by Figs. 1*a* and 1*b* are still needed in the formation of the crystal. The energy involved in the analog of Fig. 1*c* is the energy of a single molecule multiplied by a constant A which is greater than 1. The constant A, Madelung's constant, depends on the structure of the crystal.

THE COVALENT BOND

In the last section, we listed the requirements for the formation of an ionic bond and found that such bonds form between cations from the left side of the periodic chart (low I_z value) and anions from the right side of the periodic chart (large negative E_a values). What happens when the two atoms are located in the same position of the periodic chart, as for the hydrogen molecule? Since the two hydrogen atoms are identical, there is no reason why one atom should lose an electron and the other atom gain an electron. Both atoms have the same tendency to lose or gain an electron; therefore we might anticipate the compromise reached in this bond wherein the two protons, A and B, share equally the two electrons, 1 and 2. Using the particle description, we say that electron 1, originally in an orbit around proton A, now moves in an orbit circumscribing protons A and B, and electron 2 does likewise. True, there will be repulsions between electrons 1 and 2, but this repulsive force is overwhelmed by the attractive force of

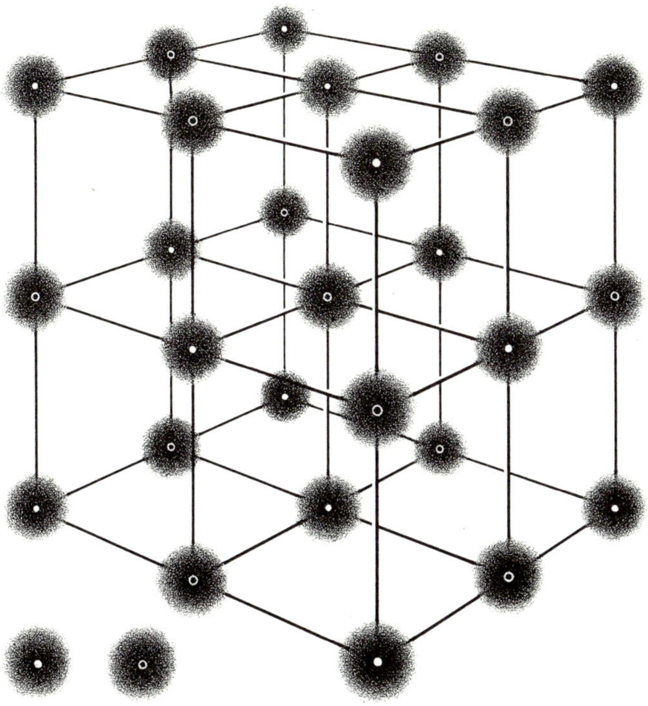

Fig. 3. Sodium chloride crystal.

the two protons acting on each electron. Thus, in this *covalent* bond, the electrons are shared by the two protons.

A correct description of the hydrogen molecule requires the wave description. In essence, the bond is the result of the sharing of electrons, now pictured as charge clouds, between the two protons. This sharing results in the concentration of the charge cloud in the region between the two protons where the coulombic attraction between the protons and the charge cloud is sufficient to bind the molecule together. In this description, the electrons lose their identity: it is not possible to divide the charge cloud and say which "electron" belongs to which atom: we can only say both electrons belong to the whole molecule. Such a stable arrangement of the charge cloud is possible only under certain conditions. It is our aim to ascertain these conditions.

The valence-bond description. In Fig. 4*a*, we have pictured what happens to the charge clouds of two hydrogen atoms approaching each other with parallel electron spins. According to the Pauli exclusion principle, if the electron spin quantum numbers are both the same, the other quantum numbers, which specify

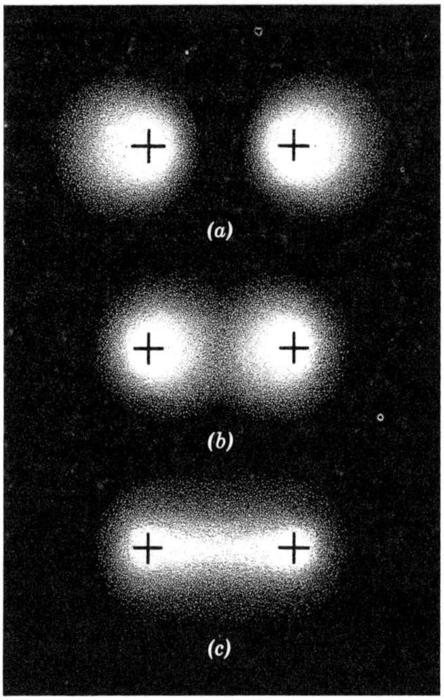

Fig. 4. Electron distribution for adjacent H atoms.

the spatial arrangement of the charge cloud, must be different. Therefore, if the electrons have the same spin, they avoid the same region of space and, in effect, the two charge clouds repel. When the charge clouds repel, this partially bares the nucleus and the repulsion between charge clouds is augmented by a repulsion between the poorly shielded nuclei. Thus, if the electrons in the two atoms have parallel spins, the atoms repel and a stable molecule cannot be formed.

In Figs. 4*b,* and 4*c* we have pictured what happens to the charge clouds of two hydrogen atoms approaching each other with antiparallel electron spins. Since the two spin quantum numbers are different, the other quantum numbers can now be the same and the charge clouds can occupy the same region of space. The most stable arrangement is that for which the charge cloud of the two electrons is concentrated between the two nuclei; that is, the electrons are shared. Under these conditions, both electrons feel an attractive force from both nuclei and the nuclei are well shielded from each other. Therefore, if the electrons in two hydrogen atoms have opposed spin, the atoms attract each other and form a stable molecule.

This sharing of pairs of electrons is conventionally represented by a line or two dots between the symbols of the elements. Thus, for a hydrogen molecule, we write

$$\mathrm{H{:}H} \quad \text{or} \quad \mathrm{H{-}H}$$

With more complex atoms, some electrons are shared and some are not. In such cases, a line or a pair of dots between the chemical symbols indicates a shared-pair bond; the other dots indicate the remaining outer shell electrons. Thus, N_2, S_2, and Cl_2 molecules contain three, two, and one shared pair of electrons. We can represent this by either of the following notations:

$$\mathrm{N{\equiv}N} \qquad \mathrm{S{=}S} \qquad \mathrm{Cl{-}Cl}$$

or

$$\mathrm{N{:}{:}{:}N} \qquad \mathrm{S{:}{:}S} \qquad \mathrm{Cl{:}Cl}$$

If we wish to account for all *outer shell* electrons in the atoms (10 electrons per N_2, 12 per S_2, and 14 per Cl_2), we write

$$:\mathrm{N{\equiv}N}: \qquad \text{or} \qquad :\mathrm{N}\vdots\vdots\mathrm{N}:$$

$$\ddot{\underset{..}{\mathrm{S}}}{=}\ddot{\underset{..}{\mathrm{S}}} \qquad \text{or} \qquad :\ddot{\mathrm{S}}::\ddot{\mathrm{S}}:$$

$$:\ddot{\underset{..}{\mathrm{Cl}}}{-}\ddot{\underset{..}{\mathrm{Cl}}}: \qquad \text{or} \qquad :\ddot{\underset{..}{\mathrm{Cl}}}:\ddot{\underset{..}{\mathrm{Cl}}}:$$

This description of the covalent bond is known as the *valence-bond* theory. It is only an approximation, but it provides a good framework for a discussion of the structure of molecules. Such covalent bond formation is found when a pair of electrons share common orbitals in both atoms, and it can occur only if the spins are opposed. On this basis, only atoms with unpaired electron spins can form molecules. Since atoms with completely filled subshells have all electrons paired, only atoms with unfilled subshells or atoms that can be readily excited to create unfilled subshells can form molecules.

Let us see if the valence-bond description can help us predict the bonding in diatomic molecules formed by the elements in the second row of the periodic chart. In Table 1 we have listed these elements, their atomic structure, and the observed structure of their diatomic molecules.

A lithium atom has one unpaired electron in its $2s$ subshell. The situation is much like that for hydrogen; we expect a single bond to be formed. For beryllium, both the $2s$ and the $1s$ subshell are filled. For atoms in such a state there are no unpaired electrons. We expect no molecules to be formed. With boron, we have one unpaired electron that can participate in bonding to form a single bond; with carbon—two unpaired electrons—a double bond; with nitrogen—three unpaired electrons—a triple bond. Oxygen is a special case. There are three p orbitals, each of which can accommodate two electrons provided they have opposite spins. In oxygen, there are four electrons in the $2p$ shell; one of these orbitals must be doubly occupied. Hence there are only two unpaired electrons and we expect a double bond. Finally, for fluorine, with one unpaired electron, we expect a single bond and, for neon, with no unpaired electrons, we expect no molecule to be formed.

Table 1

Elements	Atomic Structure	Molecular Structure	Bond Energy
Li	21	Li—Li	1 ev
Be	22	No molecule	—
B	22^{1}	B—B	3.0 ev
C	22^{2}	C=C	>3.5 ev
N	22^{3}	N≡N	9.8 ev
O	22^{4}	O=O	5.1 ev
F	22^{5}	F—F	1.6 ev
Ne	22^{6}	No molecule	—

The bond energies in Table 1 represent the amount of work we have to do to separate the two atoms completely. For single bonds, this energy is about 1 to 3 ev; for double bonds, about 5 ev; and, for triple bonds, about 10 ev. In other words, the more shared-pair bonds we have, the more tightly the two atoms are held together.

In our predictions we find one glaring discrepancy. The spin of an electron can be associated with a microscopic electromagnet. Thus we can see whether or not a molecule contains unpaired electrons by examining the spectra obtained in a magnetic field. According to the valence-bond picture, none of the molecules listed in Table 1 should have unpaired electrons. But they do. Boron, carbon, and oxygen have magnetic properties indicating the presence of two unpaired electrons. Thus the valence-bond picture is only an approximation to the truth.

A few words are needed about the molecules listed in Table 1. The second and third are observed only in electric discharges; the last three are stable molecules under ordinary conditions.

The molecular orbital description. In view of the approximate nature of the valence-bond treatment of chemical bonds, it is not surprising to find that there are alternative treatments. Of these, the most widely used is the molecular orbital treatment. This approach yields an energy level sequence for molecules similar to the energy level sequence for atoms.

To obtain the results of the molecular orbital theory, we must take a point of view different from that used in the valence-bond theory. Rather than look upon the hydrogen molecule as being formed by the approach of two hydrogen atoms, let us look upon it as being formed by the fission of one helium atom. These processes are indicated in Fig. 5. On the left-hand side, we have shown the charge cloud distribution for a neutral helium atom with the two electrons occupying various orbitals.

Assume that this helium atom contains two protons but no neutrons in the nucleus. What changes in electron distribution occur in an imaginary process in which we separate the two protons? As we separate the two protons, the charge cloud distorts and, when these protons are about 1 Å apart, the distorted helium atom becomes a hydrogen molecule. In the top figure the helium atom is in the 1*s* state. We started out with a spherical charge cloud, but after proton separation this cloud will elongate and concentrate between the two protons. If we continue with the separation, we finally create two hydrogen atoms and each of them will be in a 1*s* state.

Now consider a helium atom in which the two electrons are in a 2*p* orbital. We have to consider two cases for the *p* orbitals. For the p_x orbital, the dumbbell-shaped charge cloud points in the same direction as a line through the centers of the protons after a separation has started. The other two *p* orbitals point in a direction perpendicular to this center line. These two cases are also indicated in Fig. 5. The helium atom with electrons in the p_x orbitals will result in an unstable hydrogen molecule since, for the intermediate separation, the two protons are inadequately shielded. Let us indicate by an asterisk this hydrogen molecule with repulsion between atoms. If we continue the separation, we again form two hydrogen atoms, both in the 1*s* state.

When we start with $2p_y$ or $2p_z$ states for the helium atom, the situation is that depicted in the third line of Fig. 5. Here the bonding of the hydrogen molecule is somewhat

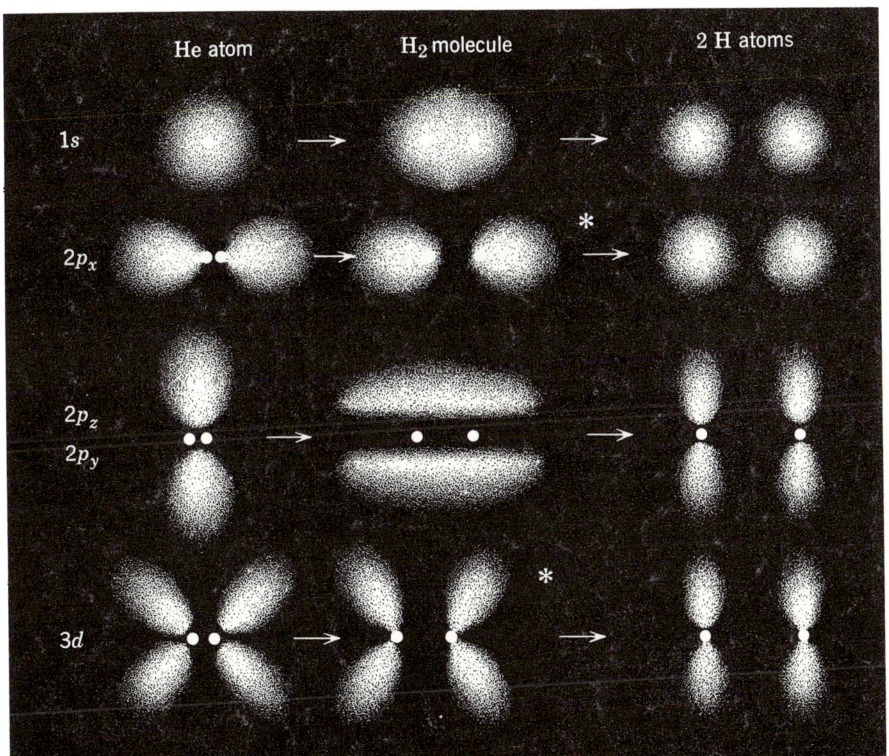

Fig. 5. Formation of molecular orbitals from atomic orbitals.

peculiar. True, we do have the charge cloud concentrated between the two protons, but there is no cloud along a line joining their centers. Since such charge distributions can be formed either by $2p_y$ helium or $2p_z$ helium, there are two arrangements like this at right angles to each other, and *in toto* these molecular orbitals accommodate four electrons, each with the same energy. If we start with such a helium atom, proton separation yields two hydrogen atoms in a p state.

Finally, suppose the helium atom has two electrons in the $3d$ state. The charge cloud distribution is as indicated. As we separate the protons, we obtain a hydrogen molecule that, once again, will be unstable (as denoted by the asterisk). At large separations, this yields two hydrogen atoms in p states.

In the molecular orbital description, the most stable states for the distorted helium atom are the most stable states for the hydrogen molecule. Just as there is a definite series of orbitals for electrons in an undistorted helium atom, there is a definite series of orbitals for electrons in this distorted atom, this hydrogen molecule. In our discussion of the building-up process of the periodic chart, we assumed that all *atoms* have states similar to those found for the hydrogen atom. It seems a logical extension to assume that all *diatomic molecules* have states similar to those found for a hydrogen molecule. Thus, just as we spoke of a series of atomic orbitals for atoms, we now speak of a series of molecular orbitals for molecules.

In Fig. 5, we showed the correlation between the state of the helium atom, the state of the hydrogen molecule, and the states of the separated hydrogen atoms. An energy diagram indicating this correlation is shown in Fig. 6. To make it general, we have indicated on the left-hand side the states of a united atom and on the right-hand side the states of the separated atoms. For the sake of clarity, we have indicated three p levels for the united atom. Similarly, on the right-hand side, since we have two atoms to consider, we have indicated two sets of $1s$ levels, two sets of $2s$ levels, and six sets of $2p$ levels. Of course, for a given atom, the three p levels do not differ in energy.

The gradual changes in electronic distribution are not always as obvious as those shown in Fig. 5. In the mathematical formulation of this treatment, however, the results are quite obvious. Since we cannot go into these aspects of the molecular orbital treatment any more than we can for the valence-bond treatment, let us accept the results as fact.

The correlation diagram describes the system of levels developed during the imaginary fission process. Stable molecule formation with different atoms occurs at different "separations." (In Fig. 6, the molecule B_2 corresponds to point *A*, whereas F_2 corresponds to point *B*.) For simplicity, however, let us assume that the first ten elements form molecules at point *B* on the correlation diagram, and discuss the expected structure in each case. These structures are summarized in Table 2.

In H_2, one stable or bonding orbital is filled, so a molecule with a two-electron single bond is formed. In He_2, two electrons occupy the σ bonding orbital, but this is offset by the fact that the repulsive antibonding orbital, σ^*, is also occupied. The repulsion due to an antibonding orbital always outweighs the attraction due to a bonding orbital; therefore the He_2 molecule is unstable.

The Li_2 molecule is stable since there are four bonding electrons compared to only two antibonding electrons. The strength of the bond is about that expected for a *net* two-electron single bond. The non-existence of Be_2 is expected on the basis of this theory, since the two filled bonding σ orbitals are cancelled by the two filled antibonding σ^* orbitals. The analysis for the net number of two-electron bonds in the next three

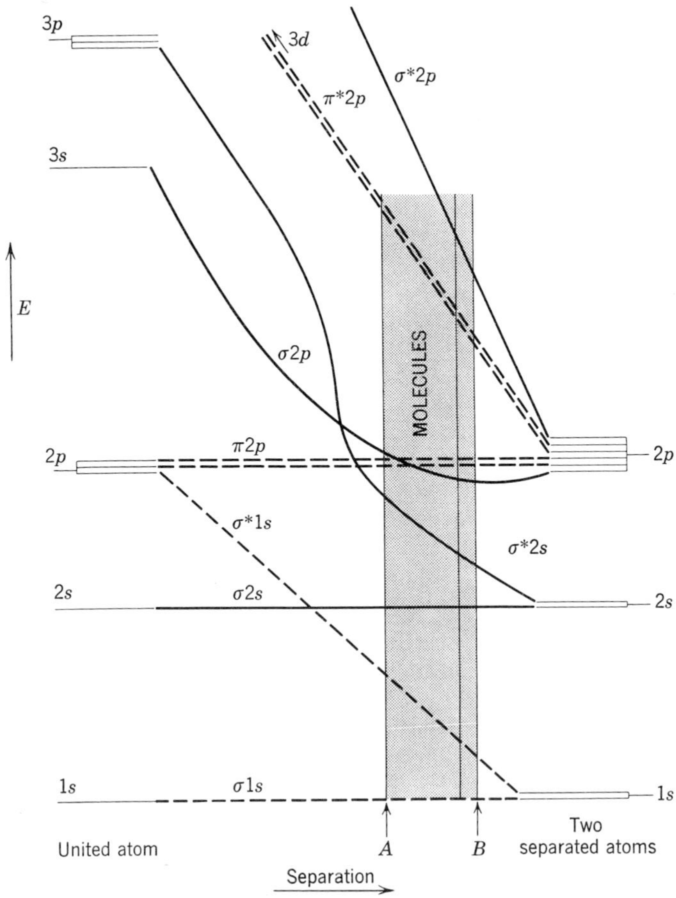

Fig. 6. Correlation diagram. The dashed lines correspond to structures shown in Fig. 5. Lines are drawn for each *m* state; hence each line will hold two electrons with opposite spins. Actually individual *m* values of the same *l* levels in an atom have the same energy. The separate lines for different *m* levels are shown to simplify reading of the diagram

molecules proceeds along similar lines: B_2, with six bonding and four antibonding electrons, has a net single bond; C_2, with eight bonding and four antibonding electrons, has a net double bond; N_2, with ten bonding and four antibonding electrons, has a net triple bond. At this point, we begin to fill the antibonding π^* orbitals. Therefore O_2 with two electrons more than N_2 has only a double bond, and F_2 with two more electrons than O_2 has only a single bond. The species Ne_2 is not stable because there are five antibonding orbitals that more than cancel the five bonding orbitals. Thus the molecular orbital description for diatomic molecules predicts the same variation in bonding for these molecules as the valence-bond description.

One superior feature of the molecular orbital treatment is that bonding does not require pairing of electrons. Whenever a π or π^* molecular orbital with a capacity for

Table 2 Molecular Orbitals

Molecule	Occupied Molecular Orbitals[a]	Comments
H_2	σ^2	Net: single bond
He_2	$\sigma^2\ \sigma^{*2}$	No molecule
Li_2	$\sigma^2\ \sigma^{*2}\ \sigma^2$	Net: single bond
Be_2	$\sigma^2\ \sigma^{*2}\ \sigma^2\ \sigma^{*2}$	No molecule
B_2	$\sigma^2\ \sigma^{*2}\ \sigma^2\ \sigma^{*2}\ \sigma^2$	Net: single bond 2 unpaired electrons
C_2	$\sigma^2\ \sigma^{*2}\ \sigma^2\ \sigma^{*2}\ \sigma^2\ \pi^2$	Net: double bond 2 unpaired electrons
N_2	$\sigma^2\ \sigma^{*2}\ \sigma^2\ \sigma^{*2}\ \sigma^2\ \pi^4$	Net: triple bond
O_2	$\sigma^2\ \sigma^{*2}\ \sigma^2\ \sigma^{*2}\ \sigma^2\ \pi^4\ \pi^{*2}$	Net: double bond 2 unpaired electrons
F_2	$\sigma^2\ \sigma^{*2}\ \sigma^2\ \sigma^{*2}\ \sigma^2\ \pi^4\ \pi^{*4}$	Net: single bond
Ne_2	$\sigma^2\ \sigma^{*2}\ \sigma^2\ \sigma^{*2}\ \sigma^2\ \pi^4\ \pi^{*4}\ \sigma^{*2}$	No molecule

[a] The assumption that the energy levels in all diatomic molecules correspond to the same "separation" on the correlation diagram is not strictly valid. If we had included this in our description, we would have found that the structure of B_2 is $\sigma^2\ \sigma^{*2}\ \sigma^2\ \sigma^{*2}\ \pi^2$, a structure that leads to two unpaired electrons per molecule. Such factors are also important in the interpretation of the spectra of C_2.

four electrons is half filled, the two electrons should be unpaired according to Hund's rule. Thus, of the molecules just discussed, C_2 and O_2 should have magnetic properties consistent with two unpaired electrons, and they do.

On the basis of the preceding discussion of diatomic molecules, it might appear that the molecular orbital description is superior to the valence-bond description. This is not true. Both treatments are approximations to the same truth. In essence, the approximate valence-bond treatment overemphasizes the atomic structure remaining after molecule formation; the approximate molecular orbital description essentially ignores the atomic structure remaining after molecule formation. Thus these two descriptions are complementary rather than antagonistic.

In one important respect the valence-bond description is more useful for us. The notion of shared-pair bonds leads naturally to the electron dot and line notation for bonds. On the other hand, it would be rather difficult to represent the structure of Li_2 in a simple way following the molecular orbital theory; here we would have to indicate two bonds and an antibond. Because of this we shall deal largely with the valence-bond theory, but, in cases where the valence-bond theory leads us astray, we shall see if the molecular orbital approach eliminates these difficulties.

POLYATOMIC MOLECULES

Simplified theory. In the simplest cases, bonding in polyatomic molecules involves *s* and *p* orbitals of the constituent atoms. The charge clouds for these orbitals are indicated in Fig. 7. The *s* orbital is spherical; the three *p* orbitals can be represented as dumbbell-shaped and pointing along the *x*, *y*, and *z* directions in the Cartesian system of coordinates.

A nitrogen atom has five electrons in the second shell, two in the 2*s* orbital, and three in the 2*p* orbitals. There is one electron in each of the three *p* orbitals, and the spins of these three electrons are uncoupled. (This is a consequence of Hund's rule.) If three hydrogen atoms approach such a nitrogen atom, it seems reasonable that they approach along the x, y, and z direction and that the bonds formed by the nitrogen will be in the same direction in the molecule as the orbitals were in the atom. In such a case, we expect the H-N-H angle in ammonia (NH_3) to be 90°.

Phosphorus and arsenic are in the same family as nitrogen, and they form hydrides similar to ammonia, that is, PH_3 and AsH_3. Since they are all in the same family, the outer electrons have the same arrangement; hence NH_3, PH_3, AsH_3 would be expected to have the same geometric arrangement. Similar considerations should hold for the hydrides of oxygen, sulfur, and selenium. In Table 3, we have indicated the formulas of the hydrides formed by these six atoms and the observed H-X-H angle. The heavier members of these two groups have the expected bond angles; the lighter members do not.

The bond angles in water and ammonia are 15° to 18° greater than we would expect. Initially, this discrepancy was given a rather simple explanation. The heavier atoms of a group are larger in size; consequently, in the arsine molecule, the hydrogen atoms are more widely separated than in the ammonia molecule.

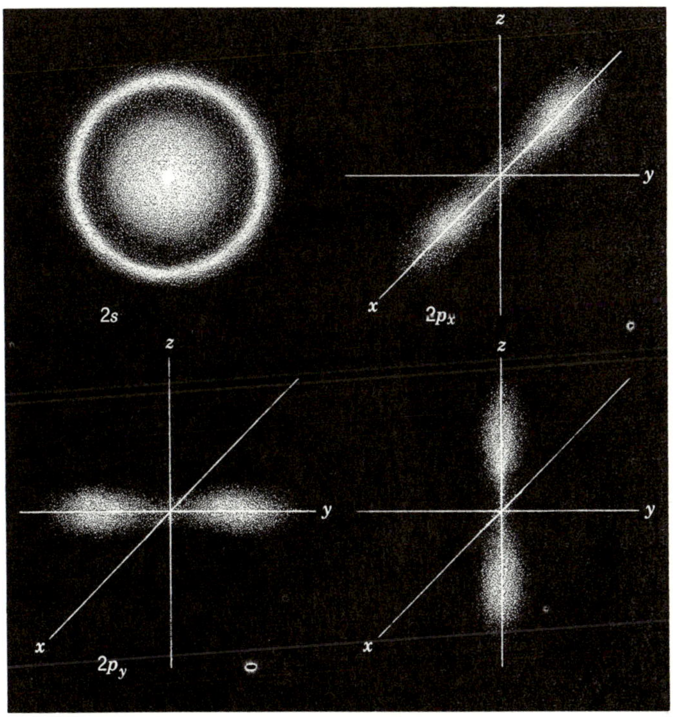

Fig. 7. Atomic orbitals.

Table 3 Structure of Hydrides

Element	Hydride	Bond Angle H-X-H	Element	Hydride	Bond Angle H-X-H
O	H_2O	105°	N	NH_3	108°
S	H_2S	92°	P	PH_3	94°
Se	H_2Se	90°	As	AsH_3	90°

(It will be seen, shortly, that the sharing of electrons between nitrogen and hydrogen atoms is not altogether equal; the nitrogen atom gets more than its share, so that the hydrogen atom has a slight positive charge.) If these hydrogen atoms are fairly close together in a molecule, the repulsive force between them will tend to increase the H-N-H bond angle, or for similar reasons the H-O-H bond angle. For the heavier hydrides, the separation of hydrogen atoms is greater and the distortion of bond angles is less.

The structure of the compounds of carbon is more alarming. Carbon will combine with hydrogen to form methane, which has the formula CH_4. The electronic structure of the carbon atom is ${}_{2\,2}^{\;\;2}$, and, since this means there are only two unpaired electrons, it is distressing to find carbon bound to four instead of two hydrogen atoms. (A similar situation holds with boron. A boron atom has the ${}_{2\,2}^{\;\;1}$ structure and has only one unpaired electron; yet, compounds like BF_3 and BCl_3 are quite common.)

One way out of this difficulty is to assume that the bond formation occurs with an excited carbon atom. Suppose that each bond that the carbon atom forms with hydrogen lowers the energy by 3 ev. Then, if a carbon atom should form two bonds, the energy would decrease by 6 ev. Now we can create four unpaired electrons in carbon by the following excitation:

$$C_{2\,2}^{\;\;2} \rightarrow C_{2\,1}^{\;\;3}$$

This "preparation" of a carbon atom would cost 4 ev, but the carbon atom can now form four bonds and give off 12 ev. Thus this over-all process of bond formation would lower the energy by 8 ev, compared to 6 ev for the direct formation of two bonds.

The above picture only partially resolves the difficulty. If the bonds in CH_4 involved one *s* electron and three *p* electrons, three of the carbon-hydrogen bonds would be different from the remaining bond. Actually, in methane, all four bonds are the same with H-C-H bond angles of 109°.

Hybridization. The final answer to this problem comes from a consideration of the wave properties of the electron. If two different states, indicated by ψ_1

and ψ_2, interact with each other, the resulting wave function is made up of a sum of ψ_1 and ψ_2. In the simple case where the resultant is $\psi_1 + \psi_2$, the charge cloud distribution is given by $(\psi_1 + \psi_2)^2$. Thus, if two electron waves mix, the resulting charge cloud is not simply the sum of the two charge clouds, i.e., $\psi_1^2 + \psi_2^2$. The wave nature of the electron makes interference (cf. pp. 85–86) possible, and, to see the result of mixing, we must consider the wave functions of the electrons rather than the charge cloud distribution.

In Fig. 8, we have sketched the $2s$ wave function (a) and the $2p$ wave function (b) together with the corresponding charge cloud distributions. The plot in Fig. 8c represents the wave function resulting from the superposition of the s and p wave function together with the resulting charge cloud distribution. Note that the charge cloud distribution for the combined sp wave function cannot be derived by combining the charge clouds for the s and p functions; the wave nature of the electrons must be considered to obtain the correct description.

In the ground state, the $_{2}{}_{2}^{2}$ carbon atom can be described by a charge cloud

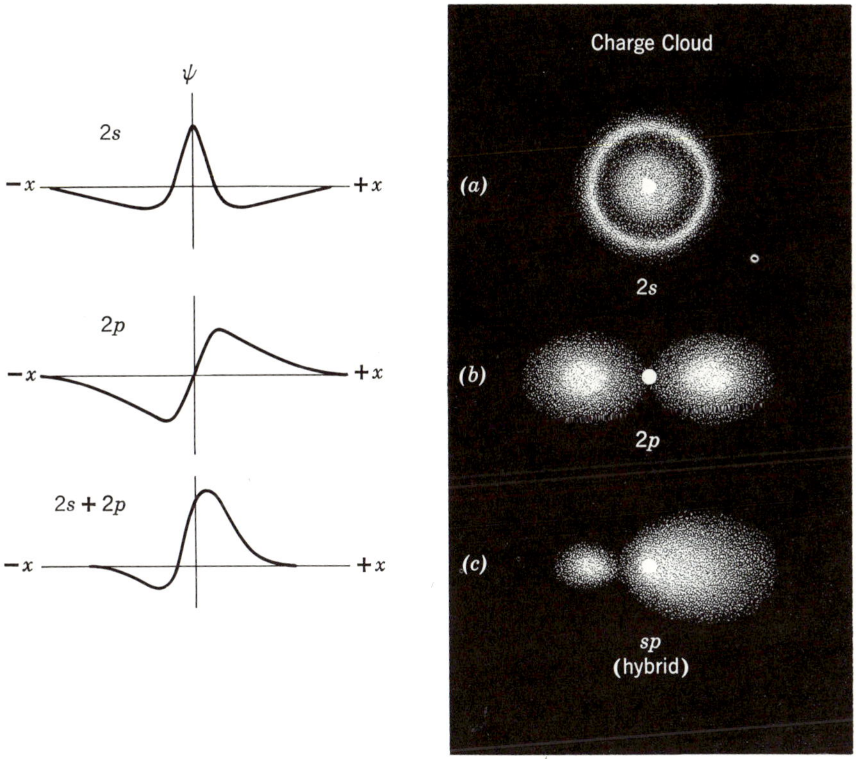

Fig. 8. Hybridization.

distribution arising from two 1*s*, two 2*s*, and two 2*p* wave functions. Similarly the $_{2\,1}^{\;\;3}$ excited carbon atom can be described by two 1*s*, one 2*s*, and three 2*p* wave functions. Both of these bear a close resemblance to states that actually exist for isolated carbon atoms. Using quantum mechanics, we can calculate the energy to these states and, for that matter, the energy of any state whether it is found or not found in nature. If we want to consider a carbon atom in an imaginary state corresponding to two 1*s* wave functions plus four new wave functions consisting of a $\frac{1}{4}s$ and a $\frac{3}{4}p$ function, we can estimate the stability from quantum mechanics; we find this energy is about 8 ev above the ground state. If hydrogen atoms are bonded to the electrons in such *hybridized* orbitals, all the H-C-H angles in the resulting CH_4 molecule are 109°.

It is customary to describe much of the bonding of the first-period elements in terms of hybridized orbitals. Such states are not found in isolated atoms, either excited or in the ground state, but we can view them as forming during the approach of a reacting atom. Several types of hybridization are possible; not all of the outer shell electrons need to occupy hybridized orbitals, but some may exist in almost pure atomic orbitals. Thus, for a carbon atom, the following types of orbitals can be imagined:

(*a*) Each of the four electrons in a $\frac{1}{4}s\ \frac{3}{4}p$ orbital, sp^3 hybridization.

(*b*) Three of the electrons in a $\frac{1}{3}s\ \frac{2}{3}p$ orbital, sp^2 hybridization; the fourth electron in a pure *p* orbital.

(*c*) Two of the electrons in a $\frac{1}{2}s\ \frac{1}{2}p$ orbital, *sp* hybridization; the other two electrons in *p* orbitals at right angles.

The properties of each of these types of orbitals is indicated in Table 4.

Why does sp^3 hybridization occur with methane? To "prepare" for compound formation with sp^3 hybridization requires 8 ev; to "prepare" for bond formation without hybridization requires only 4 ev. But, by hybridization, we have a more directed charge distribution than with pure *p* and *s* functions; therefore, with hybridization, the charge cloud is more concentrated between the two nuclei and a stronger bond is formed. The additional energy released by this stronger bond formation is sufficient to counteract the unfavorable requirement of more energy for bond "preparation."

In Fig. 9, we have sketched the charge distribution for the hybridized bonds that would be expected for CH_4, C_2H_4, and C_2H_2. The sp^3 hybrids are four foot-

Table 4

Waves	Symbol	Hybrid Bond Angle
$\frac{1}{4}s + \frac{3}{4}p$	(4) sp^3	109°
$\frac{1}{3}s + \frac{2}{3}p$; p	(3) sp^2; (1) p	120°
$\frac{1}{2}s + \frac{1}{2}p$; p	(2) sp; (2) p	180°

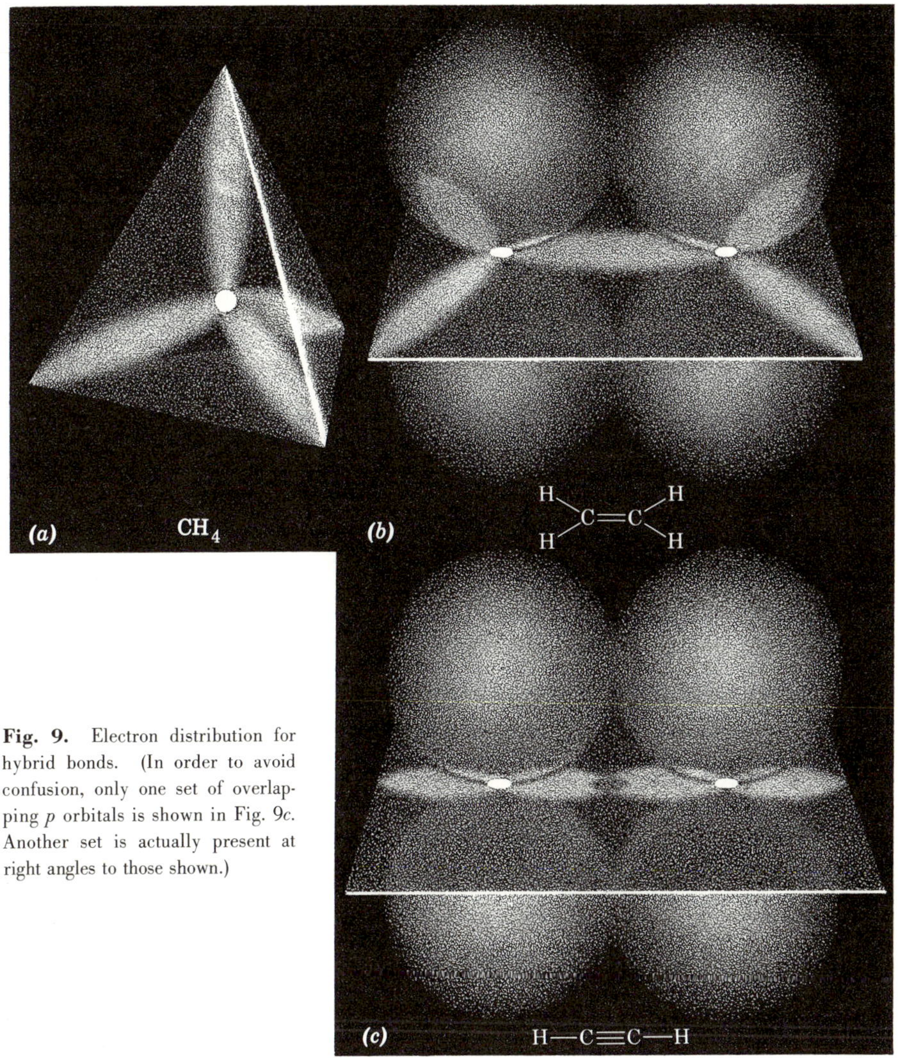

Fig. 9. Electron distribution for hybrid bonds. (In order to avoid confusion, only one set of overlapping *p* orbitals is shown in Fig. 9*c*. Another set is actually present at right angles to those shown.)

ball-shaped charge clouds, each of which is directed from a centrally located carbon atom toward hydrogen atoms located at the corners of a tetrahedron. The sp^2 hybrids are three charge clouds that are similar in shape to the sp^3 hybrids but are directed from the carbon atom toward the corners of a circumscribed equilateral triangle. In ethylene, C_2H_4, the two carbon atoms are joined together by connecting two sp^2 bonds. It must be recalled, however, that we have a pure, unhybridized *p* orbital left over. This extends above and below the plane of the three hybrid bonds, and there is an overlapping of the *p* charge clouds on the two atoms. Thus there is a total of four electrons in the bond between these carbon atoms; this corresponds to a double bond. In view of the exposed positions of these *p* bonds, we would expect such molecules to be readily susceptible to chemical attack.

The *sp* hybrid, as found in acetylene, C_2H_2, is indicated in Fig. 9*c*. In addition to the *sp* hybrid, *two* pure *p* bonds contribute to the bonding between the two carbon atoms. Two of the *sp* hybrids form one bond; the two sets of *p* orbitals form the two bonds shown in Fig. 9*c*. This results in six electrons participating in this bond, i.e., effectively a triple bond. Once again, the *p* electrons are exposed and one might expect that compounds such as these would be chemically reactive.

Bond descriptions in terms of hybridization are applicable to molecules containing atoms other than carbon. Thus, for chemical bonding in all molecules, double bonds are similar to the double bond in ethylene, and triple bonds are similar to the triple bond in acetylene.

The results of hybridization offer a partial answer for the bond angles of NH_3 and H_2O. The resulting angle is a net result of two opposing factors. Thus, for NH_3, we have an unshared pair of electrons which tends to remain in an *s* state since by so doing they have the lowest energy; on the other hand, the bonding electrons have the lowest energy for hybrid bonds that are highly directed. Since there are three bonds and one unshared pair in NH_3, the bonding electrons win out, and the resulting angle is very close to tetrahedral. In H_2O there are two unshared pairs and two bonding electrons; hence we expect the H-O-H angle to be lower than the H-N-H angle, the data in Table 3 confirm these conclusions.

THE OCTET RULE

In the discussion of methane, ammonia, and water, we noted that carbon forms four, nitrogen forms three, and oxygen forms two bonds with hydrogen. All outer shell electrons need not participate in the bonding. In methane, no electrons are unshared, but, in ammonia, two electrons are unshared and, in water, four electrons are unshared. These structures are indicated below:

$$
\begin{array}{ccccc}
\mathrm{H} & \qquad & \mathrm{H} & \qquad & \\
\mathrm{H:\underset{\cdot\cdot}{\overset{\cdot\cdot}{C}}:H} & & \mathrm{:\underset{\cdot\cdot}{\overset{\cdot\cdot}{N}}:H} & & \mathrm{:\underset{\cdot\cdot}{\overset{\cdot\cdot}{O}}:H} \\
\mathrm{H} & & \mathrm{H} & & \mathrm{H}
\end{array}
$$

In each case the pairs of electrons that the atom originally had plus those shared in bonding add up to eight. It is almost as if there are eight slots in these atoms that can accommodate electrons; the most stable arrangement corresponds to these slots being filled, if necessary, by electrons from other atoms. Thus an atom with the electronic arrangement of a rare gas appears to be specially stable. This *octet rule* may be stated as follows: *In all compounds, atoms, one way or another, attempt to obtain a completed octet in the outer shell.*

Let us now consider the structure of compounds formed by elements in the second period with fluorine. Structural formulas for these are

$\angle FBF = 120°$ $\angle FCF = 109°$ $\angle FNF = 106°$ $\angle FOF = 100°$

If, to an atom, sharing electrons is just as satisfying as having them completely, all the atoms in these compounds, including fluorine, have the rare gas electronic structure.

In spite of its usefulness in describing the compounds formed by elements in the periodic chart, the octet rule should not be taken too seriously. It is a useful rule, but eight is not a magic number. Many compounds exist in violation of the octet rule; this is especially true of the third- and fourth-period elements. Therefore, in dealing with an unknown structure; it is a reasonably good guess that it will have completed octets, but one should not be shocked if this arrangement is not achieved. The octet rule is associated with the fact that in the second period we have one *s* orbital and three *p* orbitals. Maximum bonding will be achieved when all these orbitals are filled with their quota of eight electrons. More bonds could be formed if we put more than eight electrons around such an atom in a compound, but this would involve putting electrons into the next shell, which is farther from the nucleus; hence this would be unlikely. On the other hand, in the third-period, elements we have *s, p,* and *d* orbitals in the same shell. Therefore it should not surprise us to see occasionally five, six, or seven bonds to these atoms.

Let us take a concrete example. Nitrogen, a second-period element, forms the chloride, NCl_3 with the structure

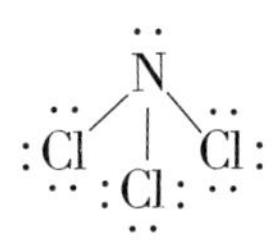

Molecules like NCl_5 *do not* exist. This is consistent with the usefulness of the octet rule for these elements. If the compound NCl_5 existed, it would probably have the structure

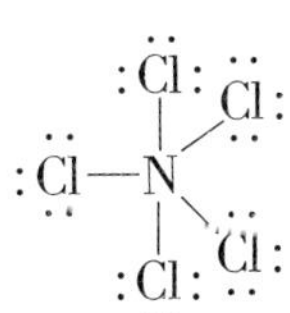

in which the nitrogen atom would be surrounded by five pairs of electrons in violation of the octet rule. Phosphorus is in the same family of the periodic chart as nitrogen, but it is a third-period element. Both PCl_3 and PCl_5 are known compounds; PCl_3, like NCl_3, has a structure consistent with the octet rule; PCl_5, like the imaginary NCl_5, has a structure incompatible with the octet rule. Presumably, in PCl_5, $3d$ levels participate in the bonding.

RESONANCE

In benzene, C_6H_6, the carbon atoms are linked together to form a ring in which each carbon atom has a hydrogen atom attached to it. The geometric arrangement is pictured in Fig. 10*a*. If we have sp^2 hybridization, the six carbon and the six hydrogen atoms must lie in a plane. Figure 10*a* represents the sp^2 hybrid bonds. In addition to these, however, there are *p* orbitals that lie above and below the plane of the ring and are described by the charge cloud distribution depicted in Fig. 10*b*. Thus we have a sandwich in which the carbon atoms and the sp^2 bonds constitute the meat, and the bread is the charge cloud of the overlapping *p* electrons.

How do we describe this molecule using the conventional bond notation? Each carbon atom contributes half of an electron pair (in addition to those in the sp^2 hybrid) to the bond linking it to other carbon atoms. This means that, on the average, there are three electrons or one and a half bonds between the carbon atoms. We can indicate the structure as follows:

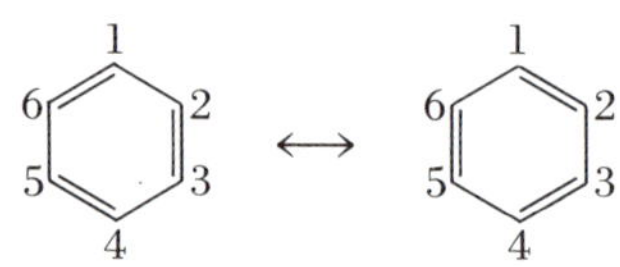

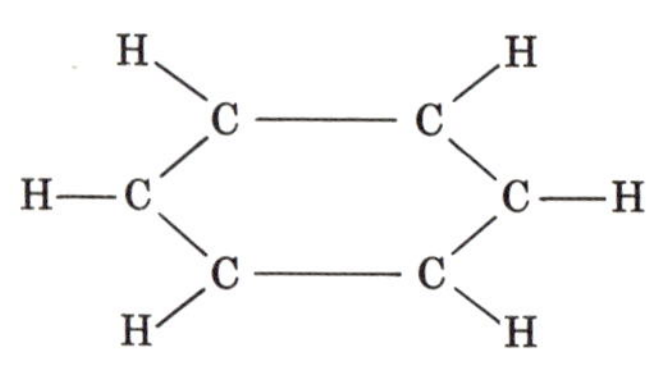

(*a*) sp^2 bonds

(*b*) *p* bonds

Fig. 10. Benzene.

where the two-headed arrow means that the actual structure is a superposition of both of the indicated structures: one-half of each. In principle, we could prepare benzene containing several isotopes of carbon so that each of the six positions is "labeled." *Such a beaker of benzene would not have half the 1-6 carbon atoms with double bonds and half the 1-6 carbon atoms with single bonds; all carbon-carbon bonds would be halfway between a single and a double bond.* This way of drawing a *resonance* structure is necessary *only* because we are too lazy to give the charge cloud distribution, which more truly represents the bonds.

We can realize from the conventional bond representation that resonance must occur. There is nothing about the 1-6 position which is energetically different from the 1-2 position; consequently, the electron spends as much time at one position as the other.

Similar considerations apply to the structure BF_3 pictured earlier. In reality, all B—F bonds in BF_3 must be the same. The true structure of BF_3 is a superposition of three structures like that pictured earlier, each of which has the double bond in a different B—F position. Structures with three single B—F bonds may also contribute.

It must be recognized that structures indicated by the line and dot notation only approximate the truth. In an atom bonded to one or more other atoms, the electrons belong to the whole molecule. Sometimes it *is* twice as likely that we will find electrons in one bond than in another and we then have a double bond. With ethylene, for example, it is reasonable to expect the position between the two carbon atoms to be decidedly different from a position between a carbon and a hydrogen atom; the environment is different. On the other hand, for the polyatomic nitrate ion, the octet structure is often written

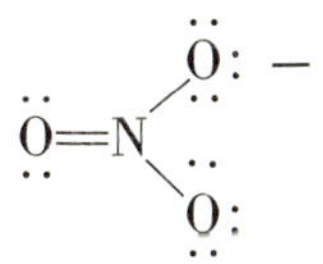

where the minus sign indicates an extra electron is present in the molecule. In this case there is no reason why it should be twice as probable to find an electron between a *particular* N—O bond and, indeed, the probability of finding an electron between any of the N—O bonds is the same. It would be closer to the truth to say that the probability of finding two electrons between any N—O bond is 1.33 times as great as the probability of finding two electrons between an N—H bond. To indicate this we write the resonance structures

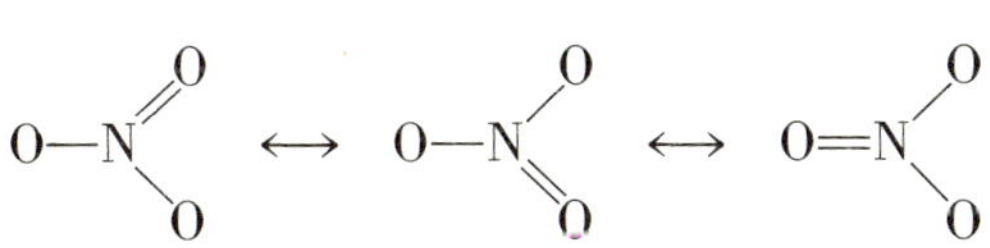

In conclusion, let us again emphasize that, if we describe a molecular structure by several contributing resonance forms, *we do not mean that a beaker full of these molecules contains several different forms. A given molecule must possess a single structure; otherwise we are dealing with more than one type of molecule.*

POLARITY OF BONDS

Ionic versus covalent bonds. Ionic bonds are expected for compounds formed between an element on the left side of the periodic chart and one on the right side. But molecules such as H_2, wherein both atoms are the same, are best described by the shared-pair bond. For all such diatomic molecules this shared pair is *equally* shared between the two atoms, but, when the atoms are not the same, this cannot be true. No doubt, LiF forms an ionic bond; presumably, BaF_2 also forms an ionic bond; but a compound like OF_2 with similar atoms probably forms a covalent bond. What about intermediate cases, like BF_3 and CF_4? These molecules are certainly not held together by ionic bonds; on the other hand, it would seem that, if a shared-pair bond is formed, the pair is not equally shared. Thus, if we consider the fluorides formed by the elements in the second period of the periodic chart, the bonding must gradually change from ionic for LiF to covalent for F_2. For a complete description of bonding, therefore, we need some criterion for determining which bonds are ionic and which are covalent. We also need a criterion for determining in which cases a completely covalent description *or* a completely ionic description is incomplete. The necessary criterion is provided by the concept of electronegativity.

Electronegativity. Consider the bond formed between two atoms A and B as depicted below:

$$\mathrm{A{:}B}$$

Is this bond perfectly covalent with the indicated pair of electrons equally shared between the atoms A and B? Let us approach this question by considering the possibility of ionic bonds such as

$$\overset{+}{\mathrm{A}}\ \mathrm{:}\overset{-}{\mathrm{B}} \qquad \Delta E = I_z\mathrm{A} + E_a\mathrm{B}$$

and

$$\overset{-}{\mathrm{A}}\mathrm{:}\ \overset{+}{\mathrm{B}} \qquad \Delta E = I_z\mathrm{B} + E_a\mathrm{A}$$

To create the first of these, we must ionize atom A and place the electron on atom B; this process requires the indicated quantity of energy. The reverse process requires a similar quantity of energy. If the energies involved are precisely the same, then neither A nor B acquires the lion's share of the electron pair. In such a case, we have a perfect shared-pair bond. Thus the condition for a perfect covalent bond is

$$(I_z\mathrm{A} + E_a\mathrm{B}) = (I_z\mathrm{B} + E_a\mathrm{A})$$

With rearrangement, this condition becomes

$$(I_z\text{A} - E_a\text{A}) - (I_z\text{B} - E_a\text{B}) = 0$$

If one of the quantities in parentheses is appreciably larger than the other, one of the two structures, A^+B^- or A^-B^+, is the more stable. Similarly, if the difference between the ionization energy and the electron affinity of A is greater than a similar difference for B, the atom A has the stronger "pull" for the shared pair in the A—B bond. This difference divided by 5.6 is designated by X and called the *electronegativity*. In other words,

$$X_\text{A} = \frac{I_z\text{A} - E\text{A}}{5.6} \quad \text{and} \quad X_\text{B} = \frac{I_z\text{B} - E\text{B}}{5.6}$$

On such a basis, if $X_\text{B} > X_\text{A}$, then the A—B bond can be represented by A^+B^-; and, if $X_\text{B} < X_\text{A}$, then the bond can be represented by A^-B^+.

Let us apply this concept of electronegativity to a specific example. Consider the compound HCl, a compound held together by a single covalent bond between the hydrogen and chlorine atoms. The chlorine atom has a high ionization potential and a large electron affinity. (Cf. Table 3, p. 111.) On both counts, therefore, we expect the chlorine atom to take more than an equal share of the bonding electrons. This, of course, is the result we get when we consider the electronegativities of hydrogen and chlorine. We can calculate the electronegativities as follows:

$$X_\text{H} = \frac{13.6 + .7}{5.6} = 2.5$$

and

$$X_\text{Cl} = \frac{13.0 + 4.0}{5.6} = 3.0$$

Thus the electronegativity of the chlorine, as expected, is greater than that of the hydrogen; hence we may represent this compound by $H^+:Cl^-$ where the + and − above the symbols do not mean that the bond is ionic but do mean that the chlorine end of the molecule has an excess of negative charge; i.e., the bond is *polar.*

We can show that HCl is polar by simple experiments involving Coulomb's Law. The force acting between two point charges is given by the equation

$$F = \frac{Qq}{Kr^2}$$

where Q and q represent the magnitude and sign of the two charges and K is the *dielectric constant.* Suppose that these two charges are separated in a polar liquid such as HCl; then we have the situation depicted in Fig. 11. The HCl

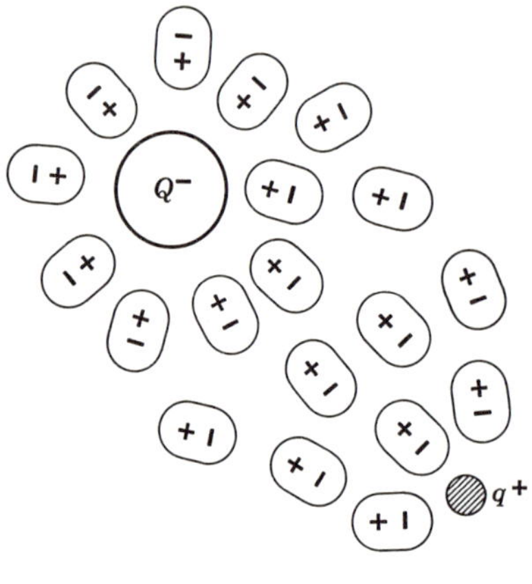

Fig. 11. Dipole orientation.

molecules orient with the positive ends directed toward Q and the negative ends directed toward q; hence the two charges are partially neutralized by the polar liquid, and the force between these charges is reduced. In Coulomb's equation, this reduction of force appears as an increase in the dielectric constant, K. Thus K for a given medium is a measure of the polarity of the molecules that make up the medium. Experiments of this type have demonstrated the validity of the concept of electronegativity.

The variation of the electronegativity in the periodic chart is predictable. For alkali metals, which have low ionization potentials and low electron affinities, the electronegativity is low. With increasing atomic weight in a given period two things happen. There is an increase in the effective nuclear charge, and a decrease in atomic size. Both the smaller size and the higher effective charge increase the tendency of these atoms to pull electrons to them; hence the electronegativity increases. With increasing atomic weight in a given family, the effective nuclear charge changes little, but the size of the atom increases. As judged by both the ionization energy and the electron affinity (cf. Table 3, p. 111), the electronegativity decreases with increasing atomic weight within a given family.

The concept of electronegativity is useful in systematizing chemistry. The electronegativities of atoms can be estimated in a variety of ways. Values vary slightly depending on how they are determined, but they are all reasonably consistent. A list of values compiled by Pauling is given in Table 5.

Table 5 Electronegativities of Atoms

Li	Be	B		C		N	O	F
1.0	1.5	2.0		2.5		3.0	3.5	4.0
Na	Mg	Al		Si		P	S	Cl
0.9	1.2	1.5		1.8		2.1	2.5	3.0
K	Ca	Sc	Ti		Ge	As	Se	Br
0.8	1.0	1.3	1.6		1.7	2.0	2.4	2.8
Rb	Sr	Y	Zr		Sn	Sb	Te	I
0.8	1.0	1.3	1.6		1.7	1.8	2.1	2.4
Cs	Ba			↑				
0.7	0.9			V→Ga				
		H 2.1						

BOND DISTANCES AND BOND ENERGIES

Bond distances. The bond distance is the distance between the centers of two nuclei connected by a bond. Let us examine a simple reaction and see what predictions we can make about bond distances. Consider the reaction

$$H_2 + Cl_2 \rightarrow 2HCl$$

Schematically, we can represent this as in Fig. 12. In the hydrogen molecule, the two atoms are bound together by a perfect shared-pair bond. If we could slice this molecule perpendicular to the bond axis midway between the two nuclei, each of the resulting "atoms" would have one electron. We could do a similar imaginary experiment with the chlorine molecule; each of the resulting chlorine "atoms" would have seven electrons in addition to those in the core. Now suppose we fit such a chlorine "atom" and hydrogen "atom" together. We expect the bond distance of the resulting molecule to be equal to $\frac{1}{2}d_{\text{H-H}} + \frac{1}{2}d_{\text{Cl-Cl}}$, where $d_{\text{H-H}}$ and $d_{\text{Cl-Cl}}$ are the bond distances in H_2 and Cl_2, respectively. The process we have described is an oversimplified version of bond formation involving no rearrangement of the electron distribution around the two atoms. This artificial situation results in a shared-pair bond in which the bond distance is a sum of the effective radii of the component atoms.

In view of the simplicity of the description above, it is surprising that it works. In Table 6, we have listed the observed radii for carbon and halogen atoms. Each radius was assumed to be half the bond distance between the atoms in the common elementary form. We have also listed in Table 6 the compounds formed by the bonding of halogen to carbon atoms. The bonds formed in these compounds are all single bonds; hence it should be possible to calculate the bond dis-

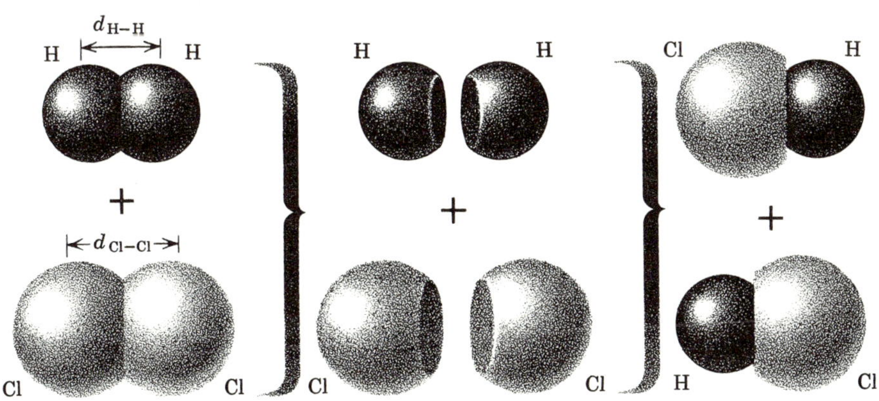

Fig. 12. $H_2 + Cl_2 \rightarrow 2HCl$ (idealized).

Table 6

Element	Observed Radius (Å)	Compound	C-X Distance (Å) Calc.	Observed
Diamond	0.77	—	—	—
I_2 gas	1.33	CI_4	2.10	2.10
Br_2 gas	1.14	CBr_4	1.91	1.94
Cl_2 gas	0.99	CCl_4	1.76	1.77
F_2 gas	0.73	CF_4	1.50	1.36

tances from the covalent radii found for the elements. In the last two columns of Table 6, we have listed the C-X distance obtained by observation and the values obtained by calculation. The agreement except for the carbon tetrafluoride is satisfactory. Thus, in dealing with covalent *single* bonds, it is approximately correct to say that the effective radius of the atom in its elementary state is the effective radius of that atom in any covalent compound.

Bond distances in compounds with double bonds cannot be described by the same approach. For example, the carbon-carbon distance in an ethylene molecule is smaller than that found in an ethane molecule; that is,

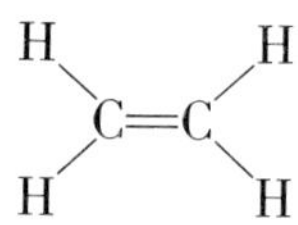

$$d_{C=C} = 1.33 \text{ Å}$$

and

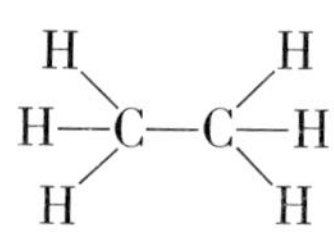

$$d_{C-C} = 1.54 \text{ Å}$$

In general, the double-bond distance is smaller than the single-bond distance.

For molecules with several resonance structures, the result is predictable. For benzene, we might guess that the carbon-carbon distance would be intermediate between the double and single bond and, indeed, it is 1.39 Å. Thus the observed bond distance indicates whether resonance is an important consideration in a given structure.

We found in Table 6 that the estimates of bond lengths are in good agreement with the experimental results except for CF_4. Such discrepancies occur whenever we have a highly polar bond. The difference in electronegativity between carbon and the halogens decreases as the atomic weight of the halogen increases. For the carbon-fluorine bond, this difference is so great that we are dealing with a bond that is more ionic than covalent. In such cases, it is not surprising that the simple approximation breaks down. Whenever this occurs, the bond distance is shorter than that predicted, and the strength of the bond is somewhat greater than that predicted.

Bond energies. Bond energies are the average energies required to separate the component atoms of a covalent bond. The bond energy for hydrogen is the energy required for the reaction $H_2 \rightarrow 2H$; the C—H bond energy is one-fourth the energy required for the reaction $CH_4 \rightarrow C + 4H$. Let us utilize the simple picture, indicated in Fig. 12, to estimate bond energies. Suppose Fig. 12 represents the formation of a bond between hydrogen and iodine. The H—H bond energy depends on the location of the two electrons relative to the two nuclei. If we have a perfect shared pair, the electronic environment for the bonding electron of the hydrogen atom in H_2 might be the same as in HI. Thus the contribution of the *hydrogen atom* to the bonding energy could be one-half the bonding energy found for the H—H bond. Similar arguments hold for the iodine atom in HI. Thus it seems reasonable (although crude) to assume that the bond energy of the HI molecule would be given by the formula

$$E_{\mathrm{HI}} = \tfrac{1}{2}E_{\mathrm{I-I}} + \tfrac{1}{2}E_{\mathrm{H-H}}$$

The bond energies of the elements are well known; that for the H—H bond is 104 kcal; that for the I—I bond is about 34 kcal. Thus the estimated bond energy of the HI molecule is

$$E_{\mathrm{HI}} = \tfrac{1}{2}(104) + \tfrac{1}{2}(34) = 69 \text{ kcal}$$

The observed value is 71 kcal.

In the calculations above, we loaded the dice. We chose the HI molecule because the difference in electronegativity of hydrogen and iodine is small. If we perform a similar set of calculations for the HCl molecule, we find that the calculated bond energy is 81 kcal, whereas the observed bond energy is 103 kcal. This poor agreement may be ascribed to the polarity, i.e., imperfect electron sharing, of the HCl bond.

How can we apply these concepts to a *polar* covalent bond? Let us carry out an imaginary process in which we first form a molecule of HCl with perfect sharing. The energy of such a bond should be 81 kcal, according to the calculations above. But we do not form a bond with perfect sharing; the electrons redistribute with the chlorine atom getting more than an equal share. This redistribution occurs only because it results in a lowering of the energy to make a stronger bond. The greater the inequality of electron-pair sharing, the greater is this bond strengthening; hence the energy of a bond is given by the equation

$$E_{\mathrm{A-B}} = \tfrac{1}{2}E_{\mathrm{A-A}} + \tfrac{1}{2}E_{\mathrm{B-B}} + \Delta_{\mathrm{A-B}}$$

where $E_{\mathrm{A-A}}$ and $E_{\mathrm{B-B}}$ are the bond energies, in kilocalories, of the constituent atoms of the A—B bond and Δ_{A-B} is the extra energy, in kilocalories, due to inequality of sharing.

Whenever there is an inequality of sharing, the bond energy increases; hence the term Δ_{A-B} must always be positive. This means that Δ_{A-B} depends only on the inequality of sharing, not on the direction in which the electron pair is shifted. Since $X_A - X_B$ is a measure of the inequality of sharing, it seems reasonable to assume that Δ_{A-B} is related to the *square* of this difference. Pauling has suggested that the relation is

$$\Delta_{A-B} = 23(X_A - X_B)^2$$

If we use the bond energy equation to calculate the bond energy of an HCl molecule, we obtain an extra 19 kcal for the unequal sharing. This makes the calculated bond energy of HCl 100 kcal, quite close to the experimentally determined value of 103 kcal.

In Table 7, we have listed values for the *single*-bond energies and bond distances of the more common elements, together with several values of bond energies and bond distances observed for compounds. We shall use this table often in our later discussions of the chemistry of the elements.

Before concluding this section, a few words of caution are necessary. First of all, the statements made regarding bond energies and bond distances apply to compounds with single bonds. For compounds with multiple bonds, bond energy calculations are more complicated. For compounds like benzene, resonance is important and we find smaller bond lengths and larger bond energies than predicted. Even under ideal conditions, it should be apparent that these calculations are *approximate*. We should always expect 1 or 2 per cent difference between the calculated and observed values; if the structure is not well known, the error may be larger.

Table 7

Bond	Bond Energy (kcal)	Bond Radii (Å)	Bond	Bond Energy (kcal)	Bond Distance (Å)
F—F	36	0.73	C—F	105	1.42
Cl—Cl	58	0.99	N—Cl	48	1.77
Br—Br	46	1.14	C—Br	66	1.94
I—I	34	1.33	I—Cl	50	2.32
H—H	104	0.37	C—H	99	1.05
C—C	83	0.77	S—Cl	60	2.00
O—O	33	0.74	H—O	111	1.38
N—N	38	0.74	N—H	93	1.01
S—S	51	1.04	C—Cl	78	1.77
Si—Si	42	1.16	Si—I	51	2.43
P—P	51	1.10	P—Br	65	2.23

EXPERIMENTAL DETERMINATION OF STRUCTURES

The ultimate test of generalizations about the structures of various molecules is, of course, the agreement with experimental results. Measurements of bond angles and bond distances can be made by a variety of methods. Studies of the spectra of molecules provide information not only about the electronic structure of the molecule but also about the geometric arrangement in space. Such information is also provided by studies of electron diffraction in gases and X-ray diffraction in solids. In recent years, the use of radiowave spectroscopy has provided a different approach and has considerably expanded our knowledge of the structure of molecules. Justice to these topics cannot be done with a short elementary discussion. Let us, therefore, simply accept the results of such techniques in the form of determined molecular and crystal structures.

OXIDATION NUMBER

One of the most vexing problems of a student taking a course in chemistry is: What constitutes a reasonable formula for a compound? For example, you might know that aluminum and hydrochloric acid react to form hydrogen and a chloride of aluminum. If this is all you know, is there any reasonable way of guessing the formula for this aluminum chloride? Fortunately, there are some relatively simple rules which will help you to solve this type of problem, but *the rules are not foolproof* and will occasionally lead you astray.

Ionic compounds. It is not a difficult task to guess the formulas for purely ionic compounds. Consider NaCl, CaF_2, Na_2O, BaO, and LaF_3. In each compound, both the cations and the anions have achieved a rare gas electronic structure by the gain or loss of the electrons. Moreover, when we take into account the charges on the cations and anions, we find that the compound has no residual charge. Ionic compounds pose few problems provided we restrict our discussion to the compounds formed by elements in the first three rows of the periodic chart.

Covalent and ionic compounds. It would be helpful if we had some way of carrying over the simplicity of ionic compounds to covalent compounds. One approach to this ideal is through the concept of an *oxidation number.* This rather cumbersome name is used because this quantity, although similar to the charge in ionic compounds, is a charge *formally* arrived at to simplify the chemical bookkeeping. If both atoms in a bond have the same electronegativity, e.g., H_2, we split the bond in the middle and assign one of the electrons to each atom; if the atoms in a bond have different electronegativities, the shared pair of electrons is assigned to the more electronegative atom; *the resulting formal charges on the atoms are their oxidation numbers. In most cases these charges are not real ones.*

Once we have determined the oxidation number of each atom in the compound, we can treat the compound as if it were purely ionic. There are, however, still difficulties; the formal approach outlined above often yields oxidation numbers that, if they were real charges, would not correspond to a rare gas electronic structure. This is discomforting but does not make the concept useless. A more important difficulty is that to specify accurately the oxidation number we should know the structure of the molecule. If this were *always* true, the concept of oxidation number would be useful only to those who did not need it, that is, to those who are fully acquainted with the structure of inorganic compounds. We must, therefore, develop a rational set of rules that will limit the oxidation numbers we can assign to a given element. These are listed below.

Oxidation number rules

1. *Fluorine.* Fluorine is the most electronegative element and always has the oxidation number -1, except when it is bonded to itself. Then the oxidation number is zero.
2. *Oxygen.* Oxygen, next to fluorine, is the most electronegative element and, therefore, has the oxidation number -2, except when it is bonded to fluorine or to itself.
3. For Groups I and II of the periodic table, the oxidation number in compounds is the group number, $+1$ and $+2$, respectively. (This rule also holds for the Group IIIB elements but does not hold for the Group IIIA elements.)
4. The oxidation number of an element always lies between N and $N - 8$, where N stands for the group number of the element in the periodic chart. Of these possible values, the most likely are the highest and the lowest.
5. The sum of the apparent charges, i.e., oxidation numbers, is equal to the charge on the molecule or ion. If the molecule is an ion, this will be a plus or a minus charge; if it is a neutral molecule, this will be a zero.

To see the usefulness of these rules, let us apply them in several examples.

What is the formula for a compound between lithium and oxygen? According to rules 2 and 3, the oxidation numbers of lithium and oxygen in a compound are $+1$ and -2, respectively. To have a neutral compound, the formula has to be Li_2O.

What is the formula of the neutral compound of aluminum and chlorine? Aluminum, since it is in Group IIIB, must have an oxidation number of $+3$. Chlorine must, therefore, have a negative oxidation number, and the only one allowed (rule 4) is -1. Thus the formula is $AlCl_3$.

What is the formula of a compound of aluminum and carbon? Aluminum, once again, will have an oxidation number of $+3$; carbon must, therefore, have a negative oxidation number. According to rule 4, this could be -4, -3, -2, or -1. The most likely of these is -4, and, in this case, the formula for the compound is Al_4C_3.

Which of the following formulas are reasonable: Al_5O_2, CaC_2, CaO_3, P_2O_7, BaCl, PCl_5, and N_2O_3? Al_5O_2 and CaO_3 are unreasonable in view of rules 2 and 3. BaCl is not reasonable; in view of rule 3, we get -2 for the oxidation number of chlorine, and this is inconsistent with rule 4. P_2O_7 can be ruled out by the application of rule 4 and rule 2. CaC_2, PCl_5, and N_2O_3 are consistent with the rules for oxidation numbers and are compounds that actually exist.

What are the possible formulas of oxides of nitrogen? According to rule 4, the possible positive oxidation numbers for nitrogen are $+1$, $+2$, $+3$, $+4$, and $+5$. If oxygen is to have an oxidation number of -2, the following compounds would be reasonable: N_2O, NO, N_2O_3, NO_2, N_2O_5. These are all known compounds.

What are the formulas for reasonable compounds between hydrogen and oxygen? The most obvious of these is H_2O, which is consistent with rule 2 and rule 4. There is, however, another compound formed between oxygen and hydrogen; this is hydrogen peroxide with the formula H_2O_2. The only reasonable interpretation of this is that the oxygen in H_2O_2 must be bound to itself and, in fact, the structure of the H_2O_2 is

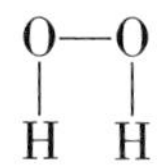

SUMMARY

In this chapter, we have discussed the ionic and the covalent bond and have shown that there is a continuous gradation in bonding between these two extremes. Sufficient discussion was provided that you should be able to estimate the bond distances, the bond energies, and bond angles for many compounds. Moreover, with the help of the "rule of eight" and the concept of an oxidation number, you should be able to make reasonable guesses for the formula and structure of unfamiliar molecules. The concepts discussed in this chapter are the result of an *approximate* quantum mechanical description of bonds. Because of this, there is no reason to be unduly shocked if our predictions regarding bond angles, energies, and lengths do not always agree with experimental results.

Blind acceptance of the "rule of eight" as a magic number will lead to many difficulties. It is a useful working rule, no more. It should also be remembered that the concept of oxidation numbers is only a useful device for seeing which compounds are reasonable. The *assigned* charge is often not a physically real quantity, but merely a bookkeeping device. If this is clearly understood, much grief will be avoided.

With such emphasis on the approximate nature of the theory of the chemical bond, you may be disturbed by the (apparent) fact that modern-day science knows so little. This is not really true. The difficulty is that, at this stage of your studies, you are not yet sophisticated enough to carry out a more exact treatment. It may comfort you to know that as early as 1933 James and Coolidge applied quantum mechanics to obtain an approximate solution for the hydrogen molecule. From the fundamental equation, they could calculate the bond energies and bond lengths. The values they obtained, 4.72 ev and 0.740 Å, are quite close to the experimentally determined values of 4.74 ev and 0.740 Å.

By calculation, one can, in principle, determine any molecular structure. In practice, the mathematical problem is too complex to be handled exactly, and approximations are needed. The real trick is deciding which approximations are reasonable; this is where a "feel" for chemistry or scientific "intuition" is needed. The latter is a talent, the growth of which is stimulated by acquaintance with a host of experimental facts.

PROBLEMS

1. With the procedure outlined in the description of the NaCl ionic molecule calculate how much energy is required or given off in the formation of the following ionic compounds.

(*a*) H^+Cl^- *Ans.* 1.7 ev. (*b*) H^+H^-
(*c*) Cs^+Cl^- ($I_{\frac{1}{2}}$ Cs = 3.87 ev) (*d*) Li^+Cl^-

Ionic radii: $H^+ = 10^{-5}$ Å; $H^- = 1.27$ Å
$Cs^+ = 1.65$ Å
$Li^+ = 0.78$ Å

See Table 3, p. 111 for other data.

2. Indicate, with the line and dot notation, reasonable structures for the following.

(*a*) $SiCl_4$ (*b*) BCl_3
(*c*) SO_2 (*d*) CO_2
(*e*) PO_4^{-3} (*f*) SO_4^{-2}
(*g*) ClO_4^- (*h*) CO_3^{-2}
(*i*) O_3

In each case, the atom written first is bonded to all the other atoms.

3. For which of the structures in the previous problem would resonance be important?

4. Germanium forms the compounds $GeCl_2$ and $GeCl_4$. Are there any features of its electronic structure that would suggest the existence of these two oxidation states?

5. Thallium forms two varieties of positive ions. On the basis of its atomic structure, what do you think they would be? *Ans.* Tl^+; Tl^{+3}.

6. Reactions with water are termed hydrolysis. In these reactions the oxidation number of the individual atoms often remains the same and the more electronegative elements combine with the hydrogen in water and the less electronegative atoms combine with the oxygen. An example of such a reaction is

$$2AlCl_3 + 3HOH \rightarrow 6HCl + Al_2O_3$$

Write reasonable equations for the hydrolysis of the following:

(*a*) NF_3 (*b*) NI_3
(*c*) $SiCl_4$ (*d*) PCl_5
(*e*) Mg_3N_2

(See Table 5.) *Ans.* (*a*) $2NF_3 + 3HOH \rightarrow 6HF + N_2O_3$.

7. Reactions analagous to hydrolysis occur with liquid ammonia. Write a reasonable equation for the reaction that is analogous to hydrolysis when $AlCl_3$ is added to liquid NH_3.

8. Which of the following formulas are likely to represent true compounds?

(*a*) $MgCl_3$ (*b*) LiCN
(*c*) NaO (*d*) ClO_4
(*e*) Cl_2O_7 (*f*) IF_5
(*g*) $BeCl_3$ (*h*) PO_6
(*i*) S_2O_7

Ans. LiCN; Cl_2O_7, IF_5.

9. Indicate the oxidation number of all the atoms in the species indicated in Problem 2.

★ **10.** In the polyatomic anion SO_4^{-2}, the structure is like methane with the sulfur atom at the center and the oxygen atoms at the corners of a tetrahedron. Compare the observed S-O bond distance, 1.50 Å, to those computed on the basis of the data in Table 7. (If the effect of the ionic nature is taken into account, the calculated bond distance is reduced by 0.09 Å.) Explain any discrepancy between the calculated and observed distance on the basis of electronic structure. (*Hint:* Sulfur is in the third period.)

11. Compute the energy change in the following reactions.

(*a*) $H_2S + O \rightarrow H_2O + S$

(*b*) $CH_4 + Cl{-}Cl \rightarrow$ 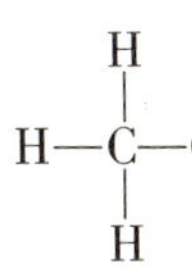 $+ H{-}Cl$

(*c*) $I_2 + 5F_2 \rightarrow 2$ 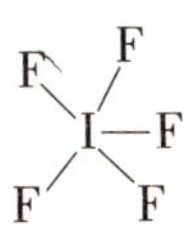

(*d*) $O + H{-}H \rightarrow H_2O$

(*e*) $F_2 + H_2 \rightarrow 2HF$

Ans. (*a*) −60 kcal; (*e*) −166 kcal.

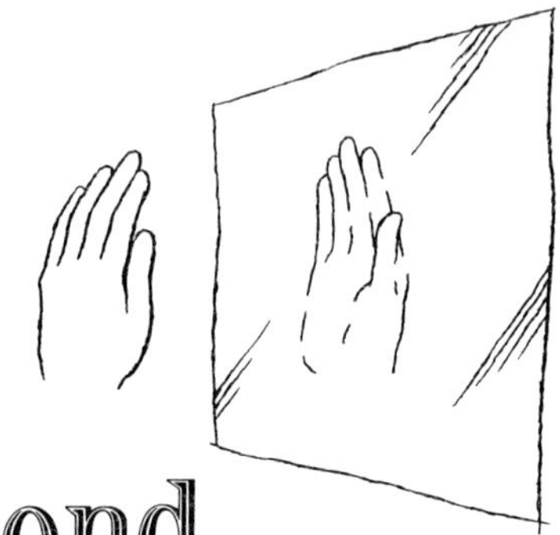

7

The covalent bond

As first used in chemical literature, the adjective *organic* applied only to compounds obtained from living matter. Synthesis of such organic compounds, which were thought to possess a "vital force," was believed to be impossible until in 1828 Wöhler prepared urea, an organic compound, from inorganic (non-vital) compounds. Today, when we speak of an organic compound, we refer to a large class of substances made by the combination of carbon with other elements.

In the four decades following Wöhler's discovery, chemists unearthed a staggering number of organic compounds; in fact, today, over a million organic compounds have been prepared and characterized. But in the early years of organic chemistry little could be done to systematize this embarrassing variety because the pattern of linkage of the carbon atoms was not understood; it was not until Kekulé and Van't Hoff developed the concept of valence and a directed chemical bond that order emerged out of this near chaos. It is from their original ideas that chemists have devised the concept of the directed *covalent* bond as we know it today; and, even today, the most convincing applications of covalent bond theory are found in organic chemistry. Therefore, in order to amplify the discussion in the previous chapter, we now discuss some organic chemistry.

TETRAVALENT CARBON

Bonding. Most of the bonds formed by carbon originate from an excited carbon atom with the $1s^22s2p^3$ electronic structure that has four unpaired electrons and forms four shared-pair bonds. These can be four single bonds, sp^3 hybridization; they can be two single and one double bond, sp^2 hybridization; or they can be one triple and one single bond, sp hybridization. The number of shared-pair bonds formed by a given atom is referred to as the *valence* of the atom; hence we say that carbon is *tetravalent,* i.e., four valent.

Carbon forms bonds with a variety of atoms, but in this section we want to consider only those formed with the halogens (Group VII Elements) and hydrogen, oxygen, and nitrogen. All such compounds are consistent with the octet rule. Neutral unattached halogen atoms have an electronic structure that is one electron short of the rare gas structure, i.e., $:\ddot{\underset{\cdot\cdot}{X}}\cdot$; hence they are monovalent. A neutral hydrogen atom has one electron less than the rare gas, helium, and is also monovalent. By way of contrast, the electronic structures of neutral oxygen and nitrogen are $1s^2 2s^2 2p^4$ and $1s^2 2s^2 2p^3$, since the rare gas, neon, has the electronic structure $1s^2 2s^2 2p^6$, oxygen and nitrogen are divalent and trivalent, respectively.

The geometry of organic compounds can be approximated if we assume that the compounds have a configuration in which all pairs of electrons (both shared and unshared) extend toward the corners of a tetrahedron with the nucleus of the carbon atom at the center. Figure 1 represents the structure of CH_4 in which the H-C-H bond angle is 109.5° as required for sp^3 hybrid bonds. A compound called formaldehyde is also known where two of the monovalent hydrogen atoms are supplanted by one divalent oxygen atom. In this compound, H_2CO, the carbon atom has three sp^2 hybrid bonds, as in ethylene, and a p-type bond between the carbon and the oxygen atom. All atoms are in the same plane and the H-C-H and H-C-O angles are about 120°. We can again visualize the geometry in terms of tetrahedra with electron pairs directed toward the corners. The

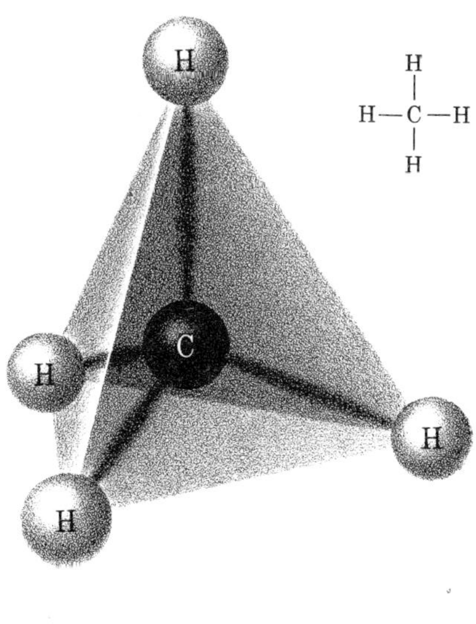

Fig. 1. Methane.

tetrahedron circumscribing the oxygen atom is now sharing an edge with the one circumscribing the carbon atom (Fig. 2). This results in all atoms lying in the same plane, but for the bond angles to be 120° the tetrahedra must be slightly distorted. The compound called hydrogen cyanide, HCN, in which the three monovalent hydrogen atoms of methane are supplanted by one trivalent nitrogen atom can be viewed as the result of the carbon and nitrogen tetrahedra sharing a base (Fig. 3). The geometry is the same as one would predict for an *sp* hybrid C—N bond with the other two *p* electrons from the carbon atom forming *in toto* a triple bond to the nitrogen.

Methane derivatives. Any group forming one bond can be substituted for a hydrogen atom in methane, CH_4. These groups may be atoms, like the halogens, or polyatomic groups, like $-\ddot{\underset{\cdot\cdot}{\mathrm{O}}}-\mathrm{H}$ and $-\ddot{\mathrm{N}}\begin{smallmatrix}\diagup\mathrm{H}\\ \diagdown\mathrm{H}\end{smallmatrix}$. If all four groups attached to the carbon atom are the same, we expect the bond angle to be 109.5°, the tetrahedral angle; if there are three groups of one kind, say hydrogen atoms, and the fourth group is different, say a bromine atom, we expect some distortion of the tetrahedron and departures from the 109.5° bond angle. The data given for compounds (*a*), (*b*), (*c*), and (*d*) in Table 1 do suggest variations in the bond angle but they are small. Therefore we assume that, whenever a carbon atom forms four single bonds, the bond angles are 109.5°.

The OH and NH_2 groups indicated in Table 1 (*c* and *d*) function as mono-

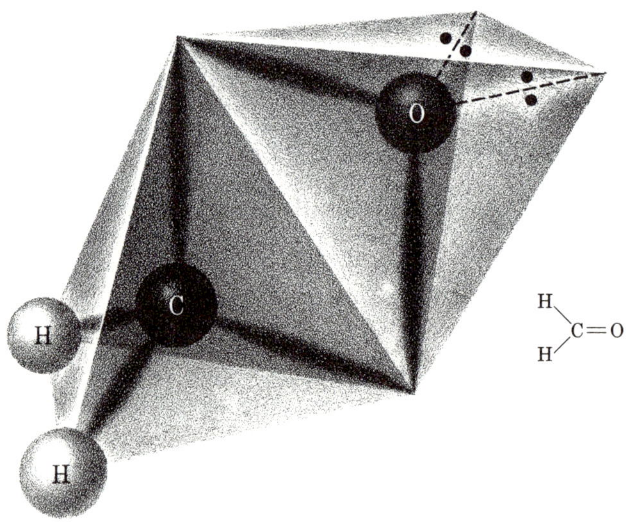

Fig. 2. Formaldehyde.

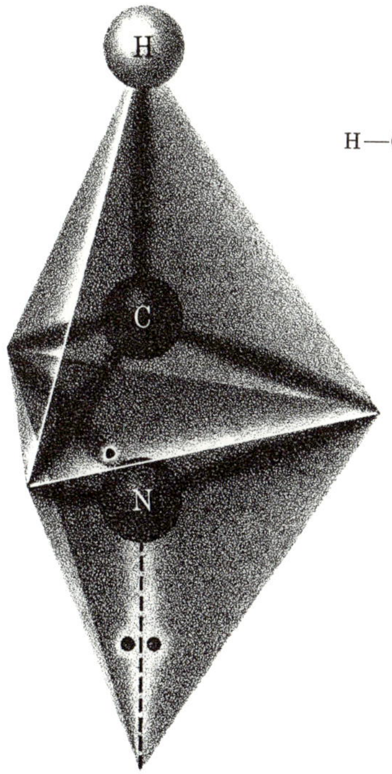

$H—C{\equiv}N$

Fig. 3. Hydrogen cyanide.

valent groups in so far as they replace one hydrogen atom in the methane molecule. In view of Figs. 2 and 3, we expect the four pairs of electrons around the nitrogen (or oxygen) atom to extend toward the corners of a tetrahedron circumscribing the atom. Thus CH_3OH and CH_3NH_2 have the structures shown in Fig. 4. Data on bond angles given in Table 1 are consistent with this picture in so far as the H-N-H and C-O-H bond angles are close to 109.5°.

Since an oxygen atom is divalent, it can replace a hydrogen atom on each of two methane molecules to form the bridged structure (Table 1*e*). Likewise the divalent $\overset{\text{H}}{\underset{|}{}}$ —N— group can form a similar structure (Table 1*f*) and the trivalent nitrogen atom can link together three carbon atoms (Table 1*g*). As expected, for all three molecules the bond angles approach the tetrahedral angle.

The last three compounds listed in Table 1 are typical examples of methane derivatives with carbon having double and triple bonds. The bond angle for compounds with double bonds (H_2CO and H_2CO_2) is found to be much closer to the theoretical 120° than the tetrahedral angle, whereas the angle for compounds with a triple bond is, as expected, 180°. The C-O-H bond angles are nearly the same for the compounds in Table 1, parts (*d*) and (*i*), even though the bonds to the carbon atom are sp^3 in part (*d*) and sp^2 in part (*i*). This is a simplifying fea-

Table 1 Structure of Simple Organic Compounds

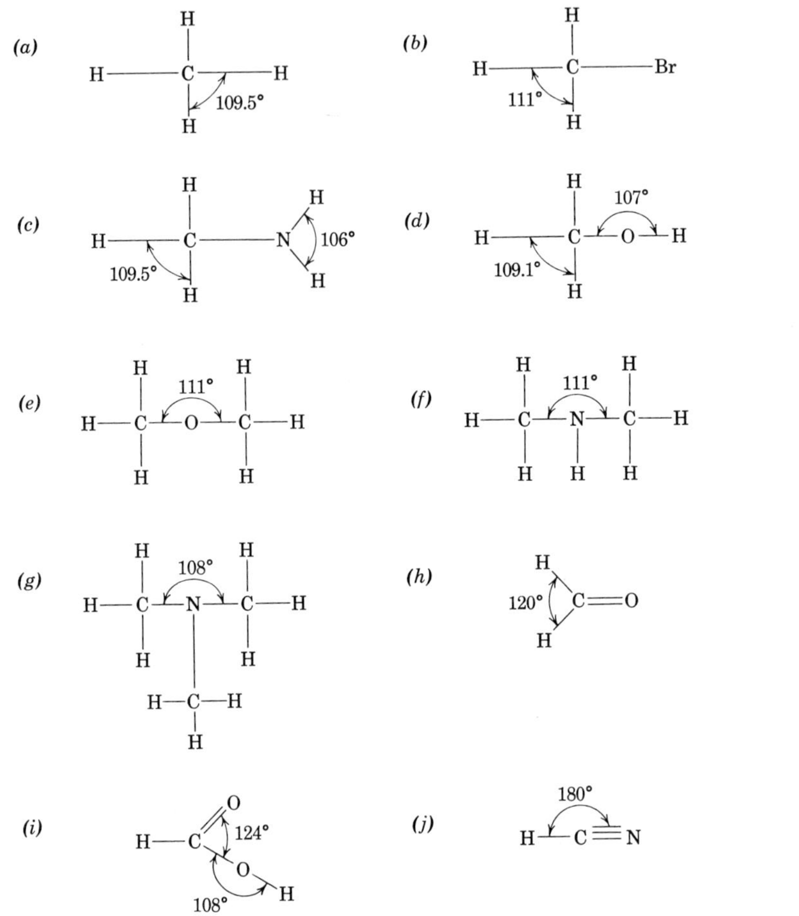

ture of the structural chemistry of these compounds: the bond angle for C-X-Y is almost independent of the types of other bonds formed by the carbon atom.

Monovalent groups such as the halogens can substitute for more than one hydrogen atom in a methane molecule. For example, when CH_4 is treated with chlorine in the presence of ultraviolet light, the following compounds are formed:

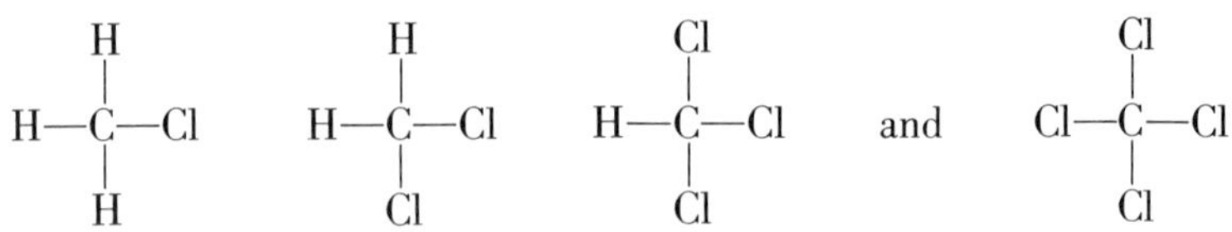

Moreover, several different groups can exist in the same molecule; for example,

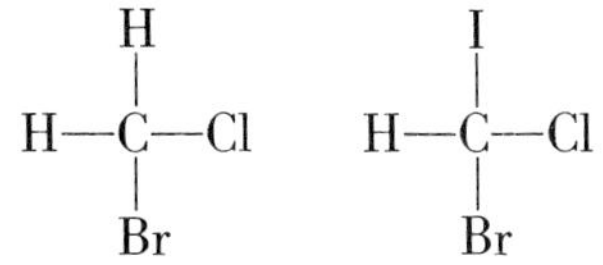

or even

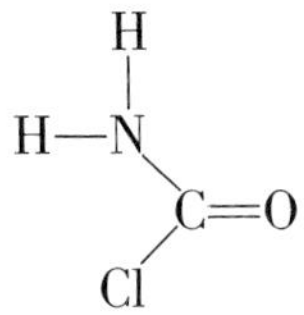 and 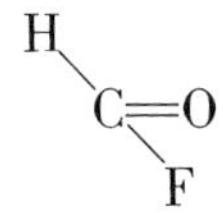

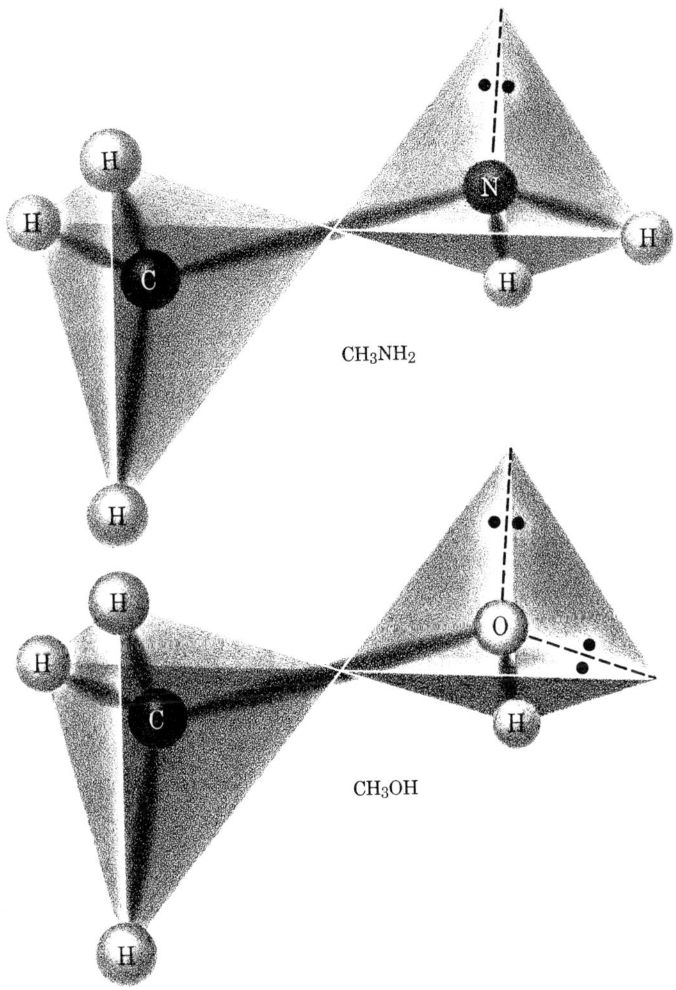

Fig. 4. Structures of methyl alcohol and methylamine

All these compounds follow the bonding rules we set down earlier for the tetravalent carbon atom.

One can use the principles of covalent bond theory to predict structures of new compounds hitherto not found in nature or synthesized in the laboratory. For example, the octet structures of sulfur analogs of the oxygen-containing compounds in Table 1 are

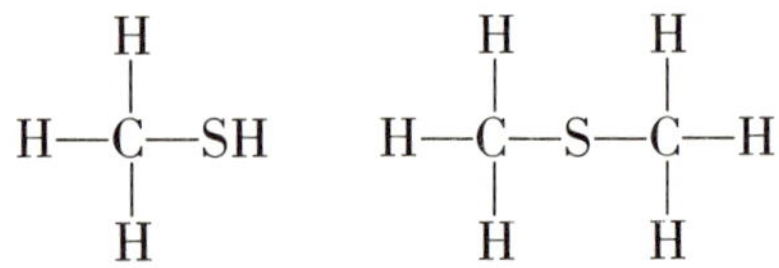

These compounds have been found in nature or have been synthesized. Occasionally, a theoretically predicted compound is unstable and changes spontaneously to form a more stable molecule. Thus, when more than one OH group is attached to a single carbon atom, the following reactions occur.

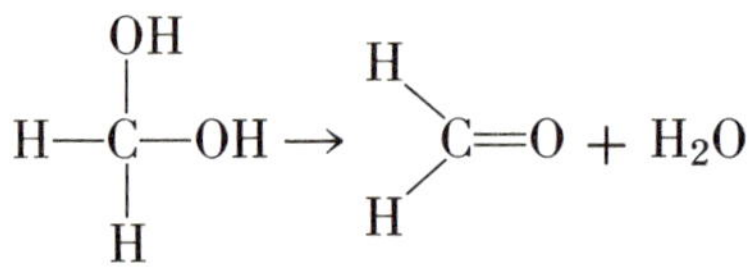

unstable stable

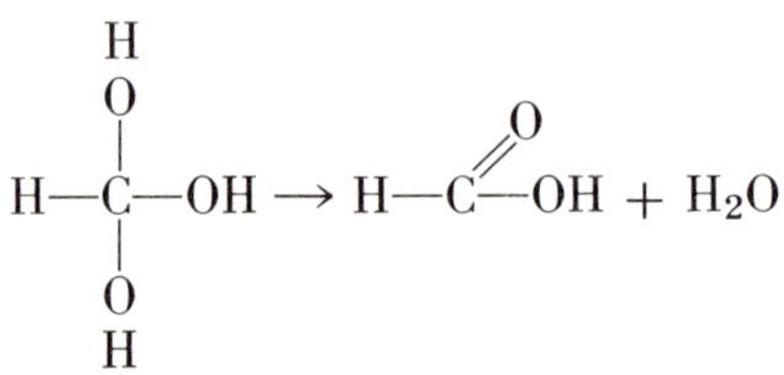

unstable stable

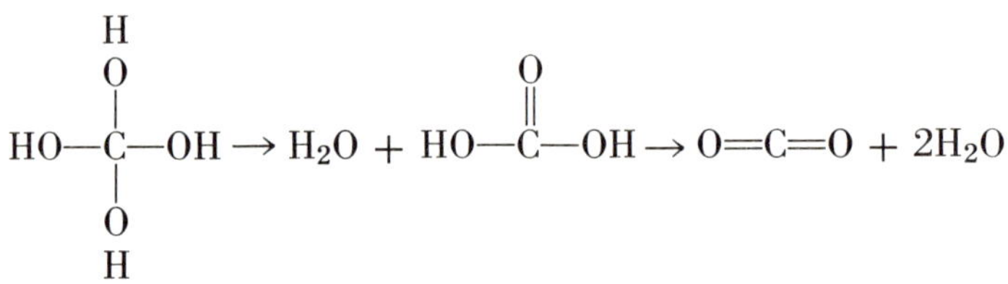

unstable more stable stable

Graphic formulas. We usually represent structures by *graphic* formulas. In these, the three-dimensional nature of the molecule is not shown. For example, we write CH_4 and CH_2Cl_2, respectively, as

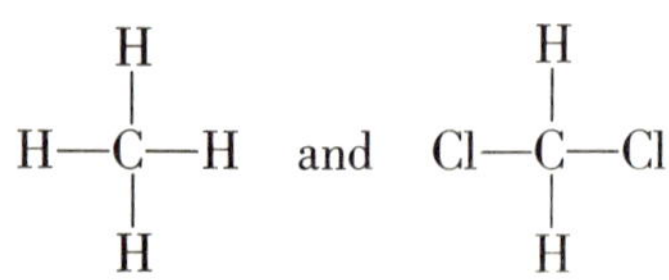

Actually, these molecules are tetrahedral as indicated in Fig. 1. If we forget this, misconceptions arise. For example, if we imagine CH_2Cl_2 to be planar as in the graphic formula,

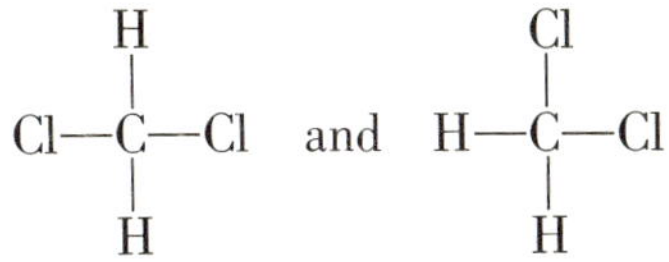

would be different molecules; Fig. 5*a*, a three-dimensional view, shows them to be the same. On the other hand,

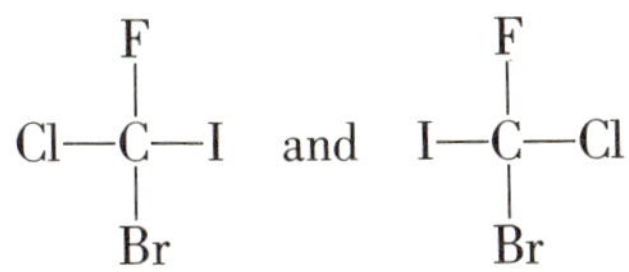

appear to be a front and back view of identical molecules; Fig. 5*b* shows they are *mirror images* of each other and as non-identical as your right and left hand.

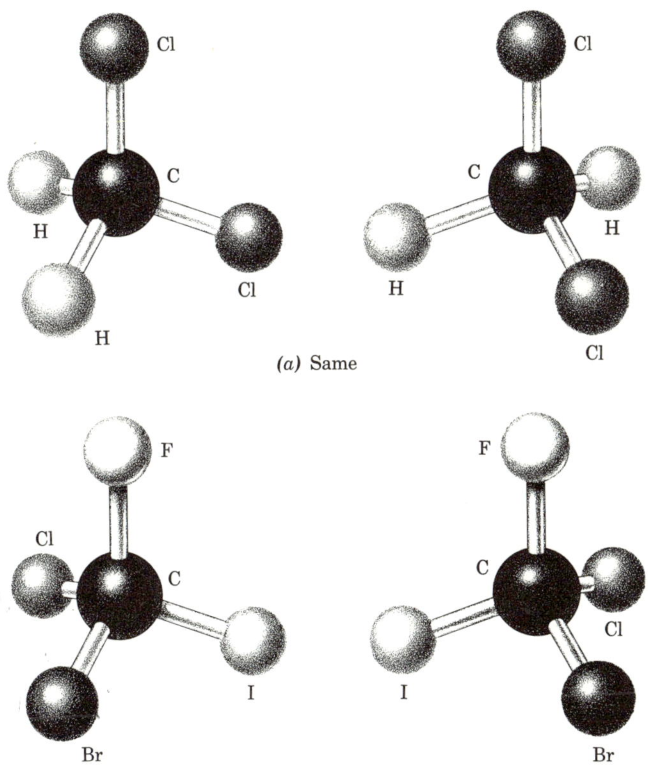

(*a*) Same

(*b*) Mirror images

Fig. 5. Geometric relations.

CARBON-CARBON BONDS

Saturated hydrocarbons. The group

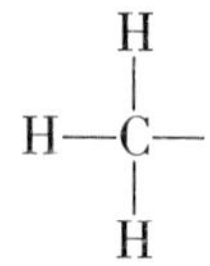

called a *methyl* group, is monovalent; hence this group can substitute for a hydrogen atom in methane to form the compound with two carbon atoms called ethane:

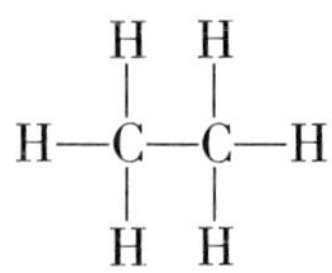

The spatial arrangement of the molecule is shown in Fig. 6. Since it corresponds to two tetrahedra linked through a corner, the tetrahedral bond angles are maintained.

Further substitution of methyl groups can result in giant molecules with a variety of structures, but all of them contain only single carbon-carbon bonds and have the general formula C_nH_{2n+2}, where n is the number of carbon atoms in the molecule. Compounds made only of hydrogen and carbon are called *hydrocarbons.* The C_nH_{2n+2} hydrocarbons are referred to as *paraffin* hydrocarbons. We shall discuss their structure in more detail shortly.

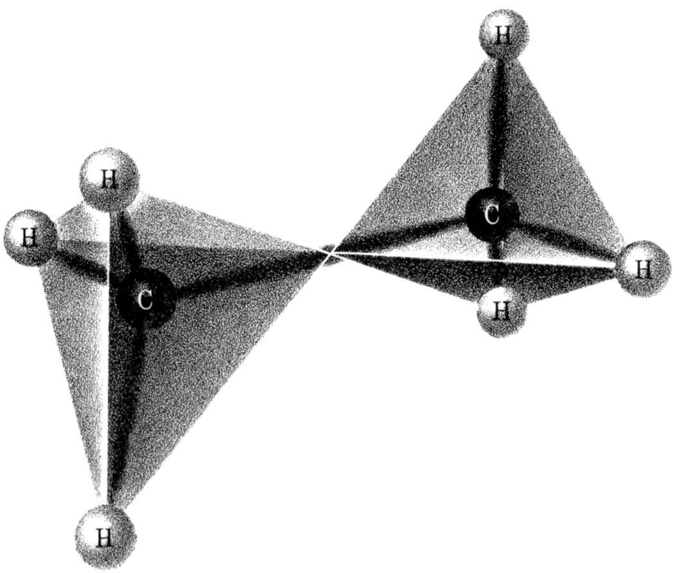

Fig. 6. Ethane.

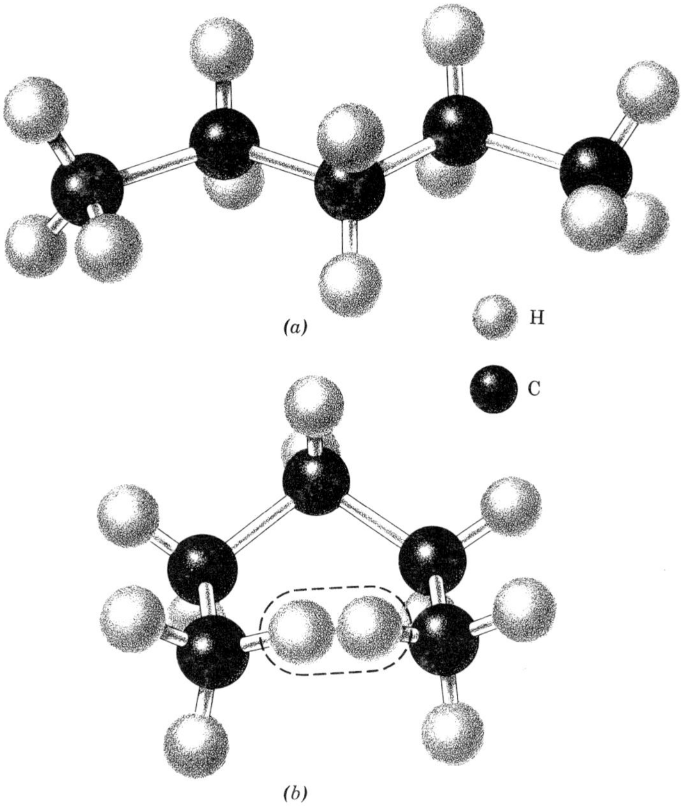

Fig. 7. Pentane—two configurations.

A given paraffin molecule can occur in any configuration in which the C-C-C, C-C-H, and H-C-H bond angles remain 109°. For example, pentane, C_5H_{12}, might momentarily have the configuration shown in Fig. 7*a*, or it could have the configuration shown in Fig. 7*b*. Either is consistent with tetrahedral bond angles. For the form shown in Fig. 7*b*, the end hydrogen atoms (*circled* by a dash line) almost touch. If these hydrogen atoms are removed, a bond can be formed between the end carbon atoms with little distortion of the tetrahedral bonds. The carbon bonds form a ring (cycle); the resulting compound is called cyclopentane, C_5H_{10}.

The preferred tetrahedral bond angles are best approximated by cyclic hydrocarbons with five carbon atoms, but cyclic hydrocarbons are also known that contain three, four, and six (or more) carbon atoms. The structures of these ring compounds are shown in Fig. 8 together with their bond angles. For simplicity, we have indicated the carbon atoms without the two attached hydrogen atoms.

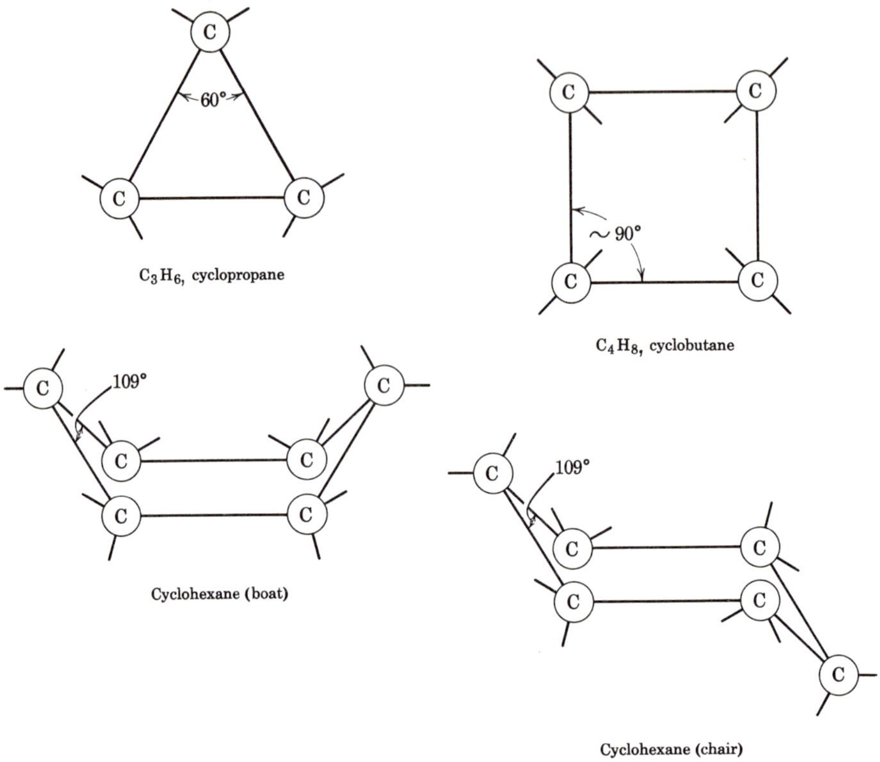

Fig. 8. Cyclic hydrocarbons.

Large deviations from the ideal tetrahedral bond angle for example, in cyclopropane and cyclobutane, make the molecule relatively unstable.

The structure of cyclohexane is peculiar. If all six carbon atoms were in the same plane, the C-C-C bond angle would be 120°; but the preferred 109.5° angle can be approached if the carbon skeleton is non-planar. Two such non-planar forms are possible (Fig. 8), and both are known to exist. The *chair* form is the more stable at ordinary temperatures.

In addition to simple cyclic hydrocarbons containing only one ring, there are also hydrocarbons, such as decalin, containing several rings. The carbon skeleton of decalin is

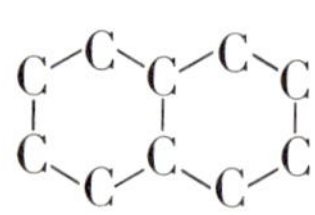

On the basis of bonding principles, you should be able to indicate the number and position of the hydrogen atoms in this molecule.

Alkenes and alkynes. If two hydrogen atoms are removed from the methane molecule, the divalent group CH_2 is formed. When two of these groups are joined together, they form the planar *ethylene* molecule (Fig. 9*a*). In this molecule the two carbon tetrahedra share an edge and the unshared corners are occupied by hydrogen atoms. Such an arrangement results in a planar structure, but to account for the observed H-C-H bond angle (120°) we must assume that the tetrahedra are distorted. An alternative description in terms of sp^2 hybridization yields the same result.

When three hydrogen atoms are removed from the methane molecule, the trivalent CH group results. Two of these join together to form *acetylene.* In this case, the two carbon tetrahedra share a face (Fig. 9*b*) to form a linear molecule. Description in terms of *sp* hybridization yields the same result.

Hydrocarbons containing carbon-carbon double bonds are called *alkenes* or *olefins.* We can imagine that they are formed by removing hydrogen atoms from two adjacent carbon atoms in a paraffin hydrocarbon; for example,

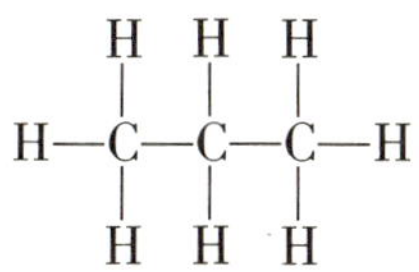

paraffin

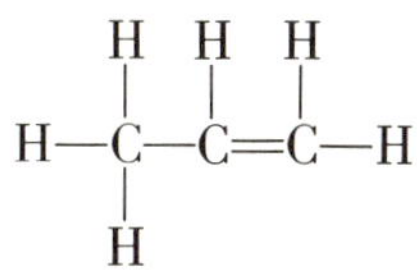

olefin

Olefins that have only one double bond have the general formula C_nH_{2n}. In theory, compounds of this class are more numerous than compounds of the par-

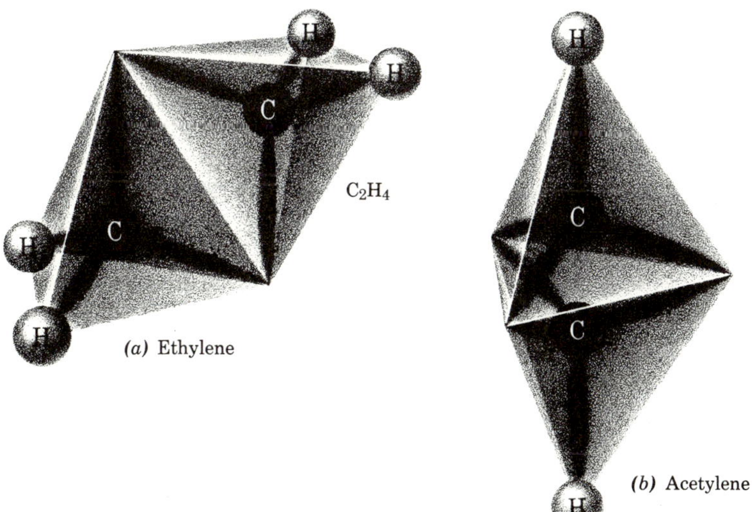

Fig. 9. Structure of olefins and olefynes.

affin class because in each parent paraffin the double bond may form in a number of different places.

Hydrocarbons containing triple bonds can arise from the removal of the two hydrogen atoms bound to carbon atoms forming the double bond in an olefin; for example,

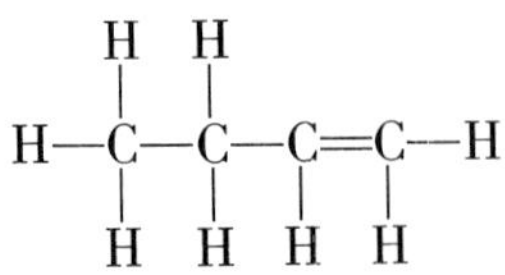 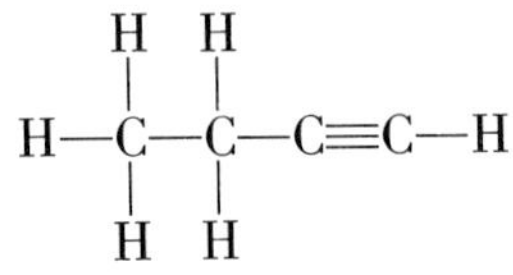

The general formula for these *alkynes* is C_nH_{2n-2}.

In general, hydrocarbons containing only single bonds are known as *saturated* hydrocarbons; those containing double or triple bonds are known as *unsaturated* hydrocarbons.

Other unsaturated hydrocarbons. Alkenes also occur with more than one double bond. For example, consider the compound

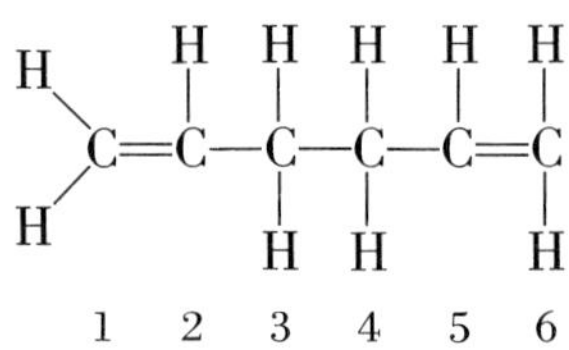

This molecule is wholly consistent with bonding principles discussed earlier. The second double bond could also be between carbon atoms 4 and 5, or 3 and 4, or 2 and 3; each of these compounds would be chemically and physically different.

In principle, a molecule with six carbon atoms can contain up to five double bonds. The structure

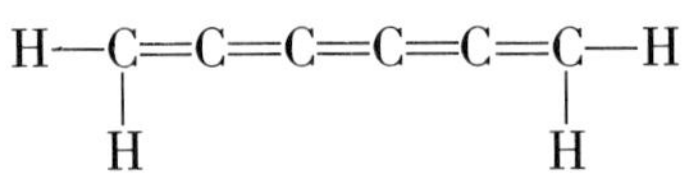

violates none of the bonding rules set forth earlier. But hydrocarbons wherein a single carbon atom forms two double bonds are usually unstable. A more common structure is one containing alternating single and double bonds; that is,

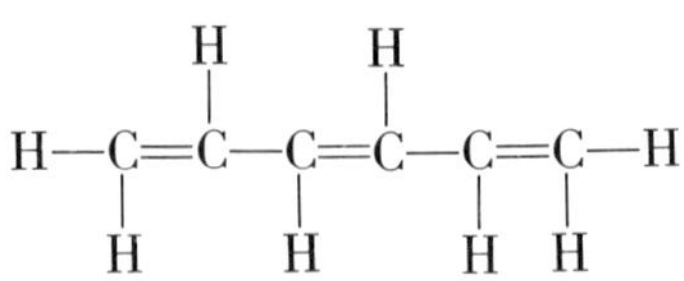

Alkynes containing more than one triple bond also exist.

MOLECULAR MOTION

Temperature. Suppose that two isolated bars of iron are identical except that one is at 100°C and the other is at −100°C. We can tell which bar is hot and which is cold by feeling them. If we touch the two together, the temperatures of both bars will approach a common value near 0°C because heat always flows from a hot to a cold body. These are common qualities of temperature with which we are all familiar. Now we want to discuss the meaning of temperature on a molecular scale, and, to do so, we must first discuss the lesser known qualities of temperature.

In theory, we can heat a substance as hot as we wish, to 10,000° or even to 10,000,000°C, but, no matter how hard we try, we can never cool a substance below −273.15°C. Since this is the lowest possible temperature, it is the zero of temperature on an *absolute* temperature scale. On this scale, −100°C becomes 173°K and +100°C becomes 373°K. To differentiate this absolute scale from the centigrade scale, we add the letter K to the number (the letter K stands for Kelvin, the man who suggested the absolute scale). Why is there this lower limit to the temperature scale?

Suppose we have two flasks of helium gas, one at 1000°K, flask A, and one at 100°K, flask B. They differ in that one body of gas is hotter than the other. This must mean that the gas in the two flasks is different; the atoms must "know" whether they are hot or cold. If we could watch the behavior of atoms in each of the two flasks, we would see the nature of this difference. Helium atoms in a gas are not touching and they are not still. They are flying around in haphazard directions at speeds of thousands of miles per hour. The atoms occasionally collide with each other and carom off the walls of the flask, but, if the temperature of the flask is kept constant, the average speed of the atoms stays constant. If we clock the average speed of the atoms in both flasks we find that the average speed in flask A is about 6300 mi/hr, whereas that in flask B is about 2000 mi/hr.

Suppose flasks A and B touch each other; the temperature in A drops and that in B increases. When both flasks reach a common temperature, the average speed of the molecules in both flasks is about 4700 mi/hr.

Let us cool flask A and simultaneously measure the average speed of the atoms. At 100°K the average speed is about 2000 mi/hr; at 25°K the average speed is 1000 mi/hr; at 0.01°K the average speed is 20 mi/hr. Thus it appears that as we approach absolute zero the average speed approaches zero, and as we increase the temperature the average speed increases. *In general, the absolute temperature is proportional to the violence of molecular motion.* The least motion an atom can have is *no motion* or *zero motion;* this corresponds to zero temperature on the *absolute scale.* (The quantum mechanical treatment of molecular motion modifies slightly our picture of Absolute Zero by introducing the concept of *zero-point* energy, but the statement above is essentially correct.)

The type of motion discussed above, in which the atom or molecule moves as a whole, is called translation. We shall have more to say about this type of motion in Chapter 8. Now we want to talk about internal motions of molecules. Broadly speaking, there are two types. The first involves a periodic stretching or bending of chemical bonds; this is called vibration. The second involves a spinning of the molecule; this is called rotation. Both types of motion increase with increasing temperature.

Vibration. The simplest type of vibrational motion is that found for a hydrogen molecule. Chemical bonds are not absolutely stiff but are elastic; for small amounts of stretching or bending, chemical bonds behave like coiled springs. When a hydrogen molecule is at 0° K, there is almost no motion.† As the temperature increases, the molecule vibrates with an amplitude that increases with temperature (Fig. 10*a*). For an average hydrogen molecule, at 1000° K, the bond distance varies periodically from about 0.67 Å to 0.81 Å. Whenever we speak of a bond distance, therefore, we mean the distance between nuclei averaged over the complete period of vibration; in this instance the *average* bond distance is 0.74 Å.

For molecules containing more than two atoms, the types of vibration are more complex. For example, it can be shown that the possible vibrations in a linear CO_2 molecule consist of the three different types of motion indicated in Fig. 10*b* singly or in combination. The first of these is a symmetric stretching in which the two oxygen atoms move along the bond axis away from the central carbon atom. The second involves a deformation of the 180° bond angle in which the molecule is bent out of a straight line. In this vibration, the carbon atom moves in a direction opposite to the two oxygen atoms. Two motions of this type occur in mutually perpendicular planes. The last type of vibration is an unsymmetric one in which one C—O bond is compressed and the other is stretched. As a molecule of CO_2 is heated, all three of these types of motion or combinations of them increase in amplitude.

We have already seen that in a molecule the bond distance is the average distance between nuclei during a vibration. In the deformation vibration of CO_2, the bond *angle* changes periodically. Thus, when we say the O-C-O bond angle in CO_2 is 180°, we are talking about the average bond angle during vibrations in which the actual bond angle may vary between 175° and 185°.

Six types of vibration are associated with an NH_3 molecule. One of these, shown in Fig. 10*c*, is particularly complex. When relatively still, the NH_3 molecule has the three hydrogen atoms at the base and the nitrogen atom at the peak of the pyramid. In one type of vibrational motion, the hydrogen atoms stay in the same plane and move up, whereas the nitrogen atom moves down. The amplitude of the vibration increases with temperature, until finally the inversion vibration takes place as indicated in Fig. 10*c*. In this mode of motion, the molecule turns inside out during the vibration, much like an umbrella turning inside out. This type of inversion vibration is found for many pyramidal molecules, e.g., PH_3.

† As in the case of gases at Absolute Zero, some motion is required if the Heisenberg uncertainty principle is to hold. Thus the molecule is slightly quivering even at 0° K; the energy associated with this motion is called the *zero-point* energy.

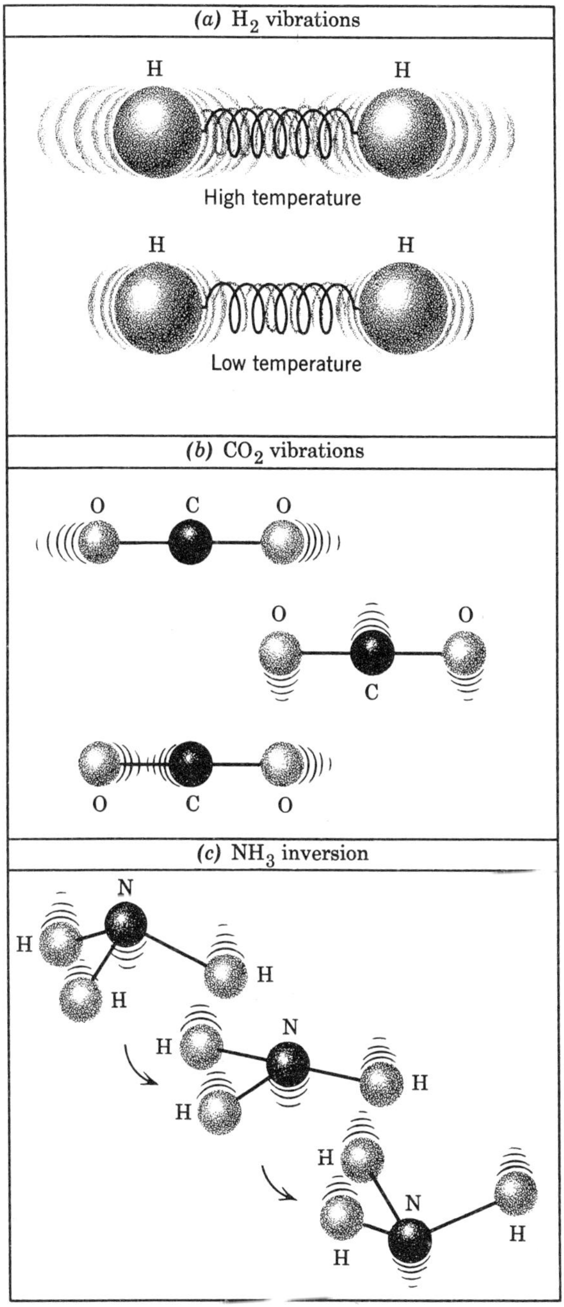

Fig. 10. Vibrations.

Rotation. Rotation is the other type of molecular motion that increases in violence with temperature. For a linear molecule such as H_2 or CO_2, only rotations about axes perpendicular to the bond absorb energy as temperature is increased. For molecules such as NH_3, rotations about three different axes absorb energy. Unlike vibration, rotation causes little change in bond distance or bond angles; hence studies of molecular rotation provide precise values for bond distances and bond angles in simple molecules.

Occasionally, rotations occur in which one part of the molecule rotates with respect to the other part of the molecule. These types of rotation have particularly important chemical consequences in providing information about the forces of attraction and repulsion acting between the two parts.

Let us consider the torsional vibration about the carbon-carbon double bond in ethylene. This type of motion, called *libration,* shown in Fig. 11, involves a twisting of one planar CH_2 group relative to the other; at the extremes of this type of motion the molecule is no longer planar. Assume that we can heat this molecule without decomposition to very high temperatures. As the temperature increases, the amplitude of this libration increases. When the twist becomes greater than 90° out of the plane, one CH_2 group can flip over and turn 180°

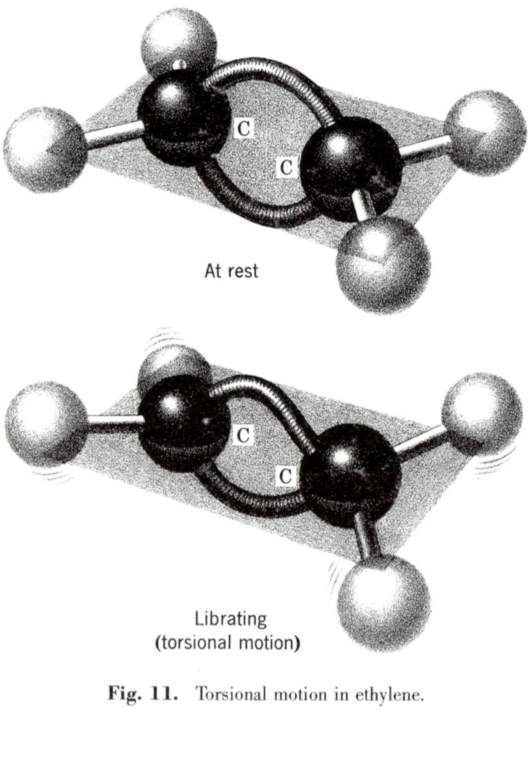

Fig. 11. Torsional motion in ethylene.

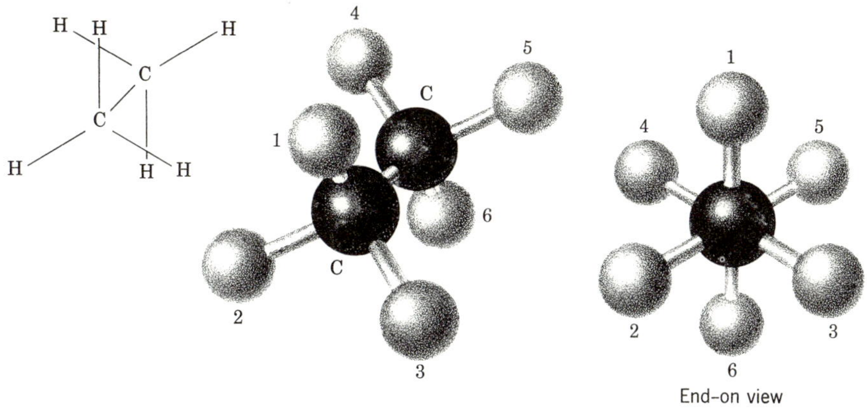

Fig. 12. Ethane—stable configuration.

around the carbon-carbon double bond. At still higher temperatures, we reach the point where this libration is no longer a true vibration but the independent rotation of the CH_2 groups around the double bond. This is called *free rotation* around a bond.

As an example, consider the molecules formed by substituting a chlorine atom for a hydrogen atom on each carbon atom in ethylene. Two different molecules result.

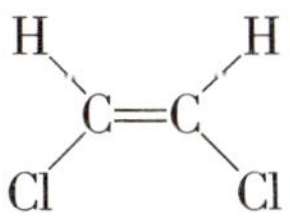

cis

and

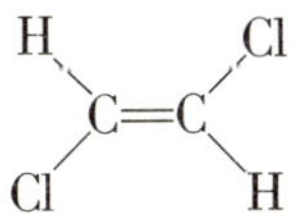

trans

These molecules have the same formula but different structures and are called *geometric isomers;* the *cis* label indicates that like atoms are on the same side of the double bond; the *trans* label indicates that they are on opposite sides of the bond. If free rotation occurs, these two compounds are indistinguishable. It is found in practice that *rotation around a double bond does not occur* except at high temperatures. Because of this, *cis* and *trans* isomers can be separated.

In a molecule like ethane, C_2H_6, the hydrogen atoms attached to different carbon atoms tend to stay as far away as possible from each other. Thus the most stable arrangement for ethane is shown in the end-on view given in Fig. 12. If we were to twist one CH_3 group relative to the other, we would find that the most *un*stable position is the one in which the 1-4, 2-6, and 3-5 hydrogen atoms are lined up when we look at the molecule end-on. Thus, here also, we find libration, motion in which the 1-2-3 CH_3 group of Fig. 12 rocks back and forth relative to the 4-5-6 CH_3 group. If we view pure librational motion end-on, we find that hydrogen atom 1 does not pass over hydrogen atom 4 or 5 but rocks between them. Since a CH_3 group is stiff, corresponding motion also occurs with hydrogen atoms 2 and 3.

At low temperatures, there is librational motion associated with the carbon-carbon single bond; at high enough temperatures, there is essentially free rotation around the carbon-carbon bond. At temperatures so low that free rotation does not occur, there are two isomers of

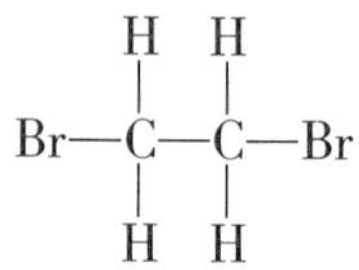

one in which the bromine atoms are close together, e.g., at the 1 and 5 positions of Fig. 12, and the other in which they are far apart, e.g., at the 1 and 6 positions of Fig. 12. At ordinary temperatures, however, *free rotation around single bonds occurs* and this molecule has the properties that would be expected if the two CH_2Br groups spin relative to each other.

At high temperatures, limited rotation around double bonds occurs, and compounds like

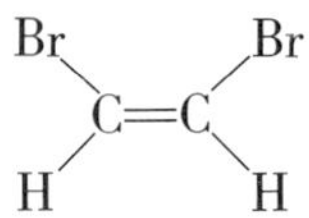

slowly change to the *trans* isomer. But, even at very high temperatures, rotation around double bonds is not completely free in the sense that a mixture of *cis* and *trans* forms has the properties of a single compound. Before the compound can be heated to a temperature high enough for this to occur, the molecule decomposes.

SATURATED HYDROCARBONS

Straight-chain paraffins. In Table 2, we have listed the names and formulas of the first ten paraffin hydrocarbons. Such a series of compounds similar in structure and representable by a general formula is called a *homologous* series. The names of the first four in this series are non-systematic with a common *ane* suffix. The compounds with five or more carbon atoms have names made up with a Latin or Greek prefix to indicate the number of carbon atoms and the *ane* suffix to show that they are members of the paraffin homologous series. The student should know the names and formulas of these compounds.

Remember that these molecules are not flat as suggested by the graphic formula. Figure 13 shows the three-dimensional structure of butane. Since free rotation occurs, the configuration is not fixed; a butane molecule can have either of the extreme configurations represented in Fig. 13 without distorting tetrahedral angles. Thus, at ordinary temperatures, a flask of butane contains both coiled and extended molecules; every second, a given butane molecule coils and uncoils millions of times because of free rotation around single bonds. Although the compounds indicated in Table 2 are not really straight, they are called *straight-chain hydrocarbons.*

Branched hydrocarbons. None of the carbon atoms in the molecules indicated in Table 2 is bound to more than two other carbon atoms. Each of these molecules could be formed by substituting a $—CH_3$ group for a hydrogen atom on the

Table 2 Straight-Chain Hydrocarbons

Name	No. of C Atoms	Formula[a]	Isomers
Methane	1	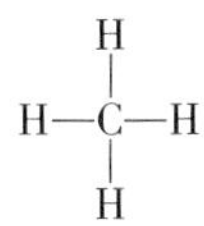	1
Ethane	2	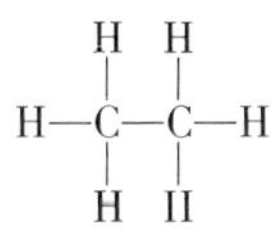	1
Propane	3	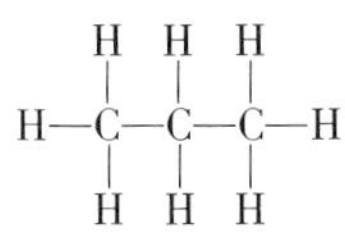	1
Butane	4	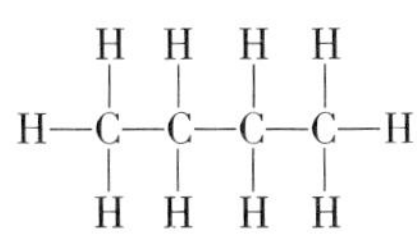	2
Pentane	5	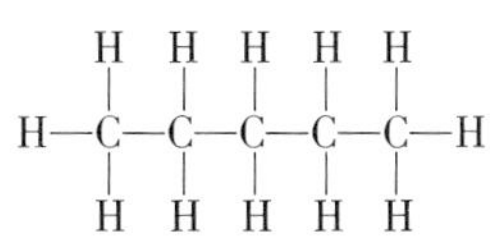	3
Hexane	6	C_6H_{14}	5
Heptane	7	C_7H_{16}	9
Octane	8	C_8H_{18}	18
Nonane	9	C_9H_{20}	35
Decane	10	$C_{10}H_{22}$	75

[a] General formula C_nH_{2n+2}.

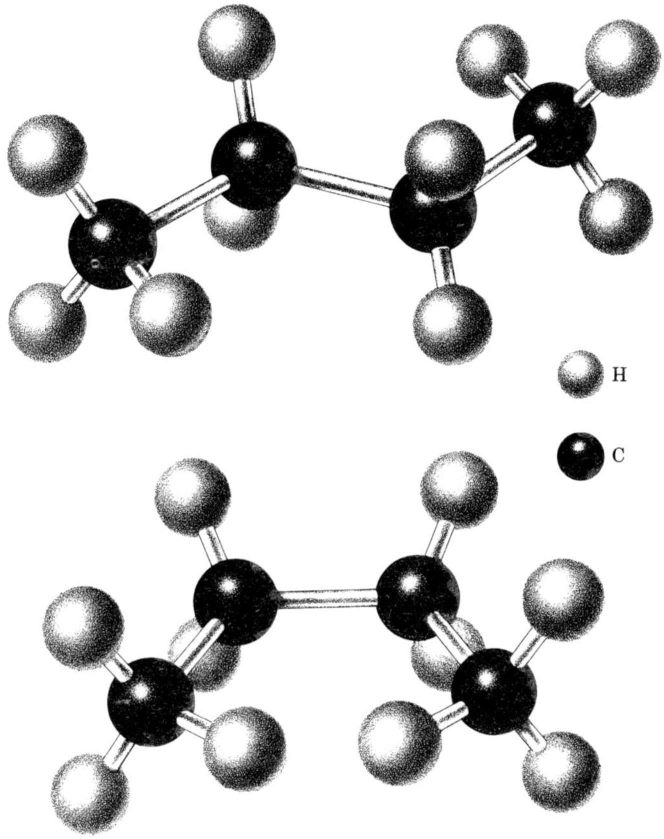

Fig. 13. Butane configurations.

end carbon atom of the preceding member of the series. In other words, butane can be formed by the following (imaginary) steps.

$$
\begin{array}{ccccccc}
 & H & H & H & \\
 & | & | & | & \\
H- & C- & C- & C- & H
\end{array}
\!\!\!\!\!\begin{array}{c} \\ \\ \end{array}
$$

$$
\begin{array}{ccccccc}
 & H & & H & & H & \\
 & | & & | & & | & \\
H- & C & - & C & - & C & -H \\
 & | & & | & & | & \\
 & H & & H & & H &
\end{array}
\rightarrow
\begin{array}{ccccccc}
 & H & & H & & H & \\
 & | & & | & & | & \\
H- & C & - & C & - & C & - \\
 & | & & | & & | & \\
 & H & & H & & H &
\end{array}
+ H-
$$

$$
\begin{array}{ccccccc}
 & H & & H & & H & \\
 & | & & | & & | & \\
H- & C & - & C & - & C & - \\
 & | & & | & & | & \\
 & H & & H & & H &
\end{array}
+
\begin{array}{ccc}
 & H & \\
 & | & \\
- & C & -H \\
 & | & \\
 & H &
\end{array}
\rightarrow
\begin{array}{ccccccccc}
 & H & & H & & H & & H & \\
 & | & & | & & | & & | & \\
H- & C & - & C & - & C & - & C & -H \\
 & | & & | & & | & & | & \\
 & H & & H & & H & & H &
\end{array}
$$

But none of the bonding principles cited earlier is violated if a four-carbon compound is formed with one carbon atom bound to *three* carbon atoms:

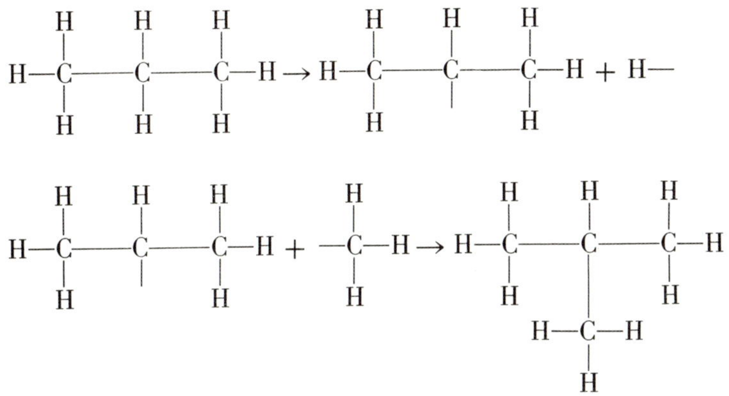

Such compounds are referred to as *branched-chain hydrocarbons.*

Both of the butanes above have the same formula, C_4H_{10}. They are, however, different molecules and have different properties; they are *structural isomers.* The straight-chain isomer is called *normal* butane, or *n*-butane; the branched-chain isomer is called *isobutane* or *i*-butane.

All paraffins with more than three carbon atoms have several isomeric forms. As an example, consider pentane. The straight-chain pentane is shown in Table 2. One branched-chain isomer can be formed by shifting the methyl group (CH_3) from the end of the chain to the next-to-the-first carbon atom in order to form

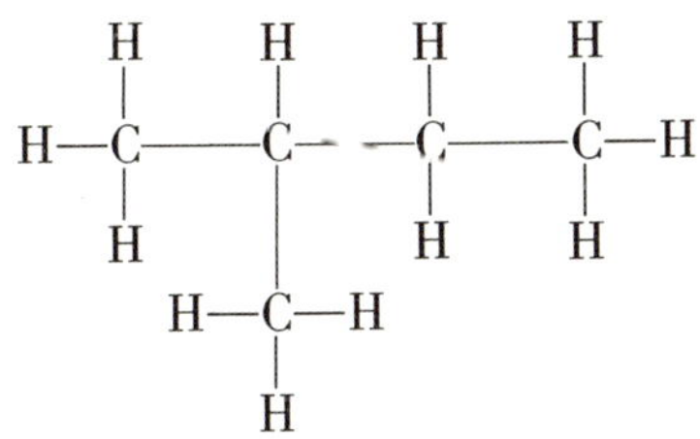

This is the same molecule as

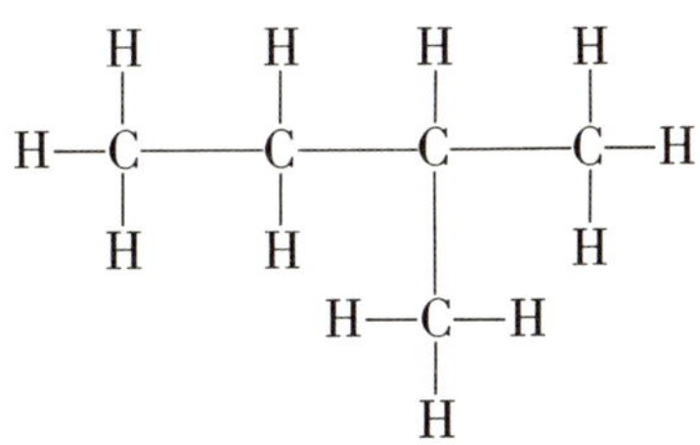

Still another isomer can be formed by removing a methyl group from both ends and attaching them to the central carbon atom of the remaining three-carbon chain; that is,

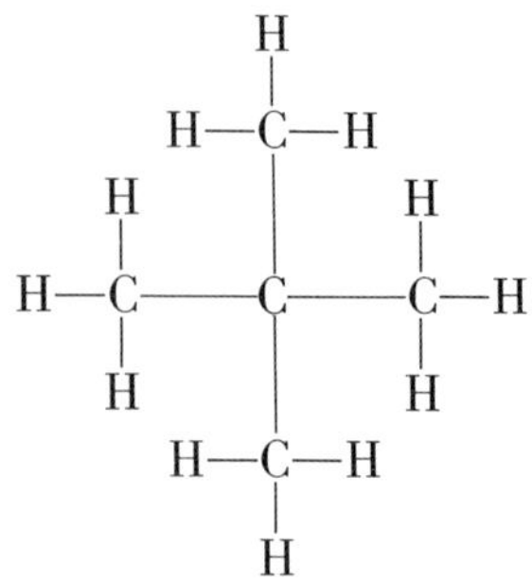

Thus there are three different molecules with the formula C_5H_{12}.

Table 2 lists the number of isomers expected for each of these paraffin hydrocarbons. The number of possible isomers increases rapidly as the number of carbon atoms in the molecule increases. There is more than one paraffin isomer of triacontane, $C_{30}H_{62}$, for each man, woman, and child in the world.

Differences in isomers. Isomers are different both chemically and physically. Table 3 lists the isomers of pentane and hexane together with their melting points and boiling points. The difference between isomers is quite pronounced. In general, the boiling points are highest for the straight-chain isomers because there is a larger area of contact between adjacent molecules. The melting points follow no simple trend, but usually symmetrical molecules have higher melting points, explainable thermodynamically. It is interesting to note that the variation in melting points of pentane isomers is greater than the difference in melting points of *n*-pentane and *n*-hexane.

Four types of carbon atoms are found in paraffin hydrocarbons. End carbon atoms that are attached to only one other carbon atom are called primary carbon atoms. Carbon atoms that are attached to two other carbon atoms are called secondary carbon atoms. Carbon atoms that are bound to three and four other carbon atoms are called tertiary and quaternary carbon atoms, respectively. Although all four types of carbon atoms form similar bonds, they differ slightly; this small difference gives rise to the different chemical properties of isomers.

The chemical difference in the carbon-hydrogen and carbon-carbon bonds in two isomers is evidenced by the heat released when the isomers, pentane and 2,2-dimethylpropane (Table 3) form from the constituent atoms. In forming either isomer, four carbon-carbon and twelve carbon-hydrogen bonds are created. If the bonds were truly identical, the heat released in their formation would be identical; in actual fact, the heat released in forming 2,2-dimethylpropane from

Table 3 Physical Properties of Isomers

Formula	Graphic Formula	Melting Point (°C)	Boiling Point (°C)	Name
C_5H_{12}	C—C—C—C—C	−130	36	Pentane
C_5H_{12}	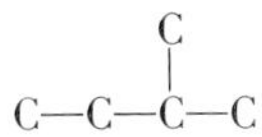	−160	28	2-Methylbutane
C_5H_{12}	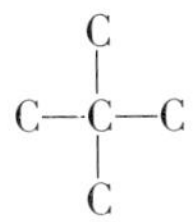	−17	10	2,2-Dimethylpropane
C_6H_{14}	C—C—C—C—C—C	−95	69	Hexane
C_6H_{14}	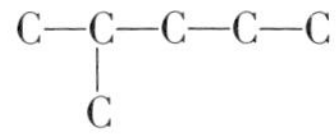	−154	60	2-Methylpentane
C_6H_{14}	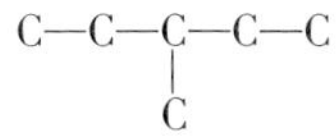	−118	63	3-Methylpentane
C_6H_{14}	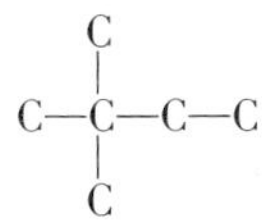	−100	50	2,2-Dimethylbutane
C_6H_{14}	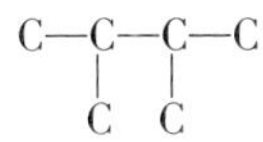	−129	58	2,3-Dimethylbutane

its constituent atoms is 4 kcal greater than that released when *n*-pentane is formed.

Chemical differences in isomers can be enormous. For example, the efficiency of these compounds as automobile fuels depends on the rate of their reaction with air in the engine cylinder. Fuels are rated on the basis of the amount of knock or noise due to explosive burning under standard operating conditions. A scale called the *octane number* has been set up for measuring the fuel performance. The fuel to be tested is compared to a mixture of two standard fuels; one of these fuels, 2,2,4-trimethylpentane, has a very low knock, the other, *n*-heptane, has a very high knock. An octane rating of 100 for a fuel means the performance is the same as that of pure 2,2,4-trimethylpentane (or isoöctane); a rating of 65 means that the fuel performance is the same as that of a mixture of 65% isoöctane and 35% *n*-heptane. Such tests reveal striking difference between isomers. For example, *n*-hexane has an octane rating of 26, whereas the hexane isomer 2,2-dimethylbutane (Table 3) has a rating of 94. In general, the more branched the isomer, the higher is the octane number. This is but one of many examples of chemical differences found for various isomers.

In any given hydrocarbon, there are several different types of hydrogen atoms; for example, in 2-methylbutane (Table 3), hydrogen atoms are attached to primary, secondary, and tertiary carbon atoms. The chemical reactivity differs for these different hydrogen atoms. As an example, consider the following reaction:

$$\underset{\text{2-methylbutane}}{C_5H_{12}} + Br_2 \rightarrow C_5H_{11}Br + HBr$$

in which a bromine atom is substituted for a hydrogen atom. In practice, it is not easy to stop this substitution after the addition of only one bromine atom, but, if we could, we would find that the following types of products are formed.

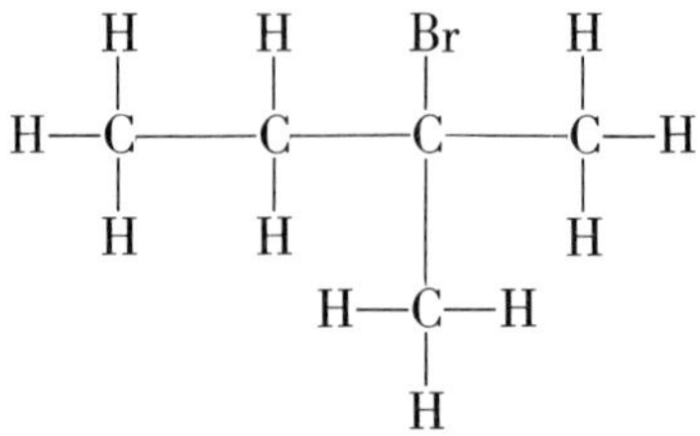

A

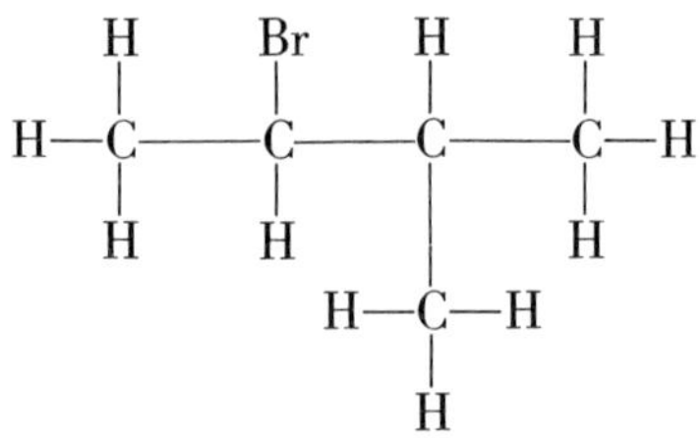

B

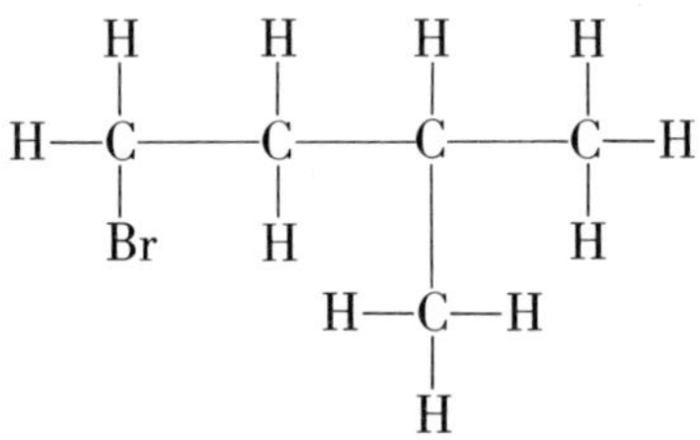

C

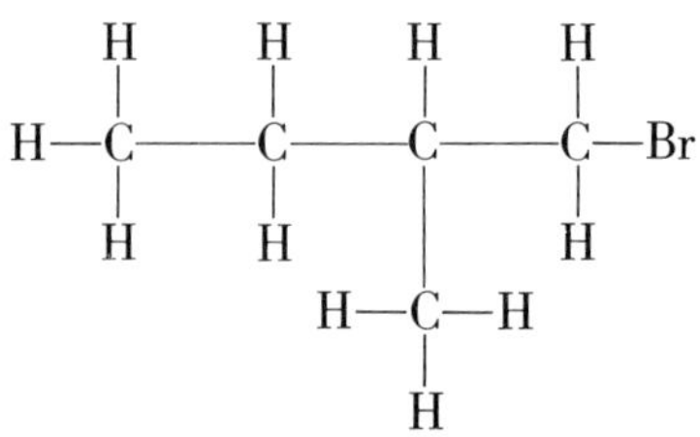

D

To produce compound A, substitution for a tertiary hydrogen atom occurs; to produce B, substitution for a secondary hydrogen atom occurs; to produce C or

D, substitution for a primary hydrogen atom occurs. In the products we find that the amount of A is greater than the amount of B and the amount of B is greater than the amount of C or D. Thus, in such substitutions, tertiary hydrogen atoms are more reactive than secondary hydrogen atoms, and secondary hydrogen atoms are more reactive than primary hydrogen atoms.

NOMENCLATURE

Paraffins. In view of the great variety of paraffin hydrocarbons, it was necessary to develop a systematic system of nomenclature. This was initiated in 1895 by an international committee at Geneva.

All hydrocarbons contain a "straight" chain of tetravalent carbon atoms. Atoms in this chain are bonded to hydrogen atoms or hydrocarbon fragments of the types

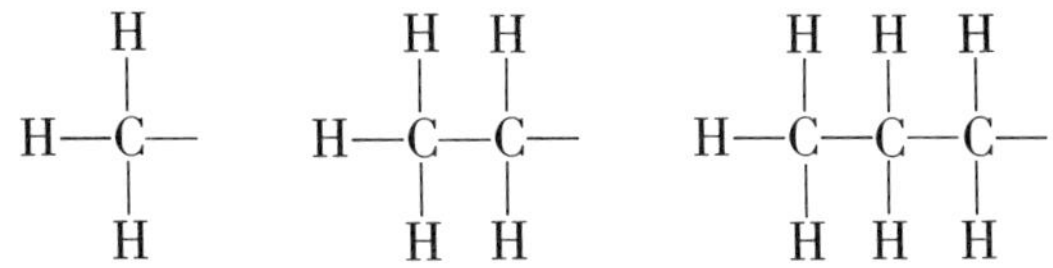

These fragments are named by substituting the suffix *yl* for the *ane* suffix of the parent hydrocarbon. Thus the above are called *methyl, ethyl* and *propyl*, respectively. Three steps are necessary to assign a name to a compound:

1. *Pick out the longest continuous chain of carbon atoms.* This step is not always easy. In the compound

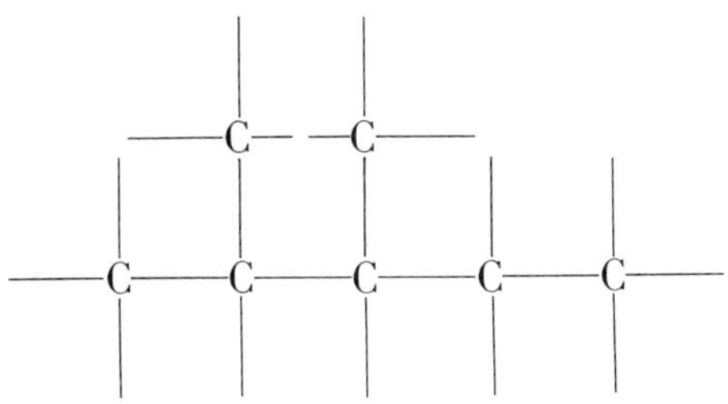

(where we have not indicated hydrogen atoms) the longest chain contains five carbon atoms, but in the compound

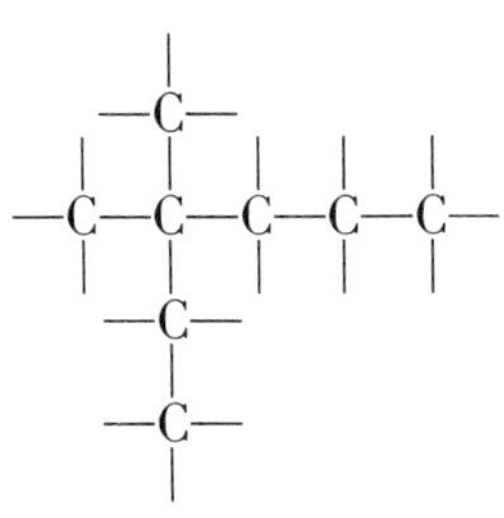

the longest chain contains six carbon atoms.

2. *Number the carbon atoms in the longest chain.* This numbering should be done in such a way that the carbon atoms bonded to groups other than H have the lowest possible numbers. Thus, in the two examples given above, the numbering is

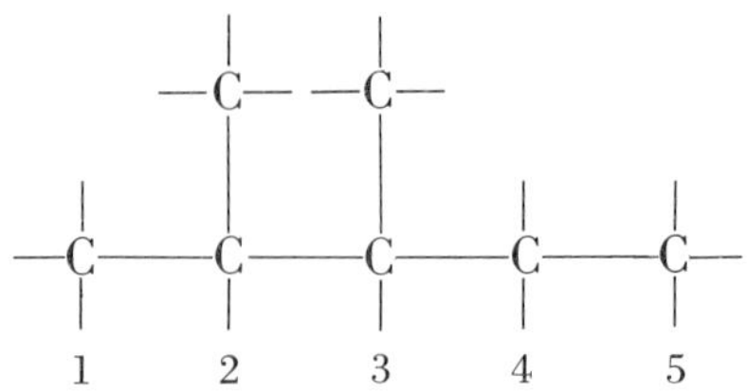

and

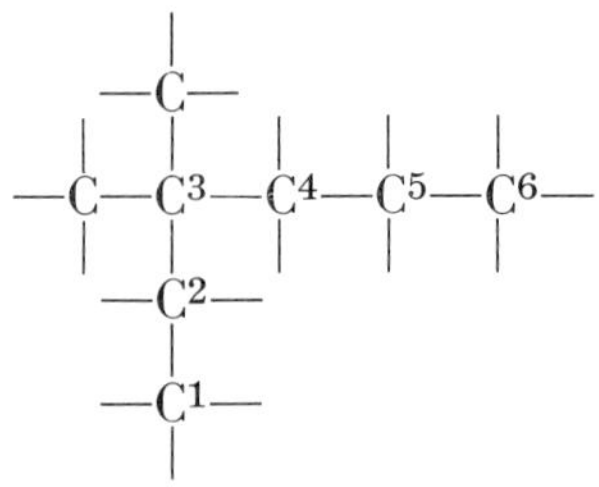

If the numbering were started at the other end of the chain, the number assigned to carbon atoms with attached groups would not be as small as possible.

3. *Note the groups attached to the chain and specify their location by number.* In the first example, there are two methyl groups located at 2 and 3; in the second example, there are two methyl groups located at position 3.

The principal name comes from the longest chain; this is written last. Before this, we give the numbers and types of groups attached to the chain. Thus the first example discussed is called 2,3-dimethylpentane, and the second is called 3,3-dimethylhexane.

Let us consider two other examples of this nomenclature. The compound

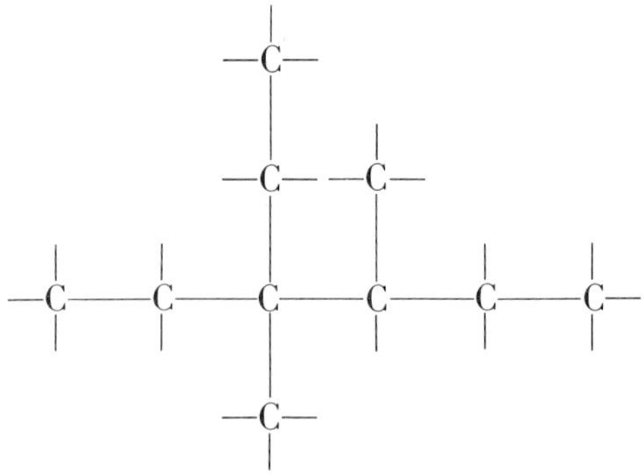

is called 3,4-dimethyl-3-ethylhexane, rather than 3-ethyl-3,4-dimethylhexane because it is customary when more than one type of group is present to name the least complicated one first. The compound shown at the top of the facing page is called, 3,3,4,5,7,-pentamethyl-4,6,7-triethyl-5-propyldecane.

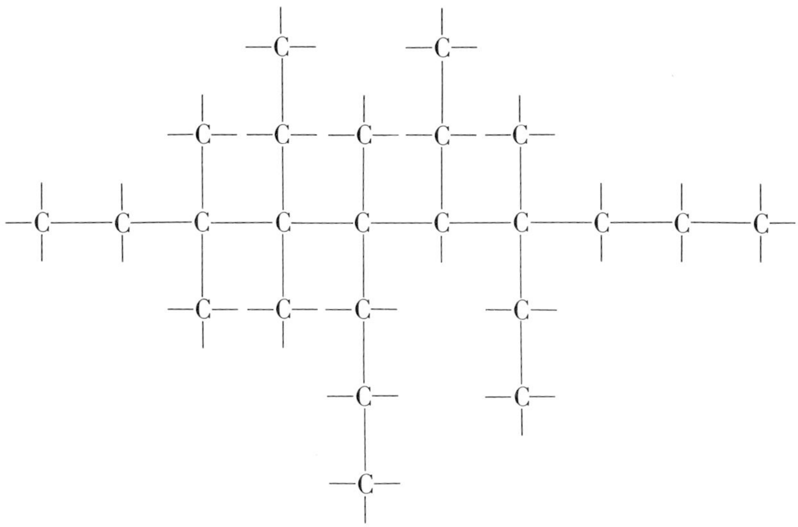

Cyclic paraffins are named by a similar procedure. If no groups are attached to the ring, the prefix *cyclo* is added to the name of the corresponding paraffin. Thus the cyclic hydrocarbons C_3H_6, C_4H_8, C_5H_{10}, and C_6H_{12} are called cyclopropane, cyclobutane, cyclopentane, and cyclohexane, respectively. Compounds with one attached group are prefixed with the name of the group; for example,

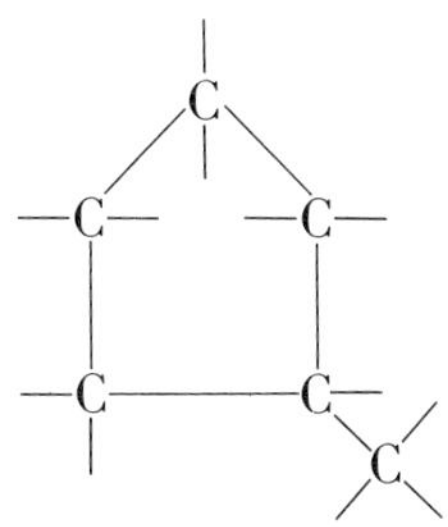

is called methylcyclopentane. If more than one group is attached, the ring is numbered starting at one of the groups and named in a manner similar to that used for the paraffins. Thus

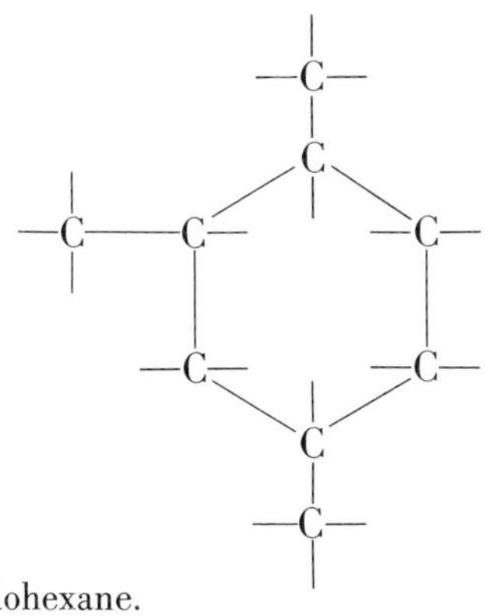

is called 1,2,4-trimethylcyclohexane.

Table 4 Mono-olefins

Name	No. of C Atoms	Formula	Isomers[a]
Ethene-1 (ethylene)	2	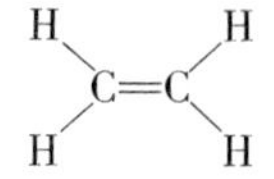	1
Propene-1 (propylene)	3	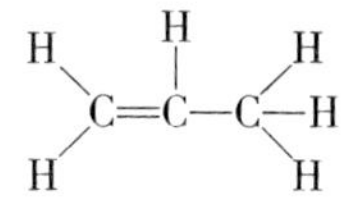	1
Butene-1 (butylene)	4	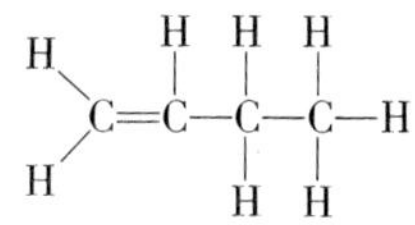	3
Pentene-1	5	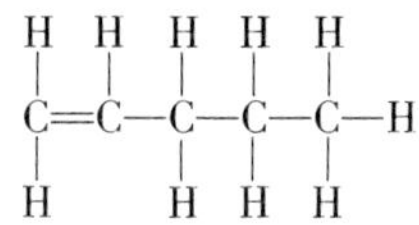	5
Hexene-1	6	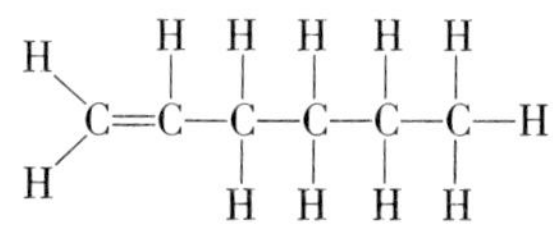	13
Decene-1	10	$C_{10}H_{20}$	377

[a] This listing does not include *cis* and *trans* isomers.

Mono-olefins. The names and formulas of a few of the members of the mono-olefin homologous series, C_nH_{2n}, are shown in Table 4. According to the Geneva system, the names of these mono-olefins have the same stem as the corresponding paraffins with the *ene* rather than the *ane* suffix.

The predicted number of isomers of a given member of this series is greater than that of the corresponding member of the paraffin series (cf. Table 2); isomers occur as the result not only of branching but also of the position of the double bond. Thus there are two straight-chain isomers of pentene:

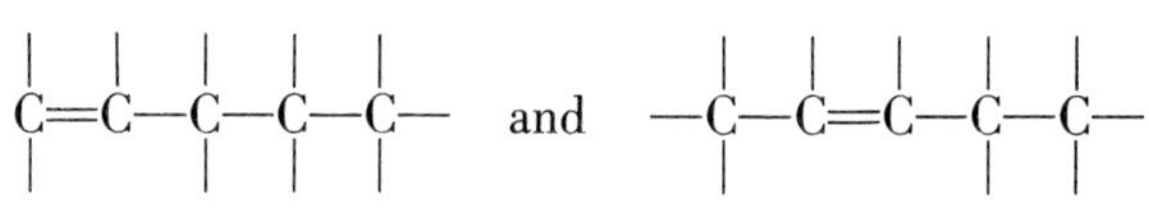

In order to identify the position of the double bond a number is added to the name. Thus these isomers are called pentene-1 and pentene-2.†

The situation is rendered more complex by the existence of *cis* and *trans* isomers of pentene-2. These are shown in Fig. 14. The physical properties of these isomers are

† These are also called 1-pentene and 2-pentene.

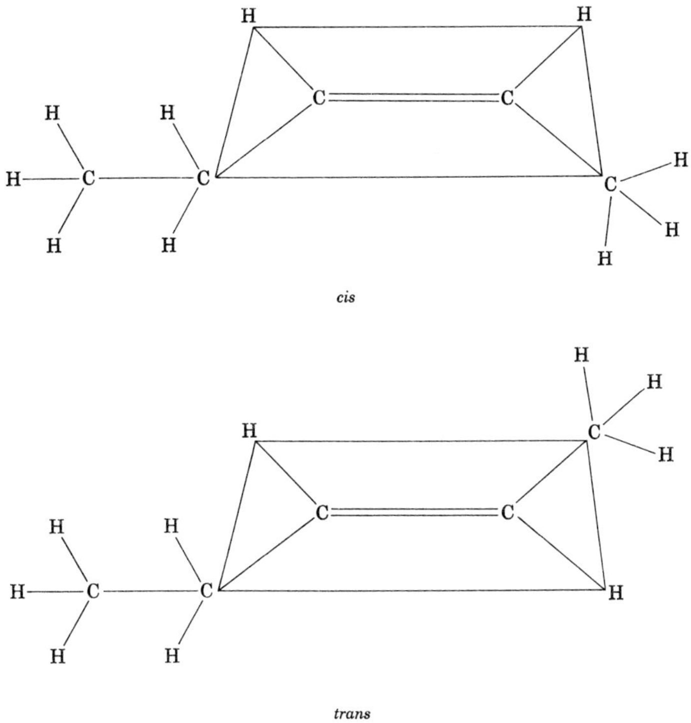

Fig. 14. Pentene-2.

listed in Table 5. The difference in properties is almost as great as that due to branching.

Carbon skeletons of the other isomers of C_5H_{10} (ignoring *cis* and *trans* isomers) and their names are given below:

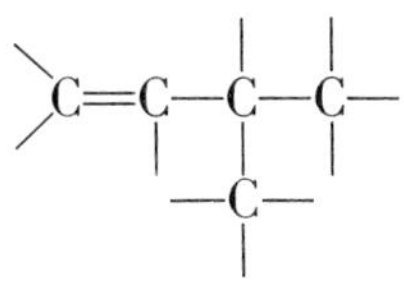

3-methylbutene-1

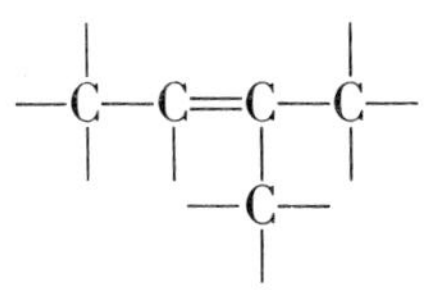

2-methylbutene-2

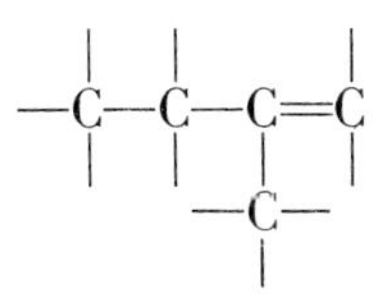

2-methylbutene-1

Table 5 Physical Properties of the Isomers of Straight-Chain Pentenes

	Pentene-1	*cis*-Pentene-2	*trans*-Pentene-2
Melting point (°C)	−165	−140	−134
Boiling point (°C)	30	36	39

Diolefins. All diolefins have the same stem as the corresponding paraffin with the suffix *diene* followed by two numbers to indicate the positions of the double bonds. The names and skeletal formulas of several diolefins are given below:

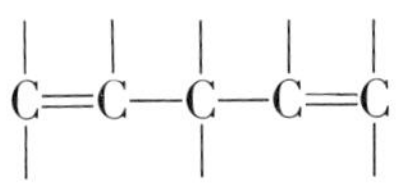

pentadiene-1,4

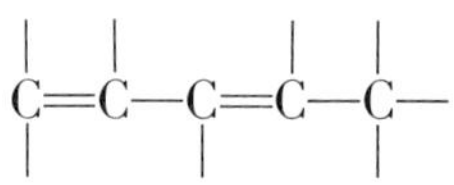

pentadiene-1,3

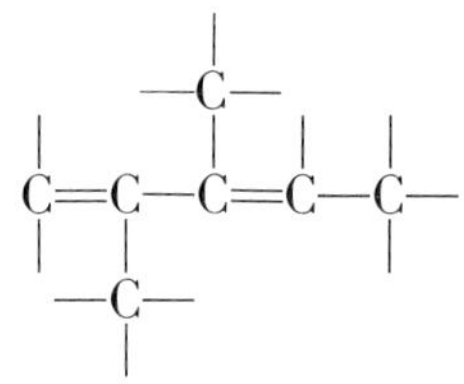

2,3-dimethylpentadiene-1,3

Alkynes. The nomenclature for compounds with a triple bond is similar to that for the olefins. All such compounds have the suffix *yne* rather than *ene*. Below we have indicated the structure and name of the first four members of this series.

H—C≡C—H	ethyne
H—C≡C—CH_3	propyne
H—C≡C—CH_2—CH_3	butyne-1
H—C≡C—CH_2—CH_2—CH_3	pentyne-1

RESONANCE IN UNSATURATED HYDROCARBONS

Conjugation. In a molecule of the type

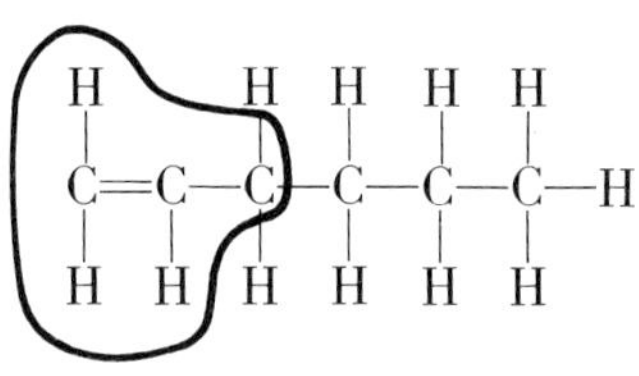

the encircled atoms must lie in the same plane. If a hydrocarbon molecule has alternating double and single carbon-carbon bonds, as in butadiene-1,3,

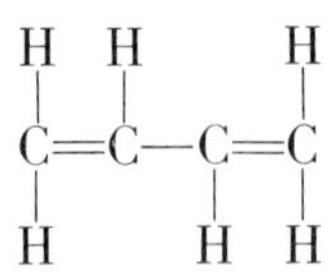

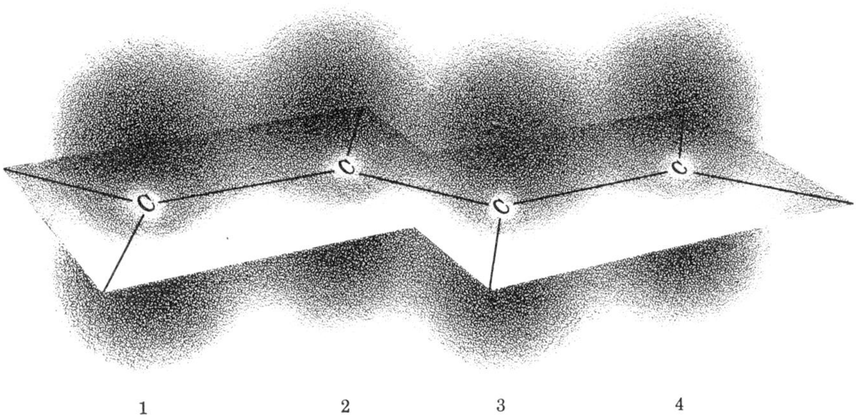

Fig. 15. Butadiene-1,3.

it turns out that *all* the atoms in the molecule lie in the same plane. This type of bonding in which single and double bonds alternate is called *conjugation*. It leads to an especially stable structure.

Resonance. The Lewis structure of butadiene, given above, suggests that the electron cloud is twice as dense between carbon atoms 1 and 2 as it is between carbon atoms 2 and 3. If we consider the electron cloud distribution for sp^2 hybridization with overlapping p orbitals, a different conclusion is reached. In Fig. 15, we have indicated the sp^2 bonds in butadiene-1,3 by lines and have sketched the cloud distribution for the p orbitals. This picture suggests that the electron cloud concentration between carbon atoms 1 and 2 is not much greater than that between atoms 2 and 3. In order to include this aspect of the bonding in a Lewis structure, we would have to indicate the structure as the resonance hybrid,

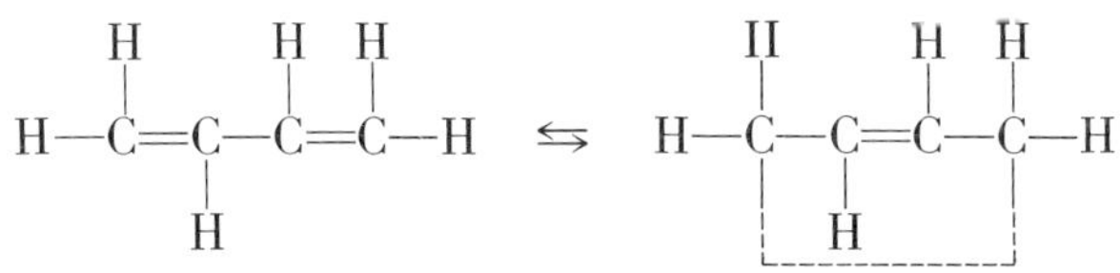

I II

The resulting molecule is a superposition of these two forms. They do not contribute equally to the structure; the actual molecule closely resembles the more stable form I.

Benzene forms a conjugated molecule in which two contributing resonance forms (shown at the top of the following page) are of equal importance. This is to be expected since the bonding in the two forms is identical.

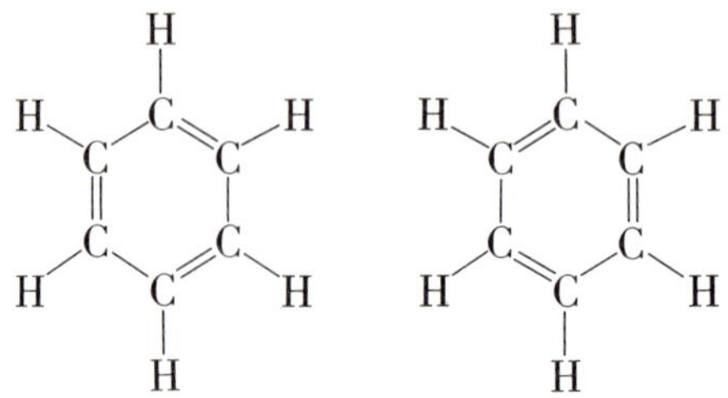

The importance of resonance stems from the fact that *whenever resonance occurs the bonds in the actual molecule are stronger than we would expect on the basis of either of the contributing Lewis structures. The more similar the contributing forms and the greater their number, the stronger will be the actual bonds and the greater will be the stability of the molecule.*

It should be emphasized that contributing forms to a resonance hybrid differ only in the location of electrons; the location of nuclei is the same in all contributing forms. As we said earlier, the concept of resonance is introduced to provide a truer representation of charge cloud distribution with the preservation of the concept of a shared-pair bond; hence the number of paired electrons is the same in all contributing forms.

Resonance energy. In the bond energy calculations in the previous chapter, we found that the energy required to break a carbon-carbon single bond is about 83 kcal. As a guess, we might expect that the energy of a double bond would be twice as great, or 166 kcal. Actually, studies of a variety of reactions show that the ordinary double-bond energy is about 147 kcal. Thus, if we imagine the rupture of a double bond to occur in two consecutive steps, the rupture of the first half of the bond requires 64 kcal; and the rupture of the second half requires 83 kcal, the carbon-carbon single-bond energy.

All carbon-carbon double bonds undergo the following reaction:

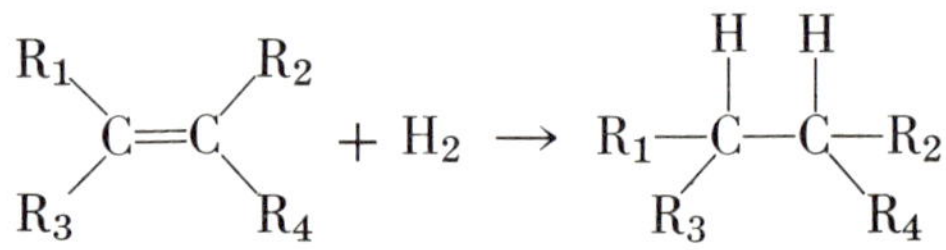

in the presence of a catalyst. In the reaction, R_1, R_2, R_3, R_4 indicate any group. We can calculate the energy† released in this reaction as follows:

$$-\Delta E_b = 2E_{\mathrm{C-H}} - E_{\mathrm{H-H}} - E_{\frac{1}{2}\mathrm{C=C}}$$

where $E_{1/2\mathrm{C=C}}$ stands for the bond energy of the first half of the double bond. From the table of bond energies (p. 158) we find that about 30 kcal of energy should be released. In Table 6, we have listed heats† of hydrogenation for

† Energy manifests itself as work *and* heat; for reactions discussed in this chapter almost all of the energy appears as heat.

Table 6 Heats of Hydrogenation

Olefin	Product	ΔH[a] (kcal)
propene	propane	−29.7
butene-1	butane	−30.0
pentene-1	pentane	−30.0
hexene-1	hexane	−30.0
butadiene-1,3	1-butene	−27.0
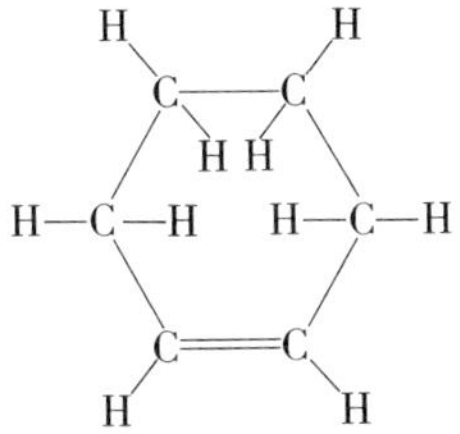 cyclohexene	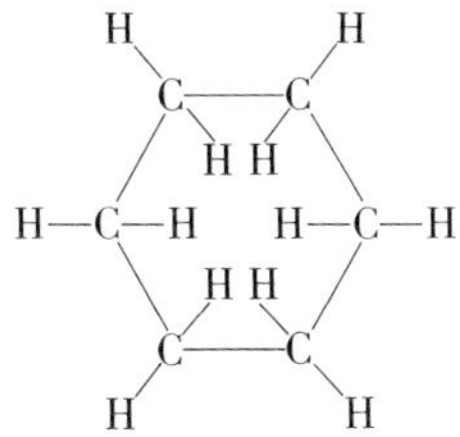 cyclohexane	−28.6
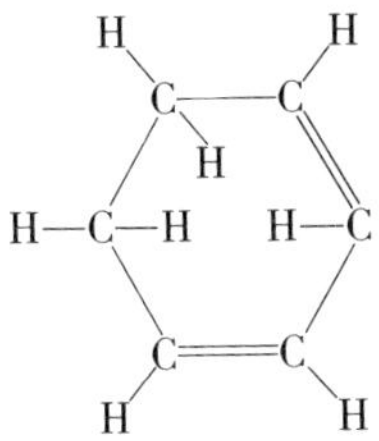 cyclohexadiene-1,3	cyclohexane	−55.4
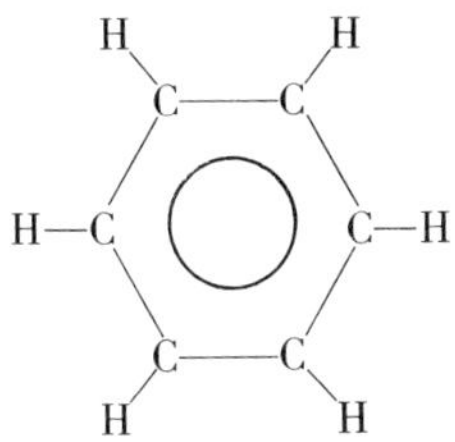 benzene	cyclohexane	−49.8

[a] A negative sign indicates that heat is evolved.

olefinic hydrocarbons. Only the first four, C_3H_6, C_4H_8, C_5H_{10}, and C_6H_{12}, contain simple double bonds; for these, the agreement with experiment is excellent.

The heat released when one of the double bonds of butadiene is hydrogenated is 3 kcal less than the value calculated on the basis of the structure of the principal contributing form. This means that the double bond in butadiene is 3 kcal stronger than expected if the possibility of resonance is ignored. This extra energy is called the *resonance energy*.

The strength of a double bond in a cyclic olefin like cyclohexene is different from that for a straight-chain olefin. Thus, in Table 6, we find the heat released by the hydrogenation of cyclohexene is only 28.6 kcal. The heat released when twice as much hydrogen reacts with the two double bonds of cyclohexadiene-1,3 should be about twice as great, or 57.2 kcal. The measured heat release is about 2 kcal less than this. On the basis of bond energy calculations, complete hydrogenation of one of the contributing forms (cyclohexatriene-1,3,5) to the benzene hybrid with three moles of hydrogen should liberate 85.8 kcal. In this case (Table 6), there is a discrepancy of 36 kcal; hence resonance makes benzene 36 kcal more stable than we might expect.

The resonance stabilization of benzene and butadiene is represented by the diagram in Fig. 16. In the diagram, butane has a lower heat content than butene-1 plus hydrogen. If butene-1 plus hydrogen reacts, heat is evolved in an amount equal to the difference in the two levels drawn in Fig. 16. Thus, to a first approximation, the higher a compound is in this diagram, the less stable it is. The heavy lines in Fig. 16 correspond to the heat content we would expect for the principal contributing resonance forms; the dotted lines represent values actually observed. From this diagram, we see that the heat of hydrogenation of butadiene is 57 kcal rather than 60 kcal because resonance makes butadiene 3 kcal more stable than the principal contributing form. On the other hand, the heat of hydrogenation of benzene is 49.8 kcal rather than 85.8 kcal because resonance makes the molecule 36 kcal more stable than the principal contributing form.

Bond distances and resonances. Whenever resonance is an important factor it affects not only the bond energy but also the bond distance. For example, in most hydrocarbons, the carbon-carbon bond distances are 1.54 Å for single bonds and 1.34 Å for double bonds. If the structure of butadiene is truly a superposition of the two forms discussed earlier, the 1-2 bond distances should be greater than 1.34 Å and the 2-3 bond distance should be less than 1.54 Å. The distances observed are 1.35 Å and 1.46 Å, respectively. For benzene, resonance

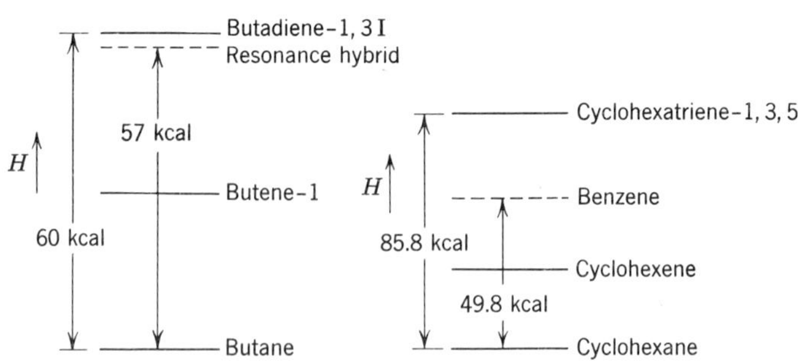

Fig. 16. Resonance.

would be expected to make all carbon-carbon distances the same and intermediate to double and single bonds. The observed bond distances are 1.40 Å.

Applications of resonance. In an earlier discussion (p. 153), we found that a perfect shared-pair bond does not occur for a polar molecule such as HCl. In other words, in the formula (pp. 157, 158)

$$E_{\mathrm{HCl}} = \tfrac{1}{2}E_{\mathrm{H-H}} + \tfrac{1}{2}E_{\mathrm{Cl-Cl}} + 23(\mathrm{X_H} - \mathrm{X_{Cl}})^2$$

the first two terms represent the bond energy for a perfect covalent bond of the type

$$\mathrm{H:\ddot{\underset{..}{Cl}}:}$$

The third term is needed because perfect sharing does not occur and the chlorine atom has more than an even share of the bonding electrons. In the language of resonance, we say that HCl is a resonance hybrid with the contributing forms

$$\underset{\text{I}}{\mathrm{H:\ddot{\underset{..}{Cl}}:}} \quad \text{and} \quad \underset{\text{II}}{\mathrm{H^+:\ddot{\underset{..}{Cl}}:^-}}$$

In the wave mechanical formulation of the problem, we assume that the wave function of the molecule is the sum of two terms containing different amounts of the wave function for the imaginary forms I and II. The resulting electron distribution is a superposition of that for form I and form II; in this particular case it is roughly 85% I and 15% II. This unequal contribution of I and II might have been anticipated because calculations show that form I alone is far more stable than form II alone. Thus the concept of resonance produces a better approximation to the charge cloud distribution in a molecule than that given by a single Lewis structure. The effects of this particular type of resonance are provided for by the formula for bond energies. The contribution of the ionic form to the resonance hybrid increases the bond energy by the amount indicated by the third term of the formula. Thus the resulting bond energy for a resonance hybrid is greater than that of either imaginary contributing form. The more stable the ionic form, i.e., the greater the electronegativity difference in the bond, the more the bond will be strengthened by the last term.

The effect of the stability of contributing forms on the resonance energy is represented in Fig. 17. The higher the energy on this diagram, the less is the stability. In such a diagram, the energy for a perfect shared-pair bond can be estimated by the first two terms of the bond energy equation; the energy for an ionic molecule, H^+Cl^- for example, can be obtained by the methods discussed in the last chapter; the resonance energy Δ is the third term in the bond energy equation. In HCl, the stability of the shared-pair structure is far greater than that of the ionic structure; hence the difference in energy Δ between the contributing form and the resonance hybrid, RH, is not very large. For HF, the stability of the ionic form is more comparable to that of the covalent form and the resonance energy is much greater.

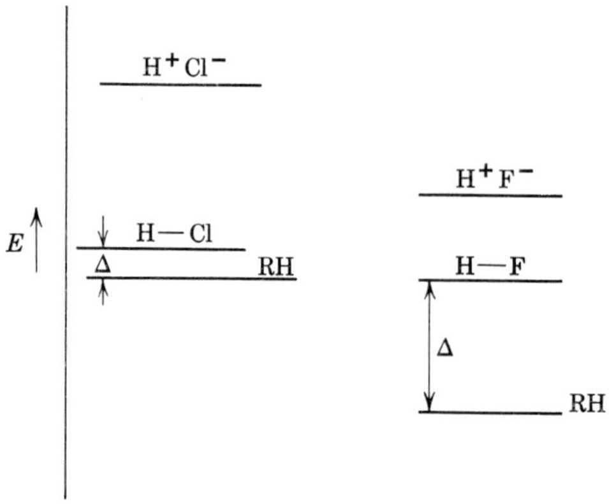

Fig. 17. Resonance effects in polar compounds.

In Fig. 18, we have indicated the energy of contributing forms and the resonance hybrid for butadiene, benzene, and the nitrate ion. The first has two contributing forms not equal in stability; the second has two contributing forms identical in stability; the third has three forms (p. 151), also identical in stability. We see from this figure that the closer together the stability of contributing forms, the greater will be the resonance stabilization. In addition, the greater the number of contributing forms the greater the stabilization.

The usefulness of the concept of resonance can be best illustrated by examples. There are two general classes of organic compounds containing an OH group:

$$\mathrm{R{-}C({=}O){-}OH} \qquad \text{and} \qquad \mathrm{R{-}O{-}H}$$

where R is any paraffin group minus a hydrogen atom. In solution, the first class of compounds (known as acids) readily yield H^+ and the $\left(\mathrm{RC({=}O){-}O}\right)^-$ anion; the second class of compounds (known as alcohols) have little tendency to yield H^+ and the $(R—O)^-$ anion. In other words, the reaction

$$\mathrm{R{-}C({=}O){-}O{-}H} \rightarrow \left(\mathrm{R{-}C({=}O){-}O}\right)^- + H^+$$

occurs readily, whereas the reaction

$$\mathrm{R{-}O{-}H} \rightarrow [\mathrm{R{-}O}]^- + H^+$$

has little tendency to occur. The O—H bond is broken in both cases; hence the only reason for the first reaction to occur more readily is that the

$$\left(\mathrm{R{-}C}\genfrac{}{}{0pt}{}{{=}O}{{-}O}\right)^-$$

ion is more stable than the $[R—O]^-$ ion. This occurs because there is no reason why the electron charge cloud in one C—O bond should be more dense than in the other; therefore the 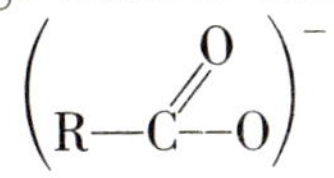 ion is a resonance hybrid of

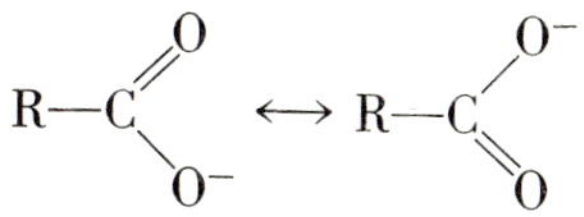

In other words, half the excess negative charge resides equally on each oxygen atom. It is this stabilization that causes 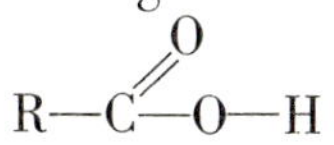 to be an acid even though R—O—H is not.

Another example is provided by HNO_2 and HNO_3, both of which are acids; that is, they undergo the following reactions.

$$H—O—N{=}O \rightarrow H^+ + (O—N{=}O)^-$$

and

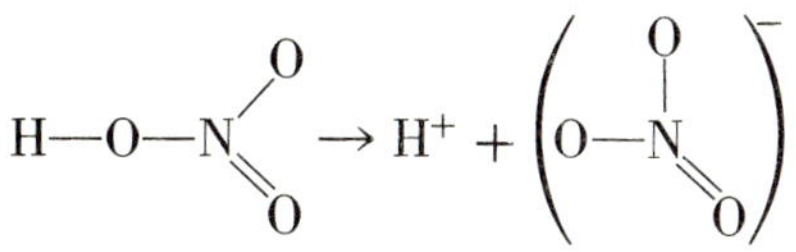

In this instance both anions are stabilized by resonance; the contributing forms are

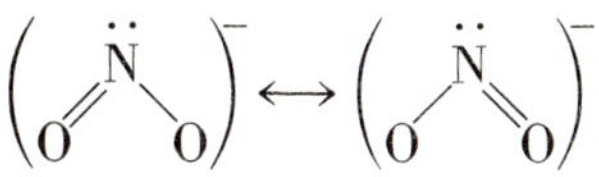

and

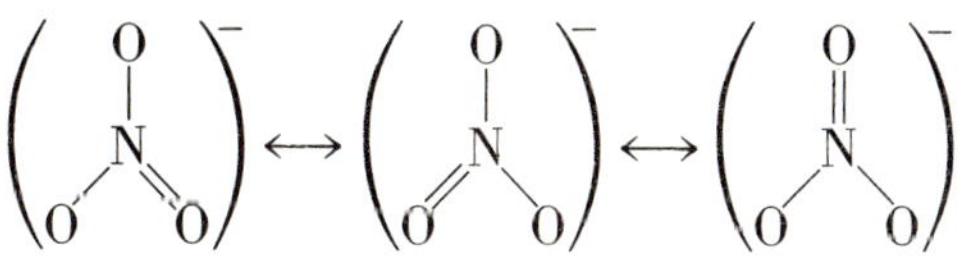

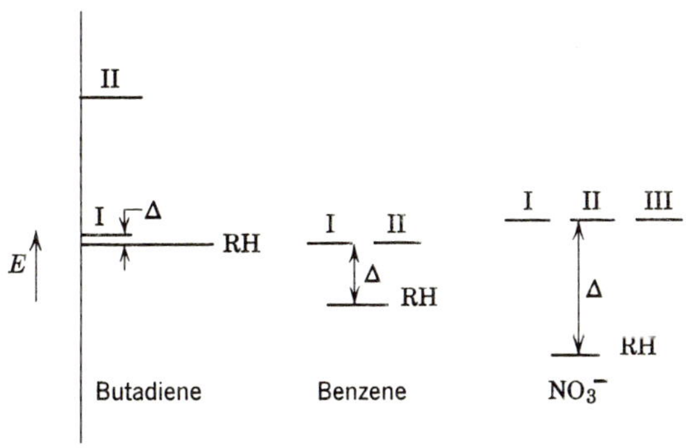

Fig. 18. Resonance effects and contributing structures to the resonance hybrid (RH).

Since the NO_3^- anion has the larger number of stable contributing forms, the resonance stabilization is greater. Thus we conclude that HNO_3 should function as a better H^+ source, a better acid, than HNO_2. This conclusion is borne out by experiment.

REACTIONS OF SATURATED HYDROCARBONS

The bond strengths of the carbon-carbon and carbon-hydrogen single bonds are 83 and 99 kcal, respectively. These represent fairly strong chemical bonds; hence chemical reactions of paraffins, which involve the breaking of these bonds and the formation of new ones, are not expected to liberate sizable quantities of energy unless the bonds formed are especially strong. Thus, in general, paraffins undergo few violent chemical reactions.

Chain-breaking reactions. One of the most violent reactions of hydrocarbons is *burning in oxygen—combustion.* If sufficient oxygen is available, the products of this reaction are carbon dioxide and water; for example, for methane,

$$CH_4 + 2O{=}O \rightarrow 2H{-}O{-}H + O{=}C{=}O$$

When all species are present as a gas, the energy change in this reaction can be calculated as follows:

$$-\Delta E = 2E_{C=O} + 4E_{O-H} - 4E_{C-H} - 2E_{O=O}$$

The bond energies for single bonds have been listed earlier (p. 158); the bond energy for O=O is 117 kcal, and that for C=O in CO_2 is 189 kcal. With these data, we find

$$-\Delta E = (2 \times 189) + (4 \times 111) - (4 \times 99) - (2 \times 117) = 192 \text{ kcal}$$

Thus this reaction occurs with the release of large amounts of energy because the O—H bond energy is far greater than the C—H bond energy and the C=O bond energy is far greater than the O=O bond energy.

The measured amount of heat released in the above reaction is 189.2 kcal. The agreement between calculated and observed values is generally good as can be seen from the comparisons given in Table 7.

Values of *heats of combustion* provide a measure of the strengths of carbon-carbon and carbon-hydrogen bonds. Large differences between the calculated and determined heats of combustion are indicative of peculiarities in the bonding. Thus the discrepancy for cyclopropane (Table 7) is explained by the fact that the 60° C-C-C bond angle distorts the carbon-carbon bond and thereby reduces the bond energy. This less stable bond reacts more readily with oxygen and yields a higher heat of combustion than that calculated.

Table 7 Calculated and Observed Heats of Combustion

Compound	Calculated Heat of Combustion (kcal)†	Observed Heat of Combustion (kcal)†
Ethane	339	336
Propane	484	483
Butane	630	640
Cyclopropane	436	465

† If these values were expressed as ΔH, they would all be negative quantities since heat is evolved.

The properties of a chemical equation plus the common products in combustion processes make heat of combustion data especially useful. For example, suppose we want to know the heat released in the reaction

$$C_2H_6 + H_2 = 2CH_4 \qquad \Delta H = ?$$

Heats of combustion are known for the following reactions.

$$(a) \qquad CH_4 + 2O_2 = CO_2 + 2H_2O \qquad \Delta H_a = -189$$

$$(b) \qquad C_2H_6 + \tfrac{7}{2}O_2 = 2CO_2 + 3H_2O \qquad \Delta H_b = -336$$

$$(c) \qquad H_2 + \tfrac{1}{2}O_2 = H_2O \qquad \Delta H_c = -58$$

In these equations, a minus ΔH means that heat is evolved. We can combine these equations any way we wish. If we take the combination $b + c - 2a$ and combine ΔH values the same way, we obtain

$$C_2H_6 + H_2 + 4O_2 - 2CH_4 - 4O_2 = 2CO_2 + 4H_2O - 2CO_2 - 4H_2O$$

$$\Delta H = \Delta H_b + \Delta H_c - 2\Delta H_a = -16 \text{ kcal}$$

This can be rearranged to yield

$$C_2H_6 + H_2 = 2CH_4 \qquad \Delta H = -16 \text{ kcal}$$

Thus heats of combustion can be used to obtain the same type of data that we calculate from bond energies. The difference in these methods is that calculations from bond energies are approximate, whereas calculations from heats of combustion are as exact as the heats of combustion, about ± 0.1 kcal.

At very high temperatures, decomposition of paraffins occurs. Typical examples of this are

$$CH_4 \xrightarrow{1000°C} C + 2H_2$$

$$CH_3{-}CH_2{-}CH_3 \rightarrow CH_2{=}CH_2 + CH_4$$

Such chain-breaking reactions absorb rather than release heat.

Substitution. In many reactions of hydrocarbons, an atom or group is substituted for a hydrogen atom. A typical reaction of this type is

$$CH_4 + X_2 \rightarrow CH_3X + HX$$

Such simple reactions (wherein the number of moles of products is equal to the number of moles of reactant) normally occur at ordinary temperatures only if energy is released. Therefore let us calculate the energy released in such reactions for X_2 equal to Cl_2, Br_2, and I_2. The general expression for the energy released is

$$E_{H-X} + E_{C-X} - E_{C-H} - E_{X-X}$$

where E_{H-X}, etc., represent bond energies. These calculations show that if X indicates chlorine or bromine the reaction is exothermic, i.e., gives off energy; whereas if X represents iodine the reaction is endothermic, i.e., requires energy. Accordingly, we predict that such substitutions occur with chlorine or bromine but not with iodine. This is correct. (Usually a catalyst such as sunlight is needed to start the reaction.)

It is interesting to note that, in principle, the energy released in the simple substitution reaction above is the same as that released for the following reactions.

$$CH_3X + X_2 \rightarrow CH_2X_2 + HX$$
$$CH_2X_2 + X_2 \rightarrow CHX_3 + HX$$
$$CHX_3 + X_2 \rightarrow CX_4 + HX$$

Thus one expects that, if the simple one-atom substitution occurs, further substitution also occurs.

If we carry out these calculations for paraffins other than methane, we find that the same equations apply. Thus we conclude (correctly) that single and multiple substitutions occur (in the presence of a catalyst) whenever a hydrocarbon is mixed with bromine or chlorine, but that no reaction occurs with iodine.

Bond energy calculations do not provide all the answers. Such calculations show that the reaction

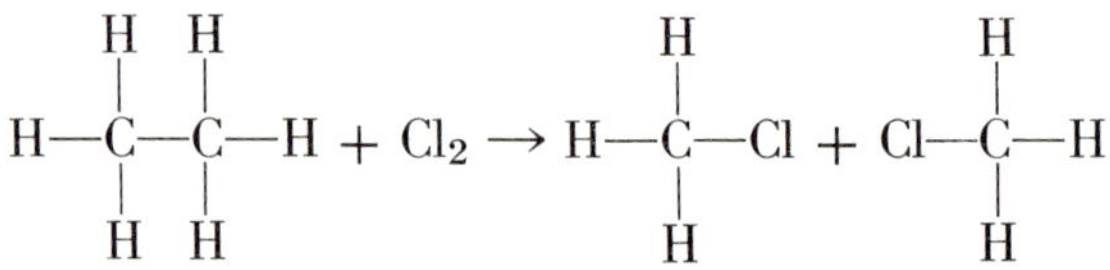

should occur with the release of about 12 kcal, but such a reaction does not occur readily. On the other hand, the reaction with fluorine is even more exothermic (~100 kcal), and in this case chain breaking is so pronounced that direct fluorine substitution without chain breaking is rare.

Isomerization. One of the mildest reactions that occur with hydrocarbons is isomerization, the change-over from one isomer to another. The heat released in such reactions is generally quite small; for example, the reaction

$$n\text{-butane} \rightarrow \text{isobutane}$$

occurs with the release of 1.6 kcal. Nevertheless, such reactions are quite important in petroleum refining because the octane number usually increases with increased branching. For example, the octane number of heptane is 0, whereas that of the heptane isomer 2,3-dimethylpentane is 89. Thus, by the catalytic transformation from heptane to this isomer, the octane rating increases from an unusable to a usable value. The octane ratings of fuels can be further enhanced by adding small amounts, about 1 part per 1000, of a knock inhibitor such as tetraethyl lead.

Cyclic hydrocarbons. Cyclic hydrocarbons undergo all the types of reaction discussed above. In addition to this, however, the strain in the ring weakens the carbon-carbon bond so that this linkage is often broken by reactions of the type

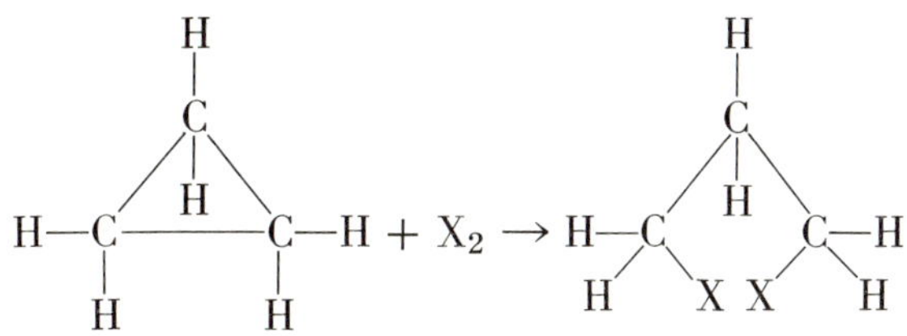

Such reactions are called *addition reactions* since the atom X is added to the compound rather than substituted for a hydrogen atom.

The heats of combustion of cyclic compounds show the weakening of the C—C bond. The decreases in C—C bond energies are 12, 9, 3, and 2 kcal for cyclopropane, cyclobutane, cyclopentane, and cyclohexane, respectively. In line with this, cyclopropane and cyclobutane will add hydrogen through catalysis more readily than cyclopentane or cyclohexane. Reaction occurs below about 100°C for the three- and four-membered rings, but temperatures in excess of 300°C are required for five- and six-membered rings.

REACTIONS OF UNSATURATED HYDROCARBONS

The reactions of unsaturated hydrocarbons include all of those found for saturated compounds. In addition to these, many reactions occur in which half the double bond opens and adds new groups. Thus a large part of the chemistry of olefins is ascribable to the fact that the second bond between carbon atoms is more readily broken than a single bond.

Addition reactions. The general equation for an addition reaction can be written

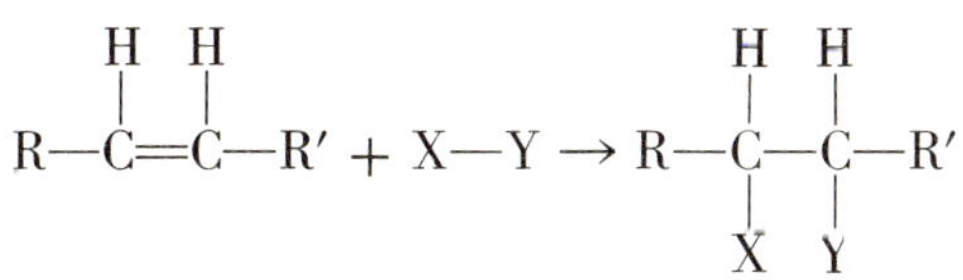

Table 8 Addition Reactants

X—Y	X	Y
H_2	H—	—H
Cl_2	Cl—	—Cl
Br_2	Br—	—Br
I_2	I—	—I
HI	H—	—I
HBr	H—	—Br
HOCl	H—O—	—Cl
H_2SO_4	H—	$-O-\overset{O}{\overset{\|}{\underset{\underset{O}{\|}}{S}}}-O-H$

wherein R and R′ stand for a hydrogen atom or hydrocarbon group and X—Y stands for the group adding. Typical X—Y groups that add to unsaturated hydrocarbons are listed in Table 8. (The list is by no means complete.) These types of addition reactions also occur somewhat less readily with cyclic hydrocarbons.

In some of the cases listed, X—Y = Br_2 for example, it is possible not only to have an addition reaction, but also to have substitution of the type

$$R-\overset{H}{\overset{|}{C}}=\overset{H}{\overset{|}{C}}-R' + Br_2 \rightarrow R-\overset{H}{\overset{|}{C}}=\overset{Br}{\overset{|}{C}}-R' + HBr$$

In general, the addition occurs more readily than substitution.

When R and R′ differ and X and Y differ, two isomers can be formed by an addition reaction. Thus, when R is CH_3 and R′ is H, the following reactions with HBr can occur:

$$CH_3-CH=CH_2 + HBr$$

$$\xrightarrow{a} CH_3-\underset{Br}{\underset{|}{CH}}-\underset{H}{\underset{|}{CH_2}}$$

$$\xrightarrow{b} CH_3-\underset{H}{\underset{|}{CH}}-\underset{Br}{\underset{|}{CH_2}}$$

On the basis of bond energies, both products should have the same stability and the two reactions should occur to the same extent; however, the rates of the two reactions differ enormously and product *a* is formed almost exclusively. In other words, the slight difference in the structural environment of the bromine atom affects the stability of the two products trivially but affects the reaction rate enormously. This type of preferential addition is common; in general, the more electronegative group in X—Y adds to the carbon having the smaller number of hydrogen atoms.†

In the presence of a catalyst, intramolecular additions to double bonds can occur; for example,

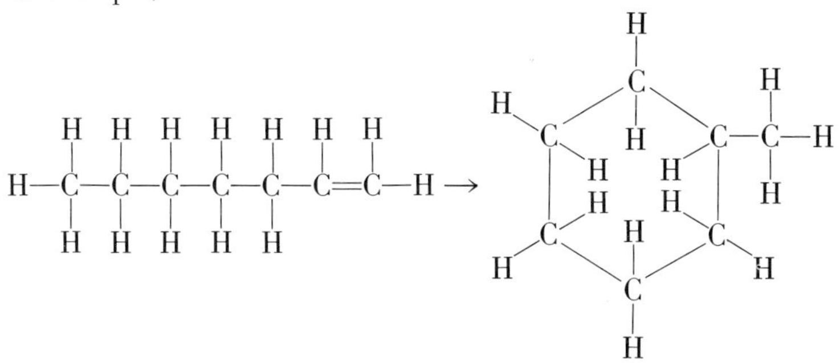

We can represent this schematically, as follows.

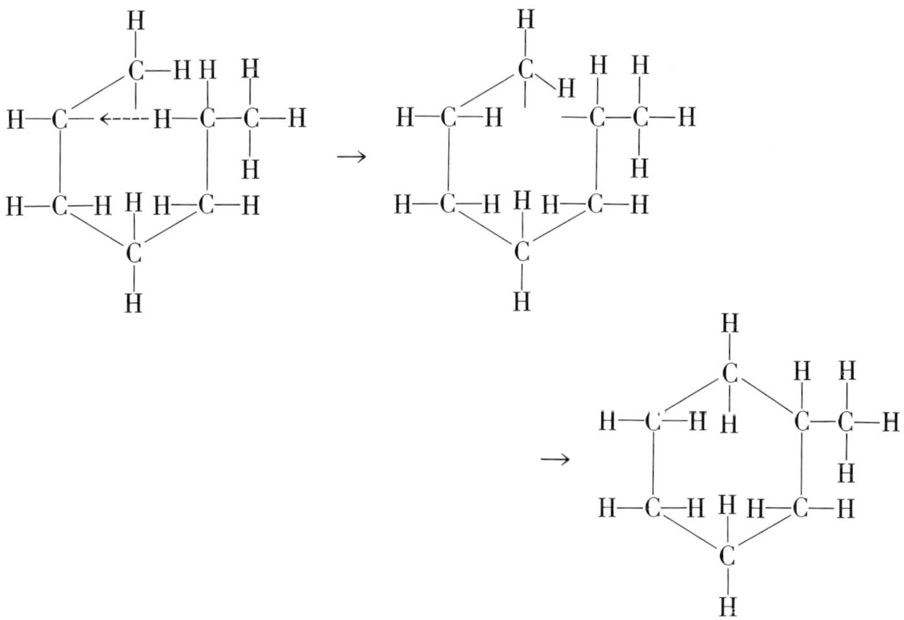

Polymerization. Intermolecular addition of olefins also occurs. For example, in the presence of a catalyst, ethylene will undergo the following reaction.

† An illustration of the sensitivity of rates of reaction to seemingly trivial changes is encountered in the foregoing reaction. Reaction *a* predominates if the reactants are pure, but a trace of hydrogen peroxide makes reaction *b* the predominant one.

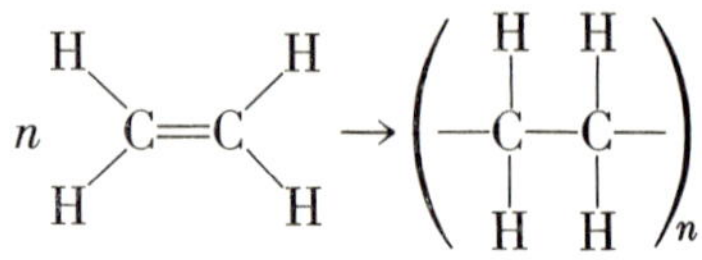

where the symbol on the right indicates a chain made up of links of the type indicated. We can imagine this reaction takes place in a stepwise fashion as follows:

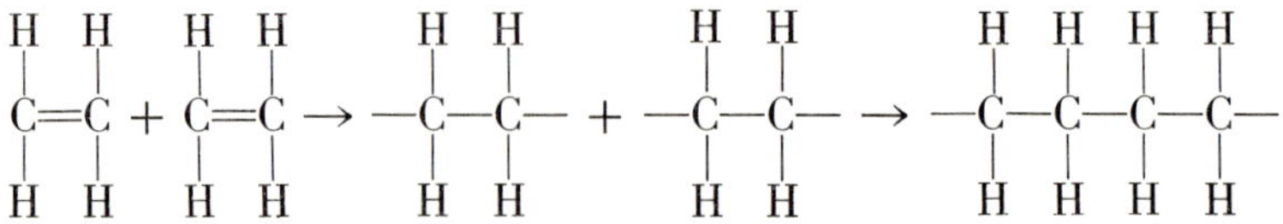

This reaction is stopped by the addition of trace additives to the ends of the chain. The compound that yields the link is called a *monomer;* a two-link compound is called a *dimer;* a three-link compound is called a *trimer;* a multilinked compound is called a *polymer.* Polymers are known containing as many as a million links in an individual molecule. Such giant molecules are often visible, individually, under an electron microscope.

Polymerization is typical of unsaturated hydrocarbons. The resulting polymers, however, are often strikingly different in properties. Polyethylene is a hard solid with a waxy feel; polyisobutylene is a milky rubber-like polymer; polystyrene, monomer

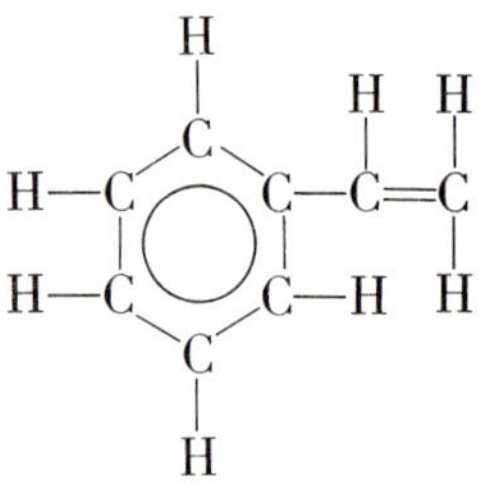

is a hard brittle polymer and clear as glass. In general, commercial polymers are called *plastics.*

Isomerization. Skeletal isomerization of olefins similar to that described for paraffins occurs in the presence of a catalyst. In addition to skeletal isomerization, double-bond isomerization of the type

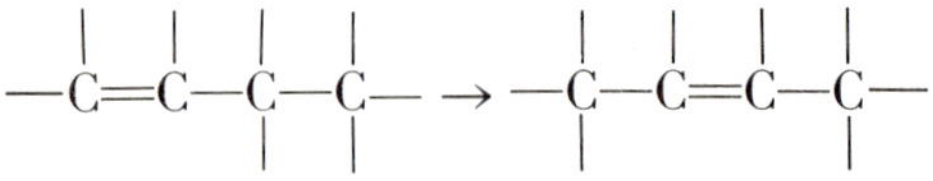

also occurs. For the most part, double-bond isomerization occurs more readily than skeletal isomerization and skeletal isomerization of olefins occurs more readily than isomerization of paraffins.

Reactions of alkynes parallel those of olefins. Addition of bromine to acetylene proceeds as follows:

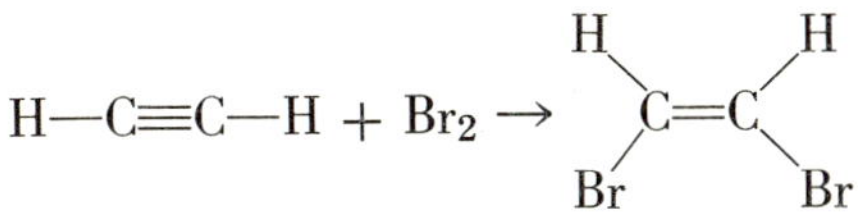

The product can then undergo the addition reaction characteristic of olefins.

As expected, 1,2-dibromoethylene occurs in *cis* and *trans* isomers. Both are formed by addition.

CONJUGATED DIOLEFINS

Conjugated diolefins such as butadiene have chemical properties similar to those of ordinary olefins in so far as they undergo polymerization and addition reactions. The character of these reactions, however, is definitely affected by conjugation. For example, if butadiene were a typical olefin, we would expect bromine addition to proceed as follows:

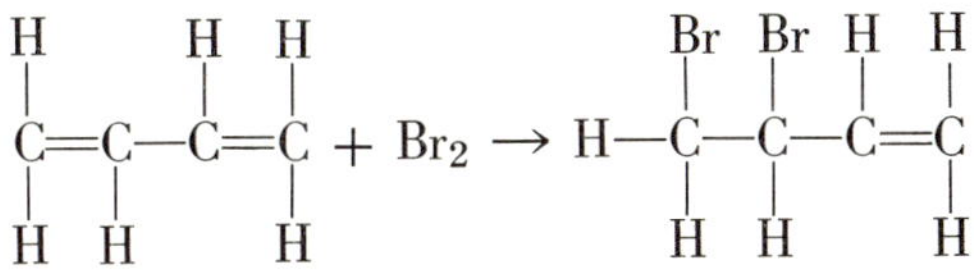

I

But we know that the molecule is a hybrid of I and II.

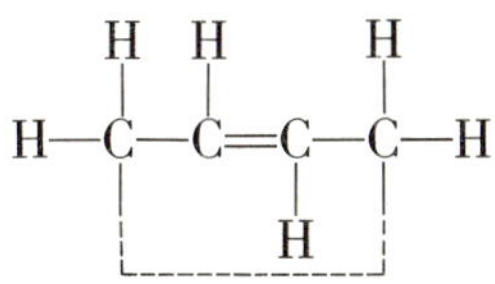

II

If II is an important contributing form, we expect addition to the 1-4 carbon atoms as well; that is,

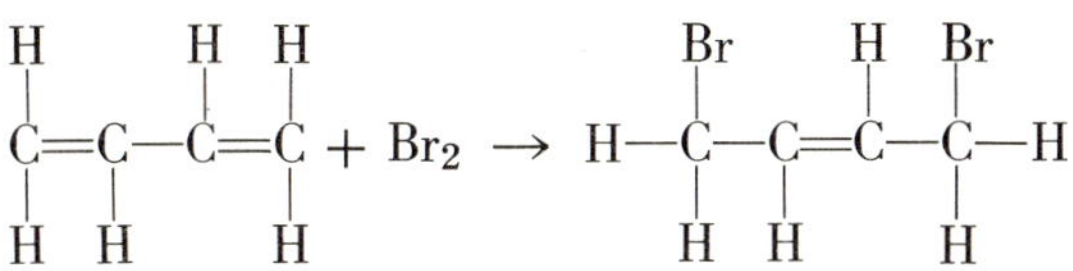

Both 1-4 and 1-2 addition are observed.

AROMATIC HYDROCARBONS

Whenever there is conjugation with perfect alternation of single and double bonds, resonance effects become so great that the character of both bonds is drastically altered; hence such compounds are no longer called olefinic hydrocarbons, they are called *aromatic* hydrocarbons. The most important example is benzene, a resonance hybrid with the two contributing forms:

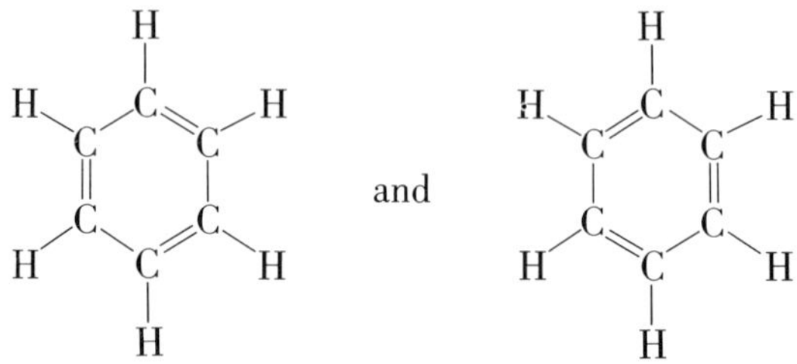

Often the formula for benzene is abbreviated by not indicating the carbon and hydrogen atoms; that is,

Since this represents only one of the contributing forms, it has become common in recent years to use the symbol

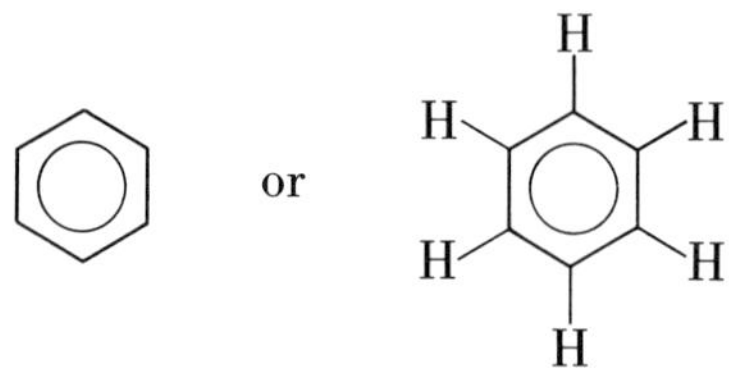

which is closer to the charge cloud distribution in the actual molecule (Fig. 10, p. 150).

Benzene definitely does not contain alternating single and double bonds. If the structure were that of one of the contributing forms, there would be five isomers formed by substituting two groups, say A and B, for hydrogen atoms. These are

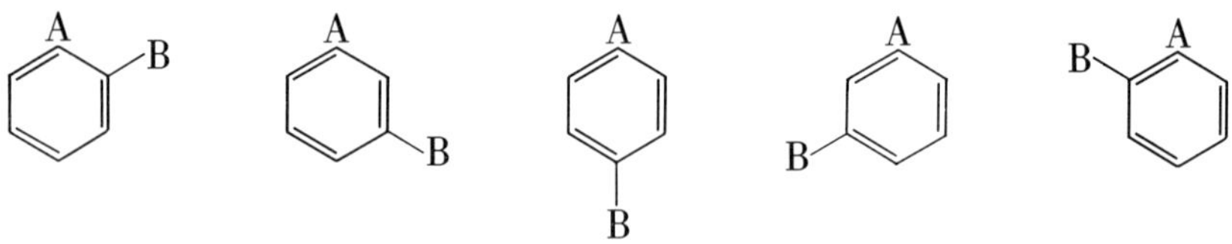

(1) (2) (3) (4) (5)

The fact that only three isomers are found is consistent with the occurrence of resonance, which makes all six bonds equivalent and makes compounds 1 and 2 identical with 5 and 4, respectively.

The three isomers formed by the above substitution are labeled *ortho* if the groups are on adjacent carbon atoms (1 and 5); *meta* if the bonded carbon atoms are separated by one carbon atom (2 and 4); and *para* if the bonded carbon atoms are separated by two carbon atoms (3). Thus dichlorobenzene forms the three isomers:

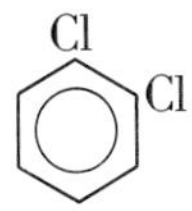

ortho-dichlorobenzene or *o*-dichlorobenzene

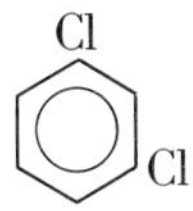

meta-dichlorobenzene or *m*-dichlorobenzene

and

para-dichlorobenzene or *p*-dichlorobenzene

If the number system is preferred, these can also be named 1,2-, 1,3-, and 1,4-dichlorobenzene, respectively. These isomers differ significantly in physical properties and chemical properties. The melting and boiling points are listed in Table 9.

The nomenclature of benzene derivatives is complicated by the fact that special names are often used for specific benzene derivatives; for example, methylbenzene,

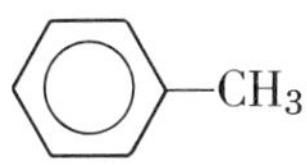

is often called *toluene*. The correct nomenclature, however, follows rules similar to those used for other hydrocarbons. Below are the names and formulas of a few compounds.

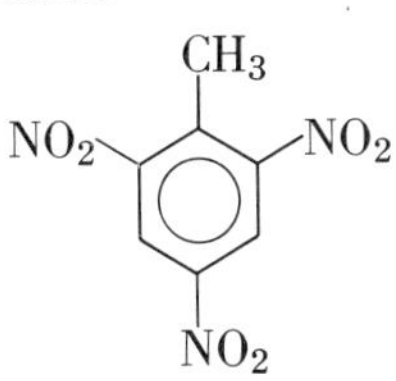

2,4,6-trinitrotoluene (TNT) or 2,4,6-trinitromethylbenzene

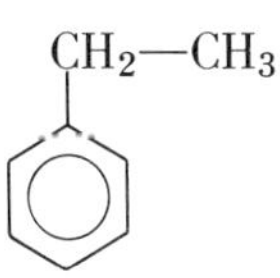

ethylbenzene

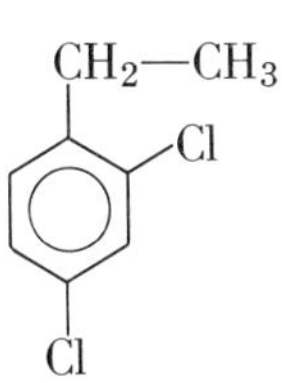

2,4-dichloroethylbenzene

Addition. As might be expected, addition reactions do occur with benzene. However, because of the occurrence of resonance the benzene structure is much more stable than that of common olefins. Therefore addition reactions with benzene occur less readily than with olefins. For example, catalytic hydrogenation

Table 9 Isomers of Dichlorobenzene

	Melting Point (°C)	Boiling Point (°C)
ortho	−17	180
meta	−25	170
para	53.1	174

of propylene to propane occurs easily below −80°C; with the same catalyst, hydrogenation of benzene to cyclohexane does not occur below +60°C.

Substitution. Substitution reactions of benzene are far more common than addition reactions. If bromine is mixed with benzene, the following is the predominant reaction.

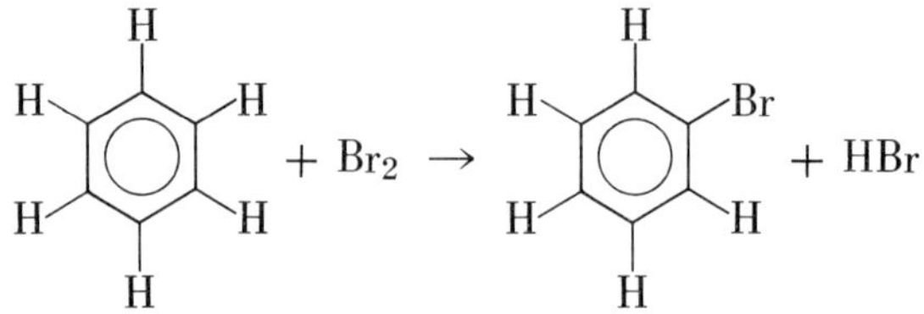

These substitution reactions occur not only with halogens but also with oxygen acids. The reactions with sulfuric and nitric acids are indicated below:

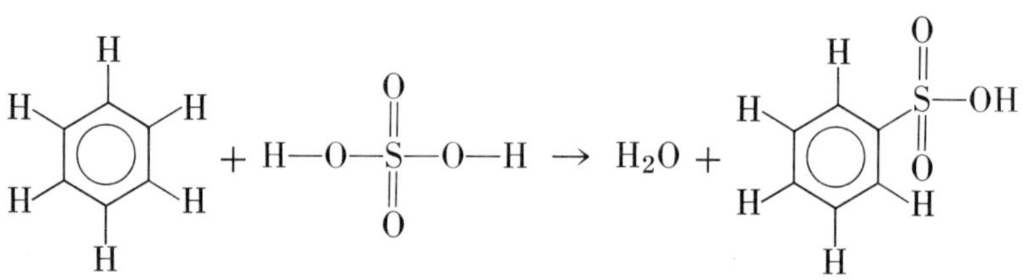

benzenesulfonic acid

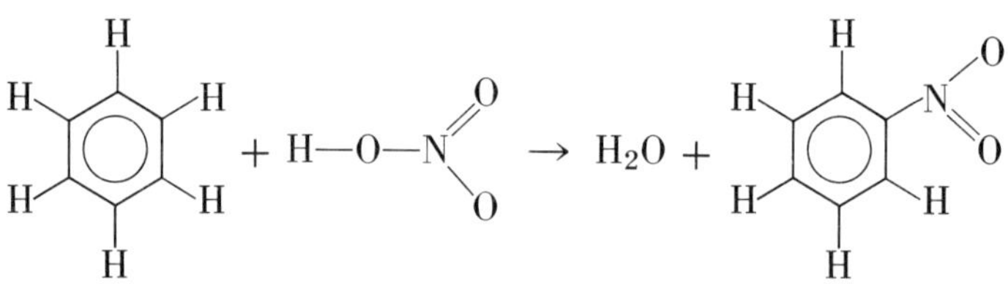

nitrobenzene

Such reactions proceed most readily in the presence of concentrated sulfuric acid.

Directional effects. Substitution reactions of benzene derivatives of the type

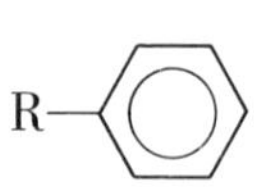

can result in a variety of products: *ortho, meta,* and *para* derivatives. The structure of the group R determines the particular isomers formed. If the group R contains only single bonds, the substitution is in the *ortho* and *para* positions; if the group R contains double bonds, the substitution is *meta.*† Thus, in the bromination of toluene, the reaction is

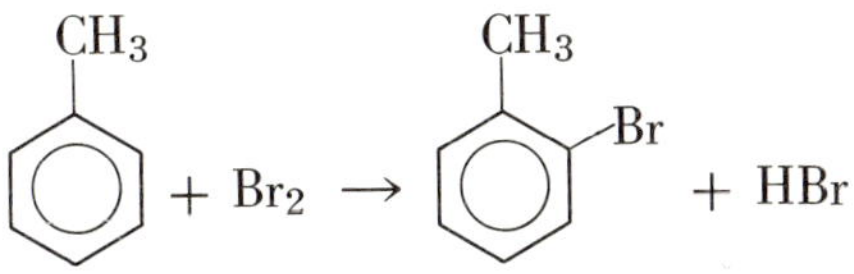

(*ortho*)

or

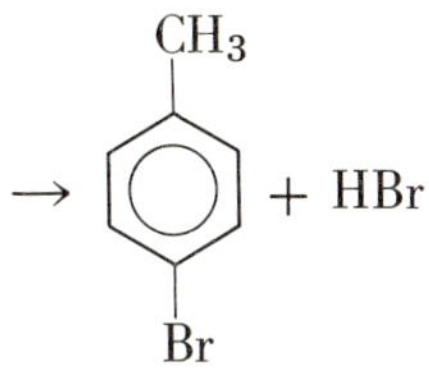

(*para*)

Only trivial amounts of *m*-bromotoluene are formed. On the other hand, in the bromination of benzoic acid, the following occurs.

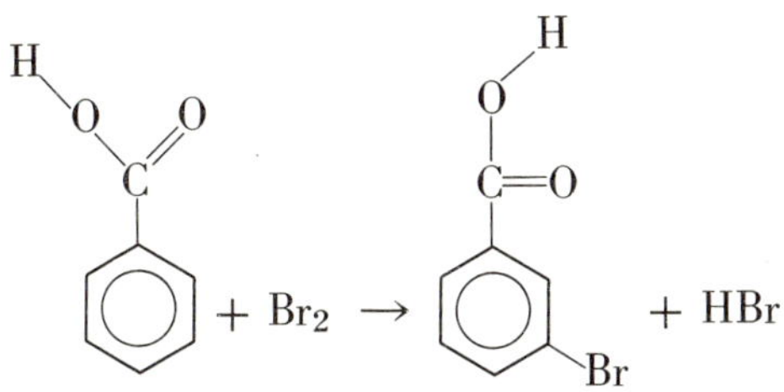

benzoic acid (*meta*)

Very little *o*- and *p*-bromobenzoic acid is formed.

These directional properties must be considered in devising a synthesis scheme. Suppose we wish to prepare *m*-nitrobromobenzene,

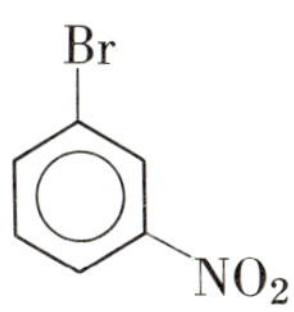

If we attempt to prepare this by nitration of bromobenzene, we will obtain *o*- and *p*-nitrobromobenzene because the bromine atom is an *ortho*-and-*para* directing group. On the other hand, the bromination of nitrobenzene would yield *m*-nitrobromobenzene because the nitro group is *meta* directing.

† Exceptions occur.

In general, *ortho*-and-*para* directing groups enhance the reactivity of the benzene ring; *meta* directing groups decrease the reactivity. Thus phenol,

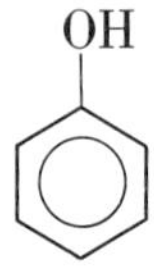

compared to benzene, undergoes an addition reaction with nitric acid more readily, whereas nitrobenzene undergoes an addition reaction with nitric acid less readily. The *ortho-para* directing groups activate the whole molecule but activate the *ortho* and *para* positions more than the *meta* position; the *meta* directing groups deactivate the whole molecule but deactivate the *meta* position least.

PROBLEMS

1. Assume that the octet rule is universally valid and write reasonable structures for the following.

(*a*) CH_5P (*b*) CH_4Se
(*c*) CH_6Si (*d*) CH_3P
(*e*) CH_4Si (*f*) CH_2Se
(*g*) CH_2Se_2 (*h*) CH_2OSe

Assume in each case that the element other than hydrogen is bonded to the carbon atom. Indicate the expected bond H-C-X bond angles.

2. Give the structures of three compounds containing two carbon atoms that could conceivably be formed by the action of oxygen on ethane. Indicate the expected bond angles in these compounds.

3. Indicate the structures of all straight-chain olefins with the formula C_5H_{10} (*Ans.* Pentene-1; *cis*-pentene-2; trans-pentene-2.) Indicate only different structures and specify all bond angles.

4. Indicate the structures of all stable straight-chain diolefins with the formula C_5H_8 Indicate only different molecules and specify all bond angles.

5. A compound is formed by the substitution of two bromine atoms for two hydrogen atoms in propane. Give the structures of the possible molecules resulting.

Ans.

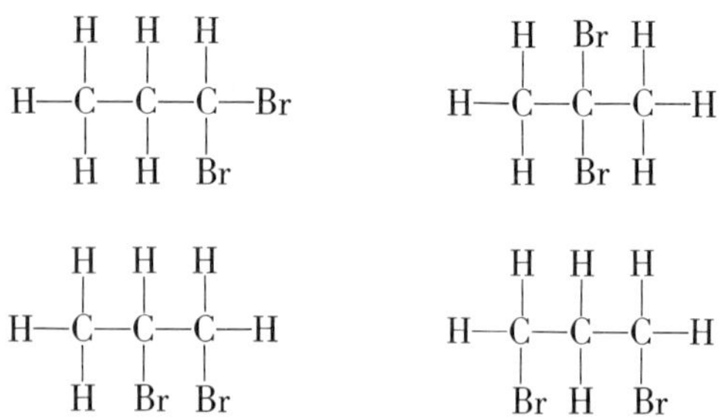

6. If we indicate the structure of a molecule as

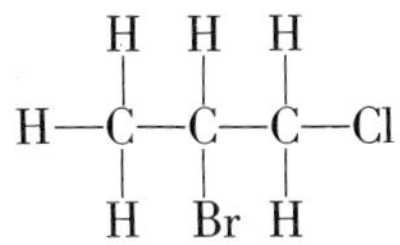

the specification is incomplete. Whereas the specification is complete for the molecule

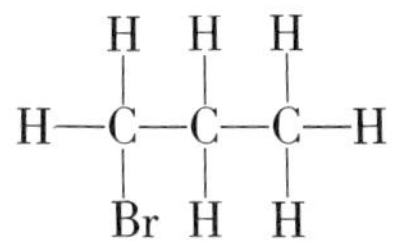

Explain in detail. *Hint:* Draw a three-dimensional view of the central carbon atoms.

7. If the temperature were sufficiently low that free rotation did not occur around single bonds, which of the following compounds would exist in more than one form?

(*a*) 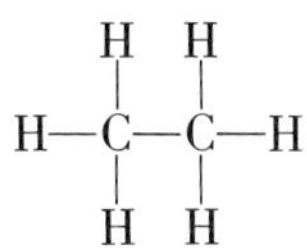 (*b*) 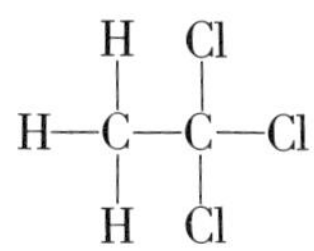

(*c*) 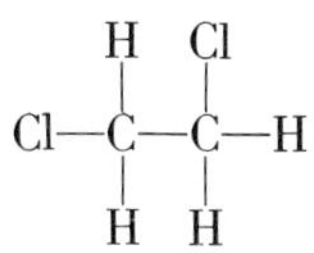 (*d*) 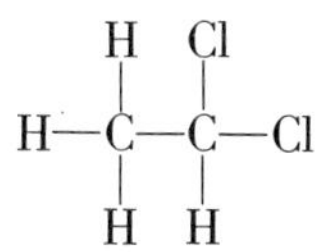

(*e*) 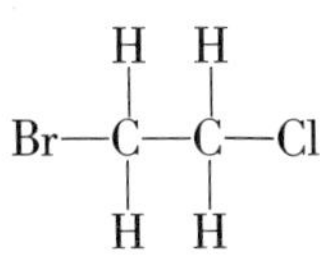 (*f*) 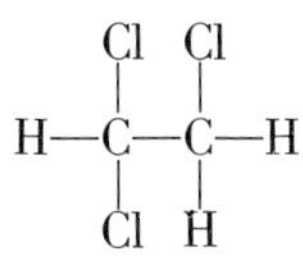

Ans. (*c*), (*e*), (*f*).

Indicate the number of forms in each case.

8. At ordinary temperatures, which of the isomers of the straight-chain olefin C_6H_{12} would exist in *cis* and *trans* forms?

9. Give the names of the following hydrocarbons.

(*a*) 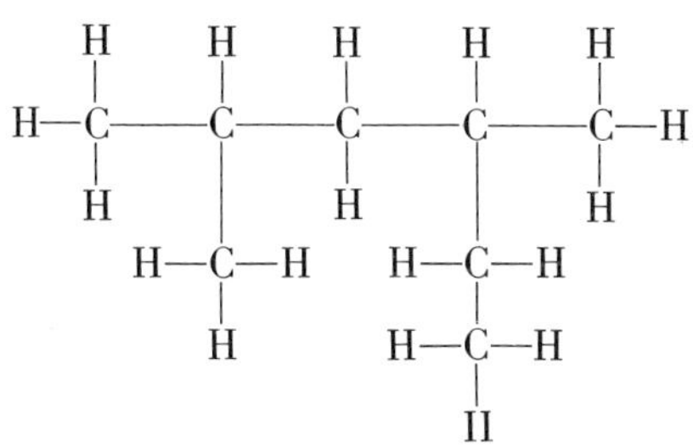

(*b*)

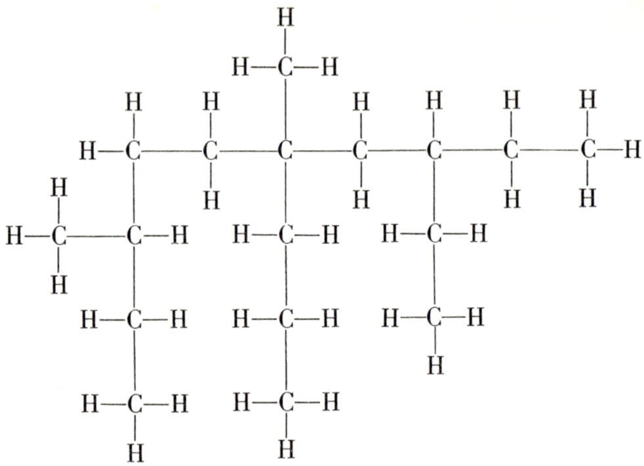

(*c*)

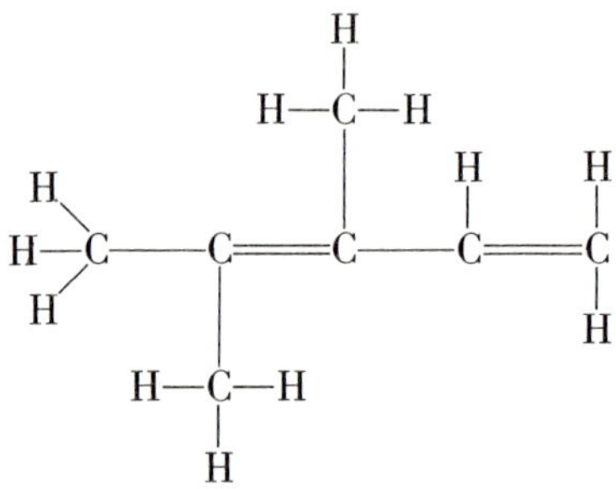

(*d*)

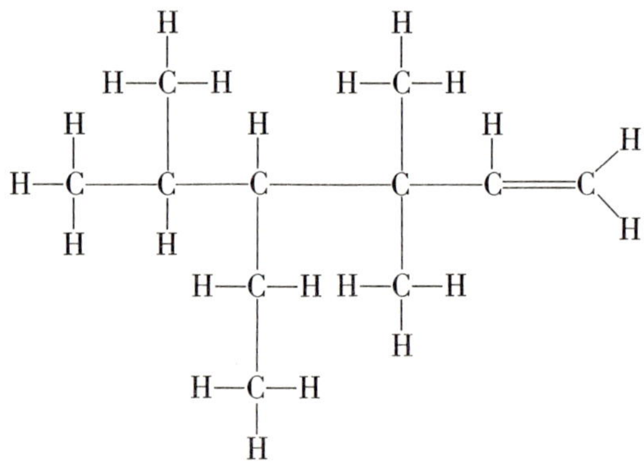

(*e*)

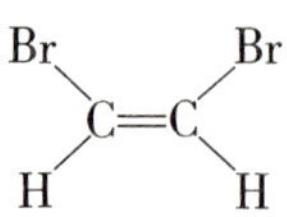

Ans. (*a*) 2,4-Dimethylhexane; (*e*) *cis*-1,2-dibromoethylene.

10. Give the structural formula for: (*a*) *trans*-1-bromo-2-chloroethene; (*b*) 1,3-dimethylcyclopentane; (*c*) 1,3-cycloheptadiene; (*d*) 3,3,4-trimethyl,4-ethylheptyne-1; (*e*) 5,5-diethylheptadiene-1,3

11. Indicate the names and formulas of all isomers of hexene and heptane.

★ **12.** On the basis of the accompanying data, estimate the energy in kilocalories per mole (1 ev = 23 kcal/mole) required to form the ionic molecules H^+F^- and H^+Cl^-.

	I_z, ev	E_a, ev	Ionic Radius
F	17.4	3.6	1.33
Cl	13.0	3.8	1.81
H	13.6		0

(See the calculation for the Na^+Cl^- molecule in the preceding chapter.) From the first two terms in the bond energy equation estimate the bond energy expected for a perfect shared-pair bond in HCl and HF. On the basis of these two estimates, is resonance involving the contributing ionic form more important in HF or in HCl? Explain.

13. The compounds $HClO_4$ and H_2SO_4 are both acids in so far as they undergo the following reactions in solution.

$$H_2SO_4 \rightarrow HSO_4 + H^+$$
$$HClO_4 \rightarrow ClO_4 + H^+$$

Their structures are often written

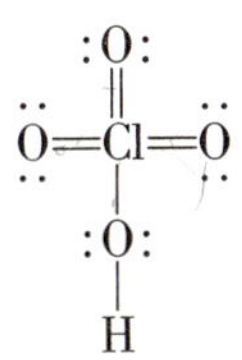

and

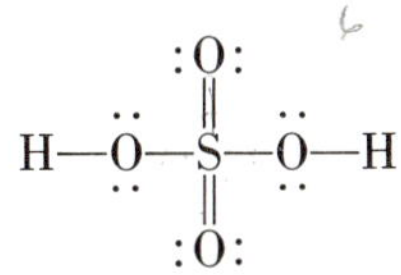

If these are their structures, can you predict, on the basis of resonance, which would dissociate more readily to form a hydrogen ion (i.e., which would be the stronger acid)? *Ans.* $HClO_4$.

14. When 1 mole of gaseous ethylene is burned to form gaseous water and carbon dioxide, 310 kcal of heat is evolved. From this and the data given on p. 203, compute the heat given off in the reaction

$$C_2H_4 + H_2 \rightarrow C_2H_6$$

15. When 1 mole of gaseous ethylene is burned to form liquid water and gaseous carbon dioxide, 332 kcal of heat is liberated. Compare this to the data in Problem 14. On the basis of this comparison, determine the amount of heat required to change water into steam.

16. Estimate the heat of the reaction

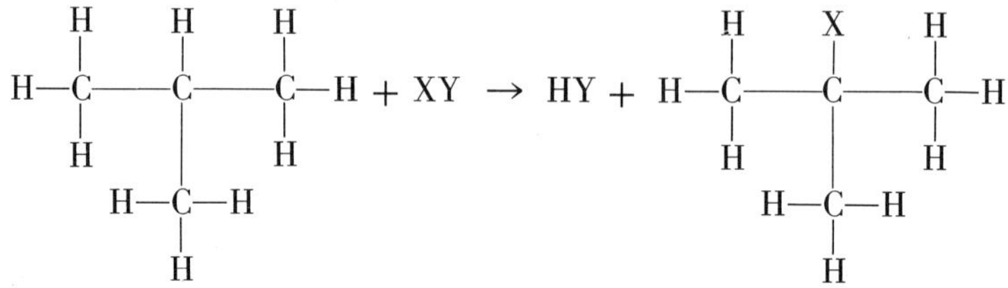

for $XY = Cl_2$, Br_2, and I_2.

17. Repeat Problem 16 for the substitution reaction with H_2SO_4.

18. Complete the following equations.

(*a*) 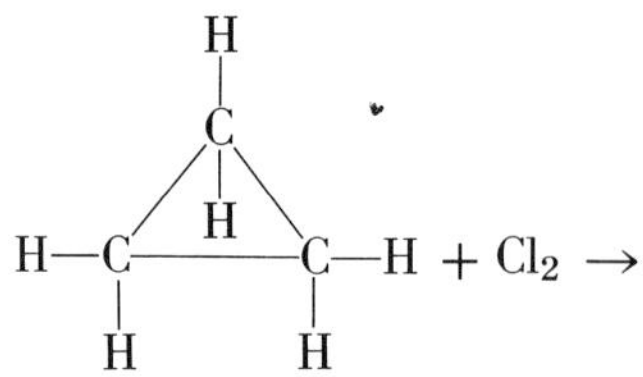

(*e*) 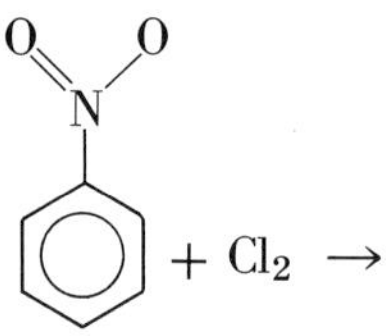

(*b*) 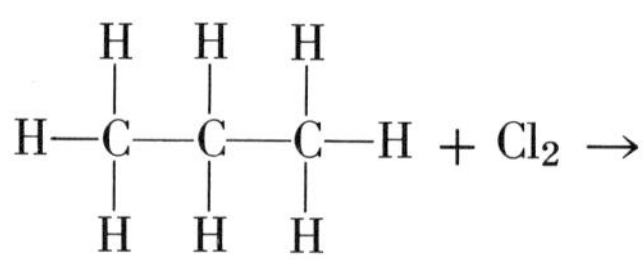

(*f*) 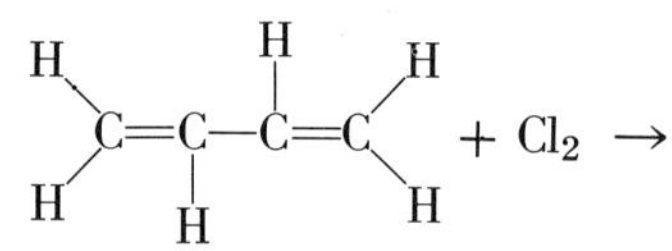

(*g*) $H—C \equiv C—H + Cl_2 \rightarrow$

(*c*) 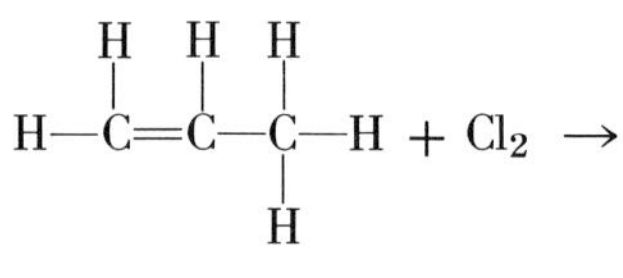

(*h*) 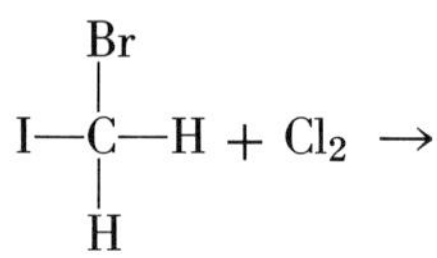

(*d*) 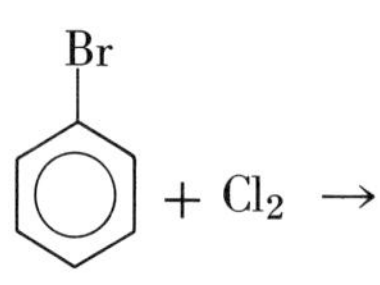

(*i*) 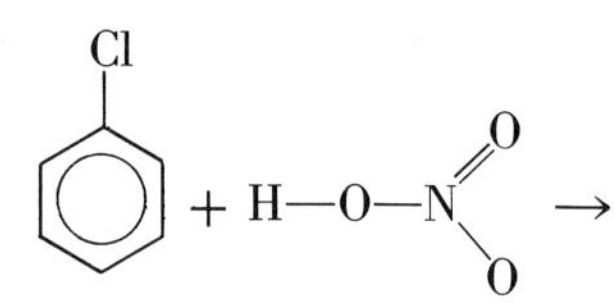

If more than one product is formed, indicate all products.

Ans. (*a*) 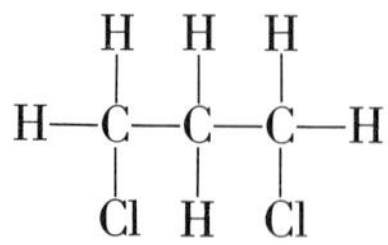 (*e*) 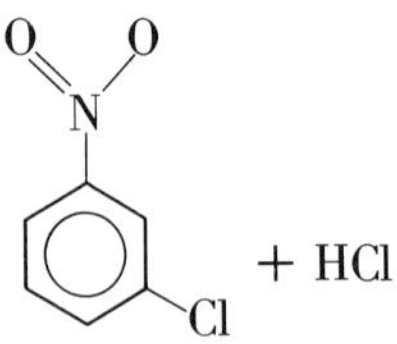

19. Write the formula for the following compounds. (*a*) *o*-dichlorobenzene; (*b*) 2,4-dinitrotoluene; (*c*) 2,4-dichlorophenol; (*d*) *m*-bromochlorobenzene; (*e*) 2,3-dichloro-bromobenzene.

Ans. (*e*) 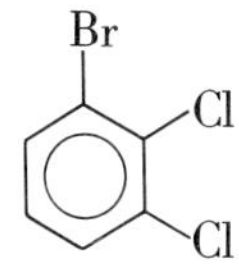

20. Write the equation for the addition of chlorine to benzene.

★**21.** Give equations showing the steps necessary to prepare the following compounds.

(*a*) 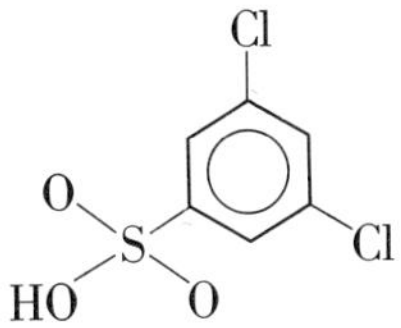

(*b*) 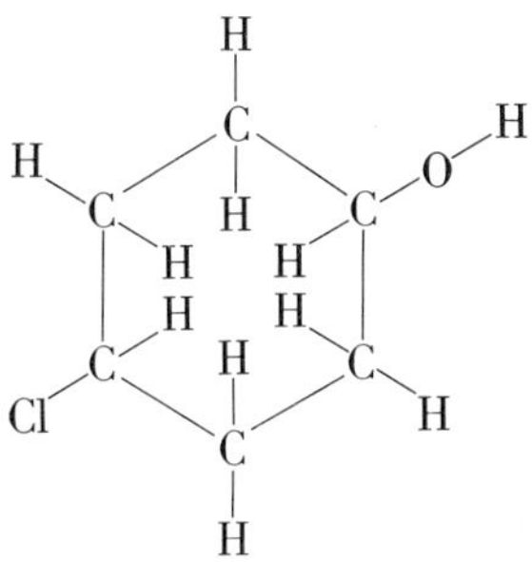 or 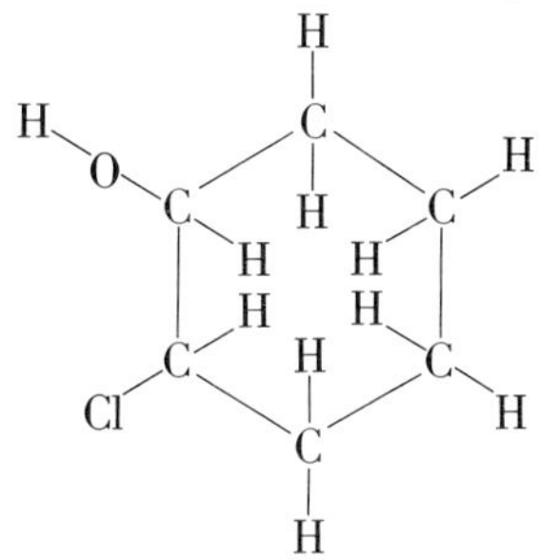

Hint: Start with benzene and chlorine, then hydrogenate.

(*c*) 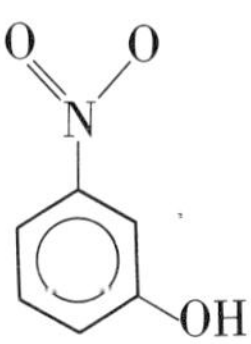

22. List the following in the order of decreasing reactivity with bromine.

(*a*) 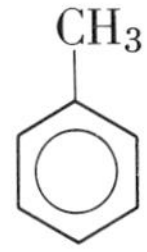 (*b*) 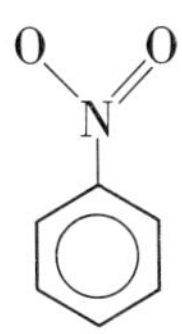 (*c*) 

Explain.

Ans. 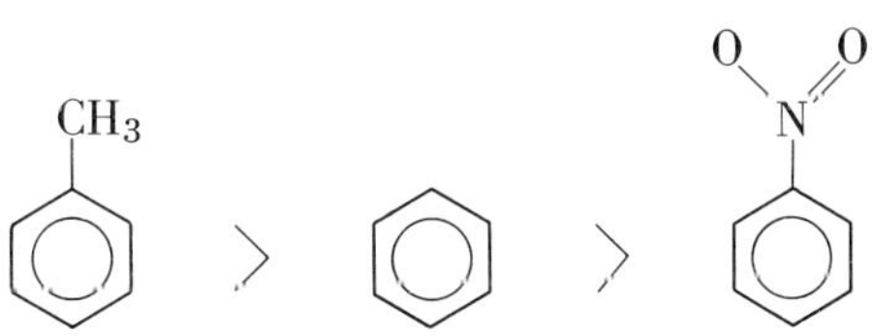

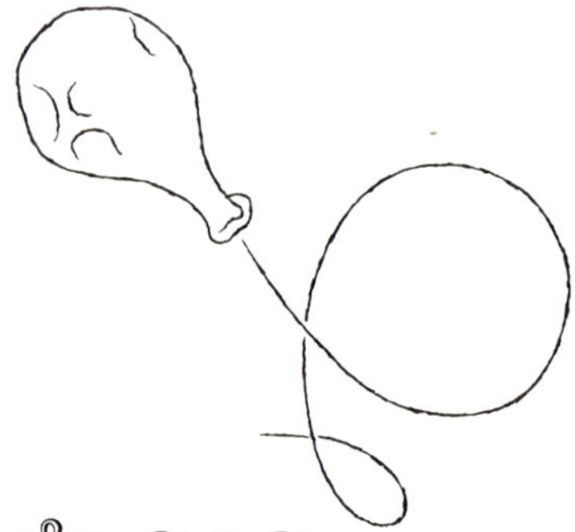

8 The nature of gases

STATES OF MATTER

All of us are familiar with the three common states of matter—solids, liquids, and gases. Solids have their own specific shape, and a definite weight of solid matter always occupies a definite volume regardless of its shape. Liquids, on the other hand, conform to the shape of the bottom of their container, even though a given weight of liquid always occupies the same volume. In contrast to both of these, gases always completely fill a container; hence, both the shape and the volume of a given weight of gas change with the container.

Examples of each of the three states of matter are a part of our everyday experience. Iron, ice, stones are solids; water, alcohol, milk are liquids. These states of matter are the most evident because they are visible. Gases, like air and steam, are not so evident because they are invisible and less substantial. But steam, like water or ice, has mass and, like water or ice, is composed of H_2O molecules.

Most substances exist in all three physical states. For example, at $-210°C$, a closed container filled with nitrogen (the principal constituent of air) contains a gas, a liquid, and a solid. Since the solid is the densest, it is at the bottom of the container and is covered by the less dense liquid. Above the liquid, there is gaseous nitrogen at a pressure corresponding to air pressure at an altitude of about 10 miles. The masses of 1 cm^3 of each of these three forms of matter differ enormously; for solid, liquid, and gas, the mass is 1.03 gm, 0.81 gm, and 0.0007 gm, respectively. From the atomic weight table, we can estimate the average volume per molecule for each state. If we assume that each molecule is confined to a cube, the edge-length of these cubes is 3.56 Å for the solid, 3.86 Å for the liquid, and 40 Å for the gas. It has been established by other methods that the effective diameter of the N_2 molecule (which is actually dumbbell-shaped rather than spherical) is 3.6 Å; hence, in a solid or liquid, individual molecules nearly touch, whereas, in a gas, only 0.1% of the volume is occupied by molecules, 99.9% of the volume is empty space.

Finally, let us look at the elasticity of these three forms of nitrogen. Solids are fairly stiff. If we put a 1000-cc chunk of solid nitrogen in a cylinder and raise the pressure to 15,000 lb/in.^{2}, the volume would drop to about 990 cc. A similar value would be obtained for a liquid. But, for a gas, the final volume would be 1 cc! The change would be one thousand-fold! Thus gases, which are mostly empty space, are far more compressible than liquids or solids, which are mostly filled space.

In the last two chapters, we discussed the structure and reactions of molecules without regard to the physical states actually observed for aggregates of molecules. Often the physical state has a significant effect on the reactivity of a compound. In solids and liquids, where a molecule is actually in contact with its neighbors, the molecule is no longer independent. In gases, on the other hand, distances between molecules are large, and independent molecular behavior is closely approximated. (This is the reason why bond energy calculations are always carried out for compounds in the gaseous state.) Thus the molecular behavior in gases is the closest to that discussed earlier; for this and other reasons, we discuss the properties of this state of matter first.

ENERGY

Energy levels. In the hydrogen atom, energy is quantized. By this we mean that the electron can exist *only* in orbits associated with particular values of energy. In all chemical systems, this situation prevails not only for the electrons in an atom but also for the atom as a whole, and not only for the atoms in a molecule but also for the molecules as a whole. Thus there is always a definite series of permitted energy levels, and all other levels are forbidden. Sometimes, because the methods of measuring energy are not sufficiently sensitive to detect these discrete energy levels, it may appear that all energies are permitted, but, whether apparent or not, discrete levels are always present.

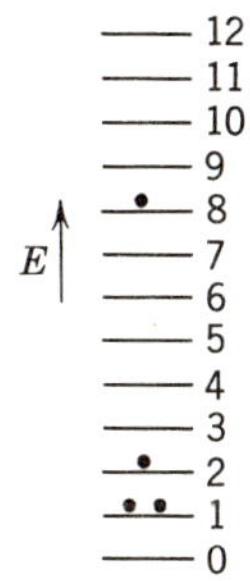

Fig. 1. Energy diagram.

Quantization is depicted in Fig. 1 for a system consisting of four molecules; here the energy levels are shown evenly spaced with the permitted values of

$E = 1, 2, 3$, etc. This might correspond to four molecules of helium inside a box. Two molecules are moving slowly, and each has only one unit of energy; another is moving a little faster and has two units; the fourth is moving much faster and has eight units.†

If four molecules have a fixed total energy, they can only occupy energy levels consistent with the total energy. If the total energy, E_T, is equal to 12 (in ergs or in arbitrary units), there are many distributions consistent with E_T, such as one molecule with twelve units of energy and the other three with none, or two molecules with six units of energy and the others with none. There are, in fact, 455 distributions consistent with $E_T = 12$. Thus it appears that, from the total energy of molecules, we cannot specify the condition of each molecule. If we know the total energy of a collection of molecules, the best we can do is specify the probability of various types of behavior. In other words, to discuss energy on a molecular scale, we must talk in terms of the probability of individual modes of molecular behavior.

Since probability controls the distribution of a group of molecules, true understanding of the meaning of energy involves the calculation of the most probable distribution of N molecules among their permitted energy levels such that the total energy E_T is fixed. Thus this calculation is subject to two conditions. The first of these is that the total number of molecules is N; that is,

$$\Sigma\, n_i = N$$

where the symbol $\Sigma\, n_i$ indicates the sum over all energy states of the number of molecules n_i in each energy state, i. The second condition is that the energies of individual molecules must add up to E_T; that is,

$$\Sigma\, n_i E_i = E_T$$

which is to be read, the sum of the number of molecules in the energy state i times E_i, the energy of the state i.

(In this mathematical shorthand, i stands for the first, second, third, fourth, etc., permitted energy levels. Thus, in the example illustrated in Fig. 1, the equation would read

$$\Sigma\, n_i E_i =$$

$$n_0E_0 + n_1E_1 + n_2E_2 + n_3E_3 + n_4E_4 + n_5E_5 + n_6E_6 + n_7E_7 + n_8E_8 + n_9E_9 + \cdots$$

or

$$0\times0+2\times1+1\times2+0\times3+0\times4+0\times5+0\times6+0\times7+1\times8+0\times9+\cdots$$

† For molecules confined to a box, the spacing between energy levels depends on the mass of the molecules and the dimensions of the box. For a box 1 cm on a side, the spacing is about 10^{-30} erg, an extremely small quantity. Throughout this discussion, we shall leave the unit of energy unspecified so that the discussion can apply to a box of any size. We shall also assume that the spacing of energy levels is even. Actually, the type of spacing in the energy level sequence depends on the particular problem and need not be even (cf. the energy level diagram for the hydrogen atom, p. 90).

The meaning of the Σ can be better understood if we recognize that we would calculate the *average* energy E of the molecules from the formula

$$\frac{\Sigma n_i E_i}{N} = \overline{E}$$

Probability. In order to calculate the average of a large number of observations of molecules in their various energy states, we make the assumption that any single distribution is equally likely. Then the procedure is similar to calculations dealing with games of chance. As an example, consider an experiment in which we cast two dice 36×10^3 times and count the number of times that we cast a two, three, four, etc. If this number of tries is large enough, the frequency of the occurrence of six will be determined by probability. There are a total of thirty-six combinations that *could* come up with two labeled dice; any one of six values could come up with die A, and, for each of these values, die B could have one of six values. The ways we can get a six are: A, B = 1, 5; 2, 4; 3, 3; 4, 2; and 5, 1. Thus there are five ways out of thirty-six of casting a six, and in 36×10^3 tries six is likely to turn up about 5×10^3 times. If, instead of casting the dice 36×10^3 times, we cast them 36×10^9 times, the number of times we get a six would be far closer to 5×10^9 percentagewise than it would be to 5×10^3 in the first case. This is because the laws of probability give us the best approximation to average behavior with large numbers. In the corresponding chemical calculations, the samples contain about 10^{23} molecules; hence the average behavior is rather precisely governed by the predictions of probability.

At this stage in our studies, we cannot provide the solution for the general problem with a large number of molecules, but we can work with the example provided in Fig. 1. Suppose we have four molecules with the total energy fixed at 12 units; how many ways can we distribute these four molecules among 13 "energy boxes" (Fig. 2)? For the permitted arrangements, how many times will a molecule turn up in the energy state 0, the energy state 1, etc.? We shall show that this average behavior is very closely specified by *Boltzmann's equation,*

$$\frac{n_i}{n_0} = e^{-E_i/kT}$$

where the subscript i indicates the ith energy level, the subscript 0 indicates the lowest energy state, T is the absolute temperature, and k is a universal constant called Boltzmann's constant. Clearly, the analysis we are outlining will make Boltzmann's equation reasonable, but it will *not* be a proof. Proof of this equation requires advanced and complicated mathematical reasoning.

In how many ways can we put the four molecules, A, B, C, and D, into the numbered boxes in Fig. 2 so that the sums of the numbers of the occupied boxes is 12?

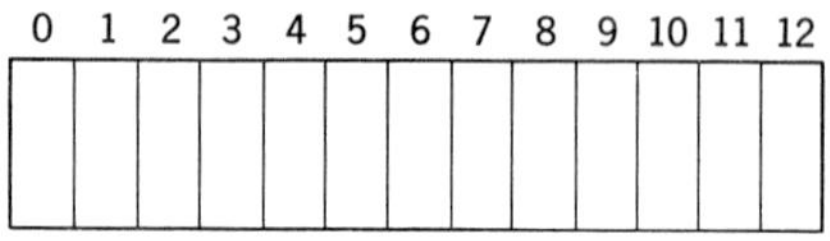

Fig. 2. Energy boxes.

In other words, in how many different ways can we have the molecules moving with different amounts of energy so that the sum of their total energy equals twelve units? A compact tabulation can be effected if we list the numbers of the boxes occupied by A, B, C, and D. In Table 1, we have listed each type of allowed combination. On the left-hand side of a particular listing, we have shown the number of different ways in which we can have the same four levels filled by the four labeled molecules. Consider, for example, the number of ways one molecule has $E = 12$ and the rest have $E = 0$. Since we have four molecules, any one molecule can be the favored molecule with 12 units of energy, while the rest have no energy at all; and this means that there are four different ways to form this arrangement, namely, A with 12 units, or B with 12 units, or C with 12 units, or D with 12 units.

Table 1

Ways	n_A	n_B	n_C	n_D	Ways	n_A	n_B	n_C	n_D
4	12	0	0	0					
12	11	1	0	0	--------	~~5~~	~~7~~	~~0~~	~~0~~
12	10	2	0	0	--------	~~5~~	~~6~~	~~1~~	~~0~~
12	10	1	1	0	12	5	5	2	0
12	9	3	0	0	6	5	5	1	1
24	9	2	1	0	24	5	4	3	0
4	9	1	1	1	24	5	4	2	1
12	8	4	0	0	12	5	3	2	2
24	8	3	1	0	12	5	3	3	1
12	8	2	2	0	4	4	4	4	0
12	8	2	1	1	12	4	4	3	1
12	7	5	0	0	6	4	4	2	2
24	7	4	1	0	12	4	3	3	2
24	7	3	2	0	1	3	3	3	3
12	7	3	1	1					
12	7	2	2	1					
6	6	6	0	0					
24	6	5	1	0					
24	6	4	2	0					
12	6	4	1	1					
12	6	3	3	0					
24	6	3	2	1					
4	6	2	2	2					
					455 Total				

The situation becomes slightly more complex when we consider the second combination listed in Table 1. The combination 11-1-0-0 can be achieved by having any one of the four molecules occupy box 11; for *each* of these four possibilities, it is possible that any one of the three remaining molecules can be in box 1; those left must go into box 0. This gives us a total of twelve ways of getting the combination 11-1-0-0; these are listed in Table 2.

Lastly, let us consider the combination 8-3-1-0. How many ways can we have the boxes 8-3-1-0 occupied singly, each by one of the four labeled molecules? Any one of four molecules can occupy box 8; for *each* of these arrangements any one of three molecules can occupy box 3. Thus far there are twelve arrangements and for *each* of these any one of two molecules can occupy box 1, and the last molecule must occupy box 0. All together this yields the 24 possibilities listed in Table 2.

(The listing of all the combinations that add up to 12 and the number of ways of achieving each of these combinations (Table 1) is tedious, and care must be taken not to repeat arrangements. For example, when we start with molecule A in the fifth box, it is tempting to list 5-7-0-0 as one of the possible combinations, but this has already been taken care of by permutations of the combination 7-5-0-0. It will be amusing and educational for you to verify some of the combinations listed in Table 1.)

Upon completion of Table 1, we find that there is a total of 455 different possible

Table 2

n_A	n_B	n_C	n_D	n_A	n_B	n_C	n_D
12	0	0	0	8	3	1	0
0	12	0	0	8	1	3	0
0	0	12	0	8	1	0	3
0	0	0	12	8	3	0	1
				8	0	3	1
11	1	0	0	8	0	1	3
11	0	1	0	3	8	1	0
11	0	0	1	1	8	0	3
0	11	0	1	1	8	3	0
0	11	1	0	3	8	0	1
1	11	0	0	0	8	1	3
0	0	11	1	0	8	3	1
0	1	11	0	3	1	8	0
1	0	11	0	1	0	8	3
0	0	1	11	1	3	8	0
0	1	0	11	0	1	8	3
1	0	0	11	3	0	8	1
				0	3	8	1
				3	1	0	8
				0	1	3	8
				1	3	0	8
				3	0	1	8
				0	3	1	8
				1	0	3	8

Table 3

	0	1	2	3	4	5	6	7	8	9	10	11	12
W	364	312	264	220	180	144	112	84	60	40	24	12	4

ways of putting these four molecules into the 13 boxes. Since each way corresponds to putting four molecules in the boxes, these 455 ways correspond to 1820 positions for the molecules. Now, how many times does a molecule turn up in box 6? If we refer to Table 1, we find that there are six ways in which we can have two molecules in box 6; hence, for these six ways, a molecule appears 12 times in box 6. There are also other combinations corresponding to a molecule in energy box 6; these are such that we find box 6 occupied 24, 24, 12, 12, 24, and 4 times, respectively. All told, box 6 is occupied

$$(6 \times 2) + 24 + 24 + 12 + 12 + 24 + 4 = 112$$

or 112 times out of the total of 1820. We have tabulated such data in Table 3, wherein W indicates the number of times the indicated level is occupied.

In Fig. 3, we have presented the over-all results as a bar graph. The dotted lines represent the numbers of molecules counted in each level for all possible combinations. Just as with dice, if we perform the hypothetical experiment of counting the numbers of molecules in each energy level for billions of similar samples, this bar graph represents closely the average behavior. In Fig. 3, we have also indicated the values calculated from Boltzmann's equation. The agreement is not perfect, but we do find that the equation approximates the average behavior for this very small sample. It can be shown that, if we perform the experiment with 10^{16} molecules distributed among 10^{20} energy levels, Boltzmann's equation is nearly exact.

We have said nothing about the state of the molecules, i.e., gases, liquids, or solids. Thus these results are quite general.

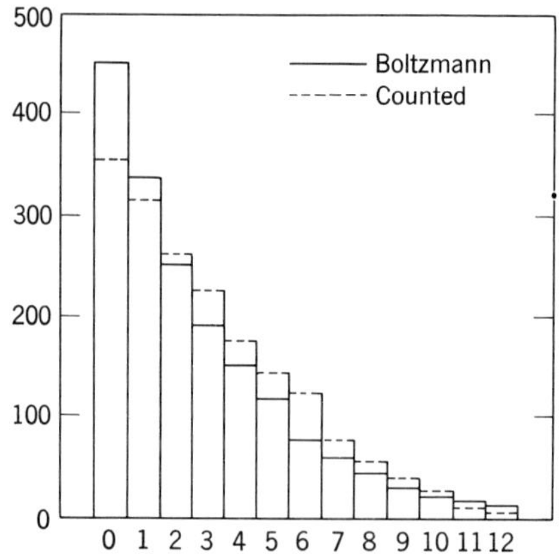

Fig. 3. Boltzmann distribution.

THE IDEAL GAS

The model. Let us devise an idealized model of a real gas. This model must be conditioned by known experimental facts. We know that a gas consists mostly of empty space, that the "shape" of a gas is that of the container, that a gas fills every part of its container, and that a gas is elastic and exerts a force on the walls of its container. What properties must we assign to this imaginary ideal gas so that it will behave, qualitatively, like a real gas? We shall show that we need only two assumptions: (1) the molecules of an ideal gas are point masses, i.e., although they have mass, they do not occupy space; and (2) all of the energy of a gas molecule is translational kinetic energy. The first of these assumptions recognizes that a gas is mostly empty space; hence the actual volume of the molecules is so small compared to the total volume of the gas that we assume that the volume of individual molecules is effectively zero. The second assumption specifies that there is no potential energy and, hence, no intermolecular forces. This approximation stems from the fact that gas molecules are usually so far apart that they seldom feel the influence, i.e., attraction or repulsion, of other gas molecules. Because of this all their energy is a self-contained energy of motion, kinetic energy.

Boltzmann distribution. We now want to formulate as precisely as possible the properties associated with this gas. The total energy of the gas is given by the equation

$$\Sigma\, n_i \tfrac{1}{2} m v_i^2 = E_T \tag{1}$$

where m is the mass of the molecules and v_i is the value of the velocity for a molecule in the ith energy level. This summation indicates the sum of the kinetic energy for each allowed value of the velocity times the number of molecules that have that allowed value of velocity. Equation 1 coupled with Boltzmann's equation describes completely the ideal gas. To show this, we must discuss some aspects of velocity and state some of the consequences of the Boltzmann distribution as applied to gases.

First, we need a way to represent velocity that includes not only magnitude (speed) but also direction. This has been done in Fig. 4 with the usual three-dimensional Cartesian coordinates. In this representation, the length of the line representing velocity indicates the magnitude; its relation to the three coordinates determines the direction. Thus, if we want to represent the velocity of a molecule, we draw a line starting from the zero point of coordinates fixed in space; the direction of this line indicates the *direction* in which the molecule is moving; the length of the line shows the *speed.* It is important to realize that this drawing does *not* show where the molecule is. In fact, we do not care where it is; it may be in the upper left-hand corner of the box, in the lower right-hand corner,

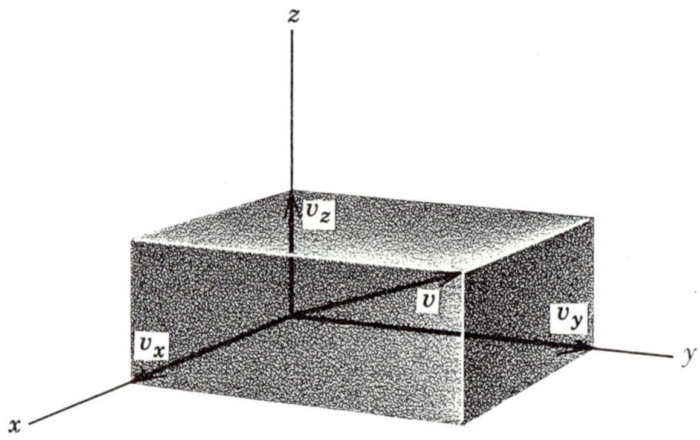

Fig. 4. Velocity components.

or in the middle. No matter where the molecule is, all lines that indicate its direction of motion and speed start at the zero of the velocity coordinates.

We can also specify the direction and magnitude of velocity by the values v_x, v_y, and v_z. For the moment, let us think of x and y as being south and east and of z as being up; then, to specify the velocity of a body, we could say that in one second it went v_x distance south, v_y distance east, and v_z distance up. The laws of solid geometry show us that the relation between these four quantities is

$$v^2 = v_x^2 + v_y^2 + v_z^2$$

where v^2 is the square of the magnitude of the velocity and is the same as the square of the speed. The quantities v_x, v_y, and v_z are referred to as the x, y, and z components of the velocity.

With these new definitions, we can write Eq. 1 in terms of the components of velocity:

$$\Sigma\, n_i \tfrac{1}{2} m(v_{ix}^2 + v_{iy}^2 + v_{iz}^2) = E_T$$

Since the x, y, and z directions are arbitrarily chosen, there is no physical reason why the contribution from the three terms should be different; hence it appears that one-third of the total energy comes from the x component of velocity, and the remaining two-thirds is divided equally between the y and z components. Thus there is an equal division of E_T among the three components of velocity; this is known as *equipartition* of energy.

Suppose we make a very detailed observation of an ideal gas. We will not determine exactly what the velocity of each molecule is, but we will see how many molecules move with v_x between 0 and 1 cm/sec, how many move with v_x between 1 and 2 cm/sec, and so forth. If we plot these results as a bar graph, the plot looks like that in Fig. 5, and a smooth curve through the top center of each bar is a plot of Boltzmann's equation in which E is the kinetic energy of the molecules *in the x direction.*

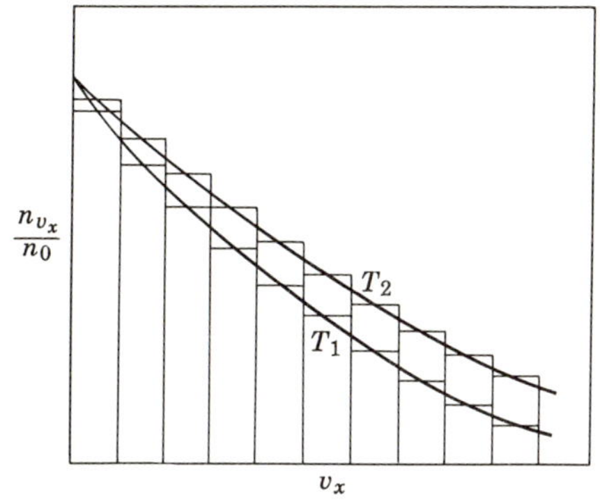

Fig. 5. Boltzmann distribution—one dimension, where $T_2 > T_1$.

Ultimately, we are interested in the distribution of molecules among *all* the energy states. This question can be answered if we know how many molecules have a velocity of magnitude v within an uncertainty, Δv. Each point in Fig. 6 represents a velocity, but what we want to know is how many such allowed points are a certain distance from the origin. Figure 5 gives us the values along any radial line; now we want the number of such points in a thin spherical shell. We do not care about the individual values of v_x, v_y, and v_z; we are interested only in a given magnitude of the resultant velocity. For higher and higher speeds there are more and more combinations of the v_x, v_y, and v_z which can give us

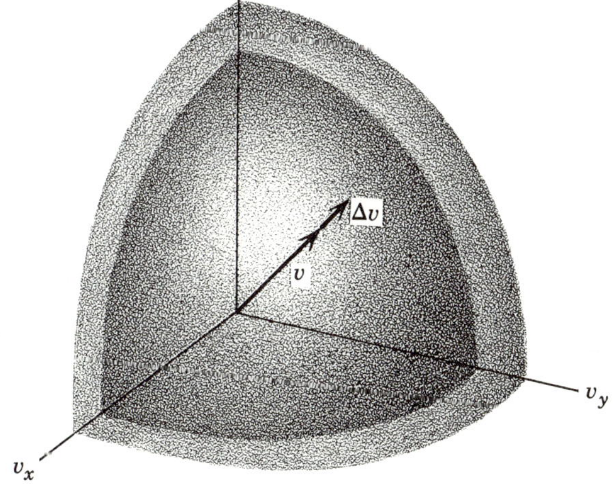

Fig. 6. Velocity distribution.

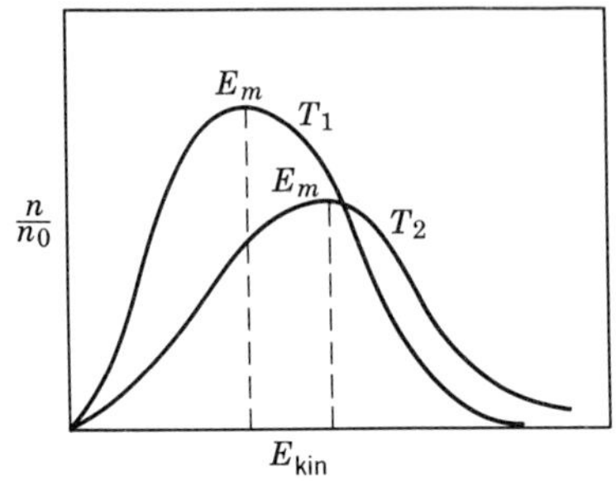

Fig. 7. Boltzmann distribution—three dimensions, where $T_2 > T_1$.

the magnitude v. This effect, therefore, acts in the opposite sense of the decrease noted in Fig. 5. The net result of these opposing factors is the energy distribution pictured in Fig. 7.† The most probable value for the energy, E_m, is not 0 as was the case when we considered only one dimension, and the character of the curve is strikingly different from that in Fig. 5.

From the Boltzmann distribution in the three dimensions, the total energy can be related to the temperature as follows:

$$E_T = \tfrac{3}{2}NkT$$

where N is the total number of molecules. With the help of this relation and Eq. 1, we can write

$$\tfrac{1}{2}m\,\frac{\Sigma\, n_i v_i^2}{N} = \overline{E} = \tfrac{3}{2}kT$$

or (2)

$$\tfrac{1}{2}m\overline{v^2} = \tfrac{3}{2}kT$$

where $\overline{E}$ represents the average energy of a molecule and $\overline{v^2}$ represents the average of the square of the velocity of the molecule. (This is *not* the same as $\overline{v}^2$.) It also follows that we can write

$$\tfrac{1}{2}m(\overline{v_x^2} + \overline{v_y^2} + \overline{v_z^2}) = \tfrac{3}{2}kT$$

and, because of the equipartition of energy,

$$\tfrac{1}{2}m\overline{v_x^2} = \tfrac{1}{2}m\overline{v_y^2} = \tfrac{1}{2}m\overline{v_z^2} = \tfrac{1}{2}kT \tag{3}$$

The above provides the background we need for the discussion of gases. A gas is a group of very small particles spread throughout a container. These particles are not standing still but are in continual motion. Some move very fast,

† The situation is analogous to that we discussed in connection with the hydrogen atom. It will be recalled that along any radial line the probability density decreased much like the curve in Fig. 5, but, if we consider successive spherical shells corresponding to the values of the atomic radius, we find that the shell most likely to be occupied by an electron is *not* at the origin.

some move very slowly. Some have very high kinetic energies and some have zero kinetic energy. The apportionment of velocities is completely random in direction, so that the average of all velocities is zero. By this we mean that it is likely that for every molecule moving north at 500 mi/hr there will be one moving south at the same speed, so that in averaging the velocity a cancellation occurs. This is comforting; the average velocity could be other than zero only if the box of gas as a whole were moving. The speed of the molecules varies as indicated by Boltzmann's equation. The average speed of these molecules is such that the average kinetic energy is always $\frac{3}{2}kT$, *independent of the mass of the molecule.* On the other hand, since the kinetic energy includes the mass, we find that at a particular temperature the average speed of a light gas molecule is greater than that of a heavier gas molecule.

Finally, let us note that an increase in temperature increases the average speed and energy of the gas molecules. This is done by decreasing the population of slow-moving molecules and increasing the population of fast-moving molecules (Fig. 7). Suppose, for example, that we want to know the number of molecules that have ten times the average energy. This turns out to be about one in a million. If we double the temperature, the average energy doubles and we expect the number of these high energy molecules to increase somewhat. Actually, the number increases a thousand-fold; now, one out of a thousand molecules has this higher energy. Thus the percentage increase in the population of high-energy molecules far outweighs that for molecules of nearly average energy.

The ideal gas law. We can now describe the behavior of ideal gases. Suppose that there are N molecules in a box of volume V (cf. Fig. 8). What is the origin of the pressure on the small section of the wall labeled A with an area of A square centimeters? Pressure is the result of the continual bombardment of the wall by gas molecules that hit the wall and then rebound elastically without change in speed. In such a collision, the path of the molecule is very much like that of light on a mirror in so far as the z and y components of velocity remain unchanged, but the x component of velocity changes signs. This change in x momentum is the origin of the force or the pressure on A.

Consider only those molecules having a velocity, v_{ix}, in the $+x$ direction. For a molecule with this velocity to collide with the wall at A in the next second, it must be located within the prism indicated in Fig. 8. If it is farther away than this, the molecule will not travel far enough in one second to reach A; but, if it is within the prism, it will reach A within one second. In other words, all these molecules that have a velocity of v_{ix} and are within a prism of volume Av_{ix} collide with A in one second. This is stated in the equation

$$R_i = \frac{n_i}{V} Av_{ix} \tag{4}$$

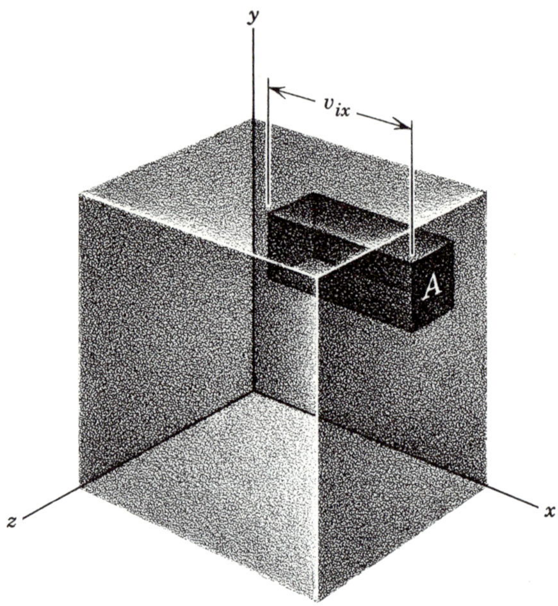

Fig. 8. Ideal gas container.

where R_i is the number of i-type molecules striking A per second. In this equation, n_i/V is the number of molecules per cubic centimeter that have the positive velocity v_{ix}. If we multiply this by Av_{ix}, the volume of the prism, we get the number of i-type molecules in the prism with a positive velocity.

Each molecule that collides with the wall changes the x component of velocity by

$$v_{ix} - (-v_{ix}) = 2v_{ix}$$

Suppose that this change takes place over a period of one second; then this is the acceleration, the rate of change of velocity, of one i-type molecule of mass m when it hits the wall. Each i-type molecule colliding with the wall exerts the force

$$F = ma = m(2v_{ix})$$

If we multiply this by the number of i-type molecules striking A per second, we find that the force due to these molecules is

$$F_i = \frac{n_i}{V} Av_{ix} \cdot 2mv_{ix}$$

The total force F_T is the sum of the force due to all types of molecules with a $+x$ component of velocity. This is given by the summation

$$F_T = \sum_{+} F_i = 2\frac{Am}{V}\sum_{+} n_i v_{ix}^2$$

This summation includes only the $\frac{1}{2}N$ molecules that have $+v_x$ velocity. We can rewrite this as

$$F_T = \frac{2Am}{V}\left(\frac{N}{2}\right)\sum_{+}\frac{n_i(v_{ix})^2}{(N/2)}$$

In the last-written form the summation proper is the average value of the square of the velocity in the x direction; hence we can write

$$F_T = \frac{AmN}{V}\,\overline{v_x^2} = \frac{AN}{V}\,kT$$

provided we make use of Eq. 3. By definition, pressure is the force per unit area; that is,

$$P = \frac{F_T}{A}$$

If we substitute this into the preceding equation we obtain

$$PV = NkT \tag{5a}$$

Since N is the total number of molecules, it is also equal to the number of moles (n) of gas multiplied by the number of molecules in a mole, *Avogadro's number*, $\mathfrak{N}_0$. If we substitute $n\mathfrak{N}_0$ for N, we get

$$PV = n\mathfrak{N}_0kT$$

The quantity $\mathfrak{N}_0k$ is called the gas constant R, so the equation becomes

$$PV = nRT \tag{5b}$$

This equation is known as the *ideal* gas law.

According to this equation, the pressure is determined by the *number* of gas molecules, their *temperature*, and *volume*. *Nothing is said about the mass of the gas molecules.* Thus the relation tells us nothing about the mass of the molecules, but it does tell us how many molecules are present; in other words, the behavior of a gas is determined by the number of molecules present; structure and mass do not enter the picture.

Effusion. Suppose the shaded section A in Fig. 8 represents a hole in the wall of the container. How fast would gas molecules stream through this opening? For i-type molecules this is given by Eq. 4:

$$R_i = \frac{n_i}{V}Av_{ix} \tag{4}$$

For streaming of all types of molecules, we need the summation

$$R_T = \frac{A}{V}\sum_{+} n_i v_{ix}$$

where once again the summation extends only over the $N/2$ molecules with $+x$ components of velocity. We can rewrite this as follows:

$$R_T = \frac{N}{2}\frac{A}{V}\frac{\Sigma\, n_i v_{ix}}{N/2} = \frac{N}{2}\frac{A}{V}v_x^+$$

where v_x^+ stands for the average v_x when we consider only those molecules that have a $+x$ velocity component.

Average properties can be calculated in a number of ways. Suppose we considered only eight molecules moving in the $+x$ and $-x$ directions with velocities of -4, -3, -2, -1, $+1$, $+2$, $+3$, $+4$ cm/sec. The average velocity v is given by

$$v = \frac{\Sigma\, n_i v_i}{N} = \frac{-4 + (-3) + (-2) + (-1) + (1) + (2) + (3) + (4)}{8}\ \text{cm/sec}$$

and is zero. The average of the square of the velocity v^2 is given by

$$v^2 = \frac{\Sigma\, n_i v_i^2}{N} = \frac{16 + 9 + 4 + 1 + 1 + 4 + 9 + 16}{8}\ (\text{cm/sec})^2$$

or

$$v^2 = \frac{60}{8} = 7.50\ (\text{cm/sec})^2 \quad \text{and} \quad (v^2)^{1/2} = 2.74\ \text{cm/sec}$$

The average plus component of velocity v^+ is not zero but is given by

$$v^+ = \frac{\sum_+ n_i v_i}{4} = 2.50\ \text{cm/sec}$$

where the summation includes only the plus velocities.

The average of the square of the $+$ component of velocity is given by

$$(v^+)^2 = \frac{\Sigma\, n_i (v_i^+)^2}{4} = \frac{1 + 4 + 9 + 16}{4} = 7.50\ (\text{cm/sec})^2$$

$$[(v^+)^2]^{1/2} = 2.74\ \text{cm/sec}$$

Thus we can conclude that

$$v \neq v^+ \neq (v^2)^{1/2} = [(v^+)^2]^{1/2}$$

In particular for an ideal gas,

$$(v_x^2)^{1/2} = [(v_x^+)^2]^{1/2} = \sqrt{\frac{kT}{m}}$$

$$v_x^+ = \sqrt{\frac{2kT}{\pi m}} = 0.80\sqrt{\frac{kT}{m}}; \qquad v_x = 0$$

Note that, although $(v_x^2)^{1/2}$ and (v_x^+) differ slightly numerically, they change the same way with temperature. The physical picture of an ideal gas makes this reasonable.

Since $\overline{v_x^+} = \sqrt{\dfrac{2kT}{\pi m}}$ we can write

$$R_T = \frac{NA}{2V}\sqrt{\frac{2kT}{\pi m}}$$

or, substituting from the ideal gas law,

$$R_T = \frac{(N)A}{2V}\sqrt{\frac{2kT}{\pi m}} = \frac{PV}{kT}\frac{A}{2V}\sqrt{\frac{2kT}{\pi m}}$$

or

$$R_T = \frac{PA}{2}\sqrt{\frac{2}{\pi mkT}} = \frac{PA}{\sqrt{2\pi mkT}}$$

or

$$R_T = \frac{PA}{\sqrt{2\pi kT}}\frac{1}{\sqrt{m}}\dagger$$

The last form written above reveals that *at a given temperature and pressure the rate of streaming through an orifice varies inversely as the square root of the molecular mass.* This conclusion was first reached by Thomas Graham in 1830. Note that, although the ideal gas law depends only on the *numbers* of gas molecules at a given temperature, pressure, and volume, the rate of effusion depends on the *mass* of the gas molecules. Thus these two equations are complementary and provide us with methods for the complete characterization of gases. For reasons not obvious from this derivation, Graham's Law applies only for very small holes at ordinary pressures.

If we compare the rates of effusion at standard conditions of two different gases, we find

$$\frac{R_T^{\mathrm{I}}}{R_T^{\mathrm{II}}} = \left(\frac{m^{\mathrm{II}}}{m^{\mathrm{I}}}\right)^{1/2}$$

where I and II indicate the different gases. Since, in an ideal gas, molecules act independently, the equation above represents the relative rates of effusion of the components of a mixture of gases from a hole in a box. This aspect of the kinetic theory of gases has been utilized for the large-scale separation of the isotopes. For example, naturally occurring uranium is 99.3% U^{238} and only 0.71% U^{235}. The isotope present in the smaller amount is a fissionable material. In order to isolate this material for the atomic bomb, scientists converted the uranium to UF_6, which is a gas at moderately low pressures. This gas was then allowed to effuse through

† In the finite difference notation, $R_T = \Delta N/\Delta t$ it is the change in the number of gas molecules per unit of time.

a small orifice and, since the lighter isotope effused faster, enrichment of U^{235} occurred. When the effusions were repeated many times, nearly complete separation was achieved.

APPLICATIONS OF THE IDEAL GAS LAW

Pressure. The quantities appearing in the ideal gas law are well defined by the mathematical equations. Moreover, such quantities as volume, numbers of molecules, and temperature are now familiar enough that little additional comment is needed. It is extremely useful, however, to picture changes of pressure, volume, and temperature in terms of the actual equipment needed to make such changes. A simple device for making such measurements is pictured in Fig. 9*a*. We have shown here a cylinder containing a frictionless, weightless, airtight piston. Below the piston is the gas to be measured; above the piston there is no gas at all, a vacuum. The weight on the piston can be adjusted so that the piston does not move. When this is done, the pressure of the gas pushing up is just equal to the pressure (weight ÷ piston area) pushing down. The volume of the gas is the volume of the cylinder below the piston. The temperature can be adjusted by putting the cylinder in a pail of water at the desired temperature. To increase the pressure on the gas, weights are added to the piston; to decrease the pressure on the gas, weights are removed from the piston.

It is difficult to construct a weightless, frictionless, airtight piston and it is also difficult to add or subtract weights inside a vacuum. For these reasons, gas pressures are usually measured by the device shown in Fig. 9*b*, a *manometer*. It con-

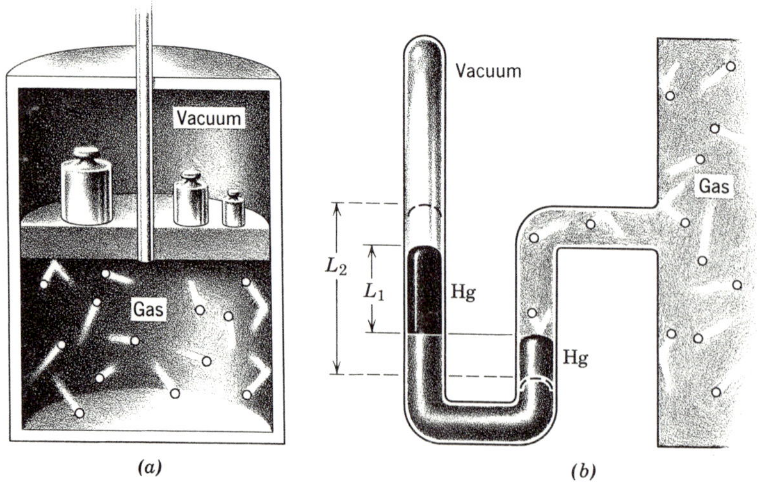

Fig. 9. Manometer.

sists of a U-shaped tube, partially filled with a non-volatile liquid. One side of the U-tube is connected to the gas; the other side is sealed off and evacuated. Figure 9*b* shows the level of the liquid in the U-tube. The lightly shaded portion of the liquid exerts no force on the gas since the weights of liquid pushing down on either side of the U-tube are equal; it functions as a weightless, frictionless, air-tight piston. The darkly shaded part exerts a force on the "piston" equal to its mass; this part functions like the weights in Fig. 9*a*. If the pressure on the gas is increased, the level in the U-tube readjusts to the levels indicated by the dotted lines and the effective weight automatically changes from that corresponding to L_1 to that corresponding to L_2. Thus with this device "weights" on the "piston" are added automatically. This is the usual method used to measure pressure.

If the U-tube is a cylinder of cross-sectional area A, the weight corresponding to L_1 is AL_1d, where d is the weight per unit volume of the liquid; the pressure (force/unit area) is L_1d. Thus the pressure is directly proportional to the net height of the column of liquid supported by the gas. Normal air pressure is 14.7 lb/in.^{2}; this corresponds to a column of mercury 760 mm high. Ordinarily, pressure of a gas is expressed as multiples of normal atmospheric pressure in atmospheres (atm), or else in terms of the height of the column of mercury it will support, in mm Hg. Thus a pressure of 29.4 lb/in.^{2} is a pressure of 2 atm or 1520 mm Hg.

A careful distinction between weight (a force) and mass is necessary. The density of mercury is 13.6 gm/cm^3. Thus 1 cm^3 of mercury has a mass of 13.6 gm. The weight, force exerted by gravity, is expressed in dynes and is obtained by multiplying the mass by the standard acceleration of gravity, 980 cm sec^{-2}. The force exerted by 1 cc of mercury is $980 \times 13.6 = 13{,}300$ dynes; this is the weight of 1 cc of mercury.

The gas constant. It is apparent from Eq. 2 that Boltzmann's constant k has the dimensions of energy/°K molecule. In Eq. 5*b*, we have introduced the alternative constant R, and, in view of the discussion leading up to Eq. 5*b*, we find that R applies to moles of gas whereas k applies to molecules of gas. Thus, whereas k is expressed in energy/°K *molecule*, R is expressed in energy/°K *mole*. In order to distinguish between these two constants, R is referred to as the *gas constant* although it is really Boltzmann's constant expressed on a per mole basis.

The gas constant R has the dimensions of energy per mole per degree but there are many ways of expressing energy. For example, we can express energy in electron volts or ergs; moreover, according to Eq. 5*b*, we can express it as a product of pressure and volume. Thus the numerical values of R depend on the units in which it is expressed. In Table 4 are some of the commonly used values. If we make calculations involving pressure expressed in atmospheres and volume

Table 4

R	Units
0.082054	$\frac{\text{atm liter}}{\text{mole °K}}$
82.055	$\frac{\text{atm cm}^3}{\text{mole °K}}$
8.314	$\frac{\text{joule}}{\text{mole °K}}$
8.314×10^7	$\frac{\text{ergs}}{\text{mole °K}}$
1.987	$\frac{\text{cal}}{\text{mole °K}}$

in liters, we must use the value 0.082054. On the other hand, if we keep the same units for pressure but express our volume in cubic centimeters we use the value 82.055.

Boyle's law. Consider what happens in a box containing a definite number of molecules when the temperature is held constant but the volume of the box is varied. The relation between P and V is given by Eq. 5*b*. When both T and n are constant,

$$PV = (nRT) = \text{constant}$$

This is a mathematical statement of Boyle's law. If we increase the volume, we must decrease the pressure, so that the product of the two remains constant. Graphically, this is shown in Fig. 10, where we have plotted P versus V for two different temperatures, T_1 and T_2.

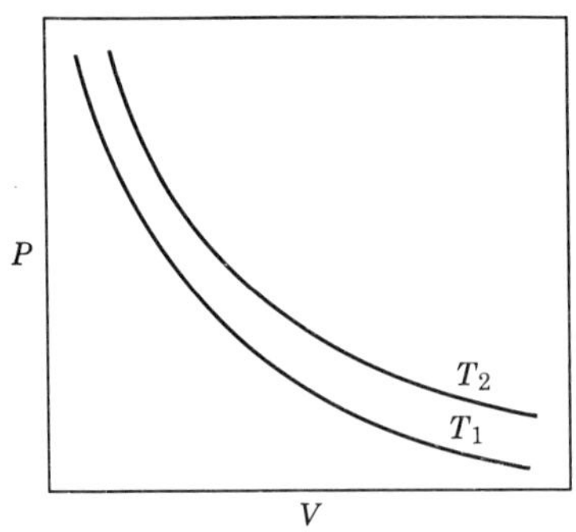

Fig. 10. Boyle's law.

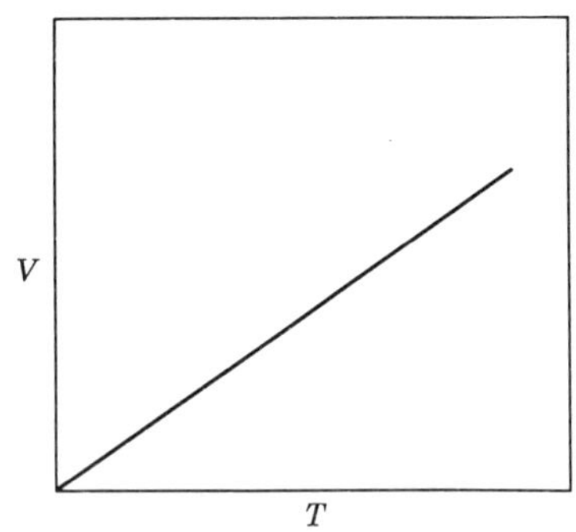

Fig. 11. Charles' law.

Charles' law. Charles' law states that, if we keep the pressure constant and vary the temperature, the volume of gas is proportional to the absolute temperature, or mathematically

$$V = \left(\frac{nR}{P}\right)T = \text{constant} \times T$$

The quantity in the parentheses is again constant. This is the basis for the ideal gas thermometer. In Fig. 11, we have given the plot of V versus T. The volume is directly proportional to temperature and becomes zero at 0°K.

Dalton's law. Suppose a sample of an ideal gas contains two kinds of molecules, A and B. The A molecules have no effect on the number of B molecules striking the wall per second; likewise, the B molecules have no effect on the number of A molecules striking the wall per second. The total effective force on the walls is the sum of the force due to A molecules and B molecules. In general, in a gas containing more than one type of molecule, the total pressure is the sum of the pressures that the individual components would exert if each were present in the container alone. This is Dalton's law, and it can be written as the equation

$$P_T = P_{\text{A}} + P_{\text{B}} + P_{\text{C}}$$

where P_T is total pressure and P_{A}, P_{B}, and P_{C} are the pressures A, B, and C would exert in the absence of the other two components. P_{A}, P_{B}, and P_{C} are referred to as *partial pressures* since each represents that part of the total pressure due to a particular kind of molecule.

Avogadro's principle. In 1811, Avogadro hypothesized that equal volumes of gas at the same pressure and temperature contain equal numbers of molecules. We can rearrange Eq. 5*a* to read

$$N = \frac{PV}{kT}$$

This, of course, contains Avogadro's hypothesis; once the volume, pressure, and temperature are fixed, the number of gas molecules must be fixed.

According to Avogadro's hypothesis, at a standard temperature and pressure 1 mole (6.023×10^{23} molecules) of any ideal gas occupies a definite volume. A convenient choice for a standard temperature and pressure (STP) is the temperature of melting ice, 0°C, and the normal pressure of the atmosphere. The volume occupied by 1 mole of gas at STP can be calculated from Eq. 5*b*. We obtain

$$V = 1.000\,\frac{(0.082054)273.15}{1.000} = 22.413 \text{ liters (STP)}$$

It is easier to remember that a mole of an ideal gas at STP occupies 22.4 liters than to remember various values of R. This is the only fact needed to apply the ideal gas law.

IDEAL GAS LAW CALCULATIONS

All ideal gas law calculations involve Eq. 5*b*. The gas law can be written in the form

$$\frac{P_1V_1}{n_1T_1} = \frac{P_2V_2}{n_2T_2}$$

where the subscripts 1 and 2 refer to initial and final conditions. Normally, we deal with problems in which $n_1 = n_2$, so we can rearrange this equation to yield

$$V_2 = V_1 \times \frac{T_2}{T_1} \times \frac{P_1}{P_2}$$

Thus, to calculate the volume V_2, we multiply V_1 by a ratio of pressures and temperatures. In this form, the correct ratios, i.e., P_1/P_2 versus P_2/P_1 are evident from the physical situation. If the units in the ratios are absolute and consistent, the units cancel; hence the particular units chosen do not affect the answer.

The way in which the appropriate ratios are chosen can be illustrated by an example. Suppose a gas with the initial volume V_1 is *cooled* to T_2 and *expanded* to P_2; what is the final volume? The cooling decreases the volume; hence we must multiply by T_2/T_1. (A factor of T_1/T_2 would be incorrect since it would *increase* the volume.) The expansion increases the volume; hence we must multiply by P_1/P_2. (A factor of P_2/P_1 would be incorrect since it would *decrease* the volume.)

Example Problem 1: A sample of gas with a volume of 20 cc† at 1.0 atm and 300°K is compressed to 2.0 atm and heated to 400°K. What is the final volume?

We summarize the problem in the following way:

V	P	T
20 cc	1.0 atm	300°K
?	2.0 atm	400°K

Imagine that the final volume is obtained by an increase in the pressure from 1.0 to 2.0 atm followed by the increase in temperature. The compression decreases the volume by an amount specified by the multiplication of 20 cc by the appropriate ratio of pressures, viz:

$$20 \text{ cc} \times \frac{1.0}{2.0}$$

† For cm^3 we often write cc (cubic centimeters) to avoid the use of exponents.

When the gas is then heated from 300° to 400°K, the volume increases; hence the final volume is

$$20 \text{ cc} \times \frac{1.0}{2.0} \times \frac{400}{300} = 13 \text{ cc}$$

Example Problem 2: If the initial volume of a gas is 3200 cc at 100 mm, and 232°K, what will be the volume at 800 mm and 116°K?

This calculation, without comment, is given below.

V	P	T
3200 cc	100 mm	232°
?	800 mm	116°

$$3200 \text{ cc} \times \frac{100}{800} \times \frac{116}{232} = 200 \text{ cc}$$

According to Dalton's law, if a gas consists of components 1, 2, and 3, and P_1, P_2, and P_3 are the pressures each gas would exert if it were the only one present, the total pressure is $P_1 + P_2 + P_3$. Alternatively, we can say that, if we have n_1 moles of gas 1, n_2 moles of gas 2, and n_3 moles of gas 3 all in the same container, the over-all pressure depends only on the number of moles of gas present; hence

$$PV = (n_1 + n_2 + n_3)RT$$

Example Problem 3: A mixture of helium and oxygen at a total pressure of 272 mm at room temperature and a small strip of magnesium are in a 1-liter container. The container is heated to 1000°C and all the oxygen is removed from the gas by reaction with the magnesium. The flask is cooled to room temperature and the pressure is found to be 136 mm. What was the composition of the original gas?

From the data above, we conclude that

$$P_{\text{He}} + P_{\text{O}_2} = 272 \text{ mm}$$

After we have removed the oxygen, the pressure is due solely to helium; hence

$$P_{\text{He}} = 136 \text{ mm}$$

and P_{O_2} must be

$$P_{\text{O}_2} = 272 - P_{\text{He}} = 136 \text{ mm}$$

Since the relative partial pressures of the gases must be the same as the relative number of moles, the original gas was half oxygen and half helium.

An important type of calculation is that in which the gas has been collected in the presence of liquid water. Such a gas contains a certain definite pressure of gaseous water at a given temperature; this pressure is called the vapor pressure. (We shall discuss this in detail later. For the present time, accept it as a fact.)

Example Problem 4: A 30-cc sample of gas is collected at 60°C over water. The total pressure is 529.2 mm. What is the volume of dry gas at 30°C and 760 mm pressure?

The fact that the gas is collected over water indicates that the gas is a mixture of dry gas and water vapor. According to the tables in Appendix H, the vapor pressure of water at 60° is 149.2 mm. Thus the pressure of the dry gas is (529.2 − 149.2) mm, or 380 mm. With this correction, the calculation is straightforward, viz:

V	P	T
30 cc	529.2 − 149.2 mm	(273 + 60)°K
?	760	(273 + 30)°K

$$30 \text{ cc} \times \frac{380}{760} \times \frac{303}{333} = 14 \text{ cc}$$

Example Problem 5: If 1000 cc of dry gas at 760 mm of pressure are bubbled through water at a given temperature, the volume of the effluent gas at the same pressure is 1100 cc. The temperature of the water and the gas are the same throughout the experiment. What is the temperature of the water?

Initially the pressure is due wholly to the dry gas, but the final pressure is the sum of the partial pressures of the water and the dry gas. The partial pressure P_g of the dry gas after it has been bubbled through the water is given by

V	P	T
1000 cc	760 mm	T
1100 cc	P_g mm	T

$$1000 \times 760 = 1100 \times P_g$$
$$P_g = 691 \text{ mm}$$

If the partial pressure of the dry gas is 691 mm, the vapor pressure of the water is 69 mm. From vapor pressure tables, we find that this corresponds to a temperature slightly below 45°C.

The number of moles of gas, n, is related to the weight, w, and molecular weight, M, as follows:

$$n = \frac{w}{M}$$

If we substitute this into Eq. 5*b*, we obtain

$$PV = \frac{w}{M} RT$$

This is an equation involving five unknowns: if we know any four of them, we can calculate the fifth. We avoid the use of R if we remember that each mole of gas at STP occupies a volume of 22,413 cc or

$$\frac{22{,}413 \text{ cc STP}}{1 \text{ mole gas}}$$

STP is a part of the units in this factor.

Example Problem 6: A sample of 37 cc of gas at a pressure of 760 mm and 273°C weighs 4.0 mg. What is the molecular weight of this gas?

Such problems are a two-step process. The first requires calculation of the number of cc STP of gas. From this and the conversion factor, we can calculate the number of moles and the molecular weight. The procedure, without comment, is given below:

(1)

V	T	P
37	546	760
?	273	760

$$37 \times \frac{273}{546} = 18.5 \text{ cc STP}$$

(2)

$$\frac{4.0 \times 10^{-3} \text{ gm}}{18.5 \text{ cc STP}} \times \frac{22.4 \times 10^{3} \text{ cc STP}}{1 \text{ mole}} = \frac{4.8 \text{ gm}}{\text{mole}}$$

Example Problem 7: Calculate the density of methane at 100°C and 500 mm.

The density of methane will not depend on the size of the sample, so, for convenience, let us work with one mole, that is, 16.0 gm of methane. At STP, this will occupy a volume of 22.4 liters. If we heat the 22.4 liters to 100°C and expand it to a pressure of 500 mm, the volume will change but the mass *must stay the same;* hence for this problem we need the volume at 100°C and 500 mm pressure.

V	T	P
22.4 liters	273°K	760 mm
?	373°K	500 mm

$$22.4 \times \frac{373}{273} \times \frac{760}{500} = 46.5 \text{ liters}$$

Then

$$\text{Density} = \frac{16.0 \text{ gm}}{46.5 \text{ liters}} = 0.344 \frac{\text{gm}}{\text{liter}}$$

Avogadro's principle can sometimes be used as a short-cut in weight calculations. An example of this is provided by the following problem.

Example Problem 8: A chemist wishes to determine the atomic weight of a new element, X. He prepares a number of different X-containing molecules and analyzes them for X. Then he fills a standard-sized flask at 213°C with each gas until the pressure reaches 1 atm. From the weight of each sample of gas and his analysis, he computes the weight of X in each case. These data, together with similar data for pure oxygen, are given in the accompanying table.

Gas	Total Wt. of Gas	Wt. of X in Gas
O_2	0.70 gm	—
Compound A	130.20 gm	3.90 gm
Compound B	41.70 gm	1.56 gm
Compound C	32.66 gm	5.46 gm
Compound D	71.23 gm	2.34 gm
Compound E	18.10 gm	3.12 gm

What is the likeliest atomic weight of X?

In a compound containing X, each molecule must contain at least one atom of X; there can be more than one atom of X but never less. Similarly, a mole of compound containing X will contain one atomic weight of X or an integral multiple of the atomic weight. Thus the weight of X in one mole of any X-containing compound is a whole number multiple of the atomic weight of X. If we had a series of such compounds, the greatest common divisor of the weights of X per mole for the whole series would be likely to be the atomic weight. Thus our task is to determine this greatest common divisor.

The volume of the flask is not needed in this calculation. The volume of gas corresponds to 0.70 gm of O_2, or

$$0.70 \text{ gm } O_2 \times \frac{1 \text{ mole}}{32.00 \text{ gm } O_2} = 0.022 \text{ mole of } O_2$$

If we had a flask containing 1 mole of O_2 at this T and P, the volume would be larger by the factor $1.00/0.022 = 46$. A flask this size at 213°C and 1 atm would contain 1 mole of any gas, according to Avogadro's principle. Thus, immediately we see that 46 times the weights in the second column of our table are the weights of 1 mole of compound A, B, C, D, and E, respectively. Correspondingly, the weights of X in 1 mole of compounds A, B, C, D, and E are 179, 72, 251, 108, and 143, respectively. The greatest divisor is 36; this corresponds closely to the atomic weight of chlorine.

This method was first applied by Cannizzaro, a student of Avogadro, in 1858. His forceful presentation of the above logic, together with its application, was instrumental in the acceptance of atomic theory by chemists. Even today, thorough understanding of the method provides a student with a firmer understanding of atomic and molecular weights as well as the properties of gases.

Thus far, we have not needed the principles of stoichiometry developed earlier, but often the problems involve stoichiometric calculations. Since these deal with weights, we proceed via the conversion factor at STP.

Example Problem 9: The decomposition of $KClO_3$,

$$2KClO_3 \rightarrow 2KCl + 3O_2$$

is often used as a source of O_2 in the laboratory. If some $KClO_3$ decomposed to form 50 cc of O_2 (collected over water at 40°C and 715 mm), how much $KClO_3$ decomposed?

From the equation, we know

$$\frac{3 \text{ moles } O_2}{2 \text{ moles } KClO_3}$$

From the cc STP of pure O_2 and the factor 22,413 cc STP O_2/1 mole O_2, we can calculate the number of moles of O_2 formed, and then, with the help of the preceding factors, the amount of reacting $KClO_3$ can be calculated.

As a first step, we calculate the number of moles of O_2 formed. Since the gas was collected over water at 40°C,

$$P_{O_2} + \text{V.P.}^{40^\circ\text{C}}_{H_2O} = 715 \text{ mm}$$

From the vapor pressure tables, we find

$$V.P._{H_2O}^{40°C} = 55 \text{ mm}$$

Hence

$$P_{O_2} = 660 \text{ mm}$$

Thus, for the first part of this problem, we write

V	P	T
50 cc	660 mm	313°K
?	760 mm	273°K

$$50 \text{ cc} \times \frac{660}{760} \times \frac{273}{313} = 38 \text{ cc STP}$$

$$38 \text{ cc STP } O_2 \times \frac{1 \text{ mole } O_2}{22.4 \times 10^3 \text{ cc STP } O_2} = 1.70 \times 10^{-3} \text{ mole } O_2$$

With this information, the last part of the problem becomes

$$1.70 \times 10^{-3} \text{ mole of } O_2 \times \frac{2 \text{ moles } KClO_3}{3 \text{ moles } O_2} \times \frac{122.6 \text{ gm } KClO_3}{1 \text{ mole } KClO_3} = 0.139 \text{ gm } KClO_3$$

Note that units are used to guide our thinking.

A particularly simple type of calculation dealing with chemical equations is encountered when the reactants and products are gases.

Example Problem 10: If 100 cc of N_2 at 862 atm and 500°C reacts completely with hydrogen to form NH_3, what volume of NH_3 is produced?

The equation is

$$N_2 + 3H_2 \rightarrow 2NH_3$$

It tells us that 1 mole of N_2 produces 2 moles of NH_3. But Avogadro's principle tells us that, if this is true, then at the same P and T 1 *volume* of N_2 produces 2 *volumes* of NH_3. (At the same P and T, equal volumes of gas contain equal numbers of molecules, or moles.) Thus, for every 100 cc of N_2 that reacts, 200 cc of NH_3 is produced *at the same P and T*, namely, 862 atm and 500°C.

INTERMOLECULAR FORCES

The gas law was derived for an idealized model of a gas, in which the molecules had no volume and in which the energy was wholly kinetic energy. The time we have devoted to this law is justifiable only if real gases approach this behavior. The two assumptions are approximations; the real question is: How good are they? One way to answer this is by comparison of 22.413 liters to the values observed for the volumes of real gases at STP. In Table 5, this is done. Note that all the values are fairly close to the ideal value. There are, however,

Table 5

Gas	V_{STP}	Boiling Point
Ideal	22,413 cc	
H_2	22,424	20°K
N_2	22,401	77°K
O_2	22,393	90°K
CO_2	22,263	195°K[a]
NH_3	22,083	240°K
CH_3Cl	21,878	249°K

[a] Sublimes.

deviations, and these deviations are most pronounced for gases with the highest boiling points.

A more detailed summary of the deviations of real gases from Eq. 5*b* is obtained by a plot of PV/nRT versus pressure. In Fig. 12, we show such a plot for 1 mole of nitrogen gas at 0°C. For low pressures the quotient is close to unity, in line with Eq. 5*b*, but for large pressures the value becomes first smaller, then larger than unity. Thus, at high pressures, the assumptions made in deriving the ideal gas law do not apply to this (or any other) real gas.

What are the causes of this departure from ideal behavior? A clue is furnished by the data in Table 5 and Fig. 12. In Table 5, deviations from the ideal value of 22,413 cc STP/mole are most evident for gases with higher boiling points. We shall see that, the higher the boiling point, the stronger are the intermolecular forces. Thus, for the gases listed in Table 5, the departures from ideality are greatest when the intermolecular forces are significant. When would this ideal gas approximation that there are no intermolecular forces be most likely to be valid? The approximation is most valid at low pressures and far above the boiling point. Intermolecular forces are short-range forces—though they are large at 5 Å, they are negligible at 10 or 15 Å. Therefore, at low pressures when there are relatively large distances between molecules, the intermolecular

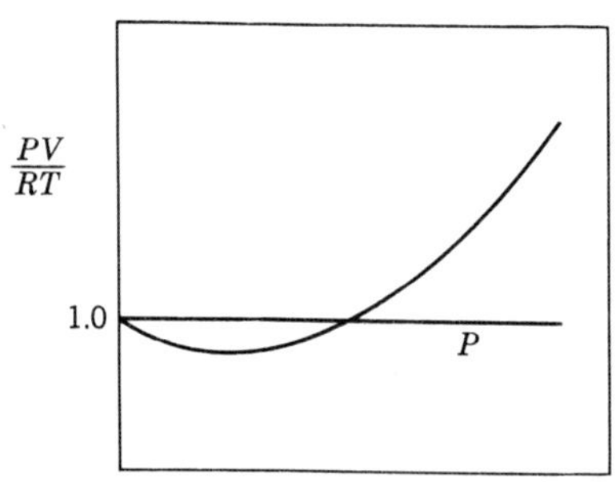

Fig. 12. Ideal gas law and real gases.

forces are negligible. In addition, at high temperatures, where the kinetic energy is much greater than the potential energy due to intermolecular attractions, i.e., at temperatures far above the boiling point, these forces are unimportant.

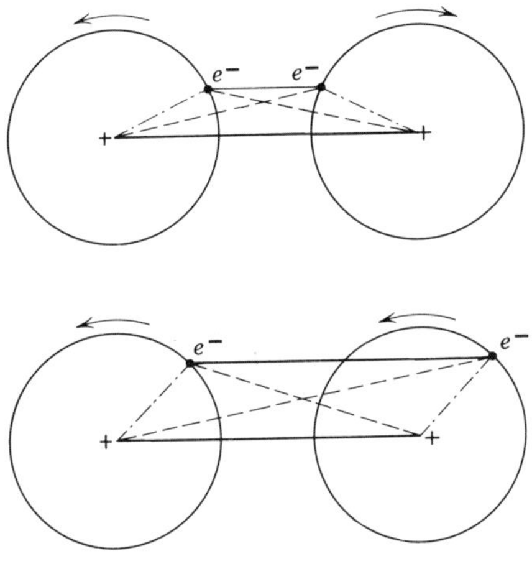

Fig. 13. Electron correlation—intermolecular forces.

What are the origins of these intermolecular forces? Attractive forces operate between all molecules as the result of correlation of electronic motion. Even for rare gases, such forces exist although the electronic structure is such that no overlap of charge clouds occurs, and the forces are far weaker than those found in chemical bonding. The origin of these non-specific forces can be represented qualitatively by the drawing in Fig. 13. As the molecules approach, there may be no correlation between the electronic motion as indicated in the upper part of the picture. There are repulsive forces between the particles, indicated by solid lines, and attractive forces between particles, indicated by dotted lines. In the situation in the upper figure, the net effect will be repulsion between the two molecules. On the other hand, the energy would be a minimum if the motions of the electrons on these different molecules were correlated, as suggested in the lower part of the picture. Then the over-all effect would be attraction between the molecules. An attraction due to correlation of the electronic motion of two separate molecules is called *van der Waals'* attraction, in honor of the Dutch physicist who first developed a theory of real gases.

If we push these molecules close together, the force becomes repulsive since the electron-charge clouds repel by virtue of the Pauli exclusion principle. Thus, *in toto,* there is an attractive force a and a repulsive force r that vary with the intermolecular distance r, as shown in Fig. 14. The repulsive force is not significant at distances larger than about 5 Å, and, at these larger distances, the attractive force predominates. At smaller distances, the repulsive force overwhelms the attractive force. Thus, as two gas molecules approach each other, there is a net attractive force at about 8 Å; this attractive force increases at first and then decreases when the intermolecular distance falls to 5 Å. For distances less than 5 Å, the force between the two molecules is repulsive. This sudden onset of repulsive force means that the molecules resemble hard billiard balls with the diameter indicated by D. In the real gas a molecule cannot move around freely within the whole of the container; part of the volume of the container will

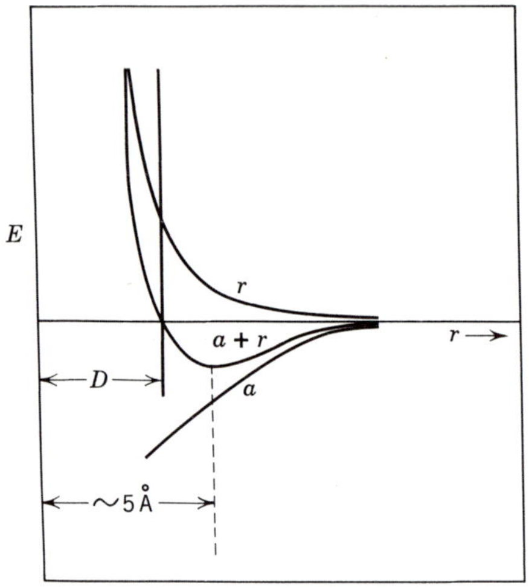

Fig. 14. Potential energy for various intermolecular distances.

be occupied by other molecules. The deviations from ideal behavior at high pressures (Fig. 12) occur simply because individual molecules now use up much of the "empty" space in the box of gas.

Joule's experiment. One of the most convincing experimental evidences of intermolecular forces is that of Joule. In an ideal gas, the energy depends only on the temperature and is independent of the pressure. If we expand such a gas into a vacuum and, thereby, increase the intermolecular distance, this has no effect on the energy of the gas. On the other hand, with real gases, the closer the molecules are to one another, the lower the energy. It requires work against intermolecular forces to separate these molecules; by so doing, we increase the potential energy. In Joule's experiment, a gas in a thermally insulated container is expanded into a vacuum. When this is done, it is found that the gas cools.† The reason is that the energy required for the expansion comes from the gas itself and is obtained at the expense of the kinetic energy; hence the temperature of the gas decreases. This principle is utilized for the liquefaction of gases.

REAL GASES

Owing to the nature of the intermolecular potential (Fig. 14): (*a*) all the container volume will not be available to the gas molecules; (*b*) there will be, *on the average*, association of gas molecules. Association of gas molecules is not long-lived since intermolecular forces are weak, but, at any instant, a definite fraction of the molecules are within the sphere of influence of other molecules

† This is not always true in the Joule-Thomson experiment.

and are weakly bonded. The molecules in these groups will constantly change, but, at any instant, the average number of groups of each type is the same.

Now we can rationalize the behavior of real gases. Let us assume that the ideal gas law does hold for real gases provided we use "ideal" pressures (P_I) and "ideal" volumes (V_I); that is,

$$P_I V_I = RT$$

for a mole of real gas. To estimate what the ideal volume is, we must consider the volume needed by the molecules themselves. In 1 mole of gas, there will be a certain inaccessible volume because the molecules themselves take up space. Call this inaccessible volume, which varies with the molecular volume, b. Then the relation between the ideal volume and V the volume of the container is

$$V_I = V - b$$

In an ideal gas, the pressure is a product of the number of molecules striking the wall and the momentum transferred. In a real gas, the number of "molecules" is reduced because, at any instant, some of the molecules are associated. Therefore,

$$P_I > P$$

where P is the measured pressure. The difference between the ideal and the real pressure becomes most pronounced as the density of the gas increases; for, when the gas is densest, the intermolecular forces increase the association. An equation which predicts such behavior is

$$P_I = P + \frac{a}{V^2}$$

At very low pressures, where the intermolecular forces are ineffective, association is trivial and the ideal pressure is equal to the real pressure; also at very low pressures, where b is small compared to V, the ideal volume is very close to the real volume. Thus, for dilute real gases, the ideal gas law is a good approximation.

By combining the equations above, we obtain

$$(P + a/V^2)(V - b) = RT$$

This is van der Waals' equation of state. In Fig. 15, we have sketched P versus V for this equation at a series of temperatures. At low pressures and high temperatures, the behavior is that predicted by the ideal gas law. As we decrease temperature, we notice pronounced deviation from ideal behavior (cf. Fig. 10).

Suppose for specificity that the data in Fig. 15 represent the behavior of steam. Then T_4 is about 500°C. At these temperatures, steam is a gas regard-

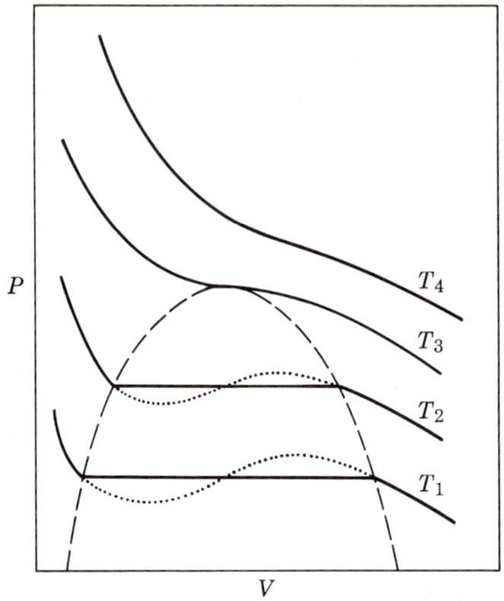

Fig. 15. Isotherm for a real gas.

less of the pressure. If we cool steam to 100° and slowly increase the pressure, it will follow the behavior indicated by T_2. At pressures near 1 atm, pronounced deviations from the ideal gas are apparent. Finally, when we attempt to increase the pressure above 1, we get a rather drastic decrease in volume, but no change in pressure! The pressure remains at 1 atm! This represents condensation of steam to liquid water. If we continue to apply the pressure, the volume will continue to change until the volume is roughly one one-thousandth of the volume at the start; then we find that the condensed steam is virtually incompressible, i.e., there is little volume change with an increase in pressure. At this stage, only liquid water is present.

Whenever we increase the pressure on a real gas, condensation occurs at a definite pressure provided the gas is below a critical temperature. The pressure where condensation begins is called the *vapor pressure.* The vapor pressure increases with temperature until, at a certain temperature, we reach the point where, no matter how much the gas is compressed, no condensation occurs. This point is found on the curve labeled T_3. The temperature T_3 is called the *critical temperature;* it is the temperature above which we cannot condense a gas to a liquid. The value of this critical temperature is determined by the values of the van der Waals constants, a and b and, like a and b, it varies from one gas to another.

van der Waals' equation does not yield the horizontal straight lines shown in the graph; it yields the dotted curve. The dotted curve corresponds to unstable conditions. At pressure and volumes corresponding to this portion of Fig. 15, we observe the behavior indicated by the horizontal lines.

SUMMARY

The behavior of gases forms the cornerstone of our theory of chemical reactions. Of the three states of matter, the gaseous state is the best understood. This is rather surprising for this is the most chaotic state. The fact that gases are the best understood chemically provides us with a glimmer of a vital factor in the theory of chemical reactions. Gases are understood because they are truly chaotic and are governed almost exactly by laws of chance. Chemical reactions are also governed by these laws. We shall find it convenient to discuss energy and probability factors in dealing with reactions, but it should be recalled that the former, via Boltzmann's law, is also described by laws of chance. Thus we shall find ourselves in a very curious position as we continue our chemical studies; when we deal with the gaseous state, where we know the least about the actual location and motion of individual molecules, we shall be able to describe most accurately the over-all macroscopic physical and chemical behavior. The liquid and solid states, for which we know most about the location of the molecules, will be the most difficult to describe chemically.

PROBLEMS

1. Calculate the weight, in pounds, of a column of mercury 1 in.^{2} in cross section and 760 mm high. The density of mercury is about 13.5 gm/cc. Would you have known the answer to this without making the calculation?

2. Two cylindrical columns of mercury, 3.000 ft high, have diameters of 5.000 cm and 10.000 cm, respectively. Calculate the pressure in dynes per square centimeter at the bottom of each column.

3. A tank of helium has a volume of about 2.00 ft^3. If it is filled with helium at 2000 psi (lb/in.^{2}), what volume would the gas fill at a pressure of 1 atm?

Ans. 272 ft^3.

4. An iron tank of helium contains the gas at a pressure of 2000 psi at 25°C. The tank will hold up to a pressure of about 7000 psi. If this tank is in a building that catches fire, will the tank blow up before it melts? The melting point of iron is 1535°C.

5. The average size of a breath taken by an adult doing no exercise is 250 cc STP. An average person takes about 15 breaths per minute. To get the same amount of air, how many breaths per minute must a diver take 300 ft below the surface of the ocean? (Pressure about 150 psi.)

6. What would be the volume of 100 cc STP of carbon at its boiling point and 200 mm pressure? The boiling point of carbon is 4200°C.

7. What is the volume of 100 cc of helium STP at its boiling point (4°K) and at the boiling point of carbon?

8. Oxygen can be made by heating potassium chlorate at 368°C. If, after the reaction, the oxygen in the *heated vessel* (volume 1000 cc) exerted a pressure of 7 mm, what would the volume of this oxygen be at STP? *Ans.* 3.92 cc STP.

9. A chemist is determining the atomic weight of X. To do so he first prepares a number of gaseous, X-containing molecules and analyzes them for X. Then he fills a 324 cc flask with each gas at a pressure of 3 atm and a temperature of 302°C. From the weight of these gases and his analysis he computes the weight of X in each sample of gas. These data, together with those for pure O_2, are given in the accompanying table. What is the most likely atomic weight of X?

Gas	Total Wt. of Gas	Wt. of X in Gas
O_2	0.66 gm	0.00 gm
Compound A	———	0.78 gm
Compound B	———	1.17 gm
Compound C	———	2.34 gm
Compound D	———	1.95 gm
Compound E	———	1.17 gm

10. A chemist prepares a series of gaseous compounds, each of which contains the new element X. To determine the atomic weight of X, he fills a single flask with each compound in turn, adjusts the pressure to 500 mm at a temperature of 127°C, and weighs each flask. He then fills the flask with pure Ar at a pressure of 500 mm at 127°C. From these results and the chemical analysis of the compound, he obtains the data in the accompanying table. What is the most likely atomic weight of X?

Gas	Total Weight of Gas	Weight % X
Ar	14.5 mg	0.0%
Compound A	232 mg	14.8%
Compound B	517 mg	4.99%
Compound C	42 mg	41.0%
Compound D	75 mg	80.3%
Compound E	118 mg	36.4%

11. A volume of 17 cc of oxygen is collected at 23° over water. If the total pressure was 540 mm, how many cubic centimeters of dry gas would this be at STP? *Ans.* 10.7 cc STP.

12. Hydrogen is produced electrolytically and is essentially saturated with water at about 30°. If this is compressed to 2000 psi and put into a tank, what will be the composition of the gas at 30°C?

13. What is the density of C_2H_6 at 200°C and 100 atm? *Ans.* 77.3 gm/liter.

14. Oxygen at atmospheric pressure (760 mm) is bubbled through water at 10°C. What is the per cent of water in the gas? How many cc STP of water would 1 liter of the *resulting* gas mixture contain?

15. A mixture of gas at 300 mm pressure contains C_2H_4 and H_2. A catalyst is added so that the following reaction takes place, as completely as possible.

$$C_2H_4 + H_2 \rightarrow C_2H_6$$

The resulting mixture has a pressure of 200 mm. What could the initial composition of this gas be?

16. If 100 cc of hydrogen at 300°C and 400 mm pressure is mixed with 200 cc of oxygen at the same temperature and pressure, what will be the STP volume of the dry gas left after reaction to form water? What will be the composition of the gas after complete reaction?

17. The density of gas X is 1 gm/cc at −135°C and 500 atm. What is the density at STP?

18. A volume of 227 cc of gas at 25° and 300 mm pressure weighs 0.100 gm. What is the molecular weight of the gas? *Ans.* 27.4 gm/mole.

19. A molecule contains carbon and hydrogen in the ratio of 1 carbon for every 1 hydrogen. At 273°C and 2.0 atm, 1 liter of this gas weighs 4.06 gm. Determine the correct formula.

20. A gas consisting of molecules containing only oxygen atoms has a density of 4.28 gm/liter at 273°C and 4 atm. What is the formula for this gas?

21. A mixture of helium and X in a 1-liter flask weighs 4 gm. The pressure is 872 mm and the temperature 17°C. X freezes out at −7°C. If the mixture is cooled to −263°C, the pressure in the vessel will be 10 mm. What is the molecular weight of X?

22. Calculate the ratio of the rate of effusion of hydrogen to the rate of effusion of deuterium under the same conditions.

★ **23.** A gas consists of 99% H_2 and 1% O_2. (*a*) Calculate the rate of effusion of each through a 10^{-10} cm^2 orifice at a total pressure of 760 mm at 25°C. (*b*) What would be the concentration of the first amount of gas that comes through the orifice? (*c*) If the effusion was repeated five times with only the first traces of gas, what would be the final concentration? (*d*) If all the original gas effused through the orifice, what would be the final concentrations of H_2 and O_2?

24. Using the ideal gas law, with P expressed in the proper cgs unit, calculate the value of R in ergs per degree.

25. Calculate the $(\overline{v^2})^{1/2}$ of a hydrogen atom at 25°C. *Ans.* 2.7×10^5 cm/sec.

26. Calculate the $(\overline{v^2})^{1/2}$ of a radon atom at 25°C.

27. Calculate the $(\overline{v^2})^{1/2}$ of an "air atom" at 25°C. Air is about 80% N_2 and 20% O_2. Compute this in feet per second and compare it to the speed of sound, 1100 ft/sec.

28. Compute the average kinetic energy of 1 molecule of oxygen at 25°C.

29. If all the energy we put into heating a gas goes into the gas in the form of kinetic energy, how much energy is required to raise the temperature of 1 mole of gas one degree centigrade?

30. Convert the answer to Problem 29 to a different type of energy, called a calorie, using the conversion factor:

$$1 \text{ erg} = 2.39 \times 10^{-8} \text{ calorie}$$

31. CO_2 can be produced by the reaction

$$Na_2CO_3 + 2HCl \rightarrow H_2O + CO_2 + 2NaCl$$

If we have a large amount of the sodium carbonate, how many grams of HCl will be needed to produce 100 cc of dry CO_2 at 200°C and 450 mm pressure?

Ans. 0.109 gm HCl.

32. Zinc reacts with sulfuric acid according to the equation

$$Zn + H_2SO_4 \rightarrow ZnSO_4 + H_2$$

If enough zinc was added to a large amount of sulfuric acid to produce 50 cc of hydrogen, collected over water at 30°C and 760 mm pressure, how much zinc was added?

33. Limestone (calcium carbonate) decomposes according to the equation

$$CaCO_3 \rightarrow CO_2 + CaO$$

If 100 cc of CO_2 is collected at 223°C and 482 mm pressure, how many grams of calcium oxide must have been formed in the reaction?

34. In the Mond process for the purification of nickel, the impure metallic nickel is first treated with carbon monoxide gas to form a volatile compound called nickel carbonyl. This reaction can be represented by the equation

$$4CO + Ni \rightarrow Ni(CO)_4$$

What volume of carbon monoxide will be needed at 100°C and atmospheric pressure in order to purify 1 lb of a mixture containing 73% nickel?

35. Lithium hydride is a convenient source of hydrogen. It reacts with water according to the equation

$$LiH + H_2O \rightarrow LiOH + H_2$$

How many grams of lithium hydride will be needed to fill a balloon 100 ft in diameter if the internal pressure in the balloon is 800 mm and the temperature is 30°C? (Assume the hydrogen is saturated with water.)

36. According to Archimedes' principle, a body immersed in a fluid is pushed up by a force equal to the weight of the displaced fluid. In air, at normal atmospheric pressure and 30°C, the balloon mentioned in Problem 35 could just lift a certain amount of weight. If the balloon itself (but not its contents) was weightless, how much weight could it lift? Assume that air has an effective molecular weight of 28.8. If the balloon were filled with helium rather than hydrogen, how much weight could it lift?

★ **37.** Pure CO_2 can be produced in the laboratory according to the reaction

$$2NaHCO_3 \rightarrow Na_2CO_3 + H_2O + CO_2$$

Usually, this CO_2 is wet and must be dried. If we produce this CO_2 without drying it, and we obtain a yield of 500 cc of gas at 760 mm pressure and 30°C, how many cubic centimeters of pure CO_2 at STP are produced and how many grams of sodium bicarbonate were used?

38. The production of sulfuric acid involves the oxidation of sulfur dioxide to sulfur trioxide, according to the equation

$$2SO_2 + O_2 \rightarrow 2SO_3$$

In order to do this in an experiment, $KClO_3$ is used as a source of oxygen. How many grams of $KClO_3$ would have to be decomposed in order to produce 10 gm of SO_3?

Ans. 5.11 gm $KClO_3$.

39. If we produce nitrogen by the decomposition of ammonium nitrite,

$$NH_4NO_2 \rightarrow 2H_2O + N_2$$

and collect 1 liter of gas over water at 100°C, and 800 mm pressure, how many grams of nitrogen gas have we produced?

40. Calculate the volume of O_2 required to burn 100 cc of CH_4, C_2H_6, C_2H_4, and C_6H_6. Assume that the final products of combustion are CO_2 and water.

Ans. For CH_4, 200 cc O_2.

9

Crystal structure and oxides

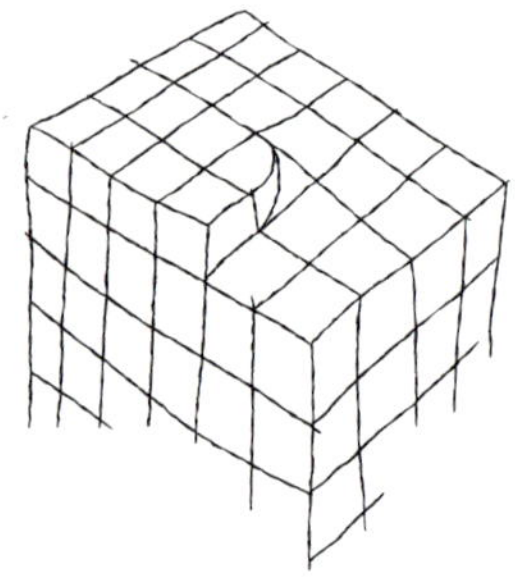

The obvious characteristics of solids are a definite shape and size, and a substantiality far greater than that of gases. Definite shape and size require that the atoms or molecules of a solid be more or less fixed in position; the substantiality requires that the packing be efficient in the sense that most of the space in a solid is occupied by the atoms themselves. All solids have these properties. But, in its most stable form called *crystal*, a solid has another property, regularity. Such crystalline solids are characterized by a regularly repeated arrangement much like the arrangement of bricks in a wall. This order is so perfect (ideally) that the position of each atom in a crystal containing billions of atoms can be specified if the positions of a few atoms are known.

Four bond types are found in crystals: *ionic, metallic, molecular,* and *covalent.* The last three types are associated with elements in specific regions of the periodic table. For example, metallic crystals are formed by the elements located left of and below a line starting with B in the first row and extending diagonally through Si, As, Te, and At. The elements to the right and above this line form mostly molecular crystals. The elements along the line are called metalloids, substances having partly metallic and partly non-metallic character. Many of these form covalent crystals.

When a metal and non-metal combine, transfer of electrons often occurs from the metal to the non-metal with the formation of an ionic crystal, wherein ions are the building blocks. Examples of these are NaCl, BaF_2, Li_2O, and CaS. They do not conduct electricity appreciably, are usually transparent, and cleave easily along definite directions when subjected to shock. Generalizations, however, must be accepted with caution since some ionic crystals are deeply colored and others exhibit a fiber structure like asbestos, or a layer structure like mica. The properties of ionic crystals are more varied than those of metals.

Metallic crystals are formed by pure elements or by combinations of two or more elements in definite or indefinite atomic proportions. All these crystals exhibit metallic luster, conduct electricity, and are malleable and ductile.

When two elements do not differ much in electronegativity, they form molecules like CH_4, PBr_3, and SO_2. In the crystalline form of these compounds the fundamental building unit is the molecule rather than atoms or ions. Such a molecular crystal does not conduct electricity and vaporizes far more readily than either metallic or ionic crystals. The forces between molecules in such a crystal are far weaker than the forces between ions or metal atoms.

Finally, we have the covalent crystal wherein shared-pair bonds connect all the atoms of a crystal into one gigantic molecule. In diamond, every carbon atom is linked to all its nearest neighbors by a covalent bond so that the bonding extends unbroken throughout the whole crystal. This bonding gives the crystal great hardness and low volatility. We may also have two or more different elements linked together into a giant molecule that extends throughout the crystal: in quartz the Si and O atoms form such a three-dimensional network.

METALS

Atomic structure. Metals are characterized by well-shielded nuclei; i.e., the core electrons reduce the effective nuclear charge seen by the outer electrons to a relatively low value. Because of this, the outer electrons are loosely held; hence the electron cloud around metal atoms is extensive compared to that around non-metals. A picture of a metal atom would reveal an ion core surrounded by a fuzz of electrons.

A picture of a single neutral iron atom would reveal the nucleus with 26 + charges at the center of a relatively dense, roughly spherical electron cloud consisting of the atomic inner core of 24 electrons $_{2}{}_{2}^{6}{}_{2}^{6\,6}$. This inner core, with a net charge of +2, is surrounded by a far more tenuous spherical charge cloud, consisting of the two remaining electrons in the 4*s* state.

When 10^{23} iron atoms come together to form a crystal, the packing arrangement is largely the result of two kinds of force. First of all, there is the mutual repulsion of like charges. Thus the cores of the atoms, each with a +2 charge, tend to repel each other and would rush apart if they alone were present. The tenuous clouds of 4*s* electrons if present alone would also repel each other. But, because of the attraction of unlike charges, the 4*s* electron clouds tend to hold the cores within them and the cores keep the clouds from dissipating. This balance of force produces two striking results. First, the clouds of the 4*s* electrons from individual atoms merge together into a common cloud of negative electricity spread throughout the crystal in which no particular 4*s* electron can be said to be associated with any particular atom. Then, within this common cloud, the cores move together into an orderly close-packed arrangement much like tennis balls tightly packed in regular rows in a box. The arrangement favored is one in which the almost empty spaces between the ion cores add up to the smallest possible volume; this is called *closest packing*. This arrangement of the atom cores is the origin of the characteristic properties of iron and other metal crystals.

Close packing. Consider the closest possible packing of spheres. Spheres packed in a box would take up a minimum of space in the arrangement shown in Fig. 1. Here, there is a straight row of spheres with adjacent rows staggered, so that those in the second row fit into the depressions between those in the first row. If we add another layer in order to form a three-dimensional crystal, we find that, with closest packing, the added spheres fit into the depressions between groups of three spheres in the first layer.

Looking at the darker center sphere C (Fig. 1) we see that between it and the neighboring spheres, there are six depressions, marked A B A B A B. Suppose that we start to build a second layer by placing a sphere in the depression A_1 (Fig. 2*a*); we cannot place a second sphere in depression B_1 because this is so close to A_1 that the spheres would have to penetrate each other. We have to place our second sphere in A_2, and, for the same reason, our third sphere must go in A_3. If originally we had placed the first sphere in depression B_1, then we would

Fig. 1. Two-dimensional closest packing.

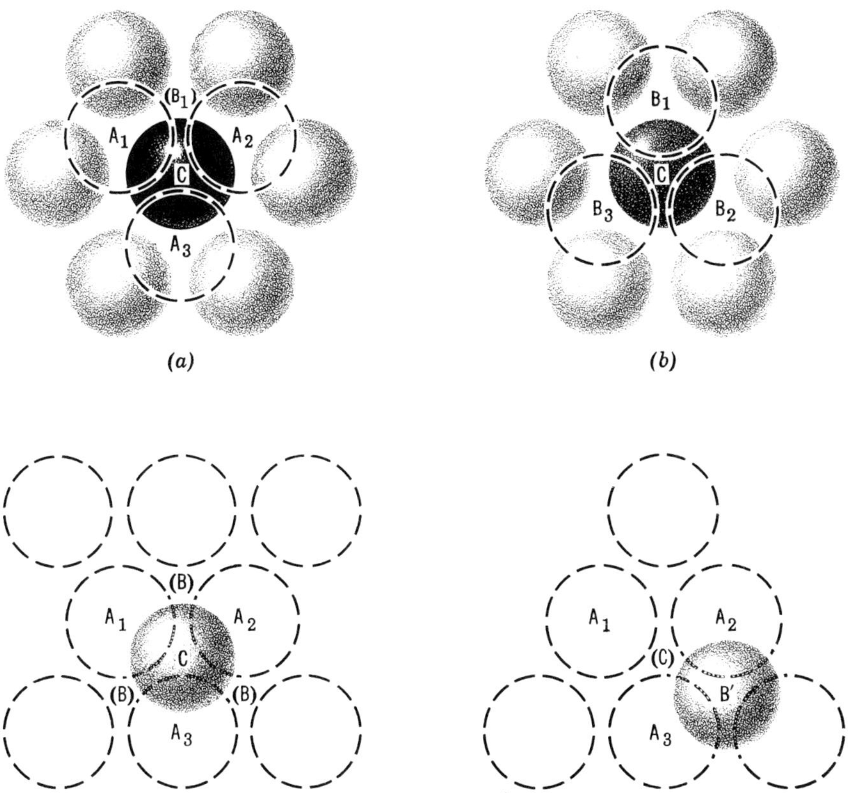

(*a*) (*b*) (*c*) (*d*)

Fig. 2. Closest packing.

have had to place the second in B_2 and the third in B_3 (Fig. 2*b*) Thus we have two choices for this second layer, and one excludes the other.

Now suppose that we pick the A depressions for the second layer and fill them all with spheres. What are the possibilities for the third layer? These are shown in Figs. 2*c* and 2*d*. We can place the first ball in the third layer directly over the C ball in the first layer. This excludes the use of the depression over B. If we follow this A C choice consistently, we get a series of layers A C A C A C, and so on, throughout the entire crystal. Such an arrangement is called *hexagonal* close packing.

Now consider the other choice illustrated in Fig. 2*d*. Here we place the sphere over a B position (labeled B′) instead of over the C position. Then positions over C are ruled out in this layer. Now the first layer has C spheres, the second has A spheres, and the third has B spheres. If we build consistently in this fashion, we get the pattern C A B C A B C A B C A B, and so on. This type of arrangement is known as *cubic* close packing.

Both cubic and hexagonal close packing effect the maximum economy of space. A crystal with *either* of these structures has 74% of the space filled by the spherical atoms, and only 26% of the space is empty.

The perfect crystal. Discussions of crystal structure are usually based on an abstraction called a "perfect crystal." A "perfect crystal" is one in which the structure is perfectly periodic; but real crystals are never perfectly periodic. Two reasons for deviations from this ideal are the finite extent of crystals and the existence of structural defects.

For the perfect crystal, each repeat unit (one or more atoms) has an environment identical with that of every other building unit. Real crystals have a surface since they are finite. Atoms within the crystal have neighbors surrounding them in all directions, but atoms on the surface have no neighbors above them. To avoid considering the distortions from perfection that must occur near the surface, we shall always assume that we are discussing the properties of a region of nearly perfect periodicity well inside the crystal.

Crystals grown in the laboratory are never free of trace impurities that destroy their perfect periodicity. Even if they were, it can be shown that, above absolute zero, we would expect that certain atoms would be out of position. Stacking faults are also to be expected; for example, in the A B C A B C . . . arrangement characteristic of cubic close packing, several layers with the hexagonal A C A C A C arrangement may occur. Thus we see that a perfect crystal, like an ideal gas, is only a model and not a physical reality. The success of the model gives us confidence that under ordinary conditions it is close to the truth.

The unit cell. Consider the hypothetical two-dimensional crystal shown in Fig. 3*a*. Assume that this crystal extends indefinitely in the plane of the paper. To describe such a crystal, we need only describe the repeat unit, a unit from

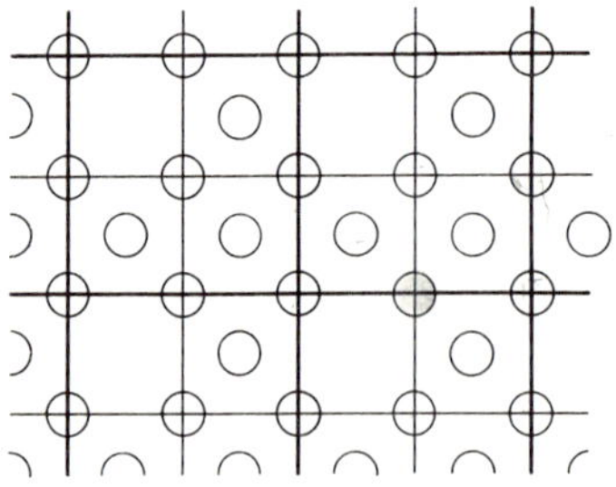

(*a*) Two-dimensional crystal

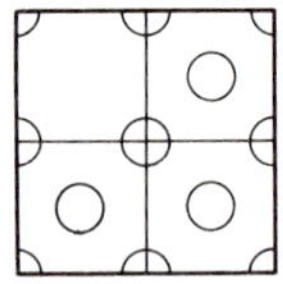

(*b*) Repeat unit (content of one unit cell)

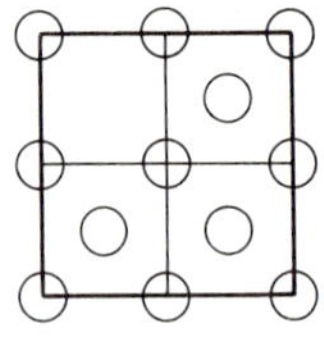

(*c*) Conventional drawing of unit cell

Fig. 3. The unit cell.

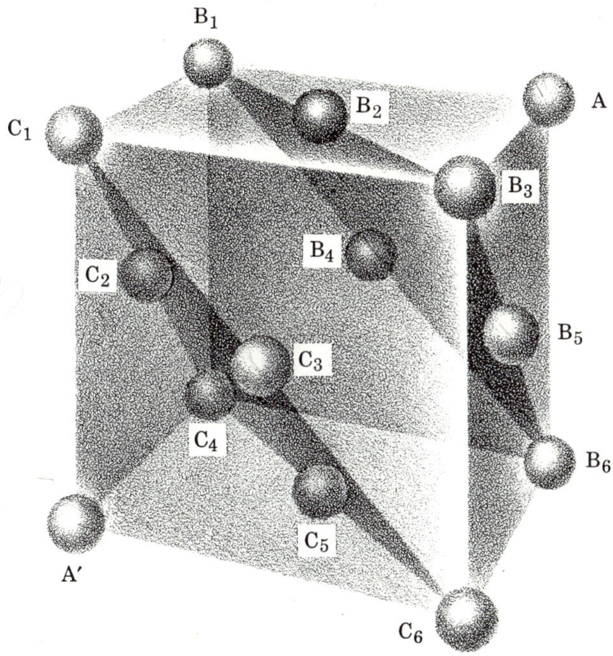

Fig. 4. Face-centered unit cell.

which the whole crystal can be constructed, just as a long chain can be constructed by connecting individual links. Thus a group of repeat units (like those shown in Fig. 3*b*) joined together reproduce the two-dimensional crystal shown in Fig. 3*a*.

If the circles in Fig. 3*b* represent atoms, this picture would horrify chemists; this unit of structure has four quarters of atoms, four halves of atoms, and four whole atoms, a total of seven atoms in all; to draw it, we had to split the atom. Chemists represent the repeat unit by a drawing, as seen in Fig. 3*c*. In this drawing, the square circumscribes the repeat unit. Even though a total of twelve atoms is shown, only four of them are *completely* within the unit cell. The four atoms on the sides of the square are shared between two unit cells, whereas those on the corners are shared between four unit cells; hence the total number of atoms divided by the total number of unit cells, i.e., the average number of atoms per unit cell, is seven, the same as before.

Extension of this description to three dimensions involves no new concepts. Figures 4 and 5 show the unit cells for two common structures. The first, the *face-centered cubic* unit cell, has an atom at each corner of a cube and one in the center of each face. The total number of atoms shown is fourteen. In the *body-centered cubic* unit cell (Fig. 5), the total number of atoms shown is nine; eight are at the corners and one is at the center of the cube.

To produce a "perfect crystal" we join unit cells together so that they fill the whole of space. When we do so with face-centered cubic unit cells, we find that

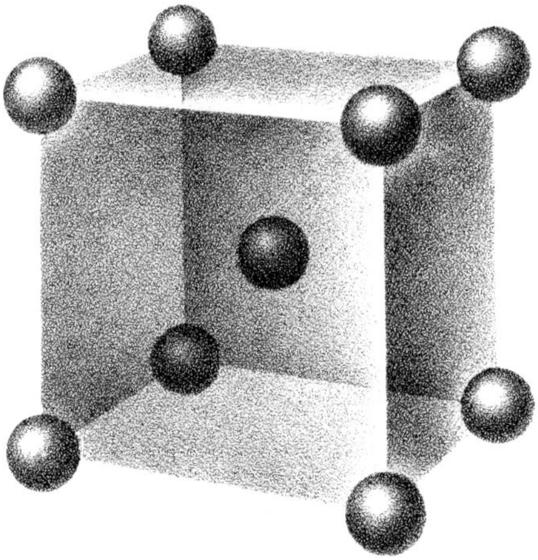

Fig. 5. Body-centered unit cell.

each of the atoms in the faces is shared between two unit cells, so that only half of the six atoms in the faces belong to the pictured unit cell; moreover, we find that the corner atoms are shared between eight unit cells, so that only one-eighth of the corner atoms belong to this unit cell. Thus a total of four atoms belongs entirely to each unit cell.

$$\begin{aligned}\tfrac{1}{2}(C_3 + B_4 + B_2 + B_5 + C_5 + C_2) + \tfrac{1}{8}(A' &+ C_6 + B_6 + C_4 \\ &+ B_1 + A + B_3 + C_1) \\ &= 3 + 1 = 4\end{aligned}$$

In the body-centered cell, each corner atom serves as a corner for eight unit cells; hence each one contributes one-eighth of an atom to the pictured unit cell. The atom at the center, on the other hand, belongs wholly to the unit cell. All together this yields two atoms per unit cell.

We can calculate the fraction of empty space in the unit cell if the atoms pictured in Figs. 4 and 5 are spheres that just touch. For the body-centered unit cell, 32% of the space is empty; for the face-centered unit cell, 26% of the space is empty. The latter value is that found for the close packing in Fig. 2. In Fig. 4, the indicated layers are in the same plane and have this close packing. There are parts of four such layers perpendicular to the diagonal AA′. These layers are such that the A atom is over neither the atoms in layer B nor over layer C, but directly over A′. This is the A B C arrangement shown in Fig. 2*d;* hence face-centered cubic packing is the same as cubic close packing.

Coordination number. The purpose of the unit cell is to indicate the environment of an atom (or molecule) in a crystal. To see this environment we must not only be familiar with the unit cell, but must also visualize the structure resulting when unit cells are joined together. A useful aid in this is the *coordi-*

nation number, the *number* of closest neighbors of the atom in question. As an example, consider the close-packing arrangement in *two dimensions* (Fig. 1). Here the coordination number is 6, since there are six neighbors nearest to any one atom. But, now, consider an atom in either three-dimensional close-packing. Six atoms in the same plane touch it; three atoms from the plane below touch it, and three atoms from the plane above touch it; hence, in all, an atom has twelve nearest neighbors, so the coördination number is 12.

The coordination number (12) determined for close packing is the same as that for the face-centered unit cell. Consider the atom labeled B_2 in Fig. 4. The atoms closest to it are the four located in the same face, plus the four atoms located in the center of the adjoining faces, i.e., those labeled A B_1 B_3 C_1 and B_4 B_5 C_2 C_3. To find the coordination number, we consider not only this unit cell but also the unit cell joining at the plane $AB_1B_3C_1$. When we do, we find that B_2 has four more nearest neighbors in this adjacent unit cell located at the face-centered positions corresponding to C_2, C_3, B_4, and B_5. Thus, for the face-centered cell, the coordination number is again 12.

Verify that the coordination number of the corner atoms in Fig. 4 is also 12 but that of the atoms in Fig. 5 is only 8.

The structure of metals. The common structure of metals is close packing, but the body-centered structure is also found. Table 1 lists the structures of elements in the left-hand corner of the periodic chart. Note that, if we draw a line dividing metallic structures from the others, we have drawn the line dividing metals from non-metals.

The metals that we regard as most metallic, namely, the alkali metals, have the body-centered cubic arrangement rather than the close-packed structure expected for metals. Reasons for this are not obvious, but they are connected with the large size of these atoms. All metals, even the body-centered cubic ones, should be close-packed at low temperatures or high pressures; lithium, below 190°C, does assume a close-packed structure, whereas cesium achieves this structure at high pressures.

Of the thirty transition metals, twenty-three exhibit close packing, five have body-centered structures, and two show neither. The close-packed arrangement is

Table 1 Structure of Metals

Li (b, c)	Be (c)		B (o)		C (d)
Na (b)	Mg (c)		Al (c)		Si (d)
K (b)	Ca (c)	Sc (c)		Ga (o)	Ge (d)
Rb (b)	Sr (c)	Y (c)		In (c)	Sn (d, c)
Cs (b, c)	Ba (b)	La (c)		Tl (c)	Pb (c)

d = diamond; b = body-centered cubic; c = close-packed; o = other.

not always the most stable at ordinary temperatures, nor is it perfect. In other words, at room temperature, iron has the body-centered structure; only at high temperatures does it have the face-centered structure. Moreover, with borderline metals like zinc, the close-packed structure is distorted.

IONIC CRYSTALS

In most ionic crystals, cations have only anions for nearest neighbors and vice versa. Thus the spatial arrangement of ions in ionic crystals is basically different from that in metals. With metals, we are concerned with the packing of spheres of the same size with the same properties; in an ionic crystal, we are still concerned with the packing of spheres, but now the spheres are of different sizes and have different properties.

Atomic structure of cations and anions. In the discussion of the chemical bond, we noted that the processes involved in the formation of an ionic crystal from gaseous atoms are as follows for NaCl:

$$\begin{array}{ll} Na_g \rightarrow Na^+ + e^- & I_z \\ Cl_g + e^- \rightarrow Cl^- & E_a \\ Na^+ + Cl^- \rightarrow NaCl_{xtal} & \text{L.E.} \end{array}$$

The first step involves ionization, which is easiest for metals; the second involves electron capture, which is most favored for non-metals; the third involves the interionic attraction and repulsion of ions packed in a definite lattice, that is, the *lattice energy*. The first two terms are determined by the atomic structures of the parent atoms of the cations and anions. They are most favorable when the atoms differ greatly in electronegativity. The last term determines the arrangement of ions in the crystal that supplies the balance of attractive forces between unlike atoms and repulsive forces between like atoms. The resulting structure is influenced greatly by simple geometric factors such as the size of the ions.

In Table 2, we have listed the ionic radii of a number of cations and compared these to that of the O^{-2} anion. (The indefiniteness of the charge cloud of the cations renders the radius somewhat indefinite. These values are estimated from the spacing in ionic crystals.) Several trends are evident: first, in any family, the size of the ion increases with atomic weight; second, in a given period, the radius decreases with increasing positive charge; third, most cations are smaller than anions. These results are expected on obvious grounds.

Ionic and covalent bonds simply represent two extremes of the chemical bond. Nevertheless, it is convenient to have a basis for classifying crystals. Therefore let us say (with the full realization that the division is artificial) that, if the elec-

Table 2 Properties of Cations

	r Å	$\frac{r_m}{r_{O^{-2}}}$	$x_O - x_m$		r Å	$\frac{r_m}{r_{O^{-2}}}$	$x_O - x_m$		r Å	$\frac{r_m}{r_{O^{-2}}}$	$x_O - x_m$
Li^{+1}	0.60	0.43	2.5	Na^{+1}	0.95	0.68	2.6	K^{+1}	1.33	0.95	2.7
Be^{+2}	0.32	0.23	2.0	Mg^{+2}	0.65	0.47	2.3	Ca^{+2}	0.99	0.71	2.5
B^{+3}	0.20	0.14	1.5	Al^{+3}	0.50	0.36	2.0	Sc^{+3}	0.81	0.58	2.2
C^{+4}	0.12	0.11	1.0	Si^{+4}	0.41	0.29	1.7	Ge^{+4}	0.53	0.38	1.7

In this table r Å stands for the effective radius of the ion, and $r_m/r_{O^{-2}}$ indicates the ratio of this ionic radius to that of the O^{-2} anion. The column labeled $x_O - x_m$ stands for the difference in electronegativity between oxygen and the atom in question.

tronegativity difference is greater than 1.7, the bond is ionic; if it is less than 1.7, the bond is covalent.

Geometric considerations. Geometric requirements of ionic crystals are similar to those of metals in so far as both structures effect a balance of repulsion and attraction. It might seem that the easiest way to obtain this balance is through close packing. Let us see what happens when we pack cations around the anions in two-dimensional close packing (Fig. 1). We can pack six cations around the black anion. But, once we do this, each cation will have two nearest neighbors that are also cations. This is unlikely; hence the close packing found for metals is not found in ionic crystals.

The coordination numbers commonly encountered in ionic crystals are 8, 6, and 4. When a cation has a coordination number of 8, the arrangement (called cubic) is like that in Fig. 5, with the central sphere a cation and the peripheral spheres anions. A coordination number of 6 involves the octahedral configuration shown in Fig. 6, with the anions at the corners of an octahedron and a cation at the center. A coordination number of 4 involves the tetrahedral arrangement in Fig. 8. In all these configurations, the anions are arranged so that the distances between them are the maximum consistent with the bond distance and coordination number to the cation.

A top view of octahedral coordination is shown in Fig. 7*a*. In this drawing, the cation is fairly large so that anions can touch the cation and still maintain their distance from the other anions. As the cation gets smaller, the interanionic distance decreases with a corresponding increase in repulsion. These repulsions become quite large when the anions just touch; hence, in the case shown in Fig. 7*b*, the cation has shrunk to a critical size. If the cation gets still smaller, close approach of anions to cations requires interpenetration of anion charge clouds and an enormous increase in the repulsion energy. Thus the octahedral configuration becomes so unstable when the cation reaches this critical size that the tetrahedral configuration with fewer cation-anion "bonds" but larger anion-anion distances becomes the stable form.

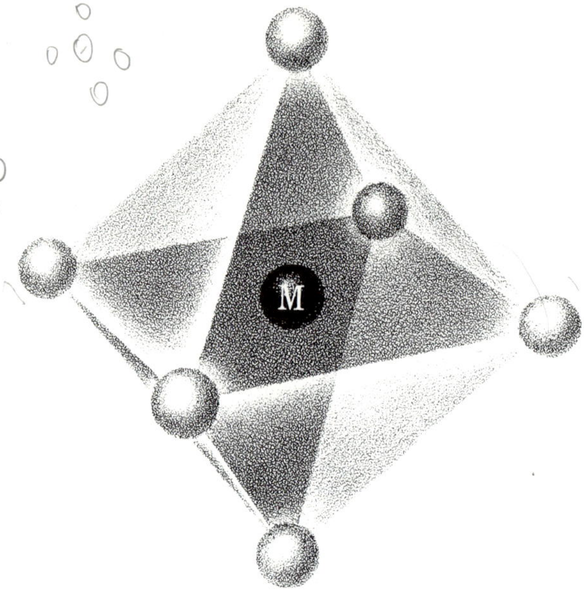

Fig. 6. Octahedral coordination.

A geometric relation must be fulfilled between the radius of the anion, r_a, and the critical radius of the metal ion, r_m. The leg and hypotenuse of the indicated right triangle have the following dimensions:

$$\text{Hypotenuse} = 2r_A + 2r_m$$
$$\text{Leg} = 2r_A$$

Therefore

$$(2r_A)^2 + (2r_A)^2 = (2r_A + 2r_m)^2$$

If we solve for the ratio of r_m to r_A, we find

$$r_m/r_A = 0.414$$

Thus, whenever the ratio of cation to anion radii falls below 0.414, transition from an octahedral to a tetrahedral configuration is likely.

Similar arguments apply to other configurations. If we repeat these arguments, we find that the critical ratio for cubic coordination is 0.732, whereas for the tetrahedral coordination this ratio is 0.225. These critical ratios are

$$\text{cubic (8)} \overset{0.732}{\rightleftarrows} \text{octahedral (6)} \overset{0.414}{\rightleftarrows} \text{tetrahedral (4)} \overset{0.225}{\rightleftarrows} \text{triangular (3)} \overset{0.155}{\rightleftarrows} \text{linear (2)}$$

Now suppose we have a compound that, on the basis of electronegativity, is ionic. If the radius ratio is 0.60, what can we say? It is clear from the above that a coordination number of 6 is likely, although 4 is also possible. On the other hand, a coordination number of 8 is unlikely, for, if we have this coordination number, we have either interpenetration of anion charge clouds or the cation rattling around in the interstice of the cubic arrangement of anions. If the radius ratio were 0.70, could we definitely rule out a coordination number of 8? NO! Ionic radii are not sufficiently precise to differentiate between 0.70 and 0.73.

Ionic crystals and closest packing. It is impossible for both anions *and* cations to have one close-packed structure, but, if one of the ions is much larger than the other, it is common for the larger ions alone to form a close-packed structure and the smaller ions to fit into holes in this structure. Let us see what the holes between two close-packed layers look like. In close packing, the minimum empty space is achieved when one sphere fits into the depression formed by three other spheres close-packed in two dimensions (Fig. 1). This fit is not perfect since the depression is not entirely filled. If a smaller sphere were placed in this remaining empty space, it would lie above the center of the triangle formed by three atoms in the close-packed layer; it would lie below the one atom in the upper close-packed layer. This is the tetrahedral configuration shown in Fig. 8. If we refer to Fig. 9*a*, we see that in two close-packed layers there is such a hole above each atom in the first layer and below each atom in the second layer. In a multilayered, close-packed structure, there is a tetrahedral hole above and below each atom; hence there are twice as many tetrahedral holes as there are close-packed atoms.

In Fig. 9*a*, the dotted lines T_2 and T_1 indicate two rows of tetrahedral holes. In the T_2 line, the base of the tetrahedron is formed by three atoms in the first

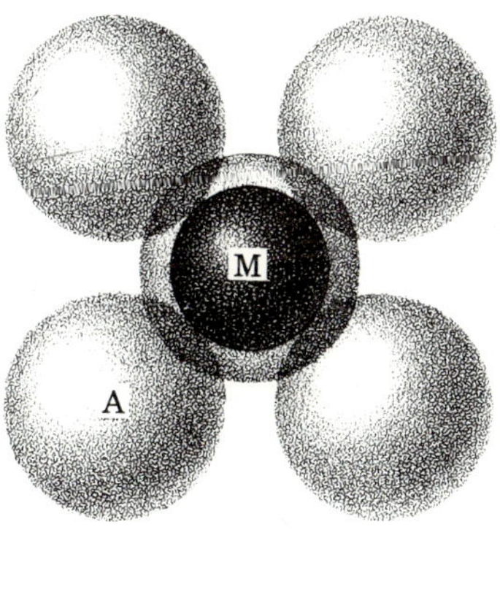

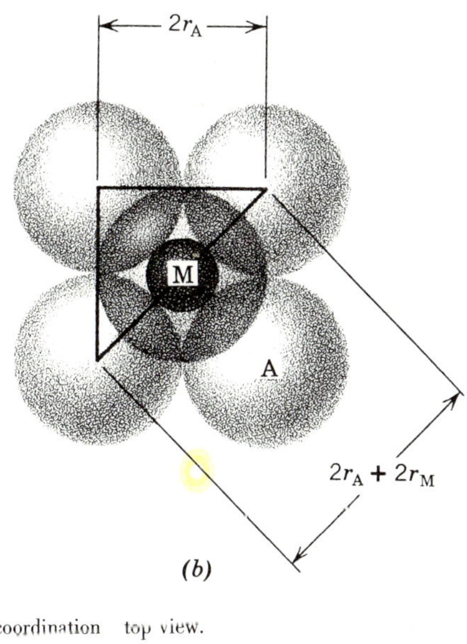

Fig. 7. Octahedral coordination top view.

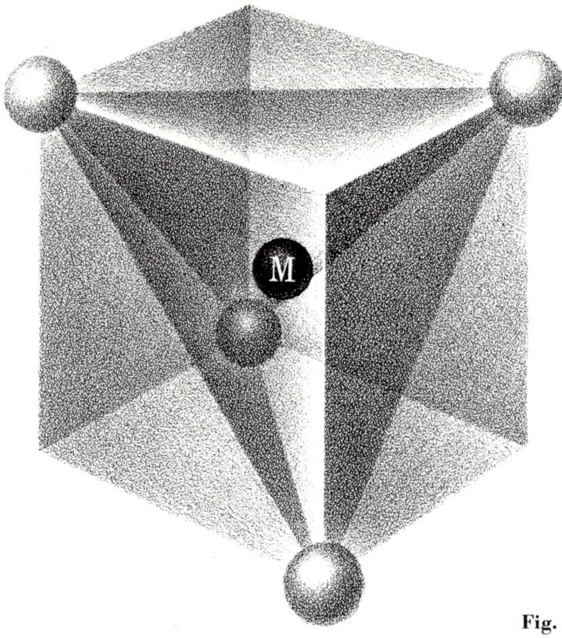

Fig. 8. Tetrahedral coordination.

layer and at the apex is a second-layer atom. In the T_1 line, the base is formed by three atoms in the upper layer and at the apex is a first-layer atom. An edge view (Fig. 9*b*) of the first two rows in both close-packed planes shows the difference more clearly. The T_1 and T_2 holes (below the crosses) are at different heights above the first close-packed layer and staggered with respect to each other.

Besides tetrahedral holes, there are octahedral holes between two close-packed layers. If we refer to Fig. 6, we see that an octahedrally coordinated cation is sandwiched between two groups of anions. Each anion group has three anions at the corners of an equilateral triangle, and the triangle on one side of the cation is inverted with respect to the triangle on the other side. Such an arrangement is shown by the interlayer hole indicated by the arrow in Fig. 9*a*. The three spheres in the lower layer that are closest to the hole form one triangle; the spheres in the upper layer closest to the hole form another triangle, inverted with respect to the lower one. A line of such octahedral holes is indicated by the dotted line O. In a close-packed structure, the number of octahedral holes is equal to the number of atoms present, and the holes are located midway between the close-packed layers. The relation of the close packed layers to their holes is shown in Fig. 9*c*.

The relation of these holes to the cubic closest packing becomes quite clear if we indicate guide lines for eight small cubelets in the face-centered unit cell, as in Fig. 10. (The close-packed layers in this unit cell are shown in Fig. 4.

For simplicity, hereafter we shall refer to the close-packed arrangement as CCP for cubic close-packed, and as HCP for hexagonal close-packed.) The atoms in any one cubelet actually touch. Since these atoms are at alternate corners of the cubelet, they form a tetrahedron (cf. Fig. 8) with a hole in the center. The unit cell contains eight tetrahedral holes, one in each cubelet. Since there are only four atoms in the unit cell, there are two tetrahedral holes per close-packed atom. When a whole crystal is made up from these unit cells, we find that each line

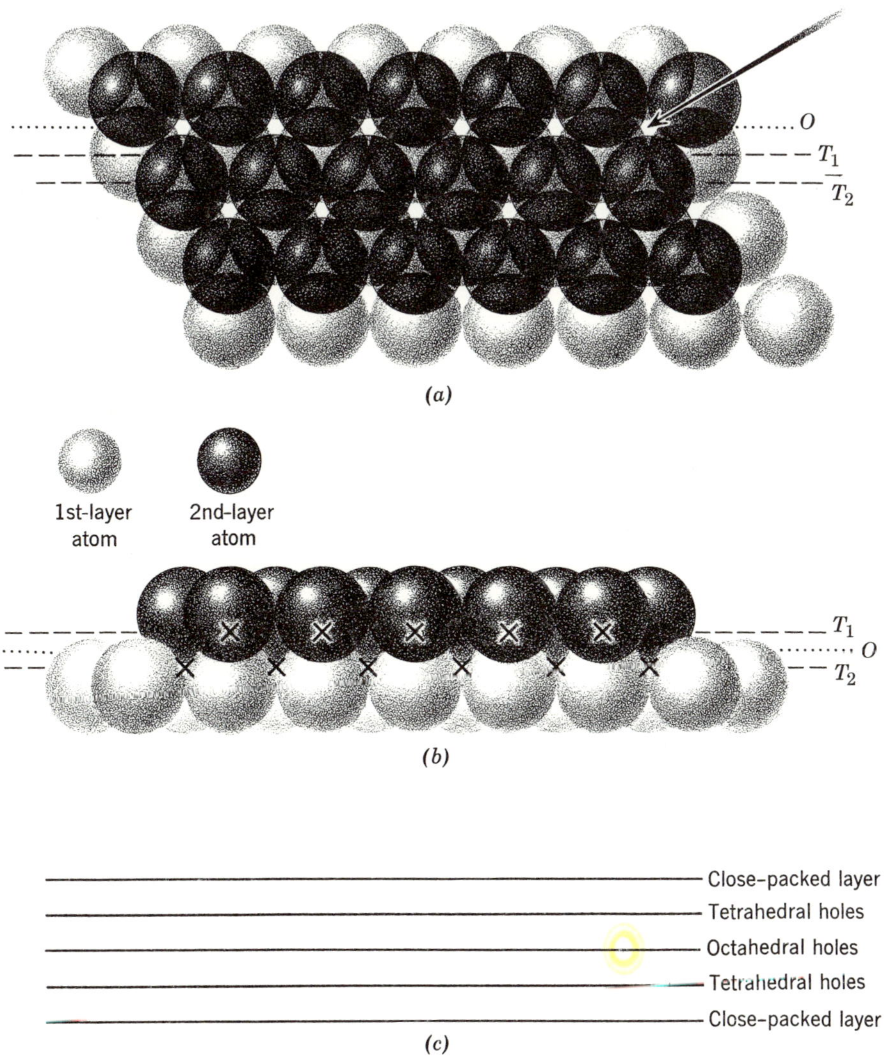

Fig. 9. Holes in close-packed layers.

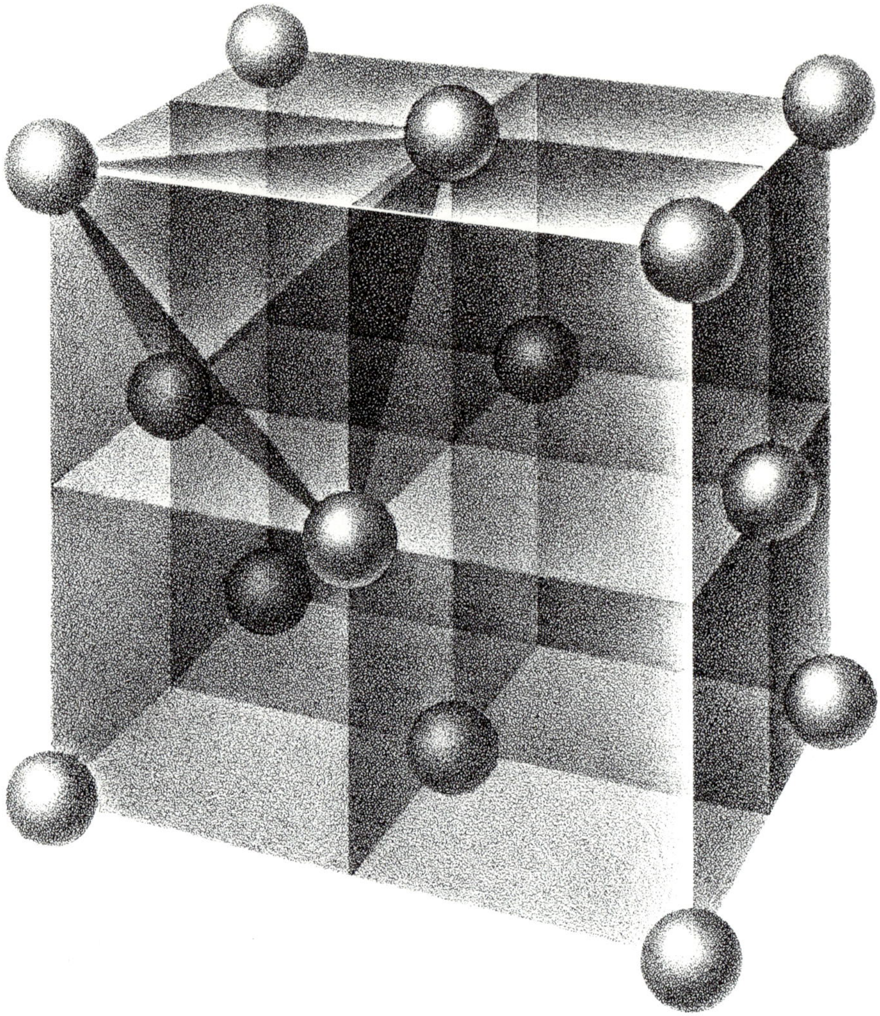

Fig. 10. Octahedral and tetrahedral holes in face-centered cubic unit cell.

intersection in Fig. 10 not occupied by atoms is surrounded by six atoms at the corners of an octahedron. These intersections become octahedral holes when the atoms just touch. There are twelve intersections at edges and one in the center. Since each edge is common to four unit cells, there are only four intersections (holes) that belong completely to one unit cell containing four atoms; hence there is one octahedral hole per close-packed atom.

The structure of common ionic oxides. Unit cells of several common ionic structures are shown in Fig. 11 together with the formula of the compound. For

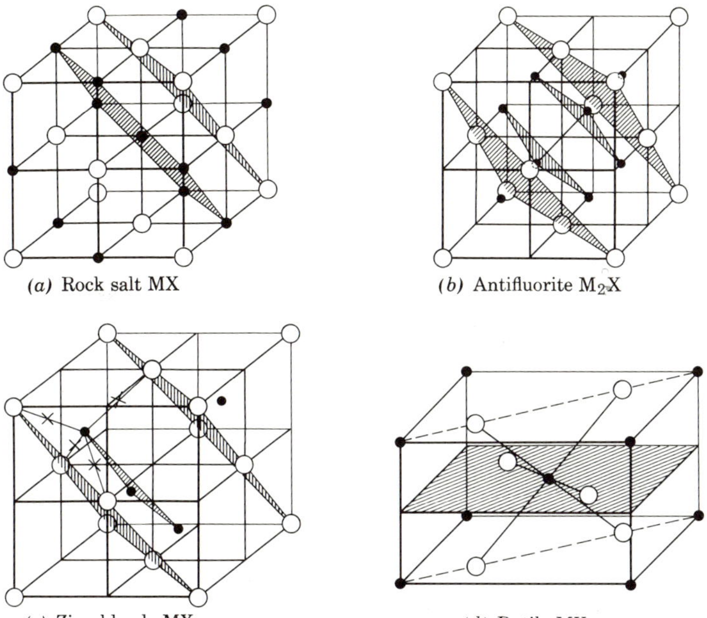

(a) Rock salt MX *(b)* Antifluorite M_2X

(c) Zinc blende MX *(d)* Rutile MX_2

Fig. 11. Common ionic structures.

the structures shown, the black dots indicate cations, the open circles indicate anions. The unit cell shown in Fig. 11*b* occurs both as shown and also with the positions of cations and anions opposite those shown; it is called the *fluorite* unit cell in the latter case.

In Table 3, we have listed the structure, cation coordination number, and radius ratio of common ionic oxides. It can be seen that most of these oxides have one of the structures represented in Fig. 11. The structural factors are best understood if we consider each case in turn.

Table 3 Ionic Oxides

Compound	Radius Ratio	Structure	Coord. No.
Li_2O, Na_2O, K_2O, Rb_2O	0.43, 0.68, 0.95, 1.05	Antifluorite	4
MgO, CaO, SrO, BaO	0.46, 0.71, 0.81, 0.97	Rock salt	6
BeO	0.22	Zinc blende	4
α-Al_2O_3, Ga_2O_3	0.36, 0.45	Corundum	6
α-Fe_2O_3, Cr_2O_3	0.43, 0.46	Corundum	6
SiO_2, GeO_2	0.29, 0.38	Cristobalite	4
GeO_2, SnO_2, PbO_2	0.38, 0.51, 0.60	Rutile	6
ThO_2, UO_2	0.68, 0.64	Fluorite	8

The rock salt structure is found for many binary ionic compounds with the formula MX. As is often true of ionic salts, the anions alone have a CCP structure. All the octahedral holes in the CCP layer of anions are filled with cations. When this occurs, the cations by themselves form a face-centered cubic (FCC) unit cell; hence the rock salt unit cell contains alternate layers of anions and cations with each type of ion in a FCC arrangement. (If the unit cell in Fig. 11*a* is split down the middle, you will see that each half is half of a unit cell containing the *cations* in FCC positions.)

It can be shown that a compound with the rock salt structure has the general formula MX, either through the concept of coordination number or through the concept of a unit cell. Consider the anion in the upper left rear corner of Fig. 11*a*; it forms a corner for eight unit cells. The closest cations will be the three shown in this drawing, plus three from the other unit cells. Thus the anion coordination number is 6. That of the cations is also 6. (Repeat this argument for an edge cation.) Therefore only one-sixth of a given cation belongs to an anion neighbor and only one-sixth of the anion neighbor belongs to this cation; hence the formula is $M_{1/6}X_{1/6}$ or, in terms of whole numbers, MX. The same conclusion is reached if we consider the number of cations and anions within the unit cell. With the FCC arrangement; there are four anions in the unit cell. The central cation is wholly within the cell; each of the twelve edge cations is one-fourth within the unit cell; hence four cations are also included. Thus the formula is M_4X_4 or, simply, MX.

According to the radius ratio rules, the octahedral coordination is stable down to a ratio of 0.414. All the alkaline earth oxides except BeO (Table 2) have a radius ratio that permits a coordination number of 6. In agreement with this, all but BeO have the rock salt structure. Note that, even though strontium oxide and barium oxide could have a coordination number of 8, 6 is actually observed. Thus the ratio rules indicate the maximum coordination number possible; the coordination number can be lower but it cannot be higher.

All the alkali halides (except CsCl, CsBr, and CsI) and ammonium halides have the rock salt structure. It is also found for the halides of silver (except the iodide) and the sulfides of the alkaline earth elements.

In the antifluorite unit cell (Fig. 11*b*), the anions are again at FCC positions. All the tetrahedral holes at the center of each cubelet are occupied by cations. In this structure, therefore, the CCP layers of anions are separated by two layers of tetrahedrally coordinated cations (cf. Figs. 9 and 11*b*). The coordination number of the cation is 4 for tetrahedral coordination. The coordination number of the anion is 8. (This can be seen if we visualize the environment of the anion in the top face when another unit cell is joined to this face. There are four cations in the four cubelets below it and four cations in the four cubelets above it.) One-fourth of each cation belongs to a neighboring anion; one-eighth of the neighboring anion belongs to this cation; the formula is $M_{1/4}X_{1/8}$, or M_2X. You can also arrive at the formula M_2X on the basis of the number of ions in the unit cell.

The preceding discussion also applies for the fluorite structure except that the roles of the anion and of the cation are interchanged. Thus the fluorite structure represents compounds with the formula MX_2, in which the cation coordination number is 8 and that of the anion is 4.

The antifluorite structure with a tetrahedrally coordinated cation is possible whenever the formula of the compound is M_2X and the radius ratio is greater than 0.225. This is true for the alkali metal oxides indicated in Table 3. Note, once again, that the observed radius ratio for these compounds would allow higher coordination numbers, but these are not observed. This structure is found not only for the oxides of the alkali metals but also for their sulfides, selenides, and tellurides.

The fluorite structure with cubically coordinated cations is not expected for MX_2 compounds unless the cation is large enough that the radius ratio is above 0.73. This will be true with oxides only if the $+4$ cations stem from heavier atoms in the lower portion of the periodic chart. The compounds UO_2 and ThO_2 (Table 3) nearly fulfill these requirements and they have the fluorite structure.

The fluoride ion is only slightly smaller than the oxide ion; hence the radius ratios for alkaline earth fluorides are roughly the same as those for the alkaline earth oxides (Table 2). Accordingly, it is no surprise to find that CaF_2, SrF_2, and BaF_2 have the fluorite structure.

In the zinc blende structure, the anions are in FCC positions and half of the tetrahedral holes (cubelet centers) are occupied. Thus the CCP anion layers are separated by only a single layer of tetrahedrally coordinated cations. Since there are four cations and four anions in a unit cell, the formula of such a compound is MX. The coordination of the cation is 4; the coordination of the anion is 4, half of that in the antifluorite structure.

The rock salt structure is possible only for MX compounds with a radius ratio greater than 0.414. When the ratio falls below this, as it does for BeO, the cation coordination drops from 6 to 4, and the zinc blende structure is found (Table 3).

The prototype of the zinc blende structure is ZnS. It might be expected that the structure of ZnO would be similar to that of zinc blende. Actually, ZnO and ZnS form different structures that are closely related. In zinc blende, half the tetrahedral holes in the CCP anion layers are filled. In the zinc oxide structure (wurtzite), half the tetrahedral holes in an HCP layer of anions are filled.

The zinc blende structure is also related to the diamond structure. In diamond, both the anion and the cation positions are occupied by carbon atoms.

The structure of β-cristobalite, a form of SiO_2, represents still another variation on the zinc blende structure. If both cation and anion positions in Fig. 11*c* are replaced by silicon atoms and oxygen atoms are placed between close pairs as suggested by the crosses, this is the cristobalite structure. The coordination number of the cations is 4; that of the anions is 2. Structures with the formula MX_2 and radius ratios less than 0.414 would be stable with this structure rather than the fluorite or rutile structure. This is true of SiO_2 and GeO_2.

Rutile differs from the other structures in Fig. 11 in that the anions are not obviously arranged in a CCP structure. This structure has two cations wholly within the unit cell; one of these is in the center and one-eighth is contributed by each of the eight cations at the corners. There are four anions within the unit cell; two are completely inside, and half of the four atoms in the upper and lower faces belong to this cell. Thus the formula is M_2O_4, or MO_2. The coordination number of the central cation is 6. That of the anions is hardly obvious, but it must be 3 in order to give the correct formula.

The rutile structure would be possible for any compound with the formula MX_2 for which the radius ratio is greater than 0.414. We have already seen that compounds with this formula and a ratio greater than 0.73 crystallize in a fluorite structure. Hence we might expect that those compounds in Table 3 with a rutile structure would have ratios between these two limits; and they do, except for GeO_2. GeO_2 is at the borderline between the rutile and cristobalite structures; both structures are known.

We have already noted that CaF_2 forms the fluorite structure; it should not surprise us to find that $CaCl_2$ and $CaBr_2$ have the rutile structure. Can you see why?

The corundum structure, which is characteristic of the sesquioxides of the Group III and transition elements, is related to the rock salt structure. There is, once again, a close packing of the anions, but here the close packing is hexagonal rather than cubic. Not all the octahedral holes are occupied; only two-thirds of them contain cations. Thus, instead of the formula MX, corundum-type compounds have the formula $M_{2/3}X$ or M_2X_3.

Many oxides have a slight excess of oxygen over the stoichiometric amount. When this occurs, some cation sites are vacant, i.e., are *holes*. If the excess is variable and the empty sites are randomly distributed, the structure is termed defective; if the excess and empty sites are fixed, it can still be geometrically a perfect crystal.

MOLECULAR CRYSTALS

Molecular versus ionic crystals. In the model ionic crystal, there is complete transfer of electrons from the cations to the anions. This ideal is *approached* in some real crystals. For example, experimental data on NaF shows that the electron cloud density is almost zero midway between the ions. As the electronegativity of the metal increases, however, it becomes increasingly difficult to remove completely electrons from the cation and place them on the anion. The more electronegative elements retain their electrons, and, instead of forming an ionic bond, they form a shared-pair bond. The directional properties of a shared-pair bond usually favor a structure with discrete molecular units.

Suppose we perform an imaginary experiment with the compound MX in which we increase the electronegativity of M. What happens? In Fig. 12*a*, a two-dimensional ionic crystal is shown; shading indicates the electron density. In this ideal ionic crystal, the electron density falls to zero between the ions. As the electronegativity of M increases, it becomes more difficult for complete transfer of electrons to occur, and there is a tendency to form a covalent bond. This can be accomplished by the slight motion of ions shown in Fig. 12*b*. Here *one* of the neighboring anions is much closer to a cation than the others; in this sense, we are justified in saying that there is an MX molecule in the crystal. Nevertheless, in addition to that anion forming a molecule with the cation, there are other anions close enough to affect the structure. Thus the molecular structure in a solid need not be identical with that found in a gas. Usually, however, gaseous molecules are similar to those found in a crystal.

Consider the oxides formed by elements in the second row of the periodic chart. Li_2O crystallizes in an antifluorite structure wherein there is no evidence of the formation of discrete molecules. Beryllium oxide crystallizes in the zinc blende structure. In the resulting BeO_4 tetrahedra, we find that one oxygen atom is closer than the other three. We could, therefore, maintain that a BeO group is distinguishable, but, since the difference is very small, BeO is still mostly ionic. B_2O_3 crystallizes to form BO_4 tetrahedra that share corners. One oxygen is significantly closer to the central metal atom than the other three, and we can regard the inequality of B-O distances as evidence of incipient covalent bond formation. Carbon is even more electronegative than boron. The CO_2 crystal consists of an array of distinct CO_2 molecules. As expected, molecular structures are found for oxides of elements more electronegative than carbon.

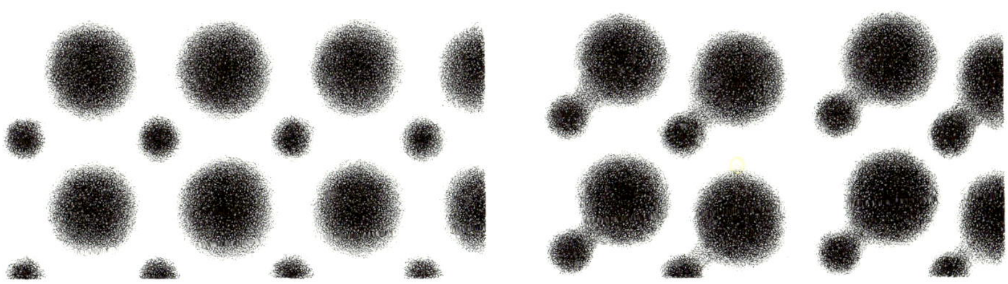

(*a*) (*b*)

Fig. 12. Transition from ionic to molecular crystal.

If we contrast the behavior of the second period elements to that of the third period elements, we find that the differences are in line with the variation in electronegativity. In the second period, evidence for departure from ionic bonding is found in all the oxides except lithium oxide. In the third period, both sodium *and* magnesium oxides are completely ionic. Recognizable departures from ideal ionic behavior appear whenever the electronegativity difference is less than 2.

Oxides of non-metals. Nitrogen forms compounds with the following formulas: N_2O, NO, N_2O_3, NO_2, N_2O_4, and N_2O_5. Two of these compounds (NO and NO_2) are unusual in that they contain an odd number of electrons. The stability of NO can be explained by the molecular orbital theory (p. 138). NO has the molecular orbital structure of an oxygen molecule in which one proton has been removed from one of the nucleii and one electron has been removed from the molecule. O_2 has five bonding and three antibonding orbitals filled and hence forms 2 shared-pair bonds. Removal of an electron from an antibonding orbital would change the bond from a double bond to a $2\frac{1}{2}$ bond; hence NO should be even more stable than O_2. It is.

Valence-bond structures of the gaseous nitrogen oxides are given below. All

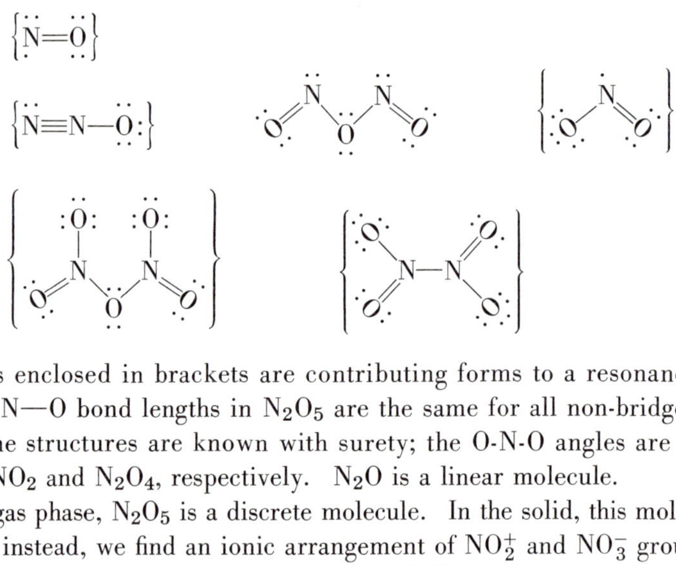

structures enclosed in brackets are contributing forms to a resonance hybrid. Thus the N—O bond lengths in N_2O_5 are the same for all non-bridge oxygens. Not all the structures are known with surety; the O-N-O angles are 133° and 108° in NO_2 and N_2O_4, respectively. N_2O is a linear molecule.

In the gas phase, N_2O_5 is a discrete molecule. In the solid, this molecule does not exist; instead, we find an ionic arrangement of NO_2^+ and NO_3^- groups. Why this should be is certainly not clear, but such differences in structure for gaseous and solid states are observed with other compounds.

There are two principal oxides of phosphorus, P_4O_6 and P_4O_{10}. Their structures and that of elementary phosphorus vapor are shown in Fig. 13. In all three, the phosphorus atoms are at the corners of a tetrahedron.

Solid P_4O_{10} forms a molecular crystal in which P_4O_{10} groups are in a body-centered array. This solid is a meta-stable form that is readily attacked by water. If P_4O_{10} is heated to 500°C for several days, it slowly changes into another crys-

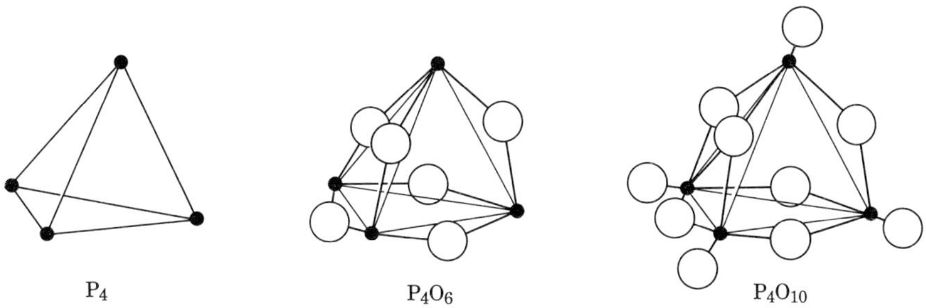

Fig. 13. Structure of P_4, P_4O_6, and P_4O_{10}.

talline form in which discrete P_4O_{10} molecules are no longer recognizable. This form is only slowly attacked by water. It consists of linked PO_4 tetrahedra with the oxygen atoms approximately close-packed.

Other members of the Group V elements are arsenic, antimony, and bismuth. They exhibit a bewildering variety of oxide structures. Bismuth, the most metallic of these, forms an oxide that is more ionic than molecular.

Two important oxides of sulfur are SO_2 and SO_3. Gaseous SO_3 has a planar structure with the sulfur at the center and the oxygens at the corners of an equilateral triangle; gaseous SO_2 has the same arrangement with a missing oxygen atom. The structure of solid SO_2 is that of a molecular crystal, but SO_3 occurs in two varieties. One of them contains closed rings of three SO_4 tetrahedra. The other form is fibrous, like asbestos, and consists of chains of SO_4 tetrahedra linked through two of their corners as shown below.

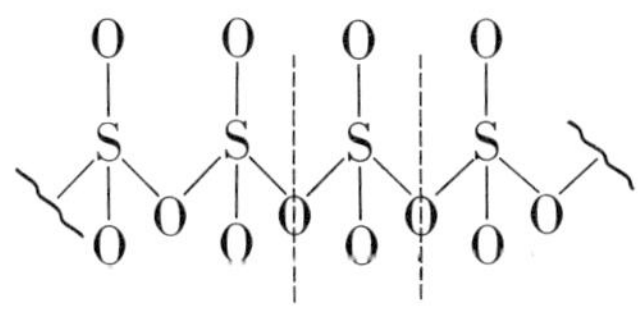

For the oxides of the halogens, only the structures of F_2O, Cl_2O, and ClO_2 are well established. They are

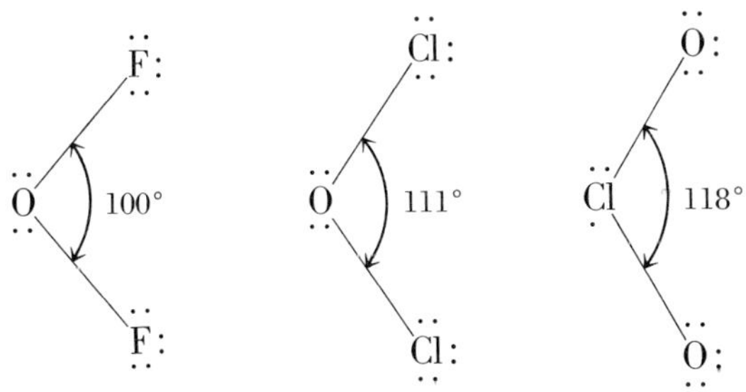

ClO_2 is another example of an odd-electron molecule.

Many other oxides of the halogens exist. For example, the known oxides of chlorine are Cl_2O, ClO_2, ClO_3, and Cl_2O_6; those of iodine are: I_2O_5, I_2O_4, and I_4O_9.

OXYANIONS

So far we have discussed only the neutral compounds of non-metals and oxygen. There are, however, a host of complex, negatively charged oxyanions which stem from these non-metal oxides. For example, SO_3 reacts with water as follows:

$$SO_3 + H_2O \rightarrow 2H^+ + SO_4^{-2}$$

The ion, SO_4^{-2}, is called the sulfate ion. Many non-metallic oxides undergo such reactions and thereby form oxyanions.

Isoelectronic molecules. The molecular orbital treatment provides a convenient basis for the discussion of oxyanions. Diatomic molecules have a set of molecular orbitals analogous to atomic orbitals. Since all diatomic molecules have the same set of molecular orbitals, the fact that both CO and N_2 have fourteen electrons, i.e., are *isoelectronic,* leads us to believe they would have similar electronic structures. This is borne out by experiment. N_2 and CO boil at about the same temperature, and the bond energies of both are in excess of 200 kcal. This principle that isoelectronic molecules are structurally similar works not only for diatomic molecules but also for polyatomic molecules. *Whenever molecules contain the same number of electrons (and the same number of nuclei) we expect them to have similar structures.*

A series of isoelectronic molecules would be: NO_3^-, CO_3^{-2}, BO_3^{-3}, and BF_3. On the basis of the preceding argument, all of these molecules should be similar in structure, and, indeed, in each molecule the oxygen (or fluorine) atoms are located at the corners of an equilateral triangle with the other atom in the center.

Another isoelectronic series is ClO_4^-, SO_4^{-2}, PO_4^{-3}, SiO_4^{-4}, SiF_4. This series, with the general formula MX_4, has structures such that M is at the center of a tetrahedron with the four X atoms at the corners.

The tetrahedral molecules formed by non-metals in the third row of the periodic chart have unusual features. The M—O distance is shorter than one would expect for a single covalent bond. Part of the reason for this can be ascribed to ionic resonance, but this is not the whole story. All the centrally located elements have 3*s* and 3*p* subshells occupied when the completed octet is formed. In addition, each has an unfilled 3*d* level. If the oxygen atoms surrounding these levels have non-bonded electrons, they can contribute to the 3*d* shell and, thereby, strengthen the bond. This could lead to multiple bonds to the central atom instead of the single bonds associated with octet binding. Even a partial contribution by these $3d$ orbitals would account for shortening of the M—O bond.

There is one more small group of isoelectronic structures to consider, i.e., SO_3^{-2} and ClO_3^-. These molecules are tetrahedral with three oxygen atoms and a lone pair of electrons at the corners and the chlorine or sulfur atom in the center.

It seems reasonable to attempt to extend this isoelectronic principle to solids. Diamond has a zinc blende structure (Fig. 11*c*) with carbon atoms replacing both anions and cations. Beryllium oxide has this structure, and the average number of electrons per atom is the same as for carbon. Boron nitride, BN, is also isoelectronic with carbon and can crystallize in the same form. If it is valid to speak of an isoelectronic series for solids, we would expect these three to be similar in properties. Beryllium oxide melts at about 2600°C, whereas diamond and boron nitride melt at about 3600. On Mohs' scale of hardness,† diamond is the hardest and is called 10; quartz, which finds many uses as an abrasive, is called 7; soft materials, such as talc and copper, are between 1 and 4. The hardness of boron nitride is close to 10, that of beryllium oxide is 9; hence isoelectronic solids appear to be similar in properties provided the structure is the same.

CONDENSED POLYANIONS

The solid oxides of many non-metals, e.g., B, Si, P, and S, have structures wherein the non-metal atom is located at the center of a tetrahedron defined by corner oxygen atoms. All these elements form complex polyanions in which the structural unit consists of tetrahedral MO_4 groups or triangular BO_3 groups linked through their corners. The subject is complicated, but in recent years the chemistry of polyanions has been studied extensively. By far the most has been done with the silicates. We shall discuss these compounds in some detail in order to show the rich variety of structures in condensed polyanion systems.

Silicates. In SiO_2 each oxygen atom has two nearest neighbors, but in the precursor of the silicates, SiO_4^{-4}, each oxygen has only one neighbor. We call those oxygens that have only one neighbor *active* oxygens, and those that have two neighbors, *inactive* oxygens.

All silicates can be viewed as the result of the combination of molecules of H_4SiO_4 with the elimination of H_2O, viz:

$$\begin{array}{c} \mathrm{OH} \\ | \\ \mathrm{HO{-}Si{-}O{-}H} \\ | \\ \mathrm{OH} \end{array} + \begin{array}{c} \mathrm{OH} \\ | \\ \mathrm{HO{-}Si{-}OH} \\ | \\ \mathrm{OH} \end{array} \rightarrow \begin{array}{c} \mathrm{OH}\quad\mathrm{OH} \\ |\qquad| \\ \mathrm{HO{-}Si{-}O{-}Si{-}OH} \\ |\qquad| \\ \mathrm{OH}\quad\mathrm{OH} \end{array} + H_2O$$

Both the reactions and the acids are mythical, but the anions corresponding to the

† Mohs' complete scale is: talc, 1; gypsum, 2; calcite, 3; fluorite, 4; apatite, 5; orthoclase, 6; quartz, 7; topaz, 8; corundum, 9; diamond, 10. A substance with a hardness of 6 will scratch any substance below it on the scale and be scratched by anything above it.

acids are real. In this simplest case, two tetrahedra share a corner to yield the $H_6Si_2O_7$ or, when protons are removed, the $Si_2O_7^{-6}$ anion. This anion has six active and one inactive oxygen, and compounds containing it are called pyrosilicates. Similar molecules are found for phosphorus, sulfur, and chlorine.

Silicates containing more than one silicon atom can also be formed, e.g.,

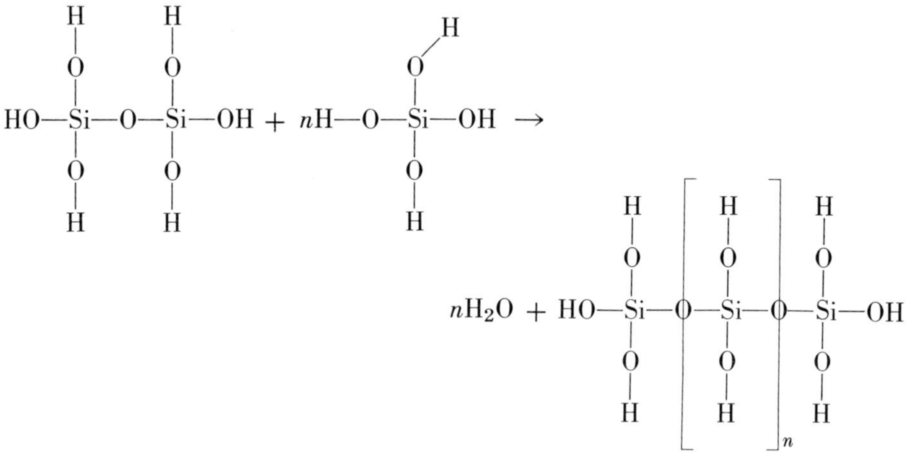

where n is very large. The resulting anion has two active and two inactive oxygens. Since the chain is very long we neglect the ends. Straight chains can be formed of indefinite length, or chain ends can join to form cyclic structures. Examples of both are shown in Fig. 14. Cyclic structures are known with three, four, or six tetrahedra in a ring. In none of these cases is it proper to speak of an SiO_3^{-2} ion, even though the general formula of the cyclic (ring) anion or the chain anion (called pyroxenes) is $(SiO_3^{-2})_n$.

A complex polyanion is formed by joining two pyroxene groups side-by-side. This structure is pictured in Fig. 15. The silicon atoms that make up this

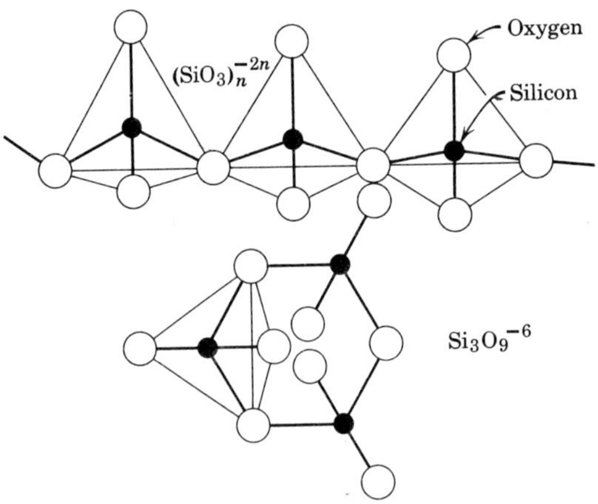

Fig. 14. $(SiO_3)_n^{-2n}$ polyanions.

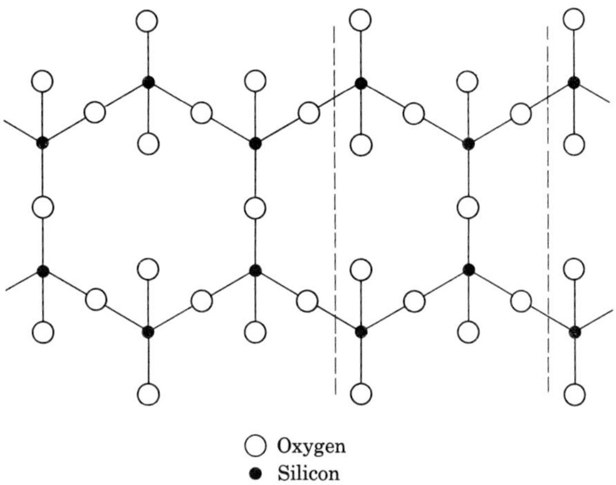

Fig. 15. The amphibole chain.

amphibole chain are of two types; one is bonded to two active oxygens, and the other is bonded to one active oxygen. The repeat unit, indicated by the dotted lines, has the formula ($Si_4O_{11}^{-6}$). Most of the commercial asbestos is an amphibole and made up of such chains. To break a long chain, it is necessary to break a chemical bond; hence individual chains are difficult to break. On the other hand, adjacent chains are held together by weak forces and it is easy to separate them. These characteristics give rise to the fibrous nature of the solid.

Sharing of three corners of the tetrahedra leads to the sheets pictured in Fig. 16. *The active oxygen above each silicon atom is not shown,* and the drawing represents only part of an infinite sheet. The general formula for this anion is

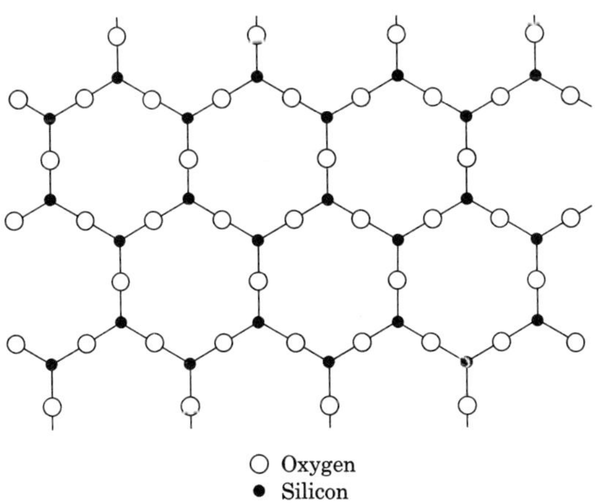

Fig. 16. $(Si_2O_5)_n^{-2n}$ polyanions.

$(Si_2O_5^{-2})_n$; it is found in many of the mica and clay minerals. The ready cleavage of mica into sheets occurs because the forces holding the crystal structure together are strong within the layer, but weak between layers.

In the cristobalite structure of SiO_2 all four corners of the SiO_4 tetrahedra are shared. An SiO_4 group, however, can be replaced by other groups. For example, we can replace a silicon atom by atoms that have nearly the same size, e.g., aluminum. If we replace a silicon atom by an aluminum atom, we must also add a cation to preserve electrical neutrality. In *orthoclase,* $KAlSi_3O_8$, charge neutrality is provided by the potassium ion; in *anorthite,* $CaAl_2Si_2O_8$, charge neutrality is provided by the calcium ion. Both of these are *feldspar* minerals. Such structures are termed "framework structures" because a few positively charged ions are now embedded in the framework of a giant anion.

Another example of a framework structure is $NaAlSi_2O_6 \cdot H_2O$, a member of the zeolite family. In this compound, *analcite,* the aluminum and silicon atoms are distributed throughout an SiO_2 type of network with one-third of the silicon positions occupied by aluminum atoms. Near each aluminum atom there is a sodium ion bound to the network by electrostatic forces.

Glasses. SiO_2 also forms an unordered solid which is called a *glass* and described as a supercooled liquid. This glass forms because the structure is not completely broken down in the molten state as it is for a purely ionic crystal. The liquid state of SiO_2 contains many tetrahedra, still linked through corners. In the liquid, the Si—O—Si bond is not restricted to any particular angle, nor does the network of tetrahedra extend indefinitely. When the liquid is cooled, the unlinked corners join to each other in a random fashion. To form crystals, these tetrahedra must rearrange into a regular pattern. But to do so requires the rupture and reformation of covalent bonds. This is a *very* slow process; ordinarily, molten SiO_2 forms an amorphous solid, a glass. The other oxides mentioned at the beginning of this section also form glasses.

CRYSTAL ENERGY

In ionic crystals, positive and negative ions arrange in space to yield the minimum crystal energy. The governing factor in the structure of covalent crystals and metallic crystals is also this minimum of energy. The details that lead to this minimum are different for these three extremes, but all types of crystals are similar in so far as the atoms or groups of atoms are fixed at certain positions in the crystal. The principal atomic motions in a crystal are vibrations in a cage formed by the neighbors.

The potential energy and the kinetic energy of an atom in a crystal change during vibration. At the extremes of its excursion about the equilibrium position, the atom stops and reverses its motion; since there is no motion at these extremes, kinetic energy is zero, but the potential energy is its greatest. As the atom moves through the equilibrium position, its speed is greatest; hence the kinetic energy is now a maximum and the potential energy is minimum. The total energy of a crystal includes both kinetic and potential energy; since the former depends on velocity and the latter depends on position, we can say that the energy of an atom in a crystal depends on its velocity and position.

Heat capacity of crystals. The energy of one mole of an ideal gas is related to temperature as follows:

$$E = \tfrac{3}{2}RT$$

The energy, which is wholly kinetic, is divided equally among the three variables that determine the instantaneous energy, i.e., velocity in the x, y, and z directions. Thus we associate with each component of velocity an amount of energy equal to $\tfrac{1}{2}RT$.

Heat capacity is defined as the change in energy per unit change in temperature for one mole of a substance; that is,

$$C_V \equiv \frac{\Delta E}{\Delta T} = \tfrac{3}{2}R$$

C_V is called the heat capacity at constant volume. (We are not taking into account any work done by the expansion of the gas as we increase the temperature.) Substituting for R, we find that $\tfrac{3}{2}R$ corresponds to a heat capacity of 3 cal/°C mole or 1 cal/°C mole for each variable needed to define the energy.

Equipartition of energy also takes place in a crystal at ordinary temperatures. In a crystal, however, the energy depends not only on the three components of velocity, but also on the three coordinates; hence we might expect the heat capacity of an ideal monatomic crystal to be twice that of an ideal monatomic gas, or 6 cal/°C gm at. This is the law of Dulong and Petit. If we have a complex structure in which there is more than one atom, e.g., CaS, each of the atoms present contributes about 6 cal/gm at. This means that for CaS the heat capacity is 12 cal/°C gm mole, whereas for $FeBr_3$ the heat capacity is 24 cal/°C gm mole.

Because of the quantization of energy first explained by Einstein, lighter elements do not contribute their full share until they reach temperatures far above room temperature; hence it is best to use empirical contributions for the lighter elements. Values for such contributions are listed in Table 4. Elements not listed contribute 6 cal/°C gm at.

The difference between the atomic environment in liquids and in solids is not so different as their appearance would indicate. Therefore we expect a rule like the one above to describe not only the heat capacity of solids but also the heat capacity of liquids. For liquids, the contribution per atom to the heat capacity

Table 4 Kopp's Rule Values

Atom	Contribution in Solid, cal/gm at. °C	Contribution in Liquid, cal/gm at. °C
H	2.3	5.0
B	2.7	4.8
C	2.0	3.0
N	3.0	4.5
O	4.0	6.0
F	5.0	7.0
All others	6.0	8.0

is about 8 cal/°C gm at. Once again, lighter elements do not contribute as much to the heat capacity at ordinary temperatures and correction factors are used. Values for liquids are included in Table 4.

Example Problem 1: Calculate the heat capacity of 1 mole of solid H_2SO_4 and 1 mole of liquid H_2SO_4.

The numbers of gram atoms of the various elements and their contributions according to Table 4 are listed below.

	Liquid	Solid
2H	10.0	4.6
1S	8.0	6.0
4O	24.0	16.0
	42.0 cal/mole °C	26.6 cal/mole °C

Example Problem 2: How much heat is required to heat 10 gm of liquid H_2SO_4 from 20° to 37°C?

Units can be used to advantage as follows:

$$\frac{42.0 \text{ cal}}{°\text{C mole } H_2SO_4} \times 17°\text{C} \times \frac{1 \text{ mole } H_2SO_4}{98 \text{ gm } H_2SO_4} \times 10 \text{ gm } H_2SO_4 = 73 \text{ cal}$$

Example Problem 3: If 10 gm of solid AgCl at 50°C is dropped into 10 gm of water at 20°C, what is the final temperature?

When the two are mixed together, the temperature of the solid drops and that of the water rises until they reach a common temperature t_c. The heat lost by the solid equals that gained by the liquid. We can calculate the heat capacity of the solid from Kopp's rule, and that of water is 1 cal/gm °C; hence we set up the equations

$$\begin{aligned} &\text{Heat gained by water} &&= 10 \text{ cal/°C} \times (t_c - 20)\text{°C} \\ &\text{Heat lost by solid} &&= C_v \, 10 \text{ gm AgCl} \times (50 - t_c)\text{°C} \\ &\text{From Kopp's rule, } C_v \text{ AgCl} &&= 12 \text{ cal/°C mole AgCl} \end{aligned}$$

$$C_v \, 10 \text{ gm AgCl} = \frac{12 \text{ cal}}{°\text{C mole AgCl}} \times \frac{1 \text{ mole AgCl}}{143 \text{ gm AgCl}} \times 10 \text{ gm AgCl} = 0.84 \text{ cal/°C}$$

Then we write

$$10(t_c - 20) \text{ cal} = 0.84(50 - t_c) \text{ cal}$$
$$10.84t_c = 242$$
$$t_c = \frac{242}{10.84} = 22°$$

PROBLEMS

1. Since the unit cell represents the behavior throughout the whole crystal, the fraction of empty space in body-centered cubic (BCC) and face-centered cubic (FCC) crystals can be calculated from the unit cell. Assume that the spherical atoms touch their nearest neighbors, and compute the fraction of the unit cell not occupied by that portion of the spheres within the unit cell. *Hint:* In the FCC structure, atoms touch along a face diagonal; in the BCC structure, they touch along body diagonals.

2. The structure of CsCl is cubic with the cesium atom at the center and chloride atoms at the corners of the cube. Assume that the compound MX has this structure. In general, the anion X will touch the cation M.

(*a*) What will be the ratio of r_m, the radius of the cation, to r_A, the radius of the anion, when the anions also just touch?

(*b*) Calculate the fraction of this structure that is empty space.

★ **3.** The effective radius of an iron atom is 1.24 Å. Iron occurs both in a BCC structure and a FCC structure. Calculate the density of each in grams per cubic centimeter.

Ans. BCC, 7.9 gm/cm^3; FCC, 8.6 gm/cm^3.

4. MgO crystallizes in a rock salt structure for which the Madelung constant is 1.75. The ionic radii of Mg^{+2} and O^{-2} are 0.70 and 1.40 Å, respectively. From this and the data in Table 3 in Chapter 4, calculate the energy released when a crystal of MgO is formed from magnesium and oxygen atoms. E_a $O \rightarrow O^{=}$ + 7.3 ev.

5. Show that the critical radius ratio is 0.225 for tetrahedral coordination and 0.155 for triangular coordination.

6. The ionic radii of the alkali metal ions are

Li^+	0.60 Å
Na^+	0.95 Å
K^+	1.33 Å
Rb^+	1.48 Å
Cs^+	1.69 Å

Those for the halides are

F^-	1.36 Å
Cl^-	1.81 Å
Br^-	1.95 Å
I^-	2.16 Å

If the radius ratio rules predict the structures exactly, give the expected cation coordination number in each case.

7. A spinel is an important class of oxides consisting of two types of metal ions with the oxide ions arranged in CCP layers. The normal spinel has one-eighth of the tetrahedral holes occupied by one type of metal ion and one-half of the octahedral holes occupied by another type of metal ion. Such a spinel is formed by Zn^{+2}, Al^{+3}, and O^{-2} with Zn^{+2} in the tetrahedral holes. Give the formula of the spinel.

Ans. $ZnAl_2O_4$.

8. Carbon, nitrogen, boron, and hydrogen form compounds with a variety of metals. These are called interstitial compounds because the added atoms fit into holes in the close-packed metallic structures. When carbon or nitrogen reacts with titanium, these atoms occupy octahedral holes; when hydrogen reacts with titanium, these atoms occupy tetrahedral holes. Suggest a reason why carbon and nitrogen do not occupy the tetrahedral holes.

9. On the basis of the molecular orbital diagram, compare the energies necessary to break the bonds in the following species.

(*a*) F_2^+ vs. F_2
(*b*) O_2^+ vs. O_2
(*c*) N_2^+ vs. N_2
(*d*) Ne_2^+ vs. Ne_2

Ans. (*a*) F_2^+ bond energy $>$ F_2 bond energy.

10. In view of the known geometry, describe the hybridization of bonds to the central atom in CO_3^{-2} and ClO_4^-.

★ **11.** When P_4O_{10} reacts with water at ordinary temperatures, H_3PO_4 is formed. If it reacts with cold water, a compound with the formula $H_4P_4O_{12}$ is formed in which the $P_4O_{12}^{-4}$ ion has a ring structure. Show by a drawing how stepwise addition of two water molecules to the P—O—P bond, i.e.,

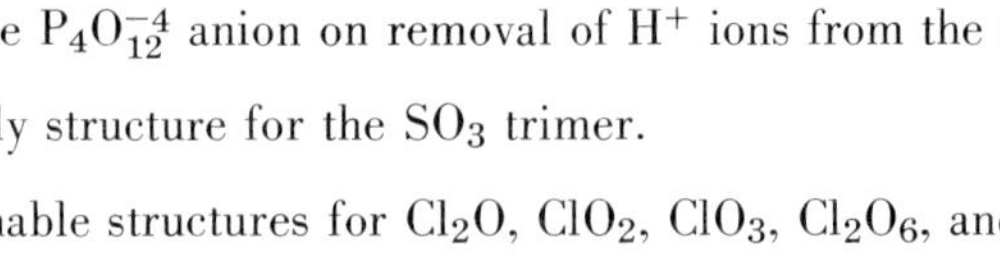

could give rise to the $P_4O_{12}^{-4}$ anion on removal of H^+ ions from the resulting structure.

12. Indicate a likely structure for the SO_3 trimer.

13. Indicate reasonable structures for Cl_2O, ClO_2, ClO_3, Cl_2O_6, and Cl_2O_7.

14. Indicate a neutral molecule isoelectronic to each of the following:

(*a*) NO_2^+ (*b*) CN^-
(*c*) NO^+ (*d*) NO_2^-

15. The solid $AlPO_4$ is isoelectronic to a familiar solid. Predict the structure.

Ans. Cristobalite.

16. In the text, it is said that phosphorus, sulfur, and chlorine form structures analogous to $Si_2O_7^{-6}$. Indicate these structures.

★ **17.** Indicate the structure of the rings formed by linking four and six tetrahedra of SiO_4 to give ions with the formulas $Si_4O_{12}^{-8}$ and $Si_6O_{18}^{-12}$.

18. Of the following, which can be "framework structures"?

(*a*) $KNa_3Al_4Si_4O_{16}$ (*b*) $KAlSiO_4$
(*c*) K_2AlSiO_6 (*d*) $Na_2KAl_4Si_4O_{16}$
(*e*) $CsAlSi_2O_6$ (*f*) $BaAl_2Si_3O_{10}$
(*g*) $K_2Al_3SiO_{10}$ (*h*) $CaAl_2Si_2O_8$

Ans. (*a*), (*b*), (*e*), (*f*), and (*h*).

19. A 100-gm crystal of $CaCO_3$ is heated to 100°C and dropped in 100 gm of water at 50°C. If no heat is lost, what is the final temperature of the water plus the crystal?

20. If 20 gm of AgCl solid at 30°C is dropped into 40 gm of water at 40°C, what is the final temperature?

21. How much heat is required to heat 20 gm of solid benzene from −20 to 5°C? How much heat is required to heat 20 gm of liquid benzene from 5° to 30°C?

10
Vapor pressure

In Chapter 8, we saw that real gases differ from ideal gases because the molecules themselves occupy space and because intermolecular forces inhibit the independent existence of gas molecules. It is these intermolecular forces that cause the condensation of a gas to a liquid or solid; hence the study of gas-liquid and gas-solid transformations provides an introduction to the nature of these forces.

CONDENSATION

Association in gases. We rationalized the behavior of real gases by the supposition that the net effect of intermolecular forces is statistical association of gas molecules. Since at very low pressures the ideal gas law is obeyed, association becomes important only at higher pressures. In order to appreciate the nature of this statistical association in the gas, it is helpful to see what happens in a real gas during compression. Imagine that, at the start, the pressure is so low that ideal behavior occurs and that, at the finish, the pressure is so high that non-ideal behavior is pronounced.

Figure 1 represents an instantaneous picture of a real gas at various pressures. We can look upon this as a representative gas in which each molecule is the symbol for trillions of molecules. The solid circles indicate the "billiard ball" size of the gas molecules; the dotted circles indicate the outer limit of effectiveness for van der Waals attraction. In the instantaneous picture taken at the lowest pressure, Fig. 1*a*, no molecule is within the sphere of influence of another.

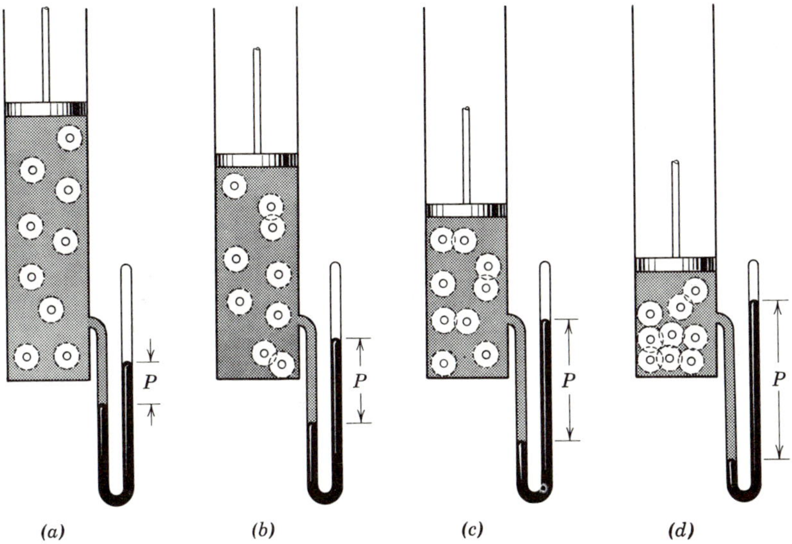

Fig. 1. Compression of a real gas.

Moreover, if this is truly representative, later instantaneous pictures will reveal that association occurs very rarely. Therefore the gas behaves ideally at this and at lower pressures.

Figure 1*b* represents another view at a higher pressure. Here a few molecules have interpenetrating spheres of influence; hence association occurs and departures from ideality are significant. A series of instantaneous pictures reveals that a given molecule is sometimes free and sometimes associated. In other words, if we see two clusters of molecules at one time, we usually see two clusters at subsequent times, but the molecules themselves are continually coming together to form a cluster and breaking free. Thus a given small cluster may hang together for only 10^{-12} second, but the formation of clusters occurs at a rate equal to the breakup of clusters, so that at any instant a look at the gas reveals numbers and types of cluster close to an average value.

As the pressure is further increased (Figs. 1*c* and 1*d*), the number and average size of these groups of associated molecules increase. Together with this, departures from ideality become more pronounced. As long as the molecules are in the form of a gas, however, and the pressure is not too high, the average number of molecules associated is only a fraction of those that remain free.

One might suppose that the average size of these clusters could increase indefinitely as the pressure is increased. This, however, is not true. Beyond a certain pressure, it is more favorable to have association of molecules into a single large cluster with only relatively few smaller clusters, rather than a great

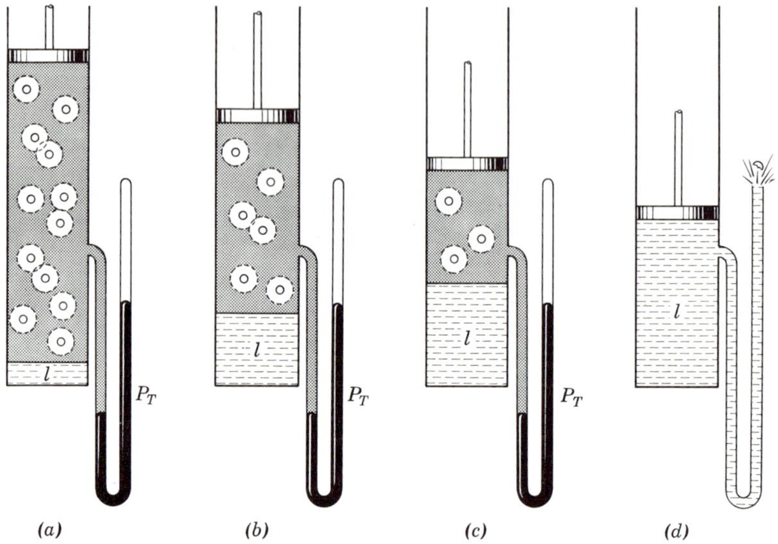

Fig. 2. Liquefaction of a real gas.

many clusters of intermediate size. This single large cluster is a droplet of liquid; the pressure at which it forms is the *vapor pressure.* Further compression of the gas does not increase the pressure; instead, condensation occurs and the liquid droplet grows.

Qualitatively, the aggregates in a real gas and the liquid droplet are similar. A liquid droplet is more firmly held together than a group of associated molecules in a gas, but nevertheless molecules continually free themselves from the liquid (evaporate) and condense from the gas. Thus, if we focus our attention on a droplet of liquid in equilibrium with its vapor, we see a dynamic condition in which molecules condense and evaporate at an equal rate so that a constant amount of liquid and vapor is always present even though there is a continual change of residence on the part of the molecules.

What happens if we compress a real gas in which condensation has already occurred? At equilibrium, the number of molecules entering the liquid surface is equal to the number of molecules leaving. We shall see shortly that the rate at which molecules evaporate depends solely on temperature. Thus, at a fixed temperature, the evaporation rate is fixed. The rate of condensation is determined by the number of molecules striking the surface. For an ideal gas at a given temperature we found (p. 233) that the number of gas molecules striking per second is proportional to the pressure. Now, consider the compression indicated in Fig. 2. At the start (Fig. 2*a*), there are twelve molecules in the gas. (Once again this is a representative picture.) If we halve the volume of the gas (Fig. 2*b*), we instantaneously double the pressure and thus double the number of molecules striking the surface per second. Since the rate of evaporation is still the same, molecules accumulate in the liquid until the number of molecules in the gas phase has dropped to a value that yields the initial pressure. At a given temperature, this occurs for any amount of compression wherein vapor remains, e.g., of Fig. 2*a*, *b*, or *c*.

What happens if we try to halve the volume of the liquid itself? Any attempt to bring the molecules closer together involves the steep repulsive region of the potential energy curve in Fig. 14 of Chapter 8. An attempt to halve the volume of the liquid usually has the result pictured in Fig. 2*d*, namely, the breakdown of the pressure-measuring device. Liquids are almost, but not quite, incompressible.

It is the purpose of this chapter to develop quantitatively our qualitative picture of vapor-liquid equilibrium. It may seem curious that we devote so much discussion to the equilibrium of a liquid and a vapor. But condensation and evaporation are actually simple chemical reactions. For, if we heat sodium metal and form a gas, we actually decompose a giant molecule; likewise, if we condense sodium gas to a crystal, we synthesize a giant molecule. In these processes, we create or destroy bonds between atoms. Thus, many of the principles developed in the following discussion apply to all *chemical* equilibria. Consequently, a discussion of vapor-liquid equilibria is really an introduction to the general principles of chemical equilibria.

EQUILIBRIUM

Let us introduce the quantitative aspects of equilibria by discussion of the two simple systems represented in Figs. 3 and 4. Figure 3 indicates a box divided by a permeable partition and filled with an ideal gas. The compartment on the left has a volume V_L; the compartment on the right has a volume V_R. The second system is also a box filled with an ideal gas (cf. Fig. 4). In this case, however, we are going to assume that the compartment V_R is in a field, perhaps an electrical field, such that the potential energy of a gas molecule in V_R is U_0 whereas the gas molecule in V_L has zero potential energy.

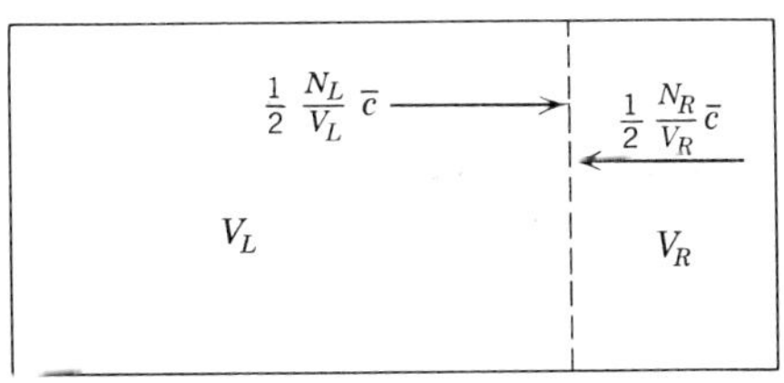

Fig. 3. Gaseous equilibrium.

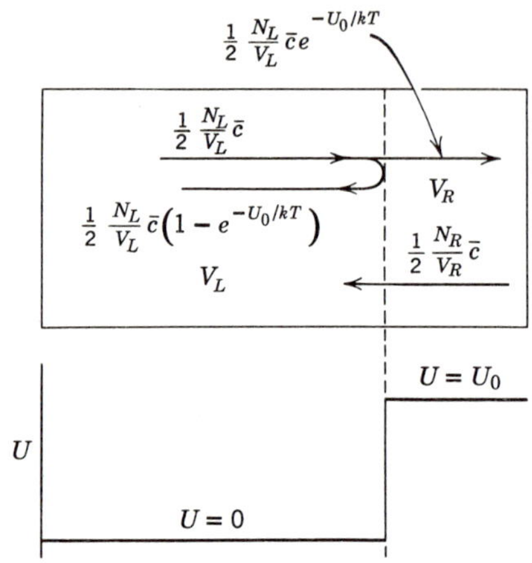

Fig. 4. Idealized gas-liquid equilibrium.

For these two simple systems, let us find the time-independent or equilibrium distribution of molecules, between the two compartments, i.e., the relative number of molecules found in each compartment when the system has reached a steady state. We do this first by consideration of the dynamic nature of the equilibrium and the necessary equality of opposing rates. In a second treatment, we shall introduce the statistical concept of equilibrium, which requires no recognition of the dynamic aspects of the equilibrium but deals only with the ultimate probability of residence in each compartment.

Dynamic equilibrium. If the gas in the two compartments of Fig. 3 is in equilibrium, the number of molecules crossing the partition from left to right must equal the number crossing from right to left. If this were not true, molecules would accumulate in one of the compartments. This condition requires

$$\frac{1}{2}\frac{N_L}{V_L}\bar{c}A = \frac{1}{2}\frac{N_R}{V_R}\bar{c}A$$

N and V are, respectively, the number of molecules in, and the volume of, the indicated partitions. As usual, $\bar{c}$ is the average speed of the molecules. This is the formula for the number of molecules striking A cm^2 of wall per second.

The two compartments are at the same temperature; consequently, $\bar{c}$ as well as A cancels and we write

$$\frac{N_L}{N_R} = \frac{V_L}{V_R}$$

The equations defining equilibrium for the system shown in Fig. 4 are more complicated. To go from the left to the right compartment, a molecule must reach the partition *with sufficient energy* to climb the potential energy hill. From Boltzmann's distribution, we know that the velocities of the molecules vary from zero to infinity; hence there will always be a certain fraction of high-energy molecules in V_L that can penetrate the partition. If the barrier is high, however, most of the molecules will have less than the required energy and they will be reflected. For dynamic equilibrium, the condition is

$$\left(\frac{1}{2}\frac{N_L}{V_L}cA\right)e^{-U_0/kT} = \frac{1}{2}\frac{N_R}{V_R}cA$$

The first factor (in parentheses) is the number striking the partition per second. The next factor is the fraction of these molecules that have sufficient energy to surmount the energy barrier. This product equals the number of molecules passing from left to right. All molecules on the right side of the partition penetrate this partition if they strike it.

Neither the average value of the *kinetic* energy nor c depends on the value of the potential energy, so we obtain

$$\frac{N_L}{N_R}e^{-U_0/kT} = \frac{V_L}{V_R}$$

Statistical equilibrium. We can arrive at these equilibrium equations by an alternative approach that considers only the probability of molecular residence in either compartment. In the system shown in Fig. 3, we are dealing with an ideal gas. Suppose we label one gas molecule in some fashion, paint it red, and then make a billion observations. There is no more reason to suppose that this molecule would be in the upper left corner of the box more than that it would be in the lower right corner. In fact, if we observe regions of equal volume, we find that the red molecule appears as often in one region as the other. On the other hand, for the unequal volumes, V_R and V_L, the probability of finding the red molecule in V_R is

$$\frac{V_R}{V_R + V_L}$$

and the probability of finding it in V_L is

$$\frac{V_L}{V_L + V_R}$$

(Note that the probability of finding the red molecule in V_L *or* V_R is 1.)

According to the principles of probability, if we make a large number of observations of a single labeled molecule, or *a single observation on a large number of independent, identical molecules,* the number of times molecules appear in V_L divided by the number of times they appear in V_R approaches the ratio of the respective probabilities. Thus

$$\frac{N_L}{N_R} = \frac{V_L}{V_L + V_R} \div \frac{V_R}{V_L + V_R} = \frac{V_L}{V_R}$$

This is the same equation that described dynamic equilibrium for the system in Fig. 3.

A statistical viewpoint can also be used to derive the equilibrium equation for the system pictured in Fig. 4. Because of the potential energy hill, not all the molecules in V_L can penetrate the partition, but all the molecules in V_R can. Hence, from the same argument used in the preceding paragraph, we say that at equilibrium

$$\frac{N_L^*}{N_R} = \frac{V_L}{V_R}$$

where N_L^* indicates those molecules in V_L that possess greater than the minimum energy required for surmounting the energy hill. N_L^* is related to N_L by Boltzmann's equation; that is,

$$N_L^* = N_L\, e^{-U_0/kT}$$

Combination of these equations yields

$$\frac{N_L}{N_R}\, e^{-U_0/kT} = \frac{V_L}{V_R}$$

the equation derived for dynamic equilibrium.

The last paragraph introduces a new feature into the statistical analysis of equilibrium. Here we find a competition between energy requirements and spatial probability requirements for equilibrium. If all molecules had the same energy, and this was less than U_0, *no* molecules would appear on the right side of the box. But Boltzmann's equation leads to a distribution of energies. Thus there will always be some molecules with sufficiently high energy to surmount the potential barrier even though U_0 is far greater than the average energy. For these energetic molecules, spatial probability arguments apply.

LIQUIDS

In Chapter 8 we found that gases are of low density with large distances between molecules, whereas liquids are compact, and solids even more so. In liquids, many modes of molecular motion are possible; molecules can vibrate in a "cage" created by their neighbors, or they can even rotate more or less freely. These motions require energy. But the energy is not divided equally among the liquid molecules; it is distributed according to Boltzmann's law; hence some liquid molecules have a great deal of energy and some have hardly any. As with gases, the average molecular energy is proportional to the absolute temperature. If we compare the average energy of a liquid to that of the corresponding solid and gas, we find that the average molecular energy in the liquid is less than that of the gas but more than that of the solid.

Liquid-gas equilibria. The equilibrium between a liquid and the corresponding gas (vapor) is dynamic with the rate of condensation equal to the rate of evaporation. Evaporation is possible because there are always some high-energy liquid molecules on the surface that may be moving in the direction of the gas phase. If their energy is high enough, they break away from the surface of the liquid and become gas molecules. For this to occur, the upward component of kinetic energy must be greater than the difference in average energy of the liquid and the gas. Only then will this liquid molecule have sufficient kinetic energy to overcome the cohesive forces in the liquid. This occurs in just the same way that a space-bound rocket leaves the surface of the earth when kinetic energy (sufficient to overcome the gravitational attraction) is channeled upward.

The requirements for liquid-vapor equilibrium are much like those pictured in Fig. 4. The left side of the box corresponds to the liquid; the right side of the box corresponds to the gas. As for the liquid, only those molecules that have more than a certain minimum kinetic energy can evaporate, i.e., pass from V_L to V_g. (We now replace the subscript R by g. This stands for "gas.") Thus the conditions for vapor-liquid equilibrium are

$$\frac{N_L}{V_L^*} e^{-U_0/kT} = \frac{N_g}{V_g} \qquad U_0 = \mathcal{E}_g^0 - \mathcal{E}_L^0 = \Delta\mathcal{E}_v^0$$

where $\mathcal{E}_g^0 - \mathcal{E}_L^0$ is the difference between the average energies of the gas and of the liquid, the energy of vaporization. In writing V_L, we now include an asterisk to indicate that this is not the entire volume of the liquid, but the volume of *empty* space available for liquid molecules in the "cage" created by its neighbors.

This equation assumes a more useful form if we substitute for N_g/V_g according to the ideal gas law:

$$\frac{P_g}{kT} = \frac{N_g}{V_g}$$

Then we can write

$$\frac{P}{kT} = \frac{N_L}{V_L^*} e^{-\Delta\mathcal{E}_v^0/kT}$$

On a *per mole* rather than a *per molecule* basis this becomes

$$P = \frac{n_L RT}{V_L^*} e^{-\Delta E_v^0/RT}$$

The quantity ΔE_v^0 represents the energy needed to change a mole of liquid to a mole of gas. In the actual vaporization process, work is done by expansion against the confining pressure of the vapor. The total heat, ΔH_v°, needed to carry out this process includes both the energy change and the work done. For the above formula we can use ΔH_v° rather than ΔE_v°; then the equation becomes

$$P = \frac{n_L RT}{V_L^*} e^{-\Delta H_v^\circ / RT}$$

or

$$P = \frac{RT}{v_f} e^{-\Delta H_v^\circ / RT}$$

where v_f is now the free volume per mole of liquid, V_L^*/n_L, and ΔH_v° is the change in H when one mole of a liquid evaporates to form a gas at constant pressure.†

A plot of this vapor pressure equation is given in Fig. 5. The vapor pressure P increases sharply with temperature, and, the greater the *enthalpy of the vaporization*, ΔH_v°, the sharper will be this increase with temperature.

Note that this vapor pressure equation is a modified form of the ideal gas law. Apparently, molecules of the liquid are essentially gas-like, but only the fraction indicated by the exponential are completely free to act as a gas within the free volume v_f.

If we take the natural logarithm‡ of both sides of the vapor pressure equation, we obtain

$$\ln P = \ln \frac{RT}{v_f} - \frac{\Delta H_v^\circ}{RT}$$

On the basis of experimental observations we may conclude that for our present purpose RT/v_f and ΔH_v° do not vary significantly with temperature. If we make the substitutions

$$\ln \frac{RT}{v_f} \equiv b, \quad \frac{\Delta H_v^\circ}{R} \equiv M, \quad Y \equiv \ln P, \quad \text{and} \quad X \equiv \frac{1}{T}$$

we can write $Y = -MX + b$. Since M and b are constants, this is the equation of a straight line. A plot of Y versus X is shown in Fig. 6. The slope, $\Delta Y/\Delta X$, is related to the heat of vaporization; the intercept, i.e., the extrapolated value of $\ln P$ when the temperature is infinite, specifies the value of RT/v_f.

The boiling point. At the boiling point of a liquid, the vapor pressure is equal to the external pressure. When liquids are heated from the bottom, the vapor pressure first becomes equal to the external pressure at the bottom of the

† The change from ΔE_v° to ΔH_v° causes a slight change in the manner in which v_f is calculated.

‡ The natural logarithm has the base of $e = 2.718$ instead of 10. Natural and common logarithms are related as follows: $2.303 \log_{10} X = \log_e X$. To differentiate between natural and common logarithms we write $\log X$ for $\log_{10} X$ and $\ln X$ for $\log_e X$. The relation between natural logarithms (ln) and common logarithms (log) is explained in the Appendix.

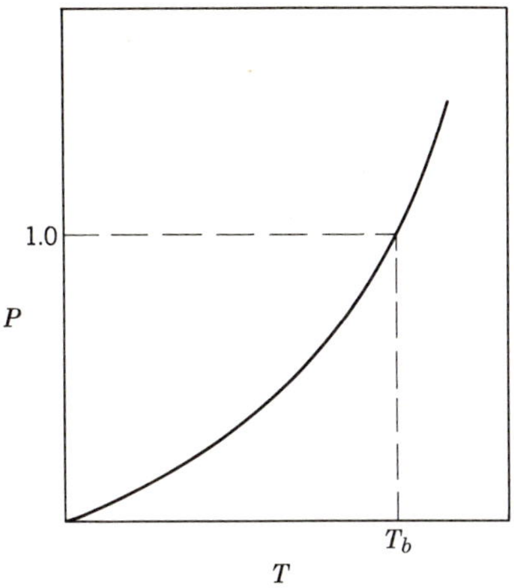

Fig. 5. Vapor pressure of liquid.

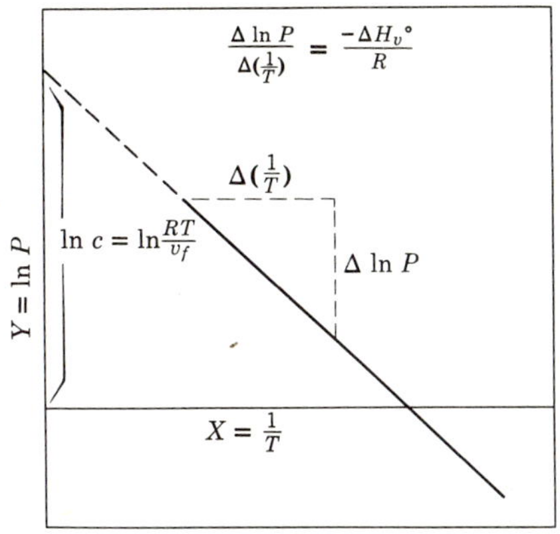

Fig. 6. The vapor pressure plot.

container. Thus the transition from liquid to gas takes place here with the formation of vapor bubbles, which rise to the top as the liquid boils.

At the normal boiling point ($P = 1$ atm), $\ln P$ is equal to zero; hence the vapor pressure equation yields the relation

$$\ln \frac{RT}{v_f} \equiv b = \frac{\Delta H_v^\circ}{RT_b}$$

where T_b is the temperature at the boiling point. If we substitute this into the vapor pressure equation we obtain

$$\ln P = \frac{\Delta H_v^\circ}{RT_b} - \frac{\Delta H_v^\circ}{RT} = -\frac{\Delta H_v^\circ}{R}\left(\frac{1}{T} - \frac{1}{T_b}\right)$$

This is known as the Clausius-Clapeyron equation (integrated form).

Trouton's rule and free volume. A hundred years ago, Trouton observed that the heat of vaporization, ΔH_v°, is related to absolute temperature at the boiling point by the equation

$$\Delta H_v^\circ = 22T_b \text{ cal/mole}$$

This relation has far-reaching consequences.

First of all, it means that a plot like that in Fig. 6 has the same intercept for *all* liquids. As an example of this, Fig. 7 indicates vapor pressure plots for nitrogen, benzene, and mercury, three compounds that differ significantly in boiling point. (Nitrogen boils at $-195°$C, benzene boils at 80°C, and mercury boils at 360°C.) Although the curves have different slopes, all three curves have about the same intercept.

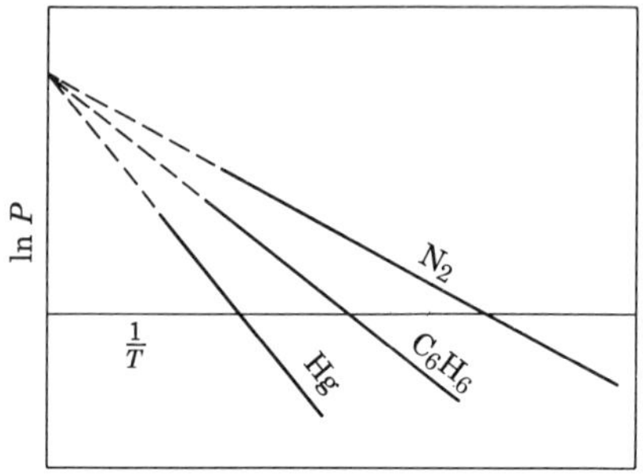

Fig. 7. The Trouton focus.

A second consequence of Trouton's Rule is that RT/v_f is the same for all liquids. Since RT/v_f does not vary with temperature, v_f is proportional to T; consequently, all liquids have the same free volume at the same temperature! In other words, liquids like mercury and benzene, which are altogether different in properties, have available the same amount of empty space at room temperature for each molecule. This conclusion is borne out approximately by the facts. Table 1 gives the free volumes for several very different liquids at 25°C. In view of the differences in the properties of these liquids, the agreement is surprisingly good.

Unstable equilibria. True equilibrium can be defined as the most stable condition of a system. Often, however, a system can exist for a long period of time in a fixed condition that is not the most stable condition. For example, diamond, which is a long-lasting form of carbon, is actually an unstable form (graphite is the stable form). When a system can persist for a long time in an unstable condition, we refer to this condition as a *metastable* or *unstable* equilibrium.

Suppose we confine steam to a cylinder at 110°C and 1 atm. What happens when we chill this cylinder to 90° while we maintain the external pressure at 1 atm? If we wait long enough, water condenses out and the pressure of the gas drops to the equilibrium vapor pressure of water at 90°. But, for a short time before condensation occurs, we may have water vapor, *supercooled,* i.e., steam at 90°C and atmospheric pressure. This situation does not persist for long, but it can occur momentarily.

Table 1

Liquid	v_f(cc)
Hg	0.23
CS_2	0.65
C_6H_6	0.21
CCl_4	0.29

Now, suppose that liquid water is under pressure, as indicated in Fig. 2*d* and we maintain this pressure at 1 atm. What happens when we heat the cylinder to 110°? At equilibrium, of course, the water vaporizes to form steam, but there is, once again, the possibility of forming an unstable state in which the water is present as *superheated* water and behaves as we would expect liquid water to behave at 110°. This unstable situation does not persist for long; in a short time, the superheated water vaporizes.

INTERMOLECULAR FORCES

According to Trouton's rule, the heat of vaporization is proportional to the boiling point. Since the heat of vaporization is a measure of the energy required to separate the molecules of the liquid, i.e., to overcome the intermolecular forces, the boiling point serves as a convenient index of the strength of these forces.

van der Waals forces. In Chapter 8 dealing with gases, we indicated that there are always non-specific attractive forces between molecules. These forces stem from the correlation of electronic motion on different atoms. In view of this, it seems reasonable that, the greater the number of electrons in a given compound, the greater will be these intermolecular forces.

In Table 2, we have listed a number of substances in the order of the number of electrons they contain. If we assume that molecules with the same number of electrons experience similar intermolecular forces, the boiling point of fluorine, a molecule with 18 electrons, should be comparable to that of argon, a molecule with 18 electrons. In Table 2, we find that the boiling points do, indeed, in-

Table 2 Boiling Points of Simple Molecules

No. of Electrons	Molecule	t (°C)	d (Å)
2	H_2	−253	0.74
2	He	−269	0.93
10	Ne	−246	1.60
14	N_2	−196	1.06
16	O_2	−183	1.20
18	F_2	−187	1.40
18	Ar	−186	1.91
34	Cl_2	− 35	1.94
36	Kr	−152	2.00
54	Xe	−109	2.20
70	Br_2	59	2.26
86	Rn	− 62	2.4

crease as the number of electrons in the molecule increases, provided the total number of electrons is less than 20.

For the heavier diatomic molecules, i.e., Cl_2 and Br_2, this oversimplified picture breaks down. These failures occur because correlation forces depend not only on the number of electrons but also on the tightness with which they are held and the symmetry of the charge cloud distribution.

Let us now examine a number of the hydrides that are isoelectronic to a corresponding rare gas. The hydrides HF, H_2O, NH_3, and CH_4 are all isoelectronic with neon. We can imagine that they are formed by distorting a neon atom; for example, except for its weight, NH_3 could be formed by pulling three protons out of the neon nucleus. The boiling points of these and other members of different isoelectronic series are compared in Table 3.

For the series with 18 electrons, the boiling points range from 87°K to 213°K. Thus our simple picture of van der Waals forces works even less satisfactorily for polyatomic molecules than for diatomic molecules. There must be additional factors that contribute to the strengths of these intermolecular bonds.

Dipole-dipole interaction. A partial interpretation of the data in Table 3 is possible when we take into account the polarity of bonds. The HCl molecule has the electron dot structure indicated below:

$$\mathrm{H}:\ddot{\underset{\cdot\cdot}{\mathrm{Cl}}}:$$

Although HCl forms a shared-pair bond, the electronegativities of hydrogen and chlorine are 2.1 and 3.0, respectively; consequently we expect unequal sharing of the bonding electrons. This leads to a surplus of negative charge on the chlorine atom and a net positive charge on the hydrogen atom, so that the molecule is *dipolar*. Similar conclusions hold for the bonds in H_2S, PH_3, and SiH_4.

In the determination of polarity of *bonds*, the molecule is regarded as an egg-shaped body containing positive and negative charges. When the center of positive charge coincides with the center of negative charge, there is no *dipole moment*, but when these two centers do not coincide the molecule is characterized by the value of the *dipole moment*. The magnitude of the dipole moment

Table 3 Boiling Points of Isoelectronic Molecules

No. of Electrons	Compounds	T (°K)	Compounds	T (°K)	Compounds	T (°K)	Compounds	T (°K)	Compounds	T (°K)
10	Ne	27	HF	292	H_2O	373	H_3N	240	H_4C	112
18	Ar	87	HCl	188	H_2S	213	H_3P	188	H_4Si	161
36	Kr	121	HBr	206	H_2Se	231	H_3As	218	H_4Ge	183
54	Xe	164	HI	237	H_2Te	271	H_3Sb	256	—	—

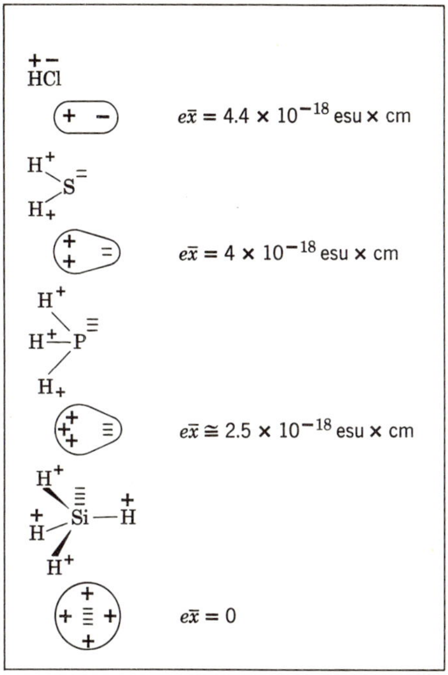

Fig. 8. Dipole moments.

is defined as follows. Consider the molecules indicated in Fig. 8. We can represent the distribution of charge in HCl by saying we have a single plus charge and a single minus charge, separated by a distance x. In such a case, the dipole moment is ex. A parallel description holds for the other compounds pictured in Fig. 8. Alongside the sketches in Fig. 8, the determined values of the dipole moments are listed.

Because of the unequal distribution of charge in an HCl molecule, each negatively charged chlorine atom tends to be surrounded by other HCl molecules with the positive end, i.e., the hydrogen atom, oriented toward the chlorine atom. To vaporize such a liquid, we have to work against not only the van der Waals but also these *dipole-dipole forces.* The greater the value of the dipole moment, the more significant will be this added contribution. For the group isoelectronic to argon, the dipole-dipole term is absent in Ar, fairly strong in HCl and H_2S, weaker in PH_3, and, again, absent for SiH_4. There is a tendency for the boiling points to change in this direction (Table 3) but the relatively small changes from hydride to hydride emphasize the fact that features subtler than dipole-dipole forces are still in evidence, e.g., the difference between SiH_4 and Ar both of which have zero dipole moments.

Dipole-dipole interactions certainly account in part for intermolecular forces. For most compounds, however, these additional forces only add a small increment to the ever-present van der Waals forces. Since the latter depend not only on the number of electrons but also on the molecular structure, it is not always easy to separate dipole-dipole and van der Waals attraction or to predict the net effect of both.

The most significant aspect of the data in Table 3 is the *very high* boiling points of HF, H_2O, and H_3N. With these three exceptions, compounds in a given family show an increase in boiling point with an increase in the number of electrons. We ascribe the unusually strong intermolecular forces in these three hydrides to the formation of a *hydrogen bond.*

The hydrogen bond. The nature of the hydrogen bond can best be described in terms of a specific example like water. The structure of water is indicated in Fig. 9. We noted earlier (p. 148) that the four pairs of electrons surrounding the oxygen atom in H_2O occupy sp^3 hybrid orbitals; consequently, these pairs of electrons extend toward the corners of a tetrahedron, and, in a single water molecule, hydrogen atoms are attached at two of the corners. The O—H bond is highly polar; the electron distribution around the hydrogen atom is distorted by the electronegativity of the oxygen atom so that there is a net positive charge on the two hydrogen atoms. Since the hydrogen atom has no core of electrons, the proton is partially bared.

In liquid water, this poorly shielded proton is surrounded by other molecules of the same type that have two pairs of unattached electrons projecting toward two corners of the tetrahedron. Understandably, there is a *very* strong attraction between these projecting unshared electrons and the poorly shielded proton of an adjacent water molecule. This results in a *super* dipole-dipole attraction.

The requirements for this very strong dipole-dipole interaction are a highly electronegative atom bonded to a hydrogen atom. The electronegative atom must be small; otherwise the hydrogen atoms on adjacent molecules are not able to approach closely enough to form these strong dipole-dipole bonds. Experimentally, it is found that only fluorine, oxygen, and nitrogen atoms fulfill these requirements. Compounds in which hydrogen atoms are bound to these atoms show evidence of hydrogen bonds.

In the lower portion of Fig. 9, we have shown a schematic structure of liquid water. For any of the oxygen atoms in water (the larger circles in the figure), there are two hydrogen atoms close by and two hydrogen atoms somewhat farther away. These four hydrogen atoms are arranged tetrahedrally around the central oxygen atom. The bonding to the two distant hydrogen atoms is the hydrogen bonding.

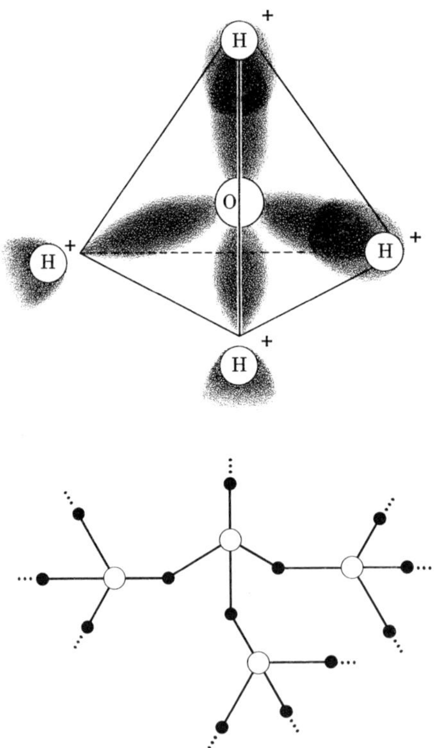

Fig. 9. The hydrogen bond in water.

In general, a hydrogen-bonded liquid is characterized by hydrogen bridge atoms located between *two* electronegative atoms. The hydrogen atom is usually closer to one of these electronegative atoms at a distance corresponding to a nearly normal bond; the other interatomic distance in the bridge is usually greater than the normal bond distance.

SOLID-GAS EQUILIBRIUM

If we compare a crystalline solid to the corresponding liquid, we find that the molecular arrangement in the solid is more regular and usually more compact. The regularity is a consequence of directed interaction between molecules that also makes it more difficult for molecules to move from place to place, so that solids are rigid and have their own shape. There are, however, gross similarities between solids and liquids. In both solids and liquids, each molecule has a certain free volume available, and, when a molecule has sufficient energy to break out of its cage, vaporization can occur. Since these similarities are the essential features needed for the derivation of the vapor pressure equation based on the model depicted in Fig. 4, we write

$$P = \frac{RT}{v_f} e^{-\Delta H_s^\circ/RT}$$

for the vapor pressure of a solid. Here v_f is the molar free volume of the solid and ΔH_s° is the difference in enthalpy between the gas and the solid, the heat required for sublimation.

The vapor pressure equation for solids can be derived on the basis of an alternative model that relies more heavily on the picture of a solid. A solid is tied together by elastic bonds between molecules and can be visualized as a regular array of spherical balls connected by springs. Once again, the thermal energy is distributed according to Boltzmann's equation; hence some of these molecules vibrate violently and others merely quiver.

For a molecule of the solid to fly off into the gas, it must be on the surface. If there are N_s atoms per square centimeter of surface, there is a total of $N_s \times A$ molecules on a crystal with A cm^2 of surface. In the simplest picture, all these molecules vibrate with the same frequency, ν_0; different energy states correspond to different amplitudes of vibration. Each second, $\nu_0 A N_s$ molecules on the surface move toward the gas. Most of the time, the bonds to other molecules prevent these molecules from sailing off into space as gas molecules and they are pulled back to the surface, but occasionally one of these vibrating molecules has enough energy to break the "spring" and fly off into space. The fraction of times this happens is $e^{-U_0/kT}$ where U_0 is the difference in average energy between the solid and the gas. Thus the total number of molecules leaving the solid per second is

$$AN_s\nu_0 \, e^{-U_0/kT}$$

At equilibrium, the number of gas molecules striking the surface and condensing is equal to the number of molecules leaving the surface. (Not all molecules that strike the surface have to condense, but we shall assume that they do.) The number of gas molecules that strike A cm^2 of wall per second (p. 235) is

$$\frac{PA}{(2\pi mkT)^{1/2}}$$

Thus, at equilibrium,

$$AN_s\nu_0\, e^{-U_0/kT} = \frac{P}{(2\pi mkT)^{1/2}}\, A$$

If we solve for P, the vapor pressure, we obtain

$$P = N_s\nu_0(2\pi mkT)^{1/2}\, e^{-U_0/kT}$$

or

$$P = N_s\nu_0(2\pi mkT)^{1/2}\, e^{-\Delta H_s^\circ/RT}$$

where ΔH_s° is the heat of sublimation. This is the same form as the earlier expression provided we identify $N_s\nu_0(2\pi mkT)^{1/2}$ with RT/v_f. This can be shown to be a reasonable identification.

If we take the logarithm of both sides of the vapor pressure equation we obtain

$$\ln P = \ln \frac{RT}{v_f} - \frac{\Delta H_s^\circ}{RT}$$

Once again, experimentally, ln (RT/v_f) is a constant and the results can be represented by a plot like that in Fig. 6. In contrast to liquids, RT/v_f varies from one solid to another; it depends on the details of the binding and the arrangement of molecules in the solid; hence there is no Trouton's rule for solids. For solids, of course, the slope is related to the heat of sublimation rather than the heat of vaporization.

Further rearrangement of this equation can be made in order to obtain the form of the Clausius-Clapeyron equation. At the temperature T_s, corresponding to a vapor pressure of 1 atm for the solid, the left-hand side of the equation is zero: hence

$$\ln \frac{RT}{v_f} = \frac{\Delta H_s^\circ}{RT_s}$$

Since RT/v_f is a constant, we can substitute this into the vapor pressure equation and obtain

$$\ln P = \frac{\Delta H_s^\circ}{RT_s} - \frac{\Delta H_s^\circ}{RT}$$

We have already noted that the average energy of gases is greater than that of liquids, and that of liquids is greater than that of solids. As a consequence, ΔH_s° is greater than ΔH_v°; hence, if we plot the vapor pressure of a solid and the corresponding liquid on the same graph, we obtain curves like those in Fig. 10.

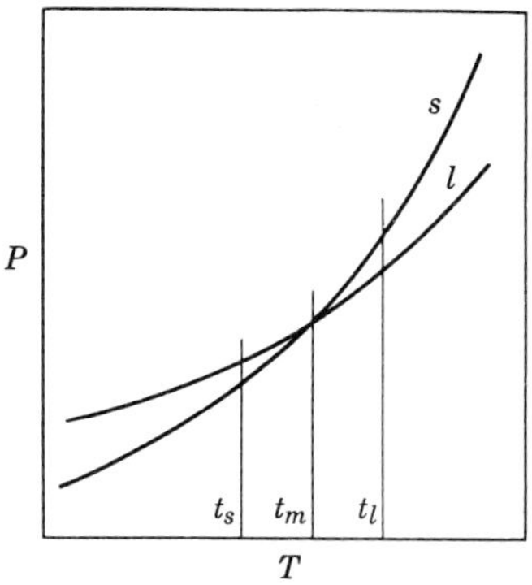

Fig. 10. Vapor pressure of solids and liquids.

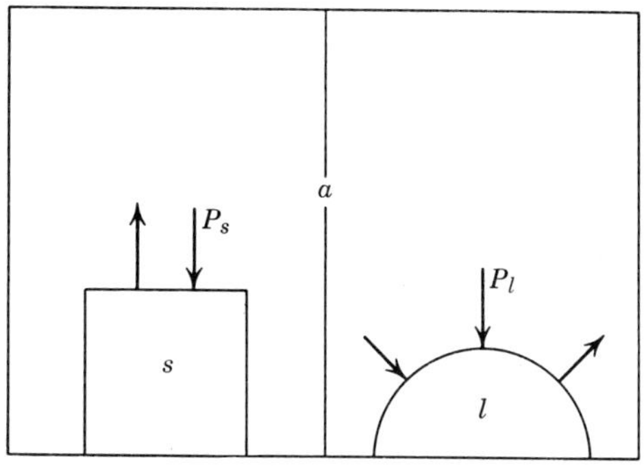

Fig. 11. Solid-liquid-vapor equilibrium.

SOLID-LIQUID EQUILIBRIUM

On the basis of the representative plots in Fig. 10, a transition must occur from the solid to the liquid at a certain temperature. Imagine an experiment carried out with a crystal on the left side of a box and a droplet of the corresponding liquid on the right side of a box (cf. Fig. 11). The two halves of the box are separated by a partition that has a small hole (cross-sectional area $= a$) to permit the flow of vapor between the two halves.

At any given temperature there is a fixed number of molecules leaving the solid per second; likewise, there is a fixed number of molecules leaving the liquid per second. At equilibrium, with the hole closed, the solid and liquid each establish an equilibrium vapor pressure such that the rate of evaporation is equal to the rate of condensation. But the vapor pressures of the solid and liquid, P_s and P_l, are only equal at a single point (cf. Fig. 10). This means that the pressures in the two halves of the box are usually different. If this is true, there will be a flow when the hole in the partition is opened. The flow to the right is given by the equation

$$N_{LR} = \frac{aP_s}{\sqrt{2\pi mkT}} - \frac{aP_l}{\sqrt{2\pi mkT}}$$

The first term in this equation represents the flow of gas to the right through a hole of cross section a when the pressure is P_s (cf. p. 238); the second term represents the flow of gas to the left when the pressure is P_l; the difference represents the net flow of gas to the right through the hole.

Now consider the conditions when the temperature is t_s (Fig. 10). At this temperature, P_l is greater than P_s; consequently, N_{LR} is negative and there is a flow of gas from the right to the left side of the box. The pressure of the vapor surrounding the solid now becomes greater than the equilibrium vapor pressure; therefore the number of molecules hitting the solid per second becomes greater than the number leaving; and the solid slowly grows in size. On the right side of this box, the situation is reversed. Because gas molecules have leaked out of this side of the box, the pressure of vapor is less than the equilibrium vapor pressure of the liquid. Therefore fewer molecules strike the liquid and condense than evaporate. Thus, at t_s, the solid increases in size and the liquid decreases in size until only solid is left. At any temperature where the vapor pressure of the liquid is greater than that of the solid, the liquid will change to solid, i.e., freeze.

At the temperature t_m (Fig. 10), P_s is equal to P_l, and there is no flow through the hole in the partition. Therefore the three phases, the liquid, the solid, and the gas, coexist. This temperature is known as the triple point.

At the temperature t_l (Fig. 10), P_s is greater than P_l. There is streaming of vapor from the left to the right side of the box, and the crystal slowly disappears while the liquid drop grows. Thus the solid becomes liquid, i.e., melts, whenever P_s is greater than P_l.

Supercooling. If a liquid at a temperature above the melting point is suddenly cooled below the melting point, it does not always freeze but sometimes forms a supercooled liquid. The vapor pressure and other properties of this liquid are those we would predict by extrapolation of the liquid properties above the melting point. This supercooled state, however, represents an unstable equilibrium; in time, the liquid freezes, to form the more stable solid. Water is one of the many liquids that readily supercool. If water is very pure, it is easy to supercool the liquid to 20° below the freezing point.

PROBLEMS

1. From the accompanying equilibrium vapor pressure data for bromine; make a plot of log P versus $1/T$.

t (°C)	P (mm Hg)	t (°C)	P (mm Hg)
−50	1.09	10	109.0
−40	2.98	20	173
−30	7.45	30	264
−20	17.1	40	392
−10	36.6	50	564
0	65.9		

(*a*) What is the melting point?

(*b*) Estimate the boiling point.

(*c*) Calculate the heat of sublimation, ΔH_s°, and the heat of vaporization, ΔH_v°.

(*d*) Some heat is released whenever a substance freezes. Compute this from these data.

(*e*) Estimate the heat of vaporization from Trouton's rule.

★ **2.** Assume for simplicity that Br_2 is a spherical molecule with a diameter of 4.0 Å. If the molecules on the surface of a bromine crystal form a two-dimensional close-packed layer, what is the value of N_s, the number of molecules per square centimeter? On the basis of the data in Problem 1 and the alternative vapor pressure equation for solids, estimate the effective frequency of vibration in the solid at $-50°C$. (Watch units!) *Ans.* 7×10^{14} at/cm^2, 5×10^9/sec.

3. Methane melts at -182.5 and boils at $-161.5°C$. At -191.8 its vapor pressure is 21 mm. Estimate the vapor pressure at $-166.8°$ and at $-195.5°C$.

4. On a mountain top the normal atmospheric pressure is 550 mm. Estimate the boiling point of water on the mountain top? *Ans.* 89°C.

5. Which of the following compounds would you expect to have a dipole moment?

(*a*) H_2	(*b*) NCl_3	(*c*) CH_3Cl	(*d*) CCl_4
(*e*) HBr	(*f*) CO_2	(*g*) NF_3	

Ans. (*b*), (*c*), (*e*), (*g*).

Note: Structural symmetry often cancels the dipole moment of a molecule even though the bonds are polar, e.g., SiH_4, p. 301.

6. In which of the following liquids would you expect hydrogen bonding? Indicate the bonding.

(*a*) CH_3F	(*b*) CH_3OH	(*c*) CH_3OCH_3
(*d*) H—S—S—H	(*e*) H—O—O—H	(*f*) CH_3NH_2
(*g*) CCl_4		

7. Hydrogen fluoride has a tendency to form associated molecules with the formulas $(HF)_2$ and $(HF)_6$. Indicate likely structures for these.

8. The ionic compound $K(HF_2)$ is well known. Indicate a reasonable structure for the HF_2^- ion. *Ans.* $[:\ddot{\underset{..}{F}}:H:\ddot{\underset{..}{F}}:]^-$

9. Many carboxylic acids, which have the general formula

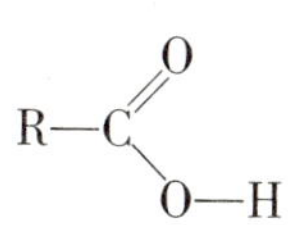

have a strong tendency to form dimers. Indicate a likely structure for the dimers.

★ **10.** Solid sulfur occurs in two forms, rhombic and monoclinic. Rhombic sulfur is the stable solid form below 96°C, monoclinic sulfur is the stable solid form above 96°C. Unstable equilibrium between the solid and the vapor exists for rhombic sulfur above 96° and for monoclinic sulfur below 96°. Sketch on the same graph a plot of vapor pressure versus temperature between 85° and 115°C for both forms of sulfur. Values of the vapor pressure are unnecessary, but the dependence on temperature and the relation of the two curves should be clearly shown. Label each curve clearly and indicate the curves corresponding to the equilibrium vapor pressure.

11. The boiling point of mercury is 360°C. Estimate the vapor pressure at 25°C.

11 Solutions

SOLUTIONS AND MIXTURES

It is not always easy to differentiate between mixtures and solutions. Many of the common materials we call solutions are merely physical mixtures in which two or more component materials retain their identities as aggregates. A true solution is a molecular mixture wherein the components lose their identity as aggregates.

Physical mixtures. Let us carry out an imaginary experiment in which we grind and mix common table salt (NaCl) with purified sand (SiO_2), melt it, and cool it, so that we have a material which seems to be the same throughout. With the unaided eye, we cannot pick out particles of sand and particles of sodium chloride; hence we might think that this is a solid solution. But, in reality, this material is not a true solution; it is a physical mixture.

If we re-examine this mixture of sodium chloride and sand under a microscope that magnifies ten thousand times, we see a material like that in Fig. 1. On this *macroscopic* scale, we can differentiate between particles of pure sand and particles of pure sodium chloride. Thus, on a macroscopic scale, this physical mixture is heterogeneous, i.e., not the same throughout. At the spot indicated by *X*, it is pure sand; at the spot indicated by *Y*, it is pure sodium chloride.

Now, let us magnify this mixture ten million times. We shall call this a *microscopic* scale. This view of the mixture is shown in the lower part of Fig. 1. Once again we see definite regions of pure NaCl and definite regions of pure SiO_2. Thus, this and all other physical mixtures are heterogeneous, both on a macroscopic scale and on a microscopic scale.

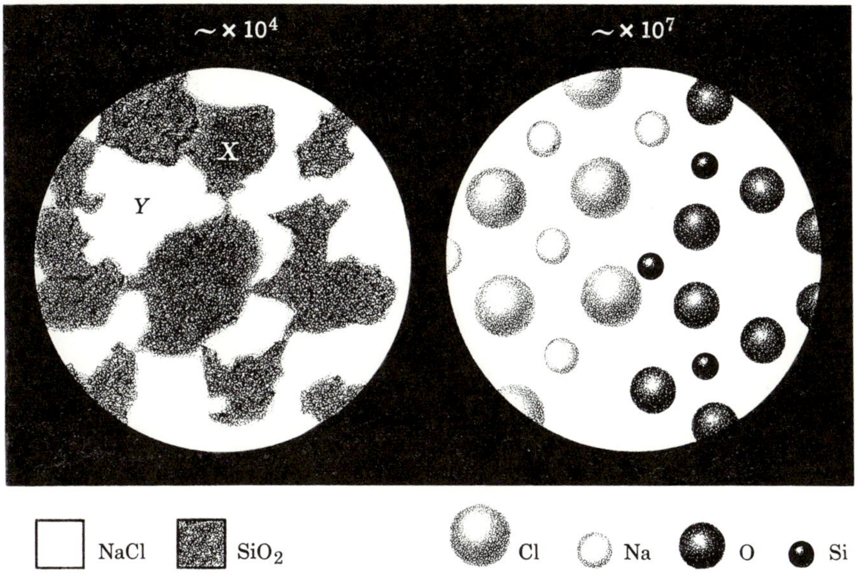

Fig. 1. Physical mixture.

Chemical solutions. If we grind together copper and silver, melt the mixed powder and cool it, we obtain a solid material that appears to be uniform just as the sodium chloride-sand mixture appears to be uniform. But this copper-silver material is a true solution, a molecular mixture. We can see this if we look at the copper-silver solution on both the macroscopic and the microscopic scale. On the macroscopic scale (Fig. 2), this material is homogeneous throughout. Only on the *microscopic* scale, where we "see" individual atoms, is this solid solution heterogeneous. Thus the difference between solutions and physical mixtures is that a true solution is homogeneous on a macroscopic scale ($\times 10^4$) and heterogeneous on a microscopic scale ($\times 10^7$), whereas a physical mixture is heterogeneous on both scales.

We have considered only *solid* mixtures and *solid* solutions, but *liquid* mixtures and *liquid* solutions are also known. For example, if benzene and water are whipped together with an eggbeater, a homogeneous-appearing, dirty-gray liquid results. If we examine this liquid at magnifications of 10^4 and 10^7, we find that it is heterogeneous on both scales and consists of globules of pure benzene mixed with globules of pure water. On the other hand, water and ethyl alcohol also form a homogeneous-appearing mixture. This mixture is homogeneous at a magnification of 10^4 and heterogeneous at a magnification of 10^7. In terms of our earlier criteria, the benzene-water is a physical mixture, but the ethyl alcohol-water is a true solution.

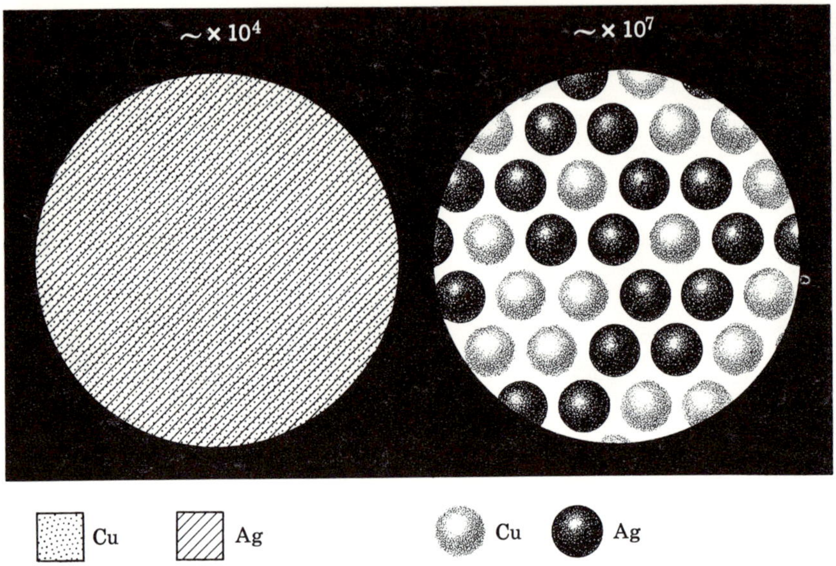

Fig. 2. Solution.

Mixtures of gases always result in a true gaseous solution. A physical mixture of gases that remains heterogeneous on a macroscopic scale is inconsistent with the physical picture of a gas.

Mixtures of different physical states, i.e., solid-gas, liquid-gas, and solid-liquid, can be true solutions only if the resulting material is in the same physical state throughout. Thus, in terms of our criteria, fog, a mixture of liquid water droplets and air, is a physical mixture, whereas humid air, air containing large amounts of gaseous water, is a true gaseous solution.

Colloids. There are cases of behavior intermediate to that of physical mixtures and solutions. If we attempt to dissolve a block of gold in water in order to form a true solution, we will not be successful. On the other hand, we can grind the gold into a very fine powder and form a suspension of the gold in water. If this is a physical mixture, the gold soon settles out. On the other hand, if we are so thorough in the grinding that the individual gold particles contain only about 100 gold atoms, then the mixture is heterogeneous on a scale somewhere between the macroscopic and the microscopic scale. This suspension might persist for years. In effect, we have dissolved not gold atoms, but relatively large gold molecules in water. Such mixtures are referred to as *colloids* or *colloidal solutions.*

THE SOLUTION PROCESS

There are three types of chemical solutions: solid solutions, liquid solutions, and gaseous solutions. All of them can be described by extensions of the models we have used earlier. Since the model for gases is the most successful in approximating physical reality, let us first consider gaseous solutions.

Gaseous solutions. Suppose that the box in Fig. 3 contains two different gases, A and B, at the same pressure but separated by a partition. For generality, we assume that the volume to the left, V_L, is not equal to the volume to the right, V_R. What happens when we remove the partition?

According to the ideal gas model, gases in a mixture are completely independent. In other words, A behaves in the presence of B just as it would in the absence of B. Because of this, we conclude that A expands from V_R and fills the whole box just as if V_L were initially evacuated. When this has occurred,

$$\frac{N_L^A}{V_L} = \frac{N_R^A}{V_R}$$

That is, the concentration becomes uniform. A similar process occurs for the B molecules.

We can arrive at the same conclusion if we consider a dynamic equilibrium in which the number of A-type gas molecules passing from right to left is balanced by an equal flow from left to right. (This approach was discussed in the preceding chapter on p. 292.) Thus the tendency for gas molecules to mix is consistent with the kinetic picture. But it is also consistent with a statistical picture in which gases mix because a mixture is more probable than a separate existence. For an ideal gas, solution always occurs, but, because intermolecular forces are nonexistent, the energy change on mixing is zero; the mixing takes place because the probability factor forces each type of gas out of its pocket into the whole box.

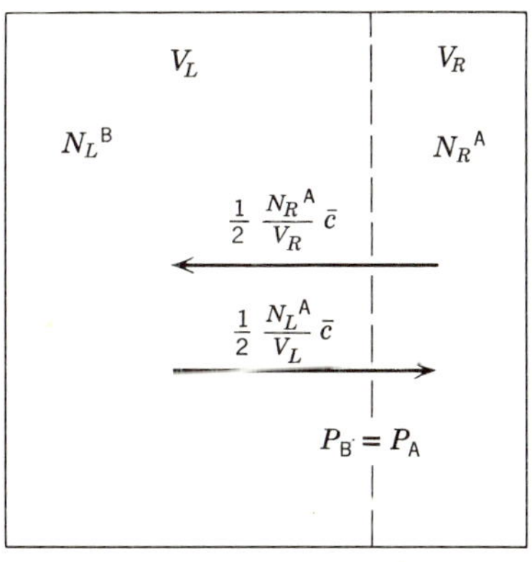

Fig. 3. Model for gaseous solution.

Solubility of gases in liquids or solids. The discussion of solution of a gas in a liquid or a solid is more complicated than discussion of solution of a gas in another gas because an energy factor must be included in the analysis. For simplicity, consider that a liquid or solid is a regular arrangement of atoms. For the solid this is true, but for the liquid it is only partly true. Nevertheless, this assumption will not lead to gross errors.

In Fig. 4*a*, we have pictured a crystal (or a droplet of liquid), and a single gas molecule. For solution to occur, the gas molecule must build into the lattice in the same way that copper builds into the silver lattice (see Fig. 2). Imagine that this process occurs stepwise. First of all, let us prepare the crystal by removing a molecule from the center and placing it on the surface as indicated in Fig. 4*b*. This requires the rupture of four solid-solid bonds, but two solid-solid bonds re-form when we place this molecule on the surface; hence the net cost of this first process will be twice the energy of the solid-solid (or liquid-liquid) bond, i.e., $2W_{ss}$. Now place the gas molecule in the hole just created (Fig. 4*c*). When this is done, four bonds are formed between the gas and the solid molecules. Therefore the second process decreases the energy by an amount equal to $-4W_{gs}$. (W_{gs} is the energy of a single gas-solid bond.)

The over-all energy change for both processes is $2W_{ss} - 4W_{gs}$. Since $2W_{ss}$ is the energy required to break all the bonds of a surface molecule and, thereby, form a gaseous molecule, $2W_{ss}$ is the energy of sublimation. Thus the energy change in the solution process can be written $\Delta E_g^s = \Delta E_s - 4W_{gs}$, where ΔE_g^s is the energy of solution and ΔE_s is the energy of sublimation of the solid. According to the expression

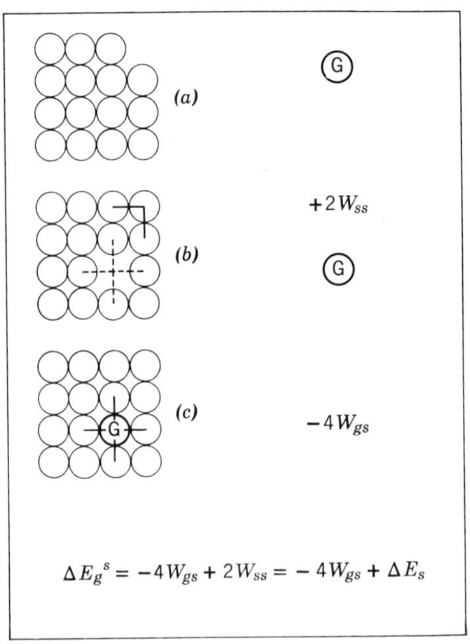

Fig. 4. The solution process—gas in crystal.

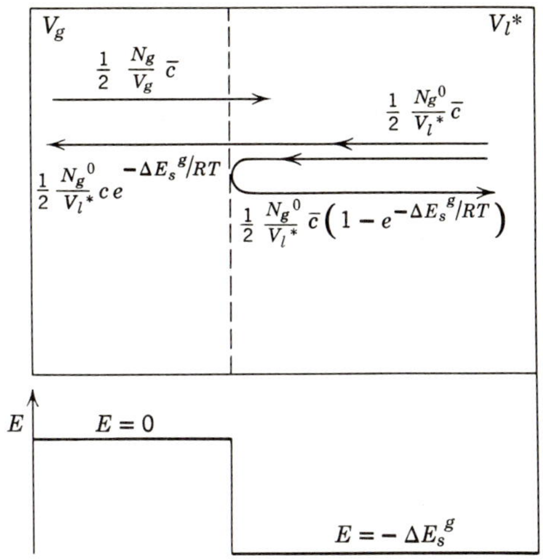

Fig. 5. Model for gas-liquid solution.

above, ΔE_g^s is negative unless the solid-solid bonds are more than twice as strong as the gas-solid bonds. Whenever ΔE_g^s is negative, the energy factor favors solution. The identical argument can be applied to liquids. ($\Delta E_g^s = -\Delta E_s^g$)

Figure 5 indicates the simple model we shall use to discuss the solution process. On the left is gas in equilibrium with liquid on the right. Assume for simplicity that the vapor pressure of the liquid is negligible. The energy of the gas molecules is different in the two regions by the amount ΔE_s^g. The number of *gas* molecules hitting the liquid surface per second and dissolving must be equal to the number of *gas* molecules leaving the surface and evaporating. Of the gas molecules that *start* to leave the liquid, only those with an energy higher than ΔE_s^g break the gas-liquid bonds; all others are turned back as indicated in Fig. 5. The condition for equilibrium is

$$\frac{n_g}{V_g} = \frac{n_g^d}{V_l^*}\,e^{-\Delta E_s^g/RT}$$

(The steps used in this derivation are identical with those used in the derivation of the vapor pressure equation; cf. p. 295.) V_l^* is the total free volume of the liquid, and n_g and n_g^d are the moles of gas in V_g and dissolved in the liquid, respectively.

We have seen earlier that a definite free volume v_f is associated with a mole of any liquid; hence, for a liquid containing n_g^d moles of dissolved gas and n_l moles of liquid, we can write

$$V_l^* = v_f(n_l + n_g^d)$$

If we substitute this into the condition for equilibrium and express n_g/V_g as P/RT, we obtain

$$\frac{P}{RT} = \frac{n_g^d}{v_f(n_l + n_g^d)}\, e^{-\Delta H_s^g/RT}$$

$$P = \left(\frac{n_g^d}{n_l + n_g^d}\right)\frac{RT}{v_f}\, e^{-\Delta H_s^g/RT}$$

Here we have made the usual change from energy to enthalpy units (see footnote on p. 296). $n_g^d/(n_l + n_g^d)$ is the fraction of moles in the liquid that are dissolved gas; this *mole fraction* is denoted by the symbol X_g.

For a given liquid-gas solution the second quantity to the right of the equality sign is a constant $1/k_{\mathrm{H}}$ at a given temperature; hence, we can write

$$P = X_g\left(\frac{1}{k_{\mathrm{H}}}\right) \quad \text{or} \quad X_g = k_{\mathrm{H}}P$$

If only a small amount of gas dissolves,

$$X_g \equiv \frac{n_g^d}{n_l + n_g^d} \approx \frac{n_g^d}{n_l}$$

Thus, under these conditions, the amount of gas that dissolves in one mole of liquid at constant temperature is directly proportional to the pressure. This is known as Henry's law. The Henry's law constants for several gases dissolved in water are listed in Table 1.

Solubility of solids in liquids. With the help of the preceding equations the discussion of the solubility of solids in liquids becomes simple. Imagine a box with three compartments containing liquid in the right-hand compartment, gas in the center, and a solid to the left (Fig. 6). The pressure of the gas, P_g, is that corresponding to the vapor pressure of the solid, *g*. We can visualize the solution of the solid *g* in the liquid *l* as taking place by the vaporization of *g*, followed by the solution of the gas in the liquid.

Table 1 Henry's Law Constants in Water at 20°C

Gas	$k_{\mathrm{H}}\left(\frac{\text{moles gas}}{\text{mole liquid 1 atm}}\right)$
N_2	1.24×10^{-5}
O_2	2.50×10^{-5}
H_2	1.46×10^{-5}
CO	1.87×10^{-5}

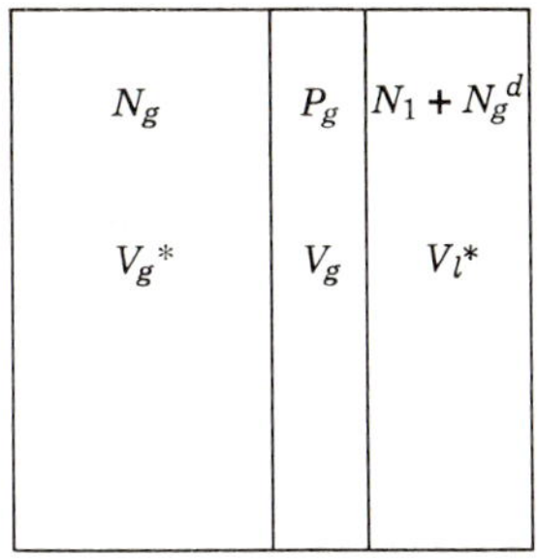

Fig. 6. Model for solution process.

The pressure of the gas in equilibrium with the solid g is given by

$$P_g = \frac{RT}{v_f^s} e^{-\Delta H_s^\circ/RT}$$

(cf. p. 303) where v_f^s is the free volume per mole of solid and ΔH_s° is the heat of sublimation, the heat needed to change a solid to a gas at 1 atm. At equilibrium, this is equal to the pressure in equilibrium with the gas dissolved in the liquid; hence P_g is also given by the equation

$$P_g = X_g \frac{RT}{v_f^l} e^{-\Delta H_s^g/RT}$$

where the various symbols apply to the solution and ΔH_s^g is the heat of solution of the gas.

If we equate these two equations and rearrange them we obtain

$$X_g = \frac{v_f^l}{v_f^s} e^{-(\Delta H_s^\circ - \Delta H_s^g)/RT}$$

or

$$X_g = \frac{v_f^l}{v_f^s} e^{-\Delta H_s^s/RT}$$

The quantity $\Delta H_s^\circ - \Delta H_s^g$ is the enthalpy change when one mole of solid dissolves in the liquid. For convenience we give this the symbol ΔH_s^s.

The last equation is quite similar to the vapor pressure equation. The pre-exponential factor, v_f^l/v_f^s, however, is not the same for all solutions since v_f^s varies from solid to solid. Thus, if ΔH_s^s is positive, the amount of solid that dissolves, X_g, increases with temperature just as the vapor pressure of a pure liquid. In contrast to the vaporization process, however, ΔH_s^s can be either positive or negative. If it is negative, the solubility decreases with temperature. In the majority of cases, ΔH_s^s is positive and small; hence, usually, the solubility increases with temperature but the relative change in solubility with temperature is considerably less than the relative change of vapor pressure with temperature.

VAPOR PRESSURE OF SOLUTIONS

Vapor pressure of solute. The pressure of gaseous solute in equilibrium with dissolved solute is proportional to the mole fraction of solute in solution.† This relation, derived in the last section, is known as Henry's law and holds only for ideal solutions. According to this law, a plot of the vapor pressure of the solute, P_B, versus mole fraction yields the straight line indicated in Fig. 7.

Vapor pressure of solvent. A similar expression for the vapor pressure of the solvent is obtained by rearrangement of equations already derived. The equation for the vapor pressure of a pure liquid is

$$P_0 = \frac{RT}{v_f} e^{-\Delta H_v^\circ/RT}$$

If the liquid is not pure but contains n_g moles of dissolved g and n_l moles of liquid, the total free volume becomes $(n_l + n_g)v_f$ rather than $n_l v_f$. Accordingly, the free volume per mole of l becomes

$$\frac{n_l + n_g}{n_l} v_f$$

and for a solution the preceding equation transforms into

$$P = \left(\frac{n_l}{n_l + n_g}\right) \frac{RT}{v_f} e^{-\Delta H_v^\circ/RT}$$

or

$$P_l = X_l P_0$$

where X_l is the mole fraction of solvent, P_l is the vapor pressure of the solvent in solution, and P_0 is the vapor pressure of the pure solvent.‡

Thus the vapor pressure of the solvent in a solution is the vapor pressure of pure solvent multiplied by the mole fraction of solvent. This is known as Raoult's law and is valid only for ideal solutions. In Fig. 7, we have indicated a plot of the vapor pressure of the solvent, P_A, versus the mole fraction. According to Raoult's law, such a plot is a straight line.

If both the solvent and the solute have an appreciable vapor pressure, the

† We shall call the major component in a solution the solvent and the minor component the solute.

‡ Pseudo-gas properties of the solution are suggested by this relation. When we increase the volume of a gas by expansion, the pressure drops. When we increase the free volume of the solvent by forming a solution, the vapor pressure drops. In both instances, the pressures are inversely proportional to the volume.

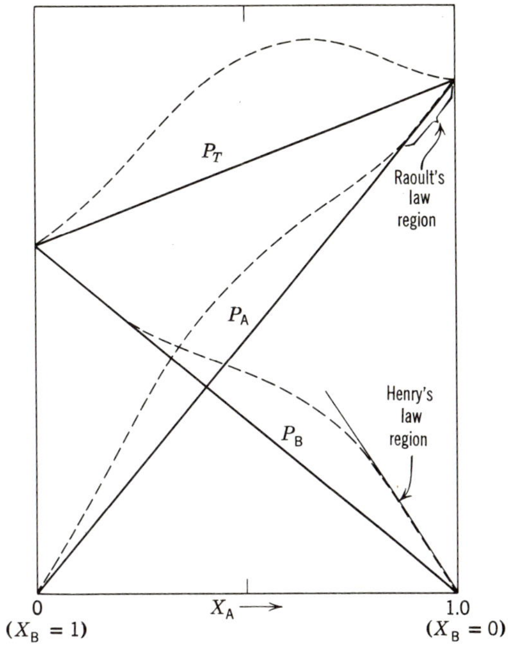

Fig. 7. Vapor pressure diagram.

total pressure of vapor is the sum of the vapor pressures of solvent and solute. This consequence of Dalton's law yields the third line, P_T, for the total pressure.

Real solutions seldom obey Raoult's law and Henry's law over the *whole* range of concentration. This is not unexpected since these laws are based on an idealized picture. For example, Raoult's law requires that the ΔH of solution be 0. This seldom occurs, and the behavior for a real solution may be like that indicated by the dotted lines in Fig. 7. Idealizations, however, become increasingly better as the solution becomes more dilute. As indicated in Fig. 7, even real solutions obey Henry's law and Raoult's law when sufficiently dilute, i.e., when X_A is close to 1.

CONCENTRATION UNITS

Before continuing the discussion of the properties of solutions, it is useful to define the standard ways of specifying concentration. To this end, we shall use the following notation:

$$\begin{matrix} n_a & W_a & d_a & M_a \\ n_b & W_b & d_b & M_b \\ V & W & d & \end{matrix}$$

The symbol n_i refers to the number of moles; W_i refers to the weights; d_i refers to the densities; M_i refers to the molecular weights. The subscripts a and b indicate the solvent and solute, respectively. Where there are no subscripts, as in the case of V, W, and d, these symbols apply to the solution as a whole. Unless otherwise specified, we always deal with volume in cubic centimeters, weight in grams, and density in grams per cubic centimeter.

Mole fractions were introduced earlier; they are defined by the equations

$$\frac{n_a}{n_a + n_b} \equiv X_a \qquad \frac{n_b}{n_a + n_b} \equiv X_b \qquad \textit{all solutions}$$

It follows, of course, that $X_a + X_b = 1$.

For *very* dilute solutions, n_b, the number of moles of dissolved material (also referred to as the number of moles of solute), is very small compared with n_a, the number of moles of solvent. Thus, as an approximation valid *only* for dilute solutions, we write

$$\frac{n_b}{n_a} \approx X_b \qquad X_a \approx 1 \qquad \textit{dilute solutions}$$

Another useful unit of concentration is that of molarity, usually indicated by M; it is defined by the equation

$$M = \text{molarity} \equiv \frac{n_b 1000}{V} \equiv \frac{\text{moles}}{\text{liter sol'n}} \qquad \textit{all solutions}$$

For dilute solutions molarity also reduces to a simpler form. The total volume of a dilute solution is approximately that of the solvent; therefore we can write, for *dilute* solutions,

$$V \approx \frac{W_a}{d_a} \approx \frac{n_a M_a}{d_a}$$

and, consequently,

$$M \approx \frac{1000 n_b d_a}{n_a M_a} \approx 1000 X_b \frac{d_a}{M_a} \qquad \textit{dilute solutions}$$

The third useful concentration unit is molality, or weight molarity, usually indicated by m_w; it is defined by the equation

$$m_w = \text{molality} = \frac{1000 n_b}{W_a} = \frac{\text{moles}}{\text{kg solvent}} \qquad \textit{all solutions}$$

Molality differs from molarity in two respects. Molarity is defined on the basis of the *volume* of *solution*, i.e., the combined solvent and solute, whereas molality is defined on the basis of the *weight* of the *solvent*.

The molality reduces to a simpler form for *dilute* solutions wherein n_a is much larger than n_b. Under these conditions,

$$m_w = \frac{1000 n_b}{M_a n_a} \approx \frac{1000 X_b}{M_a} \qquad \textit{dilute solutions}$$

Let us look at a number of illustrative problems in which we will calculate mole fraction, molarity, and molality.

Example Problem 1: A sample of solution contains 21 gm of CH_3OH and 32 gm of H_2O. What is the mole fraction of CH_3OH in the solution?

The mole fraction is defined by the number of moles of the solute and the solvent; so let us first calculate the number of moles of each present in the solution; we find

$$21 \text{ gm } CH_3OH \times \frac{1 \text{ mole } CH_3OH}{32 \text{ gm } CH_3OH} = 0.66 \text{ mole}$$

and

$$32 \text{ gm } H_2O \times \frac{1 \text{ mole } H_2O}{18 \text{ gm } H_2O} = 1.8 \text{ moles}$$

There is present a total of 2.5 moles; of these, the fraction that are CH_3OH molecules is

$$\frac{0.66}{2.5} = 0.26$$

Example Problem 2: Enough ether is added to 17 gm of CCl_4 to bring the total volume of solution to 500 cc. What is the molarity of the solution?

There are 17 gm CCl_4/500 cc sol'n. Since the molecular weight of CCl_4 is 154, we can calculate the number of moles in 500 cc of solution as follows:

$$\frac{17 \text{ gm } CCl_4}{500 \text{ cc sol'n}} \times \frac{1 \text{ mole } CCl_4}{154 \text{ gm } CCl_4} = \frac{0.11 \text{ mole } CCl_4}{500 \text{ cc sol'n}}$$

The molarity, however, is the number of moles of CCl_4 in 1 liter of solution. Therefore

$$\frac{0.11 \text{ mole } CCl_4}{500 \text{ cc sol'n}} \times \frac{1000 \text{ cc sol'n}}{1 \text{ liter sol'n}} = \frac{0.22 \text{ mole } CCl_4}{1 \text{ liter sol'n}} \equiv 0.22M$$

Example Problem 3: 45 gm of C_3H_7OH, propyl alcohol, is added to 132 gm of H_2O. What is the weight molarity of the solution?

In this solution, we have

$$\frac{45 \text{ gm } C_3H_7OH}{132 \text{ gm } H_2O \text{ (solvent)}}$$

From the molecular weight of C_3H_7OH (60), we can determine the number of moles of propyl alcohol in this amount of water; in order to determine the molality, however, we need the number of moles of propyl alcohol in 1 kg of water. In unit notation, it follows that

$$\frac{45 \text{ gm } C_3H_7OH}{132 \text{ gm } H_2O} \times \frac{1 \text{ mole } C_3H_7OH}{60 \text{ gm } C_3H_7OH} \times \frac{1000 \text{ gm } H_2O}{\text{kg } H_2O} = \frac{5.7 \text{ moles } C_3H_7OH}{\text{kg } H_2O \text{ (solvent)}}$$

but

$$\frac{5.7 \text{ moles } C_3H_7OH}{\text{kg (solvent)}} \equiv 5.7\ m_w$$

Example Problem 4: An aqueous solution contains 28.1% by weight of ethyl alcohol (C_2H_5OH). The density of this solution is 0.960 gm/cc. (*a*) What is the mole fraction? (*b*) What is the molarity? (*c*) What is the molality?

(*a*) From the information given, we know that, in 100.0 gm in solution, we have 28.1 gm of ethyl alcohol and 71.9 gm of water. If we use the molecular weight of ethyl alcohol and water, we can easily convert these weights to the number of moles in 100 gm of solution, viz:

$$28.1 \text{ gm } C_2H_5OH \times \frac{1 \text{ mole } C_2H_5OH}{46.0 \text{ gm } C_2H_5OH} = 0.610 \text{ mole } C_2H_5OH$$

$$71.9 \text{ gm } H_2O \times \frac{1 \text{ mole } H_2O}{18.0 \text{ gm } H_2O} = 3.98 \text{ moles } H_2O$$

Thus we have a total of 4.59 moles and the mole fraction of ethyl alcohol is

$$X_a = \frac{0.610}{4.59} = 0.133$$

The mole fraction of water is 1 − 0.133 or 0.867.

(*b*) In part (*a*), we found that 0.610 mole of ethyl alcohol is dissolved in enough water to make 100 gm of solution. We can express this as

$$\frac{0.610 \text{ mole } C_2H_5OH}{100 \text{ gm sol'n}}$$

Since we know the weight of 1 cc of solution, i.e., the density, we can calculate the number of moles in 1 cc of solution.

$$\frac{0.610 \text{ mole } C_2H_5OH}{100 \text{ gm sol'n}} \times \frac{0.960 \text{ gm sol'n}}{\text{cc sol'n}} = \frac{0.00586 \text{ mole } C_2H_5OH}{\text{cc sol'n}}$$

Molarity is the number of moles in 1 liter of solution; hence

$$\frac{0.00586 \text{ mole}}{\text{cc sol'n}} \times \frac{1000 \text{ cc sol'n}}{1 \text{ liter sol'n}} = \frac{5.86 \text{ moles } C_2H_5OH}{1 \text{ liter sol'n}} \equiv 5.86M$$

(*c*) In part (*a*), we found that there is 0.610 mole of ethyl alcohol dissolved in 71.9 gm of solvent; hence we can readily calculate the number of moles of ethyl alcohol in 1 kg of solvent. Thus

$$\frac{0.610 \text{ mole } C_2H_5OH}{71.9 \text{ gm solvent}} \times \frac{1000 \text{ gm solvent}}{1 \text{ kg solvent}} = \frac{8.48 \text{ moles}}{\text{kg solvent}} \equiv 8.48\ m_w$$

COLLIGATIVE PROPERTIES OF SOLUTIONS

Many of the properties of solutions depend on the concentration of the solute rather than the chemical nature of the solute. These properties are called *colligative* properties. In discussing them, we shall assume that we are dealing with an *ideal solution;* that the *solute is non-volatile;* that the *solute is insoluble in the solid solvent.*

The vapor pressure of the solvent. Figure 8, is a vapor pressure diagram for the solvent as a function of temperature. P_0 is the vapor pressure of the

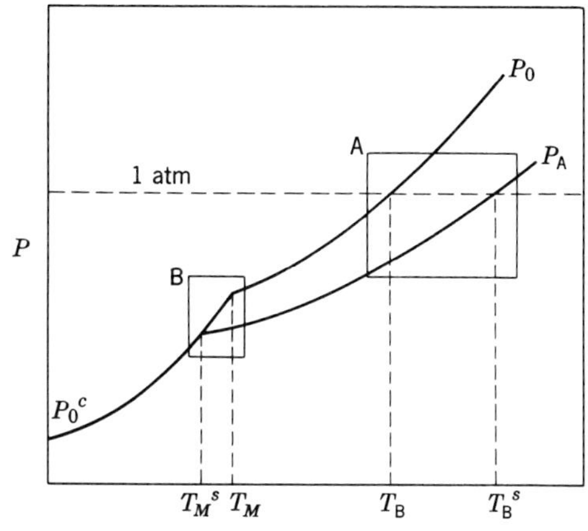

Fig. 8. Vapor pressure plot for the solvent.

pure solvent; P_A is the pressure of solvent vapor above the solution. Two things should be noted. First, since the vapor pressure of the solvent over a solution is always less than the vapor pressure of the pure solvent, the temperature at which the vapor pressure is 1 atm, i.e., the normal boiling point, is always higher for the solution. Therefore, whenever we add non-volatile solute to a solvent, the boiling point increases. If we freeze the solution, the vapor pressure of the solvent becomes equal to the vapor pressure of the solid solvent at a temperature below the normal melting point of the pure solvent. Therefore, whenever we dissolve a solute in a solvent, the freezing point decreases. In this section, we are going to deal quantitatively with these phenomena. The two regions of the vapor pressure curve in the neighborhood of the melting point and the boiling point are greatly magnified in Figs. 9 and 10.

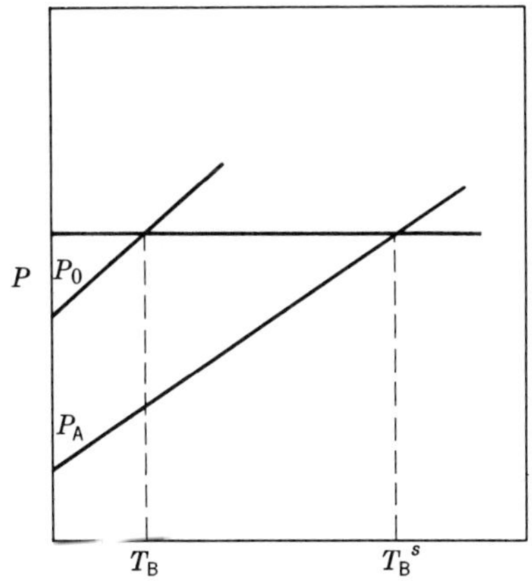

Fig. 9. Vapor pressure at boiling point.

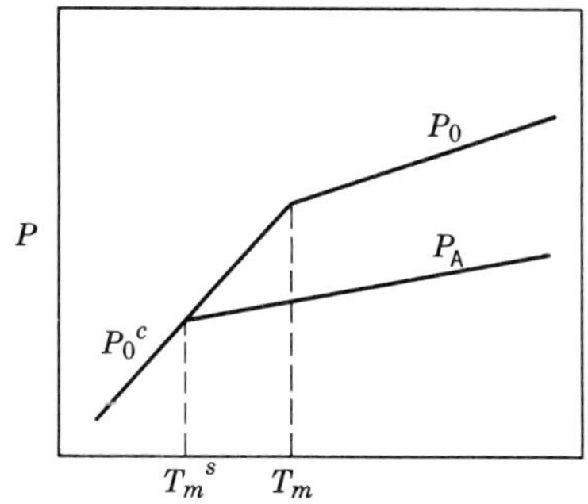

Fig. 10. Vapor pressure at melting point.

Boiling point elevation. In the limited region shown in Fig. 9, we can approximate the vapor pressure versus temperature plots by straight lines. The equations of these straight lines are

$$P_0 = \gamma(T - T_b) + 1 \text{ atm}$$

where γ is the slope. Since $P_A = X_a P_0$,

$$P_A = X_a \gamma(T - T_b) + X_a = (1 - X_b)\gamma(T - T_b) + (1 - X_b)$$

In these equations γ is the effective slope of P_0 near the normal boiling point T_b and $X_a(X_b)$ is the mole fraction of solvent (solute) in the solution. As required, when the temperature is T_b, the normal boiling point, the vapor pressure of the pure solvent is 1 atm.

At the normal boiling point of the *solution*, T_b^s, the vapor pressure of the solvent, P_A, must be equal to 1 atm; hence we write for the solution

$$P_A = 1 = \gamma\,\Delta T - \gamma X_b\,\Delta T + 1 - X_b$$

where

$$\Delta T = (T_b^s - T_b)$$

is the boiling point elevation. This equation can be rearranged to yield

$$\gamma\,\Delta T - \gamma X_b\,\Delta T = X_b$$

When X_b is very small, as it will be for dilute solutions, ΔT is also small. Thus, for dilute solutions, the first term on the left is a small quantity, but the second term is an even smaller quantity since it is the product of two small quantities; consequently, we neglect the second term with respect to the first. This yields

$$\Delta T = \frac{X_b}{\gamma}$$

For dilute solutions,

$$X_b \approx \frac{M_a}{1000} m_w$$

Hence we can rewrite this in the form

$$\Delta T = \frac{M_a}{1000\gamma} m_w = C_e\, m_w$$

Thus, for dilute solutions, the elevation of the boiling point is proportional to the weight molarity of the solute.

An important aspect of this equation is that C_e is determined only by the solvent; hence the ΔT for a one molar solution is the same regardless of the solute. In Table 2 we have listed the values of C_e for a number of solvents.

Table 2 Boiling Point Elevation and Freezing Point Depression Constants

Substance	C_e	C_d
Water	0.51	−1.86
Ethyl Alcohol	1.22	—
Ethyl ether	2.02	—
Carbon tetrachloride	5.03	—
Benzene	2.53	−4.90
Hexane	2.75	—
Cyclohexane	2.79	—
Phenol	3.56	−4.7
Chloroform	3.63	−7.3

Freezing point depression. In discussing the freezing point depression, we again make the approximation that, in the region shown in Fig. 10, all these plots of vapor pressure versus temperature are straight lines. The equations for these straight lines are

$$\begin{aligned}&\text{Pure liquid solvent} && P_0 = \beta(T - T_m) + P_M\\ &\text{Pure solid solvent} && P_0^c = \alpha(T - T_m) + P_M\\ &\text{Solvent from solution} && P_A = X_a P_0 = X_A\beta(T - T_m) + X_a P_M\end{aligned}$$

T_m is the normal melting point of the pure solvent, and P_M is the vapor pressure of the pure solid (or pure liquid) at T_m. As in the preceding case, we rewrite the last equation in terms of the mole fraction of solute, X_b, i.e.,

$$P_A = (1 - X_b)\beta(T - T_m) + (1 - X_b)P_M$$

At T_m, the melting point of the solution, the vapor pressure of the pure crystalline solvent, P_0^c, is equal to P_A, the vapor pressure of the solvent; hence

$$\alpha\,\Delta T + P_M = \beta\,\Delta T - X_b\beta\,\Delta T + P_M - X_b P_M$$

where

$$\Delta T = T_m^s - T_m$$

is the freezing point depression. Whenever X_b is very small, i.e., the solution is dilute, the second term on the right side of the preceding equation involves the product of two very small quantities and can be neglected. Then the equation rearranges to

$$\Delta T = \left(\frac{P_M}{\beta - \alpha}\right) X_b \approx \left(\frac{P_M}{\beta - \alpha}\right) \frac{M_A}{1000}\, m_w \equiv C_d m_w$$

Once again, this equation shows that C_d depends only on the solvent; for a given concentration of solute *of any kind*, the depression of the freezing point is the same. In Table 2, we have listed values of C_d for various solvents.

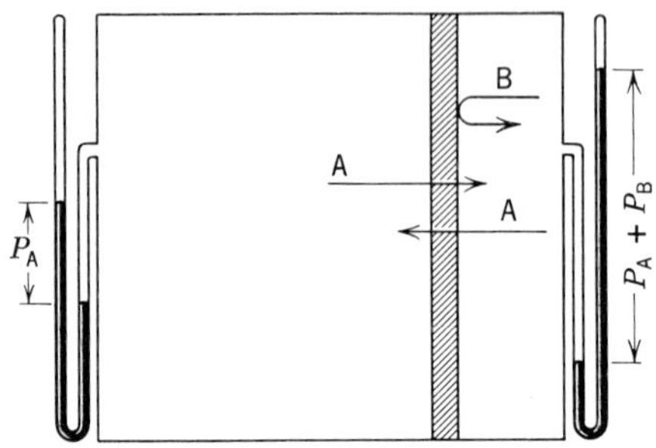

Fig. 11. Osmotic pressure apparatus.

Osmotic pressure. A physical picture of osmotic pressure in solution is difficult to formulate in a simple manner. In view of this, let us first consider osmotic pressure of gaseous solutions.

Figure 11 shows an experimental apparatus that contains two gases, A and B, on the right side of a box and only one gas, A, on the left side. The box is divided by a partition that is permeable only to A molecules; A molecules pass freely through this partition, but the B molecules do not. Suppose that at the start of the experiment the *total* pressure is the same in both compartments. Then, because the partial pressure of A on the right is less than that on the left, there is a net flow of A molecules from left to right. This flow continues until the partial pressure of A is the same on either side of the partition and equilibrium is attained. For such an equilibrium we can write

$$P_{\text{A}}^{L} = P_{\text{A}}^{R} = X_a P_{\text{B+A}}^{R} = (1 - X_b) P_{\text{B+A}}^{R}$$

where the superscripts refer to the boxes and the subscripts refer to gases and X_a is the mole fraction of A on the right side. Under these conditions there is a difference in total pressure on the two sides of the semipermeable partition. This pressure difference, π, is called the osmotic pressure and is given by the equation

$$\pi = P_{\text{A+B}}^{R} - P_{\text{A}}^{L} = X_b P_{\text{A+B}}^{R}$$

With the help of the ideal gas law this becomes

$$\pi = X_b \left(\frac{n_a + n_b}{V^R} \right) RT = MRT$$

where M is the molarity of B.

A similar experiment can be performed with liquid solutions. Suppose we fill flask I with water and flask II with a solution of sugar in water. Then we connect the two flasks by a membrane permeable only to water. If we examine the flasks from time to time, we find that water flows from flask I to flask II. This flow, of course, causes an increase in the pressure in flask II relative to that in flask I.

When there is no further flow, equilibrium is attained and the pressure difference, π, is that calculated from the equation derived for gases.

As can be seen from the fundamental equation, studies of osmotic pressures provide a means of determining molecular weights just as the gas law does. Moreover, since even dilute solutions are concentrated relative to dilute gases, the osmotic pressure developed by solutions containing relatively few molecules is measurable. For a long time one of the more pressing problems in chemistry was the determination of the molecular weights of giant molecules like proteins. Even if they were quite soluble, on a weight basis, their molecular weights were so high ($\sim 10^5$) that the number of moles present in solution was quite low. Of all the colligative properties, only osmotic pressure was sensitive enough to yield accurate results. Since for giant molecules it is easy to find fine-grained membranes permeable only to small solvent molecules, this technique has been widely utilized to study such molecules.

Although the expression for osmotic pressure of a liquid solution is identical with that for a gaseous solution, the nature of osmotic pressure is somewhat different. Imagine that Fig. 11 represents liquid water (A) and a solution of sugar (B) in water (A) which are separated by a membrane permeable only to water vapor. The vapor pressure of pure water on the left is greater than that above the solution; hence a flow of water occurs from left to right and a consequent pressure build-up occurs in the right compartment. This would continue indefinitely if the vapor pressure did not depend on the total pressure. If we exert a *large* pressure on a solution, compression occurs, the free volume is reduced, and the vapor pressure increases. Therefore, as the pressure on the solution increases, the vapor pressure increases. The pressure needed to make the vapor pressure of the solution equal to that of the pure solvent is the osmotic pressure, π.

Example Problem 5: In this problem, we are going to carry out calculations involving the colligative properties discussed in the preceding paragraphs. To show clearly the interrelation of these properties, we deal with a single solution. The molal boiling point elevation constant of water is 0.51°C. The molal freezing point depression constant is 1.86°C. An antifreeze solution containing glycerin, $C_3H_8O_3$ (molecular weight = 92.1 gm/mole) is diluted with water so that the resulting solution contains 40 weight % glycerin. If this solution is ideal:

(*a*) What is the pressure of water vapor above the solution at 35°C? At this temperature, the vapor pressure of pure water is 42.2 mm.

(*b*) What is the freezing point of this solution?

(*c*) What is the boiling point of this solution?

(*d*) What is the osmotic pressure of this solution at 35°C if the density is 1.099 gm/cc?

(*a*) This part involves direct application of Raoult's law. The first step is to calculate the mole fraction of solvent. Each 100 gm of solution contains 40 gm of glycerin and 60 gm of water; hence, in a 100-gm sample of solution, we have

$$40 \text{ gm } C_3H_8O_3 \frac{1 \text{ mole } C_3H_8O_3}{92.1 \text{ gm } C_3H_8O_3} = 0.43 \text{ mole } C_3H_8O_3$$

and

$$60 \text{ gm } H_2O \times \frac{1 \text{ mole } H_2O}{18 \text{ gm } H_2O} = 3.3(3) \text{ moles } H_2O$$

$$(\text{total} = 3.76 \text{ moles})$$

The mole fraction of water is 3.33/3.76 = 0.89 and, from Raoult's law, the pressure of water vapor in contact with this solution is

$$P_{H_2O} = 0.89 \text{ (42.2 mm)} = 38 \text{ mm}$$

(*b*) The first step in the calculation of the freezing point depression is to express the concentration in molality. From part (*a*), we know that there is 0.43 mole of glycerin in every 60 gm of water, and we can convert to molality as follows:

$$\frac{0.43 \text{ mole } C_3H_8O_3}{60 \text{ gm } H_2O} \times \frac{1000 \text{ gm } H_2O}{1 \text{ kg } H_2O} = \frac{7.2 \text{ moles } C_3H_8O_3}{1 \text{ kg } H_2O} \equiv 7.2\ m_w$$

The freezing point of water is lowered 1.86°C for a one molal solution; hence, for this solution, the freezing point depression is

$$7.2\ m_w \times \frac{1.86°\text{C dep.}}{1\ m_w} = 13.4°\text{C dep.}$$

Since the normal freezing point of pure water is 0°, the freezing point of this solution is −13.4°C.

(*c*) The calculation of the boiling point elevation is similar to the calculation above. Thus we write

$$\frac{0.51°\text{C elev.}}{1\ m_w} \times 7.2\ m_w = 3.7°\text{C elev.}$$

Since the boiling point of pure water is 100°C, the boiling point of this solution is 103.7°C.

(*d*) The formula for osmotic pressure is identical in form with the ideal gas law. Accordingly, except for the initial step, the procedure in this calculation will be similar to those we encountered in Chapter 8. First of all, we must know the number of moles of solute in a given volume of solution. We can calculate this in the following way:

$$\frac{0.43 \text{ mole } C_3H_8O_3}{100 \text{ gm sol'n}} \times \frac{1.099 \text{ gm sol'n}}{1 \text{ cc sol'n}} = \frac{0.0047 \text{ mole}}{\text{cc sol'n}}$$

For 0.0047 mole of gas at STP, the volume occupied would be

$$0.0047 \text{ mole} \times \frac{22{,}413 \text{ cc STP}}{1 \text{ mole}} = 105 \text{ cc STP}$$

The volume at 35°C can be calculated as follows:

$$105 \text{ cc} \times \frac{308}{273} = 118 \text{ cc}$$

This, then, would be the volume that 0.0047 mole of gas would occupy at 1 atm. The osmotic pressure, π, must be such that this volume is reduced to 1 cc. Hence we can write

$$118 \times \frac{1}{\pi} = 1$$

and $\pi = 118$ atm.

Example Problem 6: The colligative properties of solutions are often utilized for the determination of molecular weights. Suppose we have 5 gm of x dissolved in 100 gm of water, and the freezing point of the solution is −0.46°C. What must be the value of M_x, the molecular weight of x?

Clearly, the molality is given by

$$\frac{5/M_x}{100 \text{ gm } H_2O} \times \frac{1000 \text{ gm } H_2O}{\text{kg } H_2O} = \frac{50}{M_x} m_w$$

The freezing point depression is 0.46°C; hence

$$\left(\frac{50}{M_x}\right) m_w \times \frac{1.86°\text{C}}{m_w} = 0.46$$

and

$$M_x = 2.0 \times 10^2$$

DISSOCIATION IN SOLUTION

We have assumed in our discussion that the formation of a solution does not destroy the molecular integrity of solvent and solute. In other words, we have assumed that, when methyl alcohol is dissolved in water, the individual CH_3OH and H_2O molecules have almost the same form as in pure methyl alcohol and in pure water. This is what actually occurs for many solutions. But, if the solute is an ionic compound or the bonds in the solute molecule are largely ionic, it is possible that solution in a polar solvent, such as water, will rupture the ionic bonds and result in the *dissociation* of the solute into ions. The properties of a solution in which dissociation occurs are radically different from those of a solution in which no dissociation occurs. In this section we shall discuss these differences together with the factors that lead to dissociation.

Conductivity of solutions. An electrical current is a flow of elementary charges in just the same way that a current of water is a flow of water molecules. Figure 12*a* pictures a closed circuit in which water is pumped in a clockwise direction through a pipe; water molecules are not created or destroyed but are pushed around the circuit. We have included in the line the turbine *T*. The

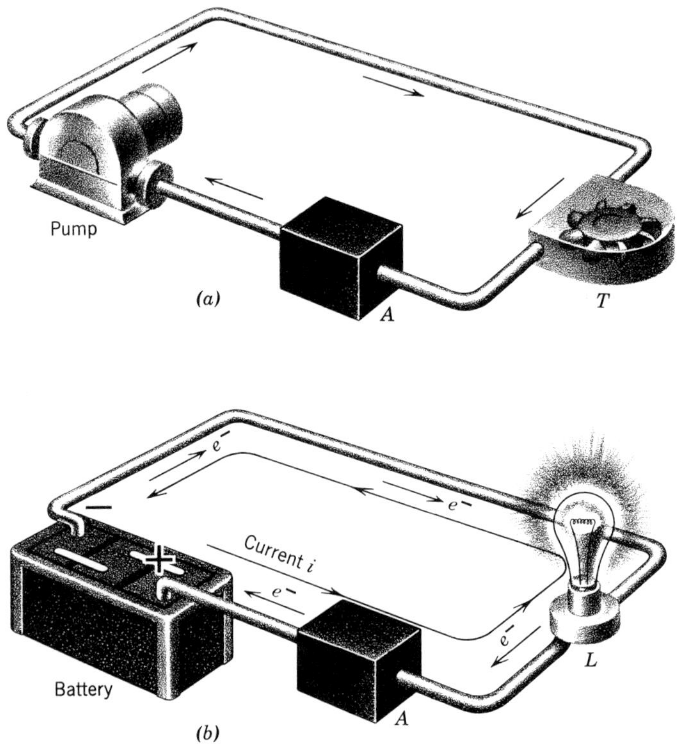

Fig. 12. Currents.

revolutions per minute of the turbine provide a measure of the current. If there is a section of the circuit, e.g., *A*, that does not contain and therefore cannot conduct water molecules, no current can exist and the turbine will not turn.

Figure 12*b* is the electrical analog of Fig. 12*a*. The battery functions as a pump and pushes electrons into the metal wire, which contains nearly free electrons, through the lamp *L* and back to the electrical pump, the battery. During operation, charges are not created or destroyed but are pushed around the circuit. The brightness of the lamp provides a measure of the current. (Note that the conventional current *i* is opposite to the flow of electrons.) If there is a section of the circuit, e.g., *A*, that does not contain and therefore cannot conduct charges, no current can exist and the lamp will not light.

Now suppose we carry out the experiments pictured in Fig. 13. In these experiments, part *A* of Fig. 12*b* is formed by immersing the two wires in a series of liquids. In Fig. 13*a* the liquid is pure water. In Figs. 13*b*, 13*c*, and 13*d*, the liquids are solutions of CH_3OH, $HgCl_2$, and NaCl in water at equal concentrations. Pure water contains no charges† and will not conduct electrons; hence the lamp in Fig. 13*a* does not light. From this experiment we see that, if the lamp lights for any of the solutions, the current flow is due to the solute rather than the water. For the CH_3OH solution the lamp does not light; for the $HgCl_2$ solution the lamp lights dimly; for the NaCl solution the lamp lights brightly. Thus we conclude that the CH_3OH solution contains no ions and is a non-conductor; the $HgCl_2$ partially dissociates to form ions and conduct electricity; and the NaCl dissociates extensively to form many ions and conduct electricity. In other words, when dissociation occurs, the solution completes the circuit with conduction in solution by the ionic current depicted in Figs. 13*c* and 13*d*.

Solutions that dissociate and conduct electricity are called *electrolytes.* On this basis we refer to CH_3OH as a non-electrolyte since its solution does not conduct electricity; we refer to $HgCl_2$ as a weak electrolyte since its solution just barely conducts electricity; and we refer to NaCl as a strong electrolyte since its solution is a good conductor.

The ions formed by an electrolyte in solution are usually obvious. For example, in both NaCl and $HgCl_2$ the chlorine atoms are the most electronegative; hence the chlorine atom tends to retain an extra electron and form the anion, Cl^-, whereas the sodium and mercury form cations. For compounds, like NaOH and H_2SO_4, the rupture usually occurs at the most ionic bond. Thus we find that these two compounds dissociate as follows:

$$Na—OH \rightarrow Na^+ + OH^-$$

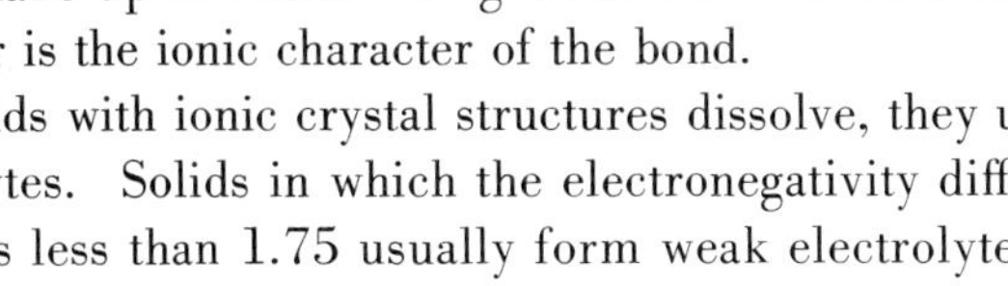

We can determine the most ionic bond from the electronegativity difference of the atoms that make up the bond: the greater the difference in the electronegativity, the greater is the ionic character of the bond.

Whenever solids with ionic crystal structures dissolve, they usually dissociate to form electrolytes. Solids in which the electronegativity difference of the cation and anion is less than 1.75 usually form weak electrolytes, whereas those for which the electronegativity difference is greater than 2.00 usually form strong electrolytes. If a compound forms a molecular crystal, the compound is usually a non-electrolyte. Exceptions occur most often for hydrogen-containing com-

† We shall see later that water does contain some H^+ and OH^- ions. In pure water there is roughly one ion per 10^8 water molecules. This is not enough to conduct an appreciable current.

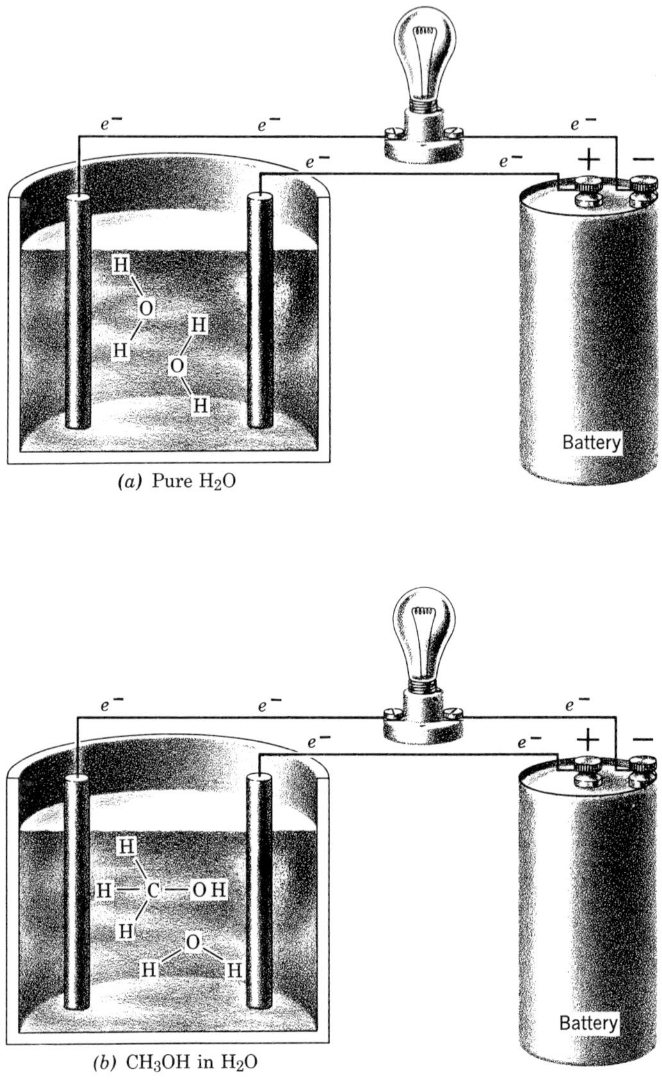

Fig. 13. Conductivity of solutions.

pounds that can form the H^+ cation. In such compounds, the tendency for dissociation is enhanced. Thus, although HCl forms a molecular crystal, it dissociates to form ions in water.

Most organic compounds form molecular crystals; hence, if they dissolve in water, they do not dissociate. The organic acids, which have the general formula

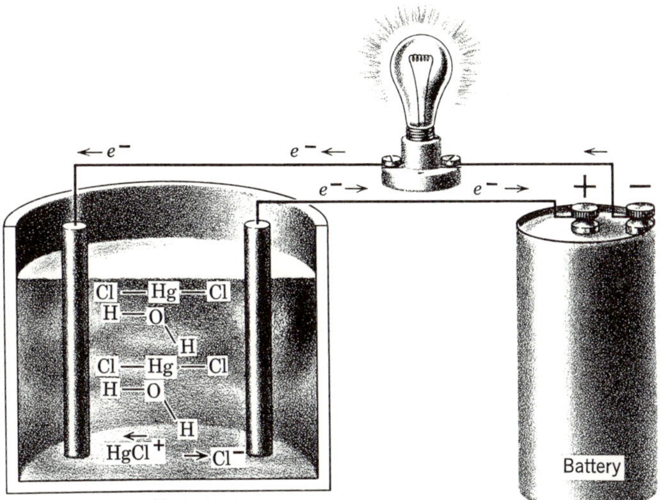

(c) $HgCl_2$ in H_2O

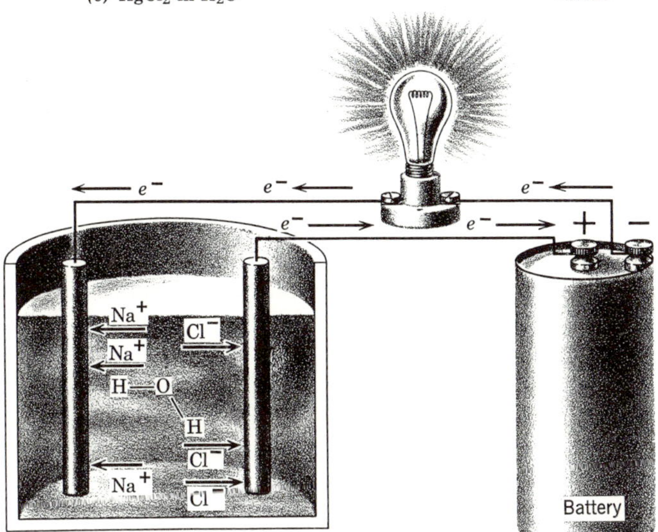

(d) Na^+Cl in H_2O

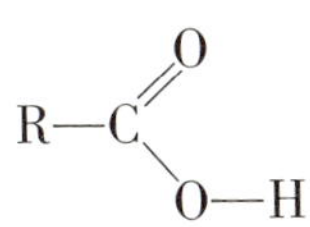

do dissociate slightly in water as follows:

$$R{-}\overset{\overset{\displaystyle O}{\|}}{C}{-}O{:}H \rightarrow [R{-}\overset{\overset{\displaystyle O}{\|}}{C}{-}O]^- + H^+$$

As we pointed out earlier (p. 201), dissociation occurs for acids but not for alcohols (general formula ROH) because of resonance stabilization of the

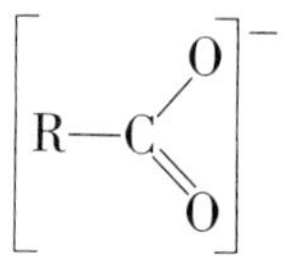

anion. Organic acids are usually weak electrolytes.

In Table 3, we have indicated some typical *strong*, *weak*, and *non*-electrolytes. In general, there is a continual gradation in behavior. For simplicity, we shall arbitrarily assume that: if a half or more of the dissolved molecules are dissociated, the compound is a strong electrolyte; if more than one out of ten thousand but less than half of the dissolved molecules dissociate, it is a weak electrolyte; if less than one out of ten thousand of the dissolved molecules dissociate, it is a non-electrolyte. Note that the division is based on the fraction of the *dissolved* molecules that dissociate; hence it is possible for a sparingly soluble salt like $BaSO_4$ to be a strong electrolyte and it is possible for a highly soluble salt like $HgCl_2$ to be a weak electrolyte, even though in a saturated solution both contain comparable numbers of ions.

Ionic equations. Chemical equations represent, as closely as possible, the changes that occur in a chemical reaction. If we are to be consistent with this ideal, the existence of strong and weak electrolytes forces us to review the method of writing chemical equations for reactions that occur in solution. For example, the following reaction occurs rapidly when solutions of $AgNO_3$ and KCl are mixed.

$$AgNO_3 + KCl \rightarrow \underline{AgCl} + KNO_3$$

(A line under a formula for a reaction in solution indicates an insoluble compound.) In view of Table 3, however, this equation is misleading. $AgNO_3$, KCl, and KNO_3 are all strong electrolytes and exist primarily as ions in solution. Thus it is closer to the truth to write

$$Ag^+ + NO_3^- + K^+ + Cl^- \rightarrow \underline{AgCl} + K^+ + NO_3^-$$

This is known as an *ionic equation.* It is redundant in so far as K^+ and NO_3^- are found in solution before and after reaction. The essential change can be represented by

$$Ag^+ + Cl^- \rightarrow \underline{AgCl}$$

This type of equation is called an *essential ionic equation.*

Essential ionic equations are far more informative than other chemical equations. The preceding equation states that, in any solution containing chloride ions and silver ions, solid silver chloride will form. The source of the chloride

Table 3 Types of Solutions Formed by Various Compounds

Compounds	Crystal Type[a]	Solutions
Halides, hydroxides, and acetates of Group I and Group II elements	Ionic	Strong electrolytes
Nitrates, chlorates, and sulfates of M^{+1} and M^{+2} cations	Ionic	Strong electrolytes
$PbBr_2$, $PbCl_2$, Pb(acetate)$_2$, $HgCl_2$, $CuCl_2$	Ionic to molecular	Weak electrolytes
HCl, HBr, HI	Molecular	Strong electrolytes
H_2SO_4, $HClO_4$, HNO_3	Molecular (H bonding)	Strong electrolytes
RC(=O)—OH, H_2CO_3	Molecular	Weak electrolytes
R—O—H (alcohols), HCN, other organic molecules	Molecular	Non-electrolytes

[a] Not all crystal structures in each group are known. This represents the structure most common in each group.

ion can be any strong electrolyte that dissociates to form, Cl^-, e.g., NaCl, KCl, RbCl, $MgCl_2$, $CaCl_2$, $BaCl_2$, HCl; the source of the silver ion can be $AgNO_3$, Ag_2SO_4, $AgClO_4$, etc. All these possibilities are contained in the single essential ionic equation; it would take twenty-one molecular equations to represent them.

In essential ionic equations, weak electrolytes and non-electrolytes are still represented by molecular formulas. For example, the essential ionic equations for the reactions

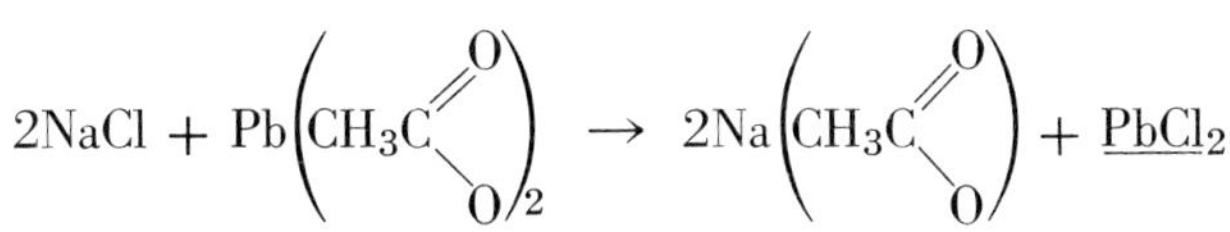

[Pb(acetate)$_2$]

and

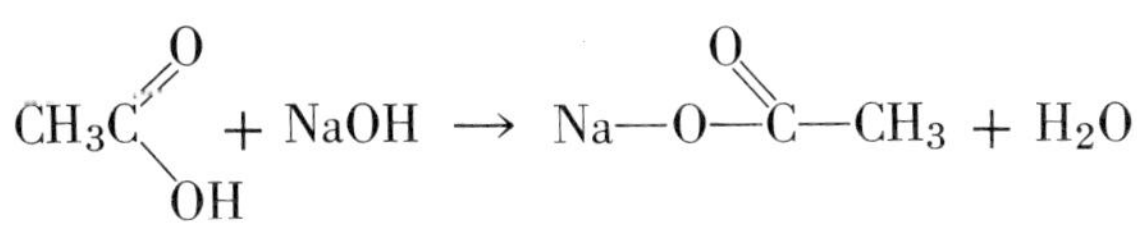

Acetic acid

are

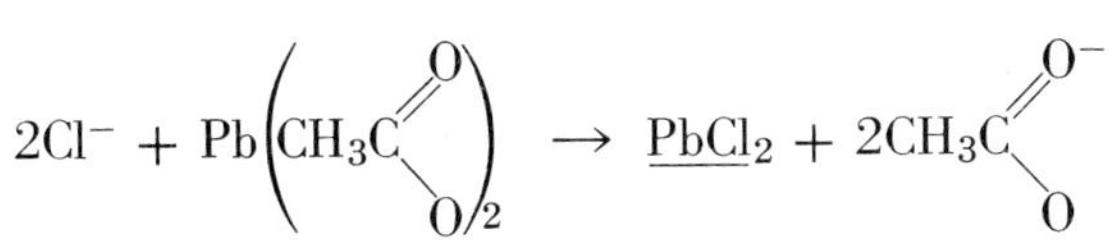

and

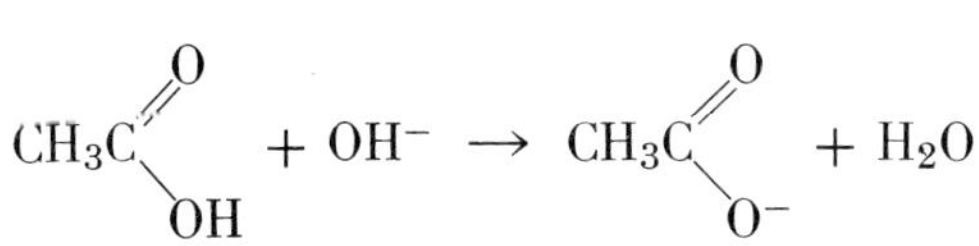

Since the essential ionic equations are the most general, we shall use them whenever possible.

Hydration. The solution of a strong electrolyte such as NaCl can be imagined to occur by the consecutive steps

$$\text{NaCl (crystal)} + \text{(water)} \rightarrow \text{NaCl (aq.)}$$

$$\text{NaCl (aq.)} \rightarrow \text{Na}^+ \text{ (aq.)} + \text{Cl}^- \text{ (aq.)}$$

wherein the symbol (aq.) indicates that the species is surrounded by water molecules. When the NaCl crystal dissolves in water, the breakdown into separate ions is complete and requires only 1 kcal of energy for each mole of NaCl dissolved in a large amount of water. On the other hand, when the NaCl crystal is separated into its constitutent ions in the absence of water, i.e., in a vacuum, about 180 kcal of energy is required; hence, in the vapor, NaCl has a strong tendency to form molecules rather than ions. Moreover, in a wide variety of organic solvents, dissolved NaCl forms NaCl molecules rather than dissociated ions. Thus there appears to be some property of water that promotes dissociation into ions.

In the simplest sense, the dissociation of an NaCl molecule involves the separation of two opposite charges. Since there is coulombic force of attraction between these two charges, work must be done and energy must be supplied to separate the charges. The general formula for this coulombic force F is

$$F = \frac{Q_1 Q_2}{Kr^2}$$

where Q_1 and Q_2 are the two charges, r is the distance between the charges, and K is called the *dielectric constant* (cf. p. 153) of the surroundings of the charges. For a vacuum, $K = 1$. On p. 131, we computed the energy required to separate an NaCl molecule into Na^+ and Cl^- ions in a vacuum and found it to be about 115 kcal/mole (~5 ev/molecule).

Water, however, consists of polar molecules oriented by hydrogen bonding; as a consequence, for water $K = 80$ at 20°C. Thus, if we compute the energy necessary to separate an NaCl molecule into its component ions in water, we find we need one-eightieth as much energy as in a vacuum, or roughly 1 kcal.

From a molecular viewpoint we can imagine that solution occurs by the steps pictured in Fig. 14. Figure 14*a* is a schematic picture of a water molecule wherein the + and − signs indicate that the electron distribution results in an excess of positive charge on the hydrogen atoms and an excess of negative charge on the oxygen atom. Figure 14*b* indicates an NaCl ion molecule wherein the two atoms have a unit positive charge and a unit negative charge, respectively.

In order to separate these ions in a vacuum we must do work against the attractive force of the two unit charges. When this molecule is in water, coulombic

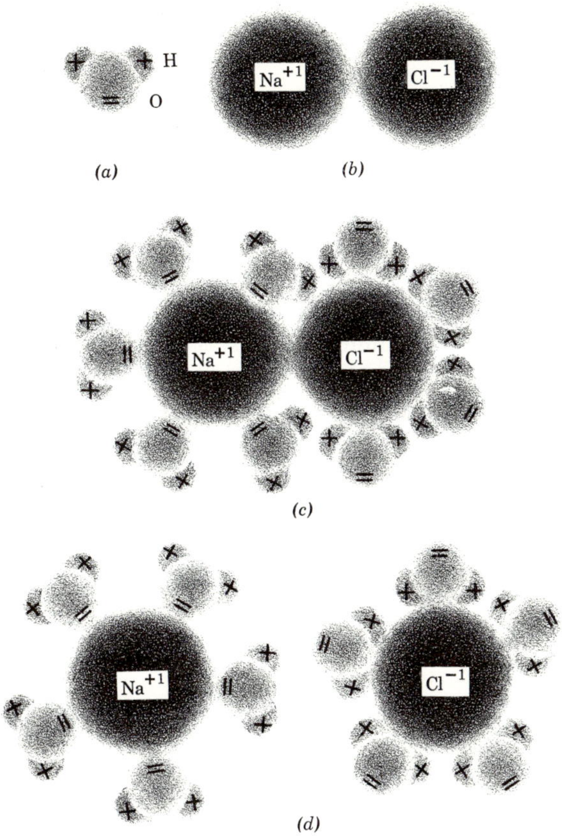

Fig. 14. Hydrated ions.

forces favor a clustering of the type shown in Fig. 14*c*, wherein the negative (oxygen) end of the water molecule is closest to the Na^+ ion and the positive (hydrogen) end of the water molecule is closest to the chloride ion. This arrangement almost neutralizes the charge and reduces the attractive force to one-eightieth of its value in a vacuum.

Figure 14*d* is a schematic picture of the separated ions in aqueous solution. The ions are not truly free but carry with them a sheath of water molecules that spread the net unit charge over a much bigger volume and thereby largely eliminate the interionic attraction. The formation of this sheath of oriented water molecules is called *hydration*. The energy given off by the approach of the positive end of a water molecule to the Cl^- ion and the approach of the negative end of the water molecules to a Na^+ ion is enough to compensate almost completely for the energy needed to separate the ions.

Table 4

Ion	Radius (Å)	Hydration Energy (kcal)	Ion	Radius (Å)	Hydration Energy (kcal)
H^+	—	263			
Li^+	0.60	125	Li^+	0.60	125
Na^+	0.95	100	Mg^{+2}	0.65	464
K^+	1.33	79	Sc^{+3}	0.81	947

The process of hydration is essentially the bringing together of unlike charges. Therefore the energy of hydration decreases as the size of the ion increases provided the charge stays constant, and it increases as the charge increases provided the size stays constant. Ions with a rare gas structure closely conform to the pattern, as may be seen from Table 4.

The hydration energy of the hydrogen ion is enormous. This is not altogether unexpected since the H^+ ion is a bare nucleus and can approach very close to a water molecule. Today we believe that a hydrogen ion in solution is bound to only one water molecule to form the *hydronium* ion,

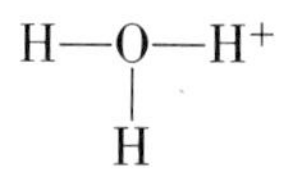

In the future, for simplicity, we shall often refer to H^+ ions in solution, but we shall do so with the full realization that they really exist as hydronium ions.

COLLIGATIVE PROPERTIES OF IONIC SOLUTIONS

Strong electrolytes. Suppose we prepare two aqueous solutions, one of them 0.01 molal in CH_3OH and the other 0.01 molal in NaCl† and determine the freezing point. Since the freezing point depression constant is 1.86°C for water, the first solution freezes at −0.0186°C. In the second solution, however, dissociation occurs. The resulting solution is 0.01 molal in Na^+ *and* 0.01 molal in Cl^-. If we re-examine the derivations dealing with colligative properties, we find

† The definition of molarity and molality involves the concept of a mole and a molecule. For ionic compounds, molecules as such do not exist. Whenever we apply the mole-molecule terminology to an ionic compound, we shall proceed as if the formula represented a molecule, i.e., as if NaCl or $CaCl_2$ represented a molecule of the compound. Often the fact that the concept of a mole is not strictly applicable to ionic compounds is emphasized by calling 58.5 gm of NaCl one gram *formula weight* rather than one molecular weight, and by calling a one molar NaCl solution a one *formal* solution or a one molal NaCl solution a one *weight formal* solution.

Dissociation gives rise to difficulties. To meet this we use the following convention: if we specify the concentration as $0.1M$ AB and AB dissociates, we mean that one liter of this solution contains 0.1 dissolved formula weight of AB; we do not mean that the concentration of undissociated AB is $0.1M$.

that only the number of solute species is important; hence the freezing point depression of 0.01 molal NaCl is that of a 0.02 molal solution, i.e., $-0.0372°C$.

Similar results obtain for other completely dissociated electrolytes. For example, if 0.01 molal solutions of KCl, $CaCl_2$, and $Al_2(SO_4)_3$ were completely dissociated in solution the freezing points would be -0.0372, -0.0558, and -0.0930, respectively. In other words, the concentration of ions for these three solutions are 0.02 molal, 0.03 molal, and 0.05 molal, respectively. This can readily be seen if we write the equations for the solution of these compounds:

$$KCl(s) \rightarrow K^+ \text{ (aq.)} + Cl^- \text{ (aq.)}$$
$$CaCl_2(s) \rightarrow Ca^{+2} \text{ (aq.)} + 2Cl^- \text{ (aq.)}$$
$$Al_2(SO_4)_3(s) \rightarrow 2Al^{+3} \text{ (aq.)} + 3SO_4^{-2} \text{ (aq.)}$$

It should be noted that the net effect depends on the number of ions formed, not on their charge. For example, 0.01 molal NaCl and 0.01 molal $MgSO_4$ both have the same melting point ideally even though NaCl forms $+1$ and -1 ions and $MgSO_4$ forms $+2$ and -2 ions.

What has been said about the freezing point depression applies *mutatis mutandis* to the other colligative properties. In other words, the boiling points of 0.01 molal KCl, $CaCl_2$, and $Al_2(SO_4)_3$ are 100.0102°, 100.0153°, and 100.0255°, respectively. The total concentration of all solute species, charged or neutral, determines the boiling point elevation. Similarly, 0.01 molal $Al_2(SO_4)_3$ will lower the vapor pressure five times as much as will 0.01 molal CH_3OH; and the osmotic pressure of 0.01 molal $Al_2(SO_4)_3$ is five times that of 0.01 molal CH_3OH. In all cases it is the total number of solute species that determines the colligative properties.

Weak electrolytes. The colligative properties of an ionic solution are often useful in determining the degree of dissociation of an ionic solution. Consider, for example, a 0.2 molal solution of MX_2. For generality, we assume that only a fraction of MX_2 dissociates. We can represent this by the equation

$$\begin{array}{ccccc} MX_2 & \leftrightarrows & M^{+2} & + & 2X^- \\ 0.2 - y & & y & & 2y \end{array}$$

where y is the number of moles of MX_2 in 1000 gm of solvent that dissociate. The molality of each species is $m_w\ (MX_2) = 0.2 - y$, $m_w\ (M^{+2}) = y$, $m_w\ (X^-) = 2y$; hence the total molality of the solute species is m_w (solute) $= (0.2 - y) + (y) + (2y) = 0.2 + 2y$. Now we can compute the freezing point depression as follows:

$$\Delta T = (0.2 + 2y) \times 1.86 = \left(\frac{0.2 + 2y}{0.2}\right)(0.2 \times 1.86) = (i) \times (0.2 \times 1.86)$$

The second factor in the last equation is what the freezing point depression would be *if* there were no dissociation; the first factor is the ratio of the freezing point depression with dissociation to that *if* there is no dissociation. This first factor provides a measure of the extent of dissociation. It is called the *van't Hoff factor, i,* after the man who first used it. This factor i is always 1 if there is no dissociation. If dissociation occurs and is complete, the factor i is equal to the moles of ions formed per mole of the initial compound.

In our specific example, when $y = 0$ (no dissociation), $i = 1$; when $y = 0.2$ (complete dissociation), $i = 3$, the moles of ions formed per mole of MX_2 present. For values of i between 1 and 3, dissociation is incomplete. For example, suppose $i = 1.05$. We can write

$$\left(\frac{0.2 + 2y}{0.2}\right) = 1.05$$

and if we solve for y we find that $y = 0.005$ or $y/0.200 = 0.025$ is the fraction of the MX_2 molecules that dissociate. This fraction is called the *degree of dissociation* and is often given the symbol α. This fraction was the basis of our classification in Table 3. On this basis, MX_2 is a weak electrolyte.

The degree of dissociation is a simple concept provided we assume that there is either complete dissociation or no dissociation. For example, the concept is clear for the salt $KHSO_4$ if we assume that the possible species in solution are those indicated by the equation

$$KHSO_4 \rightarrow K^+ + H^+ + SO_4^{-2}$$

In actual fact, solution and ionization of 1M $KHSO_4$ takes place in two steps:

$$KHSO_4 \rightarrow K^+ + HSO_4^-$$

$$HSO_4^- \rightarrow H^+ + SO_4^{-2}$$

The first step is virtually complete; for the second about 10 per cent of the HSO_4^- dissociates; hence the determined i factor would be 2.10. If we assume one-step dissociation, we calculate an α of 0.55. In other words, such a calculation says that the K^+ and SO_4^{-2} ion concentrations in a 1 molal solution are each 0.55, whereas actually they are 1.00 and 0.10, respectively. Similar conditions hold for MX_2.

In Table 5 we have listed i values for several electrolytes to indicate the range of values found.

Deviations from ideality. In the previous discussion we assumed that we were dealing with ideal solutions. Ideal behavior is always obtained for sufficiently dilute solutions, but at higher concentrations deviations from ideality occur. In effect, ideal behavior means that each solute species behaves independently, that the effect per ion is independent of the concentration. Thus, ideally, a 0.500 molal solution of KCl lowers the freezing point one hundred times that of a 0.005 molal solution or, alternatively, i will be independent of concentration.

Table 5 Van't Hoff Factor for Various Compounds

Compound	i
KCl	1.96[a]
Sulfuric acid, $H_2(SO_4)$	1.47[b]
Formic acid, $H(CHO_2)$	1.014[b]
Acetic acid, $H(CH_3CO_2)$	1.004[b]
Hydrazinium hydroxide, N_2H_5OH	1.001[b]
Hydrosulfuric acid, H_2S	1.00025[b]
Hydrocyanic acid, HCN	1.00002[b]

[a] For $0.005m_w$ solutions. [b] For $1m_w$ solutions.

In practice, this is seldom the case. For example, as m_w is changed from a very low value to 0.500, i changes from 2.00 to 1.80 for KCl and from 2.00 to 1.08 for $MgSO_4$. Thus, for 0.500 molal solutions, deviations from ideality are significant.

The origins of non-ideal behavior are the residual attraction between oppositely charged ions. Although, to a large extent, the effective ionic charge is neutralized by the hydration shell, residual effects remain. In dilute solutions, the ions are, on the average, far apart and interionic attractions are not strong enough to produce pronounced deviation from ideality. In more concentrated solutions, the ions are closer and this residual attraction is significant. Then the actions of a cation are affected by the nearby anions. The more concentrated the solution, the greater is this effect. Thus, in a concentrated solution, the independence of the cation is curtailed; when we increase the concentration one-hundred-fold, each ion becomes effectively less free and the van't Hoff factor i or the *apparent* dissociation drops to a lower value.

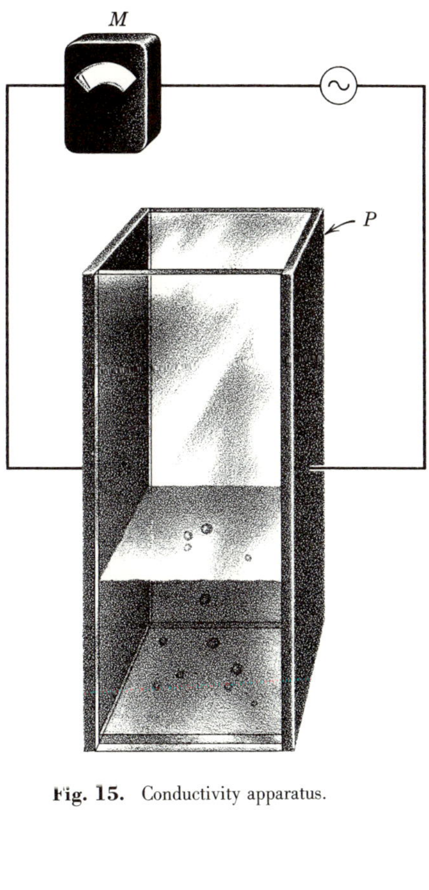

Fig. 15. Conductivity apparatus.

Interionic attractions affect not only the colligative properties but also the conductivity. Consider the apparatus shown in Fig. 15. The prism P is open at the top. The two shaded sides

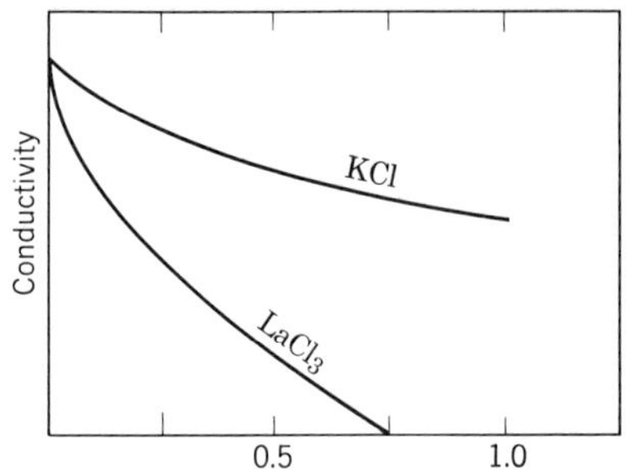

Fig. 16. Equivalent conductivity of KCl and $LaCl_3$ at various concentrations.

are made of metal; the other three sides are made of a non-conductor such as glass. The metallic sides are joined to a source of a-c voltage that produces a current that changes direction, say, 1000 times per second. The conductivity of a solution in P is directly proportional to the current measured by the meter M.

Add a 1 m_w solution of KCl to P as indicated and measure the conductivity. Then add enough water to dilute the solution to 0.5 m_w and lower concentrations, and for each concentration measure the conductivity. The results are pictured in Fig. 16; as the concentration increases, *a given amount of KCl* produces less of an effect.

A theory has been devised by Debye and Hückel that accounts for these observations. The theory assumes that substances like NaCl and $MgSO_4$ are completely dissociated even at the higher concentrations where i and the conductivity drop to low values. It also suggests that cations in solution tend to be surrounded by anions, and vice versa. When the solution is concentrated, the anions surrounding the cations are closer and exert a greater influence. Then the cations (and anions) are no longer free. When the solution conducts, cations and anions move in the opposite direction. The mutual attraction results in a mutual "drag" on the ions. The closer together the ions, i.e., the more concentrated the solution, the greater will be the drag effect; hence the conductivity (ease of ionic motion) is expected to decrease as the concentration increases.

Since the deviations from ideality arise from interionic attraction, these deviations are greatest for the more highly charged ion. Thus, in Fig. 16, the variation of conductivity with concentration is much greater for $LaCl_3$ than for KCl; in line with this, we find that the variation of i with concentration is greater for $MgSO_4$ than for KCl.

PROBLEMS

Assume that these solutions follow Raoult's law and are sufficiently dilute that the approximations we made in deriving the equations for freezing point depression, boiling point elevation, and osmotic pressure hold.

1. Calculate the molarity, molality, and mole fraction of the following solutions.

(*a*) 10.0% $C_{14}H_{10}O_9$ in H_2O; density 1.0401 gm/cc.
(*b*) 86.0% H_2SO_4 in H_2O; density 1.7872 gm/cc.
(*c*) 26% $SnCl_4$ in H_2O; density 1.233 gm/cc.
(*d*) 66% $C_6H_8O_7$ in H_2O; density 1.3071 gm/cc.
(*e*) 30% NH_3 in H_2O; density 0.8920 gm/cc.

Ans. $X_{NH_3} = 0.31$, $m_w = 25$, $M = 16$.

2. Medicinal alcohol is an aqueous solution of C_2H_5OH; 100-proof means half alcohol and half water *by volume.* If this corresponds to a solution which is 42.49% alcohol by weight, what is the density of *pure* ethyl alcohol?

3. Calculate the molarity, molality, and mole fraction of 100-proof alcohol (density 0.9344 gm/cc).

4. Calculate the osmotic pressure of 100-proof alcohol. If you utilize this pressure to support a vertical column of water, how high will the column of water be? (t = 25°C)

5. In order to see how effective it is as an antifreeze, calculate the freezing point of 100-proof alcohol. *Ans.* −30.5°C.

6. The vapor pressure of pure C_2H_5OH at room temperature is 59 mm. Assume that Raoult's law holds for 100-proof alcohol (it doesn't). The average person inhales 250 cc STP of air every 4 seconds. How long would a person have to inhale air saturated with 100-proof alcohol, in order to accumulate the same amount of alcohol as there is in 30 cc of 100-proof alcohol?

7. Temporary antifreeze is a solution of methyl alcohol. Calculate the concentration and weight per cent of an antifreeze mixture that will not freeze above −15°F. What would be the concentration of a glycerin solution that would do the same thing?

8. A 10-gm sample of polymer is dissolved in 100 gm of water in order to determine its molecular weight. The polymer is suspected to have a molecular weight of near 10^6. The resulting solution has a density of 1.00 gm/cc. Calculate the expected osmotic pressure and the freezing point depression. In the laboratory, temperatures can be measured within 10^{-4}°C, and pressures can be measured within 0.01 mm Hg. Which of the above measurements would be best for the determination of the molecular weight?

★ **9.** Normal air is about 21% O_2 and 79% N_2 on a mole basis. If the dissolved air in a beaker of water saturated with atmospheric air is released, what will be its composition; cf. Table 1.

★ **10.** The freezing point of pure water in equilibrium with its own vapor is +0.0100°C. Calculate the freezing point of water with dissolved air at a concentration corresponding to that expected for water exposed to dry air at 1 atm (cf. Table 1).

11. A chemist synthesizes a new compound X. He determines the molecular weight by dissolving 1.3 gm of X in 123 gm of water to form a solution with a density of 1.02

gm/cc. If the osmotic pressure is 143 mm Hg at 25°C, what is the molecular weight?
Ans. 1.4×10^3 gm/mole.

12. A compound Y of unknown molecular weight is dissolved in phenol. The resulting solution contains 10 gm of Y in 100 gm of phenol, and the resulting melting point is 0.306°C lower than that of pure phenol. What is the molecular weight of Y? (See Table 2.)

13. At 35°C, pure water has a vapor pressure of 42.23 mm. A chemist dissolves 10 gm of Z in 100 gm of water. The vapor pressure of the resulting solution is 41.63 mm. What is the molecular weight of Z?

★**14.** Compound A, dissolved in a solvent B, obeys Henry's law. Compound A dissolved in solvent C also obeys Henry's law. Suppose solvent C is insoluble in B. We now perform the following experiment: A solution containing A dissolved in B has the vapor pressure given by Henry's law, $P_B^A = k_H^B X_B^A$. This solution is mixed with solvent C, and some of the A dissolves in C. The vapor pressure of A in equilibrium with A dissolved in C is given by $P_C^A = K_H^C X_C^A$. The two immiscible solutions are separated, and each is analyzed for the concentration of A. Derive an expression for the ratio

$$\frac{X_C^A}{X_B^A}$$

15. In deriving an expression for the freezing point depression, we assumed that the solute did not dissolve in the solid solvent. Why? If Raoult's law held for *both* liquid solvent and solid solvent, would the freezing point law still be valid?

★**16.** Suppose the solvent and solute were both volatile and that their vapor pressures are given by the formulas $P_A = X_A P_A^0$ and $P_B = X_B P_B^0$. What relation between X_A and P_a^0 and P_B^0 must be fulfilled at the boiling point of the solution?

17. Calculate the freezing point, boiling point, and the osmotic pressure and vapor pressure at 35°C of the following solutions.

(*a*) $1.000M$ NaCl	density = 1.040 gm/cc
(*b*) $1.000M$ $CaCl_2$	density = 1.084 gm/cc
(*c*) $1.000M$ $Al_2(SO_4)_3$ (strong electrolyte)	density = 1.31 gm/cc
(*d*) $1.000M$ $CH_3C\begin{smallmatrix}=O\\ -OH\end{smallmatrix}$	density = 1.007 gm/cc

Assume that the solutions are ideal. Vapor pressure of pure water at 35°C = 42.23 mm.

Ans. (a) $T_f = -3.58$°C; $T_b = 100.98$°C; $\pi = 50.6$ atm; vapor pressure = 40.8 mm.

18. With the help of Table 5, calculate the freezing point depression for 1.00 m_w N_2H_5OH.

19. With the help of Table 5, estimate the per cent of dissociation of the following:

(*a*) $H(CHO_2)$ (*b*) $H(CH_3CO_2)$
(*c*) $N_2H_5(OH)$ (*d*) H(CN) (The parentheses indicate the anions.)

20. Write essential ionic equations for the following reactions in solution.

(*a*) $CaCl_2 + Na_2SO_4 \rightarrow \underline{CaSO_4} + 2NaCl$
(*b*) $HCl + N_2H_5OH \rightarrow N_2H_5Cl + H_2O$
(*c*) $Ca(OH)_2 + 2HCl \rightarrow CaCl_2 + 2H_2O$
(*d*) $CaCl_2 + Hg(NO_3)_2 \rightarrow HgCl_2 + Ca(NO_3)_2$
(*e*) $BaCl_2 + Hg_2(NO_3)_2 \rightarrow \underline{Hg_2Cl_2} + Ba(NO_3)_2$
(*f*) $H_2(CO_3) + BaCl_2 \rightarrow \underline{BaCO_3} + 2HCl$
(*g*) $CH_3OH + HI \rightarrow HOH + CH_3I$
(*h*) $\underline{RaCO_3} + 2HNO_3 \rightarrow Ra(NO_3)_2 + H_2CO_3$

Ans. (*a*) $Ca^{+2} + SO_4^{-2} \rightarrow \underline{CaSO_4}$
(*g*) $CH_3OH + H^+ + I^- \rightarrow HOH + CH_3I$

12

The elements of thermodynamics

Thermodynamics is a kind of chemical accounting. Just as the changes in production cost and sales price influence the movement of goods, so the changes in thermal energy and probability influence the dynamics of atoms. Just as in your daily earning and spending a "cash balance" shows whether you have some available money with which to make a purchase, so in chemistry there is for each reaction an energy-probability balance that shows whether there is some "free energy" to make the reaction go. The principle of balance is thus of paramount importance both in commerce and in chemistry. And, as in banking, it is easier to understand the principle of balance by studying it first in terms of tangible money, coins and bills, and then later in terms of credit; similarly, in chemistry it is helpful to study balance by considering first the principle of the tangible atom balance when chemical reactions take place, and then the less tangible energy and probability balances that are the core of thermodynamics.

THE ATOM-BALANCE PRINCIPLE

A correct chemical equation must be balanced. For example, the equation for the change of hydrogen and oxygen into water is

$$2H_2 + O_2 = 2H_2O$$

If 2 moles of hydrogen and 1 mole of oxygen are put in a heavy steel box and ignited by a spark, the mixture explodes; the atoms swap partners and form H_2O; and there are always exactly as many atoms of hydrogen and exactly as many atoms of oxygen in the box at the end of the reaction as there were at the beginning. In other words, atoms are never created or destroyed in any normal physical or chemical change. (We classify nuclear changes as abnormal and deal with them separately.) So in writing an equation there must always be just as many symbols for H on the right side of the equation as there are on the left side. As written above, there are 4 gram atoms of hydrogen on the left and 4 gram atoms of hydrogen on the right; there are 2 gram atoms of oxygen on the left and 2 gram atoms of oxygen on the right. This is an illustration of the principle of the *conservation of matter* discussed briefly in Chapter 1. Imagine stacking nickels and dimes in a safe-deposit box, shaking up the box, and taking the money out later. When removed, the coins are piled together differently; but there is the same number of nickels and dimes at the beginning and at the end. Coins do not appear out of nothing or vanish into nothing; and this is true because they are *matter*, they are made of atoms. In a chemical reaction, atoms are shaken up and rearranged but never created or destroyed; they never appear out of nowhere or vanish into nothing. This conservation of matter principle is a tangible example of the balance principle that we now wish to apply to the more intangible concepts of energy and probability.

THE ENERGY-BALANCE PRINCIPLE

Energy obeys the same conservation law as atoms. Energy can neither be created nor destroyed; it never can be made to appear out of nothing or vanish into nothing. This principle is known as the *first law of thermodynamics*, and it is valid during all chemical and physical changes. In violent cosmic and nuclear changes that involve huge amounts of energy, there is sometimes an apparent disappearance of mass and appearance of energy and vice versa. The great release of energy in the atomic bomb is an example of this type of change. But, within the framework of the principles of relativity, which relate energy and matter in transmutation, the balance principle still holds. (There are certain cosmologists today who maintain that in cosmic creation matter does appear out of nothing from time to time; but this conjecture is still a long way from being verified.)

The atom-balance principle is verified by weighing atoms before a chemical reaction takes place, weighing them after it has taken place, and finding that the weights are the same. But it is more difficult to verify the energy-balance principle since energy cannot be measured so directly as matter. Yet the study of energy changes has proved that the conservation of energy principle is universally valid.

A backward glance at scientific history of one hundred and fifty years ago makes it clear that one of the first and most powerful incentives to study energy was the economic motive; for the work of the world is done largely by energy.

Before the year 1800, practically all work was performed by human beings, the energy coming from the biochemical processes taking place in the muscle fibers. To supplement this labor there were a few water wheels and windmills and some draft animals, but the work day for the average man was long and hard. The necessity of having work done by human muscle was the principal factor that kept most of the world's population at a low economic level.

It was for this reason that a number of scientists and engineers started to seek ways of getting energy from other sources. The first big break-through came at the beginning of the nineteenth century with the invention of the steam engine in which the energy of heat could be changed into the energy of motion.

Perpetual motion. Now it was just at this time that scientists first began to understand how energy was transformed from kinetic to potential form, from chemical to mechanical, and so on. Since the economic importance of providing large quantities of energy was recognized, many inventors thought that there must be some way of creating energy out of nothing; they tried to invent machines that would not only run indefinitely without consuming energy but would also create energy to drive other machinery of any kind. They called this sought-for device a *perpetual motion machine.* No one was able to invent such a machine.

The fact that no one has ever devised a way of creating energy, or of destroying it, is the most convincing proof that the principle of the conservation of energy is an inviolate law of nature. No matter how many different devices one couples together to put energy through a series of transformations, one always finds that, when the cycle of transformation is complete, when the parts of the machine return to their original states, ready to repeat another series of transformations, then the balance of the energy taken in and the energy given out by the machine is always zero. It is this energy balance, always found in cyclic processes, that is the most significant proof of the conservation of energy.

Energy units. In our earlier discussion of the units of energy, we examined some of the ways of measuring energy. We learned that the *erg* is the energy contained in the motion of 2 gm with a velocity of 1 cm/sec. A *joule* is ten million ergs. It takes 4.184 joules to make a *calorie,* and a calorie is the energy required to heat 1 gm of water from 14.5° to 15.5°C. Thus it takes about 20 tons of mass moving at 1 cm/sec to heat 1 gm of water by 1 degree; a tremendous amount of motion produces only a little heat. Conversely, a little heat produces a lot of motion; when changed into heat inside the cylinders of the motor, the chemical energy in a pint of gasoline produces enough kinetic energy to propel 2 tons of automobile at 60 mi/hr for several miles.

Energy in gases. How does the energy-balance principle apply to physical and chemical changes? Using a glass bicycle pump (Fig. 1), fill the space under the piston with 1 mole (STP) of an ideal gas like helium. The piston has an area of 100 cm^2, and at the start the volume under the piston is 22.4 liters. The piston is arranged so that it can be pushed down by weights, putting energy into the gas in the form of work; or it can rise, withdrawing energy and lifting the weights. The top and sides of the pump are thermally insulated; at the bottom a metal rod is attached so that heat can be permitted to flow into or out of the gas. Thus there are two *channels* by which energy can flow into or out of the gas: the piston *work channel* at the top and the conduction *heat channel* at the bottom.

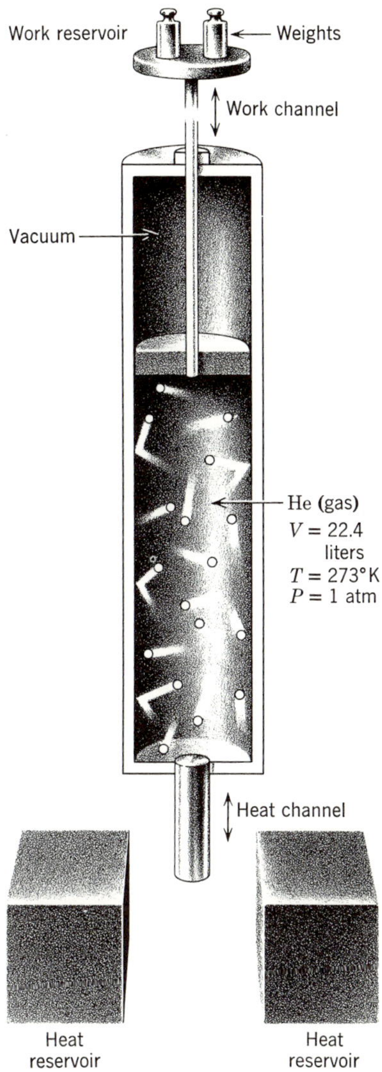

Fig. 1. Pump for expansion experiments.

Before describing energy changes in an ideal gas let us first review the properties of such a gas. According to our assumptions in deriving the gas law, an ideal gas consists of molecules behaving like point masses; and there are no interactions between gas molecules. All the energy is in the form of kinetic energy, and the energy per mole is given by the equation $E = \frac{3}{2}RT$. The interrelation of pressure (P), volume (V), the temperature (T) is specified analytically by the ideal gas law; it can also be specified graphically by isotherms as in Fig. 5*c* of Chapter 2, where pressure is plotted versus volume for several different fixed temperatures. In Fig. 2 with P and V as variables, *State 1* corresponds to the pressure, volume, and temperature at the start of these experiments.

Because the helium gas provides the pressure inside the pump and is the center of interest in studying energy changes, we shall refer to it as the *active* component of the experiment. We refer to the energy of the gas as the *internal* energy (E). The space in the cylinder above the piston is evacuated so that the upward force due to the pressure of the gas is exactly balanced by the downward force due to the pull of gravity on the weights. We shall also employ heated blocks of metal that can be pressed against the conduction rod at the bottom of the cylinder so

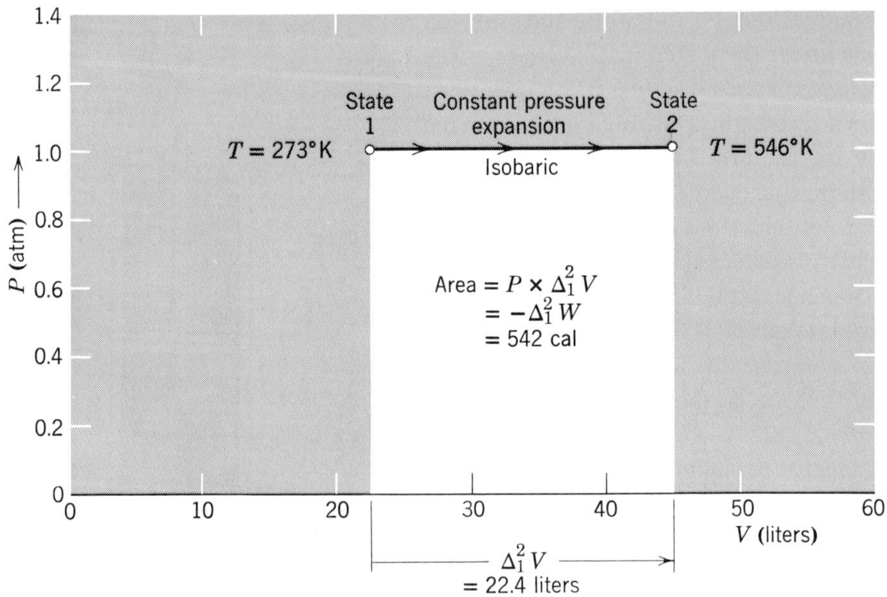

Fig. 2. Constant pressure expansion.

that heat may be made to flow into the gas. The *total thermodynamic system* consists of the gas (the active component) plus the weights and blocks of heated metal (the energy reservoirs).

Expansion at constant pressure. This experiment begins with the gas at a volume (V_1) of 22.4 liters, a pressure (P_1) of 1 atm, and a temperature (T_1) of 273°K (0°C). A block of heated metal is now placed on the rod at the bottom of the cylinder and heat flows into the gas. This increases the temperature of the gas and, consequently, its internal energy. The piston rises, increasing the volume of the gas. We permit it to rise until the volume has doubled, taking the value $V_2 = 44.8$ liters. The pressure remains constant, because it is balanced by the same weight throughout the experiment. The type of expansion with forces balanced is called *reversible* expansion. (If all the weight were suddenly removed, the gas would expand suddenly without doing any work; this is called *irreversible* expansion.)

Since the volume has doubled and the pressure has remained the same, the temperature must double, so $T_2 = 546°$K. Physically this means that, even though fewer molecules hit the bottom of the piston per second, their mean energy has increased enough that the pressure they exert has not changed. The *change of state* of our system is shown in Fig. 2 by the line extending from point

1 to point 2. The line is horizontal because the change has taken place at constant pressure.

We now examine the *energy changes*. There are two ways in which energy may be put in, or taken out of, the active system, i.e., the helium gas. At the bottom, energy can flow in or out as heat. When the system passes from state 1 to state 2, the energy flowing through the heat channel during this change is designated by $\Delta_1^2 Q$; it is a positive quantity when heat flows into the gas and a negative quantity when heat flows out. Energy may also go into or out of the gas through the piston; and the work is designated by $\Delta_1^2 W$. When the piston falls and does work *on* the gas, $\Delta_1^2 W$ is positive; when the piston rises, work is done *by* the gas and $\Delta_1^2 W$ is negative.

Energy may be taken in or out at the top as *work* and in or out at the bottom as *heat*. The first law of thermodynamics states that energy can neither be created nor destroyed. We designate the energy in the initial state of the gas by E_1, and in the final state after the change has taken place by E_2. Thus the change in the internal energy of the gas is designated $\Delta_1^2 E \equiv E_2 - E_1$. The sum of work-energy put in or out at the top, $\Delta_1^2 W$, plus heat-energy put in or taken out at the bottom, $\Delta_1^2 Q$, must equal $\Delta_1^2 E$. Thus the system is similar to a two-door safe-deposit box in which the change in the amount of money in the box is equal to the algebraic sum of what is taken in or out at the top and what is taken in or out at the bottom.

Expressing this relation as an equation,

$\Delta_1^2 E$	$=$	$\Delta_1^2 Q$	$+$	$\Delta_1^2 W$
Change in internal energy		Heat-energy passing through conduction channel at bottom		Work-energy passing through piston channel at top

We can calculate each of these quantities easily. In the movement of the piston from the initial position at 22.4 liters (V_1) to the final position at 44.8 liters (V_2), the change in volume $\Delta_1^2 V$ is

$$\begin{aligned} \Delta_1^2 V &= V_2 - V_1 \\ 22.4 \text{ liters} &= 44.8 \text{ liters} - 22.4 \text{ liters} \end{aligned}$$

As discussed in Chapter 2, the work performed, $\Delta_1^2 W$, is given by the relation

$$\begin{aligned} \Delta_1^2 W &= -P\,\Delta_1^2 V \\ -22.4 \text{ liter atm} &= -1 \text{ atm} \times 22.4 \text{ liters} \end{aligned}$$

On the diagram it may be seen that this corresponds to the white *area; P* is the height of the rectangle; $\Delta_1^2 V$ is its width; and the product of these two gives the value of the area in liter atm. Since 1 liter atm is equivalent to 24.218 cal, $\Delta_1^2 W = -542$ cal. The equivalent of 542 cal left the system as work of expansion.

The change in internal energy, $\Delta_1^2 E$, is given by the expression

$\Delta_1^2 E$	$=$	E_2	$-$	E_1	$=$	$\frac{3}{2}RT_2$	$-$	$\frac{3}{2}RT_1$
814	$=$	1628	$-$	814	$=$	$\frac{3}{2} \times 1.987 \times 546°$	$-$	$\frac{3}{2} \times 1.987 \times 273°$
cal		cal		cal		cal/deg deg		cal/deg deg

Thus the heat flowing in at the bottom, $\Delta_1^2 Q$, must provide both for the increase in internal energy, $\Delta_1^2 E$, and for the energy flowing out at the top as work. So we get

Change in internal energy		Energy flow through heat channel		Energy flow through work channel
$\Delta_1^2 E$	$=$	$\Delta_1^2 Q$	$+$	$\Delta_1^2 W$
$\Delta_1^2 E$	$=$	$\Delta_1^2 Q$	$+$	$(-P\Delta_1^2 V)$
$+814$ cal	$=$	$+1356$ cal	$+$	$(-542$ cal)
Increase in internal energy		Heat in		Work out

If the heat flow is actually measured, it will be found to have this value.

To summarize, the increase in the internal energy, $\Delta_1^2 E$, has been provided by the excess of the heat energy flowing in, $\Delta_1^2 Q$, over the work energy flowing out, $\Delta_1^2 W$. The energy of the *total* thermodynamic system has remained constant, $\Delta_1^2 E - \Delta_1^2 Q - \Delta_1^2 W = 0$. The law of conservation of energy holds.

Isothermal expansion. Consider next the experiment in which the gas expands while its *temperature* remains constant and the pressure varies. To achieve this, the thermal conduction rod at the bottom of the cylinder is packed in a mixture of ice and water so that the gas remains at the constant temperature of 0°C as it expands or contracts. When a process takes place at a constant temperature, it is called an *isothermal* process.

As contrasted with the earlier experiment, the upper part of the pump is arranged differently; there is sitting on top of the piston a box filled with small ball-bearings that can be removed one by one with a magnet manipulated from the outside. At the beginning of the experiment, the weight of these balls is 103.3 kg, a weight just sufficient to produce a downward force that exactly balances the upward thrust due to the pressure of 1 atm exerted on the under surface of the piston by the molecules of helium gas. We regard both the box and the piston itself as having negligible weight.

We now start lifting the steel balls off the top of the piston, one by one; this decreases the downward force so that the upward thrust of the gas pressure overbalances it and the piston slowly moves up. Thus the gas expands, doing work against the force of gravity that is pulling the ball-bearings down. We remove the balls slowly until the piston rises to the point where the volume of the gas has doubled. Since the temperature remains constant, the relation of pressure to

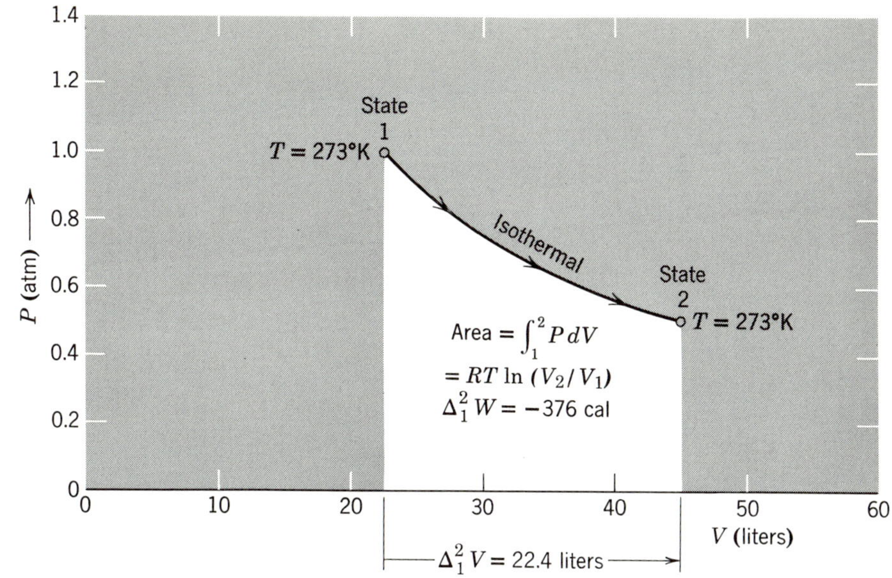

Fig. 3. Isothermal expansion.

volume during this change is given by the *isothermal* line on the diagram in Fig. 3.

As discussed in Chapter 2, it is necessary in the case of falling pressure to express the work as

$$\Delta_1^2 W = -\int_1^2 P\,dV$$

We cannot multiply P by $\Delta_1^2 V$ to calculate the work because this product is valid only when P is constant and the work area is a rectangle on the diagram. Here the work area has a curved top, as may be seen, and must be calculated by other means.

To calculate work, we first recall that for 1 mole of gas

$$P = RT/V$$

Hence, for a small change,†

$$DW = -P\,dV = -(RT/V)\,dV = -RT\,d\ln V$$

† The symbol d will be used to denote very small changes of quantities that are *state functions* of the system; the symbol D will be used to denote comparable small changes of quantities that are not state functions; the nature of this difference will be shown in the examples that follow.

(cf. p. 57) or, for a large change,

$$\begin{aligned}\Delta_1^2 W &= -RT(\ln V_2 - \ln V_1)\\ &= -RT\ln (V_2/V_1)\\ \Delta_1^2 W &= -2.303RT\log (V_2/V_1)\\ &= -2.303\underset{\text{cal/deg}}{(1.987)}\ \underset{\text{deg}}{(273°)}\ \log (\underset{\text{liters}}{44.8}/\underset{\text{liters}}{22.4})\\ &= -376 \text{ cal}\end{aligned}$$

There is no change in the internal energy of the gas; $\Delta_1^2 E = 0$ since the temperature remains constant. Thus the first law of thermodynamics may be written for this change as

$$0 = \underset{\substack{\text{No change}\\ \text{in internal energy}}}{\Delta_1^2 E} = \underset{\substack{+376 \text{ cal}\\ \text{Heat in}}}{\Delta_1^2 Q} + \underset{\substack{-376 \text{ cal}\\ \text{Work out}}}{\Delta_1^2 W}$$

Diagrammatically we may represent the energy flow in this process as 376 cal, passing in at the bottom as heat, going directly through the gas, and passing out through the top as 376 cal of work, leaving no change in the internal energy.

The last example shows how the first law of thermodynamics simplifies our description of a system. To calculate the heat flow *directly* in the problem would be extremely complicated. Instead, we calculate the work by a straightforward procedure and subtract this from $\Delta_1^2 E$ to evaluate $\Delta_1^2 Q$.

Adiabatic expansion. In this experiment the thermal conduction rod at the bottom of the cylinder is wrapped in insulating material so that the flow of heat in or out is blocked; i.e., *the heat channel* is cut off. This type of expansion is called *adiabatic* because no heat flows. Thus the only way in which energy can get in or out of the gas in adiabatic expansion is through the *work channel*, the rise or fall of the piston at the top. As in the first experiment, the gas is allowed to expand; at first we consider only a small expansion (dV); because no heat is flowing into the gas, the work of expansion must get its energy from the internal energy of the gas:

Change in internal energy		Change in heat		Change in work
dE	=	DQ	+	DW
dE	=	0 Zero because no heat flows	+	$(-P\,dV)$

The temperature falls more rapidly as the gas expands, and the pressure also drops. Because we are regarding helium as an ideal gas, its internal energy is $\frac{3}{2}RT$ and independent of the volume. For a very small change, the change in temperature is dT and the change in internal energy is given by

$$dE = \tfrac{3}{2}R\,dT$$

Thus

$$\tfrac{3}{2}R\,dT = -P\,dV = -RT(dV/V)$$

Rearranging,

$$\tfrac{3}{2}R(dT/T) = -R(dV/V)$$

or

$$\tfrac{3}{2}R\,d\ln T = -R\,d\ln V$$

For a large change,

$$\tfrac{3}{2}R\ln(T_2/T_1) = -R\ln(V_2/V_1)$$

From this expression it is possible to calculate the temperature of the gas at each volume it attains as it expands adiabatically. The temperature and volume being known, it is simple to calculate, from the ideal gas law, the pressure at each stage. Thus we find that the gas in expanding adiabatically to twice its volume follows the line labeled "Adiabatic" in Fig. 4 and the temperature drops from 273° to 172°K. The work-energy flowing out must be equal to the decrease in thermal energy, so

$$\begin{aligned}\Delta_1^2 W &= \Delta_1^2 E = \tfrac{3}{2}R(T_2 - T_1) = \tfrac{3}{2}R\,\Delta_1^2 T\\ &= \tfrac{3}{2}(1.987)\,(-101°) = -301\text{ cal}\end{aligned}$$

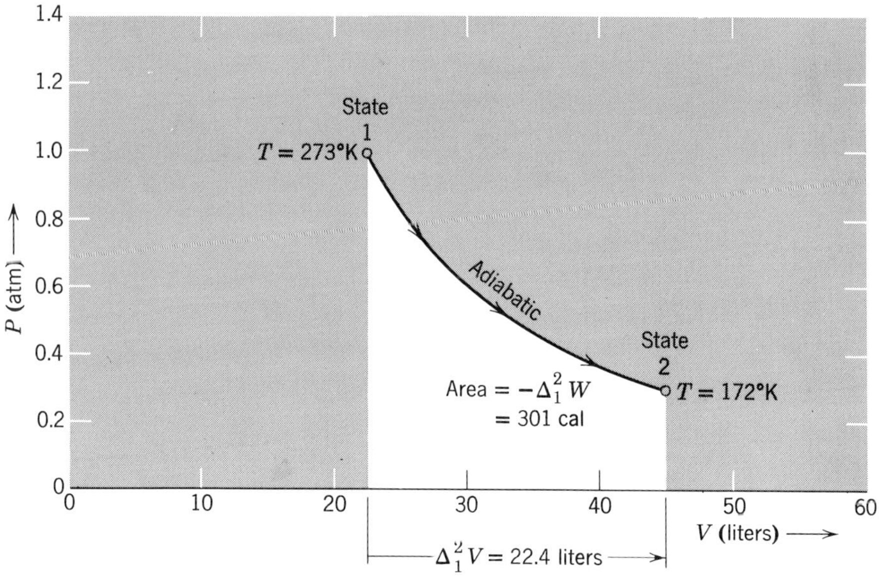

Fig. 4. Adiabatic expansion.

The expression for the first law of thermodynamics for this change is thus

$$\begin{aligned} \Delta_1^2 E \quad &= \Delta_1^2 Q + \quad \Delta_1^2 W \\ \tfrac{3}{2}R(T_2 - T_1) &= \quad 0 \quad + \quad \Delta_1^2 W \\ -301 \text{ cal} \quad &= \quad 0 \quad + (-301) \text{ cal} \end{aligned}$$

Decrease in internal energy — No heat flow — Energy flowing out as work

Summary of expansion types. In Fig. 5 are summarized the three types of expansion: (1) at constant pressure (isobaric), (2) at constant temperature (isothermal), and (3) adiabatic. In the first case enough heat must be added both to do the work and to raise the temperature of the gas in order to keep the pressure constant. In the second case, heat is required only to do the work since the temperature is constant. In the third case no heat flows in, and the energy to do the work is extracted from the internal energy, so temperature falls.

An energy cycle. Suppose that we start again with the gas at 273°K and 1 atm with $V_1 = 22.4$ liters as shown in Fig. 6*a*. We now perform three experiments in succession: (*a*) We let the gas expand adiabatically to 44.8 liters, and the temperature drops to 172°K (state 2). (*b*) We hold the volume constant, and heat the gas to 273°K (state 3). (During this process no work is done since the

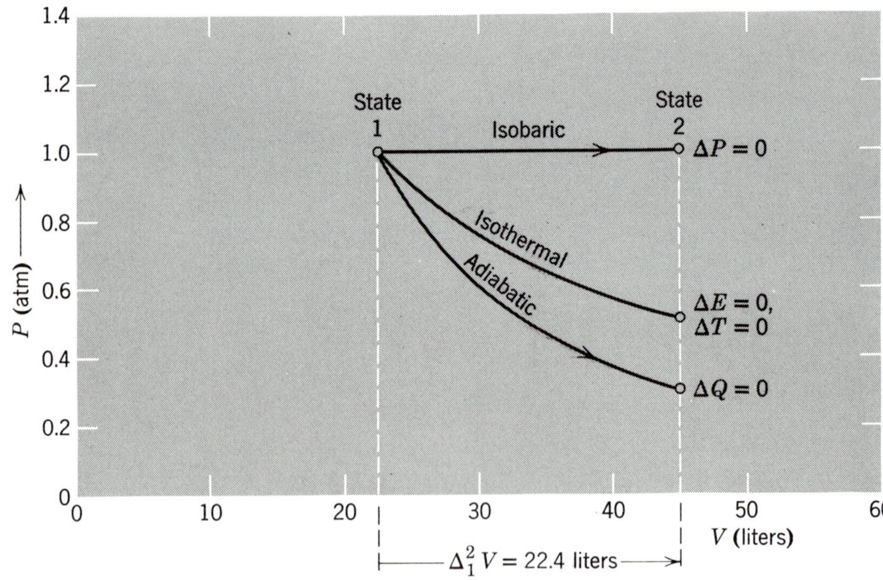

Fig. 5. Three types of expansion.

volume is constant.) (*c*) We then compress the gas isothermally back to 22.4 liters so that the final state is identical with its initial state (1). Let us summarize the energy changes:

(*a*)	$\Delta_1^2 E$	=	$\Delta_1^2 Q$	+	$\Delta_1^2 W$
	-301 cal		0 cal		-301 cal
(*b*)	$\Delta_2^3 E$	=	$\Delta_2^3 Q$	+	$\Delta_2^3 W$
	$+301$ cal		$+301$ cal		0 cal
(*c*)	$\Delta_3^1 E$	=	$\Delta_3^1 Q$	+	$\Delta_3^1 W$
	0 cal		-376 cal		$+376$ cal

We have thus carried out a cyclic change in which the gas has been carried through a series of changes that are represented by a line forming a closed figure on the state diagram. Summing all the changes in energy,

$\Delta_1^2 E$	+	$\Delta_2^3 E$	+	$\Delta_3^1 E$	=	$^\circ\Delta E$
-301 cal	+	$+301$ cal	+	0 cal	=	0 cal
$\Delta_1^2 Q$	+	$\Delta_2^3 Q$	+	$\Delta_3^1 Q$	=	$^\circ\Delta Q$
0 cal	+	$+301$ cal	+	-376 cal	=	-75 cal
$\Delta_1^2 W$	+	$\Delta_2^3 W$	+	$\Delta_3^1 W$	=	$^\circ\Delta W$
-301 cal	+	0 cal	+	$+376$ cal	=	$+75$ cal

The symbol $^\circ\Delta$ denotes the total change in a quantity as the system is moved through different states around the circuit and back to its original state; thus it is called a circuit symbol. The symbol $^\circ\Delta Q$ indicates the value of the total change in Q as the circuit is completed; it is the sum of all the changes in Q that take place. Similar meanings are attached to $^\circ\Delta E$ and $^\circ\Delta W$.

We now see that the total change in E is zero, so that it has the same value at the beginning and at the end of the experiment. This will always be the case because of the first law of thermodynamics. Since the state is the same at the beginning and at the end, E is a true *state variable. When the system is in any state, there will be one and only one value of E associated with that state.*

By way of contrast, both Q and W are not state variables. If Q were given a value at the beginning of this experiment, it would have a value 75 cal lower at the end of the experiment, since $^\circ\Delta Q = -75$ cal. Similarly W is 75 cal higher at the end than at the beginning, since $^\circ\Delta W = +75$ cal.

As another illustration of the properties of ΔQ, ΔW, and ΔE, consider the changes that would occur if, instead of going from state 1 to state 2 as shown in Fig. 6*a*, we took the alternate route indicated by Fig. 6*b*. Accordingly the 1-2 change is accomplished in two steps: by cooling at constant volume to 172°K

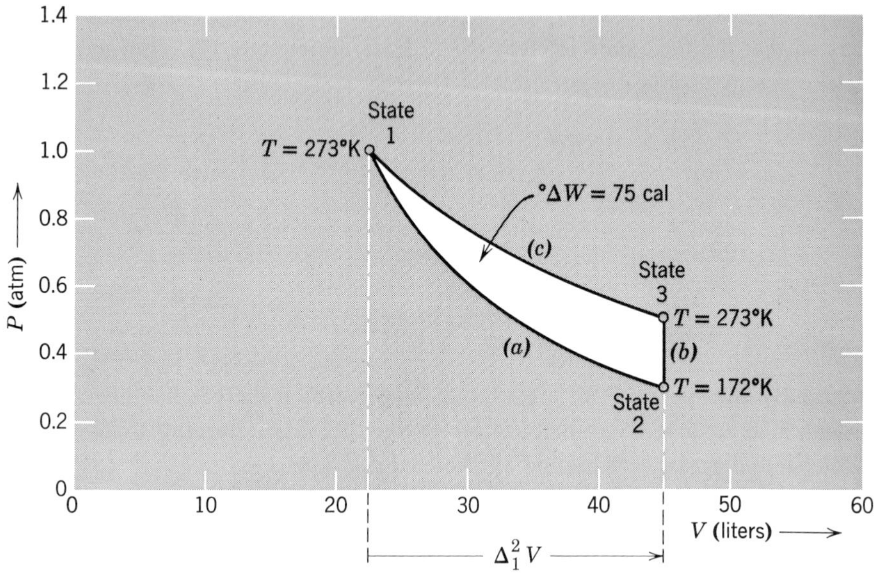

Fig. 6a. Thermodynamic cycle.

$(1 \rightarrow a)$ and by expanding isothermally to 44.8 l. Below is a summary of these changes.

		$(1 \rightarrow a)$		$(a \rightarrow 2)$			
$\Delta_1^{\overset{2}{a}} Q$	=	-301	+	237	=	-64	$\Delta_1^2 Q = 0$
$\Delta_1^{\overset{2}{a}} W$	=	0	+	-237	=	-237	$\Delta_1^2 W = -301$
$\Delta_1^{\overset{2}{a}} E$	=	-301	+	0	=	-301	$\Delta_1^2 E = -301$

Here we have used $\Delta_1^{\overset{2}{a}}$ to indicate the two-step change; for comparison, we have also listed values for the one-step change. Two things are apparent: *In a change from a given initial to a given final state, ΔQ and ΔW depend on how the change is carried out; in a change from a given initial to a given final state, ΔQ and ΔW may depend on the path but they always combine so that $\Delta Q + \Delta W = \Delta E$ is independent of path.* In practice, ΔQ and ΔW are the quantities determined by experiment. For a given over-all change their value depends on how the change occurs, i.e., on the details of the experiments. In order to express these results in a manner independent of the details of measurement we use the invented quantity ΔE, the sum of ΔQ and ΔW, which has the convenient property of being independent of path and enables us to express the results in terms independent

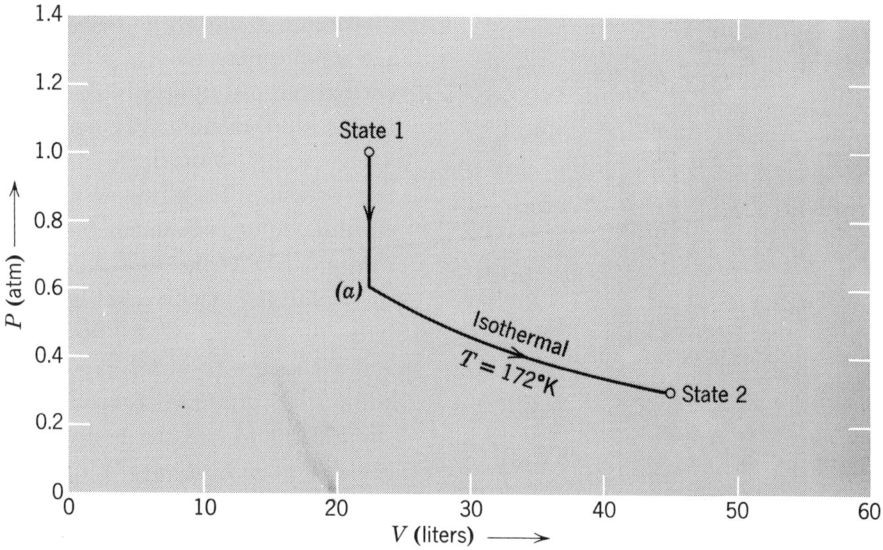

Fig. 6b. Alternate types of expansion.

of the way in which the change occurs. The first law of thermodynamics can therefore be expressed in an equivalent mathematical form as

$$^{\circ}\Delta E = 0$$

This states mathematically that the passage around any circuit, temporarily changing energy but finally restoring a system to its original state, is always found to yield a net energy change of zero; therefore energy can never be created or destroyed by any process.

Isothermal vaporization. Consider now a slightly more complicated situation wherein the pump contains both gas and a little liquid at the bottom, as shown in Fig. 7. Suppose that the ice bath is removed and in its place there is a jacket of water boiling at 1 atm pressure so that the temperature of the pump and its contents is held constant at 100°C or 373°K. The helium gas is replaced by 1 mole of water vapor at 1 atm, and at the bottom of the pump cylinder there are 2 moles of liquid water. (The volume of the liquid water, about 0.04 liter, will be neglected compared to that of the water vapor, 30.6 liters.) There is still a vacuum in the space above the piston so that the entire force bearing down on the piston is due to the weight of the ball-bearings. The water vapor is assumed to behave like an ideal gas.

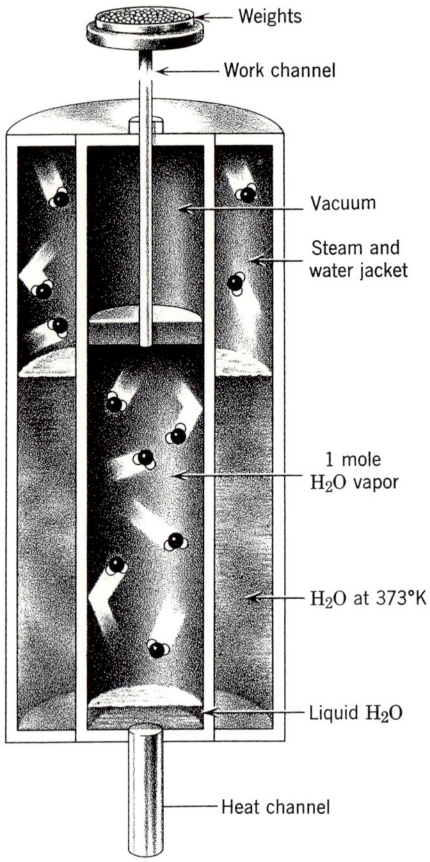

Fig. 7. Isothermal vaporization.

We now remove one ball-bearing (an infinitesimal reduction in force), so that the piston slowly rises. As the pressure momentarily decreases, some water vaporizes, bringing the pressure back up to 1 atm. As this vapor is formed, the water momentarily cools, and heat flows in from the jacket to bring it back to 100°C. Thus the liquid water changes to water vapor (gas) while the pressure remains constant at 1 atm and the temperature remains constant at 373°K or 100°C. To see the energy changes, suppose that the piston rises until 1 mole of liquid water has evaporated, forming an additional mole of water vapor which at this temperature and pressure has a volume of 30.6 liters. The piston travels upwards 306 cm, creating the additional volume $\Delta_1^2 V$ of 30.6 liters for the newly formed mole of water vapor. Work-energy will flow out of the top of the cylinder and can be calculated by

$$\Delta_1^2 W = -P\,\Delta_1^2 V$$

$$-741 \text{ cal} = -1 \text{ atm} \times 30.6 \text{ liters} = -30.6 \text{ liter atm}$$

$$= -741 \text{ cal}$$

The heat that flows in at the bottom during this change can be measured, and it is found to be 9713 cals. The energy-balance equation is

$\Delta_1^2 E$	$=$	$\Delta_1^2 Q$	$+$	$\Delta_1^2 W$
$+8972$ cal	$=$	$+9713$ cal	$+$	-741 cal
Internal energy increase		Heat in		Work out

Thus the internal energy of the system has increased by 8972 cal; 1 mole of water molecules in the gaseous or vapor state has 8972 cal more energy than it possessed in the liquid state at the same temperature. This is in contrast to the isothermal expansion of the helium where $\Delta_1^2 E = O$. There is an increase in internal energy during isothermal evaporation largely because the forces between the water molecules are stretched as these molecules move apart in going from the liquid to the vapor state. This is a process roughly resembling the stretching of the rubber band attached to a ball when you throw it from your hand. The kinetic energy of motion of the ball is transformed into the potential energy of the stretched rubber cord. In evaporating water, the heat energy that enters at the bottom of the pump is changed into potential energy in the "stretched" intermolecular forces between the gaseous molecules and the "stretched" force of gravity in raising the piston.†

You may ask why the molecules go into the gaseous state at all. If there are these invisible forces pulling these molecules together, why is not this energy factor sufficient to pull them back together and hold them in the liquid state? We shall see that the molecules move apart because of a *probability* factor. Therefore we proceed in the next section to see how this probability factor can be measured. After we have studied energy and probability as they affect physical changes of state like vaporization, we shall study the applications of these thermodynamic principles to chemical reactions.

ENTROPY

The properties of a gas are predicted by the principles of probability. We now wish to show from a somewhat different point of view how the *probability factor* influences both the physical and the chemical behavior of all aggregates of atoms and molecules that possess thermal energy.

Molecular probability. As a start, consider 1 molecule of helium confined in a box having a volume of 22.4 liters, as shown in the accompanying diagram. Suppose that it is possible to photograph the box with a magic camera that will show exactly where this molecule is at the moment the photograph is snapped. Thus the

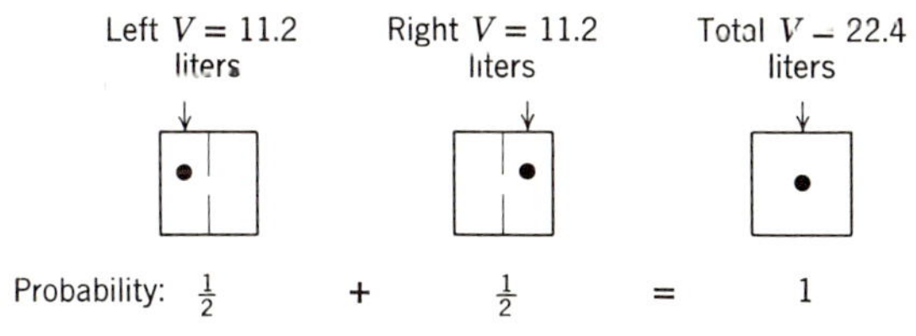

† If the water had been boiling on the stove instead of inside the pump, the effective downward force of the piston would have been replaced by the effective downward force of the atmosphere of air pressing on the water. When 1 mole of water is evaporated under these conditions, the work is done by displacing the air upward instead of the piston. Since the mole of steam occupies the same volume as in the pump (30.6 liters) and the pressure is the same (1 atm), the work done will still be $P\Delta V$ or 30.6 liter atm, equivalent to 741 cal.

photograph may show the molecule momentarily in the left half of the box; or it may show it to be in the right half. We assume that the box is uniform; there is no physical difference between the left side and the right side. What is the chance, the *probability*, of finding the molecule in the left half of the box when the picture is taken? The probability of finding the molecule in the left half of the box is ½ since that side of the box constitutes one-half of the total volume. Again, the chance of finding the molecule in the right half is also ½. The chance of finding the molecule *somewhere* in the total volume of the box is thus the sum of these two values, as shown in the diagram. Let us designate the probability of the molecule being in the left side of the box by Y_{left} and the corresponding volume by V_{left}. Then

$$Y_{\text{left}} = \frac{V_{\text{left}}}{V_{\text{total}}} = \frac{11.2 \text{ liters}}{22.4 \text{ liters}} = \frac{1}{2}$$

$$Y_{\text{right}} = \frac{V_{\text{right}}}{V_{\text{total}}} = \frac{11.2 \text{ liters}}{22.4 \text{ liters}} = \frac{1}{2}$$

Thus there are two possibilities; the molecule can be either in the left side or in the right side, and the sum of these two probabilities is unity.

Now put *two* molecules in the box and label them A and B. We have four possible states, as shown in the accompanying diagram.

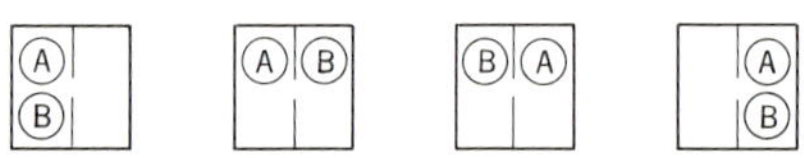

Probability: $\frac{1}{4} + \frac{1}{4} + \frac{1}{4} + \frac{1}{4} = 1$

The chance of finding A in the left side is $\frac{1}{2}$, and the chance of finding B is also $\frac{1}{2}$. The chance of finding them both there is $\frac{1}{2} \times \frac{1}{2} = \frac{1}{4}$.

We next observe that this is the same as the probability for one molecule ($\frac{1}{2}$) raised to the power (N) of the number of molecules present, which is 2:

$$\left(\frac{1}{2}\right)^N = \left(\frac{1}{2}\right)^2 = \left(\frac{1}{2}\right) \times \left(\frac{1}{2}\right) = \frac{1}{4}$$

The same reasoning applies to each of the other states, so each has the same probability and the sum is unity.

Now put a whole gram mole of helium in the box; $N = 6 \times 10^{23}$ molecules. Then

$$Y_{\text{left}} = (V_{\text{left}}/V_{\text{total}})^N$$
$$Y_{\text{left}} = (11.2 \text{ liters}/22.4 \text{ liters})^{6\times10^{23}}$$
$$Y_{\text{left}} = (1/2)^{6\times10^{23}}$$

Generalizing, the ratio of the probabilities of finding the gas in two different volumes (V_1 and V_2) is

$$\frac{Y_2}{Y_1} = \left(\frac{V_2/V_{\text{total}}}{V_1/V_{\text{total}}}\right)^N = \left(\frac{V_2}{V_1}\right)^N$$

The expansion of a gas. Suppose that a box is divided by a removable partition and the left compartment contains 1 mole of helium gas. If the partition is removed, why does the gas expand to fill the whole box? We may think of it in terms of probability. With the partition open, the chance of finding 1 molecule in the left side of the box is $\frac{1}{2}$; the chance of finding it in the whole box is 1. Thus the chance of finding all the molecules in the left half of the box is $(\frac{1}{2})^{6\times10^{23}}$, while the chance of finding the gas filling the whole box is $(1)^{6\times10^{23}}$. The ratio of the latter to the former is

$$\frac{\text{Probability of gas in whole box}}{\text{Probability of gas in left half}} = \frac{(1)^{6\times10^{23}}}{(\frac{1}{2})^{6\times10^{23}}}$$

$$= 2^{6\times10^{23}} \cong 10^{10^{23}}$$

Here, in the last term, we "round off" this large number as a power of ten. (The symbol $\cong$ means "approximately equal to.")

Thus the final state of uniform distribution is more probable than the initial state of shrinking into the left half by a factor which, if written in numbers, would be approximately 1 followed by one hundred sextillion zeros. We could start writing this number out:

1,000,000,000,000,000,000,000,000,000,000,000,000,000,000,000,000,000,-000,000,000,000,000,000,000,000,000, 000,000,000,000,000,000,000,000,000,-000,000,000 . . .

but we would need enough paper to cover the entire surface of the earth more than two hundred times in order to print the one hundred sextillion zeros necessary to express this large number. (There is no imaginable way of grasping this unthinkable magnitude. Even if we counted every electron, every proton, and every neutron in the entire universe there is evidence that we would find less than 10^{100} particles. But the chance of finding the gas in the left half of the box is only one in $10^{100,000,000,000,000,000,000,000}$!) So you see in terms of pure probability how great is the push that carries a mole of gas from the left half of the box into the entire box. By the same logic, you see what a small chance there is of the gas ever voluntarily contracting into the left side, leaving a vacuum on the right. If you waited long enough, you might *possibly* some day observe a gas shrinking into a corner of its box, but you would very *probably* have to wait a long time!

There are two ways of letting this helium gas expand from a smaller to a larger volume. We can pull out the partition suddenly and let the gas rush over into the extra volume; or we can place the gas in a cylinder with a piston as discussed in the previous section and then let the piston rise gradually so that the expansion takes place *reversibly;* in the latter case, the expanding gas does work against the piston at the top of the cylinder and absorbs heat at the bottom, where the thermal reservoir makes contact with the gas. In the former, sudden expansion, the physical reaction certainly *goes;* the gas rushes over almost explosively into the empty space because the probability factor is very large and there is no energy factor to offset it: we have essentially an ideal gas with no restraining forces operating between the molecules. Such a change *cannot* be made to go in the opposite direction by a small change in the conditions; it is *irreversible.* In the latter case, however, the work done against the piston balances the probability factor so that the "reaction" goes *reversibly.* We can readily see that the work done or the increase in potential energy of the weights balances the probability factor; if the weight is increased slightly, the piston is forced back down and restores the gas to its original volume and, therefore, to its original probability.

But how can we compare probability changes and energy changes in the same units so that we can tell from their magnitude exactly when they balance? The clue comes from the work done when 1 mole of gas expands. In the previous section, we found that

$$\Delta_1^2 Q = RT \ln (V_2/V_1)$$

Using $k \equiv R/N_0$, where N_0 is Avogadro's number, and changing from a ratio of volumes to a ratio of probabilities, we get an expression where the probability ratio is related to the expression on the left appearing in terms of energy units:

$$\Delta_1^2 Q = kT \ln (Y_2/Y_1)$$

If both sides of this equation are divided by T, we get

$$\Delta_1^2 Q/T = k \ln (Y_2/Y_1)$$

Thus, if we take the natural logarithm of the ratio of the probability of the *final* state to the probability of the *initial* state, and multiply it by the gas constant (k) for a single molecule, we have an expression that is equal to an energy factor, the heat absorbed for this reversible change, divided by the absolute temperature. This product of the natural logarithm of the probability ratio multiplied by k is so important that it is given a special name, the *entropy change;* and it is denoted by the symbol $\Delta_1^2 S = k \ln (Y_2/Y_1)$. In other words, this entropy change, $\Delta_1^2 S$, multiplied by the absolute temperature at which the change takes place, gives the measure of the "push" of the probability factor in energy units at a given temperature. So we can write

$$\text{"Probability push" in energy units} = T\Delta_1^2 S = TS_2 - TS_1$$

It is possible in many instances to calculate not only the ratio of probabilities for two different states, but also even the absolute probability for a given state. For example, a mole of helium gas in a box will be found with molecules moving at different velocities. The distribution of molecules among these different values of the velocity depends on the fact that there is a de Broglie wave associated with each molecule, just as there is a de Broglie wave associated with each electron in an atom. The nature of these waves is determined by considerations closely similar to those in a vibrating violin string; there must be an integral number of wave lengths in the box for each molecule. From the properties of these waves, the probability of this thermodynamic state of the gas can be calculated. The whole calculation is too complex to include here, but it is analogous to the calculations on p. 223. Thus we can speak of the absolute probability Y of a mole of helium at 273°K and 1 atm; or, more commonly, of its absolute entropy S. The probability is dimensionless, while the entropy has the dimensions of calories per degree. If the entropy S is multiplied by the absolute temperature T then the product TS has the dimension of calories. It is the value of change in this TS product in calories that can be compared directly with energy changes to find the relative effect of these two factors on the behavior of the system under consideration.

We shall show in the next section that, like E, this quantity S is a true state function.

THE SECOND LAW OF THERMODYNAMICS

The Carnot cycle. To see the nature of entropy, the logarithmic measure of probability, let us examine its change as an ideal gas is put through a cyclic series of changes. As shown in Fig. 8, we start with a mole of helium gas in the state where $P_1 = 1$ atm, $V_1 = 22.4$ liters, and $T = 273$°K. In Step 1, the gas expands *isothermally* until its volume is doubled. As shown in the previous section,

$$\text{(I)} \qquad \begin{aligned} \Delta_1^2 E &= \Delta_1^2 Q + \Delta_1 W \\ 0 \text{ cal} &= +376 \text{ cal} + -376 \text{ cal} \end{aligned}$$

In Step II, the gas expands *adiabatically* from state 2 to state 3 where the temperature drops to 172°K, and where $V_3 = 89.6$ liters. As shown in the previous section, the energy relations will be

$$\text{(II)} \qquad \begin{aligned} \Delta_2^3 E &= \Delta_2^3 Q + \Delta_2^3 W \\ -301 \text{ cal} &= 0 \text{ cal} + -301 \text{ cal} \end{aligned}$$

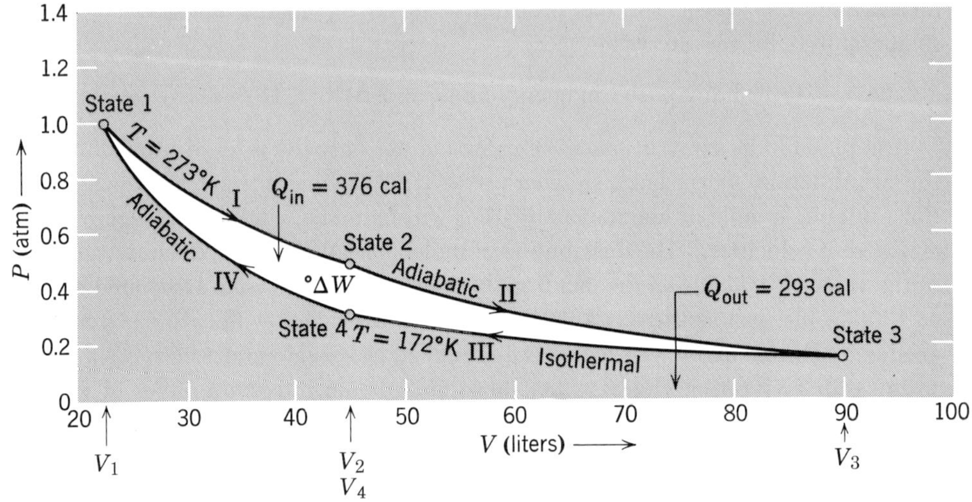

Fig. 8a. The Carnot cycle *PV* plot. (White area equals $°\Delta W$.)

Next, in Step III, the gas is compressed *isothermally* from state 3 to state 4 in which it has a volume (V_4) such that an *adiabatic* compression finally will restore it to its original state 1. To calculate V_4, recall that in Step II the gas changed adiabatically from 273° to 172°K. In the final step it will change back adiabatically from 172° to 273°. As shown in the original discussion of adiabatic expansion, the ratio of the volumes in passing adiabatically from one temperature to another is always the same; that is,

$$V_3/V_2 = V_4/V_1$$

Rearranging,

$$V_2/V_1 = V_3/V_4$$

and from this V_4 is found to be 44.8 liters. Thus

$$\Delta_3^4 W = -R(172°) \ln (V_4/V_3)$$

Substituting V_1/V_2 for V_4/V_3,

$$\Delta_3^4 W = -R(172°) \ln (V_1/V_2)$$
$$\Delta_3^4 W = +237 \text{ cal}$$

(III)
$$\Delta_3^4 E = \Delta_3^4 Q + \Delta_3^4 W$$
$$0 \text{ cal} = -237 \text{ cal} + +237 \text{ cal}$$

As shown in the previous section, the energy change in going from one temperature to another is independent of the volume and depends only on the temperatures; so

$$\text{(IV)} \qquad \begin{aligned} \Delta_4^1 E &= \Delta_4^1 Q + \Delta_4^1 W \\ +301 \text{ cal} &= 0 \text{ cal} + +301 \text{ cal} \end{aligned}$$

As is to be expected in any cycle, we find that

$$\begin{matrix} \Delta_1^2 E + & \Delta_2^3 E & + \Delta_3^4 E + & \Delta_4^1 E & \equiv {}^\circ\Delta E = 0 \\ 0 \;\; + & -301 \text{ cal} & + \;\; 0 \;\;\; + & +301 \text{ cal} & \end{matrix}$$

But our chief concern is with the entropy change. We note that there is no entropy change in the adiabatic parts of the cycle since there $\Delta Q = 0$. We thus find that in the isothermal changes

$$\begin{aligned} \Delta_1^2 S &= \Delta_1^2 Q / T_{1,2} = 376 \text{ cal}/273^\circ \\ &= 1.38 \text{ cal/deg} = 1.38 \text{ E.U.} \\ \Delta_3^4 S &= \Delta_3^4 Q / T_{3,4} = -237 \text{ cal}/172^\circ \\ &= -1.38 \text{ cal/deg} = -1.38 \text{ E.U.} \end{aligned}$$

$$\begin{matrix} \Delta_1^2 S & + \Delta_2^3 S + & \Delta_3^4 S + & \Delta_4^1 S = {}^\circ\Delta S = 0 \\ +1.38 & 0 & -1.38 & 0 \\ \text{E.U.} & & \text{E.U.} & \end{matrix}$$

(*Note:* E.U. is the abbreviation for entropy units having the dimensions of calories per degree.)

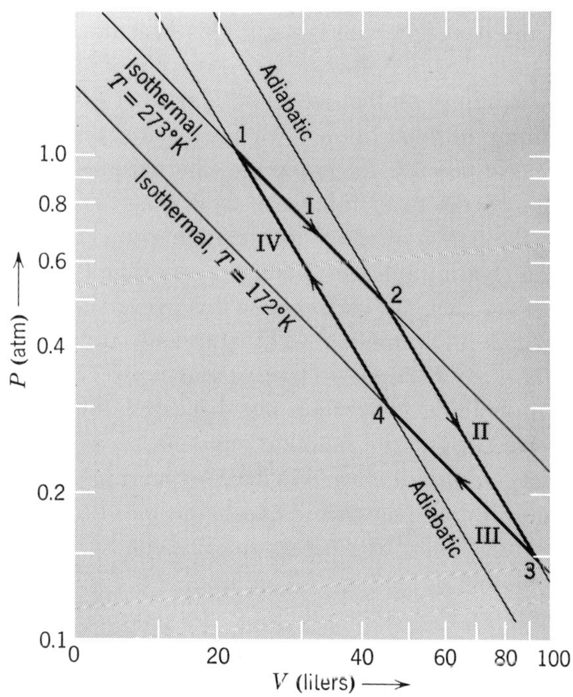

Fig. 8*b*. The Carnot cycle log-log plot. Note that all isothermals are straight lines with the same 45° slope; and all adiabatics are straight lines with a slope $\frac{5}{3}$ times steeper than the isothermals, since the gas is an ideal monatomic gas with $C_V = \frac{3}{2}R$. The linear graphs facilitate plotting.

We see that in this example, entropy S is a state function like energy E. Thus dS is an *exact* differential, $^\circ\Delta S = 0$.

This relation was first pointed out by the French scientist, Sadi Carnot, and this cycle bears his name. In examining its significance, we note first of all that each step in the cycle is carried out reversibly. The relationship we derived between heat, probability, and entropy for the reversible expansion of an ideal gas holds for all *reversible* processes in any physical or chemical system:

$$\mathrm{d}S = DQ_{\mathrm{rev}}/T$$

When a finite quantity of heat is absorbed at constant temperature, this may be written

$$\Delta_1^2 S = \Delta_1^2 Q_{\mathrm{rev}}/T$$

This relation is the *Second Law of Thermodynamics.* Its significance will be made clearer by the examples that follow.

Irreversible changes. All changes we have discussed so far have been reversible changes in which the pressure *in* the active component is substantially balanced by the pressure *on* the active component. As long as this is the case, any expansion of the gas occurs against the maximum possible force and does the maximum possible work. In other words, in a reversible expansion the gas is turning all the *probability push* into work; since this work done is the same as the heat flow into the gas, this heat flow, ΔQ_{rev}, is equal to the probability push, $T\Delta S$.

Now imagine that we allow 1 mole of gas at STP to expand isothermally to a volume of 44.8 liters. We saw earlier that, if the gas is working against the greatest possible force and the external pressure P_e is equal to the internal pressure P_i, the reversible work $\Delta_1^2 W$ is -376 cal. But suppose we do not utilize all the potential of the gas for doing work; suppose that, instead of varying P_e from 1 atm to $\frac{1}{2}$ atm in such a way that P_i is always equal to P_e, we fix P_e at $\frac{1}{2}$ atm. Then the work done in this irreversible change is $-P_e\,\Delta V = -\frac{1}{2}\,22.4 = -11.2$ liter atm $= -271$ cal, and, in accord with the first law, $\Delta_1^2 Q = \Delta_1^2 E - \Delta_1^2 W = (0) - (-271 \text{ cal}) = +271$ cal.

In both the expansions just discussed, the inital and final states were the same; hence $\Delta_1^2 S$, a state function, must be the same. This is reasonable since probability push is the same in both expansions. The difference in $\Delta_1^2 Q$ results because in the irreversible expansion we utilized only part of the push present in the gas. Thus, in general, we can write

$$T\Delta S = \Delta Q + \Delta i$$

In this equation Δi represents the *irreversibility*, the portion of the probability push that is lost when a change is carried out irreversibly. For the reversible expansion the $\Delta i = 0$; hence

$$\Delta S = (+376)/273 = 1.38 \text{ E.U.}$$

For the irreversible expansion in which P_i and P_e are only partly balanced, $\Delta i = 105$ cal. If P_e were zero, there would be complete unbalance, and Δi would be 376 cal. ΔQ would have to be zero, since no work is done in an isothermal expansion where $\Delta E = 0$.

By the very nature of Δi, this unutilized push is always positive or zero. This means that, although $T\Delta S$ can be equal to ΔQ, it is never less. For this reason, the relation above is often written without Δi:

$$T\Delta S \geqslant \Delta Q$$

where $\geqslant \Delta Q$ means equal to or greater than ΔQ. This is the most general form of the second law and applies to any process, reversible or irreversible. For a reversible change, the equality is true; for an irreversible change, the inequality is true.

Spontaneous changes. Imagine a thermodynamic system in a given state that is suddenly isolated, shut off from the outside world. What determines changes that occur spontaneously, i.e., in the absence of outside stimulations or restraints? Such a system can only change from a less probable to a more probable state. *For a spontaneous change in an isolated system the probability increases.* But the entropy is related logarithmically to the probability; if the probability increases, the entropy increases. *Therefore, for any spontaneous change in an isolated system,*

$$\Delta_1^2 S > 0$$

Suppose, however, that the initial state is such that the system is in its most probable state. Then we will never observe any change; the system is at *equilibrium.* Thus we can state our conclusion as follows:

For an isolated system,

$$\Delta_1^2 S \geqslant 0$$

The $>$ sign applies whenever a change occurs spontaneously; the equality sign applies if the system is at equilibrium.

In our discussion of the behavior of molecules in a divided box we recognize that the "gas" consisting of two molecules (a microscopic system) might sometimes be found entirely in the left side of the box with a vacuum on the right side. But, if the gas consists of 10^{10} molecules or more (a macroscopic system), we will never find all the molecules on the left side of the box; in other words, even though slight fluctuations may occur, an isolated, macroscopic system never changes spontaneously to a significantly less probable state. This distinction between microscopic and macroscopic is made on a statistical basis.

> "*Die Energie der Welt ist konstant; die Entropie der Welt strebt einem Maximum zu.*" (Clausius.)

In everyday life there are countless examples of the law of increase of probability or entropy. Take a pack of playing cards; put all the red cards in the upper half of the pack and all the black cards in the lower half; then shuffle the pack. Unless a near miracle takes place, the result of the shuffle will be an arrangement of red and black cards all mixed together in random order. There is only one state in which all red cards are on top and all black cards are on the bottom. There are about 10^{16} ways of having random red and black arrangements. Therefore the random arrangement is more probable to such a degree that it is almost a perfect certainty that a shuffle will never yield all red cards on the top and all black cards on the bottom of the pack.

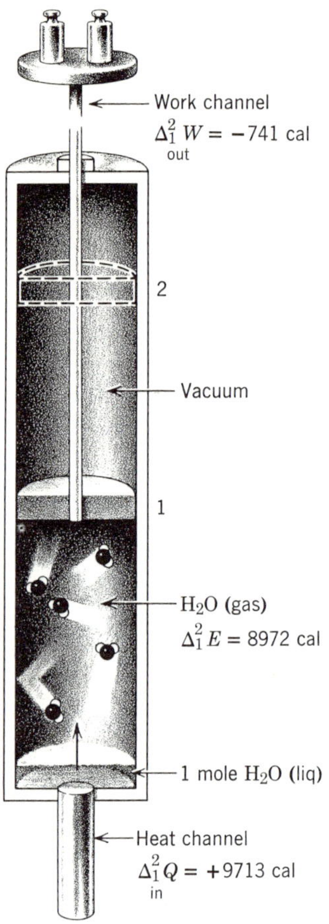

Fig. 9. Evaporation of water.

Evaporation. We now turn to the case where the molecules in a system have an attraction for one another so that when they are brought close enough together they condense and form a liquid. In Fig. 9, 1 mole of liquid H_2O is shown at the bottom of a cylinder at 1 atm pressure in the space under a piston at $100°C = 373°K$, as described in the previous section. We shall examine the nature of the changes in the energy and probability factors as we pull out the piston reversibly and let this 1 mole of H_2O pass from the liquid to the vapor state.

According to our earlier discussion,

$$\begin{aligned} \Delta_1^2 E \quad &= \quad \Delta_1^2 Q \quad + \quad \Delta_1^2 W \\ 8972 \text{ cal} &= 9713 \text{ cal} + -741 \text{ cal} \end{aligned}$$

The forces of attraction pull the water molecules together; the piston bearing down on them at the top also pushes them together. Therefore it is the sum of the 8972 calories (which as an increase in internal energy went into the forces of attraction) plus the 741 calories (which went up through the piston and out as work) that represents the *total* energy factor, the measure of the tendency of the system to contract. This sum is usually denoted by the symbol $\Delta_1^2 H$; early in the history of chemical thermodynamics, the capital H was selected because the quantity for which it stands was originally called *Heat Content;* the change in H is equal to the heat which flows in at the bottom of the cylinder. Today this quantity is more frequently referred to as the *enthalpy*. The precise definition is

$$H \equiv E + PV$$

It is important to note that the work term in the first law is generally given by $\Delta W = -P_e \Delta V$. Since P_e is the external pressure and not the pressure of the active component, the work is not a state function and it depends on the path of the change, the way in which the change is carried out. By way of contrast, the P appearing in defining the equation of the enthalpy is the pressure in the active component and V is the volume of the active component. Since E and PV are state functions, their sum H is also a state function.

For any change we can write

$$\Delta_1^2 H = \Delta_1^2 E + \Delta_1^2 (PV)$$

If this change occurs at constant pressure and temperature,

$$\underset{\text{Heat absorbed}}{\Delta_1^2 H} = \underbrace{\underset{\text{Increase in potential energy against internal forces}}{\Delta_1^2 E} + \underset{\text{Increase in potential energy due to lifting of weight}}{P\Delta_1^2 V}}_{\text{Total "squeeze" on system}} = \Delta_1^2 Q$$

When this expansion takes place under equilibrium conditions, this "squeeze" due to the pull of the molecules for each other and the force of the piston must be counterbalanced by an equal and opposite "push" outward, which is due to the probability change as the mole of water goes from the liquid to the vapor state. To a first approximation this probability push is due to the escape of these molecules into the larger volume of the vapor state. We can find the probability factor $T\Delta_1^2 S$ from the equation

$$T\Delta_1^2 S = \Delta_1^2 Q = \Delta_1^2 H$$

Thus thermodynamics supplies us with an equation which states that at equilibrium there is a perfect balance between the energy "squeeze" and the probability "push."

It is often helpful to know the change in entropy when a substance is heated from one temperature to another, for example, at constant pressure. In heating through a very small temperature increment dT over which C_p is constant, the heat added will be $C_p\,dT$ and the increment in entropy will be, as usual, the heat added divided by the temperature at which it is added, $C_p\,dT/T$. The relations discussed in Chapter 2 show that the entropy increment in going from state 1 to state 2 with C_p constant will be given by

$$\Delta_1^2 S = C_p \ln (T_2/T_1)$$

If the system is heated at constant volume,

$$\Delta_1^2 S = C_v \ln (T_2/T_1)$$

FREE ENERGY

Vaporization. At equilibrium, the energy factor is equal to the probability factor. In other words, the energy factor tending to pull the molecules together is balanced by the probability factor tending to push them apart.

$$T\,\Delta_1^2 S = \Delta_1^2 H \qquad (T,\ P \text{ constant})$$

$$0 = \underset{\text{Energy factor}}{\Delta_1^2 H} - \underset{\text{Probability factor}}{T\,\Delta_1^2 S}$$

$$0 = \underset{\text{Energy factor}}{9713 \text{ cal}} - \underset{\text{Probability factor}}{9713 \text{ cal}}$$

The zero appearing on the left side of the equality sign is the indicator that tells us the liquid phase is in *equilibrium* with the vapor phase because the two factors balance. We shall call this situation "case 2" in the following development.

We shall now use the symbol $\Delta_1^2 F$ to denote the difference between $\Delta_1^2 H$ and $T\,\Delta_1^2 S$:

$$\Delta_1^2 F \equiv \Delta_1^2 H - T\,\Delta_1^2 S \qquad (T = \text{constant})$$

The symbol F is called the *free energy* and is defined by the equation

$$F \equiv H - TS$$

Since H, T, and S are all state variables, F is also a state variable. $\Delta_1^2 F$ denotes the change in F as the system passes from state 1 to state 2:

$$\Delta_1^2 F \equiv F_2 - F_1$$

We now examine the nature of this *free energy*. Consider the situation in the evaporation of water where the pressure produced by the piston is 1.1 atm (case 1), and 0.9 atm (case 3), as well as 1 atm (case 2).†

We write an equation like a chemical equation for this change from liquid water to water vapor:

$$\begin{array}{ccc}
H_2O \text{ (liquid, P atm, 100°C)} & \rightarrow & H_2O \text{ (gas, P atm, 100°C)} \\
\text{State 1} & \rightarrow & \text{State 2} \\
-H_1 & & +\ H_2 \equiv \Delta_1^2 H \\
-S_1 & & +\ S_2 \equiv \Delta_1^2 S \\
-F_1 & & +\ F_2 \equiv \Delta_1^2 F
\end{array}$$

† Frequently when we specify the properties of states such as pressure, temperature, volume, and mole content, we will use only one or two digits, with the implication that the state is specified with precision sufficient to make the significant figures in the ensuing calculation meaningful.

where the items in parentheses specify the pressure and temperature, respectively.

If we have a more complicated change occurring, for example,

$$(a) \qquad H_2O \text{ (liquid, 1 atm, } 100°\text{C)} \rightarrow H_2O \text{ (gas, 1.1 atm, } 100°\text{C)}$$

it is often convenient to choose a path for the change consisting of several simpler steps, for example,

$$(b) \qquad H_2O \text{ (liquid, 1 atm, } 100°\text{C)} \rightarrow H_2O \text{ (gas, 1 atm, } 100°\text{C)}$$
$$(c) \qquad H_2O \text{ (gas, 1 atm, } 100°\text{C)} \rightarrow H_2O \text{ (gas, 1.1 atm, } 100°\text{C)}$$

The over-all result for the processes $b + c$ is to yield the same change as in process a. The student can verify that

$$\Delta H_a = \Delta H_b + \Delta H_c$$
$$\Delta F_a = \Delta F_b + \Delta F_c$$
$$\Delta S_a = \Delta S_b + \Delta S_c$$

Case 1: The hypothetical evaporation of water at 373°K and 1.1 atm (impossible). At 373°K and 1 atm pressure the volume of 1 mole of vapor is 30.6 liters. At 373°K and 1.1 atm the volume can be calculated from the ideal gas law:

$$\begin{array}{ccccccccc} P & \times & V & = & n & \times & R & \times & T \\ 1.1 \text{ atm} & \times & 27.8 \text{ liter} & = & 1 \text{ mole} & \times & 0.0821 & \times & 373° \\ & & & & & & \text{liter atm/mole deg} & & \end{array}$$

The entropy at 27.8 liters is less than that at 30.6 liters, and this decrease can be calculated:

$$\Delta S = R(2.303) \log (27.8/30.6) = -0.1894 \text{ cal/deg}$$

So the probability factor for evaporation at 1.1 atm is less than that at equilibrium pressure (1 atm) by

$$T\Delta S = -373° \times 0.1894 \text{ cal/deg} = -70.7 \text{ cal}$$

The enthalpy increase is the same in all three cases, since at constant temperature ΔE and ΔPV remain constant for an ideal gas. The balance equation becomes

$$\begin{array}{ccccc} \Delta_1^2 F & = & \Delta_1^2 H & - & T\Delta_1^2 S \\ +70.7 \text{ cal} & = & +9713 \text{ cal} & - & 9642 \text{ cal} \\ & & \text{"Pull" inward} & & \text{"Push" outward} \end{array}$$

Thus the pull in is greater than the push out, and the water will *not* evaporate at 373°K under a pressure of 1.1 atm; or if it is in the vapor state it will be pulled together or *condensed* into the liquid state irreversibly.

Case 2: The evaporation of water at 373°K and 1 atm (equilibrium).

$$\begin{array}{ccccc} \Delta_1^2 F & = & \Delta_1^2 H & - & T\Delta_1^2 S \\ 0 & = & 9713 \text{ cal} & - & 9713 \text{ cal} \end{array}$$

Here the reaction will go either way since $\Delta_1^2 F = 0$.

Case 3: The evaporation of water at 373°K and 0.9 atm (irreversible). Here, at the end of the reaction, the gas will have a volume calculated by

$$P \times V = n \times R \times T$$
$$0.9 \text{ atm} \times 34 \text{ liters} = 1 \text{ mole} \times 0.0821 \times 373°$$
$$\Delta S = R(2.303) \log (34.0/30.6) \text{ liter atm/deg mole}$$
$$T\Delta S = 373° \times 0.2092 = 78 \text{ cal}$$

Thus

$$\Delta_1^2 F = \Delta_1^2 H - T\Delta_1^2 S$$
$$-78 \text{ cal} = 9713 \text{ cal} - 9791 \text{ cal}$$

Net tendency to go — "Pull" in — "Push" out

Here the push out is greater than the pull in, so the reaction will go; the water will evaporate irreversibly.

We now summarize the rules for using the free energy value as a guide to show whether a given reaction will take place.

1. Write the equation for the change:

$$\underset{\text{State 1}}{H_2O \text{ (liquid)}} = \underset{\text{State 2}}{H_2O \text{ (vapor)}}$$

2. If $-F_1 + F_2 = \Delta_1^2 F$ is positive, the reaction will not go spontaneously as written from left to right.

If $-F_1 + F_2 = \Delta_1 F = 0$, the two phases or states represented by the two sides of the equation will be in equilibrium.

If $-F_1 + F_2 = \Delta_1^2 F$ is negative, the reaction will go spontaneously as written from left to right.

We should also note that if we write

$$H_2O \text{ (gas, 1.1 atm, 100°C)} \rightarrow H_2O \text{ (liquid, 1.1 atm, 100°C)}$$

then the sign of the free energy change, $\Delta F = F_{\text{liquid}} - F_{\text{gas}}$, will now be negative; the vapor will condense spontaneously at 100°C. Thus, whenever the net free energy change is not zero, the reaction can always be written in a direction which will make the net change a negative quantity and the reaction will always go spontaneously in that direction.

Free energy and equilibrium. The use of free energy as a criterion for the direction of chemical change in a simple process like vaporization is straightforward. This principle, however, applies to any change, physical or chemical. We can only justify such a broad statement by consideration of a general case wherein we quickly lose touch with the tangible, but, if we keep in mind the scope of the principle derived, the excursion into the abstract is justifiable.

Consider a thermodynamic system consisting of a cylinder with a piston that is placed in a box insulated so that no heat can flow in or out of the box and no work can be done on or by the box and its contents. The box, piston, and cylinder constitute an isolated system. For convenience, we will divide the system into two parts: (1) the cylinder containing the active component, and (2) the rest of the system. We will limit the properties of 1 to the extent that it has the same temperature and pressure as 2, but otherwise the properties are unrestricted so that it can correspond to any system. The rest of the system can have any reasonable properties. We will make 2 so large that any heat flow ΔQ_2 (reversible) in or out will not perceptibly alter the temperature, and we will also make the pressure exerted by 2 constant.

For any change, $\Delta S = \Delta S_1 + \Delta S_2 \geqslant 0$ where the equality applies only at equilibrium. Since

$$\Delta Q_2/T = \Delta S_2$$
$$\Delta S_1 + \Delta Q_2/T \geqslant 0$$

Now, since the system is of constant volume and isolated, we can write

$$\Delta V = \Delta V_1 + \Delta V_2 = 0$$
$$\Delta E = \Delta E_1 + \Delta E_2 = 0$$

or

$$\Delta V_1 = -\Delta V_2 \quad \text{and} \quad \Delta E_1 = -\Delta E_2$$

According to the first law,

$$\Delta E_2 + P\Delta V_2 = \Delta Q_2$$

but, if we substitute for ΔE_2 and ΔV_2 in terms of ΔE_1 and ΔV_1, we find

$$\Delta Q_2 = -(\Delta E_1 + P\Delta V_1)$$

For a constant pressure change this becomes

$$\Delta Q_2 = -\Delta H_1$$

We can use this relation in the inequality to obtain

$$\Delta S_1 - \Delta H_1/T \geqslant 0$$
$$-(\Delta H_1 - T\Delta S_1) \geqslant 0$$
$$-(\Delta F_1) \geqslant 0$$

If we change the sign on both sides of the equation we must change its direction; hence

$$\Delta F_1 \leqslant 0$$

The inequality shows that for a spontaneous change the change in free energy is negative; at equilibrium the change in free energy is zero.

The use of free energy may seem to be a complicated and clumsy way of stating some simple facts. When the volume of the cylinder in Fig. 9 (in which a mole of gas is contained) is such that the pressure of the gas equals the vapor pressure of the liquid, then the gas (vapor) phase is in equilibrium with the liquid; if the volume is decreased, the pressure increases and the vapor tends to condense; if the volume is increased, the pressure decreases and the liquid tends to evaporate. This relates the facts in terms of tangible quantities like pressure without recourse to the free energy concept. So why use free energy? In a physical change like vaporization, it is true that the "vapor pressure" point of view may be more direct, but, in more complicated changes, especially in many chemical changes, it is often difficult to find a direct approach. On the other hand, the free energy principles are unlimited in their applicability and present the most fundamental point of view, one in which we can almost see at work the balance of forces that determine whether physical or chemical changes will take place. As we apply these ideas in later chapters, the broad scope of these principles will become evident.

PROBLEMS

1. Five moles of an ideal gas, at 30°C, and occupying a volume of 100 liters, is expanded to 150 liters isothermally and reversibly. Calculate the work done in kilocalories and liter atmospheres. *Ans.* 1.22 kcal; 50.4 liter atm.

2. From the same initial state, the same quantity of gas is expanded at constant pressure to the same final volume as before. The heat capacity of the gas (C_p) per mole at constant pressure is $\frac{5}{2}R$. What is the final temperature of the gas? How much work is done, and how much heat is absorbed?

3. From the same initial state the same gas is expanded adiabatically and reversibly to 150 liters. What is the final temperature and the work done? ($C_v = \frac{3}{2}R$.)

4. The same gas is cooled to the same final temperature at constant volume. What is the heat abstracted? If the same temperature is reached by cooling at constant pressure, what is the heat abstracted?

5. The boiling point of methyl alcohol is 64.65°C and the heat of vaporization is 262.8 cal gm^{-1} when the liquid is vaporized at 1 atm pressure. Assuming that the vapor behaves like an ideal gas, calculate the part of the heat of vaporization due solely to the passage from liquid to vapor state and excluding any external work. Neglect the volume of the liquid. *Ans.* 7.74 kcal mole^{-1}.

6. Calculate the entropy of vaporization per mole for methyl alcohol at the normal boiling point.

7. Calculate the entropy changes corresponding to the changes of state in Problems 1, 2, 3, and 4.

8. Assume that for methyl alcohol the average heat capacity at constant pressure between 65° and 84°C can be taken as 12.0 cal deg^{-1} mole^{-1} for the vapor and 24.0 cal deg^{-1} mole^{-1} for the liquid. Calculate the heat of vaporization at 84°C where the vapor pressure is 2 atm. *Ans.* 8.2 kcal mole^{-1}.

9. Calculate the entropy of vaporization of methyl alcohol at 84°C and 2 atm. Compare with the result in Problem 6 and explain the difference.

10. Thermodynamic efficiency is defined as the net work done in a Carnot cycle divided by the heat absorbed at the higher temperature. What is the thermodynamic efficiency of a Carnot cycle with the upper temperature at 84°C and the lower temperature at 65°C?

11. The normal boiling points and corresponding heats of vaporization in calories per mole for several compounds are: $COCl_2$, 8.3°C, 5988 cal/mole; CCl_2F_2, −29.8°, 5042; SiH_3Br, 2.4°, 5798; Si_2H_6, −14.3°, 5474; NH_3, −33.6°, 5878; H_2O, 100°, 9966; CH_3SH, 6.8°, 6442; CH_3SiHCl_2, 41.9°, 6569; $ClFC{=}CCl_2$, −27.9°, 5204; $C_2H_5NH_2$, 16.6°, 6965; $(C_2H_5)_2O$, 34.6°, 7207; CH_3OH, 64.7°, 8421. Calculate the values of the entropy of vaporization in each case. Can you suggest reasons why some values lie higher than the majority of the others?

12. Calculate the free energy change in the expansion of 1 mole of CH_3OH vapor at 84°C from 2 atm to 1 atm (ideal gas). *Ans.* −0.492 kcal.

13. Calculate the change in internal energy E and the change in enthalpy H for 1 mole of methyl alcohol vapor in passing from 65°C to 84°C at a constant pressure of 1 atm. $C_p = 12.0$ cal deg^{-1} mole^{-1} (ideal gas).

14. The entropy change for the change of state in Problem 13 is given by the relation $\Delta_1^2 S = C_p \ln (T_2/T_1)$. The entropy of 1 mole of gas at 64° is 58.0 E.U. Calculate the entropy at 84°C and 1 atm. *Ans.* 58.7 cal deg^{-1} mole^{-1}.

15. The standard enthalpy of 1 mole of methyl alcohol vapor at 65°C is −46.7 kcal. Calculate the enthalpy at 84°C and 1 atm. (See Problem 13.)

16. From the results of Problems 14 and 15, calculate the free energy of 1 mole of methyl alcohol vapor at 84°C and 1 atm.

17. Calculate the free energy change in heating 1 mole of liquid methyl alcohol from 65° to 84°C at 2 atm pressure, using results in problems above. (See Problem 8.)

18. Plot the absolute values of H, TS, and F for CH_3OH as (*a*) liquid, (*b*) gas at 1 atm, and (*c*) gas at 2 atm between 65° and 84°C.

19. A gas cylinder contains 200 moles of an ideal gas that is monatomic. The gas initially at 0°C and 100 atm expands until the pressure is 1.00 atm. (a) If the expansion is adiabatic and reversible, what is the final temperature? (*b*) If the expansion is isothermal and reversible, how much heat flows into the cylinder? (*c*) If the expansion is isothermal and irreversible ($P_e = 1$ atm), how much heat flows into the cylinder?

20. List the values of ΔE, ΔW, and ΔQ in parts (*a*), (*b*), and (*c*) of Problem 19 in cal.

★**21.** The gas in Problem 19 is expanded adiabatically as far as it will go against a constant pressure of 1 atm. What is the final temperature of the gas?

★**22.** The gas in Problem 21 is compressed adiabatically and irreversibly by an external pressure of 100 atm from the final temperature and pressure in Problem 21 to a point where internal and external pressure balance. What is the temperature when balance is reached?

23. Make a 4 x 4 cycle log-log plot of pressure against volume for 1 mole of an ideal monatomic gas, covering the pressure range from 0.01 atm to 100 atm and the volume range from 0.1 liter to 1000 liters. Draw the isothermals at intervals of 10° from 10°K to 100°K, at intervals of 50° from 100°K to 500°K, and at intervals of 100° from 500°K to 1000°K. Note that the slope of all isothermal lines is -1 and of all adiabatic lines is $-\frac{5}{3}$. (See Fig. 8*b*.) (*a*) Draw the adiabatic line downward from $P = 100$ atm and $T = 273$°K and check the answer to Problem 19*a*. (*b*) By a similar process, estimate the final temperature when a gas is compressed adiabatically and reversibly from 1 atm to 100 atm with the initial temperature at 273°K.

★**24.** When an automobile strikes a bump in the road suddenly, the air in the tires is compressed by a process effectively adiabatic in character. If the pressure jumps from 2.0 atm to 2.2 atm momentarily, estimate the increase in temperature if the initial temperature is 27°C. Assume $C_v = \frac{5}{2}R$.

25. If the upper and lower temperatures of a Carnot cycle are 400° and 300°K respectively, and 1 mole of an ideal gas (monatomic) expands initially from 3 liters to 30 liters, draw the cycle on the chart prepared in Problem 23 and determine the initial and final pressures of each step from the chart. What is the efficiency?

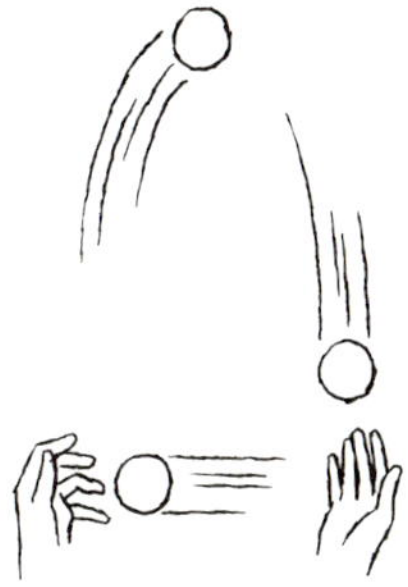

13 Chemical equilibrium

In the preceding chapters a number of examples of physical equilibrium were discussed, together with the way in which equilibrium shifts with temperature, pressure, and concentration. To illustrate equilibrium between different physical states, the example of liquid water and water vapor at 100°C in a closed box was analyzed in some detail. Now, if the temperature of this box were raised from 100°C to 4000°C, the predominant nature of the equilibrium would change from *physical* to *chemical*. At 100°C, molecules of H_2O are exchanged between the liquid and vapor state; at 4000°C, the major process is continual exchange of H and O atoms between the arrangement signified by the formula H_2O on the one hand and the arrangements signified by the formulas H_2 and O_2 on the other. When the relative amounts of H_2O, H_2, and O_2 stay constant with time, then chemical equilibrium is established. This is denoted by the equation

$$2H_2O(g) \rightleftharpoons 2H_2(g) + O_2(g)$$

These two types of equilibrium, physical and chemical, are shown in Fig. 1.

Even at room temperature there is an almost infinitesimal dissociation of water molecules to form H_2 and O_2; but, in water vapor in equilibrium with liquid water at room temperature, the calculated concentration of these molecules is so small (about 10^{-29} mole/liter) that it can be neglected. In this chapter, with the help of thermodynamic principles, we shall examine the quantitative nature of chemical equilibrium and the way in which it changes with variations in temperature, pressure, and concentration.

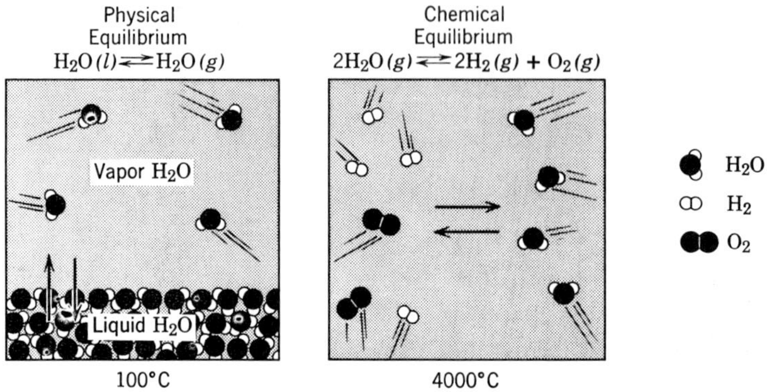

Fig. 1.

EQUILIBRIUM THERMODYNAMICS

Thermodynamic equilibrium factors. In order to make clear the role of the thermodynamic factors in influencing chemical equilibrium, we shall first compare the decomposition of water with its evaporation.

The equation for evaporation at 1 atm and 100°C may be written

State 1		State 2	
$H_2O(l)$	$\rightleftharpoons$	$H_2O(v)$	
$-H(l)$		$+H(v)$	$= \Delta_1^2 H$ $+9713$ cal
$-TS(l)$		$+TS(v)$	$= T\Delta_1^2 S$ $+9713$ cal
$-F(l)$		$+F(v)$	$= \Delta_1^2 F$ 0 cal

In this case the enthalpy (H) of the system increases as the water moves from the liquid to the gaseous state, because the forces of attraction between the molecules must be overcome as the molecules move farther apart; these forces tend to hold the molecules back in the liquid state. On the other hand, the probability factor (TS) tends to push the molecules out into the gaseous state. When these two opposite factors balance, there is physical equilibrium and

$$\Delta_1^2 F = 0$$

The analogous equation for the decomposition of water vapor at 4300°K may be written

State 1		State 2
$2H_2O(g)$	$\rightleftharpoons$	$2H_2(g) + O_2(g)$
Reactants		*Products*
$-(2H_{H_2O})$		$+(2H_{H_2} + H_{O_2}) = \Delta_1^2 H$ $+122$ kcal
$-(2TS_{H_2O})$		$+(2TS_{H_2} + TS_{O_2}) = T\Delta_1^2 S$ $+122$ kcal
$-(2F_{H_2O})$		$+(2F_{H_2} + TS_{O_2}) = \Delta_1^2 F$ 0 cal

The chemicals appearing on the left side of the equation in the initial state are called the *reactants*, and the chemicals found on the right side are called the *products*.

If H_{H_2O}, S_{H_2O}, F_{H_2O}, H_{H_2}, etc., designate the values of these quantities per mole, then each quantity must be multiplied by 2 wherever 2 moles appear in the equation. The Δ quantities designate the change in energy units when the reaction proceeds to the extent shown. For example, $\Delta_1^2 H$ is the change in enthalpy when 2 molecules of H_2O split to form 2 molecules of H_2 and 1 molecule of O_2 under the temperature and pressure specified. Actually, the energy needed to break up the H_2O molecule is considerably more than the energy released when H_2 and O_2 are formed; hence the energy factor favors the formation of H_2O; and at room temperature it is much larger than $T\,\Delta S$, so the reaction goes explosively from H_2 and O_2 to H_2O; on the other hand, at 4300°K, ΔH is exactly balanced by $T\,\Delta S$ and the reaction is in equilibrium when each constituent is at 1 atm.

The heat of reaction. The quantity of heat flowing in or out when the reaction takes place is called the *heat of reaction.* It does *not* mean the heat that flows in or out when 2 moles of water vapor are introduced into an evacuated box and then partially dissociate to establish equilibrium.

To see the correct interpretation, think of a large tank containing over a million moles of H_2O, H_2, and O_2; by some sort of control, we cause precisely 2 moles of H_2O to dissociate. The tank is so big that temperature, total pressure, and partial pressure do not change appreciably as these 2 moles undergo the reaction at equilibrium. In other words, the reaction takes place effectively at constant temperature and pressure; and the heat of reaction is the heat absorbed or given out when the number of moles written on the left side of the equation (e.g., $2H_2O$) is transformed reversibly to the moles written on the right side of the equation (e.g., $2H_2 + O_2$) under the conditions of temperature and pressure specified (e.g., 4000°C $\cong$ 4300°K and $P_{\text{total}} = 3$ atm).

We can apply the first law of thermodynamics to this reaction where water changes to H_2 and O_2. For the change in which 2 moles of H_2O form 2 moles of H_2 and 1 mole of O_2 we can write

$$\Delta_1^2 E = \Delta_1^2 Q + \Delta_1^2 W$$

Now in this case, when pressure is kept constant,

$$\Delta_1^2 W = -P\Delta_1^2 V$$

since the only work done results from the increase in volume that takes place against the constant pressure *P*. Because 2 moles of gas are initially present, and 3 moles of gas are present when the reaction is completed, we find that $\Delta_1^2 V$ equals 117.7 liters, the volume of 1 mole of gas at 3 atm and 4300°K (assuming ideal gas behavior). When one additional mole of gas is formed at 3 atm, the work done is 353 liter atm or approximately 8545 cal. Thus the equation, including actual numerical values converted to calories, is

$$\begin{array}{ccccc} \Delta_1^2 E & = & \Delta_1^2 Q & - & P\Delta_1^2 V \\ +113 \text{ kcal} & = & +122 \text{ kcal} & - & 9 \text{ kcal} \end{array}$$

or, rearranging,

$$\Delta_1^2 Q = \Delta_1^2 E + P\Delta_1^2 V$$

But, by definition,

$$H \equiv E + PV$$

and, at constant pressure,

$$\begin{array}{ccccc} \Delta_1^2 H & = & \Delta_1^2 E & + & P\Delta_1^2 V \\ +122 \text{ kcal} & = & +113 \text{ kcal} & + & +9 \text{ kcal} \end{array}$$

Therefore

$$\Delta_1^2 H = \Delta_1^2 Q$$

In words, at constant pressure the change in enthalpy equals the heat of reaction. Under these definitions, the value of $\Delta_1^2 Q$ is positive when heat is absorbed by the system, and negative when heat is given out; an opposite convention is found in many textbooks.

The free energy relations for the physical and the chemical equilibrium of water are shown in Figs. 2 and 3. *At 1 atm pressure,* physical equilibrium is established at 373°K (100°C) where the graphs for $°F_{H_2O(l)}$ and $°F_{H_2O(g)}$ intersect and $\Delta_l^g\,°F = 0$. Below this temperature the change from undercooled vapor to liquid takes place spontaneously; above this temperature the change from superheated liquid to vapor takes place spontaneously. The point representing the free energy of the system always moves downward (negative) in a spontaneous change.

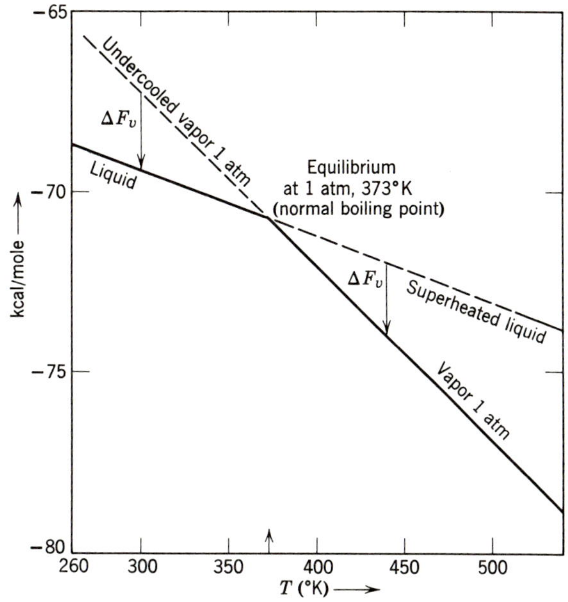

Fig. 2. Free energy of vaporization. (A spontaneous change always corresponds to a decrease in F as shown by the downward arrow.)

Similarly (Fig. 3), at 1 atm pressure for each constituent, chemical equilibrium is established at 4300°K (4027°C) where the graphs for $°F_{H_2O}$ and $°F_{H_2+O_2}$ intersect. At this temperature reactants and products have the same free energy and are in equilibrium. Below this temperature, at 1 atm each, hydrogen and oxygen combine spontaneously to form water; above this temperature water dissociates to form hydrogen and oxygen spontaneously. For each spontaneous re-

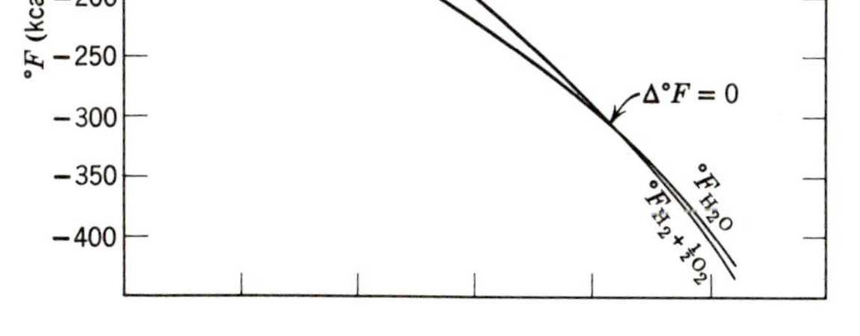

Fig. 3. Free energy. $H_2O \rightleftarrows H_2 + \frac{1}{2}O_2$.

action the point indicating the free energy of the system falls. (At 4300°K there will be side reactions involving the dissociation of the neutral molecules into atoms and ions, but these have been neglected in the discussion in order to emphasize the simple chemical principles of equilibrium.)

The law of mass action. When liquid water is placed in an evacuated box at 25°C and allowed to vaporize, the resultant pressure of the vapor (23.8 mm) is always the same; it is the vapor pressure of the water at that temperature; it is independent of the total amount of water in the box as long as there is an appreciable amount of liquid water present. But, if water vapor in a box at 3000°C dissociates, the pressures of H_2O, H_2, and O_2 at equilibrium will depend on how much H_2O vapor is present. This fact was observed early in the study of chemical reactions, and various attempts were made to find out whether there was any quantitative relation between the partial pressures of the individual substances present when the system came to equilibrium. As a result, it was shown on the basis of experimental evidence that there will be a constant value of the ratio of the pressures of the *products* to the pressures of the *reactants*, each partial pressure being raised to the power equal to the associated coefficient in the chemical equation. Thus for the reaction

$$\underset{\text{Reactants}}{2H_2O} \rightleftharpoons \underset{\text{Products}}{2H_2 + O_2}$$

it is found that

$$K_p = \frac{({}^eP_{H_2})^2({}^eP_{O_2})}{({}^eP_{H_2O})^2}$$

Each quantity in parentheses designates the partial pressure of the component under *equilibrium* conditions and is therefore marked with a superscript *e*. K_p is generally called the equilibrium constant; the subscript p indicates that it refers to the ratio of pressures. For this reaction at 4300°C the equilibrium constant has an estimated value of unity. Thus any of the following sets of pressures will result in chemical equilibrium:

$$K_p = \frac{(H_2)^2(O_2)}{(H_2O)^2} = \frac{(1\text{ atm})^2(1\text{ atm})}{(1\text{ atm})^2} = 1\text{ atm}$$

$$\frac{(0.5\text{ atm})^2(1\text{ atm})}{(0.5\text{ atm})^2} = 1\text{ atm}$$

$$\frac{(1\text{ atm})^2(0.01\text{ atm})}{(0.1\text{ atm})^2} = 1\text{ atm}$$

The units of K_p for this reaction are in atmospheres, as the units in the ratio show.

This equation is an example of the law of mass action or chemical equilibrium that applies not only to gases but also to every kind of physical and chemical change through which an equilibrium in mass exchange may be established. The general expression of this law will be discussed after we have examined this specific case a little more fully.

Before discussing the physical meaning of the equilibrium equation, it should be pointed out that the equation as applied to gases holds precisely only if they obey the ideal gas law. Similar restrictions hold for other states of matter. (Methods for dealing with non-ideal gases are developed in more advanced thermodynamics.) It should also be noted that K_p normally varies with temperature. Methods for calculating this variation are developed in a later section of this chapter.

The physical meaning of the mass action law may be shown by a comparison with the physical equilibrium between water in the liquid state and in the vapor state. Pressure has very little effect on the rate at which molecules leave liquid water. The number leaving per second depends significantly only on temperature and is constant when the temperature is constant. For an ideal gas, however, the number passing from the gas per second into the liquid is doubled if the pressure of the gas is doubled. To attain equilibrium it is necessary to have the pressure of the gas at a value where the number of molecules leaving the gas (making the reaction go from right to left) exactly equals the number leaving the liquid (left to right) according to the equation

$$\begin{array}{ccc} \text{State 1} & & \text{State 2} \\ H_2O(l) & \rightleftarrows & H_2O(g) \end{array}$$

Once the rate forward equals the reverse rate, a balance is struck wherein for every molecule of water vapor formed by evaporation another molecule disappears through condensation. The net result is that the number of water vapor molecules remains constant; hence the pressure does not change with time, and equilibrium is established.

The same balance of opposing rates is found in chemical equilibrium. Thus, in the reaction

$$\begin{array}{ccc} \text{State 1} & & \text{State 2} \\ 2H_2O & \rightleftarrows & 2H_2 + O_2 \end{array}$$

The rate of decomposition of H_2O is influenced by the partial pressures of the various gases present; similarly, the rate of formation of H_2O is influenced by the partial pressures. If there is to be no change in the amount of H_2, O_2, and H_2O, a balance of opposing rates must be reached; then equilibrium is attained. It is well established that equilibrium usually involves such a dynamic balance. Chemical equilibrium is the result of a balance of opposing rates rather than a cessation of reaction.

It is possible to devise an argument, correct in concept but fallacious in fact, that illustrates how the balance of opposing reactions can give the correct form of the mass action law. Consider the equilibrium

$$A + B \rightleftarrows AB$$

For reaction, an A and B molecule must get together. The probability that a certain small volume of the flask contains an A molecule is proportional to P_A; the probability that the *same* volume contains a B molecule is proportional to P_B. The joint probability that A and B will be in this same volume so that they can react is proportional to $P_A P_B$; hence we can write that the number of AB molecules *formed* per second $= k_f(P_A)(P_B)$, where k_f is a proportionality factor. The amount of decomposition of AB molecules per second is normally proportional to how many there are; hence we write that the number of AB molecules *decomposing* per second $= k_b P_{AB}$. At equilibrium, the two rates must be equal; hence

$$k_f(P_A)(P_B) = k_b(P_{AB})$$

or

$$\frac{k_f}{k_b} = \frac{(P_{AB})}{(P_A)(P_B)}$$

Since k_f and k_b are both constants, their ratios are also constant; consequently the pressure product is also a constant. This, of course, is the law of mass action.

The defect in the argument above is that the number of AB molecules formed per second need not be given by $k_f P_A P_B$. The same mass action law would be obtained if we found

$$\text{No. of AB molecules formed/sec} = k_f P_A^{1/2};$$

$$\text{No. of AB molecules decomposing/sec} = k_b P_{AB}/P_A^{1/2} P_B$$

Thus, although the concept of a dynamic equilibrium involving opposing reactions is correct, the added assumption that the balanced chemical equation shows how these rates depend on pressure is almost always incorrect. In the same way, the partial pressures of the chemical components influence the number of atoms passing from state 1 to state 2 (left to right) or passing from state 2 to state 1 (right to left) in a chemical reaction such as

State 1 State 2

$$2H_2O \rightleftarrows 2H_2 + O_2$$

But the dependence of these rates on partial pressure cannot be deduced from the chemical equation alone; hence the law of mass action cannot be arrived at by the equality of opposing rates without additional information. For this reason the logic of the law of chemical equilibrium is developed from the point of view of thermodynamics in this chapter; the rate approach is treated in Chapter 14.

Thermodynamics and the equilibrium constant. In essence, the law of mass action means that at a given temperature there is a fixed driving force for a forward and a reverse reaction. But the net driving force for a reaction depends on the partial pressures of reactants and products, and hence the partial pressures must have some fixed relation if they are to correspond to a given driving force; for the reaction to form water, this fixed relation of partial pressure is

$$\frac{(P_{H_2})^2(P_{O_2})}{(P_{H_2O})^2} = K$$

Now, thermodynamics gives the driving force of a reaction a concrete form, Δ_1^2F. To derive the equilibrium law involving partial pressures, we must first see the relation between the driving force, ΔF, and the pressure, P. To this end let us consider the change in free energy in the isothermal expansion of water (treated as an ideal gas).

$$\text{State 1} \quad \rightarrow \quad \text{State 2}$$

$$H_2O(g)(P_1, T) \rightarrow H_2O(g)(P_2, T)$$

We write

$$\Delta_1^2F = \Delta_1^2H - T\,\Delta_1^2S$$

Since Δ_1^2F is a state function, we can carry out this change by any convenient path. We choose a reversible path and note

(A) $\Delta_1^2E = 0$

$\Delta_1^2Q = -\Delta_1^2W$

(B) $\Delta_1^2H = \Delta_1^2E + \Delta_1^2(PV) = 0$

(C) $\Delta_1^2S = \Delta_1^2Q/T = R\ln(V_2/V_1)$

(D) $\Delta_1^2F = \Delta_1^2H - T\,\Delta_1^2S = -RT\ln(V_2/V_1)$

Steps A and B hold for any isothermal expansion of an ideal gas; step C is true for a reversible isothermal expansion; step D follows from B and C. Each of these cases has been considered in the last chapter. The last equation can also be expressed in terms of pressure:

$$V_2 = RT/P_2,\ V_1 = RT/P_1;$$

hence

$$\Delta_1^2F = -RT\ln(V_2/V_1) = +RT\ln(P_2/P_1)$$

Thus the change in the free energy in the expansion of 1 mole of an ideal gas at constant temperature is given by

$$\Delta_1^2F = RT\ln(P_2/P_1)$$

For the expansion of n moles of an ideal gas, we find

$$\Delta_1^2F = nRT\ln(P_2/P_1)$$

We can use this equation to get the relation between $\Delta_1^2 F$ and K by considering the chemical change from H_2O to H_2 and O_2 under two conditions:

Case I. Equilibrium Conditions

$$\underset{\text{gas at } {}^eP_{H_2O}}{2H_2O} \rightleftarrows \underset{\text{gas at } {}^eP_{H_2}}{2H_2} + \underset{\text{gas at } {}^eP_{O_2}}{O_2}$$

$$\underset{\text{reactants}}{-2\,{}^eF_{H_2O}} + \underset{\text{products}}{(2\,{}^eF_{H_2} + {}^eF_{O_2})} = \Delta^e F$$

Case II. Standard Conditions

$$\underset{\text{gas at } {}^\circ P = 1 \text{ atm}}{2H_2O} \rightleftarrows \underset{\text{gas at } {}^\circ P = 1 \text{ atm}}{2H_2} + \underset{\text{gas at } {}^\circ P = 1 \text{ atm}}{O_2}$$

$$\underset{\text{reactants}}{-2\,{}^\circ F_{H_2O}} + \underset{\text{products}}{(2\,{}^\circ F_{H_2} + {}^\circ F_{O_2})} = \Delta^\circ F$$

In case I we use the set of symbols ${}^eP_{H_2O}$, ${}^eP_{H_2}$, and ${}^eP_{O_2}$ to denote any set of equilibrium pressures for these gases at the temperature for which we are studying the reaction, say at 2000°K. Also, ${}^eF_{H_2O}$, ${}^eF_{H_2}$ and ${}^eF_{O_2}$ are the values of the free energy per mole for each of the components at these equilibrium pressures. By the same token, ${}^\circ P_{H_2O}$, ${}^\circ P_{H_2}$, and ${}^\circ P_{O_2}$ are each the standard pressure of 1 atm; and ${}^\circ F_{H_2O}$, ${}^\circ F_{H_2}$, and ${}^\circ F_{O_2}$ are the values of the free energy per mole when each of the component gases is at 1 atm pressure. It is customary to use a superscript circle (°) to designate standard conditions. On the right side of the equation, $\Delta^e F$ is the change in free energy when 2 moles of water dissociate under equilibrium conditions to form $2H_2$ and O_2; thus $\Delta^e F = 0$ because the change in free energy is zero when a reaction takes place at equilibrium. $\Delta^\circ F$ is the change in free energy of the system when 2 moles of H_2O at 1 atm dissociate and recombine to give 2 moles of H_2 and 1 mole of O_2, each at the pressure of 1 atm; this is the change in free energy under *standard conditions.*

Note that the expression for the change in free energy with pressure relates ${}^eF_{H_2O}$ and ${}^\circ F_{H_2O}$ by the equation

$${}^eF_{H_2O} - {}^\circ F_{H_2O} = RT \ln ({}^eP_{H_2O}/{}^\circ P_{H_2O})$$

For 2 moles of H_2O we get

$$2\,{}^eF_{H_2O} - 2\,{}^\circ F_{H_2O} = RT \ln ({}^eP_{H_2O}/{}^\circ P_{H_2O})^2$$

since multiplying the logarithm by 2 is the same as squaring the quantity in parentheses. Similarly

$$2\,{}^eF_{H_2} - 2\,{}^\circ F_{H_2} = RT \ln ({}^eP_{H_2}/{}^\circ P_{H_2})^2$$
$${}^eF_{O_2} - {}^\circ F_{O_2} = RT \ln ({}^eP_{O_2}/{}^\circ P_{O_2})$$

If we subtract the equation for the free energy under standard conditions (case II) from that for the free energy under equilibrium conditions (case I) and substitute the expressions involving pressure for each pair, like ${}^eF_{H_2O} - {}^\circ F_{H_2O}$, we get

$$RT\ln ({}^eP_{H_2}/{}^\circ P_{H_2})^2 + RT\ln ({}^eP_{O_2}/{}^\circ P_{O_2}) - RT\ln ({}^eP_{H_2O}/{}^\circ P_{H_2O})^2 = \Delta^e F - \Delta^\circ F$$

Rearranging the values of the pressure so that all ${}^eP$ are in one term and all ${}^\circ P$ in another gives

$$\Delta^e F - \Delta^\circ F = RT\ln\left(\frac{{}^eP^2_{H_2}\cdot {}^eP_{O_2}}{{}^eP^2_{H_2O}}\right) - RT\ln\left(\frac{{}^\circ P^2_{H_2}\cdot {}^\circ P_{O_2}}{{}^\circ P^2_{H_2O}}\right)$$

However, ${}^\circ P_{H_2} = {}^\circ P_{O_2} = {}^\circ P_{H_2O} \equiv 1$; hence

$$\Delta^e F - \Delta^\circ F = RT\ln\left(\frac{{}^eP^2_{H_2}\cdot {}^eP_{O_2}}{{}^eP^2_{H_2O}}\right)$$

Up to this point we have made no use of the fact that ${}^eP$ represents equilibrium pressure and $\Delta^e F$ is the free energy change at equilibrium. The free energy principle, introduced in the last chapter, assures us that $\Delta^e F = 0$. This leaves

$$-\Delta^\circ F = RT\ln\frac{({}^eP_{H_2})^2({}^eP_{O_2})}{({}^eP_{H_2O})^2}$$

In discussing equilibrium it may be taken for granted that the P symbols refer to equilibrium pressures unless otherwise specified; so we will omit the e superscripts from here on. Expressing the equation in exponential form,

$$\frac{(P_{H_2})^2(P_{O_2})}{(P_{H_2O})^2} = e^{-\Delta^\circ F/RT}$$

The quantity on the right side is a constant at any fixed value of T. Thus the thermodynamic analysis shows the theoretical justification for writing

$$\frac{(P_{H_2})^2(P_{O_2})}{(P_{H_2O})^2} = K_p$$

where

$$K_p = e^{-\Delta^\circ F/RT}$$

The latter relation frequently is written

$$-\Delta^\circ F = RT\ln K_p$$

The negative of the free energy change in a system under standard conditions is equal to the product of RT and the natural logarithm of the equilibrium constant. In order to simplify the notation used in discussing equilibrium from the standpoint of partial pressure, it is convenient to write

$$({}^eP_{H_2}) \equiv (H_2)$$

Thus the equilibrium equation in this notation is

$$K_p = \frac{(H_2)^2(O_2)}{(H_2O)^2}$$

For the reaction

$$H_2O \rightleftarrows H_2 + \tfrac{1}{2}O_2$$

the equilibrium equation is

$$K_p = \frac{(H_2)(O_2)^{1/2}}{(H_2O)}$$

As an example, the graphs of Fig. 4 show the values of $°F$ for reactants and products and also values of ^eF for reactants and products at pressures that result in equilibrium at 1400°K. By increasing the pressure of H_2O vapor from 1 atm to $^eP_{H_2O}$, its free energy at 1400° is raised from −126 kcal to −100 kcal; at this increased pressure the whole free energy graph for H_2O is displaced upward as shown on the diagram. By decreasing the pressure of H_2 and O_2 from 1 atm each to $^eP_{H_2}$ and $^eP_{O_2}$, the free energy of the products is lowered from −83 kcal to −100 kcal at 1400°K; the whole F curve for the products is correspondingly lowered. With these values of ^eP, the two graphs intersect at 1400°K and $\Delta^eF = 0$; thus equilibrium is established at this temperature.

The general equilibrium equation. Let us now expand our equations into a more general form applicable to any reaction.

Reactants				Products			
2H₂O			⇄	$2H_2$	$+ 1O_2$		A particular reaction
aA	+ bB	+ cC + ···	⇄	qQ	+ rR	+ sS + ···	A general form of reaction

The capital letters stand for any different species of molecules, single atoms, ions, and the like that enter into chemical reactions. The small letters stand for the numbers of each species taking part. Thus we get for the equilibrium equation in generalized form

$$K_p = \frac{(Q)^q(R)^r(S)^s \cdots}{(A)^a(B)^b(C)^c \cdots}$$

It is sometimes convenient to use units of concentration instead of pressure. For an ideal gas the concentration in moles per liter is related to the pressure by taking the ideal gas equation:

$$P_A V = N_A RT$$

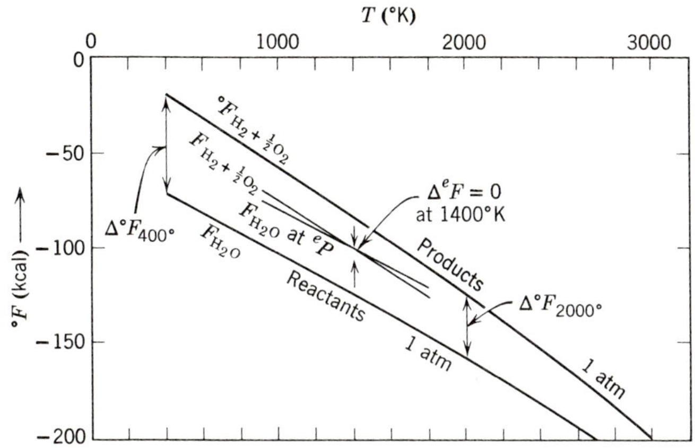

Fig. 4. H_2O (reactant) $\rightleftarrows$ $H_2 + \frac{1}{2}O_2$ (products).

where P_A is the partial pressure of component A and N_A the number of moles of A present in the system. We can also write

$$\underset{\substack{\text{Concentration}\\ \text{in moles per liter}}}{\frac{N_A}{V}} = \frac{P_A}{RT}$$

The symbol N_A/V will be written as the chemical formula in square brackets to designate concentration in moles per liter:

$$\frac{N_A}{V} \equiv [A]$$

Thus

$$\Delta_1^2 F = RT \ln \frac{P_2}{P_1} = RT \ln \frac{[A]_2}{[A]_1}$$

and, with a corresponding change in the defined standard state,

$$K_c = \frac{[Q]^q[R]^r[S]^s \cdots}{[A]^a[B]^b[C]^c \cdots} = e^{-\Delta^\circ F_c/RT}$$

In studying reactions in solution it is necessary to express the factors in the equilibrium equation in units of concentration rather than pressure. In an ideal solution for which Henry's law holds, it can be shown that the solute behaves thermodynamically like an ideal gas. Therefore

$$\Delta_1^2 F = RT \ln \frac{[A]_2}{[A]_1}$$

where $[A]_2$ is the concentration of component A in state 2 and $[A]_1$ in state 1.

In view of this, the equilibrium equation in concentration units may also be applied to ideal solutions.

For a gas the conversion from pressure units to concentration units is given by the formula

$$\frac{[\mathrm{Q}]^{\mathrm{q}}[\mathrm{R}]^{\mathrm{r}}[\mathrm{S}]^{\mathrm{s}}\cdots}{[\mathrm{A}]^{\mathrm{a}}[\mathrm{B}]^{\mathrm{b}}[\mathrm{C}]^{\mathrm{c}}\cdots}(RT)^{\Delta n} = \frac{(\mathrm{Q})^{\mathrm{q}}(\mathrm{R})^{\mathrm{r}}(\mathrm{S})^{\mathrm{s}}\cdots}{(\mathrm{A})^{\mathrm{a}}(\mathrm{B})^{\mathrm{b}}(\mathrm{C})^{\mathrm{c}}\cdots}$$

where $\Delta n = (\mathrm{q} + \mathrm{r} + \mathrm{s} + \cdots) - (\mathrm{a} + \mathrm{b} + \mathrm{c} + \cdots)$. Therefore

$$K_p = K_c(RT)^{\Delta n}$$

Also, when $\Delta^\circ F_p$ refers to the change in free energy in passing from reactants at 1 atm pressure to products at 1 atm pressure, it is related to $\Delta^\circ F_c$, the change from reactants at unit concentration, by the equation

$$\Delta^\circ F_p - \Delta^\circ F_c = -RT \ln (RT)^{\Delta n}$$

As a rule, the subscripts p and c will be omitted from the symbol $\Delta^\circ F$ unless the context does not make clear the conditions under which the standard state is specified.

The variation of the equilibrium constant with temperature. Thermodynamic analysis (see Appendix) shows that the equilibrium constant varies with temperature in a way closely resembling the variation of vapor pressure. For vapor pressure we write

$$\log \frac{P_2}{P_1} = \frac{-\Delta H_{\mathrm{vap}}}{2.303R}\left(\frac{1}{T_2} - \frac{1}{T_1}\right)$$

where P_2 and P_1 are the vapor pressures at absolute temperatures T_2 and T_1, and ΔH_{vap} is the heat of vaporization. In a similar way we write

$$\log \frac{K_2}{K_1} = \frac{-\Delta^\circ H_{\mathrm{react}}}{2.303R}\left(\frac{1}{T_2} - \frac{1}{T_1}\right)$$

where K_2 and K_1 are the equilibrium constants at absolute temperatures T_2 and T_1 and $\Delta^\circ H_{\mathrm{react}}$ is the standard heat of reaction.

In the case of vapor pressure we can equate T_1 to the boiling point T_b, where $P_1 = 1$ atm, and the equation becomes

$$\log P = \frac{-\Delta H_{\mathrm{vap}}}{2.303R}\frac{1}{T} + \frac{\Delta H_{\mathrm{vap}}}{2.303RT_b}$$

Since the heat of vaporization does not vary much with temperature, it may be regarded as a constant, and we write $-\Delta H_{\mathrm{vap}}/(2.303R) = a$; the last term on the right is also a constant (b); we let $\log P = y$, the dependent variable; and $1/T = x$, the independent variable. Thus we write

$$y = ax + b$$

and we see that the plot of log P against $1/T$ gives a graph that is approximately a straight line.

In the same way we write

$$\log \mathrm{K} = \frac{-\Delta^{\circ} H_{\text{react}}}{2.303R}\frac{1}{T} + \frac{\Delta^{\circ} H_{\text{react}}}{2.303RT_0}$$

where we select T_0 as the temperature at which $K = 1$. So this expression is also in the form of a linear equation; and, when log K is plotted against $1/T$ even for a range as large as several hundred degrees, one usually obtains a graph that is a straight line to a good approximation, as shown in Figs. 5, 6, and 7 later in this chapter. This is a useful relation to calculate K at different temperatures when one knows the value of K at a single temperature together with the value of the heat of reaction.

EQUILIBRIUM CALCULATIONS

We now apply the law of chemical equilibrium to several specific reactions in order to illustrate in detail the nature of equilibrium under a variety of conditions.

Dissociation of acetic acid. Using the symbol Ac^- to denote $COO \cdot CH_3^-$ and HAc to denote *acetic acid*, examine the equilibrium expressed by the equation for the dissociation of acetic acid:

$$HAc \rightleftarrows H^+ + Ac^-$$

The corresponding equilibrium expression is

$$\frac{[H^+][Ac^-]}{[HAc]} = K$$

At 25°C the value of K is 1.8×10^{-5}. The following sets of concentrations satisfy the equation

(*a*) $$\frac{[0.006][0.006]}{[2]} = 1.8 \times 10^{-5}$$

(*b*) $$\frac{[0.003][0.003]}{[0.5]} = 1.8 \times 10^{-5}$$

(*c*) $$\frac{[6 \times 10^{-6}][6 \times 10^{-6}]}{[2 \times 10^{-6}]} = 1.8 \times 10^{-5}$$

These figures all represent different dilutions of a pure HAc solution.

Another set of equilibrium concentrations may be obtained by adding an acid like HCl to a solution of HAc. This increases the concentration of H^+ and also adds chloride ions; but the latter do not take part in the reaction of the dissociation of HAc. Suppose that in this way x moles per liter of hydrogen ion are added to the solution. In order to simplify the equation, let the concentration of Ac^- be denoted by the symbol y; then the concentration of H^+ formed by this dissociation will also be equal to y, because one H^+ is produced for each Ac^- formed. The equilibrium equation reads

$$\frac{[x + y][y]}{[HAc]} = K$$

If $[HAc] = 2.0$ and x is 0.009, then y must be 0.003, and another set of equilibrium concentrations is

$$\frac{[0.012][0.003]}{[2]} = 1.8 \times 10^{-5}$$

A parallel situation could have been brought about if, instead of adding hydrogen chloride, we added sodium acetate, NaAc, the salt formed when sodium hydroxide and acetic acid combine according to the equation

$$NaOH + HAc \rightleftharpoons NaAc + H_2O$$

If 0.009 mole of sodium acetate is added forming 0.009 mole per liter of acetate ions (assuming that this salt is completely dissociated) and 0.003 mole per liter of acetate ion is present (coming from dissociated acetic acid), there will be a total concentration of 0.012 mole per liter of acetate ion and 0.003 mole per liter of hydrogen ion. Again it will be necessary to have 2 moles per liter of acetic acid in order to satisfy the mass action relationship:

$$\frac{[0.003][0.012]}{[2]} = 1.8 \times 10^{-5}$$

Finally, returning to the concentrations first considered with $[H^+] = [Ac^-]$, suppose that $HAc = 2.00$ moles per liter; then the equation is

$$\frac{[0.006][0.006]}{[2]} = 1.8 \times 10^{-5}$$

Degree of dissociation (d) may be defined by the relation:

$$d = \frac{\text{final concentration of ion}}{\text{initial concentration of undissociated molecule}}$$

If there were originally 2.006 moles of acetic acid added to the solution, at equilibrium $[H^+] = [Ac^-] = 0.006$ and $[HAc] = 2$; and $d = 0.006/2.006 = 0.00299$. With $[H^+]$ and $[Ac^-]$ each equal to 0.003 mole per liter, $d = [0.003]/[0.503] = 0.00596$. Decreasing the concentration increases the amount of dissociation.

Applying the same principle to the concentrations obtained when NaAc was added to the solution, $d = [0.003]/[2.003] = 0.0015$. Thus an increase in the concentration of either ion, achieved by the addition of another substance which generates that ion, decreases the amount of dissociation.

The synthesis of ammonia. The reaction of nitrogen with hydrogen to form ammonia has been studied extensively to determine the equilibrium at various temperatures. The reaction is written

$$\tfrac{1}{2}N_2 + \tfrac{3}{2}H_2 = NH_3$$

The equilibrium expression is

$$K_p = \frac{(NH_3)}{(N_2)^{1/2}(H_2)^{3/2}}$$

where the symbols in parentheses represent the partial pressures of the constituents. At 400°K, the value of K_p is 6.9 atm^{-1}. Either of the following sets of values of the pressures satisfies the equilibrium condition:

$$\frac{(NH_3)}{(N_2)^{1/2}(H_2)^{3/2}} = \frac{(6.9\ \text{atm})}{(1\ \text{atm})^{1/2}(1\ \text{atm})^{3/2}} = 6.9\ \text{atm}^{-1}$$

$$\frac{(166\ \text{atm})}{(9\ \text{atm})^{1/2}(4\ \text{atm})^{3/2}} = 6.9\ \text{atm}^{-1}$$

Suppose we compare this reaction with the formation of water from its elements at 400°K:

$$H_2 + \tfrac{1}{2}O_2 = H_2O$$

$$K_p = \frac{(H_2O)}{(H_2)(O_2)^{1/2}} = 1.73 \times 10^{29}\ \text{atm}^{\ 1/2}$$

These figures show the enormous difference in the two reactions: a mixture in which the partial pressures of NH_3, H_2, and N_2 are all 1 atm is not far from equilibrium; on the other hand, when the partial pressures are all 1 atm, there is a tremendous tendency for the reaction to go from H_2 and O_2 to H_2O, as shown by the factor 10^{29}. A comparison of the two thermodynamic factors follows.

For

$$H_2 + \tfrac{1}{2}O_2 \rightarrow H_2O$$

$$\Delta^\circ F = \Delta^\circ H - T\Delta^\circ S$$

$$-53.517 = (-58.042) - (-4.525)\ \text{kcal}$$

Large associative tendency	Energy factor associative	Probability factor dissociative

For

$$\tfrac{3}{2}H_2 + \tfrac{1}{2}N_2 \rightarrow NH_3$$

$$\begin{array}{cccc} \Delta^\circ F & = \Delta^\circ H & - & T\Delta^\circ S \\ -1.496 & = -11.552 & - & (-10.056)\ \text{kcal} \end{array}$$

Small associative tendency	Energy factor associative	Probability factor dissociative

The large energy factor of 58 kcal pulls the H_2 and O_2 together into the form H_2O. In other words, the O—H bond is formed with the release of a great deal of energy. Working in the opposite direction, the dissociative probability factor is ineffective against the overwhelming associative energy factor.

For NH_3, the associative tendency due to the energy factor is much smaller. The disassociative tendency due to the probability factor is about twice as large. This increase is due in part to the extra hydrogen. As a result, there is only a small tendency for the formation of NH_3 under standard conditions at 400°K.

The plot of log K for these two reactions shows that K for NH_3 changes from a value >1 to a value <1 at about 450°K. At higher temperatures the probability factor overcomes the small energy factor and the dissociative tendency dominates. On the other hand, K for H_2O does not become less than 1 until T is above 4300°K.

In conclusion, we see that these thermodynamic factors tell the whole story of the kind of chemical behavior to be expected when substances come into chemical equilibrium with each other. To find out why the thermodynamic factors have the values they do, i.e., to find out why $\Delta^\circ H$ is so much greater for H_2O than for NH_3 formation, we would have to turn to quantum mechanics.

The hydrogenation of benzene. If a mixture containing 1 atm partial pressure of benzene and 1 atm partial pressure of hydrogen is placed in a flask at 200°C, what will be the equilibrium concentration of cyclohexane?

	C_6H_6	$+\ 3H_2$	$\rightleftarrows C_6H_{12}$	
At the start:	1 atm	1 atm	0 atm	values of pressure
At equilibrium:	$1 - x$	$1 - 3x$	x	

In this summary, x represents the partial pressure of C_6H_{12} produced. At 200°C, $K_P = 10^{-3}$. Thus

$$K_p = \frac{x}{(1 - x)(1 - 3x)^3} = 10^{-3}$$

Because the equilibrium constant is small, the numerator is smaller than the denominator in the pressure ratio and the pressure of product formed must be small with respect to unity. So, to a first approximation, it is permissible to neglect x and $3x$ with respect to 1. Thus the resultant equilibrium pressure of cyclohexane in such a mixture is almost exactly 10^{-3} atm. Clearly, the neglect of 10^{-3} atm with respect to 1 is a legitimate approximation. On the other hand, at a higher temperature where $K_p = 1$, the same approximation gives the ridiculous answer that $x = 1$; x cannot be neglected with respect to 1; and a fourth-powered equation must be solved in order to determine the value of x.

Many times it is necessary to make a calculation of this type with the equilibrium constant far greater than 1, e.g., 10^3. If the equilibrium expression is

$$\frac{x}{(1 - x)(1 - 3x)^3} = 10^3$$

no approximation can be made, but the solution of the fourth-power equation may be avoided because the equilibrium is reversible. In other words, the same value for the equilibrium pressures is found by assuming that all the benzene reacts to form cyclohexane, and that the system then comes to the equilibrium from the cyclohexane end. Thus

	C_6H_{12}	$\rightleftarrows 3H_2$	$+ C_6H_6$	
Start:	0.33 atm	0 atm	0.66 atm	pressures
Equilibrium:	$0.33 - x$	$3x$	$0.66 + x$	

The equilibrium constant for this reaction is $1/10^3$; when we substitute these values into the new equilibrium expression, we obtain

$$\frac{(0.66 + x)(3x)^3}{(0.33 - x)} = 10^{-3}$$

Since $K = 10^{-3}$, x must be small and can be neglected with respect to 0.33 and 0.66. Then the preceding expression becomes

$$\frac{(0.66)}{(0.33)} 27x^3 = 10^{-3}$$

and

$$x = 0.026$$

Therefore the equilibrium partial pressures are

$$P_{C_6H_6} = 0.69 \text{ atm}; \quad P_{H_2} = 0.08 \text{ atm}; \quad \text{and } P_{C_6H_{12}} = 0.30 \text{ atm}$$

In general, it is possible to simplify equations by the neglect of one of the unknowns in a sum or a difference. However, this must be done with caution and, after completion of the problem, a check should be made to see if the answer is reasonable in the light of the approximations used.

Decomposition of PCl_5. If 1 mole of PCl_5 is put in a 10.0-liter flask and heated to 250°C, it will decompose partially to an equilibrium state:

$$PCl_5 \rightleftarrows PCl_3 + Cl_2$$

Both reactants and products are gases at this temperature. If at equilibrium the amount of chlorine is 0.47 mole, what is the value of K_c? At the start, there is a concentration of PCl_5 of 0.100 mole per liter; at equilibrium there is 0.047 mole of chlorine per liter. Thus

	PCl_5	$\rightleftarrows PCl_3$	$+ \; Cl_2$	
Start:	0.100	0	0	moles/liter
Equilibrium:	$0.100 - 0.047$	0.047	0.047	

$$K_c = \frac{(0.047)(0.047)}{(0.053)} = 0.042$$

What is the value of K_p in this reaction? According to the relationship we derived earlier, the relation between K_p and K_c is

$$K_p = K_c(RT)^{\Delta n}$$

In order to express the concentration in terms of moles per liter, the value of R must be given in liter atmospheres, and, as always, the value of T is given in degrees Kelvin. In this particular reaction the change in the number of moles is 1. Substituting the values of n, R, T, and K_c,

$$K_p = (0.082 \times 523)(0.042) = 1.80$$

What amount of chlorine would have been produced if the pressure were increased a hundred-fold by reducing the volume to 0.1 liter? The initial concentration of PCl_5 changes from 0.1 mole per liter to 10 moles per liter. Hence

	PCl_5	$\rightleftarrows PCl_3$	$+ \; Cl_2$	
Start:	10	0	0	moles/liter
Equilibrium:	$10 - x$	x	x	

Substituting these values into the expression for K_c, we find

$$\frac{x^2}{10 - x} = 0.042 \quad \text{or} \quad \frac{x^2}{10} \approx 0.042$$

so that $x = 0.65$, or 6.5% of the PCl_5 decomposes.

Compare this with the first reaction that was carried out at a somewhat lower pressure. In that reaction, the amount of PCl_5 decomposed was about 47%. Thus, as the pressure increases, less PCl_5 decomposes. This is an important fact. Re-examining the equation describing this reaction, we see that 1 mole of PCl_5 decomposes to give 2 moles of product. Since only gases are involved, this means that, when the reaction is carried out in a closed container, the decomposition is accompanied by an increase in pressure. Thus, as the initial pressure becomes greater, the shift in the equilibrium is in the direction that makes the change in pressure smaller; it is harder for the PCl_5 to decompose.

Le Chatelier's principle and pressure change. A generalization of this type of behavior is embodied in the principle of *Le Chatelier.* This principle states: *If by any action (pressure changes, changes of temperature, changes in concentration) a shift in the equilibrium state is produced, the nature of this shift is such that the initial action is reduced in magnitude.* In the PCl_5 system, when the total pressure is increased, the equilibrium shifts in a way that lessens the pressure increase. The same kind of action is found in the hydrogenation of benzene. If the pressure on the benzene-hydrogen mixture is increased, there is a higher yield of cyclohexane, which thus reduces the pressure increase. If only the partial pressure of P_{PCl_5} is increased, then more decomposition takes place to lessen this increase. The application of Le Chatelier's principle to effects produced by changing temperature will be discussed later.

Equilibrium under constant pressure. If a sample of PCl_5 is put into a cylinder closed at one end by a gastight piston, and pressure is maintained at 1 atm until the equilibrium is reached, what will be the partial pressure of chlorine? This problem requires a slightly different approach. Here, instead of carrying out the reaction in a constant volume, it is carried out at constant pressure. This reaction can be represented

	PCl_5	$\rightleftarrows$ PCl_3	+ Cl_2	
Initial:	$P_0 = 1$	0	0	atm
Equilibrium:	$P - x$	x	x	

where $P - x$ is the pressure of PCl_5 when equilibrium is achieved. The total pressure in the system will be the pressure of PCl_5 plus that of PCl_3 plus that of Cl_2:

$$P_{PCl_5} + P_{Cl_2} + P_{PCl_3} = 1$$
$$P - x + x + x = 1$$

and

$$K_p = \frac{(PCl_3)(Cl_2)}{(PCl_5)} = \frac{x^2}{P - x} = 1.80$$

P can be eliminated in view of the relationship between P and x. Then the equation may be written

$$x^2/(1 - 2x) = 1.80$$

In this equation, $2x$ cannot be neglected with respect to 1. If it were neglected, a solution would result where x had a value of more than 1, a ridiculous answer.

The equation must be solved exactly. Rearranging it gives a quadratic form:

$$x^2 + 3.60x - 1.80 = 0$$

so that

$$x = -1.80 \pm 2.24(5)$$

The negative value for x has no physical meaning; the partial pressure of chlorine is 0.445 atm when the total pressure is 1 atm.

Heterogeneous equilibrium. The reactions just discussed are called homogeneous because both the reactants and the products are in the same phase. If one or more of the components are in different physical states, a somewhat different approach is required. As an example, let us consider the decomposition of NOBr that takes place according to the equation

$$2NOBr \rightleftarrows 2NO + Br_2$$

If we put NOBr into a flask at 1 atm and 25°C, 23% of the compound decomposes. Let us calculate K_p and also calculate the per cent decomposition if the initial pressure is 2 atm, 4 atm, and 8 atm. At first glance, this problem looks exactly like the one we have done for PCl_5. The actual computations, however, are complicated by the fact that the vapor pressure of bromine at 25°C is 0.282 atm. We can calculate K_p from the following sets of data.

	2NOBr	⇄ 2NO +	Br_2
Start:	1	0	0 atm
Equilibrium:	1 − 0.23	0.23	0.115

The expression for K_p is

$$K_p = (Br_2)(NO_2)^2/(NOBr)^2$$

If we substitute for the pressures, we find

$$K_p = \frac{(0.115)(0.23)^2}{(0.77)^2} = 1.0 \times 10^{-2} \text{ atm}$$

Now that we know the value of K_p, we can determine what the per cent decomposition will be at 2 atm. Let us designate by x the pressure of the NO formed. Under these conditions, we can summarize what is occurring in the following equation.

	2NOBr	⇄ 2NO +	Br_2
Initially:	2	0	0 atm
Equilibrium:	$2 - x$	x	$x/2$

If we put these values into the equilibrium expression,

$$\frac{(x/2)(x)^2}{(2 - x)^2} = 1.0 \times 10^{-2}$$

The value of the various quantities are such that we cannot neglect x with respect to 2 and get a reasonably accurate answer. Therefore we must solve the resulting cubic equation. If we do this, we find that x is equal to 0.38 and at this higher pressure 19% of the NOBr will decompose, compared to the 23% decomposition at 1 atm. This trend is predicted by the principle of Le Chatelier. The overall reaction involves the change of 2 moles of reactants into 3 moles of product. If we increase the pressure, the per cent decomposition should decrease. It is also worth noting that the pressure of bromine calculated in this example is 0.16 atm. Such a pressure can be obtained without the gaseous bromine condensing to liquid bromine, and hence this problem is exactly like any other equilibrium problem which deals with a homogeneous gas-phase reaction. Let us now determine the per cent decomposition at 4 atm. The equilibrium expression will be given by

$$\frac{(x/2)(x)^2}{(4 - x)^2} = 1.0 \times 10^{-2} \text{ atm}$$

where x again represents the pressure of the NO formed. If we solve this equation, we find that $x = 0.61$. On the basis of the fact that $x/2$ is the pressure of bromine, this yields a value of 0.30 atm. Such a partial pressure cannot exist at 25°C. Bromine gas will condense to form some liquid bromine, with a maximum pressure of bromine of 0.28 atm; hence this is the value that we must use in the equilibrium expression. The correct equilibrium expression for this reaction is

$$\frac{x^2}{(4 - x)^2}(0.28) = 1.0 \times 10^{-2}$$

We can rearrange this as follows:

$$\frac{x^2}{(4 - x)^2} = \frac{1.0 \times 10^{-2}}{0.28} = 3.6 \times 10^{-2} = K'$$

This is an example of the change in the equilibrium expression whenever one of the components occurs in the liquid or solid state, because then its pressure must be the vapor pressure at the temperature in question. Because a vapor pressure does not vary significantly with the total pressure on the solid or liquid, it enters the equilibrium equation not as a variable but as a constant. Because it is a constant, it is generally incorporated in the equilibrium constant, which thus takes on a new value for the *heterogeneous* equilibrium.

Returning to the equation, if we take the square root of both sides of this equation, we find

$$\frac{x}{4 - x} = 1.9 \times 10^{-1}$$

Then we can solve for x as follows:

$$x = 0.76 - 0.19x = 0.64$$

From this value of x, we calculate that the per cent decomposition at 4 atm is 16%.

Once the liquid phase of bromine appears, the result of an application of Le Chatelier's principle to this reaction yields an answer different from the one we would obtain if both reactants and products were gases. If we go on to calculate the third part of this problem, we get for the equilibrium expression

$$\frac{x}{8 - x} = 1.9 \times 10^{-1}$$

and, if we solve for x, we find that it is equal to 1.28 atm, which corresponds once again to 16% decomposition. It is clear from this that, once the pure liquid phase of bromine comes into existence, the equilibrium yield no longer depends on how much bromine is present but only on the fact that there is a small amount of bromine in the form of liquid. To put it another way, we can say that, once the liquid bromine appears, we can write a new equilibrium expression:

$$K' = \frac{K_p}{\text{(vapor pressure of bromine)}} = \frac{(\text{NO})^2}{(\text{NOBr})^2}$$

Since Δn is now zero, $K_p = K_c$ and we can write

$$\frac{(\text{NO})^2}{(\text{NOBr})^2} = K_p = K_c = 3.6 \times 10^{-2} = \frac{[\text{NO}]^2}{[\text{NOBr}]^2}$$

The effective concentration of Br_2 in this expression is therefore unity. Whenever there is present a pure liquid phase or a pure solid phase, we can assign the value of unity for these pure phases in the equilibrium expression and still have an equilibrium constant that will describe the behavior of the system. This method of writing the equilibrium constant has become the most common one, and we shall use it in this book.

To summarize, we can say: If we are dealing with a gas, the parentheses ( ) indicate the pressure of the gas in atmospheres; if we are considering a pure liquid or a pure solid present as a separate phase, the effective concentration is taken as *one;* and, if we have a solution that is not a gaseous solution, ( ) generally indicates the concentration expressed as molarity. Henceforth we shall strictly adhere to this convention and ( ) will indicate pressure, concentration molarity or unity in accord with this convention. To see how this works out in practice, let us consider one final example. Tin oxide can be reduced to metallic tin by hydrogen gas. This reaction is a typical example of a heterogeneous chemical reaction. In the equation

$$SnO(s) + H_2(g) \rightleftarrows Sn(l) + H_2O(g)$$

the tin oxide and the tin are present as separate phases and each one is pure. At 800°C and a total pressure of 10 atm, the equilibrium concentration of hydrogen is found to be 22% of the gas phase. Hence the equilibrium expression for this reaction is

$$K = \frac{(H_2O)(Sn)}{(H_2)(SnO)} = \frac{(H_2O)}{(H_2)}$$

$$K = \frac{7.8}{2.2} = 3.5$$

since (Sn) = 1 and (SnO) = 1.

Vapor pressure and equilibrium constant. In the analysis of problems in heterogeneous equilibrium, it is helpful to note again the parallelism between equilibrium constant and vapor pressure. Consider again the equilibrium between liquid water and water vapor at temperature of 121°C (= 394°K) at which the vapor pressure is 2 atm.

We write the equilibrium equation as

$$H_2O_{\text{liq}} \rightleftarrows H_2O_{\text{vap}}$$

According to the convention for writing the equilibrium constant,

$$K_p = \frac{(H_2O_{\text{vap}})}{(H_2O_{\text{liq}})}$$

But we have just adopted the rule that the ( ) is unity for all components in the liquid and solid phases. Therefore $(H_2O_{\text{liq}}) = 1$, and the expression becomes $K_p = (H_2O_{\text{vap}})$. Now (H_2O_{vap}) is just the value of the vapor pressure. Thus we see that for heterogeneous physical equilibrium the vapor pressure (P) *is* the equilibrium constant, K. It is thus to be expected that both vary in the same way with temperature. Therefore the expression for the variation of P with temperature is simply a special example of the more general expression for the variation of K with temperature.

$$\log P = \frac{-\Delta^\circ H_{\text{vap}}}{2.303R}\left(\frac{1}{T} - \frac{1}{T_b}\right)$$

$$\log K = \frac{-\Delta^\circ H_{\text{react}}}{2.303R}\left(\frac{1}{T} - \frac{1}{T_\circ}\right)$$

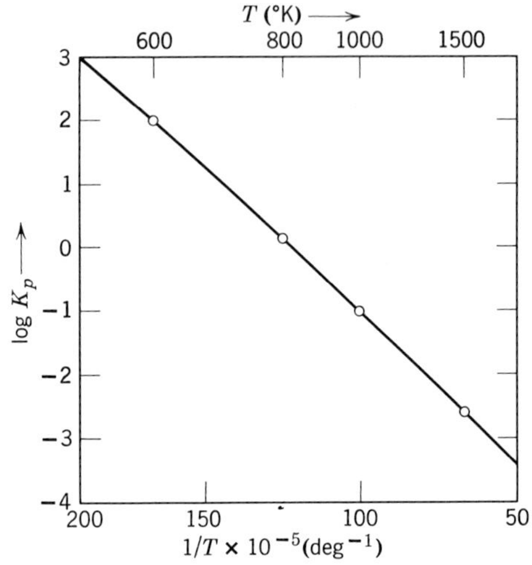

Fig. 5. log K_p for $C + 2H_2 \rightleftarrows CH_4$.

In the first equation, P is the vapor pressure, which is really the equilibrium constant for this physical equilibrium between liquid and gaseous states; $\Delta^\circ H_{vap}$ is the standard heat of the physical reaction; and T_b is the boiling point, the temperature where $P = 1$.

In the second equation K is normally the equilibrium constant for a chemical reaction, $\Delta^\circ H$ is the standard heat of that reaction, and $T_\circ$ is the temperature at which $K = 1$. Thus plots of log K against $1/T$ are almost linear, as shown in Figs. 5, 6, and 7.

Le Chatelier's principle and temperature change. The evaporation of a liquid into the gaseous state absorbs heat.

$$H_2O_{liq} \rightarrow H_2O_{vap}$$
$$-H_{liq} + H_{vap} = \underset{\text{a positive quantity}}{\Delta H}$$

If a little liquid water is in equilibrium with vapor in a closed vessel and heat is passed into the vessel (raising the temperature), then the vapor pressure increases and more liquid goes into the vapor state to increase the pressure and restore equilibrium. Thus the change in equilibrium is one that absorbs the heat that was the initial disturbing factor, just as equilibrium shifts to "absorb" or reduce pressure if pressure is the disturbing factor.

The same kind of shift will always be found also in chemical equilibria. If the value of ΔH is positive for a reaction as written, then the shift in equilibrium is from left to right, as temperature is increased, since that direction of displacement absorbs heat.

$$H_2O \rightarrow H_2 + \tfrac{1}{2}O_2$$
$$\Delta H = +$$

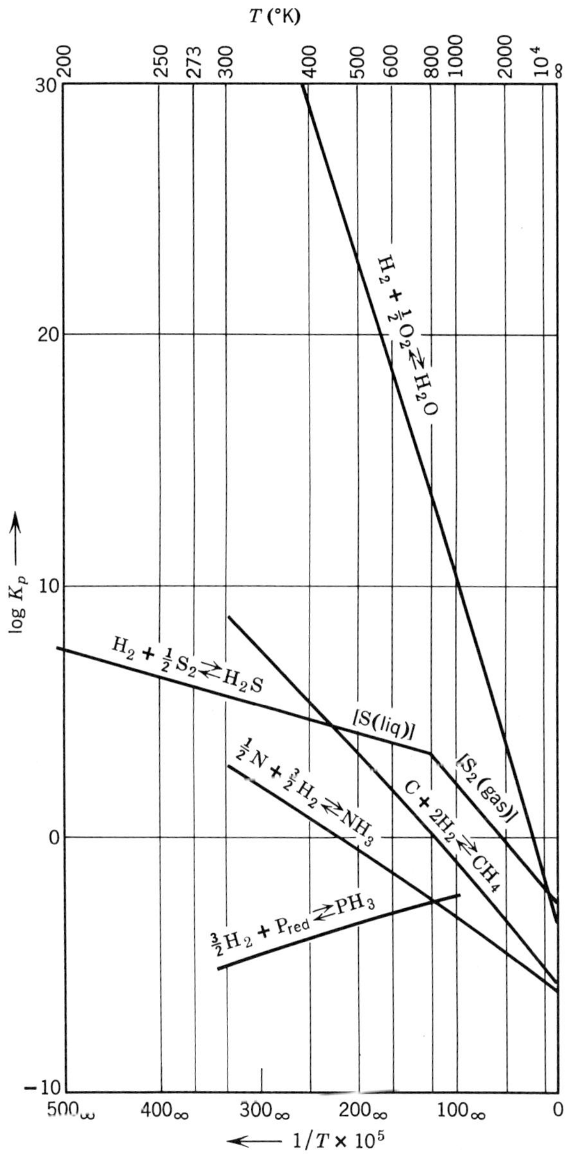

Fig. 6. Linear graphs for log K_p vs. $1/T$ (units based on atm) (Ref. 1).

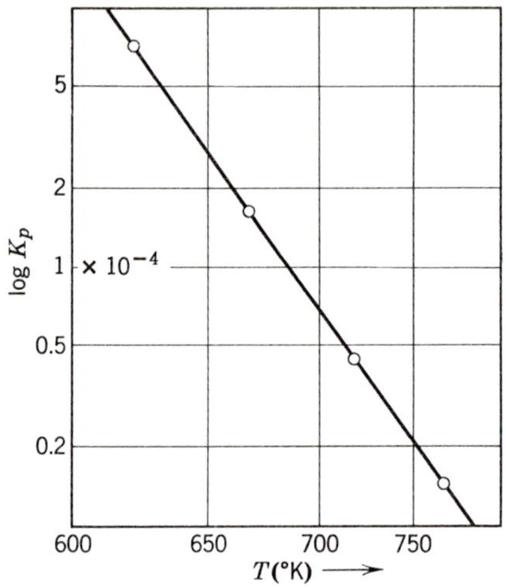

Fig. 7. Variation of K_p with T for $N_2 + 3H_2 \rightleftarrows 2NH_3$. (After A. T. Larson and R. L. Dodge.)

Increasing temperature displaces the reaction from left to right. But in reverse

$$H_2 + \tfrac{1}{2}O_2 \rightarrow H_2O$$
$$\Delta H = -$$

increasing temperature displaces the reaction from right to left.

In many of the chapters that follow, there will be a discussion of the chemistry of the various elements in the periodic table, and we shall illustrate as many phases as possible of this chemistry by discussions of equilibria and the way in which they are influenced by the thermodynamic factors of energy and probability.

PROBLEMS

(Use data in the appendix or the graphs in this chapter where required.)

1. Using the graphs in Fig. 2, plot the graphs for the free energy of water vapor at 2 atm and at 0.5 atm, assuming that the vapor behaves as an ideal gas. From the intersection of these graphs with the graph for the free energy of liquid water, estimate the temperatures at which water vapor and liquid water will be in equilibrium at these pressures. Check your results with the table for the vapor pressure of water in the Appendix.

2. From the graph in Fig. 6, estimate the value of K_p at the normal boiling point of water. What are the concentrations of H_2 and O_2 formed from the dissociation of water vapor under these conditions at equilibrium?

3. Calculate the pressure of NH_3 at 100°C such that the partial pressure of H_2, resulting from the dissociation of NH_3, equals the partial pressure of H_2 from dissociating H_2O at 1 atm and 100°C. *Ans.* 8×10^{41} atm.

4. If water vapor at 10^{-8} atm at 25°C in a sealed glass bulb ($V = 10$ cm^3) is heated to 1000°K, what will be the equilibrium concentration of O_2?

5. Calculate the total equilibrium pressure when 1 mole of CH_4 dissociates forming C and H_2 at 1000°K with $V = 1$ liter. *Ans.* 95 atm.

6. Make a plot of $\Delta°H$ and T $\Delta°S$ for the synthesis of NH_3 from its elements over the range 300° to 1000°K. $\Delta°H = 11.0$ kcal at 300° and -13.3 kcal at 1000°K.

7. Calculate the equilibrium constant for the reaction $CO + \frac{1}{2}O_2 \rightleftarrows CO_2$ at 298°K, together with values of $\Delta° H$, $T° \Delta S$, and $\Delta° F$ at this temperature.

8. If 1 mole of gaseous phosphorus (P_4) is in contact with $\frac{3}{2}$ moles of H_2 in a volume of 20.0 liters at 600°K, what will be the ratio of H_2 to PH_3 at equilibrium? *Ans.* 0.44×10^{-3}.

★**9.** Estimate the free energy change for the reaction of CH_4 with Cl_2 to form $CHCl_3$ at 500°, 1000°, and 2000°K. At what temperature will 1 atm each of H_2, Cl_2, and CH_3Cl be in equilibrium with carbon? (See Appendix for data.)

10. Make similar calculations for CCl_4.

11. Write the equilibrium constants for the following reactions and clearly state the conventional meaning of ( ) in the equilibrium expression.

(*a*) $Fe_3O_4(s) + 4H_2(g) \rightleftarrows 4H_2O(l) + 3Fe(s)$
(*b*) $2C_6H_6(l) + 15O_2(g) \rightleftarrows 12CO_2(g) + 6H_2O(l)$
(*c*) $3CO(g) + 7H_2(g) \rightleftarrows C_3H_8(g) + 3H_2O(l)$
(*d*) $Al^{+3}(aq) + 3OH^-(aq) \rightleftarrows Al(OH)_3(s)$
(*e*) $NH_4OH(aq) \rightleftarrows NH_4^+(aq) + OH^-(aq)$

12. Calculate the amount of tin that can be formed by reduction of SnO with 1 mole of H_2 in a closed vessel at 800°C if the initial hydrogen pressure is 1 atm. Repeat the calculation when the initial hydrogen pressure is 10 atm. The equilibrium constant for the reaction is 3.5.

13. The decomposition of NOBr occurs as follows at 25°C.

$$2NOBr \rightarrow 2NO + Br_2 \qquad K = 1.0 \times 10^{-2}$$

Calculate the amount of Br_2 formed when 1 mole of NOBr in a closed flask at an initial pressure of 2 atm decomposes.

14. Calculate the value of K_c for NOBr decomposition.

15. How much bromine would be formed in a 10-liter flask at 25°C if a 50-50 mixture of NO and NOBr equilibrates? The initial pressure is 10 atm. What would be the final pressure?

★**16.** A stream of hydrogen gas is saturated with water at 50°C at a total pressure of 1 atm. This gas is then passed through a chamber heated to 800°C that contains SnO. If the pressure of the inlet gas, the exit gas, and the gas in the chamber is maintained at 1 atm, what is the composition of the exit gas? Assume equilibrium is obtained. See Problem 12.

17. The equilibrium constant for the reaction

$$C_6H_{12} \rightarrow C_6H_6 + 3H_2$$

is 1.0×10^{-6} at 170°C and 1.0×10^{-3} at 200°C. Estimate $\Delta H°$ for this reaction. *Hint:* $\Delta H°$ and $\Delta S°$ usually vary little with temperature.

★**18.** The equilibrium constant is 1.0×10^{-6} for the reaction

$$C_6H_{12} \rightarrow C_6H_6 + 3H_2 \quad \text{at } 170°\text{C}$$

(*a*) If 1 mole of C_6H_{12} is placed in an evacuated bulb at 170°C and the pressure is initially 1 atm, how many moles of C_6H_6 will be formed?

(*b*) If $\frac{1}{2}$ mole of C_6H_{12} and $\frac{1}{2}$ mole of H_2 are placed in the same bulb at 170°C, how many moles of C_6H_6 will be formed at 170°C?

(*c*) If $\frac{1}{2}$ mole of C_6H_{12} and $\frac{1}{2}$ mole of C_6H_6 are placed in the same bulb, how many moles of H_2 will be formed at 170°C?

(*d*) If $\frac{1}{2}$ mole of C_6H_6 and $\frac{1}{2}$ mole of H_2 are placed in the same bulb, how many moles of C_6H_{12} will be formed?

(*e*) If 10 moles of C_6H_{12} are initially present in this bulb, how many moles of C_6H_6 will be formed?

(*f*) If 3 moles of C_6H_{12} and 6 moles of C_6H_6 are initially present in the bulb, how many moles of H_2 will be formed?

(*g*) If 3 moles of C_6H_{12} and 7 moles of H_2 are initially present in the bulb, how many moles of C_6H_6 will be formed?

★**19.** At a given temperature and a total pressure of 1 atm, a mixture of N_2 and H_2 (ratio 1:3) is placed in a cylinder. The total pressure is maintained at 1 atm and 15% of the N_2 reacts to form NH_3. What is the value of *K*?

14 Chemical kinetics

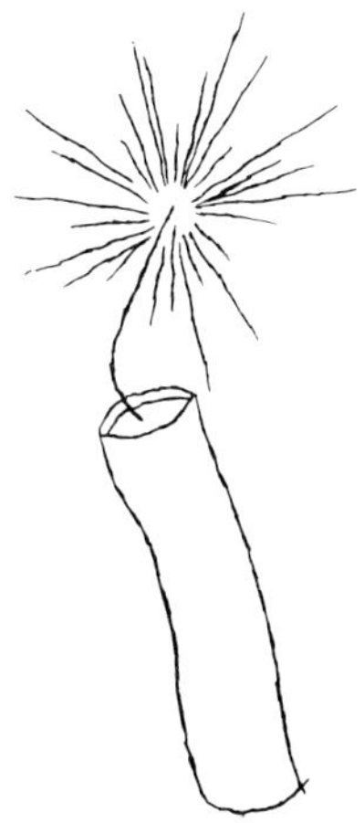

If we have the necessary thermodynamic data, we can calculate what will happen whenever two reactants are mixed. Such calculations show that, when N_2O_4 decomposes into NO_2 at 1 atm and room temperature, about one-fifth of the N_2O_4 reacts; similar calculations show that, when hydrogen and oxygen react at room temperature, essentially all of the reactants combine to form water. When we attempt to verify the N_2O_4 calculation by experiment, we find that N_2O_4 readily decomposes to form the calculated amount of NO_2. A similar investigation of the reaction between hydrogen and oxygen shows that, in direct contradiction to the thermodynamic calculations, *no* water is formed. This apparent contradiction is explained when we recall that thermodynamics tells us what must occur eventually. If we are willing to wait eons for the hydrogen and oxygen to combine, the calculated amount of water will form, but this occurs so slowly that a practical experiment reveals no water formation. In contrast, N_2O_4 decomposes almost instantaneously even though the thermodynamic tendency $(-\Delta F°)$ for this decomposition is far less than that for the combination of hydrogen and oxygen to form water. In general, *there is no obvious correlation between thermodynamic instability and the speed of a chemical reaction.*

The speed of a chemical reaction is governed by structural and energetic factors not uniquely specified by the thermodynamic quantities; hence studies of reaction rates or *chemical kinetics* provide a complimentary technique for examining the chemistry of a reaction. Chemical kinetics is not yet as complete as thermodynamics, but many questions are answered by modern theories of reaction rates. This chapter provides an introduction to these theories and shows how these techniques lead to a more detailed picture of the chemistry of the combination of the halogens with hydrogen.

RATES OF REACTION

Instantaneous rate. The rate of the reaction

$$H_2 + I_2 \rightarrow 2HI$$

was first studied near the turn of the century. The experimental procedure was simple. (The speed of this reaction is immeasurably slow at room temperature, but it increases with temperature and at 300° to 400°C it can be measured.) Known amounts of hydrogen and iodine were sealed in a flask at room temperature. The flask was then heated to a fixed temperature above 300°C and the amount of HI formed was determined by periodically withdrawing a sample of the reacting gases, chilling them, and analyzing for HI, I_2, and H_2. A typical plot of the results is shown in Fig. 1.

The only factors that affect the speed of this reaction are the pressures of HI, I_2, H_2 (which are all gases at these temperatures) and the temperature. The temperature is kept constant; only the partial pressures vary during the experiment. Since they do vary, the amount of HI produced each minute changes as the reaction proceeds. It can be seen from Fig. 1 that the amount of HI produced in the interval $t_1 - 0$ is greater than the amount produced in the equal interval $t_3 - t_2$.

For a complete description of the speed of this reaction we must specify how the rate of production of HI changes as the partial pressures of H_2, HI, and I_2 change. This rate is usually expressed as the change per unit of time; it can have such dimensions as molecules per second, molecules per hour, or moles per second. Thus over the interval $t_2 - t_3$ the rate is

$$\frac{N_3 - N_2}{t_3 - t_2}$$

where N_3 and N_2 are the molecules of HI present at t_3 and t_2. This rate depends on the partial pressures of the reactants. But does it depend on the partial pressures at t_2 or at t_3? If we make the interval small enough, it does not matter; the difference in the partial pressures at t_2 and t_3 is trivial when $t_3 - t_2$ is very small. To indicate the smallness of the N and t intervals we use the finite difference notation, i.e., $\Delta N/\Delta t$. This represents the rate at a particular instant; hence $\Delta N/\Delta t$ is called the *instantaneous rate.*

Reaction order. Usually, the instantaneous rate of a chemical reaction is proportional to the product of some power of the pressures of reactants. Specifically, the rate of formation of HI is given by an equation of the form

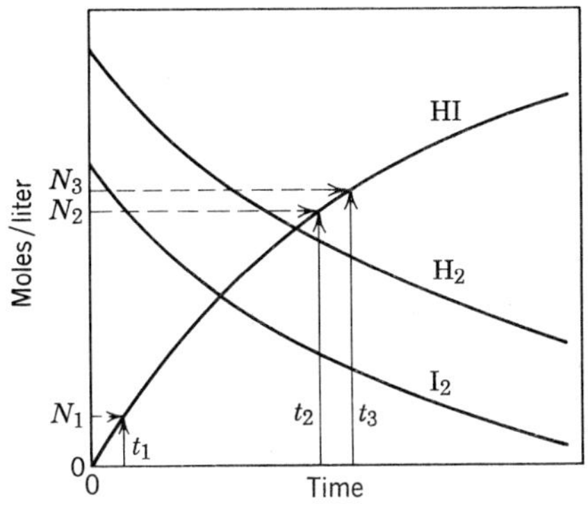

Fig. 1. Formation of HI from $H_2 + I_2$.

$$\frac{\Delta[\text{HI}]}{\Delta t} = k_{\text{HI}}(\text{H}_2)^n(\text{I}_2)^m$$

where [HI] stands for the number of HI molecules; k_{HI}, called the *rate constant,* is constant at a fixed temperature; and (H_2) and (I_2) represent pressures. The values of the exponents, n and m, determine the *order* of the reaction. Thus, if $n = 1$, the reaction is said to be "first order" in hydrogen; if $n = 2$, the reaction is said to be "second order" in hydrogen; and so forth. In order to describe the reaction, we specify both n and m. Thus, if $n = 3$ and $m = 2$, we say that the reaction is third order in hydrogen and second order in iodine.

In some reactions the instantaneous rate expression involves powers that are not whole numbers. For example, the formation of water,

$$2\text{H}_2 + \text{O}_2 \rightarrow 2\text{H}_2\text{O}$$

is described by the rate expression

$$\frac{\Delta[\text{H}_2\text{O}]}{\Delta t} = k_{\text{H}_2\text{O}}(\text{H}_2\text{O})(\text{O}_2)^{4/3}$$

This rate is four-thirds order with respect to oxygen. In addition to this type of complication, some reaction rates cannot be represented by a simple product of reactant pressures and may involve the pressures of products. For example, the rate of formation of HBr from H_2 and Br_2 is expressed by

$$\frac{\Delta[\text{HBr}]}{\Delta t} = k_{\text{HBr}}\frac{(\text{H}_2)(\text{Br}_2)^{3/2}}{(\text{Br}_2) + \frac{1}{10}(\text{HBr})}$$

Rates at equilibrium. The rate of formation of hydrogen iodide is first order in hydrogen and first order in iodine; hence this rate is represented by

$$\frac{\Delta[\text{HI}]}{\Delta t} = k_{\text{HI}}(\text{H}_2)(\text{I}_2) \equiv R_f$$

The decomposition of hydrogen iodide into hydrogen and iodine has also been studied; this rate is second order in hydrogen iodide; i.e., that is,

$$\frac{-\Delta[\text{HI}]}{\Delta t} = k_{-\text{HI}}(\text{HI})^2 \equiv R_r$$

Suppose we perform an experiment in which hydrogen and iodine are mixed in a flask and allowed to react until equilibrium is reached. At the very start, the only reaction occurring is the formation of hydrogen iodide ($R_r = 0$, $R_f \neq 0$). As soon as a trace of hydrogen iodide is formed, however, the formation and decomposition of hydrogen iodide take place simultaneously. In the early stages of the reaction, the term $(H_2)(I_2)$ is far greater than $(HI)^2$; hence, the number of hydrogen iodide molecules formed is greater than the number decomposing, so that there is a net increase in the number of hydrogen iodide molecules in the flask. As the reaction proceeds, however, (H_2) and (I_2) decrease as hydrogen iodide is formed and $(HI)^2$ increases. Consequently, R_f and R_r change as shown in Fig. 2. Once R_r and R_f become equal, the rate of production of hydrogen iodide is equal to the rate of decomposition. This means that (H_2),(I_2), and (HI) no longer change with time. Since the amount of reactants and products does not change with time, equilibrium has been attained. The type of behavior indicated in Fig. 2 is encountered for all types of reactions regardless of the complexity.

At equilibrium the following conditions are fulfilled:

$$R_f = R_r$$

or

$$k_{\text{HI}}(\text{H}_2)(\text{I}_2) = k_{-\text{HI}}(\text{HI})^2$$

This can be arranged to yield

$$\frac{(\text{HI})^2}{(\text{H}_2)(\text{I}_2)} = \frac{k_{\text{HI}}}{k_{-\text{HI}}}$$

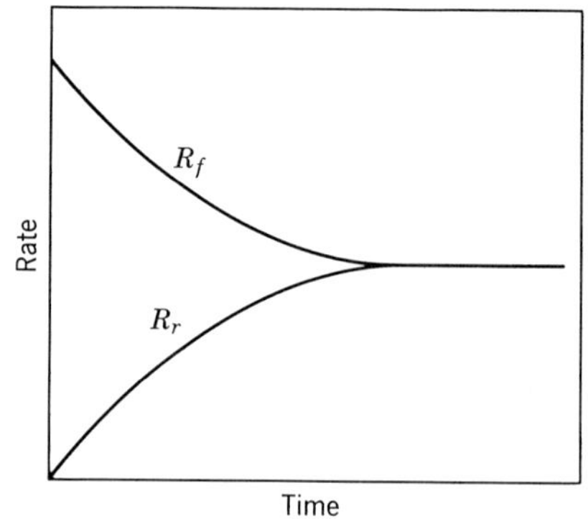

Fig. 2. Approach to equilibrium.

But this product of pressures is simply the equilibrium constant for the reaction

$$H_2 + I_2 \rightleftarrows 2HI$$

$$\frac{(HI)^2}{(H_2)(I_2)} = K_{eq.}$$

By this comparison we conclude that the rate constants, k_{HI} and k_{-HI} are related to the equilibrium constant by the equation

$$\frac{k_{HI}}{k_{-HI}} = K_{eq.} = e^{-\Delta F^\circ/RT} = e^{-\Delta H^\circ/RT + \Delta S^\circ/R}$$

where ΔF° is the thermodynamic standard free energy change for the reaction (cf. p. 386).

For the specific reaction above we have concluded:

(*a*) At equilibrium the forward and reverse instantaneous rates are equal.

(*b*) The ratio of the forward to the reverse rate constant is the equilibrium constant.

These conclusions hold for all reactions regardless of their complexity, and they often reduce the information necessary to specify a rate equation.

Example Problem 1: Studies of the reaction $C_6H_6 + 3H_2 \rightarrow C_6H_{12}$ show that the rate equation for the forward reaction is

$$R_f = \frac{\Delta[C_6H_{12}]}{\Delta t} = k_f(C_6H_6)(H_2)$$

This relation also holds at equilibrium. What is the rate expression for the reverse reaction?

According to (*b*) above,

$$\frac{k_f}{k_r} = K_{eq.} = \frac{(C_6H_{12})}{(H_2)^3(C_6H_6)}$$

This can be arranged to yield

$$k_f(C_6H_6)(H_2) = k_r\frac{(C_6H_{12})}{(H_2)^2}$$

Since at equilibrium

$$R_f = R_r$$

and R_r must be equal to k_r multiplied by a product of pressures, we conclude that

$$R_r - k_r\frac{(C_6H_{12})}{(H_2)^2}$$

COLLISION THEORY

Consider what must happen on a molecular scale for the following reaction to occur.

$$H_2 + I_2 \rightarrow 2HI$$

If we exclude the breakdown of hydrogen and iodine into atoms, the first requirement for reaction is the bringing together of H_2 and I_2 molecules. Since gas molecules are in constant motion, this "bringing together" of H_2 and I_2 molecules is a collision.

Collisions in real gases. How can we estimate the number of collisions? Since we are dealing with real gases, we must recognize that molecules themselves take up space. In order to account for this and still retain the mathematical simplicity of the ideal gas model, we consider the molecules to be mass points and assume that an H_2-I_2 collision occurs whenever the distance between them is less than σ cm. The following symbols will be used.

n_1 = No. H_2 molecules/cm^3
n_2 = No. I_2 molecules/cm^3
$\bar{c}_r$ = average relative speed of H_2 and I_2 molecules
V = volume of flask

Around each hydrogen molecule there is a sphere of influence of volume $\frac{4}{3}\pi\sigma^3$. If the center of an iodine molecule is within this sphere, a collision is taking place. At a given instant the number of collisions in progress is

$$(n_1 V \tfrac{4}{3}\pi\sigma^3) \times (n_2)$$

Since $n_1 V$ is the total number of hydrogen molecules in the flask, the first factor is the *total* volume of the hydrogen spheres of influence. If iodine molecules are distributed evenly throughout the flask, then the number undergoing collision is the product of the total volume of hydrogen spheres of influence and n_2, the number of iodine molecules per cubic centimeter.

In a very short time a given collision will end by the iodine molecule passing out of the hydrogen sphere of influence. On the average, the duration of the collisions is $\sigma/\bar{c}_r$, the time required for the iodine molecule to travel a distance σ. From these quantities we can calculate the number of collisions per second.

$$\frac{\text{No. of collisions}}{\text{sec}} = \frac{\text{No. of collisions at any instant}}{\text{average duration of collision}} = \frac{n_1 V \frac{4}{3}\pi\sigma^3 n_2}{\sigma/\bar{c}_r}$$

or $n_1 n_2 V \frac{4}{3}\pi\sigma^2 \bar{c}_r$. It is customary to refer to the number of collisions per second in 1 cm^3 of gas; then $V = 1$ in the equation above.

The average relative speed $\bar{c}_r$ changes with temperature in much the same way as does the average speed; that is,

$$\bar{c}_r \approx \sqrt{\frac{kT}{m}}$$

where m is the mass of a gas molecule. Moreover, since H_2 and I_2 are both gases, we can make use of the ideal gas law to rewrite the preceding equation as follows:

$$\frac{\text{No. coll./sec}}{\text{cm}^3\text{ of gas}} \approx \frac{(H_2)}{kT}\frac{(I_2)}{kT}\frac{4}{3}\pi\sigma^2\sqrt{\frac{kT}{m}}$$

where (H_2) and (I_2) indicate partial pressures. This can be further simplified to yield

$$\left\{\frac{\frac{4}{3}\pi\sigma^2}{(kT)^{3/2}m^{1/2}}\right\}(H_2)(I_2) \equiv Z_0(H_2)(I_2)$$

where Z_0 is a constant that depends only on temperature. ($\sigma = r_{H_2} + r_{I_2}$, where r_{H_2} and r_{I_2} are the effective radii of the molecules.) This is of the same form as the rate expression

$$\frac{\Delta[\text{HI}]}{\Delta t} = k_{\text{HI}}(H_2)(I_2)$$

For $(H_2) = (I_2) = 5.7$ atm at 410°C, the number of collisions per second in 1 cm³ is $\sim 10^{29}$; the number of HI molecules formed is $\sim 10^{16}$. Thus only a fraction of the collisions are effective in causing a reaction. Moreover, since k_{HI} trebles every 10° whereas Z_0 decreases slightly, this fraction must vary with temperature.

Steric factors. To react, molecules must not only collide; they must also *collide with the proper orientation.* Of the two collisions pictured in Fig. 3, it is evident that the one in Fig. 3*a* is more likely to result in a reaction than the one in Fig. 3*b*. Only a fraction, P_f, of all collisions have the proper orientation; hence we write

$$\frac{\text{No. collisions (proper orientation)}}{\text{sec}} = P_f Z_0(H_2)(I_2)$$

Although this modification helps, it is not the whole answer. P_f might conceivably be 10^{-3} or even 10^{-6}, but it should not vary much with temperature. Thus agreement of theory with experiment still eludes us.

Activation energy. To react, molecules must not only collide with the proper orientation; they must also collide *with enough force to disrupt the bonding in the*

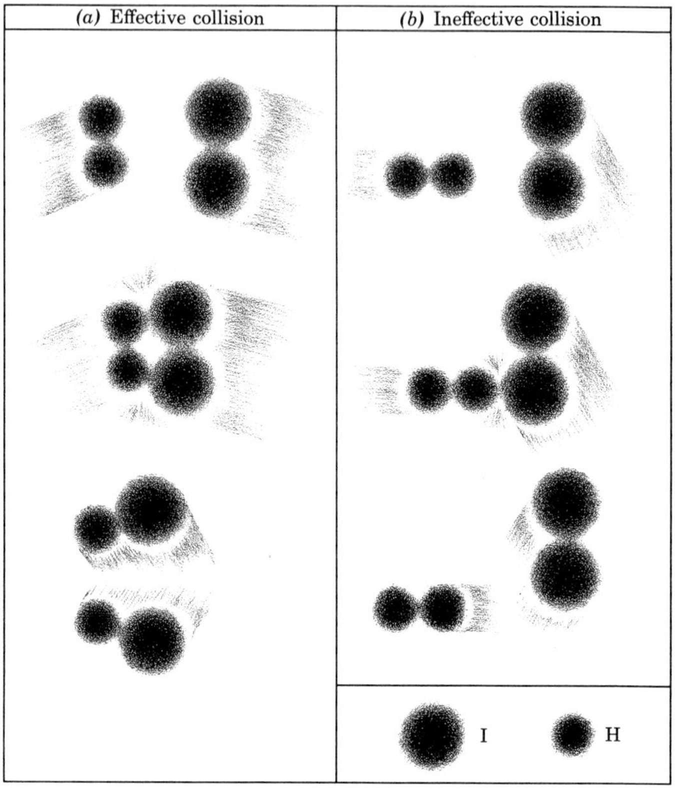

Fig. 3. Collisions of H_2 and I_2 molecules.

molecules. If the collision in Fig. 3*a* is too gentle, the HI distance is never small enough for a stable bond to form and the H_2 and I_2 molecules bounce off each other. If the collision is violent enough, interpenetration of hydrogen and iodine charge clouds occurs; and the hydrogen-iodine bond distance becomes small enough for a bond to form. In this event, the transitory H_2I_2 molecule can split either to form H_2 and I_2 or 2HI. The minimum energy required for such an effective collision is called the *activation energy*, $\Delta E^{\ddagger}$.

The fraction of the collisions in which the mean molecular energy is greater than the minimum required for reaction is approximated by the Boltzmann term $e^{-\Delta E^{\ddagger}/RT}$; hence, for the rate of formation of hydrogen iodide we write:

$$\frac{\Delta[\mathrm{HI}]}{\Delta t} = e^{-\Delta E^{\ddagger}/RT} \times P_f \times Z_0(\mathrm{H_2})(\mathrm{I_2})$$

The last factor represents the number of collisions per second; the second factor represents the fraction of these that occur with the proper orientation to form H_2I_2;

the first represents the fraction of the "proper" collisions that occur with sufficient violence for reaction to occur.

The rate constant is given by the factor

$$P_f Z_0 \, e^{-\Delta E^{\ddagger}/RT}$$

Since P_f and Z_0 change so little with temperature, the variation of rate with temperature is dominated by the Boltzmann exponential. In other words, the rate constant varies with temperature just as vapor pressure does.

In order to account for the observed data for the hydrogen iodide reaction we must assume $\Delta E^{\ddagger} \approx 40{,}000$ cal. We can interpret this in a fashion similar to that used in our discussion of liquids. Figures 4 and 5 show the Boltzmann distribution for two temperatures. Only molecules with energy greater than $\Delta E^{\ddagger}$ can react. This fraction always increases with temperature, and, the greater the value of $\Delta E^{\ddagger}$, the more rapidly, percentagewise, this fraction increases. This temperature variation is expressed precisely by the Boltzmann factor $e^{-\Delta E^{\ddagger}/RT}$.

Relation of $\Delta E_f^{\ddagger}$ and $\Delta E_r^{\ddagger}$. In the experimental study of the reaction of hydrogen iodide we have to consider not only the forward reaction but also the reverse reaction. In terms of the collision theory we can write

$$\left(\frac{\Delta[\mathrm{HI}]}{\Delta t}\right)_f = Z_{0f}\, e^{-\Delta E^{\ddagger}/RT}(\mathrm{H}_2)(\mathrm{I}_2)\, P_f$$

and

$$\left(\frac{\Delta[\mathrm{HI}]}{\Delta t}\right)_r = -Z_{0r}\, e^{-\Delta E_r^{\ddagger}/RT}(\mathrm{HI})^2\, P_r$$

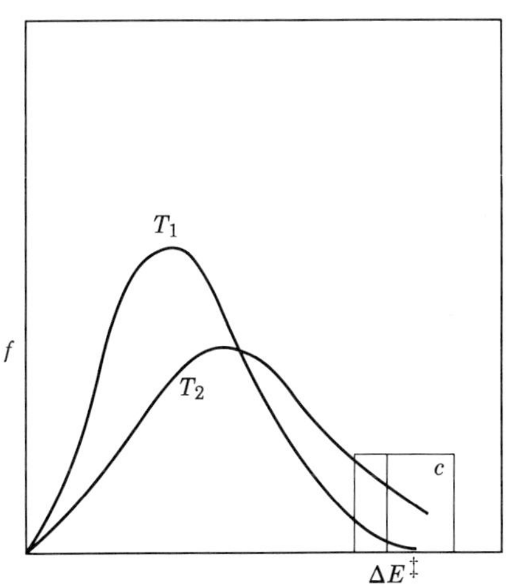

Fig. 4. Boltzmann distribution.

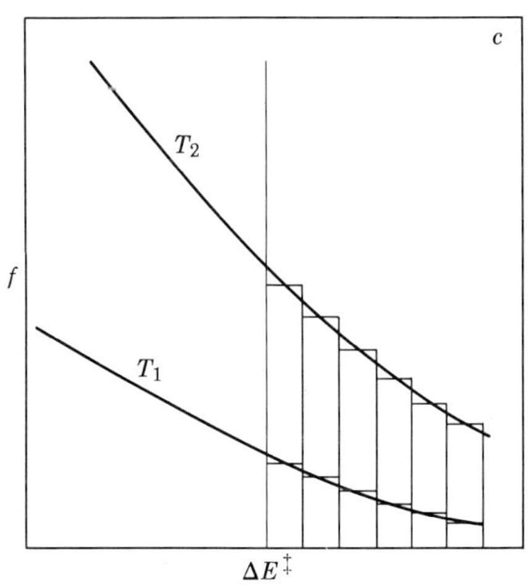

Fig. 5. High-energy molecules.

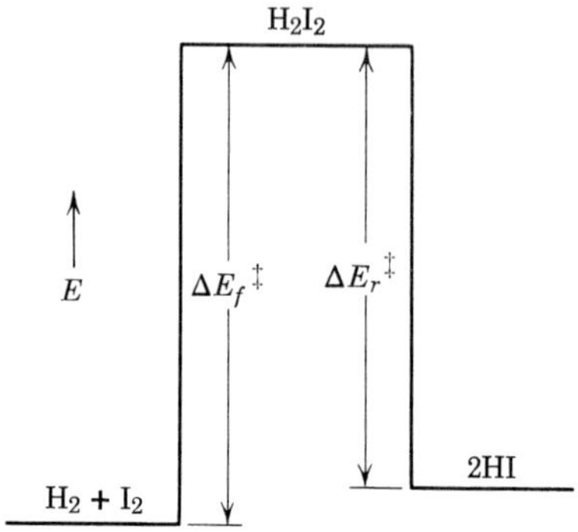

Fig. 6. Energy profile.

where the subscripts f and r refer to the forward and reverse rate for

$$H_2 + I_2 \rightleftarrows HI$$

The significance of $\Delta E_f^\ddagger$ and $\Delta E_r^\ddagger$ is shown in the energy diagram in Fig. 6. The energy of activation, $\Delta E_f^\ddagger$, represents the energy the H_2 and I_2 molecules must have in a collision if they are to collide effectively and react. To state it another way, if the formation of hydrogen iodide is to occur, the transition molecule H_2I_2 must first be formed by collision. [In general, such transition molecules are extremely short-lived ($\sim 10^{-14}$ sec); in order to emphasize this short life time, we commonly refer to such species as *transition complexes* rather than transition molecules.] All the energy requirements for the forward reaction are given by the passage from left to right in the diagram. The energy requirements for the reverse reaction, which also includes the formation of H_2I_2 as a first step, are given by passage from right to left on the diagram. In terms of this diagram, then, reaction requires the passage over an energy hill; the height of this hill relative to the initial state is the activation energy.

The energy diagram shows that the difference $\Delta E_f^\ddagger - \Delta E_r^\ddagger$ represents the energy difference between 2HI and $H_2 + I_2$. This difference is the thermodynamic energy change for the reaction

$$H_2 + I_2 \rightleftarrows 2HI$$

Reaction at equilibrium. We have seen earlier that the rate constants k_f and k_r are related to the equilibrium constants. If we now express k_f and k_r in the form predicted by the collision theory, we obtain the following relations; $k_f/k_r = K_{\text{eq.}}$ or

$$\frac{P_f Z_{0f}\, e^{-\Delta E_f^\ddagger/RT}}{P_r Z_{0r}\, e^{-\Delta E_r^\ddagger/RT}} = e^{-\Delta H^\circ/RT} e^{\Delta S^\circ/R}$$

where ΔH° and ΔS° are the standard enthalpy and entropy changes for the over-all reaction. In this reaction, $Z_{0f} \approx Z_{0r}$ and $\Delta H^\circ = \Delta E^\circ$, so we write, with rearrangement,

$$e^{-(\Delta E_f^{\ddagger}-\Delta E_r^{\ddagger})/RT}\frac{P_f}{P_r} = e^{-\Delta E^\circ/RT}e^{\Delta S^\circ/R}$$

In view of the relation $\Delta E_f^{\ddagger} - \Delta E_r^{\ddagger} = \Delta E$ and the fact that $\Delta E = \Delta E^\circ$ for ideal gases, we can cancel the factors in parentheses and obtain

$$\frac{P_f}{P_r} = e^{+\Delta S^\circ/R}$$

This gives us two relations between quantities determined by kinetics and those determined by equilibrium thermodynamics, namely;

$$\ln P_f - \ln P_r = \frac{\Delta S^\circ}{R} \text{ reaction}$$

and

$$\Delta E_f^{\ddagger} - \Delta E_r^{\ddagger} = \Delta E^\circ \text{ reaction}$$

These relations hold not only for this particular reaction but also for all reactions. Appropriately, the logarithms of the probability terms are related to entropy change for the reaction and the energy terms are related to the energy change for the reaction.

TRANSITION STATE THEORY

The transition complex. The first step in the formation of HI can be represented by

$$H_2 + I_2 \rightarrow H_2I_2^{\ddagger}$$

where $H_2I_2^{\ddagger}$ is the *transition complex* or *activated complex* formed by an effective collision. Since $H_2I_2^{\ddagger}$ is the same as the complex during the decomposition of HI, it can decompose to form either 2HI or $H_2 + I_2$; that is,

$$H_2 + I_2 \swarrow H_2I_2^{\ddagger} \searrow 2HI$$

The activated complex has a very short lifetime, but during its lifetime it behaves like any other molecule. It is not standing still but is moving, and during its motion it rotates and vibrates. In one particular vibration, two fragments of the molecule start to separate, but instead of snapping back together, as in a normal vibration, they continue the separation to form products or reactants. The mean life of a transition complex before this takes place is τ; it has been shown that this lifetime is the same for *all* types of activated complexes. It is given by the relation

$$\tau = \frac{h}{kT}$$

where h is Planck's constant, k is Boltzmann's constant, and T is the absolute temperature. Usually, the formation of products is about as likely as the formation of reactants, so that half of the time the break-up in the complex results in products and half of the time the break-up results in reactants.

In view of these statements, we can rewrite the expression for the rate of formation of HI as follows:

$$\frac{\Delta[\mathrm{HI}]}{\Delta t} = \frac{1}{2}\frac{1}{\tau}(\mathrm{H_2I_2})^{\ddagger}$$

Since τ is known, determining the rate necessitates finding an expression for $[\mathrm{H_2I_2}]^{\ddagger}$.

At the very start of the reaction between H_2 and I_2 the complex $H_2I_2^{\ddagger}$ is formed solely from reactants. Once formed, it can decompose to form products or reactants. It is often assumed that an equilibrium of the type

$$\mathrm{H_2 + I_2 \rightleftarrows H_2I_2^{\ddagger}}$$

is set up between the reactants and the complex, and that the decomposition of the complex into products depletes the reactants but does not destroy the equilibrium. If this is true, we can write

$$\frac{(\mathrm{H_2I_2})^{\ddagger}}{(\mathrm{H_2})(\mathrm{I_2})} = K_f^{\ddagger}$$

where $K_f^{\ddagger}$ is the equilibrium constant. Thus the concentration of complex is given by

$$(\mathrm{H_2I_2})^{\ddagger} = K_f^{\ddagger}(\mathrm{H_2})(\mathrm{I_2})$$

and the rate of reaction is

$$\frac{\Delta[\mathrm{HI}]}{\Delta t} = \frac{1}{2}\frac{kT}{h}K_f^{\ddagger}(\mathrm{H_2})(\mathrm{I_2})$$

Since $K_f^{\ddagger}$ is an equilibrium constant we can express it in terms of $\Delta F^{\ddagger}$, the standard free energy change for the formation of complex from reactants. In other words, since

$$K^{\ddagger} = e^{-\Delta F_f^{\ddagger}/RT} = e^{-(\Delta H_f^{\ddagger} - T\Delta S_f^{\ddagger})/RT}$$

we can write the rate as

$$\frac{\Delta[\mathrm{HI}]}{\Delta t} = \frac{1}{2}\frac{kT}{h}e^{-\Delta F^{\ddagger}/RT}(\mathrm{H_2})(\mathrm{I_2})$$

$$= \left(\frac{1}{2}\frac{kT}{h}e^{+\Delta S^{\ddagger}/R}e^{-\Delta H^{\ddagger}/RT}\right)(\mathrm{H_2})(\mathrm{I_2})$$

As before, the rate constant is strongly temperature dependent because of a Boltzmann-type exponential involving enthalpy.

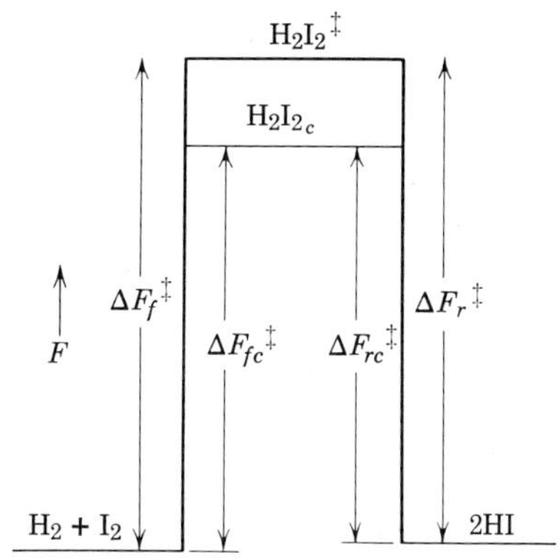

Fig. 7. Free energy profile.

The changes during reaction can again be represented by a diagram similar to that in Fig. 6. In this case, however, the ordinate is free energy (Fig. 7). Reaction occurs because the reactants establish an equilibrium with the complex. In this first reaction the standard free energy change is $\Delta F_f^\ddagger$. Once formed, a small amount of this complex is siphoned off to form products. The standard free energy change in the formation of products from the complex is $-\Delta F_r^\ddagger$. It appears from the diagram that these free energy changes are related by

$$\Delta F_f^\ddagger - \Delta F_r^\ddagger = \Delta F^\circ$$

where ΔF° is the standard free energy change for the over-all reaction

$$H_2 + I_2 \rightleftarrows 2HI$$

Rates at equilibrium. In terms of the transition state theory, the rates of the forward and reverse reaction are given by

$$\frac{\Delta[HI]}{\Delta t}\text{ forward} = \frac{1}{2}\frac{kT}{h}K_f^\ddagger[H_2][I_2]$$

and

$$\frac{\Delta[HI]}{\Delta t}\text{ reverse} = \frac{1}{2}\frac{kT}{h}K_r^\ddagger[HI]^2$$

At equilibrium these rates are equal; hence

$$\frac{1}{2}\frac{kT}{h}K_r^\ddagger[HI]^2 = \frac{1}{2}\frac{kT}{h}K_f^\ddagger[H_2][I_2]$$

and, on rearrangement, we obtain

$$\frac{[HI]^2}{[H_2][I_2]} = \frac{K_f^\ddagger}{K_r^\ddagger}$$

At equilibrium this must, of course, be the equilibrium constant. If we express $K_f^\ddagger$ and $K_r^\ddagger$ in terms of free energy, we obtain

$$\frac{K_f^\ddagger}{K_r^\ddagger} = \frac{e^{-\Delta F_f^\ddagger/RT}}{e^{-\Delta F_r^\ddagger/RT}} = e^{-(\Delta F_f^\ddagger - \Delta F_r^\ddagger)/RT} = e^{-\Delta F^\circ/RT}$$

The last relation follows from the diagram in Fig. 7 and is the thermodynamic expression for the over-all equilibrium constant.

Catalysis. One important aspect of this (and all) theories of reaction rates is that the equilibrium point is not affected by the path of the reaction. Consider, for example, that under certain conditions the intermediate in the reaction changes from $H_2I_2^{\ddagger}$ to $(H_2I_2)_C$. As indicated in Fig. 7, the path of the reaction and the free energies involved, $\Delta F_{fc}^{\ddagger}$ and $\Delta F_{rc}^{\ddagger}$, are changed. As before, however, the relation

$$\Delta F_{fc}^{\ddagger} - \Delta F_{rc}^{\ddagger} = \Delta F^{\circ}$$

still holds. As long as this is true, the results of the previous sections are applicable and the over-all equilibrium constant is unchanged.

The reaction

$$H_2 + C_2H_4 \rightarrow C_2H_6$$

should occur spontaneously at room temperature according to thermodynamic data, but a mixture of hydrogen and ethylene can be stored for years without reaction. If a small amount of finely divided platinum is added to such a mixture, reaction takes place rapidly. After the reaction occurs, the platinum can be recovered and it is apparently unchanged. *A substance that alters the rate of a reaction without itself undergoing a change is called a catalyst; the phenomenon of acceleration of the reaction rate is called catalysis.*

If a reaction is catalyzed in the strictest sense of the definition, the catalyst cannot combine permanently with either reactant or product for then the catalyst would not be recoverable unchanged. If this is true, the catalyst can only affect the short-lived transition complex. For example, the rate of a reaction could be increased if a catalyst lowered the free energy of the transition complex. Figure 7 indicates the effect of such a catalysis on the hydrogen iodide reaction. The complex, combined with catalyst, is indicated by $(H_2I_2)_C$. Since $\Delta F_{fc}^{\ddagger}$ is smaller than $\Delta F_f^{\ddagger}$, the reaction can proceed more rapidly in the presence of the catalyst. The catalyst is recovered substantially unchanged because only $(H_2I_2)_C$ combines with the catalyst and the amount of $(H_2I_2)_C$ is immeasurably small.

Two interesting consequences are evident from Fig. 7.

1. In a catalyzed reaction, $\Delta F^{\ddagger}$ is reduced for both the forward and reverse reaction so that the percentage change in forward and reverse rate is the same. Therefore "one-way" catalysis is impossible. (The laws of thermodynamics also show that "one-way" catalysis is impossible.)
2. Since $\Delta F_{fc}^{\ddagger} - \Delta F_{rc}^{\ddagger} = \Delta F^{\circ}$, the free energy for the over-all reaction, the presence of a catalyst cannot change the equilibrium constant.

COMPLEX REACTIONS

The formation of hydrogen bromide from the elements,

$$H_2 + Br_2 \rightarrow 2HBr$$

can be described by the rate expression

$$\frac{\Delta[HBr]}{\Delta t} = k \frac{(H_2)(Br_2)^{3/2}}{(Br_2) + \frac{1}{10}(HBr)}$$

This expression is a typical example of the complex kinetics found for many chemical reactions. In this section we shall discuss this reaction in some detail in order to provide examples of the theoretical and experimental techniques used to study a complex reaction.

Hydrogen bromide kinetics. The kinetic expression for the hydrogen bromide reaction can be obtained if we assume that the following steps occur:

$$Br_2 \underset{k_{-1}}{\overset{k_1}{\rightleftarrows}} 2Br \qquad \Delta H = +45.2 \text{ kcal}$$

$$Br + H_2 \xrightarrow{k_2} HBr + H \qquad \Delta H = +18 \text{ kcal}$$

$$H + Br_2 \xrightarrow{k_3} HBr + Br \qquad \Delta H = -42 \text{ kcal}$$

$$H + HBr \xrightarrow{k_{-2}} H_2 + Br \qquad \Delta H = -18 \text{ kcal}$$

$$Br + HBr \xrightarrow{k_{-3}} Br_2 + H \qquad \Delta H = +42 \text{ kcal}$$

Steps 1 and -1 involve an equilibrium between bromine molecules and atoms. Steps 2 and 3 involve atom reactions. In general, the activation energies for atom reactions are nearly zero for exothermic reactions and equal to the heat of reaction for endothermic reactions. Thus we can draw the energy diagram in Fig. 8 for this reaction. Accordingly, to form one mole of HBr we need a total energy of 40.6 kcal. This represents in effect the activation energy.

The third step in the kinetic analysis adds an interesting feature to this reaction. Each bromine atom that forms a hydrogen bromide molecule also forms a hydrogen atom. This hydrogen atom reacts via step 3 to replace the bromine atom that generated it in step 2. Thus the formation of one bromine atom triggers the reaction of $H_2 + Br_2$; each Br that is used up by step 2 is regenerated by the subsequent reaction of the hydrogen atom formed in 2. Such a reaction is called a *chain reaction.* Steps 2 and 3 are called the *chain propagation* steps.

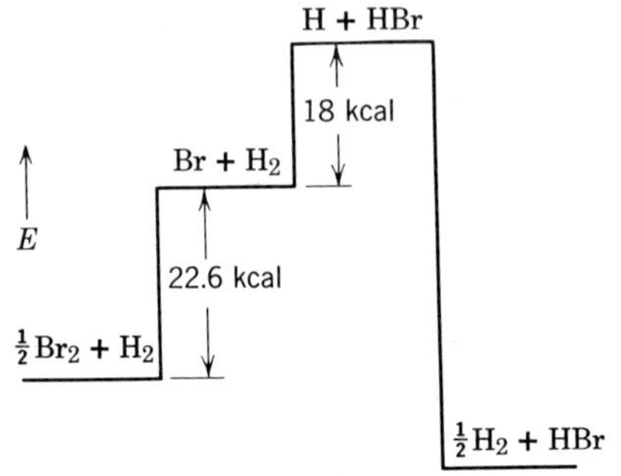

Fig. 8. Energy profile, HBr reaction.

Suppose the ΔH for step 2 were zero and instead of 3 we had the reaction

$$H^* + Br_2 \rightarrow H + 2Br$$

where H^* indicates an excited H atom and for each Br used in the second step, two Br were regenerated; then a *branched-chain reaction* would occur. After each product HBr molecule was formed, two reactive Br would be created; the rate of formation of HBr would increase and this would generate more Br. Thus, in a branched-chain reaction, the reaction takes place extremely fast after initiation. Common modes of initiation are shock, sparks, and light. Most explosions occur because of branched-chain reactions.

Rates at equilibrium. In a reaction involving a series of consecutive steps all opposing rates must be equal at equilibrium; hence we write

$$k_2(\text{Br})(\text{H}_2) = k_{-2}(\text{HBr})(\text{H})$$

$$\frac{(\text{HBr})(\text{H})}{(\text{Br})(\text{H}_2)} = \frac{k_2}{k_{-2}} = K_2$$

$$k_3(\text{H})(\text{Br}_2) = k_{-3}(\text{HBr})(\text{Br})$$

$$\frac{(\text{HBr})(\text{Br})}{(\text{H})(\text{Br}_2)} = \frac{k_3}{k_{-3}} = K_3$$

Such equilibrium expressions are not very useful since they involve (Br) or (H), quantities so small at equilibrium that they are undetectable. These equations can be combined so that they contain only measurable concentrations:

$$K_2K_3 = \frac{(\text{HBr})(\text{H})}{(\text{Br})(\text{H}_2)} \times \frac{(\text{HBr})(\text{Br})}{(\text{H})(\text{Br}_2)} = \frac{(\text{HBr})^2}{(\text{H}_2)(\text{Br}_2)}$$

This, as it should be, is the thermodynamic equilibrium expression.

Photochemistry. The most common way of exciting a molecule so that it can undergo reaction is by an energetic collision resulting from the thermal motion of the molecules, but other means can be used. Of these, the most common is *photochemical excitation.*

In photochemical excitation the energy of one quantum of light is wholly absorbed by a single molecule. Not all frequencies, ν, are suitable for such excitation; only those frequencies are adsorbed that satisfy the relation $\Delta E = h\nu$, where ΔE is a difference of energy levels in a molecule. If a particular level, E_1, must be excited for a chemical reaction to occur, the frequency of light to be used is given by $E_1 - E_0 = h\nu$, where E_0 is the lowest energy state of the molecule.

The most important type of photochemical reaction is that which occurs in plants and results in the formation of carbohydrates. This reaction, known as *photosynthesis,* can be written

$$CO_2 + H_2O + h\nu = \tfrac{1}{6}(C_6H_{12}O_6) + O_2$$

where $h\nu$ represents the absorbed (reacting) photon of light. It is catalyzed by chlorophyll, the green-colored substance in plants, and the light source is sunlight. Ultimately, it is the origin of our food.

A particularly simple photochemical reaction is the photochemical HBr reaction. In our discussion so far, we have only considered the occurrence of this reaction when the bromine atoms are produced thermally, but they can also be produced by the reaction

$$Br_2 + h\nu \rightarrow 2Br$$

The light absorbed provides a convenient measure of the number of bromine atoms present, and in this case no thermal activation energy is needed to produce bromine atoms. If the mechanism is correct, therefore, the energy of activation should be only 18 kcal (cf. Fig. 8). Studies of this type give an activation energy of 17.6 kcal, in good agreement with expectations. Often, studies of kinetics are supplemented by such photochemical studies.

Hydrogen halide reactions. The kinetics of the formation of hydrogen iodide are best represented by the formation of the H_2I_2 intermediate; the kinetics of the formation of the other hydrogen halides are best represented by a series of consecutive reactions of the type shown for hydrogen bromide formation. In view of the chemical similarities of the halides, it seems surprising to find this difference in kinetic behavior. This apparent contradiction emphasizes the fact that the chemical properties of a compound, which depend largely on equilibrium properties, are not nearly as sensitive to small differences in structure as are the kinetic properties.

The type of kinetics found for hydrogen halide formation is determined by the energetics of the two alternative paths. If X_2 represents any one of the halides, we can represent the molecular and chain mechanisms as follows.

Table 1 Activation Energies

Hydrogen Halide	Calculated Over-all Activation Energies MM	CM
HI	48	57
HBr	43	41
HCl	~50	36

Molecular Mechanism (MM):

$$X_2 + H_2 \rightarrow H_2X_2 \quad \Delta H_M$$
$$H_2X_2 \rightarrow 2HX$$

Chain Mechanism (CM):

(1) $X_2 \rightleftarrows 2X \qquad \Delta H_1$

(2) $X + H_2 \rightarrow HX + H \qquad \Delta H_2$

(3) $H + X_2 \rightarrow HX + X \qquad \Delta H_3$

The over-all activation enthalpy for MM is ΔH_M; the over-all activation energy for CM is $\Delta H_1/2 + \Delta H_2$. (For atom reactions, the activation energy is the heat of reaction for an endothermic reaction and 1 or 2 kcal for an exothermic reaction. Since step 3 is always exothermic, its activation enthalpy is ignored.) In general, whichever of these is the lower will yield the more favorable reaction path. The quantities involved in CM can be calculated from bond energies; that for MM can be estimated by other means. Such estimates are given in Table 1.

In general, a reaction occurs by the path with the lowest activation energy. Accordingly the estimates in Table 1 show that CM is less likely for HI than MM, whereas the reverse is true for HBr and HCl. This change-over from MM to CM is not ascribable to any single factor; it involves relatively subtle differences in all the reactions listed above.

REACTIONS IN SOLUTIONS

Reactions in solution are far more complex than reactions in gases; in spite of this, the pseudothermodynamic transition state theory is still applicable. Equilibrium in solution, like that in gases, is independent of the particular path of the reaction, and therefore a catalyst in solution, like one in a gas, accelerates the forward and reverse reactions by the same relative amount so that the presence of a catalyst does not change the equilibrium constant. In addition, since the general form of the theoretical rate expression is the same as that for a gas, effects of temperature on the rate constants and relations of instantaneous rate expressions at equilibrium are the same as for gaseous reactions. Thus, superficially at least, catalysis, temperature, and equilibrium play the same role in solution kinetics as they do in gas phase kinetics.

When a molecule dissolves in a solvent, a chemical reaction occurs wherein the solute molecule enters into a loose combination with a solvation shell of the surrounding molecules. The kinetics of any reaction involving this dissolved molecule depends, therefore, on the nature of this solvent shell. For this reason, the rates of many reactions in solution depend on the solvent in which they occur; in extreme cases, the mechanism of a given reaction may change when the solvent is changed.

Most solution reactions we are interested in occur in aqueous solutions. Because of hydrogen bonding, water does not form ideal solutions, and, therefore, theoretical treatment of reaction rates is likely to be complex. Certain patterns of reactivity do, however, appear. For example, reactions between ions to form ions are usually rapid in aqueous solutions if they occur at all. Those ionic reactions of most interest to us are complete as soon as the reactants are mixed. In contrast to this, reactions between molecular species in water are often sluggish; in fact, reactions in which one of the products forms an insoluble precipitate may take hours or days to occur completely. Almost all such reactions, however, take place in a reasonable period of time in a heated solution. Thus, with few exceptions, reactions in hot aqueous solutions occur reasonably fast.

HETEROGENEOUS REACTIONS

If a piece of metal is heated in oxygen (or a crystal of salt is dropped into water), the oxidation (or solution) reaction occurs between two separate phases, and the reaction is called a *heterogeneous reaction.* Such a reaction as the oxidation of a metal (Fig. 9) can occur only where the two phases meet. Therefore, in a heterogeneous reaction, *the rate increases as the contact area is increased.*

A sodium chloride crystal weighing 2.16 gm forms a cube 1 cm on an edge. If this is dropped into a beaker of water, the solution rate will be slow because the area of contact is only 6 cm^2. When the crystal is ground to a powder so that 10^{12} smaller cubes are formed and the powder is dropped into water, the contact area is 60,000 cm^2 and the solution rate becomes very fast. This increase in contact area through grinding can cause spectacular effects with metals. Nickel in massive form is quite resistant to air oxidation, but finely divided nickel powder (with a larger area) becomes incandescent when exposed to air.

Many of the properties we assign to common metals are a consequence of the nature of their oxides rather than of the metal itself. When oxygen reacts with a

Fig. 9. Oxidation of aluminum.

metal, the initial reaction takes place at the surface. Any further reaction occurs by the penetration of the surface oxide layer by oxygen, followed by reaction with the underlying metal. If the surface oxide is compact and adhesive, only the surface is oxidized even though the metal itself is very reactive. Thus it is the nature of this oxide coating that determines the corrosion resistance of the metal. Aluminum, magnesium, chromium, and nickel are actually very reactive metals; it is the adhesive, impenetrable surface oxide that makes these metals corrosion-resistant.

Metal crystals are seldom perfectly formed. Occasional flaws in the periodic structure occur; in some cases, atoms may even be missing from a normal lattice position. When this occurs, these *defects* provide a channel whereby reactants can penetrate into the interior of the metal (Fig. 10). Reactions with acids occur fastest here and corrosion also occurs more rapidly at these points. Such defects are more prevalent at higher temperatures. Thus, although many metals are corrosion-resistant at ordinary temperatures, they corrode readily at high temperatures. For example, stainless steel maintains its polished surface indefinitely in air at 300°C but at 600°C it will tarnish in a few hours.

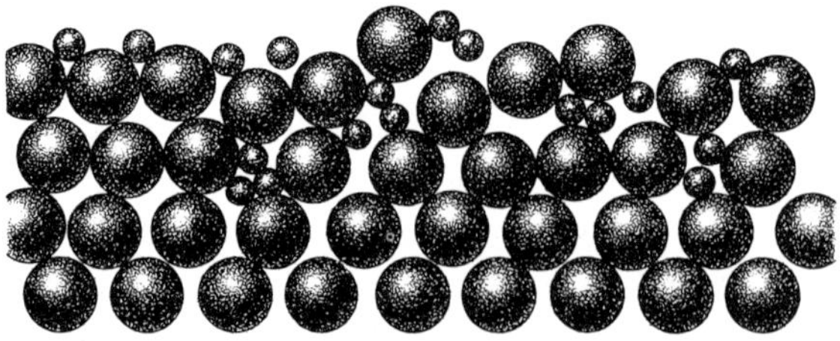

Fig. 10. Defective metal.

PROBLEMS

1. The reaction

$$2NO + O_2 \rightarrow 2NO_2$$

is first order in oxygen pressure and second order in the pressure of nitric oxide. Write the rate expression.

2. The indicated rate expressions have been obtained for each of the reactions listed below. If these rate expressions hold even at equilibrium, what is the rate expression for the reverse reaction at equilibrium?

(*a*) $C_2H_2 + H_2 \rightarrow C_2H_4$

$$R_f = k_f \frac{(H_2)}{(C_2H_2)} \qquad \textit{Ans. } R_r = k_r \frac{(C_2H_4)}{(C_2H_2)^2}$$

(*b*) $C_2H_4 + H_2 \rightarrow C_2H_6$
$R_f = k_f(H_2)$

(*c*) $2H_2 + O_2 \rightarrow 2H_2O$
$R_f = k_f(H_2)(O_2)^{4/3}$

(*d*) $N_2O_5 \rightarrow 2NO_2 + \frac{1}{2}O_2$
$R_f = k_f(N_2O_5)$

3. If σ for the HI reaction is 3.5 Å, $P_f = 1$, and $\Delta E_f^\ddagger$ is 44 kcal, what is the rate of the reaction in moles per second at $P_{H_2} = P_{I_2} = 1$ atm at 400°C in a 1-liter flask? What is the corresponding reaction rate at 300°C? What is the rate at 400°C if the activation energy is 31.5 kcal. What is the rate at 400°C if $\Delta E_f^\ddagger \approx 0$?

★ **4.** For the HI reaction, $\sigma \approx 3.5$ Å, $P_f = 1$, and $\Delta E_f^\ddagger = 44$ kcal. Calculate the rate constant at 300, 320, 340, 360, 380, and 400°C. Plot ln k versus $1/T$; draw the best straight line through the points. Determine the slope multiplied by the gas constant R. What is it? What does the value of the intercept at $1/T = 0$ correspond to (roughly)? This is how experimental data are analyzed.

5. For the decomposition of hydrogen iodide, σ is again about equal to 3.5 Å and $P_r \approx 1$. With the help of the data in Problem 4, calculate the rate constant for the decomposition reaction at 400°C. Calculate the equilibrium partial pressure of H_2 when $P_{I_2} = P_{HI} = 1$ atm. *Hint:* First calculate the ΔE for the overall reaction.

★ **6.** Radioactive decay is a kinetic process in which there is no reverse reaction. Consider the decay process in thorium:

$$_{92}U^{234} \rightarrow {}_{90}Th^{230} + {}_2He^4 \qquad t_{1/2} = 265{,}000 \text{ yr}$$
$$_{90}Th^{230} \rightarrow {}_{88}Ra^{226} + {}_2He^4 \qquad t_{1/2} = 83{,}000 \text{ yr}$$
$$_{88}Ra^{226} \rightarrow {}_{86}Rn^{222} + {}_2He^4 \qquad t_{1/2} = 1590 \text{ yr}$$

After this process has occurred for a long time, a state is reached where, for every thorium atom formed from $_{92}U^{234}$, one decomposes to form $_{88}Ra^{226}$; and, for every $_{88}Ra^{226}$ formed, one decomposes. What is the ratio of $_{90}Th^{230}$ to $_{88}Ra^{226}$ in such a case?

Ans. 52.

7. Given the following data for the reaction

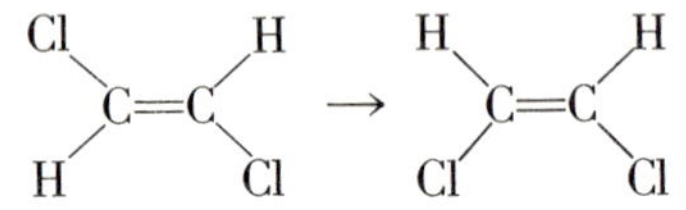

trans *cis*

t (min)	moles *trans*
0	1.00
10	0.90
20	0.81
30	0.73

what is the order of the reaction? How long will it take for half of the *trans*-$C_2H_2Cl_2$ to decompose?

★**8.** A kinetic study of the reaction

$$C_2H_4 + H_2 \rightarrow C_2H_6$$

is made by mixing H_2 with C_2H_4 in a 1-liter flask at room temperature, where there is no reaction. The mixture is then heated to the reaction temperature, 600°C. Samples (1 cc STP) are withdrawn periodically, chilled, and analyzed for hydrogen. The following data are obtained:

Run I

100 cc STP H_2	100 cc C_2H_4
time	cc H_2/sample
0	0.5
1	0.40
2	0.32
3	0.26

Run II

50 cc STP H_2	150 cc C_2H_4
time	cc H_2/sample
0	0.25
1	0.20
2	0.16
3	0.13

Write a consistent rate expression for the instantaneous rate of formation of ethane.

9. Given the following information:

$$C_2H_4 + H_2 \rightarrow C_2H_6 \qquad \Delta H^\circ = -30 \text{ kcal}$$
$$R_f = k_f(H_2) \text{ holds everywhere.}$$

(*a*) Write the expression for the reverse rate at equilibrium.

(*b*) If the temperature is increased, does the forward rate constant k_f or the reverse rate constant k_r increase more percentagewise? *Ans.* k_r.

(*c*) If the activation energy is 28.0 kcal for the forward reaction, what is the activation energy for the reverse reaction? *Ans.* 58 kcal.

(*d*) If a catalyst is added, the activation energy for the forward reaction drops to 10.7 kcal. What is the activation energy for the reverse rate of the catalyzed reaction? *Ans.* 40.7 kcal.

(*e*) The use of a catalyst speeds up the forward rate by a factor of ten million. By what factor does the reverse rate increase?

(*f*) The equilibrium constant at 500°C is 10^5. What is it (roughly) at 600°C?

(*g*) What is the value of the equilibrium constant at 500°C in the presence of the catalyst of part (*c*)?

10. The reaction

$$H_2 + Cl_2 \rightarrow 2HCl$$

occurs explosively in the presence of light. Assume that this explosion takes place by a chain reaction and is initiated by the formation of (*a*) hydrogen atoms, (*b*) chlorine atoms. From the bond energies of H_2 and Cl_2 calculate the wavelength of light needed in (*a*) and (*b*).

11. According to the drawing in Fig. 7, the $\Delta F°$ for the reaction is positive; that is,

$$H_2 + I_2 \rightarrow 2HI \qquad \Delta F° = 0.62 \text{ kcal at } 25°C.$$

Calculate the amount of HI formed at 25°C when 1 mole of H_2 and 1 mole of iodine are allowed to equilibrate in a 1-liter flask. Calculate the amount of HI formed when 1 mole of H_2 and 1 mole of iodine are allowed to equilibrate in a 10-liter flask.

12. For the formation of the activated complex

$$H_2 + I_2 \rightarrow H_2I_2^{\ddagger}$$

the $\Delta H^{\ddagger}$ is 44 kcal, the $\Delta S_f^{\ddagger} \approx -30$ E.U. Calculate the partial pressure of $H_2I_2^{\ddagger}$ in equilibrium with 1 atm of H_2 and 1 atm of I_2 at 400°C.

13. For the reaction

$$Ni + \tfrac{1}{2}O_2 \rightarrow NiO$$

$\Delta H° = -59.3$ kcal/mole. Suppose that the reaction takes place on the surface so rapidly that all the heat is used to heat up the remaining nickel. If at 25°C 1 oxygen atom reacts with each 10 Å^2 of surface, what will be the final temperature when a cube of nickel 1 cm on one edge reacts with oxygen? The density of nickel is about 9 gm/cm^3. If the 1-cm cube is ground up to form 10^{15} equal-sized cubes and reaction occurs with these, what will be the final temperature? (Ignore the change in heat capacity when nickel oxide is formed. Assume that the Dulong-Petit rule holds.)

14. Compute the amount of nickel oxide formed in each case cited in Problem 13.

★**15.** Repeat the calculations in Problem 13, but take into account the changes in heat capacity during reaction by an application of Kopp's rule.

★ **16.** The following thermodynamic data are known:

$$HI(g) \rightarrow H^+(aq.) + I^-(aq.) \qquad \Delta F^\circ = -18.6 \text{ kcal}$$

If solution involves zero free energy change for H_2, I_2, and H_2I_2, how would the rates of the reaction

$$H_2(aq.) + I_2(aq.) \rightarrow 2H^+(aq.) + 2I^-(aq.)$$

change from that in the gas phase? If you wanted to prepare pure HI, would you carry out the reaction in solution or in the gas phase. Use the data in Problem 11 to justify your choice.

17. If a cube of NaCl 1 cm on an edge is dissolved in an enormous quantity of water in a stirred tank, it takes 6 hours before solution is complete. If the cube is ground to a fine powder containing 10^{15} equal-sized spheres, what will be the time required for solution if this time is proportional to the initial area of contact between the NaCl and the water? *Ans.* 0.45×10^{-2} min.

18. Sulfur exists in two forms, rhombic and monoclinic. At room temperature, rhombic sulfur is stable and monoclinic sulfur is unstable. Which has the higher free energy? Both forms can be obtained at room temperature and both dissolve to form the same species in CS_2. If the transition complex in the solution process is the same for both forms, which form dissolves faster?

Ans. ΔF° rhombic $< \Delta F^\circ$ monoclinic; monoclinic dissolves faster.

★ **19.** If a single cube of nickel (area 6 cm^2) is dissolved in HCl, it dissolves much slower than the same amount of nickel powder with an area of 6×10^6 cm^2. If the rate is proportional to the contact area, it should be a million times faster for the finely divided nickel; the rate is actually faster by a factor of 10^8. This behavior is accounted for by the following statement: Nickel atoms on the surface are not bound to as many atoms as nickel atoms in the interior; hence the free energy per gram of nickel increases as the fraction of the nickel atoms on the surface increases. Explain the effect of this on the rate of reaction by the use of a free energy reaction profile.

20. By the proper combination of the rate data for HBr (p. 422) estimate the ΔH for the reaction

$$H_2 + Br_2 \rightarrow 2HBr$$

Hint: Write K_2K_3 in thermodynamic form and compare with the equilibrium constant. *Ans.* -24 kcal.

21. Show that for the reaction

$$H_2 + I_2 \rightarrow 2HI$$

at 1 atm, $\Delta H^\circ = \Delta E^\circ = \Delta E$ at any pressure provided H_2, I_2, and HI are ideal gases.

★**22.** Suppose it were possible (*it isn't*) to have two catalysts, A and B, and catalyst A would cause the reaction

$$H_2(g) + C_2H_4(g) \rightarrow C_2H_6(g)$$

to occur completely and catalyst B would cause the reaction

$$C_2H_6(g) \rightarrow C_2H_4(g) + H_2(g)$$

to occur completely. Sketch a perpetual motion machine utilizing these catalysts and this reaction that will convert heat entirely into work in violation of the second law of thermodynamics. Assume that $\Delta E = 0$ for both reactions.

15 Equilibrium in solution

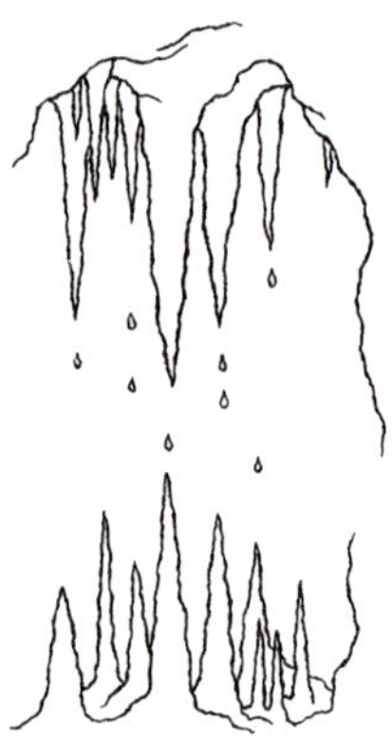

PROPERTIES OF PURE WATER

Physical properties. In Chapter 10, we found that water forms hydrogen bonds. Such bonding makes water a liquid with the remnants of crystallinity. If we choose a given water molecule and "look out" from it, the nearest water molecules have an orientation similar to that found in ice. This structure is most pronounced for the nearest neighbors and slowly vanishes for molecules farther removed. The average size of these ordered regions decreases as the temperature increases and molecular motion becomes more violent; hence liquid water is not H_2O, but $(H_2O)_n$, where n varies with temperature. Liquids that show this type of behavior are called *associated* liquids to distinguish them from ideal liquids.

The structure of ice is similar to that of SiO_2. In SiO_2, tetrahedral SiO_4 groups are linked through their corners; this results in a giant network held together by Si—O—Si bonds. The unit in ice is an OH_4 tetrahedra that is distorted so that two of the hydrogens are closer to the central oxygen than the other two. These tetrahedra are linked through their corners (i.e., by O—H—O bonds) to form a network similar to one of the structures of quartz. The resulting giant molecule has a relatively open structure as a result of directed hydrogen bonding.

When the ice melts, the average energy of the water molecule increases; this increase permits the rupture of some hydrogen bonds, but the forces that produce the tetrahedral arrangement in ice are still operative; hence remnants of the tetrahedral arrangement are preserved. In other words, we destroy the large ice crystal and form a large number of microcrystallites of ice, which, like the clusters found in the real gas, are short-lived. Since there is constant association and dissociation of water molecules to form and destroy these crystallites, they exist only in the statistical sense.

Most compounds decrease in density when they melt, but liquid water is denser than ice. Figure 1 shows a plot of the density of water above and below its melting point. The lower density of ice is associated with the open structure required for hydrogen bonding. When ice melts, some bonding is destroyed, and, along with this, the open structure associated with ice changes to a denser packing.

The density of a liquid usually decreases as the temperature increases because at the higher temperatures the molecules vibrate more violently and require more room. In water, we find two competing effects. As the temperature increases, the vibrations of the individual molecules require more space; but, at the same time the residual open structure (required for hydrogen bonding) is broken up by the increasing violence of molecular motion, and a denser packing results. Below 4° (cf. Fig. 1), the structure-breaking factor is more important, and this leads to an increase in density in spite of the fact that this is opposed by the normal expansion of the liquid. Above 4°, normal expansion of the liquid dominates, but structure breaking is still evident below 60°C. It is only near the boiling point that the density varies with the temperature as expected for a normal liquid.

Dissociation of pure water. The high dielectric constant of water makes it

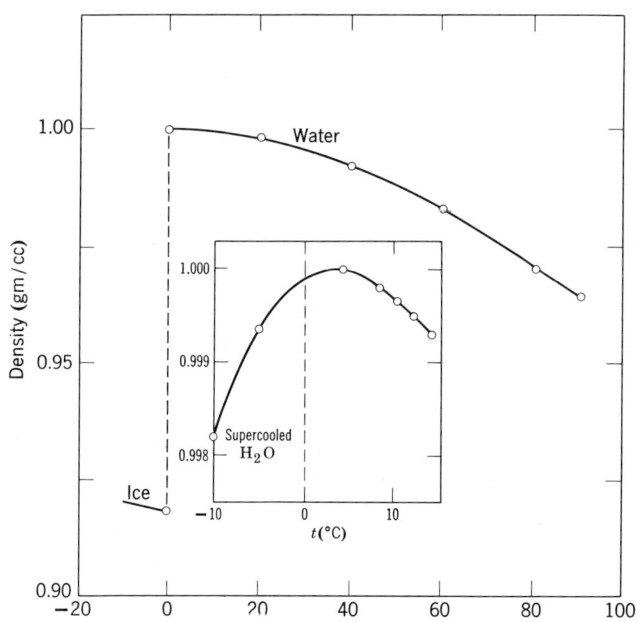

Fig. 1. Density of water.

an ideal solvent for the dissociation of electrolytes, because the work required to separate two oppositely charged particles in water is only one-eightieth that required to separate these two particles in a vacuum. Although the O—H bond in water is not ionic, it is highly polar; hence it is not surprising to find that water "dissolves" in water to form ions. Such reactions can be represented by the equations

$$HOH \rightleftarrows H^+ + OH^-$$
$$OH^- \rightleftarrows H^+ + O^{-2}$$

Since the O—H bond is not completely ionic, dissociation is not extensive. The equilibrium constant for the first reaction is 1.0×10^{-14}, and that for the second is about 10^{-37}. (We treat water as solvent.)

We write the equilibrium expressions for the dissociation of water as follows:

$$(H^+)(OH^-) = 1.0 \times 10^{-14}$$
$$\frac{(H^+)(O^{-2})}{(OH^-)} = 10^{-37}$$

If we consider only the first dissociation in pure water, the concentration of hydrogen ions is equal to the concentration of hydroxide ions,

$$(H^+) = (OH^-) = x$$
$$x^2 = 10^{-14}$$

and

$$(H^+) = (OH^-) = 10^{-7}$$

Thus, in normal water, the concentration of both hydrogen ion and hydroxide ion is 10^{-7} molar. From the second equilibrium expression, it follows that the calculated oxide ion concentration is 10^{-37} molar. In other words, if we introduce oxide ions into water they react almost completely with protons to form hydroxide ions.

Dissociation of water into hydrogen and hydroxide ions provides a convenient basis for classification of aqueous solutions. When the hydrogen ion concentration is equal to the hydroxide ion concentration, we term the solution *neutral;* when the hydrogen ion concentration is greater than the hydroxide ion concentration, we term the solution *acidic;* when the hydrogen ion concentration is less than the hydroxide ion concentration, we term the solution *basic.*

pH. In discussing hydrogen and hydroxide ions, it is common to express their concentrations in terms of the pH. The pH is defined by the equation

$$pH \equiv -\log (H^+)$$

The corresponding quantity for the hydroxide ion is

$$pOH \equiv -\log (OH^-)$$

Note that, the *higher* the value of the pH, the *lower* is the value of the hydrogen ion concentration. A neutral solution has a pH of 7; a pH *less* than this indicates an acidic solution, and a pH *higher* than this indicates a basic solution.

The pH and pOH are not independent. The equilibrium between hydrogen and hydroxide ions is satisfied by all aqueous solutions; hence

$$(H^+)(OH^-) = 10^{-14}$$

or

$$\log (H^+) + \log (OH^-) = -14$$

and, therefore,

$$pH + pOH = 14$$

DISSOCIATION IN SOLUTIONS

The mechanism of dissociation. The question of whether or not a soluble substance dissociates in water can be answered only by consideration of all the steps involved when dissociation occurs. Such a series of steps is listed below:

$MX_{2\,s} \rightarrow MX_{2\,g}$	sublimation	(*a*)
$MX_{2\,g} \rightarrow M_g + 2X_g$	bond breaking	(*b*)
$M_g \rightarrow M_g^{+2} + 2e_g^-$	ionization	(*c*)
$2(X_g + e_g^- \rightarrow X_g^-)$	electron affinity	(*d*)
$M_g^{+2} + \text{water} \rightarrow M_{aq.}^{+2}$	hydration	(*e*)
$2(X_g^- + \text{water} \rightarrow X_{aq.}^-)$	hydration	(*f*)

These add together to yield

$$MX_{2\,s} \rightarrow M_{aq.}^{+2} + 2X_{aq.}^-$$

Thus a crystalline compound MX_2 can dissolve and dissociate in water by the following steps:

(*a*) vaporization of the crystalline material to form gaseous molecules;
(*b*) dissociation of this molecule into atoms;
(*c*) and (*d*) the formation of gaseous ions from these atoms;
(*e*) and (*f*) the solvation of these gaseous ions in water.

An alternative solution process could occur via step (*a*), followed by solution of the gaseous molecules in water.

$$MX_{2\,g} + \text{water} \rightarrow MX_{2\,aq.} \quad (b')$$

Solution can occur with dissociation or no dissociation, depending on the relative free energies. If the free energies for both processes are unfavorable, there is no solution.

Of the energy terms discussed above, that for hydration is the most difficult to measure or calculate. The process of cationic solvation involves bringing the negative end of a water molecule up to the cation. The energy released is determined by the electric field just outside the cation. Thus the hydration energy depends on factors similar to those that determine the ionization energy. For a given charge and electronic configuration, the hydration energy increases as the size decreases; for a given electronic configuration and ionic size, e.g., Li^+, Mg^{+2}, and Sc^{+3}, the hydration energy increases with the *square* of the ionic charge. With ions of a given size and charge, the hydration energy parallels the ionization energy in so far as it is high when I_z is high and vice versa. Thus there is a tendency for steps (*c*) and (*e*) to cancel each other. A similar tendency is observed for other sets of terms, and it is just this that makes it very difficult to generalize about dissociation of compounds in water.

Molecular versus ionic crystals. In order to illustrate the careful balance of factors that determine whether a specific compound in solution dissociates, let us estimate the enthalpy of the various processes required for dissociation versus the enthalpy of the various processes required for molecular solution. For the sake of specificity, we choose two definite compounds:

$CaCl_2$, an ionic crystal

and

$HgCl_2$, a molecular crystal

In Table 1, we have listed estimated values of ΔH for each of the steps above. Let us compare those for $CaCl_2$ to those for $HgCl_2$. In general, the heat of sublimation is greater for ionic crystals than for molecular crystals, and, similarly, the bond energy is greater for ionic crystals; hence, if we consider the formation

Table 1 Enthalpies of the Solution Process

Step	$CaCl_2$	$HgCl_2$
	Dissociation	
a (sublimation)	50 kcal	20 kcal
b (bond breaking)	240 kcal	110 kcal
c (ionization)	410 kcal	673 kcal
d (electron affinity)	2(−86) kcal	2(−86) kcal
e (cation hydration)	−382 kcal	−441 kcal
f (anion hydration)	2(−85) kcal	2(−85) kcal
Sum	−24 kcal	+20 kcal
	Molecular Solution	
a	50 kcal	20 kcal
b′	−8 kcal	−16 kcal
Sum	+42 kcal	+4 kcal

of gaseous atoms from molecular and ionic crystals, steps (*a*) and (*b*) are easier for the molecular crystal. The third step (*c*) is harder for the molecular crystal; the higher electronegativity associated with the cation comes, in part, from the high value of the ionization potential. Since the ionization energy is high, however, the hydration energy is also high (*e*), so that the values of steps (*c*) and (*e*) partially cancel each other. The net result is that the over-all enthalpy change for $CaCl_2$ is more favorable than for $HgCl_2$; hence, from this standpoint alone, it is more likely that the $CaCl_2$ will dissociate.

The thermodynamic tendency for a reaction to occur is determined by the free energy rather than the enthalpy; hence we must also consider the entropy. Generally, this ΔS term is not large and contributes no more than 7 kcal to the over-all ΔF. Although it is true that this is enough to determine whether a given substance will dissociate in a borderline case, it is not enough to override a very unfavorable ΔH term. Thus we are safe in saying that $HgCl_2$ will not dissociate in solution.

Now look at the other possibility, i.e., molecular solution. The ΔH for the molecular solution process (Table 1) is very unfavorable for $CaCl_2$ but only slightly unfavorable for $HgCl_2$. Most of the enthalpy needed for the sublimation of $HgCl_2$ is regained when it dissolves in water, but the estimated gain in enthalpy when a $CaCl_2$ molecule dissolves in water is quite small compared with the ΔH of sublimation; hence, from the standpoint of ΔH alone, molecular solution is much more likely for $HgCl_2$ than for $CaCl_2$.

It is difficult to formulate generalizations dealing with phenomena as complicated as those discussed above. Such generalizations do, however, exist: *If* an ionic crystal dissolves, it will, in general, dissolve to form ions. *If* a molecular crystal dissolves, it will, in general, dissolve to form molecules.

THE STRUCTURE OF ACIDS AND BASES

The oxide of an electropositive metal such as sodium reacts with water as follows:

$$Na_2O + H_2O \rightarrow 2NaOH$$

The product, sodium hydroxide, is an ionic solid with the sodium atom bearing a plus charge and the OH group bearing a minus charge, but the structure does not resemble the simple ones discussed earlier. There is a strong tendency for the ionic hydroxides to crystallize with hydrogen bonding of the type

$$\begin{array}{c}\mathrm{O{-}H{-}\underset{\displaystyle\overset{|}{H}}{O}{-}H{-}O}\end{array}$$

Therefore, although sodium hydroxide crystals are ionic, they have a layer-like structure.

When a crystal of sodium chloride dissolves in water, it forms sodium ions and chloride ions; when an ionic hydroxide dissolves, it forms sodium ions and hydroxide ions. Since the concentration of hydroxide ions in such a solution is greater than that of the hydrogen ions, the resulting solution is basic. For this reason, these ionic hydroxides are referred to as bases.

If a base is completely dissociated in solution, we call it a *strong base;* those bases that are only partially dissociated in solution are called *weak* bases. For the hydroxides, we can equate ionic character to base strength. From this standpoint, the hydroxides of the alkali and alkaline earth metals are strong bases.

The oxide of a non-metal such as sulfur reacts with water as follows:

$$SO_3 + H_2O \rightarrow H_2SO_4$$

The product, a solid below 11°C, is not ionic or molecular in the usual sense of these descriptions. The structure is indicated in Fig. 2.

If we prepare a solution by adding H_2SO_4 to water, we find that the sulfuric acid dissociates to form H^+, HSO_4^-, and SO_4^{-2} ions and makes the solution acidic. Since the hydrated oxides of other non-metals behave similarly, they are called *acids.*

Both the oxygen acids and bases contain the OH group. The difference in the mode of dissociation stems from the polarity in the bonding of the OH group to the atom in question. When the bond is highly ionic, the compound forms a base; when it is highly covalent, the compound forms an acid. Accordingly, we expect a gradual change in the properties of the hydroxides of the elements across any period in the periodic chart: metals should be basic; non-metals acidic, and metalloids in between.

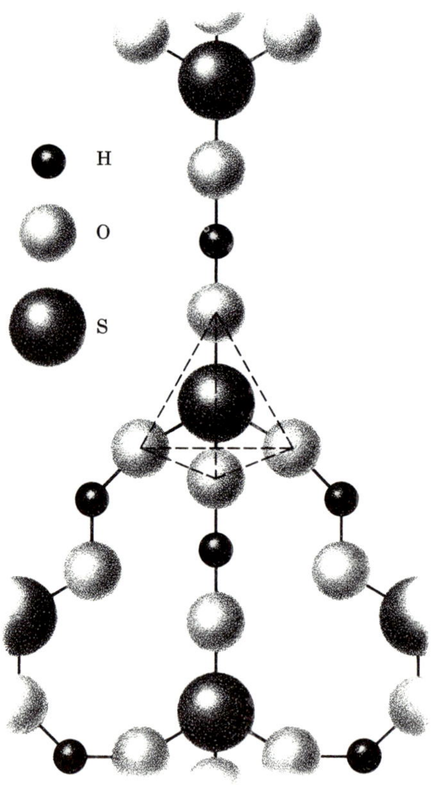

Fig. 2. Structure of sulfuric acid.

Acid strength. Even if we consider only the oxygen acids, we find that the degree of dissociation varies enormously from one acid to another.

The strength of these oxygen acids can be gauged by the value of pK_A, which is defined by

$$pK_A \equiv -\log K_A$$

where K_A represents the equilibrium constant,

$$K_A = \frac{(H^+)(A^-)}{(HA)}$$

for the acid HA. Values for the more common oxygen acids are listed in Table 2.

According to Table 2, there is a tendency for the pK_A of these acids to group around certain values. To see what factors contribute to this grouping, let us examine the structures of the oxygen acids of chlorine that appear in each of these groups.

The strength of an acid is determined by the force with which the acid anion attracts the positively charged proton. The ions formed by the dissociation of the oxygen acids of chlorine are

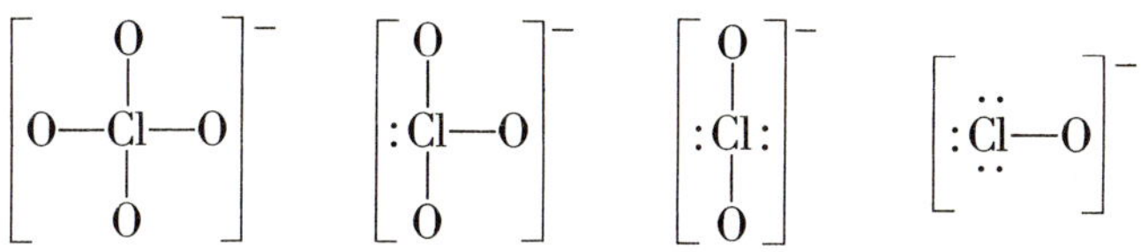

Each of these ions has an over-all charge of -1, but, since oxygen is more electronegative than the chlorine, this minus charge resides largely on the oxygen atoms. There is no reason to suppose that one of the oxygen atoms in ClO_4^- has

Table 2 Acid Strength
Oxygen Acids

pK_A	Acid	
8 to 11	$HClO$	
8 to 11	H_3AlO_3	
8 to 11	H_3BO_3	
2 to 3	$HClO_2$	HNO_2
2 to 3	H_2SO_3	H_3PO_4
2 to 3		H_3PO_3
-1 to -2	$HClO_3$	
-1 to -2	H_2SO_4	
-1 to -2	HNO_3	
-8	$HClO_4$	

more of the charge than any of the other three oxygen atoms; hence the charge is equally divided among the four oxygen atoms. This leads to an effective charge of $-\frac{1}{4}$ at each corner of the ClO_4^- tetrahedron. Similar reasoning leads to a charge of $-\frac{1}{3}$ for ClO_3^-, $-\frac{1}{2}$ for ClO_2^-, and -1 for ClO^-. Thus the proton-attracting power is greatest for ClO^- and least for ClO_4^-; hence $HClO_4$ is the strongest of these acids and HClO is the weakest (see Table 2).

If this reasoning is correct, the other acids in Table 2 should show similarities in structure. The structure of sulfuric acid we have already discussed. Upon the loss of one proton, there are three equivalent oxygen atoms on which the single negative charge can reside; hence the effective charge of each oxygen would be $-\frac{1}{3}$ as for chloric acid, and the pK_A should be comparable. A similar situation holds for nitric acid.

The reasoning is similar for the next group of acids (pK_A 2 to 3). The structure of nitrous acid, like that of chlorous acid, leads to an anion with two equivalent oxygen atoms; hence the charge on each of these oxygen atoms would be $-\frac{1}{2}$. The structure of sulfurous acid, H_2SO_3, is

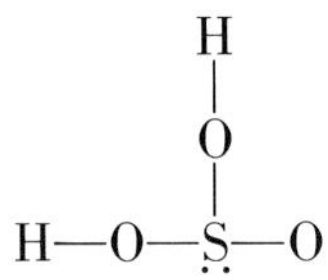

Clearly, on the loss of a proton, the resulting anion has two equivalent oxygen atoms. In phosphoric acid, the central atom is surrounded by three OH groups and one oxygen atom:

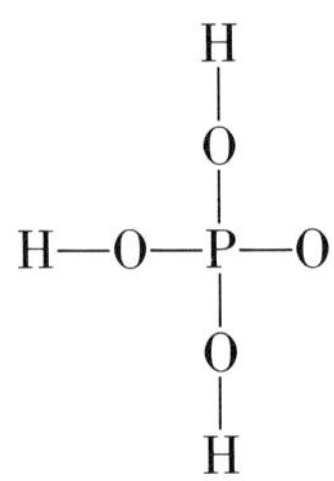

The loss of a proton leads to the two equivalent oxygen atoms found for other members of this group. If the structure of phosphorous acid is similar to those in this group, it cannot involve three OH groups on the central atom. Actually, H_3PO_3 has only two acidic hydrogens; the structure is

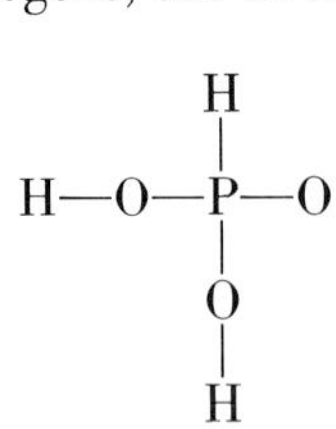

consistent with the pK.

The weakest oxygen acids are those in which only OH groups are bound to the central atom. The anion formed by a single ionization has only one negatively charged oxygen atom, and hence the proton will be strongly attracted to form a weak acid.

REACTIONS OF ACIDS AND BASES

Acid-base stoichiometry. By definition, an acid in water creates an excess of hydrogen ions, and a base in water creates an excess of hydroxide ions. Since the dissociation constant of water is small, there is a strong tendency for hydrogen ions and hydroxide ions to recombine to form water. This is the essential reaction when we combine aqueous solutions of an acid and a base.

If we mix a solution containing 1 mole of H_2SO_4 with a solution containing 1 mole of NaOH, a reaction occurs between hydroxide and hydrogen ions formed by dissociation. Evaporation of the resulting solution yields a *salt* with the formula $NaHSO_4$; consequently, we write

$$NaOH + H_2SO_4 \rightarrow NaHSO_4 + HOH$$

This equation, however, contains superfluous information. Every species but the water is dissociated; initially, there are present sodium ions, sulfate ions, bisulfate ions, hydrogen ions, and hydroxide ions, and, after reaction, there is water together with sodium ions, hydrogen ions, sulfate ions, and bisulfate ions. Thus the *essential change* that took place is indicated by the equation

$$H^+ + OH^- \rightarrow H_2O$$

If 2 moles of sodium hydroxide are added to 1 mole of sulfuric acid, the following reaction occurs:

$$2NaOH + H_2SO_4 \rightarrow 2HOH + Na_2SO_4$$

Once again, the essential change is the combination of 2 moles of hydroxide and 2 moles of hydrogen ion to form 2 moles of hydrogen hydroxide molecules, i.e., water. In all acid-base reactions, the hydroxide and hydrogen ions react as completely as possible and the ions left over after reaction constitute the salt formed.

Anhydrides. Oxygen acids are the hydrated oxides of non-metals. For example, nitric acid, HNO_3, is the product obtained when we hydrate N_2O_5. Similarly, nitrous acid is the product obtained when we hydrate N_2O_3. These reactions occur as follows:

$$N_2O_5 + II_2O \rightarrow 2IINO_3$$

and

$$N_2O_3 + H_2O \rightarrow 2HNO_2$$

Andrydrides (which means without water) can react with more than one molecule of water. For example, hydration of P_2O_5 in cold water occurs according to the equation†

$$P_2O_5 + H_2O \rightarrow 2HPO_3$$

Thus this equation is the same as that for the corresponding oxide of nitrogen, as might be expected since phosphorus and nitrogen are in the same family of the periodic chart. But, if we wait a long time, the reaction goes one step further:

$$HPO_3 + H_2O \rightarrow H_3PO_4$$

This difference between nitrogen and phosphorus occurs because in any given family the heavier elements are larger. Since phosphorus is larger than nitrogen, it has a larger coordination number and can accommodate more OH groups; hence the hydration is more extensive. The effect is even more pronounced when we contrast the behavior of N_2O_5 and Sb_2O_5. The latter is the anhydride of the acid $HSb(OH)_6$. Similar trends are observed for the halogens, e.g., $HClO_4$ compared to H_5IO_6, and the sulfur family, e.g., H_2SO_4 compared to H_6TeO_6.

In Table 3, we have indicated some of the common acids and the anhydrides from which they stem.

Bases are hydrated oxides of metals. For example, sodium hydroxide is produced by the reaction

$$Na_2O + H_2O \rightarrow 2NaOH$$

and aluminum hydroxide can be viewed as resulting from the hydration of aluminum oxide,

$$Al_2O_3 + 3H_2O \rightarrow 2Al(OH)_3$$

When an acid (a hydrated non-metallic oxide) reacts with a base (a hydrated metallic oxide), it forms water and a salt. It seems reasonable, therefore, that the dehydrated oxides of these metals and non-metals react to form the salt but no water. If we heat a mixture of calcium oxide and phosphorus pentoxide in a crucible, the following reaction occurs:

$$3CaO + P_2O_5 \rightarrow Ca_3(PO_4)_2$$

This salt is the same as that formed by the complete neutralization of H_3PO_4 by $Ca(OH)_2$. All reactions between acidic and basic oxides can be viewed from this standpoint.

† In this and the following equation, we are using the simplest formula for P_2O_5 and HPO_3. These two species actually occur as P_4O_{10} and $H_4P_4O_{12}$. The structures of P_4O_{10} and N_2O_5 are given in Chapter 9. Note the differences in the coordination numbers of nitrogen and phosphorus.

Table 3

Acid Oxide[a]	Acid
SO_3	H_2SO_4
CO_2	H_2CO_3
Cl_2O_7	$HClO_4$
SO_2	H_2SO_3
$P_2O_3(P_4O_6)$	H_3PO_3
B_2O_3	H_3BO_3
Cl_2O	$HClO$

[a] These are the simplest formulas.

Reactions also occur between an acidic oxide and a base or a basic oxide and an acid. Thus, if we expose gaseous SO_3 to sodium hydroxide, it reacts to produce a salt and some water; that is,

$$2NaOH + SO_3 \rightarrow Na_2SO_4 + H_2O$$

Equivalent weight. Since all acid-base reactions are essentially the same, it is logical to describe acids in terms of their equivalent as hydrogen ions. This leads to the definition of an *equivalent weight,* the weight of acid that yields 1 mole of hydrogen ions. Consider the acids HBr, H_2SO_4, HCl, and H_2CO_3. The equivalent weight of HBr is equal to its molecular weight, since 1 mole of the first acid yields 1 mole of hydrogen ion. For H_2SO_4, the equivalent weight is equal to one-half the molecular weight. For HCl, the equivalent weight is the molecular weight, and, for H_2CO_3, the equivalent weight is half the molecular weight. Similarly, we find that the equivalent weights of sodium hydroxide, calcium hydroxide, and aluminum hydroxide (defined in terms of the moles of H^+ they *react* with) are equal to the molecular weight, one-half the molecular weight, and one-third the molecular weight, respectively.

Normality. In dealing with acid-base reactions in solution, it is convenient to work with *normality* rather than molarity. Normality of an acid is the molarity of the *potential* hydrogen ion in the solution; normality of a base is the molarity of the *potential* hydroxide ion. This, of course, turns out to be the number of equivalents per liter of solution, so that the defining equation is,

$$\text{Normality (acid)} = \frac{\text{No. moles of available H}^+}{\text{liter sol'n}} = \frac{\text{No. of equivalents}}{\text{liter sol'n}}$$

$$\text{Normality (base)} = \frac{\text{No. moles of available OH}^-}{\text{liter sol'n}} = \frac{\text{No. of equivalents}}{\text{liter sol'n}}$$

For complete reaction between hydrogen and hydroxide ions, the number of moles of hydrogen ion added must be equal to the number of moles of hydroxide ion present; that is, the number of acid equivalents added must be equal to the number of base equivalents. According to the definition of normality, the number of equivalents is the normality multiplied by the volume of solution in liters. If we add enough acid to neutralize a given volume of base, the following equation holds:

$$N_B V_B = N_A V_A$$

where $N_B(N_A)$ and $V_B(V_A)$ refer to the normality of base (acid) and the volume of base (acid) solution.

Although the concept of normality and equivalent weight is a useful one, it is sometimes ambiguous. For example, H_3PO_4 contains potentially three equivalents per mole. It is, however, seldom possible to utilize all three of the potential hydrogen ions; hence, in practice, the *effective* equivalent weight of H_3PO_4 is half the molecular weight. Even more disturbing is the fact that the equivalent weight varies with the conditions of neutralization. Thus, in one case, the equivalent weight of H_3PO_4 may be half the molecular weight; in another case, it may be equal to the molecular weight. In spite of these drawbacks, normality is widely used in analytical chemistry.

Example Problem 1: One-half of a liter of solution is made up which contains 24.5 gm of H_3PO_4. (*a*) What is the molarity? (*b*) What is the normality? (Assume that only two of the hydrogen atoms in H_3PO_4 are available.) It requires 20 ml of this solution to neutralize 30 ml of a solution containing $Ba(OH)_2$. (*c*) What is the normality of the $Ba(OH)_2$? (*d*) What is the molarity?

(*a*)

$$\text{mol. wt. } H_3PO_4 = \frac{98 \text{ gm } H_3PO_4}{\text{mole } H_3PO_4}$$

$$\frac{24.5 \text{ gm } H_3PO_4}{0.50 \text{ liter of sol'n}} \times \frac{1 \text{ mole } H_3PO_4}{98 \text{ gm } H_3PO_4} = \frac{0.5 \text{ mole } H_3PO_4}{\text{liter of sol'n}} \equiv 0.50 \text{ molar } H_3PO_4$$

(*b*)

$$\frac{0.50 \text{ mole } H_3PO_4}{1 \text{ liter of sol'n}} \times \frac{2 \text{ moles potential } H^+}{1 \text{ mole } H_3PO_4}$$

$$= \frac{1.0 \text{ mole potential } H^+}{\text{liter of sol'n}} = \frac{1.0 \text{ eq. } H^+}{\text{liter of sol'n}} \equiv 1.0 \text{ normal } H_3PO_4$$

This is often abbreviated as 1.0N H_3PO_4.

(*c*) For complete reaction, we must have

$$\text{eq. } H^+ = \text{eq. } OH^-$$

Thus, since

$$\text{eq. } H^+ = \frac{1.0 \text{ eq. } H^+}{\text{liter of sol'n } H_3PO_4} \times 0.020 \text{ liter of sol'n}$$

and

$$\text{eq. } OH^- = \frac{X \text{ eq. } OH^-}{\text{liter of sol'n } Ba(OH)_2} \times 0.030 \text{ liter of sol'n}$$

we find

$$\frac{\text{X eq. OH}^-}{\text{liter of sol'n Ba(OH)}_2} \times 0.030 \text{ liter of sol'n} = \frac{1.0 \text{ eq. H}^+}{\text{liter of sol'n H}_3\text{PO}_4} \times 0.020 \text{ liter of sol'n}$$

or

$$\frac{\text{X eq. OH}^-}{\text{liter of sol'n Ba(OH)}_2} = \frac{2}{3} \times 1.0 = \frac{0.66 \text{ eq. OH}}{\text{liter of sol'n Ba(OH)}_2} \equiv 0.66N \text{ Ba(OH)}_2$$

(*d*) $$\frac{0.66 \text{ eq. OH}^-}{\text{liter of sol'n Ba(OH)}_2} \times \frac{1 \text{ mole Ba(OH)}_2}{2 \text{ eq. OH}^-} = \frac{0.33 \text{ mole Ba(OH)}_2}{\text{liter of sol'n Ba(OH)}_2} \equiv 0.33 \text{ molar Ba(OH)}_2$$

ACID-BASE EQUILIBRIA

General considerations. In Table 4 we have listed a number of acids. They can be divided into two groups: those for which $K_A > 1$ and those for which $K_A < 1$. The former are termed strong acids, since the value of K_A indicates that the dissociation into hydrogen ions and an anion is essentially complete. The latter are termed weak acids for analogous reasons. To determine the acidity [or, equivalently, the pH or (H^+)] of solutions of weak acids, we must know the con-

Table 4 Common Acids

	Common Strong Acids	
HNO_3	$\rightleftarrows H^+ + NO_3^-$	$K_A \sim 10^2$
HCl	$\rightleftarrows H^+ + Cl^-$	$K_A \sim 10^3$
HBr	$\rightleftarrows H^+ + Br^-$	$K_A \sim 10^9$
HI	$\rightleftarrows H^+ + I^-$	$K_A \sim 10^{11}$
$HClO_4$	$\rightleftarrows H^+ + ClO_4^-$	$K_A \sim 10^8$
H_2SO_4	$\rightleftarrows H^+ + HSO_4^-$	$K_A \sim 10$
	Common Weak Acids	
HAc (acetic)	$\rightleftarrows H^+ + Ac^-$	$K_A = 1.8 \times 10^{-5}$
HClAc	$\rightleftarrows H^+ + ClAc^-$	$K_A \sim 10^{-3}$
HCN	$\rightleftarrows H^+ + CN^-$	$K_A = 4 \times 10^{-10}$
HSO_4^-	$\rightleftarrows H^+ + SO_4^{-2}$	$K_{A\,II} = 1.2 \times 10^{-2}$
HOH	$\rightleftarrows H^+ + OH^-$	$K = 1.8 \times 10^{-12}$
H_2S	$\rightleftarrows H^+ + HS^-$	$K_{A\,I} = 1.1 \times 10^{-7}$
HS^-	$\rightleftarrows H^+ + S^{-2}$	$K_{A\,II} = 1 \times 10^{-14}$
HNO_2	$\rightleftarrows H^+ + NO_2^-$	$K_A = 4 \times 10^{-4}$
H_2CO_3	$\rightleftarrows H^+ + HCO_3^-$	$K_{A\,I} = 4.3 \times 10^{-7}$
HCO_3^-	$\rightleftarrows H^+ + CO_3^{-2}$	$K_{A\,II} = 5.6 \times 10^{-11}$
H_3PO_4	$\rightleftarrows H^+ + H_2PO_4^-$	$K_{A\,I} = 7.5 \times 10^{-3}$
$H_2PO_4^-$	$\rightleftarrows H^+ + HPO_4^{-2}$	$K_{A\,II} = 6.2 \times 10^{-8}$
HPO_4^{-2}	$\rightleftarrows H^+ + PO_4^{-3}$	$K_{A\,III} = 4.8 \times 10^{-13}$

centration and the dissociation constant, K_A. Then we can compute the concentration of various species in solution.

Consider the dissociation of a typical weak acid such as chloroacetic acid, which has the formula $H(C_2H_2O_2Cl)$ or H(ClAc). In water, we find not only chloroacetic acid molecules but also hydrogen ions and chloroacetate anions. We represent this by the equation

$$\mathrm{HClAc} \rightleftarrows \mathrm{H^+} + \mathrm{ClAc^-}$$

All three of the species occurring in the equation are in the solution, and to deal with this situation quantitatively we must make use of the equation,

$$\frac{(\mathrm{H^+})(\mathrm{ClAc^-})}{(\mathrm{HClAc})} = K_A$$

Now consider a more complicated case. H_3PO_4 is a weak acid and can yield 1, 2, or 3 hydrogen ions. Equilibrium principles will apply to *each* of these steps so that we can write the following equations:

$$\mathrm{H_3PO_4} \rightleftarrows \mathrm{H_2PO_4^-} + \mathrm{H^+} \qquad \frac{(\mathrm{H^+})(\mathrm{H_2PO_4^-})}{(\mathrm{H_3PO_4})} = K_{A\,\mathrm{I}}$$

$$\mathrm{H_2PO_4^-} \rightleftarrows \mathrm{HPO_4^{-2}} + \mathrm{H^+} \qquad \frac{(\mathrm{H^+})(\mathrm{HPO_4^{-2}})}{(\mathrm{H_2PO_4^-})} = K_{A\,\mathrm{II}}$$

$$\mathrm{HPO_4^{-2}} \rightleftarrows \mathrm{PO_4^{-3}} + \mathrm{H^+} \qquad \frac{(\mathrm{H^+})(\mathrm{PO_4^{-3}})}{(\mathrm{HPO_4^{-2}})} = K_{A\,\mathrm{III}}$$

Thus a solution of H_3PO_4 contains each of the following species.

$$\mathrm{H^+},\quad \mathrm{H_3PO_4},\quad \mathrm{H_2PO_4^-},\quad \mathrm{HPO_4^{-2}},\quad \mathrm{PO_4^{-3}}$$

Since all three equilibria occur in the same solution, the value of (H^+) must be the same in *all three* equilibrium expressions. Similar considerations hold for the other species present.

Simple bases have the general formula MOH. We write the equation for what happens in solution as follows:

$$\mathrm{MOH} \rightleftarrows \mathrm{M^+} + \mathrm{OH^-}$$

and the equilibrium constant is

$$\frac{(\mathrm{M^+})(\mathrm{OH^-})}{(\mathrm{MOH})} = K_B.$$

Whether a base is weak or strong is determined by the value of the dissociation constant; when $K_B > 1$, the base is labeled strong; when $K_B < 1$, the base is termed weak (cf. Table 5). The hydroxides of the alkali metal group, e.g., Li,

Table 5 Common Bases

Strong Bases	
Hydroxides of alkali metals and alkaline earth metals	
Weak Bases	
(Ammonium hydroxide) $NH_4OH \rightleftarrows NH_4^+ + OH^-$	$K_B = 1.8 \times 10^{-5}$
(Ethyl ammonium hydroxide) $C_2H_5NH_3OH \rightleftarrows C_2H_5NH_3^+ + OH^-$	$K_B = 5.6 \times 10^{-4}$
(Hydrazine hydroxide) $NH_2NH_3OH \rightleftarrows N_2H_5^+ + OH^-$	$K_B = 1.0 \times 10^{-6}$

Na, K, etc., and the hydroxides of the alkaline earth group, Mg, Ca, Sr, etc., are all strong bases. In solution, they completely dissociate into hydroxide and metal ions. Weak bases are generally organic.

Before proceeding with the actual solution of these problems, let us examine them from a purely mathematical viewpoint. Suppose, for example, that we are asked to find the concentration of all species present in a liter of solution made by mixing 1 mole of H_3PO_4 with water. The possible species are: H_3PO_4, $H_2PO_4^-$, HPO_4^{-2}, PO^{-3}, H^+, and OH^-. Thus we have six unknowns. According to algebra, if we wish to determine a number of unknowns, we must have as many equations as we have unknowns. Thus, in this specific case, we need six equations. Three of them are given by the equilibrium expressions for phosphoric acid indicated earlier. A fourth equation is the equilibrium expression for the dissociation of water,

$$(H^+)(OH^-) = 1.0 \times 10^{-14}$$

The two remaining equations are obtained either from the chemical equations or equivalently by a direct application of the conservation of charge and the conservation of atoms. Since the solution is electrically neutral, the total positive charge equals the total negative charge, and we write

$$(H^+) = (H_2PO_4^-) + 2(HPO_4^{-2}) + 3(PO_4^{-3}) + (OH_4^-)$$

(conservation of charge)

Since the only source of phosphorus atom is the phosphoric acid added initially, we can also write

$$(H_3PO_4)_{\text{added}} = 1 = (H_3PO_4) + (H_2PO_4^-) + (HPO_4^{-2}) + (PO_4^{-3})$$

(conservation of atoms)

Now that we have the six equations, the values of the six unknowns are specified and the problem is solved except for the computations. This does not mean that the computation is simple; in fact, solution of these equations without approximation would be a fearful prospect. In practice, we almost always make simplifying approximations. It is in making these approximations that chemistry enters the problem. For example, in the specific example we are now considering, one approximation (of many valid ones) would be to neglect (OH^-) (which is small in an acid solution) in the conservation of charge equation.

In general, the necessary equations for the solution of any equilibrium problem are obtained from the equilibrium expressions and from the equations specifying the conservation of charge and atoms. If the equilibria involved are so complex that the initial amounts of two or more substances are given, we may have two or more equations for the conservation of atoms. Regardless of the complexity, however, the problem is *in principle* solved once the equations are given; the actual values of the unknowns can then be determined by a computer.

Solutions of strong acids and bases. Calculations of the pH of solutions of a strong acid or a strong base are generally simple. The value of (H^+) or (OH^-) contributed by the acid or base is equal to the concentration of acid or base. Complications arise only when the contribution of the acid or base is comparable to that of water; then both contributions must be considered.

Example Problem 2: Calculate the pH of a solution of (*a*) $0.01M$ HCl, (*b*) $0.01M$ NaOH, (*c*) $10^{-7}M$ HCl.

(*a*) The (H^+) in pure water is 10^{-7}. In $0.01M$ HCl, complete dissociation yields (H^+) = 0.01. This overshadows the contribution from water; hence we write

$$\text{pH} = -\log (H^+) = -\log (0.01) = -(-2) = 2$$

(*b*) Since Na(OH) is a strong base, (OH^-) = 0.01. Once again, the contribution of the water is negligible. It follows, therefore, that

$$\text{pOH} = -\log (.01) = 2$$

and, since

$$\text{pOH} + \text{pH} = 14$$

$$\text{pH} = 12$$

(*c*) The problem is complicated by the fact that in pure water (H^+) = 10^{-7}, a value comparable to the contribution of the HCl; hence we must consider the equilibrium,

$$\text{HOH} \rightleftarrows H^+ + OH^-$$

where

$$(H^+)(OH^-) = 10^{-14}$$

Let us summarize the situation by

	HOH $\rightleftarrows$	H^+	$+ OH^-$
concentration (no dissociation)	s	10^{-7}	0
concentration (dissociation)	$s - x$	$(10^{-7} + x)$	(x)

In the first line, we consider only the contribution of HCl; in the second line, we recognize that a certain amount of water, x, dissociates. [The approximation in part (*a*) amounts to the neglect of x with respect to 0.01. A corresponding approximation is made in part (*b*).] We now write

$$(H^+)(OH^-) = 10^{-14} = (10^{-7} + x)(x)$$

This reduces to

$$x^2 + 10^{-7}x - 10^{-14} = 0$$

From algebra, we know that the equation

$$ax^2 + bx + c = 0$$

(where a, b, and c are constants) has two solutions, viz.,

$$x = \frac{-b + \sqrt{b^2 - 4ac}}{2a}$$

and

$$x = \frac{-b - \sqrt{b^2 - 4ac}}{2a}$$

Thus we find two solutions to our expression:

$$x = \frac{-10^{-7} + \sqrt{10^{-14} + 4 \times 10^{-14}}}{2} = \frac{-10^{-7} + 2.24 \times 10^{-7}}{2} = 0.62 \times 10^{-7}$$

and also

$$x = -1.62 \times 10^{-7}$$

The value of x is the hydroxide ion concentration; it can be small, but it can never be negative. Hence it follows that only the first solution, $x = 0.62 \times 10^{-7}$, is reasonable and

$$(\mathrm{OH^-}) = 0.62 \times 10^{-7}$$
$$(\mathrm{H^+}) = (x + 10^{-7}) = 1.62 \times 10^{-7}$$

The pH can be computed as follows:

$$\mathrm{pH} = -[\log(\mathrm{H^+})] = -\log(1.62 \times 10^{-7})$$
$$= -(\log 1.62 + \log 10^{-7}) = -(0.21 - 7) = 6.79$$

Thus we find that the contribution of the water increases the H^+ above that due to the acid alone. No matter how dilute the solution, the pH of an acid is always less than 7. By the same token, no matter how dilute the solution, the pOH of a base is always less than 7.

Solutions of weak acids and weak bases. In this section, we shall restrict our discussion to those acids which yield a single proton per molecule by dissociation. More complex equilibria involving polyprotic acids, i.e., those that yield two or more protons per molecule on dissociation, will be discussed in Chapter 22.

The law of mass action is the basis of any quantitative description of solution equilibria. In an earlier discussion, we pointed out that the equilibrium expression involving *concentrations* is only approximately correct. Just as the ideal gas law expresses the behavior of real gases in the limit of low pressures, the equilibrium expression (in concentration units) expresses the behavior of real solutions in the limit of low concentrations. In both cases, deviations from the ideal can be ascribed to intermolecular or interionic effects that result in a loss of the independence of molecules or ions. In this treatment of solutions, we *assume* ideal behavior. This is done, however, with the full realization that the treatment is an approximate one valid only at low concentrations.

Example Problem 3: (*a*) What is the pH of 2*M* acetic acid (HAc)? (*b*) What is the pH in a solution that is 2*M* in HAc and, in addition, contains 0.04*M* NaAc?

(*a*) The summarizing statement is

	HAc	$\rightleftarrows$	H^+	+	Ac
CND concentrations (no dissociation)	2		0		0
CAD concentrations (after dissociation)	$2 - x$		x		x

The equilibrium expression is (Table 4)

$$\frac{(H^+)(Ac^-)}{(HAc)} = 1.8 \times 10^{-5}$$

If we substitute the indicated values, we obtain

$$\frac{(x)(x)}{(2 - x)} = 1.8 \times 10^{-5}$$

Since the equilibrium constant is small, the value of x is small. As an approximation, we neglect x with respect to 2. On this basis, we find

$$x = 6 \times 10^{-3} = (H^+)$$

Thus our approximation is valid.†

The pH can be calculated as follows:

$$\begin{aligned} pH = -\log (H^+) &= -(\log 6 \times 10^{-3}) = -(\log 6 + \log 10^{-3}) \\ &= -(0.78 - 3) = 2.22 \end{aligned}$$

(*b*) Assume that the sodium acetate is completely dissociated. This yields a solution which is 0.04*M* in sodium ion and in acetate ion. The (Na^+) does not enter into the equilibrium, but the (Ac^-) does and must be considered in the summarizing statement:

	HAc	$\rightleftarrows$	H^+	+	Ac^-
CND	2		0		0.04
CAD	$2 - x$		x		$0.04 + x$

If we substitute these values into the equilibrium expression, we obtain

$$\frac{(x)(0.04 + x)}{(2 - x)} = 1.8 \times 10^{-5}$$

Once again, it seems reasonable to neglect x not only with respect to 2, but also with respect to 0.04. Then we obtain

$$x = 9 \times 10^{-4} = H^+$$

† In this problem, we avoided the precise algebraic treatment involving the solution of a quadratic. This is justifiable since the assumption of ideal behavior introduces errors of about 10 to 20% for most solutions. Thus, if we have a factor in the equilibrium expression of the type $(a \pm x)$, we can write $a \pm x = a$, provided $x/a < 1/10$, and still not introduce errors larger than those inherent in the method. Whenever $x/a < 1/10$, we will make such an approximation. To see if an approximation is valid, try it; if it is a valid approximation, the answer will be consistent with the approximation; if the answer is not consistent with the approximation, the approximation is not valid.

a result consistent with the simplifying assumption. The pH is given by

$$pH = -(\log 9 \times 10^{-4}) = -(\log 9 + \log 10^{-4}) = -(0.95 - 4) = 3.05$$

Example Problem 4: If the initial concentration of the hydrazine hydroxide is $3M$ (*a*) in pure water, and (*b*) in $0.5M$ N_2H_5Cl, what is the pH in both cases?

(*a*)	N_2H_5OH	$N_2H_5^+$	+	OH^-
CND	3	0		0
CAD	$3 - x$	x		x

If we substitute the indicated values in the expression for the ionization constant, this yields

$$\frac{(N_2H_5^+)(OH^-)}{(N_2H_5OH)} = \frac{(x)(x)}{(3 - x)} = 1.0 \times 10^{-6}$$

When x is neglected with respect to the 3, the equation simplifies to

$$x^2 = 3 \times 10^{-6}$$

and yields

$$x = 1.7 \times 10^{-3}$$

Thus the approximation is valid. To obtain the pH, note that x stands for (OH^-) rather than (H^+); therefore it is easier to calculate the pH via the pOH. Accordingly,

$$pOH = -(\log 1.7 + \log 10^{-3}) = -(0.23 - 3) = 2.77$$

and

$$pH = 14 - pOH = 11.23$$

(*b*)	N_2H_5OH	$\rightleftarrows$	$N_2H_5^+$	+	OH^-
CND	3		0.5		0
CAD	$3 - x$		$0.5 + x$		x

When we substitute the indicated values into the equilibrium expression, we obtain

$$\frac{(x)(0.5 + x)}{(3 - x)} = 1.0 \times 10^{-6}$$

With the x neglected with respect to 3 and 0.5, this becomes

$$x = 6 \times 10^{-6} = (OH^-)$$

and leads to a pH of 8.78.

Hydrolysis. The equilibrium constant for dissociation of acetic acid,

$$HAc \rightleftarrows H^+ + Ac^-$$

is small ($K_A = 1.8 \times 10^{-5}$); consequently, the acetate ion must have a strong affinity for the hydrogen ion. In an aqueous solution, there is always a small amount of hydrogen and hydroxide ion; hence a solution of sodium acetate, in addition to containing sodium and acetate ions, also contains hydrogen and hydroxide ions. Because of the affinity of Ac^- for H^+, Ac^- reacts with some of these H^+ ions to form HAc. Thus the (H^+) is depleted in such a solution, and, in accord with Le Chatelier's principle, this decrease is accompanied by an increase in (OH^-) as indicated by the equilibrium,

$$\begin{array}{c} HOH \rightleftarrows H^+ + OH^- \\ + \\ Ac^- \\ \updownarrow \\ HAc \end{array}$$

The net effect is that a solution of sodium acetate in water is basic. This process is known as hydrolysis.

The preceding equations are not the only way of representing what occurs. We could also say that the acetate ion reacts with water proper and forms acetic acid and hydroxide ions. Thus

$$Ac^- + HOH \rightleftarrows HAc + OH^-$$

The equilibrium expression is

$$\frac{(HAc)(OH^-)}{(Ac^-)} = K_{hyd}$$

[According to conventions for the standard state, $(HOH) \equiv 1$.] We can rearrange the equilibrium expression for hydrolysis in the following way:

$$K_{hyd} = \frac{(HAc)(OH^-)}{(Ac^-)} = \frac{(OH^-)}{(Ac^-)/(HAc)} = \frac{(H^+)(OH^-)}{(H^+)(Ac^-)/(HAc)}$$

After this rearrangement, we see that the value of K_{hyd} is given by K_w/K_A.

Hydrolysis also occurs with the cation of a weak base, e.g., N_2H_5Cl. We know from the ionization constant of hydrazine hydroxide that the $N_2H_5^+$ readily combines with hydroxide ion. This tendency is so strong that $N_2H_5^+$ can pull the hydroxide away from water, form a hydrogen ion, and thereby make a solution of N_2H_5Cl acidic. In other words, we can write

$$N_2H_5^+ + HOH \rightleftarrows H^+ + N_2H_5OH$$

and

$$K_{hyd} = \frac{(H^+)(N_2H_5OH)}{(N_2H_5^+)}$$

As before, we can rearrange this equation to give

$$K_{hyd} = \frac{(H^+)(N_2H_5OH)}{(N_2H_5^+)} = \frac{(H^+)}{(N_2H_5^+)/(N_2H_5OH)}$$

$$= \frac{(H^+)(OH^-)}{(N_2H_5^+)(OH^-)/(N_2H_5OH)} = \frac{K_w}{K_B}$$

Example Problem 5: What is the hydrogen ion concentration of a $0.1M$ NaAc solution?

Sodium hydroxide is a strong base; hence the sodium ion will not abstract a hydroxide ion from water to form an acidic solution. On the other hand, acetic acid is a weak acid; hence the acetate anion will abstract a hydrogen ion from water and make the solution basic.

$$\begin{array}{llcccc} \text{Summary:} & Ac^- & + & HOH \rightleftarrows & HAc & + \; OH^- \\ & 0.1 & & & 0 & 0 \\ & 0.1 - x & & & x & x \end{array}$$

The equilibrium expression is

$$\frac{(HAc)(OH^-)}{(Ac^-)} = \frac{10 \times 10^{-15}}{1.8 \times 10^{-5}} = 5.6 \times 10^{-10}$$

If we substitute the indicated concentrations into this expression, we obtain

$$\frac{x^2}{0.1 - x} = 5.6 \times 10^{-10}$$

Assume that the value of x is so small that we can neglect it with respect to 0.1. Then the expression above simplifies to

$$\frac{x^2}{0.1} = 5.6 \times 10^{-10}$$

and, if we solve for x, we find

$$x = 7.5 \times 10^{-6}$$

a result consistent with the approximation. This is *not* the answer to the problem. If we refer to the summary, we find that x represents (OH^-) and the problem calls for (H^+). The remainder of the solution is

$$\begin{aligned} (H^+)(OH^-) &= 10^{-14} \\ (H^+)(7.5 \times 10^{-6}) &= 10^{-14} \\ (H^+) &= 1.3 \times 10^{-9} \end{aligned}$$

Example Problem 6: What is the hydrogen ion concentration in a $0.30M$ N_2H_5Cl solution?

$$\begin{array}{llcccc} \text{Summary:} & N_2H_5^+ & + & HOH \rightleftarrows & N_2H_5OH & + \; H^+ \\ & 0.30 & & & 0 & 0 \\ & 0.30 - x & & & x & x \end{array}$$

$$K_{\text{hyd}} = \frac{K_w}{K_B} = \frac{1.0 \times 10^{-14}}{1.0 \times 10^{-6}} = 1.0 \times 10^{-8}$$

Now substitute the indicated concentrations into the expression above. This yields,

$$\frac{x^2}{0.30 - x} = 1.0 \times 10^{-8}$$

and with the approximation, $0.30 - x \approx 0.30$, we obtain

$$(x) = 5.5 \times 10^{-5} = (H^+)$$

Table 6 Titration of 1.0*M* HCl with 1.0*M* NaOH

Ml of Added Base	Total Vol. of Sol'n	Equiv. of Base Added	Equiv. of Base Left	Equiv. of Acid Left	H^+	pH
0	50	—	—	0.050	1	0
10	60	0.0100	—	0.040	0.66	0.18
30	80	0.0300	—	0.020	0.25	0.60
45	95	0.0450	—	0.005	0.05	1.3
49	99	0.0490	—	0.001	0.01	2.0
49.9	99.9	0.0499	—	0.0001	0.001	3.0
49.95	99.95	0.04995	—	0.00005	0.0005	3.3
49.99	99.99	0.04999	—	0.00001	0.0001	4.0
50.00	100.00	0.05000	—	—	10^{-7}	7
50.01	100.01	0.05001	0.00001	—	10^{-10}	10
50.05	100.05	0.05005	0.00005	—	5×10^{-11}	10.7

Titration. Many studies of acids and bases involve the controlled addition of a measured amount of base (or acid) to a measured amount of acid (or base). This process, known as *titration,* is an important analytic procedure. Therefore let us consider a specific case in detail.

Suppose we add successively larger amounts of 1*N* NaOH to 50 ml of 1*N* HCl. We can calculate the pH of the resulting solution for various amounts of base added provided we assume that the volume of the resulting solution is the sum of the volumes of the combining acid and base. In other words, we assume that the volume of the solution resulting from the addition of 10 ml of NaOH to 50 ml of HCl is 60 ml. At the start, the 50 ml of solution contain 0.050 equivalent of HCl. This yields a hydrogen ion concentration of 1, corresponding to a pH of zero. After the addition of 10 ml of base, the volume of the solution increases to 60 ml. The 10 ml of added base contains 0.01 equivalent of base that reacts with 0.01 equivalent of HCl and leaves only 0.040 equivalent of HCl remaining in the 60 ml of solution. The concentration of hydrogen ion is 0.66, corresponding to a pH of 0.18. The calculated pH for the successive additions of base are listed in Table 6.

According to Table 6, we can add 99.8% (49.9 ml) of the base required to neutralize the HCl and change the pH only 3 units. When we are within 0.01 ml of having neutralized the solution, the pH is still only 4, whereas for a neutral solution it is 7. At this stage, the pH changes very rapidly; the addition of 0.02 ml of NaOH changes the pH by 6 units. If we plot the pH versus milliliters of NaOH added, we get the curve indicated in Fig. 3.

If, instead of starting with a strong acid like HCl, we start with a weak acid like acetic acid, then the initial pH is 2.4 rather than zero. In addition, when we neutralize the acetic acid with sodium hydroxide, the sodium acetate solution formed is basic because of the hydrolysis of the acetate ion. The curve for pH versus the milliliters of base added is also shown in Fig. 3.

Indicators. An indicator is a weak acid (or a weak base) that in the undissociated form, HIn, has one color and in the dissociated form, In^-, has another color. An example of this is litmus. The undissociated form of this acid is red; the dissociated form is blue. Thus we write for the equilibrium

$$\underset{\substack{\text{Red}\\ \text{A}}}{HIn} \rightleftharpoons H^+ + \underset{\substack{\text{Blue}\\ \text{B}}}{In^-}$$

It is apparent from Le Chatelier's principle that, if we start with litmus in a basic solution, the litmus is present principally as the anion and the solution is blue. If we make this solution acidic, the anion reacts with the proton to form the undissociated acid, and the solution turns red. Qualitatively, this is the typical action of all indicators.

The best indicators are so intensely colored that only a drop is needed to color the whole solution. Since the amount of indicator present is small, very little hydroxide ion is required to change litmus from red to blue. Thus, in a titration, if we add a drop of base to the solution and a color change occurs, this indicates that all of the acid reacted. The amount of added base (or acid) required to cause this color change is but a small fraction of the total used for the titration; this can be corrected for by titrating the indicator alone, by running a *blank*.

Let us derive the general relation that describes the action of an indicator. If K is the equilibrium constant for the indicator,

$$\frac{(H^+)(In^-)}{(HIn)} = K_{eq.}$$

or

$$(H^+) = K_{eq.} \frac{(HIn)}{(In^-)}$$

The indicator will be midway between a color change when the quantity $(In^-)/(HIn)$ is 1, i.e., when there are equal amounts of the A and B. When this quantity is 10, we will have 10 times as much B as A, and the color of the solution will be that of the B form. On the other hand, when this quantity is $\frac{1}{10}$, the solution has the color characteristic of the A form. Let us rearrange this equation as follows:

$$\log (H^+) = \log K_{eq.} + \log \frac{HIn}{In^-}$$

or

$$pH = -\log K_{eq.} - \log \frac{HIn}{In^-}$$

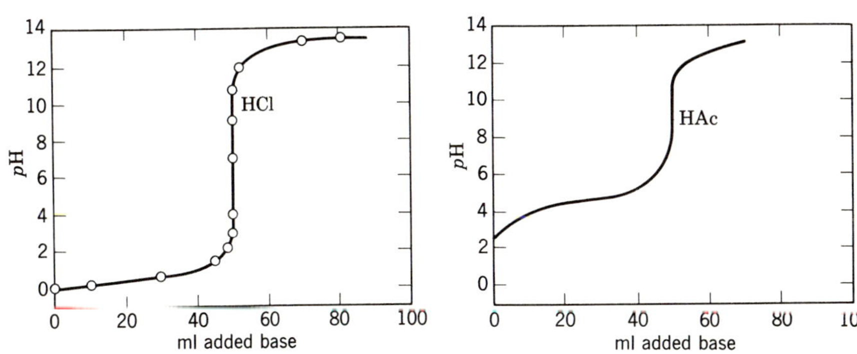

Fig. 3. Titration curve.

At the midpoint of the indicator (when the solution contains equal amounts of In^- and HIn), the pH is

$$pH = -\log K_{eq.}$$

On the other hand, if we add enough acid that there is ten times as much HIn as In^-, the pH is

$$pH = -\log K_{eq.} - 1$$

whereas, if we add enough OH^- that there is ten times as much In^- as HIn, the pH is

$$pH = -\log K_{eq.} + 1$$

Thus a solution such that the indicator is at the midpoint has a pH equal to $-\log K$. In order to change the color to that characteristic of A, we lower the pH one unit; to change the color to that characteristic of B, we increase the pH one unit.

A wide variety of indicators are available. Several of these together with the pH at their midpoint are listed in Table 7. Indicators in the pH range 6 to 8 are used in the titration of HCl with NaOH in order to tell when we have added just enough base to neutralize the HCl. Similarly, indicators with a pH range 9 to 11 are used in the titration of HAc. Indicators for a titration are always selected so that the midpoint coincides with the steepest portion of the titration curve. Then a color change is usually brought about by the addition of one drop of titrating agent, and the amount of base (or acid) needed for neutralization is determined within one drop.

The titration curve for HAc is gradual on the acid side, but still sufficiently sharp on the basic side to permit the use of an indicator. The gradual part of the titration curve on the acid side results because HAc is a weak acid. If the titration were carried out with a weak base, like NH_4OH, instead of NaOH, the titration curve at the end point would not show the large change in pH after the addition of one drop (0.05 ml) of base. When this situation prevails, indicators cannot locate the end point with a high degree of accuracy; then titration as an analytical procedure becomes unsatisfactory.

Buffers. Solutions containing a weak acid and the corresponding anion, or a weak base and the corresponding cation, are called *buffers* because they protect

Table 7 Indicators

Name	Change (Acid to Base)	pH at Midpoint
Thymol Blue	red ⟷ yellow	2.0
2,4 Dinitrophenol	colorless ⟷ yellow	3.2
Bromcresol Green	yellow ⟷ blue	4.8
Methyl Red	red ⟷ yellow	5.3
Phenol Red	yellow ⟷ red	7.2
Phenolphthalein	colorless ⟷ red	9.0
Alizarin Yellow	yellow ⟷ violet	11.0
Nitramine	colorless ⟷ orange	12.0

the solution from large changes in pH. Consider, for example, two solutions, A and B. Let A contain 10^{-5} mole of HCl per liter and B contain 1.8000 moles of sodium acetate and 1.0000 mole of acetic acid per liter. The pH of the HCl solution and that of the buffer (sodium acetate-acetic acid) is 5.0. Now, suppose we add 0.11 ml (about 2 drops) of $1M$ NaOH to 1 liter of each solution. In solution A, the 1.1×10^{-4} mole of added sodium hydroxide reacts with 0.1×10^{-4} mole of HCl to form 0.1×10^{-4} mole of NaCl with 1.0×10^{-4} mole of NaOH left over. This yields a solution with a pH of 10.0, a change of 5.0 pH units. In solution B, reaction with the sodium hydroxide leaves the solution with 1.8001 moles of sodium acetate and 1.9999 moles of acetic acid. Since in solution B the ratio $(HAc)/(Ac^-)$ is virtually unchanged and the (H^+) is determined by the equation $(H^+) = (HAc)/(Ac^-) \times 1.8 \times 10^{-5}$ ($K_A = 1.8 \times 10^{-5}$), the pH is still 5.0. Even if we added 100.0 ml of $1M$ NaOH, the pH would only change from 5.0 to 5.1. Thus buffers stabilize the pH of a solution.

OTHER SOLUTION EQUILIBRIA

Solubility. When a solid with the formula MX_2 dissolves in water, we can (in the simplest cases) describe what occurs by the equations

$$MX_2 \underset{K_1}{\rightleftharpoons} MX_{2\text{ aq.}} \underset{K_2}{\rightleftharpoons} M^{+2} + 2X^-$$

The equilibrium expressions are

$$(MX_{2\text{ aq.}}) = K_1 \quad \text{and} \quad \frac{(M^{+2})(X^-)^2}{(MX_{2\text{ aq.}})} = K_2$$

These can be combined to yield the solubility product:

$$(M^{+2})(X^-)^2 = K_1K_2 = K_{sp}$$

Solubility can occur with varying degrees of dissociation. There are cases where solubility occurs with no dissociation, or some dissociation, or nearly 100% dissociation. The last type of behavior is commonly found with ionic crystals. Then all the material that dissolves appears as ions. This is the type of solubility we discuss below; the more complicated types will be discussed in a later chapter.

Example Problem 7: The solubility of CaF_2 in water at 20°C is 16 mg per liter of solution. Calculate the K_{sp}.

Before we calculate K_{sp}, we must first express the solubility in moles per liter. This can be done as follows:

$$\frac{16 \times 10^{-3}\text{ gm } CaF_2}{\text{liter of sol'n}} \times \frac{1\text{ mole } CaF_2}{78\text{ gm } CaF_2} = \frac{2.0 \times 10^{-4}\text{ mole } CaF_2}{\text{liter of sol'n}}$$

Each mole of calcium fluoride that dissolves yields 1 mole of calcium ion and 2 moles of fluoride ion:

	CaF_{2s}	$\rightleftarrows$	Ca^{+2}	+	$2F^-$
Start	S(olid)		0		0
Eq.	$S - 2.0 \times 10^{-4}$		2.0×10^{-4}		4.0×10^{-4}

The remaining steps are straightforward.

$$\begin{aligned} K_{sp} &= (Ca^{+2})(F^-)^2 \\ &= (2.0 \times 10^{-4})(4.0 \times 10^{-4})^2 \\ &= 32 \times 10^{-12} \end{aligned}$$

Example Problem 8: What is the solubility, in moles per liter, of $Ag_2(C_2O_4)$ if the value of the K_{sp} is 1.3×10^{-11}?

Summary:	$Ag_2(C_2O_4)_s$	$\rightleftarrows$	$2Ag^+$	+	$C_2O_4^{-2}$
	S		0		0
	$S - x$		$2x$		x

The K_{sp} expression is

$$K_{sp} = (Ag^+)^2(C_2O_4^{-2}) = 1.3 \times 10^{-11}$$

Substitution for (Ag^+) and $(C_2O_4^{-2})$ in terms of x yields

$$\begin{aligned} (2x)^2(x) &= 1.3 \times 10^{-11} \\ 4x^3 &= 1.3 \times 10^{-11} \\ x^3 &= 0.32 \times 10^{-11} = 3.2 \times 10^{-12} \end{aligned}$$

In solving for x, the number of moles of oxalate ion per liter, we find

$$x = \sqrt[3]{3.2} \times \sqrt[3]{10^{-12}} = \sqrt[3]{3.2} \times 10^{-4}$$

The easiest way to extract the cube root is through the use of logarithms; that is,

$$\log (3.2)^{1/3} = \tfrac{1}{3} \log 3.2 = \tfrac{1}{3}(0.505) = 0.168$$

and therefore

$$(3.2)^{1/3} = 1.5$$

Thus the solubility of silver oxalate is 1.5×10^{-4} mole per liter.

In the problems considered so far, the crystalline solid was dissolving in pure water. Often we are interested in the solubility of a solid in a solution containing dissolved salts. If this solution contains one or more ions common to the solid dissolving, the equilibrium solubility is affected.

Example Problem 9: Assume that the K_{sp} of $Sc(OH)_3$ is 1.1×10^{-15} (it isn't). What is the solubility (*a*) in pure water, (*b*) in $0.1M$ $Sc_2(SO_4)_3$, and (*c*) in $0.1M$ NaOH?

(*a*) Summary:	$Sc(OH)_3(s)$	$\rightleftarrows$	Sc^{+3}	+	$3(OH)^-$
	S		0		0
	$S - x$		x		$3x$

The expression for K_{sp} is

$$(Sc^{+3})(OH^-)^3 = 1.1 \times 10^{-15}$$

If we substitute the indicated values, we obtain

$$(x)(3x)^3 = 11 \times 10^{-16}$$
$$27x^4 = 11 \times 10^{-16}$$
$$x^4 = 0.41 \times 10^{-16}$$

and

$$x = \sqrt[4]{0.41} \times \sqrt[4]{10^{-16}} = \sqrt[4]{0.41} \times 10^{-4}$$

Once again, to extract the root we resort to logarithms. Thus

$$\log (0.41)^{1/4} = \tfrac{1}{4} \log 0.41 = \tfrac{1}{4}(-0.387) = -0.097, = 9.903 - 10$$

or

$$(0.41)^{1/4} = \text{antilog } (9.903 - 10) = 0.80$$

This yields a solubility of 8.0×10^{-5} mole/liter.

(*b*) Let us assume that in solution scandium sulfate dissociates completely:

$$Sc_2(SO_4)_3 \rightarrow 2Sc^{+3} + 3SO_4^{-2}$$

Thus, if we have a solution that is 0.1*M* in scandium sulfate, we will have a solution that is 0.2*M* in scandium ion.

Summary:	$Sc(OH)_3(s)$	$\rightleftarrows$	Sc^{+3}	+	$3OH^-$
	S		0.2		0
	$S - x$		$0.2 + x$		$3x$

The expression for the K_{sp} is

$$(Sc^{+3})(OH^-)^3 = (0.2 + x)(3x)^3 = 1.1 \times 10^{-15}$$

The product of these two terms is very small; hence we assume that x is so much less than 0.2 that we can neglect x with respect to 0.2. Then the expression becomes

$$(0.2)(27x^3) = 1.1 \times 10^{-15}$$

or

$$x = 6 \times 10^{-6}$$

The solubility is 6×10^{-6} mole/liter.

(*c*) Sodium hydroxide is completely dissociated into sodium ions and hydroxide ions in solution; hence we now have scandium hydroxide dissolving in a solution that already contains the hydroxide ion at a concentration of 0.1*M*.

Summary:	$Sc(OH)_3(s)$	$\rightleftarrows$	Sc^{+3}	+	$3OH^-$
	S		0		0.1
	S		x		$0.1 + 3x$

and, if we substitute the indicated values into the expression for the K_{sp}, we obtain

$$(Sc^{+3})(OH^-)^3 = (x)(0.1 + 3x)^3 = 1.1 \times 10^{-15}$$

Once again, it seems quite reasonable to neglect $3x$ with respect to 0.1. If we do, we get

$$x(0.1)^3 = 1.1 \times 10^{-15}$$
$$x = 1.1 \times 10^{-12}$$

The solubility is 1.1×10^{-12} mole/liter.

Complex ions. All ions are hydrated in solution. In most cases, these hydrated ions constitute definite compounds with water. For example, the silver ion in solution is believed to be $Ag(H_2O)_2^+$, whereas the cobalt ion is believed to be $Co(H_2O)_6^{+2}$. Occasionally, ions (or molecules) present in the solution have a stronger affinity for the cation than the water itself. Such is the case for NH_3 in the presence of Ag^+ ion, and the following reactions occur:

$$Ag(H_2O)_2^+ + NH_3 \rightleftarrows Ag(NH_3)(H_2O)^+ + H_2O$$

and

$$Ag(NH_3)(H_2O)^+ + NH_3 \rightleftarrows Ag(NH_3)_2^+ + H_2O$$

Ions such as $Ag(NH_3)_2^+$ are referred to as *complex ions.*

Two factors systematize the chemistry of complex ions. First, the number of groups, water or otherwise, that pack around a given cation is usually fixed. This number is called the *coordination number.* The common coordination numbers for several cations are given in Table 8. Second, not all groups can displace the water from the hydration shell. Some of the groups commonly found in complex ions are: NH_3, F^-, Cl^-, CO, $NH_2—CH_2—CH_2—NH_2$, S^{-2}, OH^-, NCS^-.

When a molecule like $NH_2—CH_2—CH_2—NH_2$ (ethylenediamine or simply *en*) forms a complex, both NH_2 groups approach the central ion and take up two of the available coordination positions. Thus, whereas Ni^{+2} in aqueous solutions is thought to be $Ni(H_2O)_6^{+2}$, the *en* complex is $Ni(en)_3^{+2}$. Such complexing agents are said to be *polydentate* (many-teeth) or *chelates.*

The equations depicting what occurs in solution containing Ag^+ and NH_3 are

$$Ag^+ + NH_3 \rightleftarrows Ag(NH_3)^+ \qquad K_1$$
$$Ag(NH_3^+) + NH_3 \rightleftarrows Ag(NH_3)_2^+ \qquad K_2$$

(We have ignored the water of hydration.) These equilibria are represented by the equations

$$K_1 = \frac{(Ag(NH_3)^+)}{(Ag^+)(NH_3)} \quad \text{and} \quad K_2 = \frac{(Ag(NH_3)_2^+)}{(Ag(NH_3)^+)(NH_3)}$$

Just as we combined the two equations for the solubility equilibria, we can combine these two equations to yield:

$$K_1K_2 = \frac{(Ag(NH_3)_2^+)}{(NH_3)^2(Ag^+)}$$

This is generally written

$$\frac{1}{K_1K_2} = K_d = \frac{(NH_3)^2(Ag^+)}{(Ag(NH_3)_2^+)}$$

where K_d is called the dissociation or instability constant.

It must be realized that the situation is the same here as in solubility equilibria. If we use the last equation, and assume that all of the complex silver is in the form of an ammonia complex, we will be making a mistake unless K_1 is much much

Table 8 Coordination Number of Common Cations

2	4	6
Cu^+	Cu^{+2}	Ni^{+2}
Ag^+	Zn^{+2}	Fe^{+2}
Au^+	Cd^{+2}	Co^{+2}
Hg^{+2}	Hg^{+2}	Mn^{+2}
	Ni^{+2}	Fe^{+3}
	Au^{+3}	Co^{+3}
		Al^{+3}

smaller than K_2. By this we mean that the silver complex is really *two* complex ions: one with one ammonia molecule bound to the silver ion, and the other with two ammonia molecules bound to the silver ion. Thus, in order to account for all of the silver in the form of a complex ion, we must consider both $Ag(NH_3)^+$ and $Ag(NH_3)_2^+$. Only when K_1 is small compared to K_2 and, hence, $Ag(NH_3)^+$ is negligible compared to $Ag(NH_3)_2^+$, can we assume that the complexed silver ion is all in the latter form. In the case of the silver ammonia complex, K_1 is sufficiently small, so this is a good approximation. But this does not always happen. For example, in a solution containing dissolved NH_3, cupric ion forms the following complexes: $Cu(NH_3)(H_2O)_3^{+2}$, $Cu(NH_3)_2(H_2O)_2^{+2}$, $Cu(NH_3)_3(H_2O)^{+2}$, and $Cu(NH_3)_4^{+2}$. If we were to calculate how much cupric ion would complex and assume that the only complex ion formed was $Cu(NH_3)_4^{+2}$, we would make a grave error in our calculations. Significant amounts of all four of the indicated complexes are formed. When each step of a multiple dissociation must be considered, quantitative calculations are complicated. This situation will be considered later. At this time, we wish to consider the simpler case of complex ions such as $Ag(NH_3)_2^+$.

Example Problem 10: The dissociation constant of $Ag(NH_3)_2^+$ is 6×10^{-8}. How much NH_3 must be added to a liter of solution containing 0.01 mole of $AgNO_3$ and 0.01 mole of NaCl to just prevent the precipitation of AgCl ($K_{sp} = 1.6 \times 10^{-10}$)?

When the AgCl just completely dissolves, the (Cl^-) is 0.01 and the (Ag^+) is some low value. If we increase Ag^+ just a little, AgCl will precipitate and the equilibrium

$$AgCl(s) \rightleftarrows Ag^+ + Cl^-$$

will be established. This critical value of the (Ag^+) is given by

$$(Ag^+)_{crit}(0.01) = K_{sp} = 1.6 \times 10^{-10}$$

or

$$(Ag^+)_{crit} = 1.6 \times 10^{-8}$$

This determines the concentration of silver ion in the equilibrium.

$$\begin{array}{ccc} Ag(NH_3)_2^+ \rightleftarrows & Ag^+ & + 2NH_3 \\ (0.01 - 1.6 \times 10^{-8}) & (1.6 \times 10^{-8}) & (NH_3) \end{array}$$

$$\frac{(NH_3)^2(1.6 \times 10^{-8})}{(0.01 - 1.6 \times 10^{-8})} = 6 \times 10^{-8}$$

With the obvious approximation, we find

$$(NH_3)^2 = \frac{6 \times 10^{-10}}{1.6 \times 10^{-8}} = 3.7 \times 10^{-2}$$

or

$$(NH_3) = 0.19$$

This is the final concentration of NH_3, but to allow for the formation of 0.01 mole of $Ag(NH_3)_2^+$ the initial concentration of NH_3 would have to be 0.21 molar.

PROBLEMS

1. Given that the equilibrium constant at 25°C for

$$H_2O \rightarrow H^+ + OH^-$$

is 1.0×10^{-14}, calculate the standard free energy change for the reaction.

2. In the reaction above, the $\Delta S°$ at 298°K is -19.0 cal/°/mole. Calculate the $\Delta H°$.

3. Near the boiling point, the pH of a neutral solution is roughly 5. Suppose $\Delta S°$ does not vary with temperature but $\Delta H°$ does. What is the heat of ionization of water, $\Delta H°$, at the boiling point? See Problem 2.

4. The pK of H_3PO_2 is 2 to 3. On the basis of the known structure and coordination number in other phosphorus oxyacids, guess the structure of H_3PO_2. List the likely sodium salts.

5. What are the anhydrides of $HSb(OH)_6$, H_5IO_6, and H_6TeO_6?

Ans. Sb_2O_5, I_2O_7, and TeO_3.

6. In the reaction of water with P_4O_{10}, a whole series of acids are formed depending on how many water molecules react. Sketch the structure of P_4O_{10}. Assume two water molecules react with the P_4O_{10} molecule by breaking two P—O—P bonds as follows:

$$—P—O—P— + H—O—H \rightarrow —P—O—H + H—O—P—$$

If the two P—O—P bonds are as far apart as possible, what is the structure of the resulting molecule? What is the formula? What is the simplest formula? Suppose the two P—O—P bonds that react with the two water molecules are adjacent; what is the structure of the resulting molecule?

7. A liter of solution contains 20 gm of $M(OH)_3$, a strong electrolyte. The atomic weight of M is 293. What is the normality of the solution? A solution of H_2SO_4 is prepared and 20.00 ml of H_2SO_4 reacts with 30.00 ml of this solution. If the density of the H_2SO_4 solution is 1.000, what is the per cent of H_2SO_4?

8. A solution of 1.000*N* $M(OH)_3$, a strong electrolyte, is used to titrate in turn approximately 1*N* HCl, 1*N* HAc, and 1*N* HCN. On the basis of the data in Tables 4 and 7, indicate the most suitable indicators in each case.

Ans. Phenol Red, Phenolphthalein, Nitramine.

9. A 50.00-ml sample of a 1.000M solution of $M(OH)_3$, a strong electrolyte, is titrated with 0.5M H_2SO_4. How much H_2SO_4 is needed? *Ans.* 150 ml H_2SO_4.

10. A liter of solution contains 1 mole of $CuSO_4$ and 0.5 mole of NH_3. All four copper ammonia complexes are formed. List all species present in the solution. Write down the equations needed to solve for the concentration of each species.

11. Calculate the solubility (in moles per liter) of AgCl ($K_{sp} = 1.6 \times 10^{-10}$) in the following solutions: (*a*) distilled water; (*b*) 1.0M KNO_3; (*c*) 1.0M $LaCl_3$; (*d*) 0.1M Ag_2SO_4; (*e*) 1.5M KCl.

Ans. (*a*) 1.3×10^{-5} mole/liter; (*c*) 5×10^{-11} mole/liter.

12. Calculate the solubility of $Sc(OH)_3$ (assume $K_{sp} = 1.1 \times 10^{-15}$) in the following solutions: (*a*) 1.0M $La_2(SO_4)_3$; (*b*) 0.05M $Ca(OH)_2$; (*c*) 1.0M K_2SO_4; (*d*) 0.4M $Sc_2(SO_4)_3$.

13. Calculate the pH of the following solutions: (*a*) 0.1M HCl; (*b*) 0.1M HCN + 0.3M NaCN; (*c*) 0.6M NaCN; (*d*) 0.2M NH_4OH; (*e*) 0.2M NH_4OH + 0.07M NH_4Cl; (*f*) 0.14M NH_4Cl. (Cf. Tables 4 and 5 for K values.)

Ans. (*a*) pH = 1; (*b*) pH = 9.9.

14. What is the pH of the following solutions? (Assume that the final volumes are the sum of the two separate volumes.)

(*a*) 10 ml of 1M NaOH + 40 ml of 1M HAc
(*b*) 34 ml of 0.05M NaOH + 17 ml of 1M HAc
(*c*) 32 ml of 0.3M NaOH + 16 ml 1.0M HAc
(*d*) 32 ml of 0.3M NaOH + 32 ml of 0.5M HCl
(*e*) 50 ml of 0.3M NaOH + 15 ml 0.7M HAc

Ans. (*a*) 4.27; (*d*) 1.00.

15. Given the following pH, calculate the concentration of the solutions: (*a*) NaCN, pH 11.3; (*b*) HCl, pH 3; (*c*) NH_4Cl, pH 5.6; (*d*) HCN, pH 4.8. *Ans.* (*a*) 0.16 molar.

16. A 20.15-ml sample of 0.1032N NaOH is required to neutralize 32.43 ml of vinegar (aqueous acetic acid). What is the normality of the vinegar? If the density of the vinegar is 1.031, what is the per cent HAc (CH_3CO_2H)?

★ **17.** The addition of 0.20 ml of 0.010N indicator to 50 ml of water is sufficient to impart a definite color to the solution. If this indicator is used to detect the end point in the titration of 50 ml of 0.1N HCl by 0.1N NaOH, how much base must be added to bring about a color change? The ionization constant of the indicator is 4×10^{-7}.

18. The dissociation constant for $Ag(S_2O_3)_2^{-3}$ is 5.7×10^{-14}. Calculate the solubility of (*a*) AgCl ($K_{sp} = 1.6 \times 10^{-10}$), (*b*) AgBr ($K_{sp} = 7.7 \times 10^{-13}$), and (*c*) AgI ($K_{sp} = 1.5 \times 10^{-16}$) if each pure solid is immersed in three separate beakers each containing 1.6M $Na_2S_2O_3$.

★ **19.** Calculate how much AgCl and AgBr, both in the same beaker, will dissolve in

2.0 ml of solution containing 0.10 ml of 6M NH_4OH and 0.25 ml of 0.1M $AgNO_3$. See Problem 19 for K_{sp} data. The dissociation constant of $Ag(NH_3)_2^+$ is 6×10^{-8}.

20. Calculate the pH of the resulting solutions when 20 ml of 0.1N NaOH is added to 20 ml of 0.1N HCN.

21. Calculate the change in pH when 0.5 mole of NaOH is added to a liter of solution containing 1 mole of HAc and 1 mole of sodium acetate. *Ans.* 0.5.

16 Electrochemistry

Many times in the earlier chapters of this book we have discussed the properties of ions. These positively and negatively charged atoms and molecules are found in all three states of matter. They are the building blocks from which ionic crystal lattices like that of sodium chloride are made; and they play the major role in electrolytic conductivity in ionic solutions, like that of sodium chloride in water.

When we studied electrolytic conduction in Chapter 11, we discussed qualitatively how ions in solution carry an electric current. In this chapter, we shall examine in a quantitative way how this electrolytic process converts ions to neutral atoms and brings about other chemical changes; we shall also examine the opposite process by which chemical changes produce electric currents, as in a battery. We shall study the different properties associated with these processes such as the electromotive force; and we shall see that electrochemical changes are governed by the same thermodynamic factors that we have observed before—*energy* and *probability*.

ELECTROLYSIS

The circuit. As a first example of electrolysis, consider what happens when an electric current passes through a vessel or "cell" filled with molten sodium chloride, in an arrangement similar to that shown in Fig. 1*a*. Here we see the apparatus, labeled according to the standard conventions of electrical engineering. First, there is the *dynamo* which, upon rotation, provides the *force* that moves the electric current through the various media. The current emerges from the positive pole of the dynamo and passes along the wire. It next flows through an instrument called an ammeter; this is designed to measure the quantity of electricity passing through the wire per second. The standard unit quantity of electricity is the coulomb; just as a water pipe might have one cubic foot per second of water passing through it, a wire might pass one coulomb per second of electricity; under these conditions we say that the *current* of electricity is *1 ampere,* and the indicating needle on the ammeter points to the reading of *1* on the ampere scale.

The electric current then leaves the ammeter and enters the cell through the left-hand *electrode,* a metal rod that sticks down into the molten sodium chloride shown as a liquid inside the cell. This left-hand electrode is known as the *anode* because it is the place where the current passes *into* the cell. The current then passes through the cell, conducted by the ions, and goes out through the right-hand metal bar or electrode that is known as the *cathode* (*cat-* is the Greek prefix meaning *out* just as the prefix *an-* means *in*).

Before the electron had even been discovered, the decision was made to label the different parts of this circuit in the manner shown here. This choice for the direction of the current was unfortunate, because the electrons actually move in the opposite direction. The discovery of the electron with its *negative* charge made it apparent that electrons move opposite to the direction that had been used for so many years when an electric current was regarded as a positively charged fluid. Since we are interested in the way in which electrons move in and out of ions and atoms, we shall henceforth show the direction of motion of the electrons in our diagrams.

The resemblance of electrons flowing through a wire to water flowing through a pipe is surprisingly close. It is the essence of metals that the outer electrons of their atoms are loosely held. When an electrical force field pushes electrons in at one end of a wire, these in turn push neighboring electrons and the push is communicated down the wire. Thus, if we pump four electrons in one end of the wire, four are pushed out the other end; by way of analogy, if we pump four pints of water into a filled garden hose, four pints are pushed out the other end. Electrons behave like a fluid flowing around the circuit.

In the spirit of this analogy, we regard the dynamo as an electron pump; then the electrical field of force that makes the electrons move resembles the pressure originating in a real pump that forces the water through the pipe. It is in these terms that we have represented the electrolytic cell in Fig. 1*b*. The electrons are driven through the wire and into the solution at the cathode by the electrical "pressure" originating at the dynamo.

Cathode reaction. What happens to the electrons when they reach the surface of the cathode? This surface is in contact with molten sodium chloride, a liquid made up of a mixture of sodium ions with positive charges and chlorine ions with

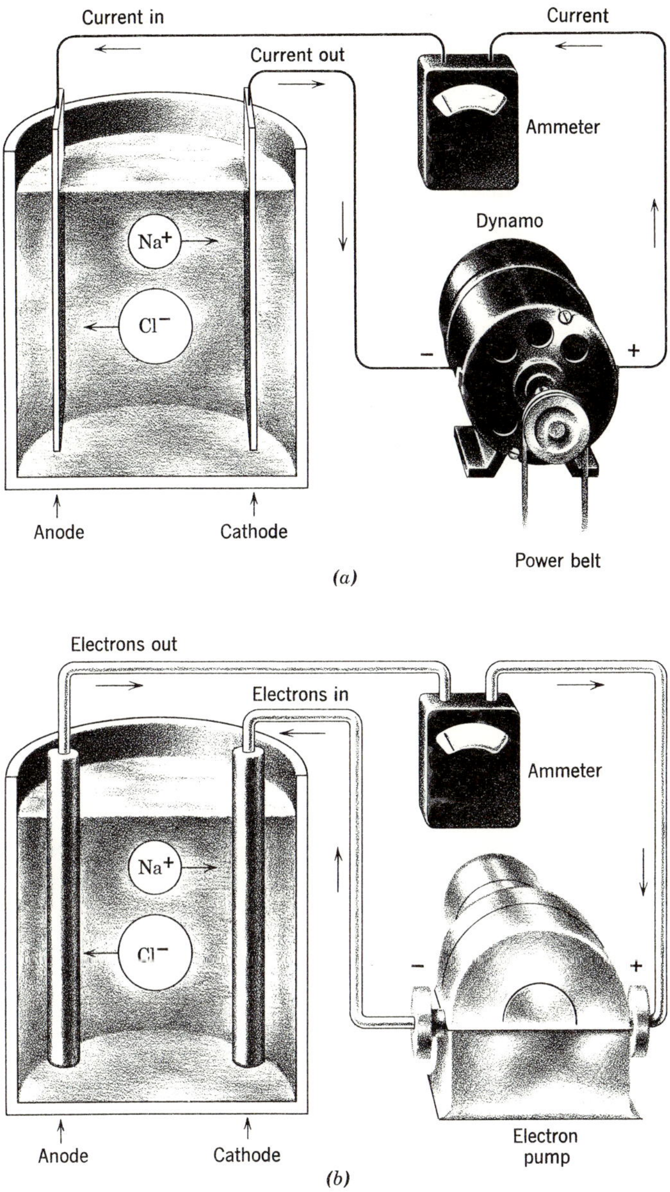

Fig. 1. Electrolysis of molten sodium chloride showing: (*a*) movement of *electric current;* (*b*) movement of *electrons.*

negative charges. Now, although this ionic state of NaCl is normally the most stable both in the crystalline state and in solution, the stability is lost when sodium ions come in contact with the surface of the cathode metal where the electrons are under pressure to escape. Here the energy conditions are sufficiently different that these "current" electrons come out and occupy the available place in the outer *s* shell of the Na, thus changing the ion back into a neutral sodium atom. This process can be written as a chemical reaction:

$$Na^+ + e^- \rightarrow Na(s)$$

The dynamo, acting as an electron pump, not only produces a pressure on one side, forcing the electrons out at the cathode, but at the same time "sucks" on the electrons in the other wire so that a force also pulls electrons into the metal at the anode. At the surface of the anode there are chloride ions each with an extra electron in the external shell held there by the tendency to form the stable shell of eight. However, when the suction at the metal surface is great enough, this extra electron can be pulled into the metal surface, leaving a neutral chlorine atom. This action can also be expressed as a chemical equation:

$$Cl^- \rightarrow e^- + Cl$$

Now recall briefly the nature of neutral sodium and chlorine atoms as shown in Fig. 2. Sodium atoms "bond" together, producing sodium metal with the loosely bound single *s* electrons moving to conduct current when force is applied. A very different situation is produced, however, when neutral chlorine atoms get together. These atoms join together in pairs, sharing two electrons that form a single covalent bond between the two Cl's; this results in a molecule of chlorine gas, Cl_2.

$$2Cl \rightarrow Cl_2(g)$$

If these three equations are added together and the symbols for electrons that appear on both sides are cancelled, the equation that is left expresses the over-all action of this electrolysis:

$$2Na^+ + 2Cl^- \rightarrow 2Na(s) + Cl_2(g)$$

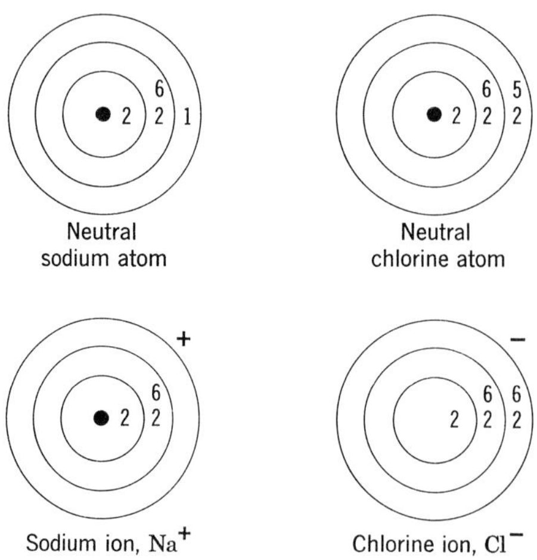

Fig. 2.

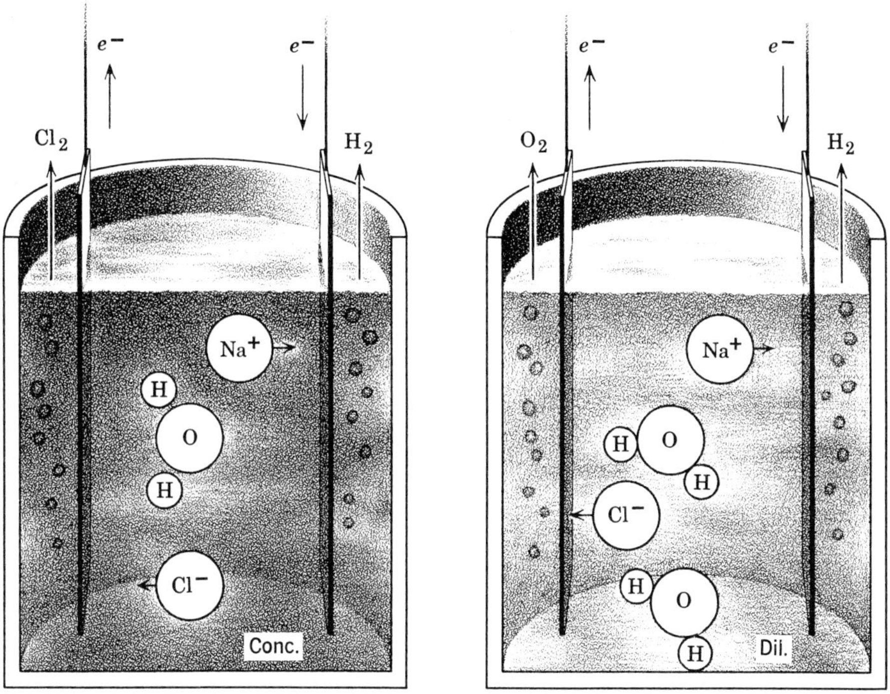

Fig. 3. Electrolysis of an aqueous solution of sodium chloride.

With electrons supplied by the cathode to form sodium metal and electrons given up at the anode as the result of the formation of chlorine gas, the circuit is complete and the dynamo can run, pumping electrons into the cathode and pulling electrons out from the anode continuously until all the original NaCl is converted to Na metal and Cl_2 gas.

Electrolysis in aqueous solution. Consider next what happens when a cell is filled with an aqueous solution of sodium chloride, with sodium and chlorine ions present in the solvent. This is shown in Fig. 3, the same apparatus with a cathode and an anode; again, electrons are pulled out of the anode and pushed into the cathode. But now there are present in the solution not only Na^+ and Cl^- ions but also neutral water molecules, H_2O, and a few H^+ and OH^- ions. If an H^+ receives an electron, this results in the formation of a neutral hydrogen atom, H, according to the reaction

$$H^+ + e^- \rightarrow H$$

The neutral hydrogen atom is reactive chemically because of the single electron on its exterior; and, as soon as such a hydrogen atom meets another hydrogen atom, the formation of gaseous hydrogen takes place according to the equation

$$H + H \rightarrow H_2$$

The resulting electron-pair bond between the two atoms yields the stable molecule H_2. Originally, the solution contained an equal number of H^+ and OH^- ions; if this reaction occurs, the number of H^+ ions becomes less than the number of OH^- ions and the solution in the neighborhood of the cathode becomes basic.

What happens if Na^+ gets an electron? There is the momentary formation of a neutral sodium atom; this atom is in contact with neighboring water molecules; and, since it is so reactive, a reaction takes place:

$$2Na(s) + 2H_2O \rightarrow 2NaOH + H_2(g)$$

Furthermore, NaOH is a strong electrolyte in aqueous solution; any molecule of this substance momentarily formed in contact with water will dissociate to give ions:

$$NaOH \rightarrow Na^+ + OH^-$$

Thus the net result of the electrolysis at the cathode by either of the preceding processes is the production of hydrogen gas:

$$2H_2O + 2e^- \rightarrow H_2(g) + 2OH^-$$

The kinetic mechanism of the electrode process is far more complicated than we have indicated here; but, regardless of the mechanism, the over-all reaction is represented by this equation.

The reaction at the anode is also complicated by the fact that chloride ions are similar in properties to the hydroxide ions formed by the dissociation of water. Therefore we might reasonably expect one or both of the following reactions to occur:

$$2Cl^- \rightarrow Cl_2 + 2e^-$$
$$4OH^- \rightarrow O_2 + 2H_2O + 4e^-$$

Actually, both do occur during the electrolysis of aqueous NaCl. The predominant reaction is controlled wholly by the relative concentrations of chloride and hydroxide ions. In concentrated sodium chloride solutions (wherein the concentration of hydroxide ion is 10^{-7}), the anode product will be largely chlorine, but in dilute salt solutions (wherein the concentration of hydroxide ion *is still* 10^{-7}), the product will be primarily oxygen. This result is a consequence of the fact that the relative probabilities of the reactions above are determined to a large degree by the relative concentrations.

The over-all reaction in a dilute salt solution is

$$2H_2O \rightarrow 2H_2 + O_2$$

This reaction occurs during the electrolysis of a number of solutions containing strong electrolytes. In these electrolyses, the salt increases the conductivity of the solution so that the current is high and a rapid electrolysis takes place. This added salt is a practical necessity. If pure water were electrolyzed, the conductivity of the liquid containing OH^- and H^+ at a concentration of 10^{-7} mole/liter would be very low and consequently the supply of electrons, i.e., the current to the cathode, would be limited to a low value. Therefore, in pure water, electrolysis is so slow as to be virtually undetectable.

Faraday's law. The first clear picture of the nature of electrolytic processes came from the work of Michael Faraday, the distinguished British scientist who made an extensive study of these reactions a hundred years ago. He pointed out that, when 1 mole of sodium atoms (6.02×10^{23}) is produced by electrolysis, the same number of electrons must flow from the dynamo into the electrode, because each Na^+ needs one electron to form a neutral Na. How much electricity is this? The common unit of electricity is the *coulomb*, defined as that amount of electricity which passes when 1 ampere flows for 1 second through a wire. In these units, the number of electrons necessary to change 1 mole of singly charged ions into neutral atoms is 96,489 coulombs. This number is known as Faraday's constant. Keep in mind that this quantity of electricity is always the equivalent of Avogadro's number (6.02×10^{23}) of electrons, the same as the number of atoms present in 1 mole. This equivalence of quantity of electricity to quantity of chemical change is known as Faraday's Law.

ELECTRIC BATTERIES

The Daniell cell. Consider a glass box divided into two parts by a porous plate as shown in Fig. 4. This plate permits ions and small amounts of water to pass through very slowly from one side to the other but prevents gross over-all mixing of the solution on either side of it. On the left side of the plate, there is a 1 molal solution of zinc sulfate, $ZnSO_4$; this is a strong electrolyte, so that there are present in the solution zinc ions, Zn^{+2}, and sulfate ions, SO_4^{-2}. On the right side, there is a 1 molal solution of cupric sulfate, which is also a strong electrolyte, so that here there are cupric ions, Cu^{+2}, and sulfate ions, SO_4^{-2}. A bar of zinc dips into the solution on the left and a bar of copper into the solution on the right. This apparatus is known as a Daniell battery or cell. The electrodes are connected with wires, and in the middle of the circuit are inserted a switch, a small electric motor, and an ammeter that will show by its reading the amount of current passing through the wire. When the switch is closed, the motor spins and an electric current flows through the wire from right to left. This is a strikingly different situation from the electrolytic cell containing the sodium and chloride ions. There, if a motor were connected with wires to the anode and cathode, nothing would happen; no electric pressure would act between the two cell electrodes to make an electric current flow, and a voltmeter (an electric

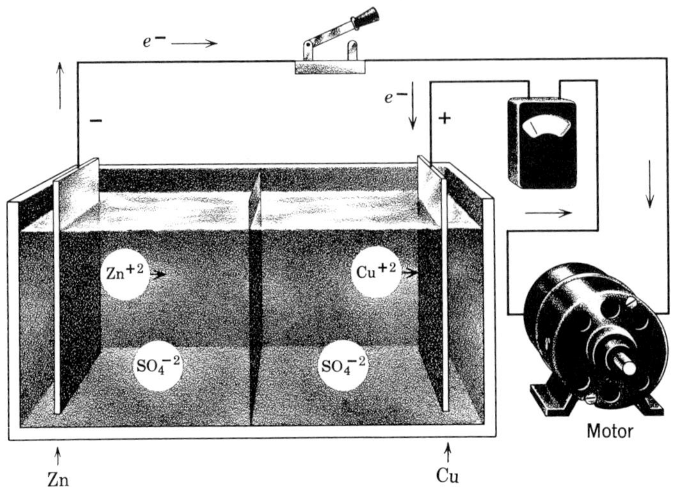

Fig. 4. The Daniell cell.

pressure meter) connected across the electrodes would give zero reading. But, with the copper-zinc cell, there *is* electric pressure, a voltage between the electrodes; when the circuit is closed, the electrons are forced to flow through the wire and make the motor spin and do work.

Where does this energy come from? We know from the first law of thermodynamics that it is not magically created. So we examine the surface of the electrodes to see how the ions of the solution impinging upon them can produce this energy.

First of all, since electrons have a negative charge, the sign of the potential shows that the electrons are flowing from left to right, out of the zinc electrode and into the copper electrode. The zinc electrode cannot provide electrons unless Zn^{+2} is formed in the solution; hence there is a chemical reaction taking place at the electrode surface. Here there are exposed zinc atoms in contact with the solution. These atoms lose two electrons to the underlying metal and pass into the solution as zinc ions:

$$Zn \longrightarrow \underset{\substack{\text{into the}\\ \text{metal}}}{2e^-} + \underset{\substack{\text{into}\\ \text{solution}}}{Zn^{+2}}$$

By passing into the metal cathode, these electrons force other electrons out at the top of the cathode and into the wire connected to it. This is the first step in setting up a current. Even though zinc is a metal, it is not easy to produce Zn^{+2} ions; it requires about 600 kcal to produce gaseous zinc ions from gaseous zinc atoms. This reaction is made possible when Zn metal is in contact with solution, because the water molecules react with and *hydrate* the zinc ions and thereby yield most of the energy required to ionize the zinc. This process is indicated schematically in Fig. 5. The dipolar water molecules are attracted to the + charge of the Zn^{+2} ion, arranged with the − pole nearest the + charge. The ions can move out into the solution like gas escaping into a vacuum, and this yields a favorable probability factor. The result is a tendency for the reaction to take place at the surface. Thus Zn atoms detach themselves from the metal, leaving behind electrons and going into the solutions as zinc ions, Zn^{+2}; and the electrons left behind are forced out of the electrode and into the wire.

Meanwhile, what is happening at the copper electrode on the right-hand side of the cell? Here, qualitatively, the situation is much the same. The electron structure diagrams for copper and copper ion and the shell structure diagrams are shown in Fig. 6. The copper atoms *might* give up their outer electrons and form ions thus:

$$Cu(s) \rightarrow Cu^{+2} + 2e^-$$

But it turns out that the tendency for Zn ions to form is considerably greater than the tendency for Cu ions to form under the conditions specified in this example; we shall show later in this chapter how we know this quantitatively. The net result is that electrons are forced out of the copper electrode and into the copper ions impinging on the electrode surface, with the result that the surface reaction takes place as shown in the following equation, the direction being reversed from that in the previous equation:

$$Cu^{+2} + 2e^- \rightarrow Cu(s)$$

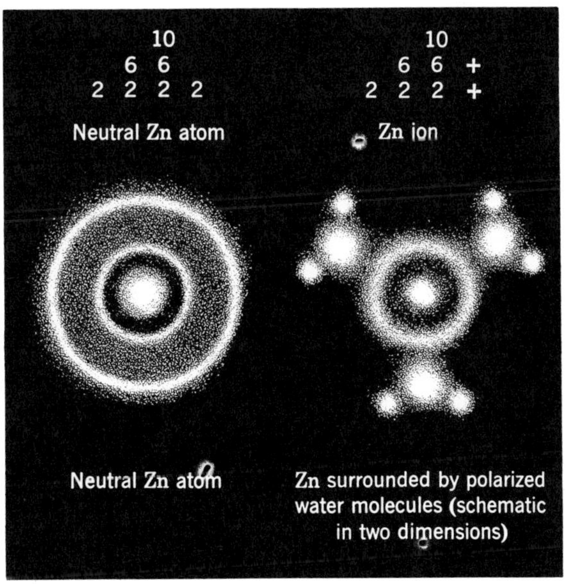

Fig. 5.

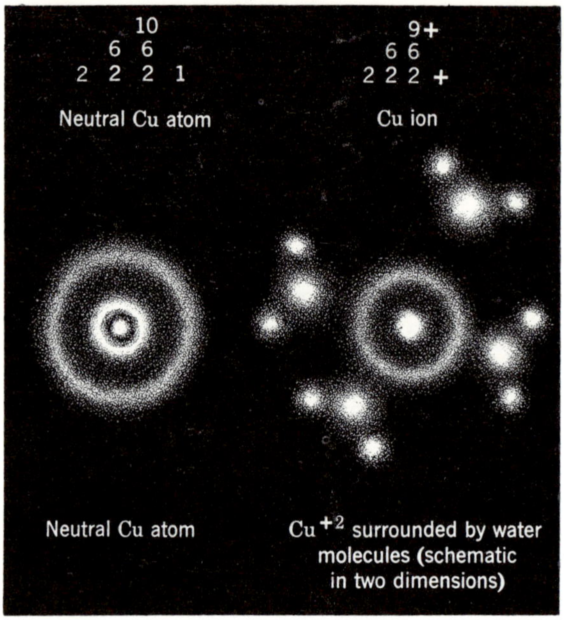

Fig. 6.

Thus the ions accept electrons, become neutral copper atoms, and stick on the electrode; this is called *plating out* of copper on the electrode.

With other pairs of metals, different results are obtained. If Ag and $AgNO_3$ are substituted for the Zn and $ZnSO_4$, the push of the copper is *greater* than the push of the silver; hence the reaction $Cu \rightarrow Cu^{+2} + 2e^-$ occurs as shown and the equivalent reaction for silver ion occurs in the reverse direction. Thus the magnitude and sign of the electron push (voltage) developed by a battery is the *net* result of two opposing tendencies.

If we were to immerse a zinc strip in a beaker containing copper sulfate, the zinc strip would dissolve and copper would plate out; that is, the following would occur:

$$Zn + Cu^{+2} \rightarrow Cu + Zn^{+2}$$

This reaction is the same as that observed in a battery. In the battery, however, the two reactants, Zn and Cu^{+2} are separated and connected with a wire so that only electrons pass between them. The greater the net tendency for this reaction, the stronger will be the net push on the electrons. It is this quality of cells that enables us to equate cell voltages to the thermodynamic functions which determine the tendency for reaction.

A combination of electrodes and solution that will force an electric current to pass is known as an *electric battery*. Since the current in passing through a wire can be made to run a motor or light a lamp or do many other useful things, such a battery is a widely used technical tool. Because the exact amount of the current can be measured, it is also a useful scientific instrument for studying the behavior of atoms and electrons; and important information comes from measurements of the push of the electrons trying to get from one electrode to the other. This push is known as the *electrical potential, voltage,* or *electromotive force* (*emf*); it is customarily measured in the electrical units called *volts.* One *volt* potential will push a current of one *ampere* through an electrical resistance of one *ohm.* Just as the flow of electrons through the wire resembles the flow of water through a pipe, this push of the electrons is similar to the pressure on the water that causes it to flow, or the pressure of a gas that either causes it to flow or makes it move a piston against a mechanical force. Thus, for many purposes, the electrons in a metal can be regarded as similar to a liquid or a gas, and it is helpful to think of a conductor as a pipe through which electron-gas flows.

The lead storage battery. One of the more familiar types of battery is the lead storage battery or accumulator used in automobiles; this is shown schematically in Fig. 7. The equations for the anode, the cathode and the complete cell reaction are

Anode:

$$Pb(s) + HSO_4^- \rightarrow PbSO_4(s) + 2e^- + H^+$$

Cathode:

$$PbO_2(s) + HSO_4^- + 3H^+ + 2e^- \rightarrow PbSO_4(s) + 2H_2O$$

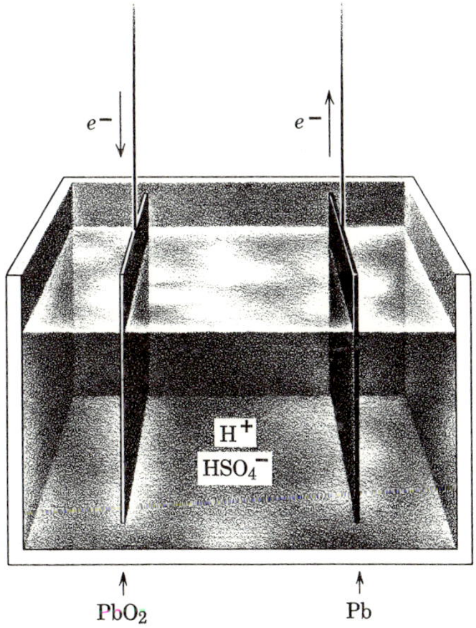

Fig. 7. The lead storage cell.

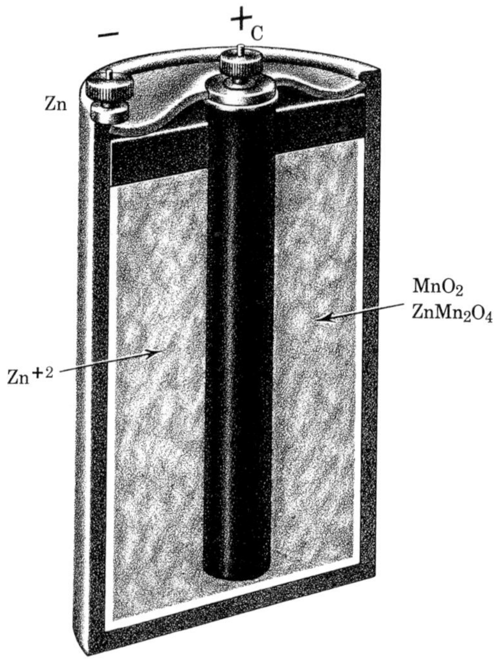

Fig. 8. The dry cell.

Complete reaction:

$$Pb(s) + 2HSO_4^- + 2H^+ + PbO_2(s) \rightarrow 2PbSO_4(s) + 2H_2O$$

When the battery discharges to drive the automobile starter, the concentration of sulfuric acid is diminished and lead sulfate is formed. As the concentration of sulfuric acid goes down, the density of the solution decreases, and consequently it is possible to tell the state of charge of a battery by measuring the density of solution. If batteries are allowed to stand for a long time, the solid lead sulfate accumulates to an undesirable degree and the batteries are said to be *sulfated.*

In charging, these reactions are reversed and the lead sulfate is changed back into lead and lead oxide.

The dry cell. Another common form of battery is the *dry cell,* which consists of a can of zinc containing a graphite rod in the center and filled with a moist paste of manganese dioxide, zinc chloride, and ammonium chloride. This is shown in Fig. 8. The equations for the cell are

Anode:

$$Zn(s) \rightarrow Zn^{+2} + 2e^-$$

Cathode:

$$2MnO_2(s) + Zn^{+2} + 2e^- \rightarrow ZnMn_2O_4(s)$$

Complete reaction:

$$Zn(s) + 2MnO_2(s) \rightarrow ZnMn_2O_4(s)$$

CELL THERMODYNAMICS

In our earlier studies of chemical reactions, we have seen that under specified conditions a reaction takes place whenever the free energy change, ΔF, is negative. ΔF is the difference between the energy factor, ΔH, and the probability factor, $T\,\Delta S$. A physical reaction like vaporization takes place if the probability factor overbalances the energy factor, making the push to send the molecules into the vapor state greater than the pull to bring them back to the liquid state. In the reaction between hydrogen and oxygen gas to form liquid water at room temperature, the energy pull to bring the atoms of hydrogen and oxygen together in the form of water far more than overbalances the probability push tending to make the atoms go from water into the form of the elements hydrogen and oxygen. Thus the energy pull produces the explosive tendency of the reaction to go from hydrogen and oxygen to water.

Electrolytic reactions, changing ions to neutral substances and vice versa, also tend to go in the direction indicated by the relative balance of the energy and probability factors. In the case of the zinc-copper battery, electrons are pushed around through the wire with considerable pressure because the energy and probability balance results in a decrease in the free energy of the whole system when the reaction occurs as indicated; but this reaction is considerably more complicated than the evaporation of water and needs a more detailed analysis.

Consider the energy factor first. Recall the energy change when a gas expands as shown in Fig. 9. One mole of ideal gas at 0°C (273°K) in a volume of 22.4 liters has a pressure of 1 atm. Let this gas expand to a volume of 44.8 liters, with heat flowing into the gas so that the pressure is kept constant. Then an amount of work is performed equal to the pressure multiplied by the change in

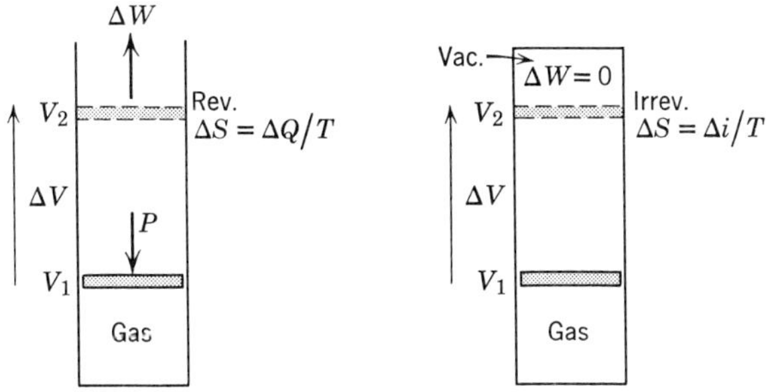

Fig. 9. Reversible and irreversible expansion of an ideal gas.

volume, 1 atm multiplied by the increase in volume of 22.4 liters, 22.4 liter-atm or 542 cal of work.

When a faraday equivalent or 1 gram mole of electron "gas" moves into a metallic conductor or wire against an electrical pressure, exactly the same kind of work is performed. Electrical pressure is normally in terms of volts. Thus, if 1 gram equivalent of electron "gas" moves into a wire against 1 volt "pressure" of electricity, the amount of work performed measured (in terms of calories) is 23,061 cal.

An expanding ideal gas:

$$-P\,\Delta V = -1 \text{ atm} \times 22.4 \text{ liters} = -22.4 \text{ liter atm} = -542 \text{ cal}$$

An "expanding" electron-gas:

$$-P_e\,\Delta X = -1 \text{ volt} \times 1 \text{ faraday} = -1 \text{ volt-faraday} = -23{,}061 \text{ cal.}$$

We use the symbol P_e to denote electrical pressure (in volts) and the symbol ΔX to denote the quantity of electricity acting against this pressure. Thus, when a reaction takes place that involves *both* a change of volume against pressure and the pushing of electrons against voltage, we denote the total work done by the expression

$$\Delta W = -(P\,\Delta V + P_e\,\Delta X)$$

The negative sign shows that energy leaves the system when it does work by expanding.

We now recall the relation between the heat absorbed or given out in a chemical reaction and the change in entropy.

$$T\,\Delta S = \Delta Q + \Delta i$$

where ΔQ is the heat absorbed in the reaction and Δi is the amount of irreversibility; the latter is equivalent numerically to the extra calories of heat that would have been absorbed if the reaction had proceeded reversibly.

The first law of thermodynamics can be written

$$\Delta E = \Delta Q + \Delta W$$

Inserting the values for the ΔQ and ΔW terms obtained from the preceding relations,

$$\Delta E = T\,\Delta S - \Delta i - P\,\Delta V - P_e\,\Delta X$$

The definition of free energy is

$$F = H - TS = E + PV - TS$$

In this discussion, reactions will be considered that take place under constant pressure and temperature. Thus

$$\Delta F = \Delta E + P\,\Delta V - T\,\Delta S$$

Substituting for the value of ΔE as given by the first law (where ΔQ and ΔW are expressed in the expanded form of $T\Delta S$, Δi, $P\Delta V$, and $P_e\,\Delta X$), we find the $P\,\Delta V$ and $T\,\Delta S$ terms cancelling out and we get

$$\Delta F = -\Delta i - P_e\,\Delta X$$

Suppose that at first the cell is connected so that it forces a current around the wire and through the motor, making the motor turn; now we force the motor to turn *backward* by spinning it in the reverse direction by a crank. A motor so turned can act as a dynamo, and it exerts a back pressure or potential against the flow of current. Let us spin the motor-dynamo at such a speed that the current is just *barely* able to flow forward. This is a situation exactly like the expansion of a gas at 1 atm under the pressure of the piston maintained at just a little under 1 atm so that the gas moves very slowly from the small volume to the large volume, expanding *reversibly*. In other words, the cell reaction now takes place *reversibly*. Under these conditions, there is no irreversibility and the term Δi is equal to zero. Thus the expression for the change in the free energy equals the negative of the voltage multiplied by the amount of electricity which is passed. If we maintain the back voltage constant and let precisely 1 mole or a Faraday equivalent† of electrons pass through the wire, then

$$\Delta F = -P_e\,\Delta X$$

In the Daniell Zn-Cu cell, when 1 mole of zinc ions is produced, 2 moles of electrons must flow out of the electrode because the Zn^{+2} ion is doubly charged; the same holds in the reverse sense for the copper ion. To make such a situation clear, the equation is usually written

$$\Delta F = -P_e n \mathcal{F}$$

where n is the moles of electrons transferred and $\mathcal{F}$ is the value of the faraday equivalent expressed as 96,489 coulombs. In the copper-zinc cell, the voltage actually exerted by the cell under reversible conditions is 1.10 volts. If we balance the pressure of electrons produced by the cell by a counterpressure and let 1 mole of zinc change reversibly into zinc ions while 1 mole of copper ions plate out as copper, the numerical values are

$$-2.2 \text{ volt-faraday} = -(1.10 \text{ volt})(2 \text{ charges})(1 \text{ faraday})$$

This may be converted to kilocalories as shown below, since one volt-Faraday is the equivalent of 23.061 kcal of energy:

$$-2.2 \text{ volt-faraday} \times 23.061 \text{ kcal/volt-faraday} = -51 \text{ kcal} = \Delta F$$

† In this ideal case, we assume that we have a very large amount of cupric sulfate and zinc sulfate solutions so that the concentration is not measurably affected by the zinc dissolving or the copper plating out.

Measurements of the heat changes show that the value of ΔH for this reaction is -21 kcal. In the equation relating ΔF to ΔH and $T\Delta S$ as given below, we thus know ΔF and ΔH; and, by taking the difference, we can calculate $T\Delta S$:

$$\begin{array}{cccc} \Delta F = & \Delta H & - & T\Delta S \\ -51 \text{ kcal} = & -21 \text{ kcal} & - & 30 \text{ kcal} \\ & \text{energy factor} & & \text{probability factor} \end{array}$$

Thus the probability factor is 30 kcal. Both the energy factor and the probability factor favor the reaction, and there is a large negative change in free energy and a relatively large electron pressure or voltage from the cell.

OXIDATION AND REDUCTION POTENTIALS

When an electric current is passed through molten sodium chloride and the electrolytic process takes place, the sodium ions present in the liquid salt combine with electrons to produce metallic sodium as follows:

$$Na^+ + e^- \rightarrow Na(s)$$

It is customary to say that in this process the sodium ions are *reduced* to form sodium metal. From the same point of view, when sodium metal is changed to sodium ions by going into solution in water and producing hydrogen, the metal is said to be *oxidized,* although it is not actually combined with oxygen:

$$2Na(s) + 2H_2O \rightarrow 2Na^+ + 2OH^- + H_2(g)$$

The reason for the choice of these terms may be seen more clearly after a consideration of analogous changes which take place with ions containing multiple charges. When iron is heated in air, the following reaction with oxygen occurs:

$$4Fe + 3O_2 \rightarrow 2Fe_2O_3$$

Fe_2O_3 is *ionic,* with a corundum structure, and it can be converted to the metal by treatment with hydrogen:

$$Fe_2O_3 + 3H_2 \rightarrow 2Fe + 3H_2O$$

In the first reaction the oxygen content is increased, and such a process is called *oxidation;* in the second reaction there is a change from a compound form of iron to the elementary form, and this is referred to as *reduction.* But we must keep in mind that, in the first reaction, neutral atoms of Fe lost electrons and became Fe^{+3} and, in the second reaction, Fe^{+3} became Fe. This same loss of electrons occurs in a variety of reactions; for example,

$$2Fe + 3F_2 \rightarrow 2FeF_3$$
$$2Fe + 3Cl_2 \rightarrow 2FeCl_3$$

For this reason, the loss of electrons by an atom in *any* chemical reaction is called *oxidation;* and the gain of electrons by an atom in any chemical reaction is called *reduction.*

Half-electrode reactions. We now examine the oxidation aspect of the electrolytic reactions in a battery or cell. In the Daniell cell, there are two reactions taking place: one at the left (the anode) where oxidation takes place, and one at the right (the cathode) where reduction takes place. The zinc metal loses electrons and goes into solution as ions; it is, therefore, oxidized. The copper ions gain electrons to become copper metal and are, therefore, reduced.

Since these two processes are taking place at different locations in the cell, it is helpful to consider them as two distinct processes:

$$\mathrm{Zn}(s) \rightarrow \mathrm{Zn}^{+2}(\mathrm{aq.}) + 2e^-$$
$$\mathrm{Cu}^{+2}(\mathrm{aq.}) + 2e^- \rightarrow \mathrm{Cu}(s)$$

instead of

$$\mathrm{Zn}(s) + \mathrm{Cu}^{+2} \rightarrow \mathrm{Zn}^{+2} + \mathrm{Cu}(s)$$

The left side of the Daniell cell is called the zinc half-cell, and the right side is called the copper half-cell. In the laboratory, it is possible to remove the right side of the cell containing the copper and replace it with other half-cells. Thus, every half-cell can be combined in turn with a large number of different half-cells, and in this way many different electrolytic relations can be studied.

Is it possible to leave out the right half of the apparatus completely and to attach a wire somehow in order to study the action of a single half-cell, not combined with anything else? In order to have a reaction take place, there must be a current circulating; in other words, if positively charged zinc ions, Zn^{+2}, go into solution, there must be some compensating way to prevent the solution from accumulating a net positive charge. Even the presence of a small positive charge would so change the energy necessary to move electrons in or out that almost instantly the reaction would cease taking place. Thus there must be some way of removing other positive charges from the solution as is done in the complete Daniell cell by the copper electrode that pulls out cupric ions, Cu^{+2}; or some negative ions must be introduced to counterbalance the positive ions. Either of these counterbalancing processes involves the removal or introduction of ions and, therefore, involves another half-cell. Thus it is impossible to study a *single* half-cell by itself; it must always be in contact with another half-cell in order to have any kind of electrolytic reaction taking place.

The hydrogen electrode. One of the most useful half-cells is one in which hydrogen ions are converted into hydrogen gas. This reaction takes place when hydrogen gas is bubbled over an unreactive metal, such as platinum, that dips into a solution where hydrogen ions are present. Under the influence of an electrical potential, the hydrogen ions move up to the platinum surface, pick up the electrons from this metallic surface, and then come together in pairs to form hydrogen gas; *or* the reverse process can take place. At equilibrium this reaction can be expressed as

$$2H^+ + 2e^- \rightleftharpoons H_2(g)$$

Standard potentials. In the Daniell cell the voltage or electrical pressure is 1.10 volts; this is the difference in electrical potential between the zinc bar and the copper bar. Since it is logical to think of this cell as made up of two half-cells, one can raise the question "How much voltage comes from the zinc half-cell and how much voltage comes from the copper half-cell?" But, since any given half-cell must always be combined with some other half-cell, it is difficult to answer this question unambiguously. Chemists have decided that the simplest way to avoid confusion is to call the voltage of the platinum-hydrogen half-cell *zero volts* when the pressure of the hydrogen gas is maintained at 1 atm and the concentration of hydrogen ion is 1 molal. This is designated as the *standard* half-cell.

Thus, if the zinc half-cell having a 1 molal concentration of zinc ions is combined with a standard platinum-hydrogen cell, the voltage observed between the zinc and the platinum is 0.76 volt. We, therefore, say arbitrarily that the voltage of the zinc half-cell is 0.76 volt and the voltage of the standard platinum-hydrogen cell is 0.00 volt. In order to save space and to represent such a combination of half-cells without drawing pictures, it is customary to write down the symbols for the left side and right side of the cells as shown below:

$$Zn(s)/Zn^{+2}(1m)//H^+(1m)/H_2(g)(1 \text{ atm})$$
$$\mathcal{E}^\circ(\text{L}) = 0.76 \text{ volt} \qquad \mathcal{E}^\circ(\text{R}) = 0.00 \text{ volt}$$

The symbol $\mathcal{E}^\circ(\text{L})$ is used for the voltages of the left-hand cell and the symbol $\mathcal{E}^\circ(\text{R})$ is used for the voltage of the right-hand cell.† Signs will be used for these voltages so that the two may be combined to give the total voltage of the cell, $\mathcal{E}^\circ_r$, as follows:

$$\begin{aligned} \mathcal{E}^\circ_r \quad &= \mathcal{E}^\circ(\text{L}) - \mathcal{E}^\circ(\text{R}) \\ 0.76 \text{ volt} &= 0.76 \text{ volt} - 0.00 \text{ volt} \end{aligned}$$

The Daniell cell has the cell symbols:

$$Zn(s)/Zn^{+2}(1m)//Cu^{+2}(1m)/Cu(s)$$

† Earlier the symbol P_e was used to emphasize the similarity between electrical pressure and mechanical pressure. In keeping with common usage, we shall henceforth use $\mathcal{E}$ to denote electrical pressure. $\mathcal{E}^\circ$ denotes the potential under standard conditions, a concentration of 1 molal and gas (if any) at a pressure of 1 atm. When deviations from the ideal gas and solution laws are taken into account, the standard state is defined with *fugacity* at 1 atm and *activity* at 1 molal, where fugacity and activity are the thermodynamic equivalents of pressure and concentration that are directly related to the free energy, enthalpy, and entropy by the simple laws for ideal systems. Ordinarily we shall neglect deviations due to non-ideality. Instead of m_w we shall use the letter without the subscript to denote molality (m) in the equations that follow.

If we write our half-cell equations, we obtain

$$\mathrm{Zn}(s) = \mathrm{Zn}^{+2} + 2e^- \qquad \mathcal{E}^\circ(\mathrm{L}) = +0.76 \text{ volt}$$
$$\mathrm{Cu}(s) = \mathrm{Cu}^{+2} + 2e^- \qquad \mathcal{E}^\circ(\mathrm{R}) = -0.34 \text{ volt}$$

Let us now subtract the equation for the right-hand electrode from that for the left-hand electrode in order to obtain the equation

$$\mathrm{Zn}(s) + \mathrm{Cu}^{+2} = \mathrm{Cu}(s) + \mathrm{Zn}^{+2} \qquad \mathcal{E}^\circ(\mathrm{L}) - \mathcal{E}^\circ(\mathrm{R}) = +1.10 \text{ volts}$$

In this case, the voltage of the copper half-cell has a negative value. These voltages are always given so that the total voltage of the cell is the voltage of the left half minus the voltage of the right half. Thus the two minus signs on the right cancel each other and the total voltage of the cell, 1.10 volts, is equal to the numerical sum of the two values, 0.76 volt + 0.34 volt. It is helpful to keep certain points in mind in connection with this conventional usage:

1. The cell potential is equal to the electrode potential of the left-hand half-cell minus that of the right-hand half-cell.
2. The sign of the total cell potential is the same as the polarity of the right-hand electrode. In the Daniell cell, the total cell potential is +1.10 volts, and, since the conventional electric current is flowing out of the right-hand electrode through the wire and in at the left-hand electrode, a voltmeter placed across the poles of the cell will show that the right-hand electrode is positive.
3. Electrons flow in the direction opposite to the conventional current; in other words, electrons flow from the negative electrode to the positive electrode.
4. The electrode where the conventional current enters the cell is called the *anode*. The electrode where the conventional current leaves the cell, flowing into the wire, is called the *cathode*. In the cell, the ions that flow toward the anode are called the anions. The ions that flow toward the cathode are called the cations.

The potentials of a number of half-cells are given in the Appendix.

Equilibrium standard potentials. In Chapter 13, a basic equation was derived that related the free energy change to the equilibrium constant. This equation had the form

$$-\Delta F^\circ = 2.303RT \log K$$

where K is the equilibrium constant, R and T have the usual meanings, and ΔF°† is the standard free energy change. The standard free energy change for the reaction

$$\mathrm{Zn}(s) + \mathrm{Cu}^{+2} = \mathrm{Cu}(s) + \mathrm{Zn}^{+2}$$

† The superscript circle will be placed at the left or at the right in the position that interferes least with the other sub- or superscript symbols.

is the change in free energy when 1 mole of Zn reacts with 1 mole of Cu^{+2} in a solution which is at unit concentration in zinc and cupric ion at the beginning and end of the experiment. This change is imagined to occur in such a large amount of solution that the loss of 1 mole of cupric ion and the gain of 1 mole of Zn^{+2} ion has no effect on the concentrations. Thus

$$n\mathfrak{F}\{\mathcal{E}^\circ(\text{L}) - \mathcal{E}^\circ(\text{R})\} = -\Delta F^\circ$$
$$n\mathfrak{F}\mathcal{E}_r^\circ = -\Delta F^\circ$$

where $\mathfrak{F}$ is the thermal equivalent of the faraday, 23,061 cal.

ΔF° is thus proportional to the negative of the standard cell potential. When the standard cell potential is large and positive, ΔF° is large and negative. According to the development in Chapter 12, this means that the reaction goes spontaneously and the yield of products is high. On the other hand, if $\mathcal{E}_r^\circ$ is negative (and ΔF° is positive), the reaction will not go spontaneously. This statement provides us with a concrete picture of the tendency of a reaction to go. When $\mathcal{E}_r^\circ$ is positive, the reaction has such a strong tendency to go that the electron transfer, necessary for the reaction, can occur even when there is an opposing voltage. The greater this tendency to react, the greater can be the opposing voltage without stopping the reaction.

We can combine the preceding relations to give a more quantitative expression of the thermodynamics of cells:

$$2.303RT \log K = n\mathfrak{F}\mathcal{E}_r^\circ$$

or

$$\log K = \frac{n\mathfrak{F}\mathcal{E}_r^\circ}{2.303RT}$$

Thus, if we know the standard cell potential, we can calculate the equilibrium constant for the corresponding reaction.

Concentration effects. It is useful to see how *cell potential* is related to the concentrations of the ions in the cell. To establish this relation, recall first some of the basic equations concerned with free energy and the laws of thermodynamics.

In Chapter 12, it was shown that, when the pressure of an ideal gas changes from P_1 to P_2, the change in the free energy of 1 mole of the gas is given by

$$\Delta_1^2 F = RT \ln (P_2/P_1)$$

If Henry's law holds and this gas is in equilibrium with dissolved gas, the vapor pressure P will be related to the concentration in C by

$$P = kC$$

where k is a constant. Thus

$$F_2 - F_1 = \Delta_1^2 F = RT \ln (P_2/P_1) = RT \ln (C_2/C_1)$$

For many cases in which we are interested, this equation may be applied with reasonable accuracy, and we shall assume its validity unless otherwise noted.

Let us now apply this equation to the change of concentration of zinc ion (Zn^{+2}) from 1 molal to some arbitrarily selected concentration, x, where there are x moles of Zn^{+2} per 1000 gm of solution; that is, the molality is x. Let $^{\circ}F$ represent the free energy in the standard state, 1 m, and ^{x}F represent the free energy at the new concentration, x. Then

$$^{x}F - {}^{\circ}F = RT \ln [{}^{x}(Zn^{+2})/{}^{\circ}(Zn^{+2})]$$

where $^{x}(Zn^{+2})$ is the concentration of Zn^{+2} arbitrarily selected and $^{\circ}(Zn^{+2}) = 1$.

Now consider the reactions

$$Zn(s) + 2H^{+}(1\,m) \rightarrow Zn^{+2}(x) + H_2(1\text{ atm}) \qquad \Delta F^{x}$$
$$Zn(s) + 2H^{+}(1\,m) \rightarrow Zn^{+2}(1\,m) + H_2(1\text{ atm}) \qquad \Delta F^{\circ}$$

where the second reaction produces Zn^{+2} at unit concentration and the first produces Zn^{+2} at the arbitrarily selected concentration x. From the difference of these two equations we get

$$Zn^{+2}(1\,m) \rightarrow Zn^{+2}(x) \qquad \Delta F^{x} - \Delta F^{\circ}$$

Since ΔF^{x} and ΔF° are both measured with respect to the same zero (the H_2-H^{+} cell), we get

$$^{x}F - {}^{\circ}F = \Delta F^{x} - \Delta F^{\circ} = RT \ln [{}^{x}(Zn^{+2})/{}^{\circ}(Zn^{+2})]$$

and, since $^{\circ}(Zn^{+2}) = 1$, we can write

$$\Delta F^{x} = \Delta F^{\circ} + RT \ln {}^{x}(Zn^{+2})$$

This expression can be related to cell potentials by means of the equation

$$\Delta F^{\circ} = -n\mathfrak{F}\mathcal{E}_r^{\circ}$$

where $\mathfrak{F}$ is the thermal equivalent of one Faraday and n is the number of moles of electrons transferred. For the reaction with the hydrogen half-cell with Zn^{+2} at unit concentration, we write

$$\Delta F^{\circ} = -2\mathfrak{F}\mathcal{E}_r^{\circ} = -2(23{,}061 \text{ cal/volt})(0.76 \text{ volt})$$

We can also write $\Delta F^{x} = -2\mathfrak{F}\mathcal{E}_r^{x}$ where $\mathcal{E}_r^{x}$ is the voltage developed by a zinc-hydrogen cell when $(Zn^{+2}) = {}^{x}(Zn^{+2})$. If we substitute for ΔF^{x} and ΔF° in the equation specifying the effect of concentration on free energy, we find

$$-n\mathfrak{F}\mathcal{E}_r^{x} = -n\mathfrak{F}\mathcal{E}_r^{\circ} + RT \ln {}^{x}(Zn^{+2})$$
$$\mathcal{E}_r^{x} = \mathcal{E}_r^{\circ} - (RT/n\mathfrak{F}) \ln {}^{x}(Zn^{+2})$$

Here we see that the potential of an electrode with ions at concentration x differs from the potential of an electrode with ions at unit concentration (1 m) by the amount, $-(RT/n\mathcal{F})\ln x$. The energy factor, in contrast to the probability factor, depends little on the concentration; variations in concentration change the cell potential mainly by way of their effect on the probability factor. For example, at unit concentration, the potential for the zinc half-cell is +0.76 volt and, compared to a hydrogen standard half-cell, the reaction as written has a great tendency to occur. But, if we increase the Zn^{+2} concentration much above unity, there will be more ions colliding with the electrode and taking electrons away from it, reversing the reaction and reducing the potential by the amount $-(RT/n\mathcal{F})\ln x$. The probability factor will thus overbalance the energy factor if the concentration of ions is sufficiently increased, and the potential can be made negative. If, on the other, we reduce the concentration, this is analogous to reducing the pressure on a gas or putting the gas in a larger volume at the same temperature. We saw in Chapter 12 how the probability of the state of the gas increases as the volume is increased. In the same way, putting a given number of ions in a larger volume of solution or, in an equivalent way decreasing the concentration, increases the probability. It is as if this decreased concentration "pulls" the ions off the electrode and, therefore, increases the tendency of ions to form and of electrons to push out of the top of the electrode with greater force into the circuit. Thus a smaller concentration makes a numerically larger logarithm with a negative sign, and this multiplied by the negative sign in front of the whole term gives a larger positive potential.

The same logic applies to the Cu^{+2} ion and we find

$$\mathcal{E}_r^x = \mathcal{E}_r^\circ - (RT/n\mathcal{F})\ln {}^x(Cu^{+2})$$

If we subtract

$$Cu(s) = Cu^{+2}(x) + 2e^-$$

from

$$Zn(s) = Zn^{+2}(x) + 2e^-$$

we get

$$Zn + Cu^{+2} = Cu + Zn^{+2}$$

where we omit the x symbols, it being understood that Zn^{+2} and Cu^{+2} are at the respective concentrations (Zn^{+2}) and (Cu^{+2}). Combining the potential equations for the two electrodes, we get

$$\mathcal{E}_r^x = \mathcal{E}_r^\circ - (RT/n\mathcal{F})\ln[(Zn^{+2})/(Cu^{+2})]$$

where $\mathcal{E}_r^x$ represents the potential of the total cell with concentration of Zn^{+2} at (Zn^{+2}) and of Cu^{+2} at (Cu^{+2}), and $\mathcal{E}_r^\circ$ is the voltage with both ions at unit concentration. If we measure the voltage of our cell when the concentration of the zinc ion is 2 m and the concentration of the copper ion is 0.2 m, the equation is

$$\mathcal{E}_r^x = \mathcal{E}_r^\circ - (RT/n\mathcal{F})\ln[(2\ m)/(0.2\ m)]$$

If we are making measurements at the customary temperature of 25° C, then the numerical values are

$$\mathcal{E}_r^{\ne} = 1.07 \text{ volts} = 1.10 \text{ volts} - \frac{2.3026 \times 1.9872 \times 298.15°}{2 \times 23{,}061} \log [10]$$

$$= 1.10 \text{ volts} - (0.05916/2) \log [10]$$

$$= 1.10 \text{ volts} - 0.0295 \text{ volt} = 1.07 \text{ volts}$$

Similar reasoning applies to any reaction, and, in general, we can write

$$\mathcal{E}_r^{\ne} = \mathcal{E}_r^{\circ} - (RT/n\mathcal{F}) \ln Q$$

or, at 25° C,

$$\mathcal{E}_r^{\ne} = \mathcal{E}_r^{\circ} - (0.05916/n) \log Q$$

where Q is the quotient of concentrations written as they would appear for the reaction as written. This is known as the Nernst equation. It is fair to say that this equation has yielded more information about aqueous solutions than any other. To obtain some feeling for its power, let us consider a few examples of its application.

Whenever a reaction is written in such a way that the symbol for the electron, e^-, appears on the right, this means that it is a reaction where the *substance* on the left-hand side is giving up an electron. In such a reaction, this substance is an *electron donor.* Conversely, if the reaction is written with the symbol for the electron appearing on the left, then, as the reaction proceeds, the ion on that side is accepting the electron and is an *electron acceptor.* Examples of this are

$$\underset{\text{electron donor}}{Fe^{+2}} \rightleftarrows Fe^{+3} + e^- \qquad \mathcal{E}° = -0.77 \text{ volt}$$

$$\underset{\text{electron acceptor}}{Cu^{+2} + 2e^-} \rightleftarrows Cu(s) \qquad \mathcal{E}° = +0.34 \text{ volt}$$

As we have seen, the potential, $\mathcal{E}$, is directly proportional to the free energy change. Since the free energy change is a measure of the tendency of the reaction to go, the potential is also a measure of the tendency of the reaction to go. A positive potential corresponds to a negative free energy; *a negative free energy* means that the reaction has a tendency to go; and, therefore, *a positive potential* means that the reaction has a tendency to go. Thus we see why the Daniell cell will push a current around a circuit; the zinc tends to go into solution as zinc ions push electrons into the wire attached to that electrode; it has a positive potential in this direction. Conversely, compared with the standard hydrogen electrode, the copper ions tend to accept electrons and plate out of solution as copper metal. This pull of the copper ions to take electrons out of the copper electrode must, therefore, be added to the push of the zinc metal trying to get rid of electrons in order to get the total push-pull to move the electrons around the wire from the left pole of the cell to the right pole of the cell. For this reason, the relative tendency of any ion or atom to lose electrons, as in the case of Zn^{+2}, or to gain electrons, as in the case of Cu^{+2}, is an important measure of chemical activity. Since the loss of electrons is referred to as *oxidation* and the gain of electrons is referred to as *reduction*, this is frequently called the *oxidation-reduction* potential. The difference in potentials at unit concentration is an expression of this push-pull of electrons *at unit concentration.* But we see that a change in the concentration will affect this push or pull. This is, in the simplest cases, a change in the probability factor. Thus, in the case of zinc tending to go into solution as Zn^{+2}, reducing the concentration of the zinc ions in the solution results in an increase in the probability factor. This is exactly like the problem of the evaporation of water from the liquid to the vapor state. If we reduce the pressure (i.e., concentration) of the water vapor, the probability factor tends to pull the water from the liquid to the vapor state. In the same way, if we reduce the concentration of zinc ion, Zn^{+2}, this probability factor tends to pull the zinc metal into solution with a consequent push on the electrons to go into the wire.

We can also have this electron donation take place directly from one ion to another without having electrodes and a cell reaction involved. Let us consider the reaction

$$Ce^{+3} = Ce^{+4} + e^-$$

where

$$\mathcal{E}^\circ = -1.61 \text{ volts}$$

This shows that cerium in the form of a triply charged ion can give up an electron and become a quadruply charged ion. Now suppose that we have in the same beaker both kinds of cerium ions and both kinds of iron ions, Fe^{+2} and Fe^{+3}.

Let us rewrite the equation with the electron appearing on the left-hand side:

$$Ce^{+4} + e^- \rightleftarrows Ce^{+3}$$

where

$$\mathcal{E}^\circ = +1.61 \text{ volts}$$

Since

$$Fe^{+2} \rightleftarrows Fe^{+3} + e^-$$

where

$$\mathcal{E}^\circ = -0.77 \text{ volt}$$

We add and we get

$$Fe^{+2} + Ce^{+4} \rightleftarrows Fe^{+3} + Ce^{+3}$$
$$\mathcal{E}_r^\circ = 1.61 - 0.77 = +0.84 \text{ volt}$$

Since we have symbols for electrons appearing on both sides, these cancel out, leaving us only with the ions. The potential will be the difference between the potentials of the two half-reactions or $+0.84$ volt. This positive potential indicates that the reaction will have a tendency to go in the direction written, from left to right when all component ions are at 1 *m* concentration. In other words, the ferrous ion, Fe^{+2}, gives up an electron to the ceric ion, and we get as a result the ferric and cerous ions. If we were to increase the concentrations of the components on the right-hand side, there would be less tendency for the reaction to go from left to right, in exactly the same way that we have seen for other non-electrolytic reactions. In fact, the concentrations of either of the components on the right could be increased to the point where we would get equilibrium, leaving no tendency for the reaction to go. Of course, under these circumstances, if we add together the potentials for our concentrations at these new values instead of at 1 *m*, we find that the potentials sum to zero when we have equilibrium. We thus get this direct and intimate relation between potentials and the tendency for chemical reactions to take place. This is why oxidation-reduction potentials are so important; they give us a measure of the tendency for the reactions to go where the essential process involved is the transfer of one or more electrons from one species to another. Many of the biochemical reactions taking place in living matter are of this sort. The study of these reactions can be carried out by placing the chemicals involved in electrolytic cells and measuring the potential. This gives a direct way of determining free energies and the way in which free energies vary with concentrations.

The equations that we have been discussing do not always hold with complete accuracy. Frequently, because of the action of the dipoles of the surrounding water molecules, the change in free energy with concentration is not what one calculates from the equations for ideal systems. A most useful set of relations has been worked out, however, in which quantities called *activities* are used instead of *concentrations*. By a careful study of solutions, one can find factors to convert concentrations to activities and, when these activities are used, the relations between activities and potentials or free energies hold with exactness.

In the Appendix are given a number of the oxidation-reduction potentials that can be used either for predicting the potentials of different cells under different conditions, or the tendency of oxidation and reduction reactions to go.

Example Problem 1: A copper-zinc battery is set up under standard conditions so that the ions are at unit concentrations. Initially, the voltage developed by this cell is 1.10 volts. When the battery is in use, the concentration of the cupric ion gradually decreases and that of the zinc ion gradually increases. Along with this, according to Le Chatelier's principle, the voltage developed by the cell drops. What will be the ratio Q of the concentrations of the zinc and the cupric ions when the voltage has dropped to 1.00 volt? The reaction we are considering is

$$Cu^{+2} + Zn(s) \rightarrow Cu(s) + Zn^{+2}$$

and the ratio equation for this reaction is

$$Q = \frac{(Cu(s))(Zn^{+2})}{(Cu^{+2})(Zn(s))} = \frac{(Zn^{+2})}{(Cu^{+2})}$$

(Note that the concentrations of metallic copper and metallic zinc are unity in this expression, in accord with conventions for the equilibrium constant.) The value of $\mathcal{E}_r$ is 1.00 volt. If we substitute the concentration and the values of the two voltages into the Nernst equation,

$$1.00 = 1.10 - (0.05916/2) \log [(Zn^{+2})/(Cu^{+2})]$$

and, if we solve for the relative concentrations of zinc and copper ions, we find

$$\log \frac{(Zn^{+2})}{(Cu^{+2})} = \frac{0.10 \times 2}{0.05916} = 3.38; \quad \frac{(Zn^{+2})}{(Cu^{+2})} = 2400$$

It is apparent from this that, if the voltage of the battery drops measurably, the concentration has changed very significantly. This is because of the logarithmic relationship between the emf developed and the concentration. It also tells us that, if we were to recharge this battery by applying an emf to reverse the reaction, the voltage would rapidly build up to 1.00 volt and, if we stopped the recharging process at this point, we would not have a very high concentration of cupric ions in solution. We would have to get very close to the $\mathcal{E}_r^\circ$ value before the battery was substantially recharged. Finally, it should be noted that the potentials of batteries can be measured very accurately. Consequently, we can trace quite precisely the extent of oxidation-reduction reactions simply by following the voltage when the reaction of interest is utilized in a battery.

Example Problem 2: A chemist constructs a battery in which one electrode is made of silver, and this is immersed in a 1 molal silver nitrate solution. The other electrode is made of platinum over which hydrogen gas (at a pressure of 1 atm) is bubbled, and this is immersed in one molal nitric acid. The emf of this cell is found to be 0.7995 volt. He then adds enough KCl to the silver cell that the *final* (Cl^-) is one molal. This, of course, precipitates the silver nitrate as silver chloride and thus reduces the silver ion concentration to a very low value. When this is accomplished, the measured value of the emf of the altered cell is 0.222 volt. Calculate the solubility product of silver chloride.

The reaction we are investigating in this cell is

$$2Ag^+ + H_2 \rightleftarrows 2H^+ + 2Ag$$

and the Q product for this reaction is

$$Q = \frac{(Ag)^2(H^+)^2}{(H_2)(Ag^+)^2}$$

The relation between the voltage developed by the cell and the concentrations of the ions is given by the Nernst equation. Therefore we can write

$$\mathcal{E}^x = \mathcal{E}_r^\circ - \frac{0.05916}{2} \log \frac{(H^+)^2}{(H_2)(Ag^+)^2}$$

In both cases, (H^+) and (H_2) are unity. In the first case, the concentration of silver ion was also unity, so the voltage developed by the cell was equal to the $\mathcal{E}_r^\circ$ value. In the second case, the silver ion concentration was that determined by the solution equilibrium of the solid silver chloride in a solution where (Cl^-) = 1. Thus, if we insert the appropriate $\mathcal{E}_r^\circ$ values and $\mathcal{E}_r^x$ values, we should be able to calculate this equilibrium concentration of silver ion. This is done by the following steps:

$$\begin{aligned}
0.2222 &= 0.7995 - (0.05916/2) \log [1/(Ag^+)^2] \\
(Ag^+)^2 &= 10^{-19.6} \\
(Ag^+) &= 10^{-9.8} = 10^{0.2} \times 10^{-10} \\
(Ag^+) &= 1.6 \times 10^{-10}
\end{aligned}$$

The K_{sp} expression is, of course, given by

$$K_{sp} = (Ag^+)(Cl^-)$$

In the case at hand, the chloride ion concentration was adjusted to 1 and the silver ion concentration is that calculated from the value of the potential developed. Thus we find, for the K_{sp},

$$K_{sp} = (1.6 \times 10^{-10})(1) = 1.6 \times 10^{-10}$$

Example Problem 3: A chemist constructs a cell similar to that described in Example Problem 2. The silver electrode is again immersed in a 1 molal solution of silver nitrate. The hydrogen electrode (i.e., the platinum electrode exposed to hydrogen gas at 1 atm) is immersed in the solution of unknown hydrogen ion concentration. The emf developed from the cell is 0.750 volt. What is the pH of the unknown solution?

Here again, the reaction we are dealing with is

$$2Ag^+ + H_2 \rightleftarrows 2H^+ + Ag$$

In the unknown solution, the hydrogen ion concentration is unknown but it can be calculated from the formula

$$\mathcal{E}_r^x = \mathcal{E}_r^\circ - \left(\frac{0.05916}{2}\right) \log \frac{(H^+)^2}{(H_2)(Ag^+)^2}$$

In this particular instance, (Ag^+) and (H_2) are both unity and only the (H^+) need be considered. Let us now insert the appropriate values of $\mathcal{E}_r^\circ$ (0.7995) and $\mathcal{E}_r^x$ and calculate what the hydrogen ion concentration or the pH must be in this particular case. This can be done in the following way:

$$\begin{aligned}
0.750 &= 0.7995 - (0.05916/2) \log (H^+)^2 \\
-0.0495 &= -(0.05916/2) \log (H^+)^2 = -0.05916 \log (H^+) \\
-\log (H^+) &= pH = -(0.0495/0.05916) = -0.837
\end{aligned}$$

Notice that, in this particular case, the pH is linearly related to the emf developed by the cell. We can determine the hydrogen ion concentration of any solution by the application of these principles. Commercial instruments utilizing a cell reaction similar to that pictured above are available for the purpose of measuring the pH of a solution. This apparatus is appropriately called a pH meter.

In the last two examples, we saw how the study of the emf developed by various cells enables us to determine the solubility products of sparingly soluble salts and the pH of any solution. Given the pH, of course, we can calculate the ionization constant of a weak acid or a weak base; moreover, we can extend measurements of this type to measurements of the instability constants of complex ions and a host of similar problems. Thus all aspects of equilibria in solution can be aided by studies of appropriate oxidation-reduction reactions.

At this point, it is probably well to remind you, once again, that the equilibrium constant, using concentrations, holds rigorously only for dilute solutions. In more concentrated solutions, the positively charged ions interact with the negatively charged ions, and, in addition, the solvation effects become more complicated. Both of these effects tend to make the equilibrium constant vary a little as concentrations change, and, for the most precise work, these factors must be taken into account either directly or indirectly. In the crudest sense, the inaccuracy of the equilibrium constant under such conditions is analogous to the departure from the ideal gas law at high pressures, i.e., at high concentrations. In most of the cases of interest to us, however, we can make fairly accurate calculations using these equilibrium constants without correcting for the interaction of these ions. This is a simple consequence of the fact that, in most cases, we are dealing with fairly dilute solutions.

Example Problem 4: A solution contains the following compounds at one molal concentrations: $Hg_2(NO_3)_2$, $AgNO_3$, $Cu(NO_3)_2$, $Co(NO_3)_2$, $Mg(NO_3)_2$. Two inert electrodes are immersed in this solution, and the voltage is gradually increased. Specify what will be evolved or plated out at the cathode and anode, and in addition the voltage at which this will begin to take place. Assume $(H^+) = 10^{-7}$ at all times.

To see clearly what is going to take place in this experiment, let us first list the potentials in the emf series for each of these metals.

$$
\begin{array}{lll}
Ag & \rightarrow Ag^+ + e^- & -0.799 \text{ volt} \\
2Hg & \rightarrow Hg_2^{+2} + 2e^- & -0.789 \\
Cu & \rightarrow Cu^{+2} + 2e^- & -0.337 \\
Co & \rightarrow Co^{+2} + 2e^- & +0.277 \\
Mg & \rightarrow Mg^{+2} + 2e^- & +2.37
\end{array}
$$

In addition to these values we must also consider the fact that these reactions are taking place in an aqueous solution. Therefore there will also be the possibility of evolving oxygen at the anode or hydrogen at the cathode. The correct $\mathcal{E}^x$ values for such reactions are: O_2 from H_2O, $\mathcal{E}^x = -0.814$; H_2 from H_2O, $\mathcal{E}^x = +0.414$. The $\mathcal{E}^x$ for oxygen from water is the most negative of all of these; therefore, in all cases, oxygen will be evolved at the anode. (This will take place in preference to decomposition of NO_3^-, since the $\mathcal{E}°$ for NO_3^- is even more negative.) At the lowest value of the voltage, silver begins to plate out. The voltage at which this occurs is given by the value of the difference between the $\mathcal{E}$ values for Ag^+ and for O_2 from water, and is found to be 0.015 volt. As soon as most of the silver has plated out, the mercury begins to plate out and the voltage climbs to 0.025 volt, the difference between the $\mathcal{E}$ value for O_2 from H_2O and the mercurous ion. Likewise, when the mercury has plated out, the copper will begin to plate out and the voltage developed will be about 0.47 volt. When substantially all the copper has plated out, cobalt will begin to plate out and the voltage will be 1.09 volts.

After the cobalt has plated out, one might at first think that magnesium would start to plate out, but on closer examination we find that before the voltage became this high hydrogen would start to evolve at the cathode. This would occur when the voltage between the two electrodes had reached a value of 1.23 volts. The electrolysis would continue until all the water was changed into hydrogen and oxygen. Only if we were dealing with a fused salt would we expect to plate out magnesium.

Molarity vs. weight molarity. Throughout this chapter we have used weight molarity (molality) as the concentration unit to define the equilibrium constants and the standard states. We followed this procedure to conform to common practice in electrochemistry. On the other hand, in chapters dealing solely with calculations involving equilibrium constants, we express the concentrations as molarity. This is done purely for convenience and introduces no significant error. In the dilute concentration region where equilibrium calculations involving concentrations are permissible, molarity and molality are virtually the same. For example, a liter of solution containing 10.05 gm of NaCl is $0.172M$ and $0.173m$.

PROBLEMS

Use the emf table in the Appendix.

1. Calculate the voltage necessary to plate out at 25°C the metal in the following solutions: (*a*) $0.01m$ $AgNO_3$; (*b*) $5m$ $Co(NO_3)_2 \cdot 6H_2O$; (*c*) $10^{-4}m$ $CuSO_4$; (*d*) $0.03m$ $Ni(NO_3)_2 \cdot 6H_2O$.

Assume in each case that oxygen is evolved at 1 atm at the anode from an aqueous neutral solution. *Ans.* (*a*) 0.132 volt.

2. How long will it take to plate out 0.100 gm of each of the metals at a current of 5.00 amp from the following solutions: (*a*) $1m$ $CuSO_4$; (*b*) $0.01m$ $Pb(NO_3)_2$; (*c*) fused $TiCl_2$; (*d*) $1m$ $NiCl_2$.

3. Calculate the equilibrium constant of the reaction of zinc with the following ions: (*a*) Cs^+; (*b*) Pb^{+2}; (*c*) Cu^{+2}; (*d*) Ag^{+1}. *Ans.* (*b*) 2×10^{21}.

4. What will be the voltage produced when a battery is made up of a hydrogen electrode immersed in one molal nitric acid and a silver electrode in a concentrated $AgI(K_{sp} = 1.0 \times 10^{-16})$ solution?

5. In Problem 4, if the hydrogen electrode were immersed in a one molal solution of acetic acid ($K_I = 1.8 \times 10^{-5}$), what would be the voltage developed? *Ans.* 0.47 volts.

6. If ferrous ion is oxidized to ferric ion by permanganate ion in acid solution, what will be the equilibrium concentration of ferrous ion if the final concentrations of ions are: $(H^+) = 0.10$; $(Mn^{+2}) = 1$; $(MnO_4^-) = 10^3$; $(Fe^{+3}) = 1$. Assume $(H_2O) = 1$, in line with our conventions.

17

Oxidation-reduction reactions

In Chapter 16 we discussed the thermodynamic aspects of oxidation-reduction equilibria and the electrochemical methods by which equilibrium constants are determined. In this chapter we take up practical applications: the stoichiometry of redox (i.e., oxidation-reduction) reactions; the uses of such reactions in analytical chemistry; and the mechanism by which such reactions occur.

REDOX EQUATIONS

It is a simple matter to balance redox equations if one has available a complete table of half-cell equations (cf. p. 786). Such tables, however, are not always available. Therefore, let us see what we must know in order to obtain a balanced redox equation.

Half-cell method. To balance redox equations, we do not need $\mathcal{E}°$ values, but we do need balanced half-cell equations. We can always determine the half-cell equations (or *couple*) if we know the principal products and reactants. For example, suppose we must balance the equation

$$H^+ + MnO_4^- + Fe^{+2} = Fe^{+3} + Mn^{+2}$$

This unbalanced equation includes two half-cell equations; in one of them, permanganate ion is converted to manganous ion; in the second, ferrous ion is converted to ferric ion. The most foolproof procedure for balancing redox equations requires that we work first with each separate half-cell equation as if it were a chemical reaction involving electrons. To obtain the final equation, we add the two half-cell equations after changing the coefficients so that the number of electrons cancel out. (This procedure seems to imply that electrons are produced in the redox reaction and travel through the solution from the reducing agent to the reduced species. This is not true; we should view the individual steps of balancing these equations as a device for obtaining the final answer rather than as a mechanism related to the physical situation.)

Let us use the preceding reaction to illustrate the standard procedure for balancing redox equations.

1. Write down the half-cell reactant and products:

$$MnO_4^- = Mn^{+2}$$

2. Balance the number of atoms involved as follows: Add hydrogen ions and water to either side of the equation if the reaction takes place in an acidic solution. Add hydroxide ions and water to either side of the equation if the reaction takes place in a basic solution.

This can be done in two steps. First, balance oxygen atoms by adding four molecules of water to the right-hand side of the equation:

$$MnO_4^- = Mn^{+2} + 4H_2O$$

This leaves us with an equation that is still unbalanced, but we can rectify this by adding protons to the left side of the equation; that is,

$$8H^+ + MnO_4^- = Mn^{+2} + 4H_2O$$

3. Effect a balance of charge by adding negatively charged electrons to the left or right side. In the equation above, there are two plus charges on the right side and a net of seven plus charges on the left side; we can correct this by adding five electrons to the left side of the equation:

$$5e^- + 8H^+ + MnO_4^- = Mn^{+2} + 4H_2O$$

We now have the balanced half-cell equation for permanganate and manganous ions.

When we apply steps 1, 2, and 3 to obtain the other half-cell equation, we find

$$Fe^{+2} = Fe^{+3} + e^-$$

If we multiply both sides of this equation by 5, we produce the same number of electrons in this half-cell equation as are needed in the reaction of permanganate ion; hence we can add the two equations and obtain

$$8H^+ + MnO_4^- + 5Fe^{+2} = Mn^{+2} + 5Fe^{+3} + 4H_2O$$

To make sure that we understand this procedure, let us do a few examples.

Example Problem 1: What is the balanced equation for the oxidation of ferrous to ferric ion by the dichromate ion ($Cr_2O_7^{-2}$) in acid solution? The dichromate ion under these conditions yields Cr^{+3}.

First, indicate the reactants and products for the dichromate-chromic ion couple,

$$Cr_2O_7^{-2} = Cr^{+3}$$

Then effect an atom balance by adding water to the right side and H^+ to the left side:

$$14H^+ + Cr_2O_7^{-2} = 2Cr^{+3} + 7H_2O$$

Finally, balance the charge. Since there are six positive charges on the right side and twelve positive charges on the left side, we add six electrons to the left side to obtain the balanced half-cell equation:

$$6e^- + 14H^+ + Cr_2O_7^{-2} = 2Cr^{+3} + 7H_2O$$

Combination of this equation with the ferrous-ferric couple (after we have multiplied it by 6) yields the balanced equation:

$$6Fe^{+2} + 14H^+ + Cr_2O_7^{-2} = 2Cr^{+3} + 6Fe^{+3} + 7H_2O$$

Example Problem 2: Balance the equation

$$H_2C_2O_4 + H_2O_2 = H_2O + CO_2$$

First, separate the two half-cell equations. One of them involves the production of CO_2 from oxalic acid, $H_2C_2O_4$; the other involves the production of water from hydrogen peroxide. The three steps needed to obtain the balanced couple for the oxalic acid are

$$H_2C_2O_4 = CO_2$$

Atom balance: $$H_2C_2O_4 = 2CO_2 + 2H^+$$

Charge balance: $$H_2C_2O_4 = 2CO_2 + 2H^+ + 2e^-$$

The steps leading to the balanced couple for the peroxide are

$$H_2O_2 = H_2O$$

Atom balance: $$2H^+ + H_2O_2 = 2H_2O$$

Charge balance: $$2e^- + 2H^+ + H_2O_2 = 2H_2O$$

Combination of these two equations yields

$$H_2C_2O_4 + H_2O_2 = 2CO_2 + 2H_2O$$

Oxidation number method. Once you have acquired skill in balancing redox equations, it is not necessary to use the step-by-step procedure indicated above to obtain a balanced equation; instead, you can start with the over-all unbalanced equation and simply add coefficients, provided you proceed in a systematic manner. The easiest systematic approach involves the use of oxidation numbers, but you *must* remember that the oxidation number is an artificial concept introduced as a bookkeeping aid. It corresponds not to a real charge on the atom, but to an assigned charge introduced to assure the conservation of charge. As long as it is used in this manner, it will not lead to misconceptions.

Let us illustrate this method by working once again with the equation that describes the oxidation of ferrous by permanganate ion in acid solutions. The unbalanced equation is

$$\overset{}{H^+} + \overset{+7}{\underset{\underbrace{\uparrow}_{5e^-}}{Mn}}O_4^- + \overset{+2}{\underset{\underbrace{\downarrow}_{1e^-}}{Fe^{+2}}} = \overset{+3}{Fe^{+3}} + \overset{+2}{Mn^{+2}}$$

In this equation, we have indicated not only the products and the reactants, but also the oxidation numbers for the atoms undergoing substantial change in the reaction. In addition, we have indicated the electrons apparently gained or lost by these atoms when the reaction occurs. If we are to produce no free electrons, we must have five Fe^{+2} ions to yield the five electrons taken up by the permanganate ion. This changes the initial equation to

$$H^+ + MnO_4^- + 5Fe^{+2} = 5Fe^{+3} + Mn^{+2}$$

After accounting for the electron balance on the left side of the equation, we make the corresponding changes on the right side. The next step is to achieve an atom balance. Generally, this works best when we consider the oxygen atoms first. There are four oxygen atoms on the left side and none on the right side. In these equations, the added oxygen always appears as hydroxide ions or water molecules. Because this reaction takes place in an acid solution, we can add only water to effect the balance for the oxygen atoms. Thus

$$H^+ + MnO_4^- + 5Fe^{+2} = 5Fe^{+3} + Mn^{+2} + 4H_2O$$

We have now balanced this equation for all the atoms except the hydrogen atom; this final balance can be achieved if we add eight hydrogen ions to the left side:

$$8H^+ + MnO_4^- + 5Fe^{+2} = 5Fe^{+3} + Mn^{+2} + 4H_2O$$

In view of the multitude of steps involved, it is always wise to recheck the final equation to see if it is consistent with the conservation of charge and the conservation of atoms.

Example Problem 3: Give the balanced equation for the oxidation of chloride ion by permanganate ion in an acid solution.

In acid solution, the permanganate ion is reduced to manganous ion. The chloride ion upon oxidation can yield a variety of products; in this particular case, the product is chlorine gas. The steps are

Known: $$H^+ + \overset{+7}{\underset{\underbrace{\uparrow}_{5e^-}}{Mn}}O_4^- + \overset{-1}{\underset{\underbrace{\downarrow}_{1e^-}}{Cl^-}} = \overset{0}{Cl_2} + \overset{+2}{Mn^{+2}}$$

e^- balance: $$H^+ + MnO_4^- + 5Cl^- = \tfrac{5}{2}\,Cl_2 + Mn^{+2}$$

e^- balance cont.: $$H^+ + 2MnO_4^- + 10Cl^- = 5Cl_2 + 2Mn^{+2}$$

Oxygen atom balance: $$H^+ + 2MnO_4^- + 10Cl^- = 5Cl_2 + 2Mn^{+2} + 8H_2O$$

Atom balance cont.: $$16H^+ + 2MnO_4^- + 10Cl^- = 5Cl_2 + 2Mn^{+2} + 8H_2O$$

The new feature of this example is that, after the electron balance, we find a fractional coefficient in the equation. We eliminate this by multiplying the equation by 2 in the third step.

Example Problem 4: Give the balanced equation for the oxidation of oxalic acid by potassium permanganate in an acid solution.

The balancing procedure is

Known: $$H^+ + \underset{\underset{5e^-}{\uparrow}}{\overset{+7}{MnO_4^-}} + \underset{\underset{2e^-}{\underbrace{\downarrow e^-}}}{\overset{+3}{H_2C_2O_4}} = \overset{+4}{CO_2} + \overset{+2}{Mn^{+2}}$$

e^- balance: $$H^+ + 2MnO_4^- + 5H_2C_2O_4 = 10CO_2 + 2Mn^{+2}$$

Atom balance: $$6H^+ + 2MnO_4^- + 5H_2C_2O_4 = 10CO_2 + 2Mn^{+2} + 8H_2O$$

The new feature of this example is that the oxalic acid molecule contains two carbon atoms that are oxidized. Thus, although each carbon atom loses only one electron, the molecule *as a whole* loses two electrons.

Example Problem 5: Give the balanced equation for the oxidation of sulfite ion by permanganate ion in a slightly basic solution. Under these conditions, the permanganate ion is reduced to manganese dioxide.

Since the sulfur atom in the sulfite ion is in the +4 oxidation state, it is reasonable to guess that the sulfur atom is oxidized to the +6 oxidation state, i.e., sulfate ion; hence we write

$$\underset{\underset{3e^-}{\uparrow}}{\overset{+7}{MnO_4^-}} + \underset{\underset{2e^-}{\downarrow}}{\overset{+4}{SO_3^{-2}}} = \overset{+6}{SO_4^{-2}} + \overset{+4}{MnO_2}$$

From the electron balance, we get

$$2MnO_4^- + 3SO_3^{-2} = 3SO_4^{-2} + 2MnO_2$$

We now have one more oxygen atom on the left than we have on the right side. This can be taken care of by adding a hydroxide ion to the right and a hydrogen to the left:

$$H^+ + 2MnO_4^- + 3SO_3^{-2} = 3SO_4^{-2} + 2MnO_2 + OH^-$$

We must remember, however, that this reaction takes place in basic solution; hence any hydrogen ion present immediately reacts with the hydroxide to form water. Therefore, we rewrite the equation as follows:

$$HOH + 2MnO_4^- + 3SO_3^{-2} = 3SO_4^{-2} + 2MnO_2 + 2OH^-$$

Example Problem 6: Balance the equation

$$FeCr_2O_4 + K_2CO_3 + O_2 = Fe_2O_3 + K_2CrO_4 + CO_2$$

The O_2 is definitely undergoing a change in oxidation number; the iron and/or the chromium atoms are also undergoing a change in oxidation number. Actually, *we can use any interpretation, provided we are consistent.* To illustrate this, let us assume that the oxidation numbers for the iron and chromium are as indicated below:

$$\overset{0}{\mathrm{Fe}}\ \overset{+4}{\mathrm{Cr_2}}\ O_4 + \overset{+4}{K_2CO_3} + \overset{0}{O_2} = \overset{+3}{Fe_2O_3} + \overset{+6}{K_2CrO_4} + \overset{+4}{CO_2}$$

$$\underbrace{\underset{3e}{\downarrow}\ \underset{(2\times 2)e^-}{\downarrow}}_{7e^-} \qquad \underset{4e^-}{\uparrow}$$

Then, in this process, we find that the "molecule" $FeCr_2O_4$ loses a total of seven electrons, whereas each oxygen molecule gains four electrons. We effect a balance of the electrons by using 4 as the coefficient for $FeCr_2O_4$, 7 for the coefficient of oxygen, and 2 and 8 for the compounds Fe_2O_3 and K_2CrO_4, respectively. These require coefficients of 8 for the K_2CO_3 and CO_2 in order to obtain an oxygen balance. The final equation is

$$4FeCr_2O_4 + 8K_2CO_3 + 7O_2 = 2Fe_2O_3 + 8K_2CrO_4 + 8CO_2$$

This assignment of oxidation numbers is not unique. We could just as well assume that the iron undergoes no change in oxidation number and is originally in the +3 oxidation state. This assumption requires that the initial oxidation state of the chromium be 2.5 and leads to the equation

$$\underbrace{\overset{+3}{\mathrm{Fe}}\ \overset{+2.5}{\mathrm{Cr_2}}\ O_4}_{\underset{7e^-}{\downarrow}} + K_2CO_3 + \overset{0}{O_2} = Fe_2O_3 + \overset{+6}{K_2CrO_4} + CO_2 \qquad (O_2:\ \underset{4e}{\uparrow})$$

Once again, for electron balance, we must have coefficients of 4 and 7 for the $FeCr_2O_4$ and O_2, respectively. The subsequent balance of atoms leads, as expected, to the same equation that we obtained when we assumed that initially the oxidation state of iron was 0 and that of chromium was 4. These results emphasize that the oxidation number is a convenient bookkeeping device rather than an actual charge.

Example Problem 7: Sulfuric acid reacts with cellulose, $C_{12}H_{22}O_{11}$, to form carbon dioxide, sulfur dioxide, and water. Below, we have indicated the unbalanced equation:

$$\overset{0}{C_{12}}\ \overset{+1}{H_{22}}\ \overset{-2}{O_{11}} + \overset{+6}{H_2SO_4} = \overset{+4}{CO_2} + \overset{+4}{SO_2} + H_2O$$

$$\underset{(12\times 4)e^-}{\downarrow} \qquad \underset{2e^-}{\uparrow}$$

We have also indicated a consistent set of oxidation numbers for the various atoms. (Here we find that the carbon atoms in cellulose have the same oxidation number as the carbon atoms in diamond or in graphite; this is ridiculous if we associate the oxidation number with a real charge on the atom.) We conclude that in the reaction each carbon atom loses four electrons and each molecule of cellulose loses a total of 48 electrons. An electron balance can be obtained only if we combine each molecule of cellulose with 24 molecules of sulfuric acid each of which gains two electrons. From here on, the procedure is standard, and we obtain for our final equation

$$C_{12}H_{22}O_{11} + 24H_2SO_4 = 12CO_2 + 24SO_2 + 35H_2O$$

In addition to the above cases, wherein all the atoms oxidized had the same oxidation state, we also find cases in which two atoms of the same element in a given molecule have different oxidation states. For example, the structure of the thiosulfate ion $S_2O_3^{-2}$ is

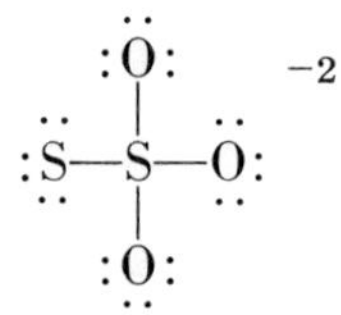

Thus $S_2O_3^{-2}$ is an analog of the sulfate ion wherein one of the oxygen atoms is replaced by a sulfur atom. For the sulfate ion, the oxidation state of sulfur is readily apparent. Oxygen is the more electronegative element in SO_4^{-2} ion; consequently, in all its bonds with sulfur, we *assign* the shared pair to the oxygen. This means that in this imaginary process the sulfur atom is denuded of its outer electrons and the oxidation state of sulfur is $+6$. For the thiosulfate ion, the situation is more complex. The electron in the sulfur-oxygen bonds is assigned to the oxygen. For sulfur-sulfur bonds, according to the rules (p. 160), we assign one electron of the shared pair to the central atom and the other electron to the outer sulfur atom. As a result, the central sulfur atom apparently has one electron in its outer shell, and has an assigned charge (oxidation number) of $+5$. The outer sulfur atom apparently has seven electrons in its outer shell; hence its oxidation number is -1. The *average oxidation number* for sulfur in this compound is

$$\frac{+5-1}{2} = 2$$

The same result is obtained even if we do not know the structure of the ion. The ion, $S_2O_3^{-2}$, has three oxygen atoms in the -2 oxidation state, contributing a "charge" of -6. For the "charge" on the whole ion to be -2, the "charge" on the sulfur atoms must be $+4$ *in toto* or $+2$ apiece.

A similar situation is found for $S_4O_6^{-2}$, the tetrathionate ion. Tetrathionate ion has the structure

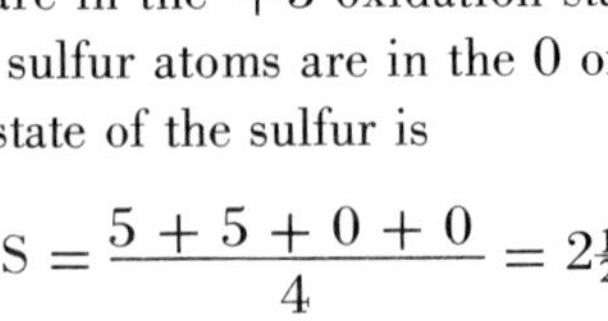

On the basis of analysis similar to that above, we find that the two tetracoordinated sulfur atoms are in the $+5$ oxidation state, whereas the two sulfur atoms bound only to other sulfur atoms are in the 0 oxidation state. This means that the average oxidation state of the sulfur is

$$S = \frac{5+5+0+0}{4} = 2\tfrac{1}{2}$$

We arrive at the same value, $2\frac{1}{2}$, from the formula alone.

Thiosulfate ion reacts with free iodine to form iodide and tetrathionate ion. The balanced equation for this reaction is

$$\underbrace{\overset{+2}{2S_2O_3^{-2}}}_{\downarrow\ 1e} + \underbrace{\overset{0}{I_2}}_{\uparrow\ 2e} = \overset{+2.5}{S_4O_6^{-2}} + \overset{-1}{2I^-}$$

The important point is that, *because the oxidation number concept is an artificial one* designed to guide us in balancing oxidation-reduction equations, we occasionally find fractional *assigned* charges for atoms.

Auto-oxidation-reduction. For many of the atoms in the periodic chart, there are more than two common oxidation states. If such an atom is in an intermediate oxidation state, it can be both oxidized and reduced; then self- or *auto-oxidation-reduction* can occur. In other words, one of these atoms can be oxidized while the other is reduced.

Hydrogen peroxide undergoes an auto-oxidation-reduction when it decomposes to form oxygen and water. From the standpoint of oxidation and reduction, we explain this reaction as follows: In hydrogen peroxide, H_2O_2, the oxygen atom is in the -1 oxidation state. In this decomposition, one hydrogen peroxide molecule reacts with the other hydrogen peroxide molecule in such a way that one set of oxygen atoms loses electrons and the other gains electrons; this yields the equations

$$\underbrace{\overset{-1}{H_2O_2}}_{\downarrow\ 2e} + \underbrace{\overset{-1}{H_2O_2}}_{\uparrow\ 2e} = \overset{-2}{2H_2O} + \overset{0}{O_2}$$

or

$$2H_2O_2 = 2H_2O + O_2$$

Auto-oxidation-reduction also occurs with the hypochlorite ion, ClO^-. In this ion, the chlorine atom is in the $+1$ oxidation state, which is intermediate between the extreme oxidation states of chlorine. *Either* oxidation or reduction is possible, and a reaction does occur to form chlorate ion and chlorine ion. The balanced equation for this reaction is

$$\underbrace{\overset{+1}{ClO^-}}_{\downarrow\ 4e} + \underbrace{\overset{+1}{2ClO^-}}_{\uparrow\ 2e} = 2Cl^- + ClO_3^-$$

or

$$3ClO^- = 2Cl^- + ClO_3^-$$

In some molecules, it is possible to have an interatomic oxidation-reduction. For potassium chlorate, $KClO_3$, the chlorine is in a $+5$ oxidation state (the oxidation number can either increase or decrease), and the oxygen is in the -2 oxidation state (the oxidation number can only increase). Thus we have the possibility of the chlorine atom gaining electrons and the oxygen atom losing electrons. This takes place as follows:

$$\underset{\substack{\uparrow\\6e}}{KCl}\ \underset{\substack{\downarrow\\6e}}{\underbrace{O_3}} = KCl + \tfrac{3}{2}O_2$$

or

$$2KClO_3 = 2KCl + 3O_2$$

ANALYTICAL APPLICATIONS OF REDOX REACTIONS

Titrations. Analytical applications of redox reactions are as extensive as those of acid-base reactions. For example, the amount of iron in an ore can be determined from the reaction

$$14H^+ + Cr_2O_7^{-2} + 6Fe^{+2} = 6Fe^{+3} + 2Cr^{+3} + 7H_2O$$

The sample to be analyzed is dissolved, converted to ferrous ion, and titrated with dichromate solution. This, of course, requires that the reaction go to completion rapidly, and that we have some way of knowing when all of the ferrous ion is converted to ferric ion.

If a solution is approximately 6 molar in ferrous ion, it requires 50 ml of 1 molar dichromate solution to titrate 50 ml of the iron solution. With each addition of dichromate to the iron solution, equilibrium is established; therefore the tendency for ferrous ions to give up electrons is balanced by the tendency of the chromic ions to give up electrons. Although the ferrous-ferric couple is balanced by the chromic-dichromate couple throughout the course of this titration, the ferrous-ferric couple may undergo a reaction with another species. The relative tendency for such a reaction to occur at any concentration of ferrous and ferric ion (or, equivalently, at any concentration of chromic and dichromate ion) is determined by the Nernst equation. In particular, for the reaction of ferrous ion with hydrogen ion to form elementary hydrogen and ferric ion, we can write

$$\mathcal{E}_T^x = -0.77 - 0.05914 \log \frac{(Fe^{+3})(H_2)^{1/2}}{(Fe^{+2})(H^+)}$$

Since $\mathcal{E}_T^x$ is related to the free energy change by the equation

$$\Delta F = -n\mathcal{F}\mathcal{E}_T^x$$

the greater the value of $\mathcal{E}_T^x$, the greater will be the driving force of the reaction.

During the course of the titration, the value of $\mathcal{E}_T^x$, which is a measure of the oxidizing power of the solution, changes. Values of $\mathcal{E}_T^x$ at various stages in the

Table 1 Titration of 50 Ml of 6*M* $FeSO_4$ with 1*M* $K_2Cr_2O_7$

ml $K_2Cr_2O_7$ added	Fe^{+3}/Fe^{+2}	$\mathcal{E}_r^x$
0	10^{-3}	−0.59
25	1	−0.77
49	50	−0.89
49.9	500	−0.95
50.1	10^{-10}	−1.33

titration are shown in Table 1. (It is assumed that initially there is a thousand times as much ferrous ion and ferric ion.) Values of $\mathcal{E}_r^x$ versus milliliters of dichromate solution added are plotted in Fig. 1. These titration curves are similar to plots of pH versus milliliters of reactants added in acid-base titrations (p. 455).

During the titration $\mathcal{E}_r^x$, the oxidizing ability of the solution, increases from that associated with the iron solution to that associated with the dichromate solution. (The $\mathcal{E}^\circ$ values for the dichromate-chromic couple is -1.34 volts.) Moreover, this increase is most rapid at the end point; here the addition of a few drops of dichromate solution changes the $\mathcal{E}_r^x$ value by about 0.4 volt.

Let us suppose that a small amount of X^{-2} is present in the iron solution and that X^{-2} can lose electrons as indicated in the equation

$$\underset{\text{blue}}{X^{-2}} = \underset{\text{red}}{X} + 2e^- \qquad \mathcal{E}_r^x = -1.1$$

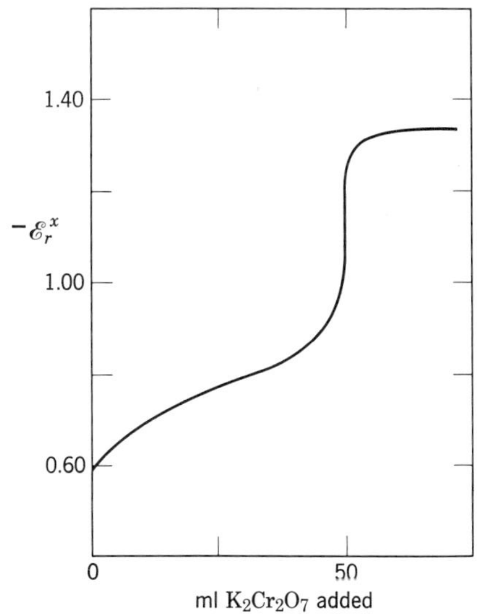

Fig. 1. Redox titration curve.

In its reduced form X^{-2} imparts to the solution a blue color, and in its oxidized form X imparts to the solution a red color. As long as the $\mathcal{E}$ value for the solution is above -1.1 volts, the principal species will be the blue variety; as soon as the $\mathcal{E}_T^\circ$ value of the solution is less than -1.1 volts, the principal species will be the red variety. Thus the X^{-2} initially present in the iron solution makes the solution blue. This blue color persists up to the end point; then, after the addition of one or two drops of dichromate solution, X^{-2} changes to X and the color of the solution changes from blue to red. A species such as X is an indicator for a redox titration.

The existence of a wide variety of redox indicators allows us to use almost any oxidizing reagent in redox analytical work. In some cases, however, the reactant itself changes color during the reaction and will act as its own indicator. This occurs in the reaction

$$8H^+ + MnO_4^- + 5Fe^{+2} = Mn^{+2} + 5Fe^{+3} + 4H_2O$$

The color of the permanganate ion itself is so intense that, as soon as an excess of permanganate is present, i.e., all the ferrous ion is oxidized, the solution turns pink; that is, permanganate acts as its own indicator.

The principles of the redox reactions, themselves, are straightforward. Often, however, a particular analytical procedure has a number of extra steps that must be followed in order to obtain accurate results. For example, in one method of analyzing for ferrous ion with permanganate solutions, a solution containing manganous ion and phosphate ion is added before the titration. This is required because the reaction is carried out with a solution containing chloride ions that permanganate can oxidize to chlorine. If this occurs, some of the permanganate ion is used for the oxidation of chloride ion instead of for the oxidation of the ferrous ion; hence the reaction is no longer specific for Fe^{+2}. In the presence of an excess of manganous ion, however, the oxidizing power of the permanganate is reduced so that it will no longer oxidize chloride ion. Of course, this also reduces the oxidizing power of permanganate ion for ferrous ion; and, if this were all that were done, the oxidation reaction would no longer be quantitative. This difficulty is overcome by the addition of phosphate ion. The resulting formation of a complex with ferric ion assures that the reaction is quantitative. Each step of any standardized determination has a rational basis; departures from the standardized procedure will lead to errors that may be quite large.

Equivalence and normality. In dealing with acids and bases, we found that the concept of an equivalent weight was useful because the common feature of all acid-base reactions is the loss of a hydrogen ion by one of the reactants, the acid, and the gain of a hydrogen ion by the other reactant, the hydroxide ion. In a similar fashion, the basis of all oxidation-reduction reactions is the *apparent* loss of electrons by one of the reactants and the *apparent* gain of electrons by the other reactant. Thus it seems natural to define an equivalent weight for an oxidation-reduction reaction. *This* equivalent weight is *that weight of material that yields or gains 1 mole of electrons in a reaction.* To make sure that this is understood, let us consider the reaction between permanganate ion and oxalic acid.

$$6H^+ + 2\overset{+7}{Mn}O_4^- + 5H_2\overset{+3}{C_2}O_4 = 10\overset{+4}{C}O_2 + 2\overset{+2}{Mn^{+2}} + 8H_2O$$
$$\underset{(2 \times 5)e^-}{\uparrow} \qquad \underset{(5 \times 2)e^-}{\downarrow}$$

We can arrange the coefficients so that only 1 mole of electrons is lost or gained by the ions or compounds involved. Then the equation becomes

$$\tfrac{6}{10}H^+ + \tfrac{1}{5}MnO_4^- + \tfrac{1}{2}H_2C_2O_4 = CO_2 + \tfrac{1}{5}Mn^{+2} + \tfrac{8}{10}H_2O$$

Thus, $\frac{1}{2}$ mole of $H_2C_2O_4$ is *equivalent* to 1 mole of electrons and reacts with $\frac{1}{5}$ mole of $KMnO_4$, the *equivalent* of 1 mole of electrons. This equation contains the same stoichiometry as the original equation, and it is an equally good representation of the facts. Just as for acids and bases, the justification for this approach lies in the fact that now one equivalent reacts with one equivalent.

The equivalent weight depends on the particular reaction involved and is not always the same for the same compound. For example, in an acidic solution, permanganate ion gains five electrons to form manganous ion; hence in such reactions the equivalent weight is one-fifth of the molecular weight. In a neutral or slightly basic solution, permanganate ion reacts to form MnO_2; hence, under these conditions, the permanganate ion gains three electrons so that the equivalent weight is one-third of the molecular weight. Finally, in a strongly basic solution, we find that permanganate ion changes to manganate ion with the apparent gain of one electron; therefore, in basic solutions, the equivalent weight of potassium permanganate is equal to its molecular weight.

We define the normality of an oxidizing or reducing agent in an aqueous solution as follows:

$$\text{Normality} = \frac{\text{No. of moles of electrons gained or lost}}{\text{1 liter of sol'n}}$$

If we multiply normality by the volume used, expressed in liters, this yields the number of equivalents of oxidizing (or reducing) agent, i.e., the number of moles of electrons gained (or lost). This must be equal to the number gained (or lost) by the other reactant. Thus, for complete reaction, the number of equivalents of oxidizing agent must equal the number of equivalents of reducing agent, or

$$V_O N_O = V_R N_R$$

where N_O is the normality of the oxidizing agent, V_O is the volume of this solution required to react completely with the volume, V_R, of reducing agent, which has a normality of N_R.

Example Problem 8: A sample of 0.2794 gm of $Na_2C_2O_4$ requires 40.17 ml of approximately 0.10 *N* $KMnO_4$ for complete reaction. What is the precise normality of $KMnO_4$?

First of all, calculate the formula weight of sodium oxalate. This turns out to be 134.00. In the reaction that occurs in acid solution, 1 mole of sodium oxalate is equivalent to 2 moles of electrons; hence the equivalent weight is 67.00 gm. Now calculate the number of equivalents of sodium oxalate present as follows:

$$? \text{ eq. } Na_2C_2O_4 = 279.4 \times 10^{-3} \text{ gm } Na_2C_2O_4 \times \frac{1 \text{ eq.}}{67.00 \text{ gm}}$$

$$= 4.170 \times 10^{-3} \text{ eq.}$$

The 40.17 ml of potassium permanganate required to react completely with the sodium oxalate contains a like number of equivalents. Thus, for the potassium permanganate solution, we write:

$$\frac{4.170 \times 10^{-3} \text{ eq. } KMnO_4}{(40.17) \times 10^{-3} \text{ liter of sol'n}}$$

From this we find out that 1 liter of solution contains 0.1038 equivalent, or the normality of the solution is 0.1038. If we made two other runs that yielded the values 0.1045 and 0.1044, the best value, the one that we believe to be closest to the truth, for the normality of potassium permanganate would be the average value of these three runs; that is,

	Deviation
0.1038	.0004
0.1045	.0003
0.1044	.0002
0.1042 Av.	.0003 Av.

Here the .0003 represents the average of the deviations.

Example Problem 9: The solution of 0.1042*N* potassium permanganate is used to determine the equivalent weight of unknown X. In this analysis, 0.2639 gm of X is titrated with 73.19 ml of potassium permanganate. What is the equivalent weight of X?

The first step in this problem is to calculate how many equivalents of potassium permanganate reacted:

$$73.19 \times 10^{-3} \text{ liter} \times \frac{0.1042 \text{ eq. } KMnO_4}{\text{liter of sol'n}} = 7.626 \times 10^{-3} \text{ eq.}$$

The sample X, therefore, contains 7.626×10^{-3} equivalent of reducing agent, or the weight of 1 equivalent is

$$\frac{263.9 \times 10^{-3} \text{ gm X}}{7.626 \times 10^{-3} \text{ eq.}} = \frac{34.60 \text{ gm}}{1 \text{ eq.}}$$

If we made two other determinations and obtained the values 34.81 and 34.76, the best value would be the average, viz:

	Deviation
34.81	.09
34.76	.04
34.60	.12
34.72 Av.	.08 Av.

THE MECHANISM OF REDOX REACTIONS

On the basis of the half-cell reactions, the most obvious mechanism of a redox reaction would involve a dissociation of the type

$$Fe^{+2} \longrightarrow e^- + Fe^{+3}$$

with the formation of an aqueous electron. This electron could then wander through the solution and attach itself to the substance to be reduced. Although such may be the case in solutions in liquid ammonia, it is definitely not the case in aqueous solutions.

Even though electron transfer in which the electron persists for a time in the aqueous solution is extremely unlikely, it does seem reasonable that in some reactions a direct transfer of electrons takes place between the oxidizing and the reducing agent. Consider, for example, the reaction

$$Fe(CN)_6^{-3} + MnO_4^{-2} \longrightarrow MnO_4^- + Fe(CN)_6^{-4}$$

Both the manganese and the iron atom are strongly attached to the oxide and cyanide groups. When a manganate ion and a ferrocyanide ion "collide" in solution, it is likely that an electron is transferred from one species to another.

Many redox reactions involve a transfer of atoms rather than a transfer of electrons. This occurs most readily when the groups attached to the anion or cation are not tightly held. Consider the reaction

$$Fe^{+3} + I^- \longrightarrow Fe^{+2} + \tfrac{1}{2}I_2$$

To be realistic about this, let us assume that sodium and chloride ions are also present. In this event, the cations in solution might contain one or more chloride ions in the hydration shell in place of water molecules. This is expected because iron forms a complex ion with chloride ion. Then we rewrite the reaction:

$$Fe:\underset{\cdot\cdot}{\overset{\cdot\cdot}{Cl}}:^{+2} + :\underset{\cdot\cdot}{\overset{\cdot\cdot}{I}}:^- \longrightarrow \cdot Fe:\underset{\cdot\cdot}{\overset{\cdot\cdot}{Cl}}:^{+1} + \tfrac{1}{2}:\underset{\cdot\cdot}{\overset{\cdot\cdot}{I}}:\underset{\cdot\cdot}{\overset{\cdot\cdot}{I}}:$$

Here we have indicated the number of electrons around each species. (For simplicity in the case of iron, we have indicated that there are no electrons around the atom in the +3 oxidation state.) Mechanistically, oxidation-reduction *could* occur by the following steps, none of which involves a transfer of an unattached electron.

$$Fe:\underset{\cdot\cdot}{\overset{\cdot\cdot}{Cl}}:^{+2} \rightarrow Fe\cdot^{+2} + \cdot\underset{\cdot\cdot}{\overset{\cdot\cdot}{Cl}}:$$

$$:\underset{\cdot\cdot}{\overset{\cdot\cdot}{Cl}}\cdot + :\underset{\cdot\cdot}{\overset{\cdot\cdot}{I}}:^- \rightarrow :\underset{\cdot\cdot}{\overset{\cdot\cdot}{Cl}}:^- + \cdot\underset{\cdot\cdot}{\overset{\cdot\cdot}{I}}:$$

$$\tfrac{1}{2}(:\underset{\cdot\cdot}{\overset{\cdot\cdot}{I}}\cdot + \cdot\underset{\cdot\cdot}{\overset{\cdot\cdot}{I}}: \rightarrow :\underset{\cdot\cdot}{\overset{\cdot\cdot}{I}}:\underset{\cdot\cdot}{\overset{\cdot\cdot}{I}}:)$$

$$:\underset{\cdot\cdot}{\overset{\cdot\cdot}{Cl}}:^- + Fe\cdot^{+2} \rightarrow \cdot Fe:\underset{\cdot\cdot}{\overset{\cdot\cdot}{Cl}}:^+$$

The first of these reactions indicates the breaking of the iron-chlorine bond with the formation of a chlorine atom. The second reaction indicates the removal of an electron from the iodide ion by the chlorine atom. The iodine atoms thus produced recombine to form iodine. The last step represents the recombination of the chloride ion, formed in the preceding reaction, with ferrous ion. The overall result is that we have transformed the iron from a +3 to a +2 oxidation state and transformed the iodide ion from the −1 to the 0 oxidation state. Thus it is not necessary to invoke an electron transfer; a change in oxidation state can be accomplished by a transfer of atoms.

PROBLEMS

1. Complete and/or balance the following equations:

(*a*) $Br_2 + CO_3^{-2} + H_2O = Br^- + BrO_3^- + HCO_3^-$
(*b*) Acid solution: $MnO_2 + Cl^- = Mn^{+2} + Cl_2$
Ans. $4H^+ + MnO_2 + 2Cl^- = Mn^{+2} + Cl_2 + 2H_2O$
(*c*) Acid solution: $Cu + NO_3^- = NO + Cu^{+2}$
Ans. $8H^+ + 3Cu + 2NO_3^- = 3Cu^{+2} + 2NO + 4H_2O$
(*d*) Acid solution: $HgS + Cl^- + NO_3^- = HgCl_4^{-2} + S + NO$
(*e*) Basic solution: $MnO_4^- + SO_3^{-2} = MnO_4^{-2} + SO_4^{-2}$
(*f*) Acid solution: $S_2O_5^{-2} + H_2O = SO_4^{-2} + H_2O_2$
(*g*) Non-aqueous medium: $H_3PO_3 = H_3PO_4 + PH_3$
(*h*) Basic solution: $P_4 = H_2PO_2^- + PH_3$
Ans. $P_4 + 3H_2O + 3OH^- = 3H_2PO_2^- + PH_3$
(*i*) Non-aqueous: $MnO_2 + KOH + O_2 = K_2MnO_4$
(*j*) Acid solution: $MnO_4^{-2} = MnO_4^- + MnO_2$
(*k*) Non-aqueous: $SiCl_4 + Na = Si + NaCl$
(*l*) Acid solution: $I_2 + NO_3^- = NO_2 + IO_3^-$

2. Compute the equivalent weight of the following substances if they are used to supply the oxidizing or reducing agent in the indicated equations: (*a*) MnO_2 in Problem 1*b*; (*b*) HNO_3 in 1*c*; (*c*) Na_2SO_3 in 1*e*; (*d*) $SiCl_4$ in 1*k*; (*e*) I_2 in 1*l*. *Ans.* (*b*) 21.0 gm/eq.

3. A solution of $KMnO_4$ was standardized in acid solution with iron wire. The iron was first dissolved in HCl, converted in $FeCl_2$, and this solution was used for the titration (0.523 gm of iron wire is required to react with 45.23 ml of $KMnO_4$). What was the normality of this $KMnO_4$?

4. A solution of $K_2Cr_2O_7$ was standardized in acid solution with $Na_2C_2O_4$ (0.3754 gm of $Na_2C_2O_4$ requires 34.32 ml of $K_2Cr_2O_7$ for complete reaction). What was the normality of the potassium dichromate? What was the molarity of the potassium dichromate?

5. 32.42 ml of 0.1047*N* $KMnO_4$ was required to titrate 1.032 gm of the reducing agent in acid solution. What was the equivalent weight of the reducing agent? In acid solution the $KMnO_4$ is converted to Mn^{+2} ion. *Ans.* 304.0 gm/eq.

6. 32.42 ml of 0.1047*M* potassium permanganate was required to titrate 1.032 gm of reducing agent in acid solution. What was the equivalent weight of the reducing agent?

7. 14.32 ml of 0.1742*N* $K_2Cr_2O_7$ was needed to titrate a 0.374-gm sample of wire containing iron. (The wire was dissolved in acid and converted to Fe^{+2}. Assume that the other material present does not react with potassium dichromate.) What is the per cent of iron in this wire?

8. If a standardization is carried out in the laboratory, what is the concentration of unoxidized $H_2C_2O_4$ when we have added 40 ml of 0.1000*N* $KMnO_4$ to 80 ml of a solution of 0.0500*N* $H_2C_2O_4$? Assume that the (H^+) is 1 and the pressure of CO_2 is 1 atm. Calculate the concentration of $H_2C_2O_4$ when we add two additional drops of $KMnO_4$ (0.05 ml apiece). The assumed half-cell equation for $H_2C_2O_4$ is

$$H_2C_2O_4 = 2CO_2 + 2H^+ + 2e^- \qquad \mathcal{E}^\circ = 0.19$$

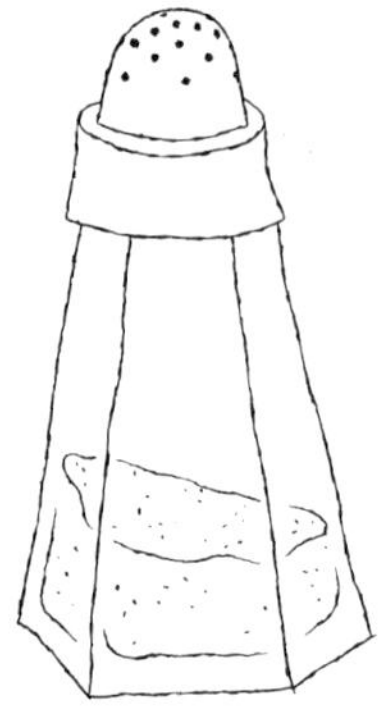

18 Chemistry of Group I, II, and III elements

Of all the varieties of chemistry found in the periodic chart of the elements, the simplest is associated with the members of Group 1. For these are the elements most closely related to the simplest atom of all, hydrogen. Just as hydrogen has but a single electron surrounding its nucleus when in the solitary neutral state, so the elements in Group I are characterized largely by the single electron in the outer atomic shell; and the inner electrons screen the bulk of the positive charge on the nucleus so effectively that this single outer electron moves under the influence of a positive electric field roughly equivalent to that of a single unit positive charge. Thus these elements of Group I may be regarded as stemming from a hydrogen prototype.

There is a similar kind of simplicity in the chemistry of the elements of Group II; it also is due to the screening of the bulk of the positive charge so that the two outer electrons move in a field roughly equivalent to that of two unit positive charges at the center of the atom; and to a lesser extent the same principle holds for the members of Group III, although there is always an increase in complexity of behavior associated with elements nearer the center of the periodic table.

ELECTRONIC STRUCTURES

In surveying the chemistry of the different groups in the periodic table, we want to relate chemical behavior to electronic structure as often as possible. To this end, we summarize in this section a number of the principles of electronic structure that have been discussed in previous chapters.

The diagrams of the electronic structures of the elements of Group I are shown in Table 1. At the bottom of the table there is a key diagram summarizing the meaning of the values and positions of the numbers as discussed in detail in Chapter 4. Each *column* (n = 1,2,3,4,5,6, or 7) corresponds to a shell in the electron cloud. An s electron with the quantum number $n = 1$ has one spherical loop or zone of maximum charge concentration in the wave cloud; for $n = 2$, there are two such spherical zones; and so on. Each *row* specifies the symmetry of the electron cloud (l = 0,1,2,3, etc.) or s, p, d, and f. The simplest symmetry is the spherical symmetry corresponding to the s row; in this electron cloud all radial directions are equivalent. As a result there can be only two electrons in an s shell, the distinction between them being the two possible spin quantum numbers usually written $+\frac{1}{2}$ and $-\frac{1}{2}$. The next in order of decreasing symmetry is the p type, represented by a dumb-bell-shaped region of charge concentration. This dumb-bell can be oriented along any of the three space axes, x, y, or z. Since to each of these 3 orientations there are the 2 possible spin orientations, there are $2 \times 3 = 6$ distinct types of p electrons. In a similar way the still less symmetrical d waves have 10 distinct types, and the waves with the still more complex f symmetry have 14 distinct types. Thus each lowering of the symmetry adds 4 new types of electrons: $s = 2$, $p = 6$, $d = 10$, $f = 14$.

The symmetries specified by the n and l quantum numbers are not independ-

Table 1 Electronic Structures of Group I Elements

Element	Row	1	2	3	4	5	6	7
Li	s	2	1					
Na	p		6					
	s	2	2	1				
K	p		6	6				
	s	2	2	2	1			
Rb	d			10				
	p		6	6	6			
	s	2	2	2	2	1		
Cs	d			10	10			
	p		6	6	6	6		
	s	2	2	2	2	2	1	
Fr	f				14			
	d			10	10	10		
	p		6	6	6	6	6	
	s	2	2	2	2	2	2	1

Key:

	l ↓	1	2	3	4	5	6	7
f	3				14			
d	2			10	10	10		
p	1		6	6	6	6	6	
s	0	2	2	2	2	2	2	2

Shell number → 1 2 3 4 5 6 7
Shell letter → *K L M N O P Q*

ent. The number of high charge density regions as we go out along a radial line is $n - l$. Thus an s function always has n dense regions, a p function has $n - 1$ dense regions, and a d function has $n - 2$ dense regions, and so on. Alternatively we could say that along a radial line (excluding the origin) there are $n - l + 1$ places where the charge density is zero. The symmetry specified by the l values is more complex. The number of surfaces (not necessarily planes) going through the nucleus on which the charge density is equal to zero is l. Thus there are zero such surfaces for an s function, one such surface for a p function, and two such surfaces for a d function.

The elements of Group II have electronic structures that are represented by *symmetrical diagrams.* Thus radium has the diagram

14
10 10 10
6 6 6 6 6
2 2 2 2 2 2 2

Since the chemistry of all the elements is so intimately connected with their electronic structures, it is helpful to have the *es* diagram (electronic structure diagram) of an element constantly in mind when discussing its chemistry. Fortunately this symmetry property of the diagrams makes them easy to remember. To make this clearer we can insert diagonal guide lines in the diagram for radium:

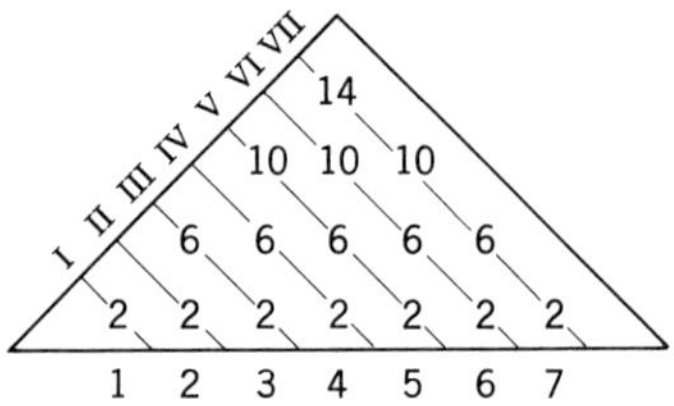

There are seven of these diagonal lines, designated by Roman numerals. Starting with line II, we write down all the diagrams that are symmetrical about a vertical central axis, thus:

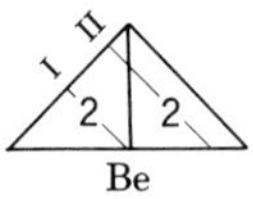

Be

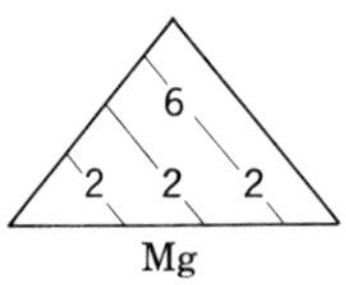

Mg

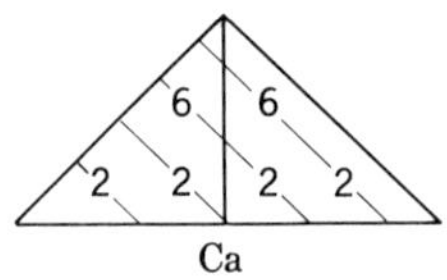

Ca

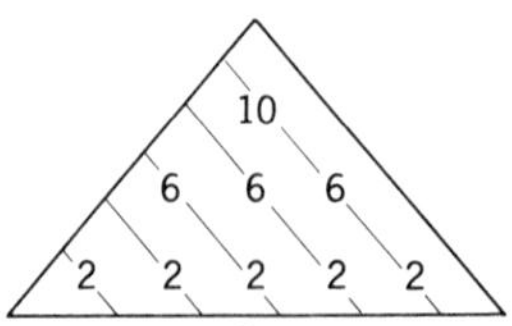

Sr

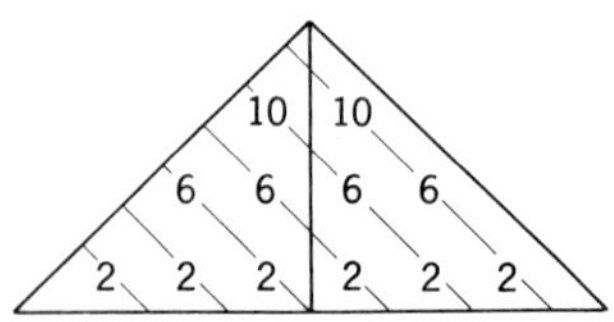

Ba

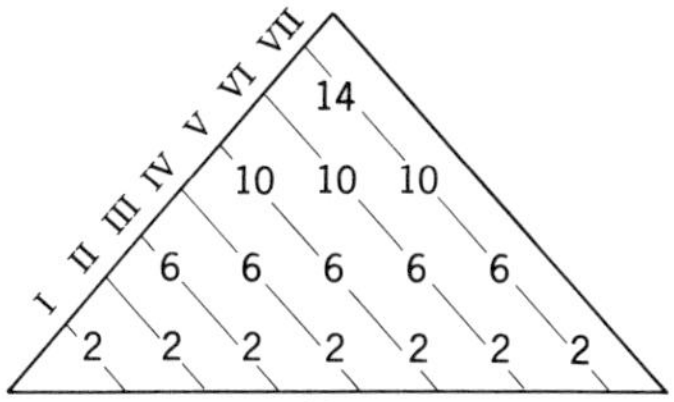

Ra

This gives us the *es* diagrams for Group II. Then recalling that the elements in Group I, have only 1 electron in the outer shell, we reduce the outer shell number at the far right from 2 to 1 and get the *es* diagrams for Group I. Thus the *es* diagrams for the elements of Group II provide a "base-line" from which the *es* diagrams of all chemical elements can be constructed with the help of a few simple rules. As an additional example, we can write the *es* diagrams for the first two elements of Group III by adding one additional *p* electron to each of the Group II diagrams:

```
  1            6 1
2 2          2 2 2
 B             Al
```

The *es* diagrams of the heavier elements of Group III are complicated by the more subtle energy relations in the larger wave types. These will be discussed at the close of this chapter.

ELEMENTS OF GROUPS I, II, AND III

The elements of Group I are called the *alkali* elements. In Table 2, we list first the atomic number of each of these elements; this is also equal to the number of units of positive charge on the atomic nucleus of each element. In the next two columns there are values for the approximate radii of the neutral atoms and for the ions. To see the meaning of these figures, we turn again to the charge clouds.

Table 2 Properties of Group I Elements

	Atomic No. Nuclear Charge	Radius (A°)		Ionization Potential	Electro-negativity
		Neutral	Ion		
H	1	0.52	10^{-5}	13.6	2.1
Li	3	1.5	0.60	5.4	1.0
Na	11	1.86	0.95	5.1	0.9
K	19	2.27	1.33	4.3	0.8
Rb	37	2.43	1.48	4.2	0.8
Cs	55	2.62	1.69	3.9	0.7

From quantum mechanics, it is possible to calculate the density of the charge cloud corresponding to a given electron as pointed out in Chapter 4. It was on this basis that the density curves and plots for the hydrogen electron were constructed. Similar calculations can be made for the other elements of Group I; and even without complicated calculations we can draw some direct qualitative conclusions. For example, we know from the Pauli exclusion principle and the quantum number rules that the electron added in passing from hydrogen to helium exhausts the places in the first shell. Therefore, when we go to lithium, this new electron must go into the second shell with $n = 2$. It will, therefore, have a distribution similar in shape to the electron in hydrogen when it is promoted to the second shell; there will be a core of charge at the center and also a maximum in shell-charge density a little way out from the nucleus. This is the result of the two "loops" in the wave for $n = 2$. Thus, in shape, there is a similarity between a 2*s* electron in hydrogen and the 2*s* electron of lithium. Also the two 1*s* electrons in lithium have wave patterns of the same shape as do electrons in hydrogen or helium. But we now recall that the nucleus of lithium has a positive charge of three units as compared with the single charge on the hydrogen nucleus. Thus the attraction of the nucleus for the electron is three times as great and the charge cloud is pulled in closer to the nucleus.

In Fig. 1, we have drawn the shell-charge plots for the electron charge for several elements. In the plot for lithium, the peak marked *k* shows the maximum due to the two 1*s* electrons.† A comparison with the peak for the 1*s* electron of hydrogen shows how much closer the lithium electrons are to the nucleus. The curve in Fig. 1 marked 2*s* for lithium shows the two peaks due to the nodal surface. This 2*s* type of wave lies much farther out from the nucleus than that for the 1*s* electron, so that the effective radius for the lithium atom is much greater than that for the hydrogen atom. The nature of this distribution of charge can also be seen graphically in the *fog* plot of Fig. 2; here the degree of whiteness of the "fog" cloud indicates roughly the density of the charge, viewing the cloud in cross section.

In the other drawings in Figs. 1 and 2 are shown similar shell charge graphs and fog plots for the other neutral atoms of this group. As the positive charge on the nucleus builds up to the value of 55 units with cesium, the electrons of the inner shells are pulled in very close to the nucleus; and these completed shells of negative electricity are very effective in neutralizing the pull of the positive charge for the last solitary electron in the outer shell. As a result this outer

† The calculated charge cloud on lithium is such that one-third of the charge due to the three electrons has 2*s* symmetry and two-thirds of the charge due to the three electrons has 1*s* symmetry. This does *not* mean that there are two electrons of one type (1*s*) and one electron of another type (2*s*). A similar interpretation applies to any charge cloud distribution involving more than one electron, such as the cloud of conduction electrons in a metallic crystal.

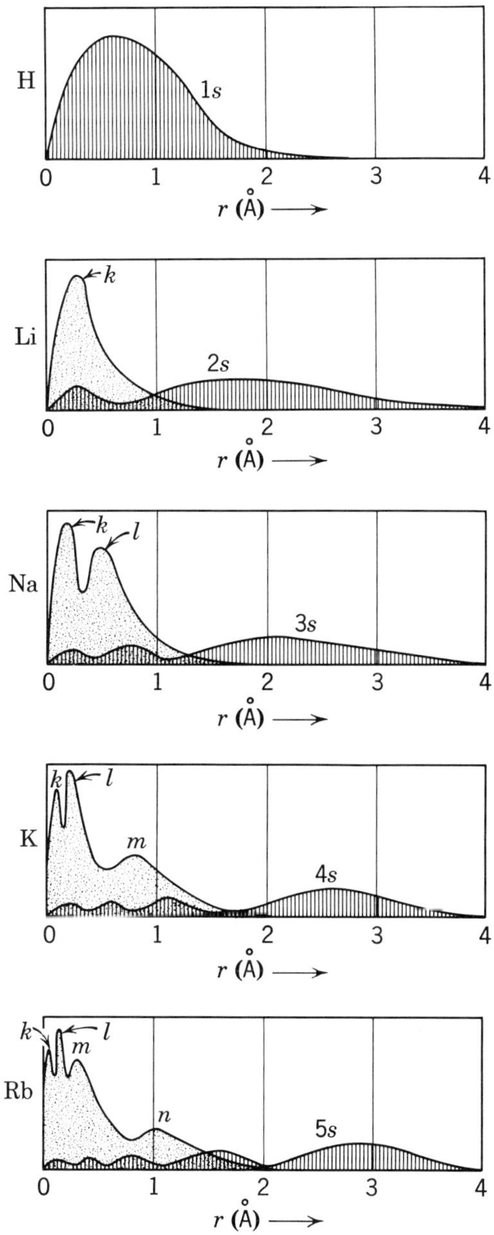

Fig. 1. Graphs of total shell charge.

cloud is quite diffuse, and the outer electron, which it represents, becomes easier to remove as we proceed down the column from Li to Cs.

It is this factor that is largely responsible for the chemistry of these elements. When packed together as neutral atoms in the elementary state, these outer shell electron clouds merge together and form a kind of sea of electrons in which the inner atomic cores are immersed. These electrons can even pass from atom to atom under the influence of an electric field, thereby producing the electrical conductivity so characteristic of the metallic state. This topic is discussed in detail in Chapter 24.

For us as chemists, the more important feature of these atoms is the ease with which this electron cloud in the outer shell is lost and transferred to another atom like a neutral halogen atom that lacks just one electron to achieve a completed shell of eight electrons on its exterior. As we have stated before, this transfer or removal of the outer electron leaves the atom in the form of the alkali *ion*, where the exterior also consists of a completed shell of eight electrons. The size of these ions can be estimated from the measurements of interionic distances in crystals with the help of X-rays. Values of ionic radii thus obtained are given in the fourth column of Table 2. Density plots of these ions are shown in Fig. 3.

We now turn to the last two columns of Table 2, which give us a measure of the ease of removing the solitary electron in the outer shell, i.e., the ionization potential and the electronegativity. It requires only 5.4 ev to pull the outer 2*s* electron from Li as compared with 13.6 ev to remove the 1*s* electron from H. As we go down the table, the ionization potential decreases; although the nuclear charge is far higher, the inner core electrons screen it so effectively that the remaining net charge acting at a greater distance on the outer shell electron resists the ionization less and less. The electronegativity is a similar measure of this tendency and reflects the same trend.

In Fig. 4, we have plotted the ionization potential for removing the first electron from the outer shell of each of the elements of Groups I, II, and III. The points for Group I show graphically the relations discussed in the preceding paragraph. In Group II, the inner core of electrons corresponds in total negative charge to *two* units less than the positive charge on the nucleus. Qualitatively speaking, an outer shell electron is thus attracted to the nucleus by roughly twice the force, as in Group I; but it is also somewhat repelled by the negative charge of its partner electron in the outer shell. Thus, in Be, the value of I_z is almost twice that for Li; but, as we move down the column, I_z falls off faster in Group II than in Group I.

When we pass to Group III, it is interesting to see that the value of I_z is now lower than in Group II for the ligher elements but slightly higher for In and Tl. This reflects the difference in ground state found in Group III that was mentioned

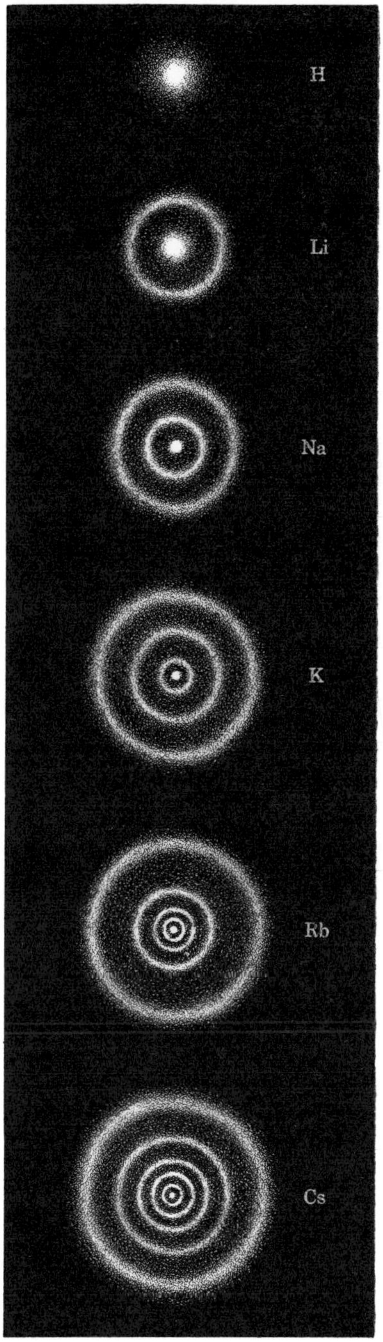

Fig. 2. Cross section of neutral atoms showing density of electron charge cloud.

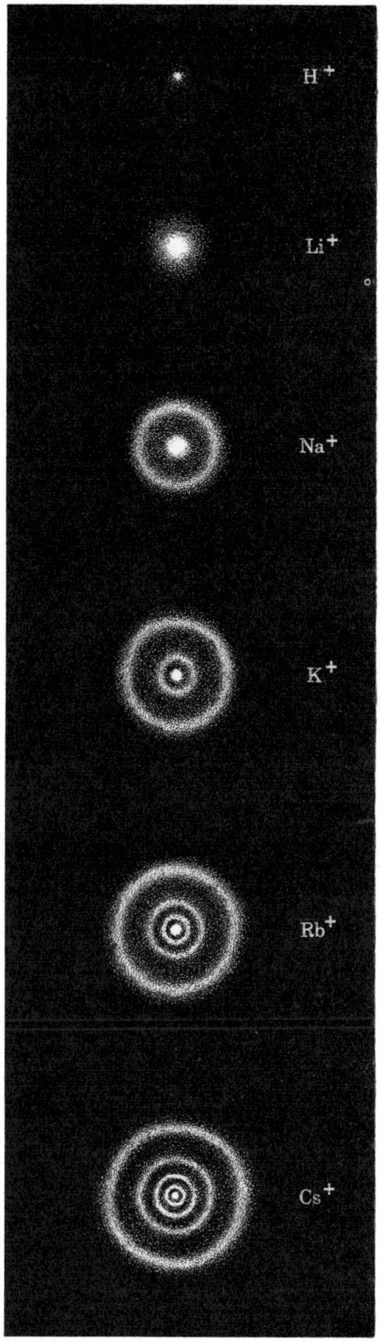

Fig. 3. Cross section of ions showing charge density of the electron cloud; approximate scale.

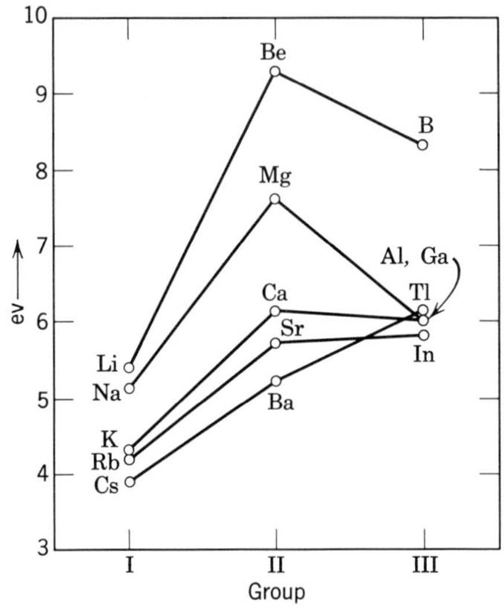

Fig. 4. Ionization potentials.

earlier. The higher values of the electronegativity for Al and B reflect the greater tendency for these elements to form covalent bonds.

As discussed in Chapter 9 on solid structures, there is a tendency among all these elements, when combined as neutral atoms, to merge the outer electron cloud with neighbors, leaving essentially ions in a close-packed arrangement immersed in a sea of conduction electrons. These merged clouds constitute an electron glue holding the ions in the close-packed arrangement in spite of the mutual repulsion between the positive charges. A measure of this attraction may be found in the melting points and boiling points. These have been plotted in Figs. 5*a* and 5*b*. In the first row (Li, Be, B) the melting point rises sharply as the number of electrons in the outer shell is increased. The boiling point jumps in going from Li to Be, but it does not increase appreciably for B. This probably is associated with the fact that boron is not a true metal. Its electrical resistivity is very high (2×10^{12} ohm cm) as compared with that of Be (19×10^{-6} ohm cm) and of Li (9×10^{-6} ohm cm). The trend indicates that the strength of binding increases from Group I to Group III; and, within a single group, the binding decreases in going from the lighter elements down to the heavier elements.

In Table 3 are shown values of the electrical resistivities for these three groups. There is great interest in attempting to correlate such data with the thermodynamic and structural features that we have been discussing; but this takes us into details beyond the scope of this book. We therefore turn to the chemistry of the different elements to illustrate some of the influences that we have just surveyed.

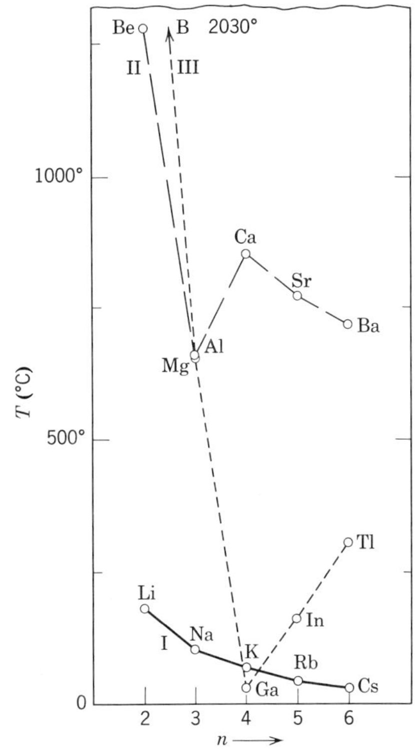

Fig. 5a. Melting points.

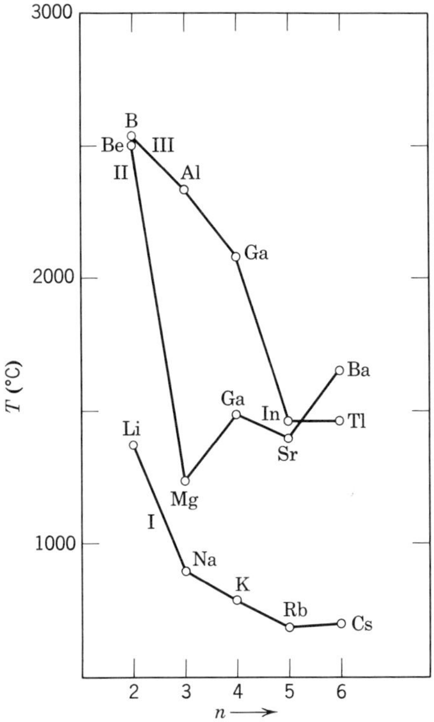

Fig. 5b. Boiling points.

CHEMISTRY OF GROUP I

Preparation. It is difficult to prepare the members of Group I of the periodic table in their elementary state. These elements lose the single electron so easily that almost any reaction results in a positively charged ion rather than in the element itself.

There is a widely used electrolytic method for the preparation of sodium. By adding sodium carbonate (Na_2CO_3) to sodium chloride (NaCl), the melting point of the mixture can be reduced to 600°C. By using iron or copper cathodes and

Table 3 Electrical Resistivities (ohm cm) at 20°C

Li	9×10^{-6}	Be	19×10^{-6}	B	2×10^{12}
Na	5×10^{-6}	Mg	4.6×10^{-6}	Al	2.8×10^{-6}
K	7×10^{-6}	Ca	4.6×10^{-6}	Ga	56×10^{-6}
Rb	12×10^{-6}	Sr	25×10^{-6}	In	9×10^{-6}
Cs	20×10^{-6}	Ba	—	Tl	18×10^{-6}

carbon anodes, this fused mixture can be electrolyzed with the formation of sodium metal and chlorine gas. Similar methods can be employed to prepare other members of this group in the elementary form.

There are also some chemical reactions by which these elements can be prepared in metallic form. For example, metallic calcium will react with rubidium chloride at high temperatures to give rubidium according to the reaction

$$Ca(s) + 2RbCl(s) \rightleftharpoons CaCl_2(s) + 2Rb(g)$$

Since rubidium is removed as a gas, Le Chatelier's principle shifts the equilibrium to the right.

Occurrence. The members of this group always occur in nature as ions with a single positive charge. Lithium is found in many rocks; sodium and potassium are quite abundant in salt deposits left by the evaporation of seas and lakes; there are also vast quantities of these elements in the oceans. Sodium is the sixth most abundant element in the earth's crust, and potassium the seventh.

Uses. In industrial plants where sodium is a by-product, it is occasionally used as a substitute for copper in bus bars to conduct electricity over short distances where currents of very high magnitude are required, as in electrolytic processes. The high thermal conductivity of sodium makes it valuable also as a medium for conducting away heat in nuclear reactors.

Because the single *s* electron of Group I elements is held so loosely, it can be ejected from the surface of the metal by a low-energy photon. For this reason, cesium is used in photoelectric cells, especially designed to be sensitive to light in the red region where the photons have low energy. When the photon strikes the surface of the cell, it is absorbed and the energy contained in it ejects the *s* electron from the surface, leaving the metal charged positively. By proper amplifiers, this charge can then be made to give signals of various kinds. Cesium oxide makes a good photoelectric surface sensitive to blue light.

Metallic sodium dissolves in liquid ammonia to form a curious type of ionic solution consisting of sodium cations and electron anions. Such solutions are useful chemical agents for many reactions.

Singly charged ions are formed not only from the elements of Group I, but also from the *coinage* groups in the middle of the periodic table in rows 4, 5, and 6. These include elements such as silver, copper, and gold. It is surprising that, although sodium, potassium, silver, and gold all give singly charged ions, sodium and potassium are among the most reactive and gold and silver are among the least reactive of metals. It is, of course, for this reason that these metals are used for coins, for which it is desirable to have the least amount of chemical reaction taking place so that the coins will remain relatively unchanged during years of use. We shall look further into the reason for this contrast in chemical behavior when we study the coinage elements later.

Compounds. The elements of Group I form a variety of oxides. All the elements form oxides of the type M_2O, but only Li combines directly with oxygen to yield this oxide. The other alkali metals form ionic peroxides, M_2O_2, wherein the anion is O_2^{-2}. The reactions can be written

$$2Li + \tfrac{1}{2}O_2 \rightleftarrows Li_2O$$
$$2M + O_2 \rightleftarrows M_2O_2 \qquad M = Na, K, Rb, Cs$$

Other oxides can also be formed:

$$M + O_2 \rightleftarrows MO_2 \qquad M = K, Rb, Cs$$

In this compound the anion is O_2^-.

The members of this group form sulfides. Selenides and tellurides are also formed. The general equation is

$$2M + X \rightleftarrows M_2X \qquad \begin{array}{l} M = Li, Na, K, Rb, Cs \\ X = Se, Te, S \end{array}$$

Lithium forms a nitride:

$$3M + \tfrac{1}{2}N_2 \rightleftarrows M_3N \qquad M = Li \text{ only}$$

The Group I elements react with heavier elements of Group V to form other compounds of this sort:

$$3M + X \rightleftarrows M_3X \qquad X = P, As, Sb$$

All elements of Group I react with the halogens to form the important compounds known as the alkali halides:

$$2M + X_2 \rightleftarrows 2MX \qquad X = F, Cl, Br, I$$

Several of these reactions will be discussed from the thermodynamic point of view at the close of this chapter.

The usual halide salts of the NaCl type are formed by all these elements; but the larger ions also form polyhalides: $NaI_3 \cdot 2H_2O$, RbI_3, CsI_3, CsI_4, $CsICl_2$, $CsBrCl_2$, $CsClBr_2$, $CsBrCl_2$, CsFIBr, RbFICl, KClIBr, CsClIBr, RbClIBr.

There are a number of reactions between the elements of Group I and hydrogen or compounds of hydrogen according to the equations

$$M + \tfrac{1}{2}H_2 \rightleftarrows MH$$
$$M + H^+ \rightleftarrows M^+ + \tfrac{1}{2}H_2$$
$$M + ROH \rightleftarrows MOR + \tfrac{1}{2}H_2 \qquad R = \text{an organic radical}$$
$$M + NH_3 \rightleftarrows MNH_2 + \tfrac{1}{2}H_2$$
$$M + H_2O \rightleftarrows MOH + \tfrac{1}{2}H_2$$

Thermodynamic data can provide considerable insight into the nature of many of these reactions. Consider, for example,

$$2Li + H_2 \rightleftarrows 2LiH$$
$$2Na + H_2 \rightleftarrows 2NaH$$

In each of these reactions the metallic bond is broken and the H_2 molecule is dissociated; then the two LiH or NaH bonds are formed. There is considerable interest in metallic hydrogen compounds as possible rocket fuels, and calculations have recently been completed on the thermodynamic properties of many compounds of this type. In Fig. 6, values of ΔH_f° and ΔF_f° for the preceding equations are plotted for LiH and NaH in the gaseous state under standard conditions. While these molecules actually solidify below 600–700°K, it is helpful to have the data for the theoretical gaseous state over the entire range of temperature. In the following discussion all Δ quantities refer to formation from the elements and the subscript ($_f$) is omitted.

Because of the relation

$$\Delta F^\circ = \Delta H^\circ - T\Delta S^\circ$$

we see that the distance between the curves for ΔH° and for ΔF° is the value of $T\,\Delta S^\circ$. We recall next that the high positive value of ΔF° indicates a strong

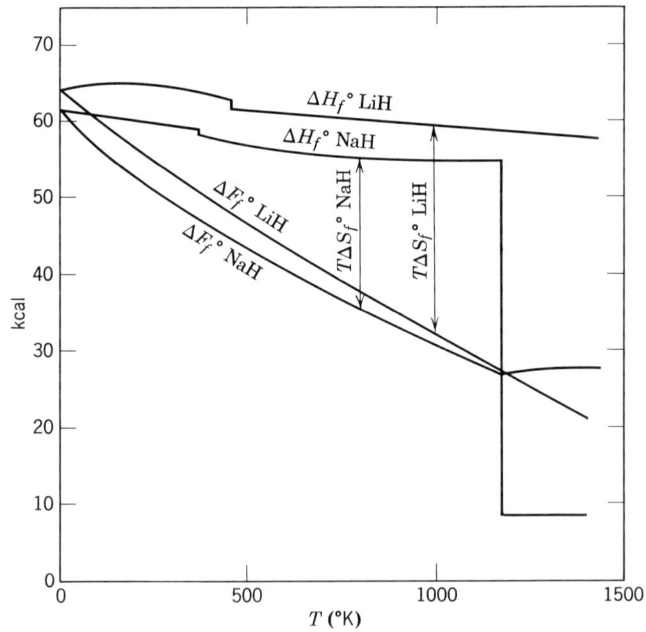

Fig. 6. ΔH_f° and ΔF_f° for LiH(g) and NaH(g) (Ref. 1).

tendency of the reaction to go in the direction that is the reverse of the way the reaction is written. We note that this high positive value of ΔF° at the low temperatures is due to high positive value of ΔH°. This tells us that, in separating the atom of Li from the metallic crystal lattice and pulling apart the H_2 molecule, we put in far more energy than we get back when the LiH bond is formed. The evidence is that in LiH we have a positive charge on the Li and a negative charge around the H. In other words, the tendency is for the outer electron of the Li to pair up with the electron of the H. It is not surprising that this compound is not particularly stable, and that there is an enormous tendency for the reaction to go backward in the direction of metallic Li and gaseous H_2 at low temperatures.

At higher temperatures, we note that $T\,\Delta S^\circ$ increases, and as a result ΔF° decreases, though even at 1000°K the tendency is still in the reverse direction to a high degree.

The two small breaks in the ΔH° curves just below and above 400°K represent the change in H° of Na and Li due to melting. Above these melting temperatures, the curves represent equilibria involving liquid metal. At 1170°K there is a sharp drop in the ΔH° curve for the NaH. This is due to the vaporization of the sodium metal, since this is its boiling point at 1 atm. Because the heat of vaporization is so large, the value of ΔH° drops abruptly. Above this temperature, the curves represent the values for the equilibrium of *gaseous* Na with H_2 and NaH. Accordingly, energy is not required for breaking the metallic bond as the sodium now consists essentially of individual atoms in the vapor state. Of course, while this *decreases* the pull on the reaction to go backward and form metallic sodium, it *increases* the tendency of the probability factor to pull the reaction backward, as Na in the vapor state at the normal boiling point has an entropy greater than liquid Na by the amount of $\Delta H_v/T_b$. As we go to higher temperatures, we see that this probability factor is greater than the energy factor and the value of ΔF° begins to rise so that high temperature favors the formation of NaH still less.

There are several reasons why at lower temperatures the effect of $T\,\Delta S^\circ$ is to push the reaction in the direction favoring the formation of NaH and LiH. We note that, as the reaction proceeds in the direction written, two atoms of metal (2Li) and one gaseous molecule (H_2) disappear and two gaseous molecules appear (2LiH). Thus the net result is more gas, and this always means greater entropy and a probability push in that direction. But, in addition, the reaction is also favored by the redistribution of energy as bonds are broken and reformed.

In Fig. 7 are plotted values of the standard free energy of formation for a series of lithium compounds. Gaseous Li_3N and LiO have positive values of ΔF_f° at room temperature and are unstable with respect to their elements for reasons much the same as in the case of LiH and NaH. When Li is linked with both an oxygen

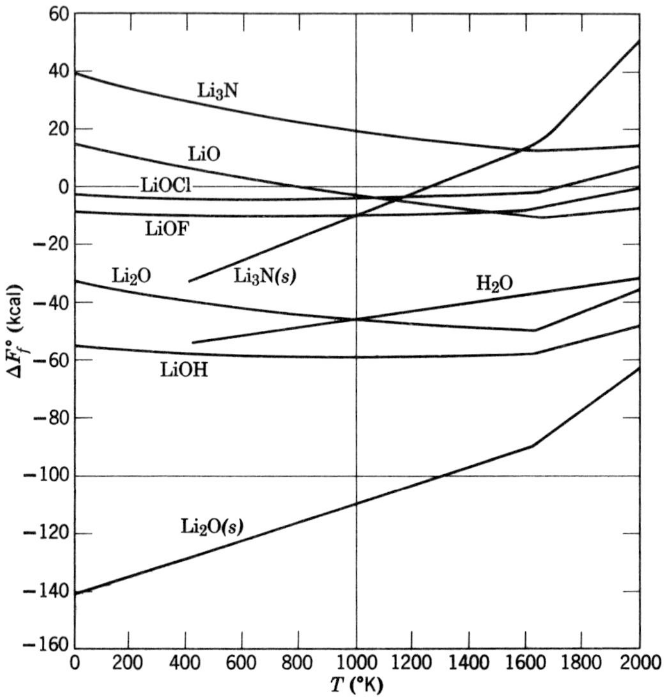

Fig. 7. ΔF_f° for some lithium compounds (Ref. 1). All compounds are in the gaseous state except where otherwise indicated.

atom and a halogen atom as in LiOCl and LiOF, the lithium compound has a small negative ΔF_f° because of the effect of the electronegativity of the F and the Cl. The former has the greater electronegativity, and its compound has the lower value of ΔF_f° and greater stability. The normal oxide, Li_2O, is shown with values for both the unstable gaseous state and the stable solid state. The almost complete transfer of the electrons from the two Li atoms to the oxygen atom makes the solid Li_2O extremely stable with respect to the elements. LiOH is also quite stable with respect to the elements, even when in the physically unstable gaseous state.

In Fig. 8, values of ΔF_f° are plotted for a series of sodium compounds. Again, the more electronegative substituents produce the greatest stability. In Fig. 9, values of ΔH_f°, $T\,\Delta S_f^\circ$, and ΔF_f° are plotted for the formation of solid NaCl from its elements. The enthalpy of formation has a large negative value and favors the formation of the compound. However, ΔS_f° is also negative and $T\,\Delta S_f^\circ$ goes to greater negative values as temperature is lowered. Thus the probability factor, $T\Delta S_f^\circ$, when coupled with the energy factor, ΔH_f°, through the equation $\Delta F_f^\circ = \Delta H_f^\circ - T\,\Delta S_f^\circ$, raises the value of ΔF_f° to a position somewhat above ΔH_f°. As temperature increases, ΔH_f° stays relatively constant, but $T\,\Delta S_f^\circ$ sinks; consequently ΔF_f° rises. This is a behavior characteristic of the vast majority of chemi-

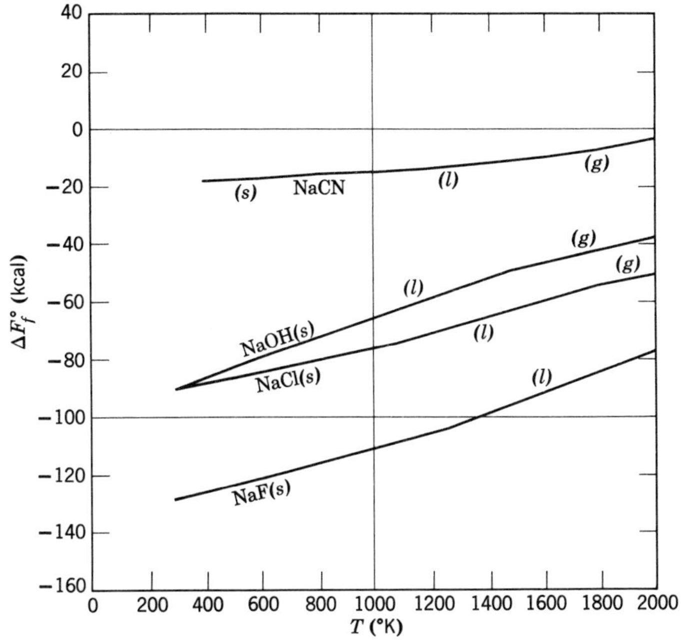

Fig. 8. ΔF_f° for compounds of sodium (Ref. 1).

cal compounds and is the thermodynamic aspect of the tendency for compounds to decompose at high temperatures. In other words, the energy factor continues to favor compound formation as temperature rises, but the probability factor ($T\,\Delta S_f^\circ$) pulls the compound apart.

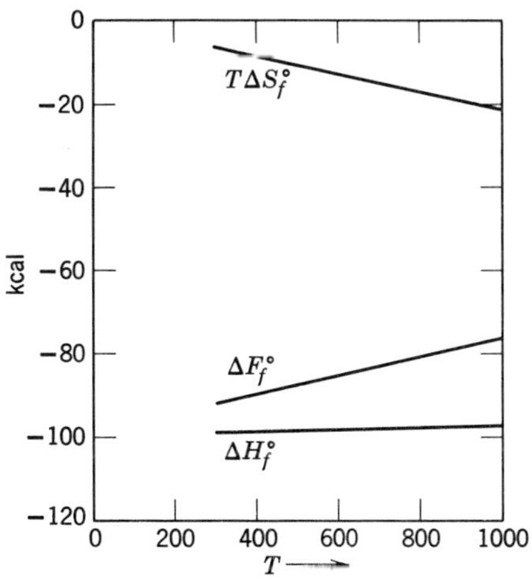

Fig. 9. ΔH_f°, $T\Delta S_f^\circ$, and ΔF_f° for NaCl (Ref. 1).

Spectra. When salts of these elements are heated even to moderate temperatures found in the flame of a Bunsen burner, they produce characteristic colors that serve as a means of identification; this emission of visible light is due to the low value of the ionization constant. Thus we find the following flames: Li = red, Na = yellow, K = violet, Rb = red, Cs = blue.

Solubility. Nearly all the salts of the alkalis with halides are very soluble. Exceptions are LiF, which is virtually insoluble (0.3 gm/100 gm H_2O), and sodium fluoride, which is only moderately soluble (4 gm/100 gm H_2O). Although the contrast of the solubility of these salts to that of the others (e.g., KF: 92 gm/100 gm H_2O) is marked, these changes are the result of relatively small changes in the thermodynamic quantities; for example, the above changes result from differences in enthalpies of solution of 2 or 3 kcal. The enthalpy of solution contains the sum and difference of many terms, only one of which is crystal energy (p. 435). Thus arguments based on crystal energy are shaky. In the present case it is tempting to argue that LiF is insoluble because the small sizes of Li^+ and F^- make the crystal exceptionally stable to solution, whereas this is not true for LiCl or KF. The same argument would convince us that AgCl is more soluble than AgF; actually AgF is by far the most soluble of the silver halides (AgCl, 9×10^{-5} gm/100 gm H_2O; AgF, 122 gm/100 gm H_2O).

The other common insoluble salts are potassium cobaltinitrite, $K_3Co(NO_2)_6$, potassium hexachloroplatinate, K_2PtCl_6, and sodium antimonate, $NaSb(OH)_6$.

Physiology. The ions of the elements of this group play an important part in physiological processes. In fluids outside the biological cell, there is a predominance of sodium ions; among other effects, these ions play a major role in governing the contraction of muscle. Inside biological cells, there is a predominance of K^+ ions. In plants, also, there is a great deal of potassium. In some cases, plants absorb more than five pounds of potassium from soil per acre; thus it is important to have potassium present in fertilizers to renew the supply in the ground.

THE CHEMISTRY OF HYDROGEN

Hydrogen has but one electron. It is a cousin of the alkali metals; but, because this electron is not enveloping a stable 2-6 shell but is directly outside the nucleus, the chemistry of hydrogen is very different from the chemistry of the alkali metals. Thus hydrogen in its crystalline state is not a metal but a molecular solid, H_2, where a covalent bond holds together the two atoms; in the vapor state the alkali metals also form molecules of the type Na_2.

Because the nucleus of the hydrogen atom consists of a single elementary particle, the proton, hydrogen plays a unique role in nuclear chemistry and in the cosmic chemistry of the universe. It is believed that hydrogen makes up about 90 per cent of the mass of the star which is our sun. It may well be that, throughout the universe, hydrogen is the most abundant element. Here on earth, there is a great deal of hydrogen near and at the surface in the fluid of our oceans, H_2O. In terms of number of atoms, hydrogen is probably the third most abundant element on earth, but because of its light weight it is only ninth in terms of the total mass of the earth's crust.

Hydrogen is rarely found in nature in its elementary state. There are occasional traces of H_2 in volcanic gases; and the upper atmosphere contains traces of atomic hydrogen. The human body is made up of water to about two-thirds of its weight, which means that it is roughly 10 per cent hydrogen by weight. Combined hydrogen is very common in coal and petroleum and is found in some minerals, such as clay.

Preparation. The purest hydrogen is formed by the electrolysis of water;

$$2H_2O \rightleftharpoons 2H_2(g) + O_2(g)$$

The oxygen gas comes off at the anode, and the hydrogen at the cathode. This process requires large amounts of electrical power. Another process for making hydrogen in a commercially practicable way is the passage of steam over hot iron:

$$3Fe + 4H_2O(g) \rightleftharpoons Fe_3O_4(s) + 4H_2(g)$$

The least expensive method for producing hydrogen is the *water-gas process:*

$$C(s) + H_2O(g) \rightleftharpoons CO(g) + H_2(g)$$

Steam is passed over hot carbon and a mixture of carbon monoxide and hydrogen is formed. However, because it is difficult to separate the carbon monoxide from the hydrogen, this is not a good process for getting pure hydrogen. The mixture of these two gases, called *producer gas,* makes an excellent fuel

Hydrogen is also obtained as a by-product in petroleum refineries. In the laboratory, it can be made through the reaction of zinc metal with acid:

$$Zn(s) + 2H^+ \rightleftharpoons Zn^{+2} + H_2(g)$$

Reactions. The reaction of hydrogen gas with oxygen is one of the first we studied in our survey of chemistry, and it serves as an excellent illustration for many chemical principles. Because of the large amount of energy released, it is a useful reaction for producing high temperatures with the oxy-hydrogen blowtorch. A similar reaction occurs with fluorine:

$$H_2(g) + F_2(g) \rightleftharpoons 2HF$$

There are many reactions in which hydrogen combines directly with other substances. We have seen in the early part of this chapter that it combines directly with the metallic elements of Group I; it also combines directly with other metals. It adds directly to many organic compounds, especially where there is unsaturation at a double bond. It also removes oxygen from many oxides.

When hydrogen is combined with elements more electronegative than itself, it is in the $+1$ oxidation state. If the difference in electronegativity is great enough, the hydrogen is present as a single positively charged proton; but one finds all degrees of ionic character in bonds thus formed. When hydrogen combines with elements like Na, less electronegative than itself, it is generally regarded as in the -1 state. Here it has two electrons associated with the nucleus and is potentially, at least, a negative ion. The hydrides of elements of Groups I and II form cubic or hexagonal crystals; they conduct electricity when melted, and they give off H_2 at the anode in electrolysis. Hydrogen also forms many covalent hydrides such as methane, CH_4, silane, SiH_4, and arsine, AsH_3.

There are also substances in which hydrogen appears to be present as a proton with a single electron; such a compound is uranium hydride, UH_3. There may be some question, in a number of combinations of metals with hydrogen, whether there is any stoichiometric relation between the hydrogen and the metal involved; apparently, the hydrogen is dissolved in the metal as elementary hydrogen.

One of the most important properties of hydrogen is its ability to form *hydrogen bonds*. We have discussed this in connection with the behavior of water, where it appears that hydrogen can bond with oxygen atoms in two different H_2O molecules at the same time. The boiling point of H_2O is abnormally high because of this type of behavior; the boiling point of HF is also abnormally high compared with those of other compounds formed by combining hydrogen with the halides and this appears to be due to hydrogen bonding. Hydrogen bonds also play a significant role in many biological systems.

Isotopes. One of the most startling discoveries of the century was the identification and isolation of *heavy hydrogen*. This is an isotope of hydrogen where, instead of a single proton, there are a proton and a neutron in the nucleus. It has been given the name deuterium (D) from the Greek root meaning "two" because its atomic weight is approximately 2. A third isotope of hydrogen with atomic weight approximately 3 has also been discovered and named tritium (T); its nucleus is made up of a proton and two neutrons. Whereas deuterium is radioactively stable, tritium decomposes and is present in nature only at a concentration of 10^{-17} times the concentration of normal hydrogen. Because of the large ratio between the masses of these isotopes, they can be separated from one another more easily than any other known isotopes. By the repeated hydrolysis of successive fractions of water, the deuterium content of the water can be increased and D_2O can be produced almost 100 per cent pure.

In physical and chemical behavior, H, D, and T are not as much alike as one might expect isotopes to be; the differences in mass make an appreciable difference in many physical and chemical properties. The use of deuterium as a tracer element has been of great help in the study of the mechanisms of many different reactions.

THE ALKALINE EARTHS (GROUP II)

In the periodic table, there are a number of chemical elements which are found with two electrons in the outer shell. These may be divided into five distinct groups. First of all, there is helium, which stands all by itself, having a nucleus with two positive charges surrounded by two electrons in the first shell. As we have seen, these two *s* electrons form such a stable configuration that helium is an unreactive gas. For this reason, helium is classed as a member of Group VIII, rather than of Group II. Continuing our search for elements with two electrons in the outer shell, we see that six atoms of this type occur in the second column of the periodic table. These are the elements beryllium (Be), magnesium (Mg), calcium (Ca), strontium (Sr), barium (Ba), and radium (Ra). They are called *Alkaline Earth Metals,* although this name originally applied only to calcium, strontium, and barium. It is this group of elements which we shall study in this section. Their *es* diagrams are shown in Table 4.

To the right of the Group II elements in the periodic chart there are thirty elements known collectively as the transition elements. Nearly all of these have two *s* electrons in the outer shell.

And there are also many other elements with two electrons in the outer shell. For example, in the twelfth column of the periodic table, counting from left to right, there are the three elements Zn, Cd, and Hg. These three elements are usually classified as members of Group IIB; and we shall consider their chemistry in Chapter 23. We also note a series of elements with atomic numbers running from 58 (Ce) to 71 (Lu). These elements have the two electrons outside of a

Table 4 ***es*** **Diagrams for Group II**

Be	Mg	Ca	Sr
2 2	6 2 2 2	6 6 2 2 2 2	10 6 6 6 2 2 2 2 2

Ba	Ra
10 10 6 6 6 6 2 2 2 2 2 2	14 10 10 10 6 6 6 6 6 2 2 2 2 2 2 2

stable 6-2 shell, but inside this there is a partially filled fourth shell containing *f* electrons. For this reason, their chemical properties are quite different from those of the elements of Group II and they are grouped in a class by themselves, called the *Lanthanides.* Finally, we note that there is a group of elements running from atomic number 89 (Ac) to atomic number 102 (No). These are the atoms with high atomic number and high atomic weight, and with large wave functions. For this reason, their chemistry again becomes quite different from that of the elements of Group II and they really form a class by themselves. As may be seen from the *es* diagrams on the inside of the back cover, they have both incomplete *f* shells and incomplete *d* shells. We now turn to a more detailed consideration of the chemistry of the elements which, in the strictest sense, belong to Group II.

In elementary form, the elements of Group II are metals somewhat similar to the elements of Group I. The two *s* electrons in the outer shell resonate and thus form covalent bonds between each atom and its nearest neighbors (roughly twelve in number) in the crystal lattice. These electrons can move through the metal lattice under the influence of an electrical potential, thus producing a current of electricity. Consequently, these elements are good conductors, both for heat and for electricity. They reflect light for the same reason and have a silvery sheen. Because there are two resonating electrons instead of one, the bonds are stronger between neighboring atoms, and the metals are, therefore, harder and more dense. In Table 5 are given values for the ionic radii, ionization potentials, and electronegativities.

Occurrence and preparation. The lightest of these elements, beryllium (Be), is a light silvery-white metal. The principal source is the naturally occurring mineral, beryl, an aluminosilicate with the formula $(Be_3Al_2)(SiO_3)_6$. This is found in only a few places on the earth, but there are some relatively large deposits. In the state of New Hampshire, single crystals weighing as much as five tons each have been quarried. Crystalline beryl with a green color produced by traces of chromium is known as *emerald* and is one of the most valuable gem stones.

Table 5 Properties of Group II Elements

	Atomic No. (Nuclear Charge)	Ionic Radius (Å)	Ionization Potential (ev)	Electro-negativity
Be	4	0.31	9.32	1.5
Mg	12	0.65	7.6	1.2
Ca	20	0.99	6.1	1.0
Sr	38	1.13	5.7	1.0
Ba	56	1.35	5.2	0.9

Beryllium can be prepared by the electrolysis of a fused mixture of beryllium chloride, $BeCl_2$, and sodium chloride. In metallic form, beryllium is used for making windows for X-ray tubes, since elements with low atomic number have very slight absorption for X-rays. Beryllium is also useful in the production of alloys with special properties for springs. BeO is extremely poisonous, and this limits somewhat the uses of Be in industry.

Magnesium (Mg) is found in nature as the carbonate, *magnesite*, and most abundantly in the form of silicates, including *talc, soapstone, asbestos,* and *mica.* Next to sodium, magnesium is the most abundant metal in sea water.

Magnesium metal can be produced by the electrolysis of fused magnesium chloride or by the reduction of magnesium oxide by carbon or by an alloy of iron and silicon:

$$MgO + C \rightleftarrows Mg + CO$$
$$3MgO + FeSi \rightleftarrows 3Mg + FeO + SiO_2$$

It is extensively used in light-weight alloys for the construction of airplanes and rockets, but it has the disadvantage of oxidizing so readily that it must be regarded as an inflammable substance. In pure form, it burns very rapidly even in air and is one of the principal constituents of flashlight powder.

Calcium (Ca) is found widely in nature. *Limestone* consists of a mixture of calcium carbonate and siliceous compounds; *calcite* is a purer form of calcium carbonate. One of the most familiar forms of calcium carbonate is marble; it is also found in the shells of marine animals. Other sources of calcium are dolomite, $MgCO_3 \cdot CaCO_3$; gypsum, $CaSO_4 \cdot 2H_2O$; anhydrite, $CaSO_4$; *fluorspar* or *fluorite,* CaF_2; phosphate rock, principally $Ca_3(PO_4)_2$. Bones of animals consist largely of calcium phosphate. There is considerable calcium present in sea water in the form of the doubly charged ions, Ca^{+2}. Metallic calcium can be prepared by the electrolysis of fused calcium chloride, $CaCl_2$. It is used as a deoxidizer for iron and steel as well as copper; it is a constituent of some lead alloys.

Strontium (Sr) is found in a mineral called *celestite* ($SrSO_4$) and in strontianite ($SrCO_3$).

Barium (Ba) is found in a mineral *barite* ($BaSO_4$). Because of its large atomic number, barium absorbs X-rays strongly. When studies of the gastrointestinal tract are made, patients usually swallow a mixture of barium sulfate and water in order to make the organs stand out in the X-ray photograph or the fluoroscopic view. Although barium compounds are highly poisonous, the sulfate is so insoluble that the body does not absorb significant amounts. A green color is produced in fireworks by barium nitrate, $Ba(NO_3)_2$, and by barium chlorate, $Ba(ClO_3)_2$.

Compounds. The elements of this group have two electrons in the outer shell;

the removal of the second electron is much more difficult than the removal of the first because the presence of the positive charge left by the departure of the first electron produces a much stronger force to hold the second electron. For example, $I_z^{\rm I}$ to form Be^+ is 9.28 ev; $I_z^{\rm II}$ to form Be^{+2} is 18.1 ev. The singly charged magnesium ion, Mg^+, does exist in the gaseous form; but the only ions found in solution are all doubly charged. Although the removal of the second electron requires about 800 kcal, the attraction of the two positive charges for water molecules also releases a great deal of energy when the doubly charged ion is formed in aqueous solution surrounded by oriented H_2O dipoles. The result is that the atoms always revert to the ionic form with two positive charges when given an opportunity to go into aqueous solution.

The same principles operate when these elements combine with halogens. There is so much energy released when the electrons join a halogen to form a rare gas shell that the compound MX_2 always results according to equation

$$M + X_2 \rightleftarrows MX_2$$

All metals of this group form oxides according to the equation

$$M + \tfrac{1}{2}O_2 = MO$$

Because of its large size, barium can form a peroxide, and even strontium under pressure can form a similar compound:

$$M + O_2 \rightleftarrows MO_2$$

All elements of this group form sulfides, selenides, and tellurides according to the equations

$$M + S \rightleftarrows MS$$
$$M + Se \rightleftarrows MSe$$
$$M + Te \rightleftarrows MTe$$

Nitrides can be formed by heating the elements in the presence of nitrogen:

$$3M + N_2 \rightleftarrows M_3N_2$$

The values of ΔF_f° are plotted in Fig. 10 for the oxides, chlorides, and carbonates of Group II metals at 25°C. All these compounds are in the crystalline form. There is a sharp drop in ΔF_f° in passing from Mg to Ca, and a corresponding increase in stability. There is a similar break in the ionization potentials and in the ionic radii. After this break, the compounds of Ca, Sr, and Ba resemble each other with only a slight increase in stability as the atomic weight increases.

An interesting variation in the behavior of the elements of this group is found as we go down the column of the periodic table. When exposed to air, Be and Mg metal corrode owing primarily to oxide formation; and Ba and Ra, on the other hand, corrode primarily because of nitride formation.

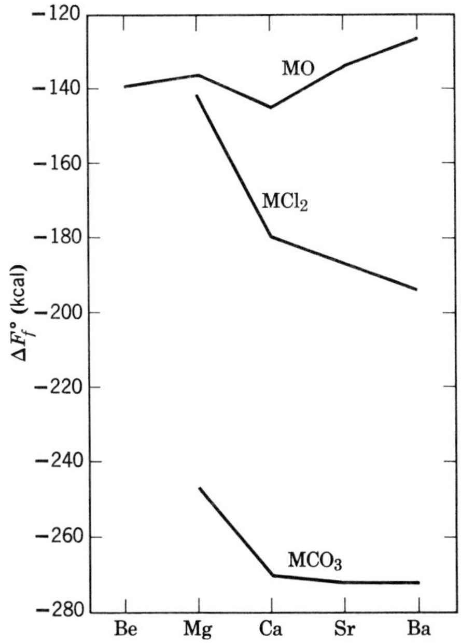

Fig. 10. Free energy of formation for some crystalline compounds of Group II elements at 25°C (Ref. 2).

All the elements of this group form hydrides:

$$M + H_2 \rightleftarrows MH_2$$

The thermodynamic data for the formation of BeH_2 (gaseous) and MgH_2 (solid) are given in Fig. 11. The tendency is to form MgH_2 in the solid state at low temperatures largely because of the strong binding between these atoms in the crystal. As temperature increases, however, the sign of ΔF_f changes from negative to positive and the solid dissociates under 1 atm pressure to give Mg and H_2.

All of these elements dissolve in acids to form gaseous hydrogen, and all except beryllium react with neutral water as follows:

$$M + 2HOH \rightarrow H_2 + M^{+2} + 2OH^-$$

As might be expected, the reaction is very slow for magnesium, but the rate increases as the atomic weight increases and the element becomes less electronegative.

The halogen compounds of this group differ considerably in bond character. In beryllium, the covalency is high and, although its halogen compounds are soluble, there is considerable hydrolysis; on the other hand, there is almost no covalency in the halogen compounds of Ba and Ra. The covalency of the beryllium compounds also is shown by the fact that $BeCl_2$, $BeBr_2$, and BeI_2 dissolve in

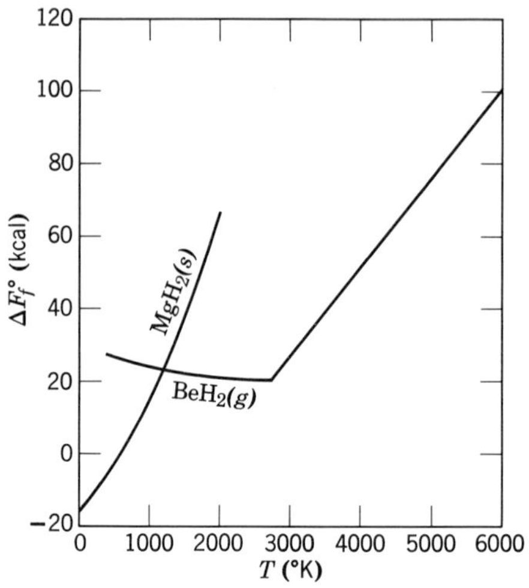

Fig. 11. Free energy of formation for BeH_2 and MgH_2 (Ref. 1).

organic solvents. The same is true of $MgBr_2$ and MgI_2, but MgF_2 is only slightly soluble.

Beryllium hydroxide is amphoteric, dissolving in acids to give Be^{+2} and in strong bases to give the beryllate ion. Magnesium hydroxide has a low solubility (1×10^{-4} mole per 1000 gm); it is not amphoteric. *Milk of magnesia* is a suspension of magnesium hydroxide in water and is used to neutralize stomach acidity.

Calcium hydroxide also has a low solubility in water (0.02 mole per 1000 gm). The aqueous solution is known as *lime water* and is useful for the detection of carbon dioxide, which forms a white precipitate of calcium carbonate:

$$Ca(OH)_2 + CO_2 \rightleftarrows \underline{CaCO_3} + H_2O$$

Whitewash is a suspension of calcium hydroxide in water. Calcium oxide (like all the Group II oxides except for beryllium) is prepared by thermal decomposition of the carbonate. In dry form it is known as *quicklime;* when combined with water to form $Ca(OH)_2$ it is called *slaked lime.*

Because of the presence of calcium as well as magnesium and iron in natural waters, there is a tendency for precipitation to take place when such water is used with soap for washing purposes. Calcium ions in this *hard water* will unite with the stearate and oleate ions from dissolved soap to form insoluble salts; this process wastes the soap and produces an undesirable curd-like solid that remains in the clothes. It is, therefore, desirable to remove the calcium ions from laundry water.

A similar undesirable action takes place when *hard water* is used in steam boilers; the carbon dioxide comes off at high temperature, and a deposit of calcium or magnesium salts is produced inside the boiler. This can clog the tubes and also reduce the thermal conductivity. There has been a long search to discover means to prevent these effects in hard water. It has been discovered that the addition of a small amount of certain phosphate compounds prevents the calcium from precipitating out; but the action is still not completely understood. It is also possible to replace the calcium ions with other ions by an ion-exchange process that holds the calcium ions on the surfaces of certain natural or artificially prepared crystals and releases sodium or similar ions into the solution.

THE BORON FAMILY (GROUP III)

As shown in Table 6, the elements of this family are characterized by three electrons in the outer shell, two of these being *s* electrons and one being a *p* electron. Because *s* levels are somewhat lower in energy than *p* levels, the third electron is relatively loosely bound. We therefore find that the ionization potentials for the elements in this group are a little bit less than for the corresponding elements of Group II, where the outer shell consists of the two *s* electrons. On the other hand, the elements of Group IV which lie to the right with two *p* electrons again have correspondingly higher ionization potentials. Values of I_z and of ionic radii and electronegativities are given in Table 7.

In the elements of Group II, all the atoms in their elementary state exhibit metallic properties because of the two electrons in the outer shell. In Group III, the first element, boron, has the three electrons in the outer shell but is distinctly not a typical metal. As we mentioned earlier, its electrical conductivity is almost non-existent (resistivity, $\rho = 2 \times 10^{12}$ ohm cm; cf. Na: $\rho = 5 \times 10^{-6}$ ohm cm) and its properties place it definitely in that diagonal row of the elements including silicon, arsenic, and tellurium that lies midway between the true metals and the completely non-metallic elements. As we go down the

Table 6 ***es*** **Diagrams for Group III**

B	Al	Ga	In	Tl
				14
		10	10 10	10 10 10
1	6 1	6 6 1	6 6 6 1	6 6 6 6 1
2 2	2 2 2	2 2 2 2	2 2 2 2 2	2 2 2 2 2 2

Table 7 Properties of Group III Elements

	Atomic No. (Nuclear Charge)	Ionic Radii (Å)	Ionization Potential (ev)	Electro-negativity
B	5	0.20	8.3	2.0
Al	13	0.50	6.0	1.5
Ga	31	0.62	6.0	—
In	49	0.81	5.8	—
Tl	81	0.95	6.1	—

column of the elements of Group III, aluminum is a true metal with good electrical conductivity and all the other properties characteristic of this kind of an element. In fact, it is used in making wire cables for power transmission lines. Because aluminum has a very low density it is important as a structural material. It is the third most abundant element in the earth's crust. By way of contrast, boron is relatively scarce; this may be due to the fact that its nucleus is readily transmuted to other elements by bombardment with cosmic rays. Such bombardments have occurred naturally throughout billions of years during the changes that established the relative abundance of the elements observed today. Gallium, indium, and thallium are metallic elements; they are not found in concentrated deposits and have not been used commercially to any extent.

Boron. The small size of the boron atom, coupled with its greater nuclear charge as contrasted with beryllium, results in an element that does not have true metallic properties. Its chemistry resembles silicon chemistry much more closely than the chemistries of the other members of its own family. Its melting point is extremely high, and this suggests strong bonding between the individual atoms in the solid state. Like all Group III elements it has a +3 oxidation state, but its tendency to form covalent bonds results in a large number of unusual compounds, particularly with hydrogen. Because these compounds offer possibilities for rocket fuels, they have been examined in great detail during the last few years.

In the elementary state, boron exists both in a crystalline and an amorphous form. The latter was obtained by Moissan by reducing boric oxide with magnesium, but the product contains considerable Mg as impurity. A pure boron metal can be prepared by the electrolysis of fused oxide; when melted and cooled, the liquid solidifies into the crystalline form, a material characterized by extreme hardness, opacity with a slight metallic luster, and, as mentioned above, with almost no electrical conductivity at room temperature. When heated to 600°C, the conductivity increases a hundred-fold; with normal metals, electrical conductivity decreases with temperature; thus, the behavior of boron is not typical of metals.

Boron in the elementary state can be oxidized only under the most extreme conditions at room temperature. It is necessary to use concentrated nitric acid, fluorine, or other agents of this sort. When fused with NaOH and $NaNO_3$, it reacts to form borates. It forms compounds such as AlB_{12} and Mg_3B_2 with metals.

One might expect that boron would form a compound like BH_3, but this molecule is unstable at room temperature. Only more complex compounds like B_2H_6 (diborane), $B_{10}H_{14}$ (decaborane), and others of this sort are known. The structure of these compounds has been determined only recently; they will be discussed in some detail in Chapter 24.

These boron hydrides oxidize to give complex compounds many of which still have not been completely identified. When heated alone, they decompose to boron and hydrogen and, with water, give hydrogen and boric acid. This is a compound with the formula H_3BO_3 and is the starting point for the other much more complex compounds of the borate family. It is an extremely weak acid with the value $K_I = 6.0 \times 10^{-10}$. In solution, it has antiseptic properties and, because of its low acidity, it can be used to bathe such delicate tissues as are found in the human eye.

Borax is the most common of the borates (formula: $Na_2B_4O_7 \cdot 10H_2O$); the boron and oxygen atoms combine to form rings suggestive of some of those which occur with carbon compounds. Borax reacts with Ca^{+2} to form insoluble calcium borate. It also dissolves many metal oxides to form easily fusible borates, and, for this reason, it is widely employed as a flux in welding operations. A flux is useful because most metals have more or less thin coatings of oxides on their surfaces, and, when one wishes to link two metal surfaces closely together by other metal atoms such as those in solder, it is necessary to get through this oxide coating so that the metal atoms to be linked will be in direct contact. By removing the coating with some flux that can be washed away, the bare metallic lattice is exposed and it will bond directly with another metallic lattice.

Boron forms trihalides because of the three electrons in the outer shell. These are BF_3, BCl_3, BBr_3, BI_3, all molecular substances. When one writes the formula for one of these compounds, however, it is clear that there are only six electrons surrounding the boron atom. Thus, when some other compound can approach one of these halides and provide electrons to give a total of eight around boron, there is a tendency for a link to form. This is illustrated in the formation of the compound BF_3NH_3:

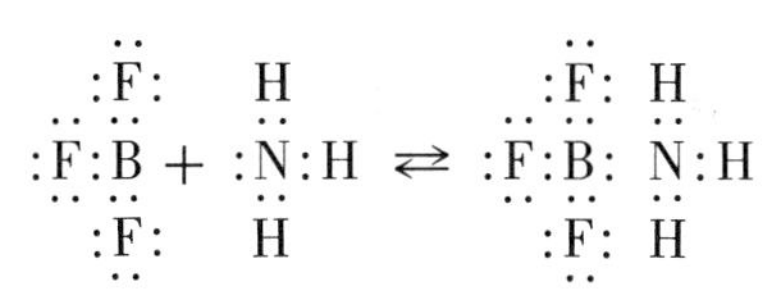

Aluminum. In terms of industrial use, aluminum is the most important element of Group III. It is the third most abundant element in the earth's crust and, therefore, readily available. It is stable when exposed to the earth's atmosphere, forming a skin of aluminum oxide that protects it against further oxidation. Aluminum is an extremely useful building material where light weight and high tensile strength are desirable. It is also used extensively for household utensils.

For many years, aluminum was not widely used in industry because the method of preparation was too expensive. Originally, it was prepared by reducing aluminum chloride with potassium according to the reaction

$$AlCl_3 + 3K \rightleftarrows Al + 3KCl$$

The cost of the metal thus produced was about $150 per pound. During the nineteenth century, extensive research was carried out to develop a better method for producing aluminum; finally, in 1886, Charles M. Hall in America and P. L. T. Héroult in France perfected an electrolytic method. Shortly the price had dropped to 20 cents per pound.

The most important ore of aluminum is *bauxite*, $Al_2O_3 \cdot 2H_2O$. Aluminum also occurs in a number of other forms such as the mineral cryolite, Na_3AlF_6, in complex aluminosilicate minerals such as clays and feldspar, and in many gem stones. The ruby and the sapphire consist largely of aluminum oxide, Al_2O_3.

In the commercial preparation of aluminum, the bauxite is purified and converted to sodium aluminate where it exists in the form of the aluminate ion, $Al(OH)_4^-$. It is precipitated as aluminum hydroxide and ignited to the oxide which is then dissolved in fused cryolite. It is then electrolyzed, the change taking place according to the equation

$$2Al_2O_3 \rightleftarrows 4Al + 3O_2$$

Metallic aluminum reacts with ferric oxide to produce iron and aluminum oxide (the *thermite* reaction):

$$2Al + Fe_2O_3 \longrightarrow 2Fe + Al_2O_3$$

The temperature rises so high that the iron and the aluminum oxide become liquid, and for this reason the process is useful in welding large pieces of iron or steel.

The equation for the cell reaction which gives the oxidation potential is

$$Al(s) \rightleftarrows Al^{+3}(aq) + 3e^- \qquad \mathcal{E}° = +1.66 \text{ v}$$

Because of this tendency of aluminum to give up electrons readily to other substances, it is a good reducing agent. It does not yield these electrons readily because they are easily removed from a neutral aluminum atom ($I_z^{I} + I_z^{II} + I_z^{III} =$ 53 ev). The high value of $\mathcal{E}°$ is possible only because the hydration energy of Al^{+3} is so high, over 1000 kcal.

When aluminum combines directly with oxygen, we get

$$4Al(s) + 3O_2(g) \rightleftarrows 2Al_2O_3 + 800 \text{ kcal}$$

The formation of this oxide on the surface of the metal provides the inert skin that makes aluminum a stable material for building purposes or for utensils. Because of the oxide skin, aluminum does not dissolve readily in concentrated nitric acid, although it does so readily in hydrochloric acid according to the equation

$$2Al(s) + 6H^+ \rightleftharpoons 2Al^{+3} + 3H_2(g)$$

Aluminum also goes into solution under basic conditions according to the equation

$$2Al(s) + 2OH^- + 6H_2O \rightleftharpoons 2Al(OH)_4^- + 3H_2(g)$$

It is this dual capacity to form solutions in either acidic or basic media which makes aluminum amphoteric. The relation between aluminum hydroxide and the hydroxyl and hydrogen ions may also be written

$$Al(OH)_4^- \rightleftharpoons Al(OH)_3 + (OH^-)$$
$$3H^+ + Al(OH)_3 \rightleftharpoons Al^{+3} + 3H_2O$$

An alternative way of writing these reactions is

$$Al(H_2O)_2(OH)_4^- \rightleftharpoons Al(H_2O)_2(OH)_3 + OH^-$$
$$3H^+ + Al(H_2O)_3(OH)_3 \rightleftharpoons Al(H_2O)_6^{+3}$$

On standing, aluminum hydroxide becomes more and more insoluble, apparently because of oxygen bridges which are formed between the neighboring molecules. This is an example of a compound where the ions can form many links with each other and, thus, behave like very large and complex ions, somewhat analogous to the structure of complex organic molecules with many bonds between the atoms. However, a "bond" between ions is relatively weak compared with the carbon-carbon bond.

The factors responsible for the behavior of the ion with respect to water also influence its behavior with respect to the halogens. For example, it is possible to form in solution various ions which combine aluminum and fluorine, such as AlF^{+2}, AlF_2^+, AlF_3, AlF_4^-, AlF_5^{-2}. In the mineral cryolite, aluminum is found combined with fluorine in the form of the ion AlF_6^{-3}. With chlorine, aluminum also forms interesting and important compounds. The ratio is normally one aluminum atom to three chlorine atoms but, in the vapor state, there are definite molecules with the formula Al_2Cl_6. It is believed that the substance exists with molecules of $AlCl_3$ in the crystalline state or at high temperatures in the vapor. By way of contrast, the bromide forms the Al_2Br_6 molecule both in the vapor and in the solid.

Another important compound is aluminum sulfate, $Al_2(SO_4)_3 \cdot 18H_2O$; in solution, this yields two triply charged aluminum ions, Al^{+3}, and three of the ordinary sulfate ions, SO_4^{-2}. This is the cheapest soluble salt of aluminum and is used in great quantities in the dye industry, in the manufacture of paper, and in the purification of water through reactions which form aluminum hydroxide.

One of the more important minerals containing aluminum is feldspar, $KAlSi_3O_8$. Aluminum is also found in the form of clay, $Al_2Si_2O_5(OH)_4$.

Thermodynamic relations. In Fig. 12 are plotted some of the values of the free energy of formation for a few compounds of boron and aluminum. B_2H_6 has a high free energy content relative to its elements, and an even higher one with respect to its products of oxidation like B_2O_3 and H_2O. It is easy to see why it is a promising rocket fuel. The two oxides have the lowest values of ΔF_f°, with curves roughly parallel. It is interesting that B_2O_3 falls below Al_2O_3 at the

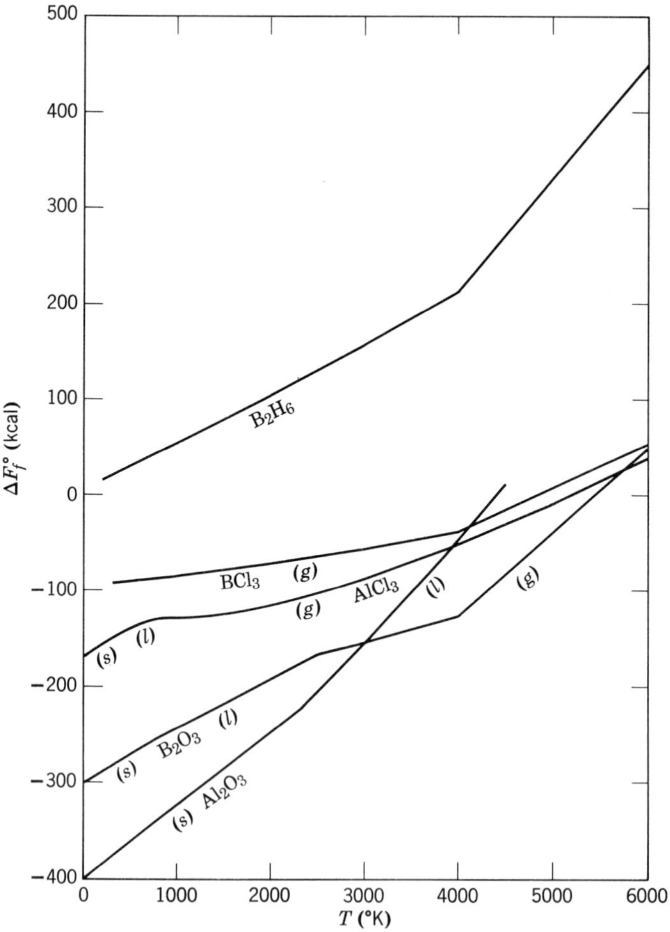

Fig. 12. ΔF_f° for some B and Al compounds (Ref. 1).

higher temperatures, though one should keep in mind that the values in this upper region of temperature are based largely on theoretical calculations.

Gallium, indium, and thallium. The chemistry of these elements will be discussed in Chapter 19, which deals with all the metals having an inner core of electrons of the 2-6-10 type.

QUESTIONS AND PROBLEMS

(Data necessary for the solution of these problems will be found in the tables and figures of this chapter and in the Appendix.)

1. Calculate the pressure of H_2 that will be in equilibrium with 1 atm of gaseous LiH and liquid Li at 1000°K.

2. If Li_2O is placed in a vacuum inside a closed vessel and heated to 1000°K, what will be the pressure of O_2 when equilibrium is reached? *Ans.* 10^{-48} atm.

3. From the curve in Fig. 8, make an estimate of the value of ΔS_f° and of ΔH_f° at 700°K for the formation of NaCl from its elements, and compare with the values in Fig. 9.

4. Calculate ΔF_f° at 1500°K for the reaction

$$Li_2O(s) + H_2O(g) \rightleftarrows 2LiOH(g)$$ *Ans.* +18 kcal.

5. Discuss the relation between the electronegativity and the free energy of formation of carbonates for the elements in this group.

★ **6.** Calculate values and make a plot of the curve for ΔF° for the reaction

$$B_2H_6 + 3O_2 \rightarrow B_2O_3 + 3H_2O$$

7. Calculate the pressure of O_2 above Al_2O_3 at 1000° and at 4500°K.
Ans. 10^{-47} atm at 1000°K.

★ **8.** Calculate the values and make a plot of ΔS_f° for BCl_3 and $AlCl_3$ from 2000° to 6000°K. Comment on the comparison of the two curves.

19

Chemistry of copper, zinc, gallium, and germanium subgroups

In the preceding chapter, we surveyed Groups I, II, and III in the periodic table. In each of these elements, there was an inner core of electrons with a rare gas configuration, and the number of valence electrons corresponded to the number of the group, one valence electron for lithium in Group I, two valence electrons for beryllium in Group II, and so on. Because of the nature of this electronic structure these metals have the simplest chemistry of all the elements.

As we continue our study of other metallic elements in this chapter, it is natural to turn to the elements in successive columns that have one or more valence electrons, but with these electrons lying just above an inner core of two *s* electrons, six *p* electrons, and ten *d* electrons. This set of twelve metallic elements is found in the lower right-hand corner of the periodic table; and, comparing them with the 2-6 core metals, we find striking similarities and differences in both physical and chemical behavior. These sets of metals are called Subgroups IB, IIB, IIIB, and IVB, the Roman numerals indicating the number of electrons normally lying outside the core. This similarity in the valence electrons and difference in the core can be seen at a glance in the electronic structure diagrams in Table 1.

PROPERTIES IN THE METALLIC STATE

Figure 1 is a plot of the melting points and boiling points for this group of twelve metallic elements. In the values for the first three (Cu, Ag, Au), we see immediately the great difference between the elements with this 2-6-10 core and the alkali metals with the 2-6 core. Of the alkali metals, lithium has the highest

Table 1 Electronic Structure Diagrams of the 2-6-10 Core Metals

Cu	Zn	Ga	Ge
10 6 6 2 2 2 1	10 6 6 2 2 2 2	10 6 6 1 2 2 2 2	10 6 6 2 2 2 2 2
Ag	**Cd**	**In**	**Sn**
10 10 6 6 6 2 2 2 2 1	10 10 6 6 6 2 2 2 2 2	10 10 6 6 6 1 2 2 2 2 2	10 10 6 6 6 2 2 2 2 2 2
Au	**Hg**	**Tl**	**Pb**
14 10 10 10 6 6 6 6 2 2 2 2 2 1	14 10 10 10 6 6 6 6 2 2 2 2 2 2	14 10 10 10 6 6 6 6 1 2 2 2 2 2 2	14 10 10 10 6 6 6 6 2 2 2 2 2 2 2

melting point, 179° C; and, as we go down the column to cesium, the melting point falls close to room temperature at 28.4° C; but the melting points of Cu, Ag, and Au lie near 1000° C. And, similarly, the boiling points of the 2-6-10 core metals range 1000° C or more higher than those of the 2-6 core metals.

A striking difference is also observed in chemical behavior. The alkali metals are so active that they are never found in the elemental metallic state in nature. They react vigorously both with oxygen and with water and cannot survive long

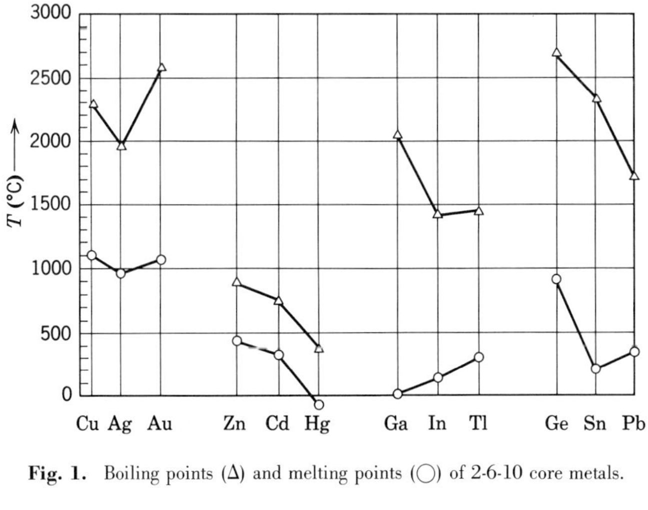

Fig. 1. Boiling points (Δ) and melting points (○) of 2-6-10 core metals.

under atmospheric conditions. On the other hand, copper, silver, and gold are found all over the world in the elementary metallic state; they are so inert that they can exist for millions of years without reacting with either oxygen or water.

Because of this lack of reactivity with oxygen and the moisture in the air, these three metals have been used during the centuries to make coins that must withstand wear and exposure to the air and moisture; and, for this reason, this group of metals is frequently called the "coinage" group. As we go down the column, the ease of oxidation decreases; copper and silver both undergo surface oxidation (tarnish) readily, but gold maintains its brilliance for centuries. As we proceed to the right across the periodic table, the activity increases; thus we find that zinc is oxidized easily by dilute acid; the rate of air oxidation of many of these metals is retarded, however, by the formation of a protective coating; in the case of zinc, this coating appears to be $Zn_2(CO_3)(OH)_2$.

Some of the most interesting physical properties of these atoms appear in their alloys or solutions with other metals. These alloys frequently have valuable properties; for example, copper and zinc form an alloy that is commonly known as brass; it is much harder than either of the metals taken separately, and it also possesses greater structural strength. Mercury is interesting because it has such a low melting point ($-39°$C), the lowest of all known elemental metals.

It is interesting to note that copper and gold are the only colored metals in this group. Whenever one finds a color in a chemical compound, this is evidence of a low-lying energy level that can be excited by a low-energy photon of visible light corresponding to an energy of about 1 to 3 ev. If we examine the electronic structure of copper, silver, and gold, we find that all three ions undergo the transition $d^{10} \rightarrow d^9s^1$. The energy needed for this transition is 2.7 ev for copper, 4.8 ev for silver, and 1.9 ev for gold. Thus copper and gold absorb visible light and are colored, whereas silver absorbs in the invisible ultraviolet and, hence, is a white silvery metal.

In Fig. 2 are plotted the ionization potentials for these 12 metals, with values given for removing the second and third electrons for some elements. It is interesting to compare these with the values for the alkali metals. For example, potassium bears the closest resemblance to copper, since it is exactly like copper except for having the 2-6 core instead of the 2-6-10 core. The ionization potential for potassium is 4.32 ev as compared to 7.68 for copper. As may be seen from the figure, the ionization potentials for zinc, cadmium, and mercury are higher; the values then drop for gallium, indium, and thallium; and they rise again back to values close to those for the coinage metals when we reach germanium, tin, and lead.

As a consequence of these ionization potentials, the alkali metals normally occur in compounds only in the $+1$ state; for it is easy to remove one outer s electron, but almost impossible to remove an electron from the inner 2-6 core. On the

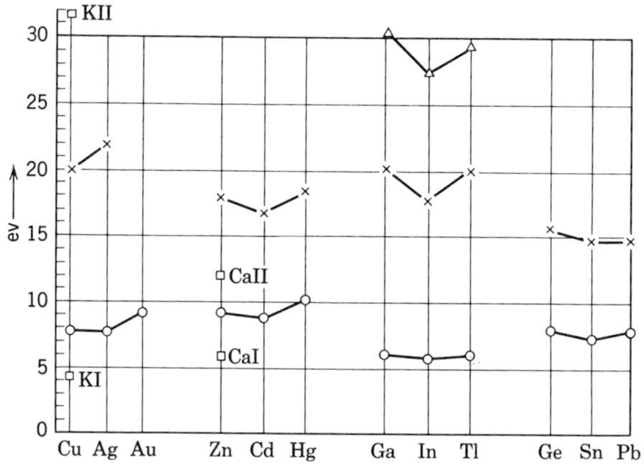

Fig. 2. Ionization potentials of 2-6-10 core metals: ○ I; × II; Δ III.

other hand, electrons can be removed from the inner 2-6-10 core far more easily with Cu, Ag, and Au; and these core electrons also participate in covalent binding. Thus we find that the states Cu^{+2} and Au^{+3} are common, the extra energy being provided by the attachment of other atoms to the central copper atom in processes that are essentially complex formation.

In Table 2 there are listed the oxidation-reduction potentials for a number of these metals, together with the values for potassium and rubidium by way of comparison. Unless otherwise specified, the value is shown for the reaction in which the metal forms the ion by giving up one or more electrons; the number of charges on the ion is indicated. The oxidation potentials give a clue to the inertness of these subgroup elements in the metallic state, the values lying far lower than for the elements in Group I. In this survey, we shall discuss each element, first considering it in the metallic state and then discussing the chemistry of its compounds.

Copper. This element is found with high purity in its native state, and it also occurs in concentrated deposits of sulfides and oxides. When copper of extreme purity is desired, it is refined by electrolysis from the sulfate, $CuSO_4$. The impure copper is placed in the cell as an anode, and the pure copper appears as a deposit on the cathode. The principal impurities to be removed are silver and iron. If the voltage applied to the electrolytic cell is kept just below the voltage necessary for silver deposition, the silver impurity can be kept from going into solution at the anode; and fortunately, at this voltage, other impurities like iron are prevented from coming out of the solution at the cathode. Thus the copper is deposited in an extremely pure form when the voltage is kept at precisely the right value.

Table 2 Oxidation-Reduction Potentials (Ref. 3)

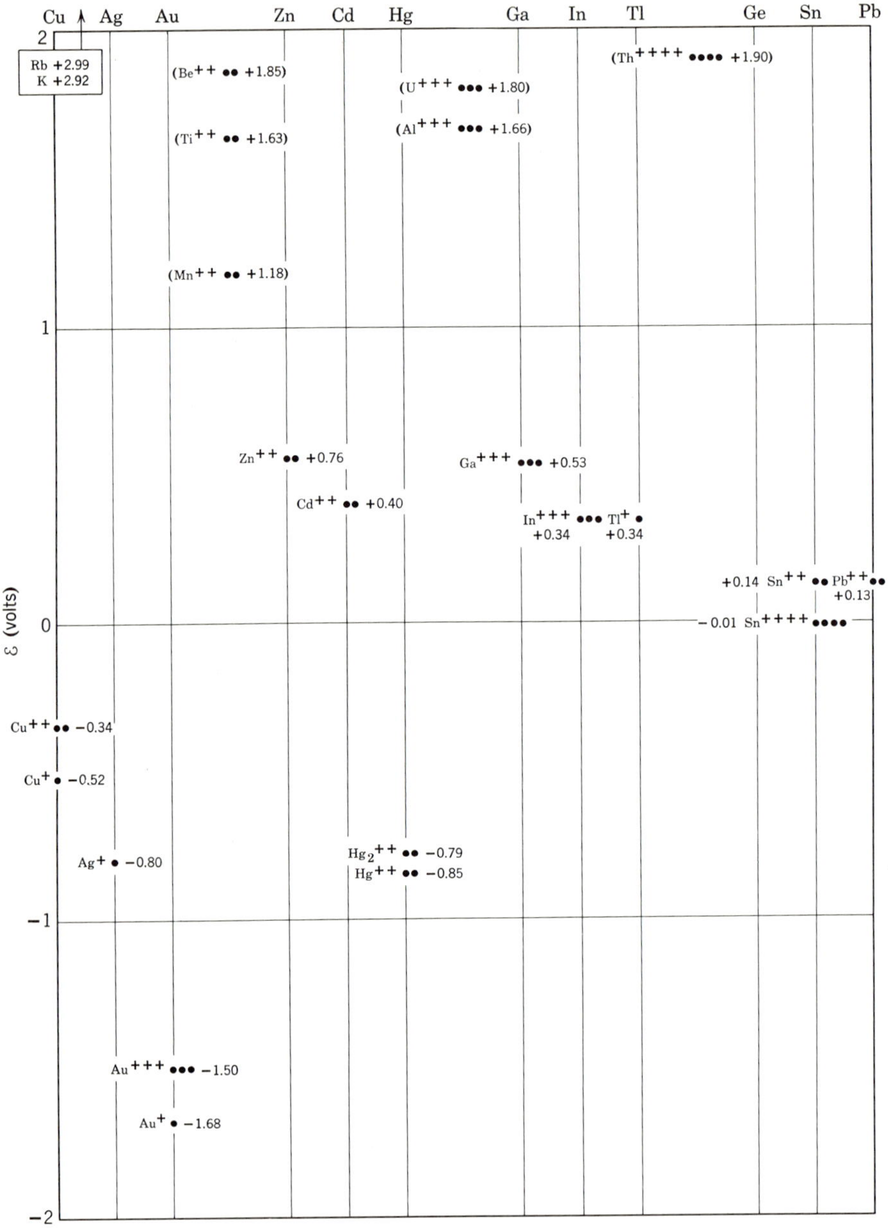

Although the fairly high ionization potential would not suggest high electrical conductivity, actually copper is next to the top in its ability to conduct electricity, being surpassed only by silver. Because it is more generally available than silver, it is the most widely used metal for electrical conduction; three to four million tons per year are produced for this purpose. Copper also alloys with many other metals to make substances that have special uses. The addition of zinc hardens it, forming alloys generally classified as brass; with tin, alloys are formed that are called bronzes; with manganese, nickel, and iron, alloys are formed that are particularly resistant to corrosion and are known as *Monel* metals. In its native state, copper oxidizes only slowly, so that bars and wires and other useful forms of copper can be exposed to air for years without serious deterioration.

Silver. A large part of the silver produced in the world comes from the residues obtained in the production of copper. However, like copper, silver is also found in nature as the sulfide, Ag_2S. In its elementary form, silver has a high luster, but it is a soft metal and this makes it impractical for coins or even tableware and jewelry unless alloyed to some extent with another metal like copper that increases the hardness and durability. The pure metal oxidizes only slowly, but the formation of a thin coating of oxide or of sulfide impairs the luster and consequently the beauty of tableware and jewelry, so that "cleaning" silver is a frequent task for the neat housewife.

Gold. Elemental gold is found in nature as fine particles and nuggets; it is also found combined with elements of the sixth group of the periodic table (e.g., $AuTe_2$). Gold has much less tendency to form coatings of oxides and sulfides and is, therefore, far more valuable than silver, for both jewelry and other ornamental purposes. It has been a favorite metal for coins for thousands of years and much prized because of its beautiful color and luster. Like silver, it is often alloyed with copper to increase its hardness.

Zinc. The principal source of zinc is a mineral, ZnS, called *zinc blende* or *sphalerite.* The steps in the process of refining the ore are similar to those for copper: heating in the air, which converts the sulfur to sulfur dioxide and leaves zinc oxide, ZnO; this is then heated with carbon and reduced to metallic zinc. The equations are

$$2ZnS + 3O_2 \rightleftarrows 2ZnO + 2SO_2$$
$$ZnO + C \rightleftarrows Zn + CO$$

Like most metals, zinc forms a coating of oxide but does not corrode like iron. For this reason, a light plating of zinc on iron protects the iron; this product is called galvanized iron. As mentioned before, zinc alloyed with copper gives the important structural material, brass.

Cadmium. Cadmium occurs principally in nature as an impurity in zinc ores with a ratio of about 200 to 1. It also occurs like zinc as a fairly pure sulfide called *greenockite.*

Mercury. Like its congeners, zinc and cadmium, mercury is found as the sulfide, HgS, called *cinnabar.* It is also found occasionally alloyed with gold and silver, or as the element.

Mercury has many important uses. Because it is a liquid, freezing only at a low temperature (−39°C) and expanding uniformly as temperature is increased over a wide range, it is used as a fluid for thermometers. In the vapor state, it conducts electricity when sufficient voltage is applied and an arc is started. In this conduction process, it emits a characteristic light that is green in the visible part of the spectrum and contains also much ultraviolet; so mercury arcs are used to give a winter sun-tan. The alloys of mercury are called amalgams and some of these are used for filling cavities in teeth.

Gallium. This is a rare element; it is found throughout nature but only in the most minute quantities. The same is true of the elements indium and thallium, which are found in the ores of lead, zinc, and iron. The key problem in the preparation of any of these three metals is the separation of these minute quantities to obtain the metal in pure form. Gallium has an abnormally low melting point, somewhat like mercury but with a far higher boiling point; it has the longest liquid range of any known substance; it is quite hard but possesses so little tensile strength that it is brittle. On the other hand, both indium and thallium are soft.

Germanium, tin, and lead. These elements are also found largely as sulfides; and tin occurs extensively as the oxide, SnO_2; from these sources both tin and lead have been prepared in metallic form since the earliest beginnings of civilization. Tin has been regarded as particularly valuable because of its resistance to corrosion; and lead has been used because of its high density and its malleability. Until recently, there was little demand for germanium; but within the last few years a type of electronic component, called a *transistor*, has been developed. It is made of germanium and takes advantage of this element's electrical properties, which classify it more as a semiconductor than a true conducting metal.

Tin exists in three different types of crystal lattice; below 13°C, it changes to a non-metallic form which has a gray color and is extremely brittle. (When tin organ pipes developed spots of gray tin during a cold winter, the diagnosis was "tin disease.")

COMPOUNDS AND CHEMISTRY

In order to show the relationships among these twelve metals, both horizontally and vertically in the periodic table, there are listed in Table 3 some of the more common ions and compounds formed from these elements. The electronic structure diagrams are also given at the head of the column for each element.

Ions. For 2-6 core metals, we found that the ions formed were almost exclusively those where one electron was removed in Group I, two electrons were removed in Group II, and three electrons in Group III. For these 2-6-10 metals, the situation is more complicated. As shown by the points marked with little squares at the left of Fig. 2, it requires only about 4 ev to remove the first electron from the *s* shell in potassium; but it requires about 32 ev to remove an electron (II) from the underlying 2-6 shell. By way of contrast, it requires about 7.5 ev to remove the one valence electron from the outer shell of copper, but it takes only 20 ev to remove the *d* electron from the underlying shell. The result is that we find copper ions both with a single positive charge and with two positive charges where two electrons have been lost from the atom. Actually, in solution, because of the other energy factors such as the heat of hydration, it turns out that the Cu^{+2} ion is the more stable; this is known as the *cupric ion.* The Cu^{+} ion (cuprous) is generally unstable in solution and has a tendency to disproportionate and oxidize to the doubly charged state:

$$2Cu^{+} \rightarrow Cu^{+2} + Cu(s)$$

Compounds in the cuprous state are Cu_3N, Cu_2SO_4, and CuSCN; Cu_2O is formed only above 1000°C with normal pressure of O_2.

With the ions of the alkali metals, we found that there was a grouping of water molecules around the ion in aqueous solution, but presumably no actual covalent bonding. By way of contrast, with copper it is believed that in aqueous solution the cupric ion generally exists as $Cu(H_2O)_4^{+2}$. When ammonia is present, this ion can also exist as a complex ion, $Cu(NH_3)_4^{+2}$. There is, thus, a tendency not only to lose electrons from the group of ten in the *d* shell, but also to form covalent bonds with these electrons. Moreover, when copper combines with an element like chlorine, it appears that the bonds have a far greater covalent character than in the case of NaCl.

The structural chemistry of the complexes of copper is rich in variety. To a first approximation, $CuCl_2 \cdot 2H_2O$ and $Cu(NH_3)_4^{+2}$ are square coplanar structures. But there is increasing evidence that the bonding in solution of the blue ammonia complex (often used in the test for copper) is actually a distorted octahedron with ammonia or water molecules at the corners and copper at the center. In these complexes, four of the attached groups lie at the corners of a square with the cupric ion at the center. The remaining two groups are somewhat farther away, above and below the plane of the molecule, and lead to a structure that is square coplanar if we include only nearest neighbors; but it is octahedral if we include all near neighbors. Similar behavior is found for other complexes of copper. This behavior has been predicted theoretically by extensions of the crystal field theory of complex ions discussed briefly in Chapter 25.

Divalent copper chloride complexes have a tendency to form chains. For example, in solid $CuCl_2$ we have infinite chains of the type

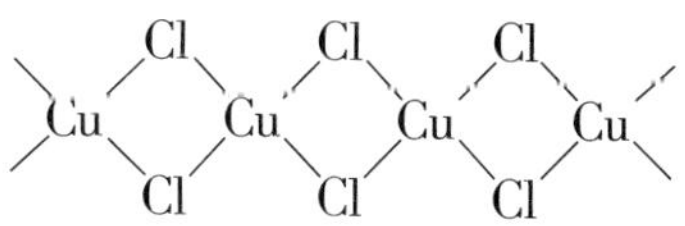

Table 3 Important Ions and Compounds

(Parentheses around the *whole* ion or compound denote instability.)

	Cu 10 6 6 2 2 2 1	Ag 10 10 6 6 6 2 2 2 2 1	Au 14 10 10 10 6 6 6 6 2 2 2 2 2 1
+ 3			(Au^{+3}) $AuCl_4^-$
+ 2	$Cu(H_2O)_4^{+2}$ $Cu(NH_3)_4^{+2}$ $CuCl_2(H_2O)_2$ CuO $CuCl_3(H_2O)^-$ $Cu(OH)_2$ CuS		
+ 1	(Cu^+) CuCl CuCN Cu_2O Cu_2S	Ag^+ $Ag(NH_3)_2^+$ AgCl $Ag(CN)_2^-$ Ag_2O Ag_2S	(Au^+) $AuCl_2^-$ $Au(CN)_2^-$ AuCl
	Ga 10 6 6 1 2 2 2 2	In 10 10 6 6 6 1 2 2 2 2 2	Tl 14 10 10 10 6 6 6 6 1 2 2 2 2 2 2
+ 4			
+ 3	Ga^{+3} $GaCl_3$ $Ga(OH)_3$ Ga_2O_3 Ga_2S_3	In^{+3} $InCl_3$ $In(OH)_3$ In_2O_3 In_2S_3	(Tl^{+3}) $TlTlCl_4$ $TlCl_3$ $Tl(OH)_3$ Tl_2O_3 Tl_2S_3
+ 2	$(GaCl_2)$	$(InCl_2)$	
+ 1	(Ga_2O)	(InCl)	Tl^+ Tl_2S TlOH Tl_2O

In this structure the chains are packed together so that each copper has above and below it one more chlorine atom slightly farther away than the ones in the chain.

In the compound $CsCuCl_3$, the $CuCl_3^-$ complex does not form a discrete complex ion but forms an infinite spiral chain of the type

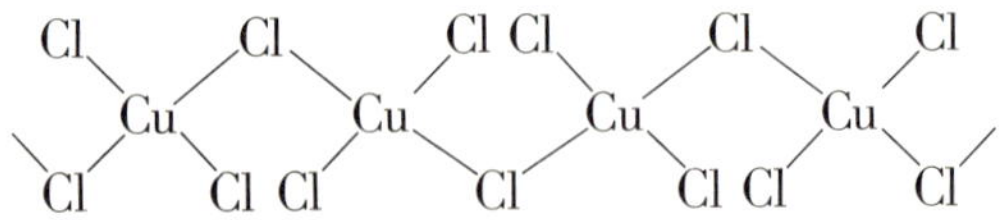

Table 3 (Continued)

	Zn 10 6 6 2 2 2 2	Cd 10 10 6 6 6 2 2 2 2 2	Hg 14 10 10 10 6 6 6 6 2 2 2 2 2 2
+ 3			
+ 2	$Zn(H_2O)_4^{+2}$ $Zn(NH_3)_4^{+2}$ $ZnCl_2$ $Zn(OH)_2$ $Zn(OH)_4^{-2}$ ZnO ZnS	$Cd(H_2O)_4^{+2}$ $Cd(NH_3)_4^{+2}$ $CdCl_2$, CdI_4^{-2} $Cd(OH)_2$ CdO CdS	Hg^{+2} $HgCl_2$ $HgNH_2Cl$ HgI_4^{-2} HgO HgS
+ 1			Hg_2^{+2} Hg_2Cl_2 (Hg_2S)
	Ge 10 6 6 2 2 2 2 2	Sn 10 10 6 6 6 2 2 2 2 2 2	Pb 14 10 10 10 6 6 6 6 2 2 2 2 2 2 2
+ 4	GeH_4 H_2GeF_6 $GeCl_4$ Na_2GeO_3	Sn^{+4} SnH_4 $SnCl_4$ Na_2SnO_3 SnO_2 SnS_2	Pb^{+4} H_2PbCl_6 ($PbCl_4$) H_2PbF_6 PbO_2
+ 3			Pb_2O_3
+ 2	Ge^{+2} GeO	Sn^{+2} $SnCl_2$ SnO SnS	Pb^{+2} $HPbO_2^-$ $PbCrO_4$ $PbCO_3$ $PbSO_4$ PbO PbS
+ 1			Pb_2O

An interesting example of the effect of environment on bonding is found in the anhydrous solid Cs_2CuCl_4. In the solid compound the $CuCl_4^{-2}$ unit is a distorted tetrahedron, but the $CuCl_4^{-2}$ group reverts to the planar (or distorted octahedral) structure when the solid dissolves. The electron structures of some copper complexes are shown in Table 4.

With few exceptions (like the above) cupric complexes are square-coplanar and cuprous complexes are tetrahedral or linear depending on the coordination. Recently it has been found that polynuclear complexes may involve bonding between the copper atoms; this is a striking example of copper covalency.

A large number of cupric compounds contain water of hydration; these water molecules frequently group themselves around the Cu^{+2} ion, and their linkage

Table 4 Some Complexes of Copper

d			10	
p		6	6	
s	2	2	2	1

Cu (neutral atom)

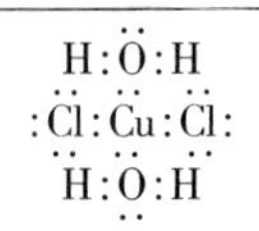

$CuCl_2(H_2O)_2$

$CuCl_3(H_2O)^-$

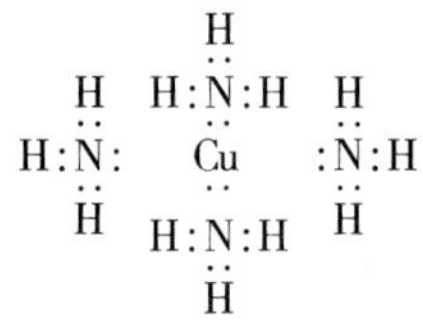

$Cu(NH_3)_4^{+2}$

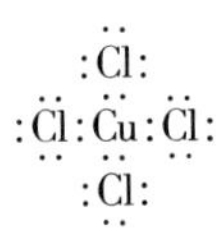

$CuCl_4^{-2}$

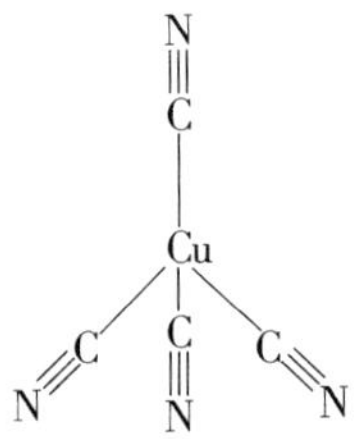

$(Cu(CN)_4)^{-3}$

with it affects the energy levels of the electrons so that the resulting color is blue. The familiar compound cupric sulfate, $CuSO_4(5H_2O)$ is an example of this coloration; dehydrated, it has a whitish color slightly tinged with green.

With hydrogen sulfide gas, cupric compounds precipitate cupric sulfide, CuS. The usual compounds with oxygen are formed, Cu_2O above and CuO below 1000°C under normal pressure of O_2. Copper also forms the normal cupric halides with the exception of CuI_2, which is unstable. These compounds are far more covalent in character than the corresponding alkali halides.

Silver. As contrasted with copper, the only stable ion of silver is the singly charged Ag^+. The compound formed between silver and nitric acid, $AgNO_3$, is highly soluble and provides one of the most common solutions of this ion. The compounds between silver and the halogens also illustrate this monovalence. Curiously, silver fluoride is quite soluble in water, but the other three halogens salts are extremely insoluble, having the solubility-product constants with the values: AgCl, 12×10^{-10}; AgBr, 5×10^{-13}; AgI, 9×10^{-17}. As in the case of copper, there is a strong tendency to form covalent bonds, and this, plus the attractions in the lattice, produces the insolubility.

Silver, like copper, forms complexes with ammonia and cyanide ion, $Ag(NH_3)_2^+$ and $Ag(CN)_2^-$; at appropriate concentrations, the ammonia complex is just strong enough to dissolve silver chloride without dissolving the bromide or the iodide. This is often made use of for separating the halides of silver. Silver also forms a complex with thiosulfate, $Ag(S_2O_3)_2^{-3}$; this complex is strong enough to dissolve silver bromide.

When the silver halides (with the exception of AgF) are exposed to light, an activation process takes place in which some of the electrons are displaced. An exposed silver compound can remain in this activated state for days; when it is then treated with a mild reducing agent, metallic silver can be deposited as a black coating. This deposit is most intense wherever electrons are displaced by light. This is the basis of photographic film. Unactivated silver bromide is removed by reaction with thiosulfate.

In both the oxide and the sulfide, we have the formulas to be expected on the basis of the loss of one electron, Ag_2O and Ag_2S. Like the other metals with 2-6-10 cores, silver forms many interesting complexes of the coordination type. There is the linear $Ag(CN)_2^-$ ion as well as the ammonia complex, $Ag(NH_3)_2^+$. The former is the basis of the cyanide process used in leaching silver from its ores:

$$O_2 + 4Ag + 8CN^- + 2H_2O \rightleftarrows 4Ag(CN)_2^- + 4OH^-$$
$$AgCl + 2CN^- \rightleftarrows Ag(CN)_2^- + Cl^-$$

We also find $AgCl_2^-$ and $AgCl_3^{-2}$.

Gold. With the third member of this family, gold, we find still a different situation in the formation of ions. Although there is nothing in the form of the *es* diagram to suggest it, there are many compounds of gold with three positive charges called *auric*, Au^{+3}. Like its neighboring element, silver, gold also exists in the form of a singly charged ion, the *aurous* state, Au^+; but this disproportionates like copper to go to the +3 oxidation state as Au^{+3} and metallic gold:

$$3Au^+ \rightarrow Au^{+3} + 2Au$$

Covalent bonds are frequently formed by sharing some of the ten *d* electrons, giving the complexes shown in Table 3.

Zinc. We now turn to the first member of Group II*B*, zinc, which bears much the same relation to calcium that copper does to potassium; zinc is characterized almost exclusively by ions and compounds produced by the removal of the two valence electrons. But we see, from Fig. 2, that the second ionization potential for zinc is higher than that for calcium.

In aqueous solution, zinc has a tendency to form covalent complexes with water molecules to give the complex ion $Zn(H_2O)_4^{+2}$; like copper, it also forms a complex with ammonia as shown in Table 3. Its compounds with the halogens, such as $ZnCl_2$, are normal; its hydroxide also has the normal formula, and so do the oxide and sulfide as shown in Table 3.

The Zn^{+2} ion has no color and is not paramagnetic; it does hydrolyze slightly to give an acid solution according to the equation

$$Zn^{+2} + H_2O \rightarrow H^+ + Zn(OH)^+$$
$$K_h = 2 \times 10^{-10}$$

When the solution is made basic, zinc hydroxide precipitates as $Zn(OH)_2$; and in strongly basic solution this dissolves, giving the zincate ion that can exist as $HZnO_2^-$ and ZnO_2^{-2}. ZnO is an oxide with a high sublimation point around 1800°C and a lattice structure that is much closer to quartz than to the true covalent molecular oxides like OsO_4.

ZnS finds an important use in fluorescent screens because it can accept the photons of light in the ultraviolet X-ray region, absorb part of the energy, and then re-emit the photon in the visible. In this way, the X-ray image can be perceived by the eye. Again, when ZnS is bombarded with electrons of sufficiently high voltage, the electron image becomes visible; many television screens are coated with this compound.

Cadmium. Cadmium and zinc resemble each other closely, though the quantity of cadmium found in nature is only about $\frac{1}{200}$ that of zinc. Cadmium exists primarily in the +2 oxidation state, as shown in Table 3.

Mercury. When we go from cadmium to mercury, there is a change in properties that is quite different from that found between Ag and Au. In going from Ag to Au, the melting points and boiling points rise; in going from Cd to Hg, they both fall sharply. And though, as we expect from the electronic structure diagram, the mercuric ion, Hg^{+2}, is stable, we also find two mercury atoms combining with a *covalent* bond to form the mercurous ion with the formula Hg_2^{+2}. Mercury forms the divalent oxide and divalent sulfide as shown in Table 3; it also forms compounds of the type of mercurous chloride; the mercurous sulfide is not stable but disproportionates into HgS and Hg.

All the metals we have discussed so far in this chapter have a strong tendency to form complex ions with ammonia. Mercury forms a particularly strong bond with the nitrogen in ammonia; the resulting electronic rearrangement is accompanied by a weakening of the N—H bond with the following result:

$$HgCl_2 + 2NH_3 \rightarrow \underset{\text{unstable}}{Hg(NH_3)_2Cl_2} \rightarrow HgNH_2Cl + NH_4Cl$$

$HgNH_2Cl$ is so stable that if insoluble mercurous chloride, Hg_2Cl_2, is treated with ammonia the following occurs:

$$Hg_2Cl_2(s) + 2NH_3 \rightarrow HgNH_2Cl(s) + Hg + NH_4Cl$$

Thus the driving force for the formation of $HgNH_2Cl$ is sufficient to cause an auto-oxidation-reduction of the mercury. The change from the white Hg_2Cl_2 on treatment with NH_3 to the gray mixture of Hg and white $HgNH_2Cl$ is used as a qualitative test for mercury. Mercury salts, like zinc salts, have a good deal of covalent character. For example, mercuric chloride, $HgCl_2$, is predominantly covalent. It melts at 277°C and boils at 305°C; although it dissolves in water readily, it forms solutions that are poor conductors, and hence it is a weak electrolyte. Mercurous chloride dissolves with dissociation.

Gallium, indium, and thallium. These three elements are rare; their chemical behavior resembles that of aluminum. The ionization potential for aluminum is 5.98 ev and, as may be seen from Fig. 2, the three values for gallium, indium, and thallium lie very close to this. In the case of gallium, there is some indication of monovalent and divalent atoms that can be stable under restricted conditions only; but the chemistry is largely concerned with the gallic ion, Ga^{+3}. This is illustrated by the compounds $GaCl_3$, $Ga(OH)_3$, Ga_2O_3, Ga_2S_3. Gallates such as Na_3GaO_3 are formed when $Ga(OH)_3$ is treated with excess base.

The chemistry of indium resembles that of gallium, being characterized by the indic ion, In^{+3}. Typical compounds are given in Table 3.

The chemistry of thallium, like that of the other heavier atoms, gold and mercury, exhibits at least two different valence states. The thallous ion, Tl^{+}, is stable, and the compounds TlOH, Tl_2O, and Tl_2S are formed. Univalent thallium is quite similar to silver in chemical properties. The thallic ion, Tl^{+3}, is not stable in solution, but compounds are found as indicated in Table 3. Like some of the other elements, thallium can occur in the same molecule in two different ways, as in the thallous-thallate, $TlTlCl_4$.

The metallic elements of Group IV. The first two members of Group IV in the periodic table, carbon and silicon, are among the most important of all chemical elements, and we include their diagrams for comparison with the metallic elements of this group (Table 5). All these elements have four elec-

Table 5 Electronic Structure of Group IV Elements

Carbon	Silicon	Germanium	Tin	Lead
				14
		10	10 10	10 10 10
2	6 2	6 6 2	6 6 6 2	6 6 6 6 2
2 2	2 2 2	2 2 2 2	2 2 2 2 2	2 2 2 2 2 2

trons in the outer shell; and, because 4 lies midway between 0 and 8, these elements have a peculiarly flexible electronic structure. Thus the outer shell can gain 4 electrons to form the stable shell of 8, or lose 4 electrons, leaving a stable configuration of 2 or 8.

All members of this group form hydrides similar to the hydrocarbons formed by carbon (Table 6). The stability of these hydrides decreases with increasing atomic weight.

The diagrams show that there are two *s* electrons and two *p* electrons in the outer shell of each element of this group. If the atom loses the *p* electrons, then there is the partially stable configuration of two electrons left, formed by the two *s* electrons. When this happens, there is left a +2 oxidation state. However, the two 2*s* electrons also can be removed, leaving a +4 oxidation state. The +2 oxidation state is uncommon in silicon and, when found in germanium, it has a tendency to revert to +4 state; but, in tin and lead, the +2 state is quite common. The following equation shows the relations between the two for tin:

$$Sn^{+2} \rightleftarrows Sn^{+4} + 2e^- \qquad \mathcal{E}^\circ = -0.15 \text{ volt}$$

Germanium. Like silicon, germanium forms compounds with hydrogen, although somewhat less readily. Only GeH_4, Ge_2H_6, and Ge_3H_8 have been identified. These compounds oxidize readily to form GeO_2. Evidence has been found for the formation of compounds of germanium that are similar to the zincates and are called germanates. Germanium also forms all the tetrahalides, like $GeCl_4$, and dihalides, like $GeCl_2$, obtained by heating the tetrahalide with Ge.

Table 6 Hydrides of Group IV Elements

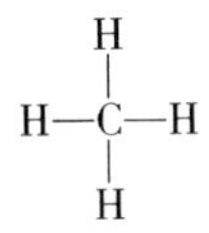 Methane	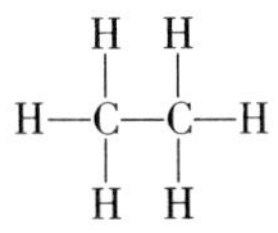 Ethane	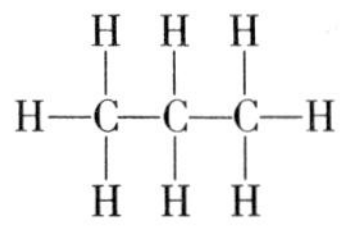 Propane
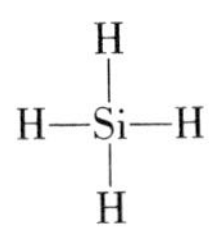 Silane	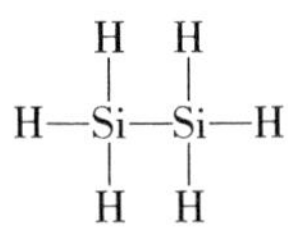 Disilane	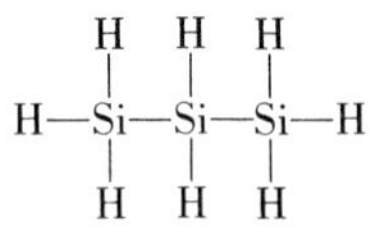 Trisilane
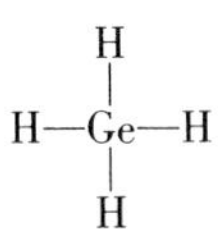 Germane	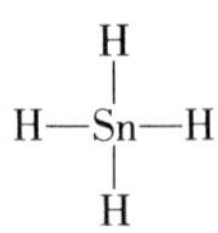 Stannane	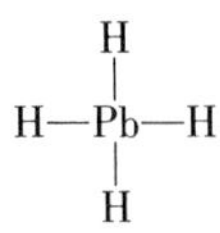 Plumbane

Table 7 Electronic Structures of Tin

Sn	Sn^{+2} Stannous ion	Sn^{+4} Stannic ion
10 10 6 6 6 2 2 2 2 2 2	10 10 6 6 6 + + 2 2 2 2 2	10 10 6 6 6 + + 2 2 2 2 + +

Tin. As shown in Table 3, tin is found generally either in the +2 or +4 oxidation state.

When tin is heated with warm hydrochloric acid, it forms stannous chloride, $SnCl_2$, and hydrogen according to the equation

$$Sn + 2HCl \rightarrow SnCl_2 + H_2$$

It also forms stannous sulfate, $SnSO_4$, when heated with hot concentrated sulfuric acid, and stannous nitrate, $Sn(NO_3)_2$, when heated with cold dilute nitric acid.

In neutral solution, stannous chloride hydrolyzes, forming a precipitate of stannous hydroxychloride, Sn(OH)Cl. This hydrolysis involves the equilibrium

$$SnCl_2 + HOH \rightleftarrows H^+ + Cl^- + Sn(OH)Cl(s)$$

and can be reversed by the addition of an acid; hence hydrochloric acid is often added to solutions of stannous chloride to prevent precipitation of the basic salt. In the presence of chloride ion, the precipitation is further inhibited by the formation of the weak complex ion as indicated by

$$SnCl_2 + 2Cl^- \rightleftarrows SnCl_4^{-2}$$

The diagrams for the stannous ion and stannic ion along with that for tin in the elementary state are given in Table 7. The stannous ion readily loses its two electrons in the outer shell to form the stannic ion and is, therefore, an active reducing agent. It forms a chlorostannate ion in the presence of excess chloride ion, $SnCl_6^{-2}$.

Stannic chloride, $SnCl_4$, is a liquid boiling at 114°C, a clear sign of its covalent nature. In the gaseous state, it combines with water vapor to produce hydrochloric acid and stannic acid, $H_2Sn(OH)_6$. The hexahydroxy stannate ion is similar in structure to the chlorostannate ion, $SnCl_6^{-2}$, there being in both cases six groups octahedrally disposed around the central tin atom.

When combined with water, the stannous ion forms $SnOH^+$ and $H^+ \cdot Sn(OH)_2$. Under slightly basic conditions, solutions of stannous salts yield a precipitate of a white solid known as stannous hydroxide, $Sn(OH)_2$. As the solution gets more basic, this dissolves to form the stannite ion that is sometimes written $Sn(OH)_3^-$

or $HSnO_2^-$. This ion is unstable and disproportionates according to the equation

$$2Sn(OH)_3^- \rightarrow Sn(s) + Sn(OH)_6^{-2}$$

It is possible to use hydrogen sulfide, H_2S, to precipitate either stannous sulfide, SnS, or stannic sulfide, SnS_2. In high concentration of the sulfide ion, S^{-2}, the solid stannic sulfide dissolves to give the thiostannate ion, SnS_3^{-2}. When the solution is acidified, the stannic sulfide is reprecipitated.

Lead. The Latin name for lead is *plumbum.* It is from this root that we get the term *plumber,* because years ago so many pipes were made of lead. This is also the origin of the symbol for lead, Pb, and for the name of the ions which are referred to as *plumbous,* Pb^{+2}, and *plumbic,* Pb^{+4}. There are three common oxides of lead: litharge, or lead monoxide, PbO; red lead, Pb_3O_4; and lead dioxide, PbO_2. In the +2 oxidation state, lead forms many compounds of the normal type such as lead nitrate, $Pb(NO_3)_2$, and lead carbonate, $PbCO_3$, which occurs in nature as the mineral *cerussite.* Other common compounds are lead sulfate, $PbSO_4$, and lead chromate, $PbCrO_4$, which is used as a pigment and called *chrome yellow.* Another common pigment is *white lead,* $Pb_3(OH)_2(CO_3)_2$. These lead compounds are poisonous.

Because of the large size of the atom, it is possible to form many complex ions. With the halogens, we find:

$$PbCl^+ \rightleftarrows Pb^{+2} + Cl^- \qquad K = 0.8$$
$$PbBr^+ \rightleftarrows Pb^{+2} + Br^- \qquad K = 0.7$$
$$PbI^+ \rightleftarrows Pb^{+2} + I^- \qquad K = 0.03$$

It is also possible to form $PbCl_3^-$ and $PbBr_4^{-2}$. Plumbite ions, $Pb(OH)_3^-$, are also formed. As contrasted with the stannite ion, the plumbite ion is stable in solution. It is interesting to compare the behavior of lead in acid and basic solution through the following equations:

$$Pb(s) \rightleftarrows Pb^{+2} + 2e^- \qquad \mathcal{E}^\circ = 0.13 \text{ volt}$$
$$3OH^- + Pb(s) \rightleftarrows Pb(OH)_3^- + 2e^- \qquad \mathcal{E}^\circ = 0.54 \text{ volt}$$

This is all in accord with Le Chatelier's principle. In the plumbic form, lead dioxide is a strong oxidizing agent, as shown by the equation

$$PbO_2(s) + 4H^+ + 2e^- \rightleftarrows Pb^{+2} + 2H_2O \qquad \mathcal{E}^\circ = +1.46 \text{ volts}$$

The halides of lead as well as those of tin are largely covalent in character; hence, although lead and tin salts are electrolytes, they are only partially dissociated in solution.

In the plumbous form, most of the salts of lead are relatively insoluble: $PbHPO_4$, $PbCO_3$, $PbSO_4$, PbS. In solution with sulfide ion, the other salts will revert to lead sulfide since this is the least soluble.

Another common use of lead is in the form of *lead tetraethyl,* $Pb(C_2H_5)_4$, a relatively volatile compound, where four ethyl radicals are attached to a central lead atom. This compound prevents the sudden oxidation of gasoline in the internal combustion engine, the phenomenon of "knock" that causes serious loss of power. To attain the increased efficiency of operation under high compression, it is necessary to use some anti-knock agent and, as such, lead tetraethyl has been extremely important. Lead, like many other heavy metals, is a powerful inhibitor for enzyme reactions, and this may be the reason why it is poisonous. It has a tendency to accumulate in the body, and it affects the central nervous system.

Thermodynamic relations. Unfortunately, the compounds of the elements studied in this chapter have not been the subject of as extensive thermodynamic study as the compounds of the elements in Groups I, II, and III. The available data refer almost exclusively to room temperature. In Fig. 3 are plotted values of ΔH_f°, and in Fig. 4 a more restricted set of values of ΔF_f°, all for 25.00°C or 298.15°K. These values do permit some interesting comparisons of certain aspects of thermodynamic behavior at this temperature.

Turning first to the upper left corner of Fig. 3, it is interesting to note how many of the compounds of the coinage elements have values of ΔH_f° lying close together just below the zero axis. In this respect, the halogen compounds of Cu^+ and Ag^+ closely resemble one another; the values for analogous compounds for Au lie about 20 kcal higher. By way of contrast, the value of ΔH_f° for Ag_2O lies 30 kcal above that for Cu_2O.

With Zn, Cd, and Hg, there is a more even progression of values of ΔH_f° through this triad; this is true for the oxides, sulfides, halides, and nitrates. Turning to the compounds containing anions with oxygen—the carbonates, sulfates, and silicates—the values of ΔH_f° fall sharply, indicating greatly increasing stability. A comparison of these values with Fig. 4 shows that ΔH_f° is by far the dominant factor in determining the value of ΔF_f°; in these compounds, the energy factor determines the stability and the entropy factor plays a minor role, raising the value of ΔF_f° above that of ΔH_f° by only a few kilocalories in most instances.

In the columns for Ga, In, and Tl, we observe that the values of ΔH_f° for the halides fall off slightly from Ga to In and then rise sharply for Tl. Note that the thermodynamic properties of the halides and oxides of Tl^+ resemble closely those of silver. This probably accounts for the similarity in chemical properties.

Viewing Fig. 3 as a whole, we note that compounds with common anions lie in about the same range. The values of ΔH_f° for nearly all the halogen compounds lie between 0 and -100 kcal; on the right-hand side, a few of the compounds with three and four halogen atoms lie between -100 and -200 kcal, and PbF_4 has the largest negative value, as we might expect. The carbonates are grouped just above the −200 kcal line, with the sulfates close by, $HgSO_4$ lying a little higher, and $ZnSO_4$ a little lower. The silicates are grouped around the −300 kcal line.

In conclusion, it may be said that the chemistry of these twelve elements is fairly regular. A knowledge of the trends in the periodic table, interpreted in

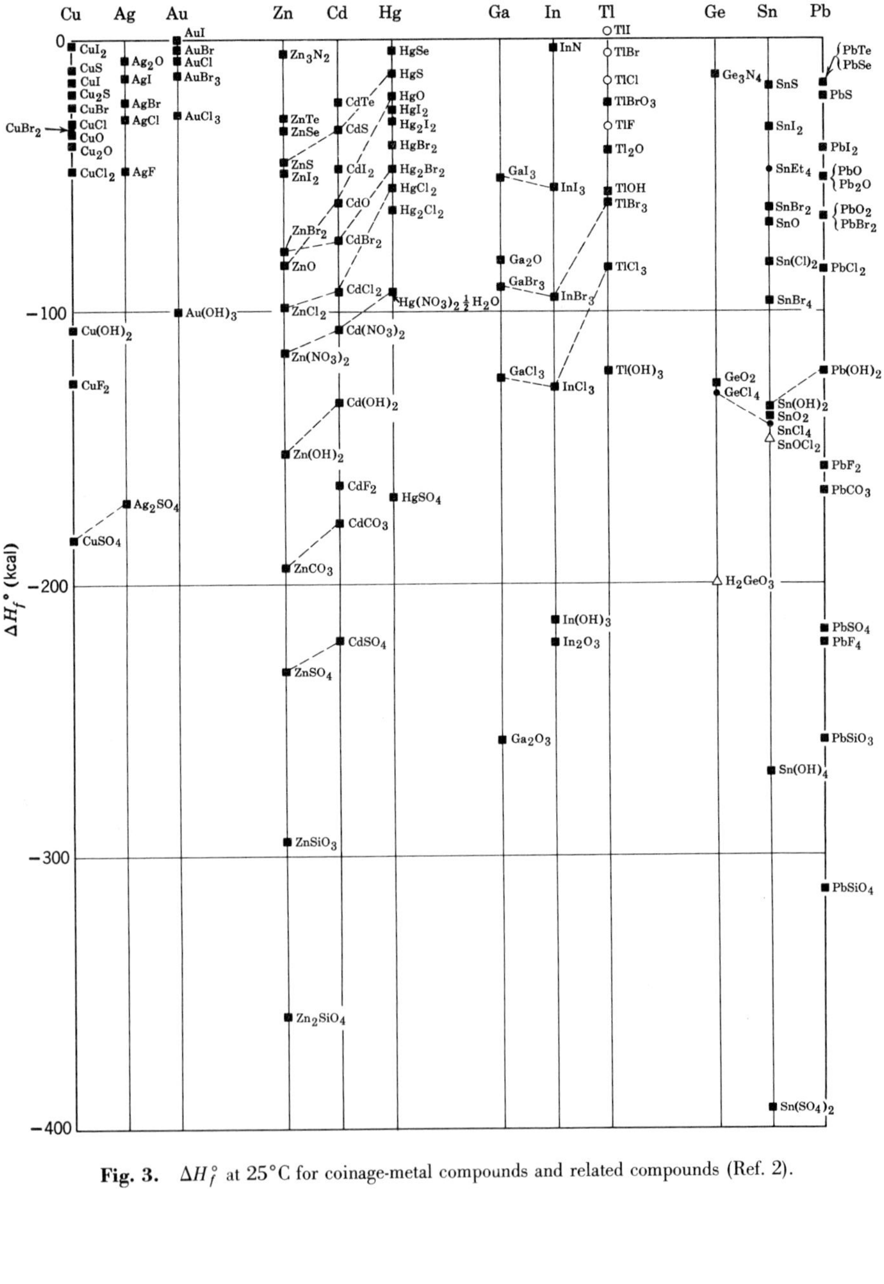

Fig. 3. ΔH_f° at 25°C for coinage-metal compounds and related compounds (Ref. 2).

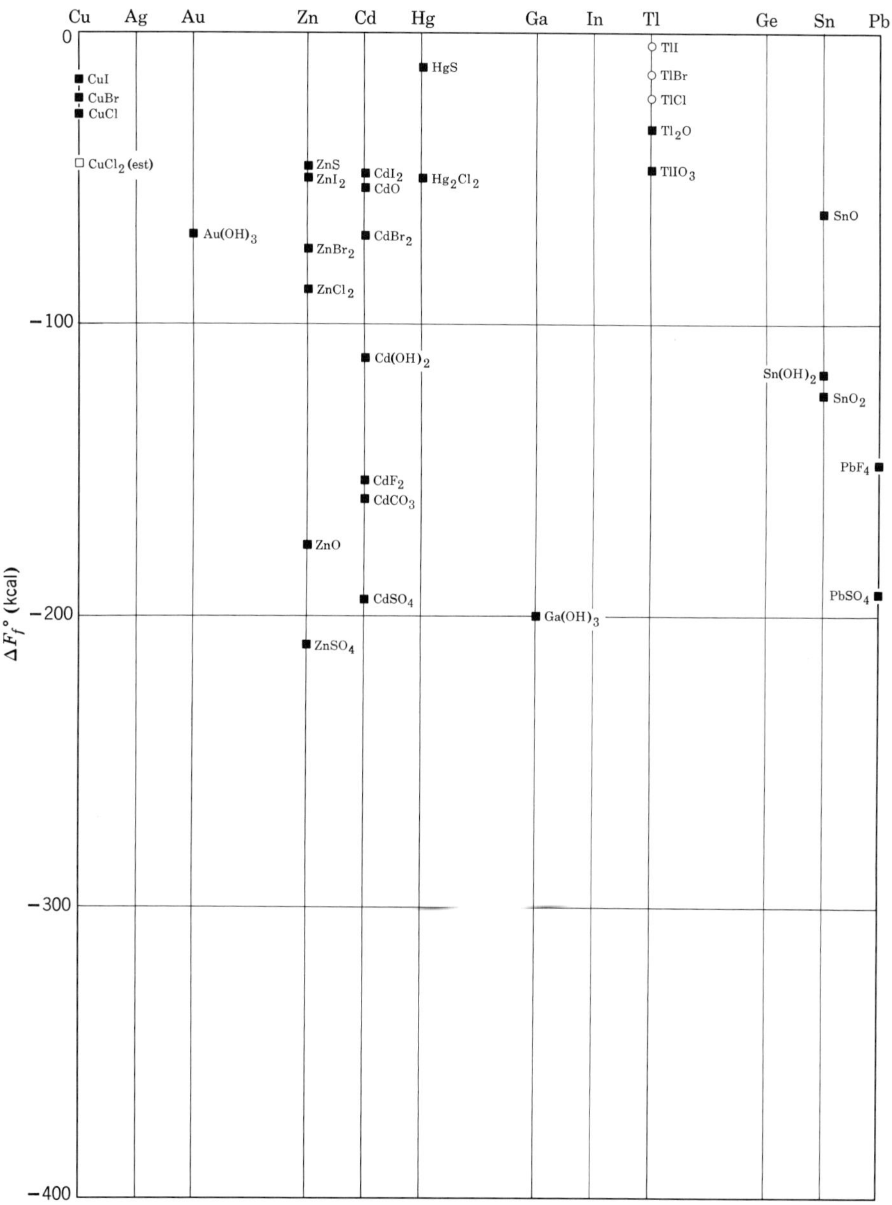

Fig. 4. ΔF_f° for coinage-metal compounds and related compounds at 25°C (Ref. 2).

terms of electronic structure and of thermodynamic relations, permits reasonably good predictions about chemical behavior.

PROBLEMS

1. What is the principal reason for the stability of Ag^+, Zn^{+2}, and Ga^{+3} ions? Why is the stability of Cu^{+2} surprising?

★ **2.** From the data in Table 2, calculate the voltage of a half-cell for each of the following reactions:

(a) $Cu^+ \rightleftarrows Cu^{+2} + e^-$ *Ans.* −0.16 v.

(b) $Au^+ \rightleftarrows Au^{+3} + 2e^-$

(c) $Sn^{+2} \rightleftarrows Sn^{+4} + 2e^-$

3. Why is the half-cell potential for the formation of Cu^+ so much lower than the potential for the formation of K^+? Why is the potential of Ga^{+3} not so far below the potential of Al^{+3}?

4. Calculate the pressure of H_2 in equilibrium with Zn metal, Zn^{+2} at 1 *m*, and H^+ at 1 *m*.

5. From the values in Figs. 3 and 4 estimate the pressure of iodine in equilibrium with crystalline CuI over the range from 0° to 500°C at each 100° interval.†

6. Calculate the equilibrium constant for the reaction $CuCl + \frac{1}{2}Cl_2 \rightleftarrows CuCl_2$ at 25°C. What pressure of chlorine will be in equilibrium with a mixture of these two crystals at this temperature? *Ans.* 1.2×10^{-25} atm.

7. Calculate the entropy factor for each of the reactions involving the formation from elements of $CdSO_4$ and $PbSO_4$. Would you expect the results to be similar or different?

8. Calculate the equilibrium constant for the oxidation of ZnS to $ZnSO_4$ at room temperature. Make an estimate of this equilibrium constant at 1000°K.

★ **9.** Estimate the pressure of CO_2 over $CaCO_3$ and over $CdCO_3$ at 1000°K.

10. Calculate the heat of formation of $ZnCO_3$ from ZnO and CO_2 at 25°C. *Ans.* −18 kcal.

† At 25°C, $\Delta F°$ and $\Delta H°$ for $I_2(g)$ at 1 atm are 4.63 kcal and 14.9 kcal respectively.

20

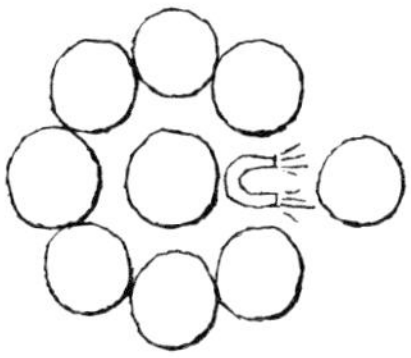

Chemistry of Group V, VI, and VII elements

Elements on the left side of the periodic chart are usually metallic; those on the right side are usually non-metallic. In the last two chapters we discussed the properties of the elements that make up the first three rows of the periodic chart; in the next two chapters, we shall discuss the properties of the elements that make up the last three rows of the periodic chart. In the study of Group I, II, and III elements, we found that these elements are primarily metallic with a trend toward non-metallic properties for the lighter Group III elements, e.g., boron. In Groups V, VI, and VII, we shall find that these elements are primarily non-metallic with a trend toward metallic properties for the heavier members of Group V. Thus the study of these elements completes a perspective view of the properties of the elements.

HORIZONTAL AND VERTICAL RELATIONS

The properties of the Group V, VI, and VII elements are listed in Table 1. In the discussion that follows we survey each of the properties in turn.

Oxidation state. According to the rules developed in Chapter 6, the oxidation states of the Group VII elements, the *halogens,* range from $+7$ to -1; similarly, the oxidation states of the Group VI elements range from $+6$ to -2; those of the Group V elements range from $+5$ to -3.

Only for nitrogen are all possible oxidation states found. For the other elements only half of the expected oxidation states are stable. Thus, for the halogens, oxidation states occur in jumps of 2; the most stable states for chlorine, for ex-

Table 1 Properties of Elements

	F	Cl	Br	I
Oxidation states				
Common	−1, 0	−1, 0, 1, 3, 5, 7	−1, 0, 1, 3, 5	−1, 0, 1, 5, 7
Also observed		+4, +6	+4	+4
Covalent radius, Å	0.73	0.98	1.13	1.35
X^{-1}, ionic radius, Å	1.33	1.81	1.96	2.20
Ionization energy, ev	17.4	13.0	11.8	10.4
Melting point, °C	−223	−102	−7	114
Boiling point, °C	−188	−34	59	185
Resistivity, ohm cm	—	—	10^{14}	10^{12}
Single-bond energy, kcal	37	58	46	36
Electronegativity	4.0	3.0	2.8	2.5
Electron affinity, ev	3.7	4.0	3.8	3.4
$\mathcal{E}°$, $X^- \rightarrow \frac{1}{2}X_2 + e^-$	−2.8	−1.4	−1.1	−0.53
Hydration energy, kcal/mole	117	85	74	61
K_i HX	10^{-4}	10^{7}	10^{9}	10^{11}

	O	S	Se	Te
Oxidation states				
Common	+2, 0, −2	−2, 0, 2, 4, 6	−2, 0, +4, +6	−2, 0, +4, +6
Also observed	−1	−1, +7, +8, +9		+9
Covalent radius, Å	0.74	1.06	1.16	1.44
X^{-2}, ionic radius, Å	1.32	1.74	1.91	2.03
Ionization energy, ev	13.6	10.4	9.8	9.0
Melting point, °C	−219	119	220	450
Boiling point, °C	−183	445	688	1390
Resistivity, ohm cm	—	10^{23}	12	0.03
Single-bond energy, kcal	33	51	44	33
Electronegativity	3.5	2.5	2.4	2.1
$\mathcal{E}°$, $X^{-2} \rightarrow X + 2e^-$		+0.51	+0.77	+0.91
K_i^I H_2X	10^{-14}	10^{-7}	10^{-4}	10^{-3}

	N	P	As	Sb	Bi
Oxidation states	$-3 \longleftrightarrow +5$	−3, 0, 1, 3, 4, 5	0, 3, 5	0, 3, 5	0, 3, 5
Covalent radius, Å	0.74	1.1	1.25	1.44	1.45
X^{-3}, ionic radius, Å	1.7	2.1	—	—	—
Ionization energy, ev	14.5	11.0	10	8.6	8
Melting point, °C	−210	44	817	630	271
Boiling point, °C	−196	280	633	1640	1560
Resistivity, ohm cm		10^{17}	35	40	115
Single bond energy, kcal	38	51	32	30	25
Electronegativity	3.0	2.1	2.0	1.8	1.7
K_i^I H_3X	10^{-23}	10^{-16}	—	—	—

ample, are -1, 1, 3, 5, and 7. (Other oxidation states are either unstable or uncommon.) This behavior is explicable. Chlorine has seven electrons in its outer shell. It can form electron-pair bonds by contributing one, three, five, or seven electrons to other atoms. It can form an ion with the rare gas configuration by gaining an electron to achieve the -1 oxidation state. Thus, these jumps of two are to be expected for *all* elements if the electron pair is the most stable arrangement. Exceptional compounds, wherein the oxidation state does not conform to these jumps of two, contain either an odd number of electrons, e.g., ClO_2, or bonds between identical atoms, e.g., H_2S_2.

Compounds with odd numbers of electrons are called free radicals. Often, two free radicals combine to pair their electrons; thus we find NO_2 reacting to form N_2O_4 and ClO_3 reacting to form Cl_2O_6. Sometimes the bonds formed in this association are quite weak. NO does not combine to form N_2O_2 in the gaseous state, but such combination does occur in the pure liquid. Similarly, oxygen, which has two unshared electrons, forms the species O_4 in liquid oxygen.

Covalent radius. According to the arguments advanced in Chapter 6, the covalent radius should increase with increasing atomic weight within a given family. This is borne out by data in Table 1. By similar arguments (based on effective nuclear change), the covalent radius should decrease with increasing atomic weight within a given period. This expectation is realized (Table 1) for $n > 2$, but, for the $n = 2$ elements this is not found; that is, the radius stays constant with increasing atomic weight. This discrepancy is the result of interelectronic repulsions that increase as the shell fills with electrons just as the effective nuclear charge increases. This increase in repulsion tends to expand the atom and oppose the contraction produced by the increasing effective nuclear charge. For larger atoms (where electrons are well separated) repulsions are subordinate, but for smaller atoms the effect of repulsion can be significant. When the atoms are as small as N, O, and F atoms, the repulsion term balances the contraction due to increasing nuclear charge, and the covalent radius is fixed for these second period elements (Table 1) even though it decreases for elements in periods where $n > 2$.

Ionization energy. Theoretically, the ionization energy decreases with increasing atomic weight in a given group and increases with increasing atomic weight in a given period. According to the data in Table 1, there is a strong tendency in this direction, but certain irregularities appear. If we compare the ionization energies of nitrogen, oxygen, and fluorine, we find that the ionization energy of oxygen is lower than that of nitrogen whereas that of fluorine is, as expected, higher than that of oxygen. This irregularity is observed for all the periods in Table 1 and emphasizes that the simple concept of effective nuclear charge does not completely predict the properties of atoms.

To rationalize these irregularities, consider the fine points of the atomic structure of these elements. The last three electrons added to nitrogen go into the $2p$ subshell, and all three of these electrons have the same spin. Because of the Pauli exclusion principle, electrons with the same spin cannot come close to each other; that is, their other quantum numbers cannot all be the same. This results in a lower interelectron repulsion and a more stable electronic arrangement than the nitrogen atom would have if the electrons were paired and approached each other. When we add an additional electron to form an oxygen atom, this electron pairs up with one of the three electrons in the p subshell, and at the same time the effective nuclear charge increases. The increase in the effective nuclear charge makes it more difficult to remove the electron; the increase in the interelectron repulsions, because of pairing, makes it easier to remove the electron. Since it is easier to remove an electron from oxygen than it is to remove an electron from nitrogen, the interelectron repulsion overrides the effect of increasing nuclear charge. The smaller the atom, the more important should be this electron repulsion; hence we find the biggest jump for the second period elements and a progressively smaller jump for the third, fourth, and fifth periods.

Melting points and boiling points. The binding energy of metallic and molecular crystals is quite different. In a molecular crystal, the forces are weak, and a small increase in intermolecular distance results in a rapid fall-off in these forces. In a metallic crystal, the forces are strong and not so definitely oriented as they are in a molecular crystal. Thus the behavior of these two types of crystals on melting is quite different. With a molecular crystal, less energy or a lower temperature is required to melt the crystal. Once melted, it takes only a little more energy (increase in temperature) to vaporize the molecules, since in the disordered liquid the intermolecular forces are less effective than in an ordered solid. Lastly, in view of this fall-off in the effectiveness of intermolecular forces when the solid is melted, a rather large increase in volume takes place when a molecular solid is melted. For metals, the melting point is high, and, since the interatomic forces do not depend much on order, the change of volume on melting is small and the boiling point is far above the melting point.

Let us illustrate this behavior by actual examples. Nitrogen is a typical molecular crystal. It melts at $-209°$C; the change in volume on melting is about 20 per cent, and it boils at $-196°$C. On the other hand, sodium (a metal) melts at about 98°C with a volume change of only 3 per cent and does not boil until the temperature is nearly 900°C.

Consider the melting points and boiling points of the elements listed in Table 1. Those of the halogens are low and fairly close together. With increasing atomic weight, there is no sudden change that indicates that these elements are other than molecular crystals. This conclusion is strengthened by the high values of the resistivities of the solid halogens. (Metals have resistivities less than 1.)

By the same criteria, we find that oxygen and sulfur are primarily molecular solids, whereas tellurium and perhaps selenium have metallic properties. This, of course, is expected; on the basis of electronegativity, the heavier elements in a group should be the more metallic. If we examine the resistivity of these elements, we find that our conclusions are in line with the resistivity data in so far as selenium and tellurium have the low resistance characteristic of metals.

This transition from non-metallic to metallic properties is even more pronounced when we consider the Group V elements. Nitrogen and phosphorus are primarily molecular compounds, whereas antimony and bismuth show metallic behavior. (Arsenic is difficult to discuss; it exists in several forms and sublimes before it boils.)

Bond energy. The variations in single-bond energy within a group are surprising. For example, the single-bond energy of oxygen is only 33 kcal, but for sulfur it increases to 51 kcal and then drops again for selenium. Similar behavior is observed for the halogens and the Group V elements. This type of behavior does not lend itself to simple explanation. The changes are more regular for the energies of multiple bonds. Thus the double bond of oxygen requires about 120 kcal to break, whereas that of sulfur requires only 80, and that of selenium requires 65. The triple bond of nitrogen requires over 200 kcal to break, whereas the triple bond of phosphorus requires only about 120 kcal to break. The fact that the multiple-bond energy decreases within a given group, whereas the single-bond energy increases, at least initially, suggests that the $n = 2$ elements are likely to form multiple bonds, whereas the $n = 3$ elements, i.e., phosphorus, sulfur, and chlorine are likely to form single bonds. This is the case.

Reactions of elements with water. Reactions of these elements with water are determined by their $\mathcal{E}^\circ$ values relative to those for the reduction or the oxidation of water. For practical reasons, we are concerned with the relative ease of oxidation and reduction in pure water for which the hydrogen-ion concentration is not unity but is 10^{-7}; hence we cannot gain the required information by consideration of $\mathcal{E}^\circ$ values but must consider concentration effects. Therefore we list below the $\mathcal{E}$ values (not $\mathcal{E}^\circ$ values) for the oxidation and reduction of neutral water wherein the hydrogen-ion concentration is 10^{-7}:

$$H_2 = 2H^+\ (10^{-7}M) + 2e^- \qquad \mathcal{E} = 0.414$$
$$2H_2O = O_2 + 4H^+\ (10^{-7}M) + 4e^- \qquad \mathcal{E} = -0.815$$

From these data, we see that an anion at unit concentration can react with neutral water to produce hydrogen provided its $\mathcal{E}^\circ$ value exceeds 0.414; on the other hand, if the $\mathcal{E}^\circ$ value is less than $-.815$, the element can react with water to form the anion and gaseous oxygen; finally, if its $\mathcal{E}^\circ$ value is greater than -0.815, the ion in aqueous solution can, in principle, be oxidized by atmospheric oxygen.

According to the $\mathcal{E}^\circ$ values (Table 1) fluorine, chlorine, and bromine should react with water to form oxygen. Actually, chlorine and bromine do react with water but not to form oxygen, whereas fluorine does react with water to form elementary oxygen. We may be sure that oxygen will be formed by chlorine and bromine *in time;* but the kinetics are such that the reaction proceeds rapidly part way toward the formation of oxygen, and then further reaction proceeds too slowly to be observed conveniently.

At first glance it appears that all of the Group VI anions should be reduced to the elements by reaction with water to form hydrogen. For kinetic reasons, this does not occur. The more favorable reaction with atmospheric oxygen, however, *does* occur with S^{-2}, Se^{-2}, and Te^{-2}. As might be expected, telluride ion is most readily oxidized by air to the element. A similar reaction occurs fairly rapidly with selenium but only slowly with sulfide ions.

The group V anions are unstable in water; they react to form hydrides, e.g., $N^{-3} + 3H_2O \rightarrow NH_3 + 3OH^-$.

Hydrides. Aqueous solutions of Group VI hydrides are weak acids; those of the halogens, except HF, are strong acids, and those of the Group V elements are basic or nearly neutral. In each of these groups the dissociation constant of the hydride increases with increasing atomic weight. In a given group, the increase in the degree of ionization is more pronounced between the $n = 2$ and $n = 3$ elements than between the $n = 3$ and $n = 4$ elements. This is related to the formation of hydrogen bonds by the hydrides of nitrogen, oxygen, and fluorine. Across a given period, the ionization constant of the hydride increases. Thus we find that PH_3 is nearly neutral, H_2S is a weak acid, and HCl is a strong acid. Since phosphorus, sulfur, and chlorine atoms are comparable in size, this effect is associated with increasing electronegativity.

In each group, the stability of the hydride decreases with increasing molecular weight. For example, at 1000°C, in a vessel containing HCl at 1 atm, only one molecule in 10,000 will decompose; if the gas were HI, one-third of the molecules would decompose.

All hydrogen halides are stable in water, whereas hydrogen sulfide is stable for a limited period of time, and hydrogen selenide and telluride readily decompose. Of the Group V elements, only ammonia is stable in water. It reacts via hydrogen bonding to form a hydrated species called ammonium hydroxide, a moderately weak base. Phosphine, the phosphorus analog of ammonia, decomposes slowly in water, and the hydrides of the heavier members of this group, i.e., AsH_3, SbH_3, and BiH_3, decompose readily in water.

Many complex hydrides are known for Group V and Group VI elements. One of the better known is H_2O_2, hydrogen peroxide. In this compound, the two oxygen atoms are linked together with hydrogens on either side to give the structure

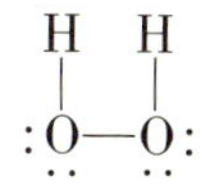

This non-planar molecule forms hydrogen bonds, and consequently the boiling point is high, i.e., about 151°C. Again, like water, it is a weak acid with a dissociation constant of about 10^{-12}. The dielectric constant of hydrogen peroxide is higher than that of water, and for this reason some salts are more soluble in pure hydrogen peroxide than in water. Many salts crystallize from this solvent with bound molecules of H_2O_2, just as many salts form hydrates when crystallized from water. The use of undiluted H_2O_2 as a solvent is limited because it is violently reactive and hard to prepare pure. Products of the decomposition are water and oxygen.

The sulfur and selenium analogs of H_2O_2 are also known. In general, they are more stable than hydrogen peroxide because the sulfur-sulfur and selenium-selenium single bonds are stronger than the oxygen-oxygen single bond. Thus one can form sulfur bridges containing two, three, four, and even more sulfur atoms; the corresponding hydrides, H_2S_2, H_2S_3, and H_2S_5, are quite stable.

Nitrogen and phosphorus also form hydrides analogous to hydrogen peroxide. Imagine that we pull a proton from each of the oxygen nuclei in a hydrogen peroxide molecule; the result is a compound with the formula $NH_2—NH_2$, hydrazine. This compound bears some resemblance to hydrogen peroxide. It decomposes violently when heated, and the boiling point, 114°C, is high because of hydrogen bonding. Like hydrogen peroxide, it has a high dielectric constant and dissolves many salts. Unlike hydrogen peroxide, however, it is a weak base in the presence of water, and it hydrates to form the compounds, $N_2H_4 \cdot H_2O$ and $N_2H_4 \cdot 2H_2O$. This hydrate has a dissociation constant slightly less than that of ammonia. The unstable phosphorus analog P_2H_4 is also formed. Phosphorus is similar to sulfur in so far as it forms long-chain hydrides of the general formula $(P_2H)_n$. The chemistry of P_2H_4 is complex.

The oxygen acids. The common oxygen acids for these three families of elements are shown in Table 2. Since we discussed acids in some detail in Chapter 15, we now concentrate on the unusual structural features of these compounds. In a given group, the strongest acids with a given structure are those formed by the elements of the lowest atomic weight. (This is the opposite of the behavior found for the hydrides.) Thus we can say that, the more metallic the element, the weaker is the acid. This tendency is dramatically illustrated by the oxygen acids for the +1 oxidation state of the halogens. Of the halogens, iodine is the most alike metal. The "acid" HIO is best represented by the formula IOH, since in an aqueous solution it is not an acid but a base.

Several unusual structural features are found among these acids. H_3PO_3 and H_3PO_2, for example, have hydrogen atoms directly bound to the central atom.

Table 2 Oxygen Acids[a]

Oxidation State[b]	V	VI	VII
N	HNO_3		
	H_3PO_4	H_2SO_4	$HClO_4$
	H_3AsO_4	H_2SeO_4	
	$HSb(OH)_6$	H_6TeO_6	H_5IO_6
$N-2$	HNO_2		
	H_3PO_3	H_2SO_3	$HClO_3$
	H_3AsO_3	H_2SeO_3	$HBrO_3$
		H_2TeO_3	HIO_3
$N-4$	$H_2N_2O_2$	(H_2SO_2)	
	H_3PO_2		$HClO_2$
			$(HBrO_2)$
$N-6$			$HClO$
			$HBrO$
			HIO

[a] Parentheses indicate that the compounds are not well characterized.
[b] N = group number.

Such hydrogen atoms are not replaceable, so we find only two salts formed from H_3PO_3 and only one salt formed for H_3PO_2, i.e., Na_2HPO_3 and NaH_2PO_3, and NaH_2PO_2, respectively.

CHEMISTRY OF THE HALOGENS

Solution chemistry. The solution chemistry of the halogens is summarized, in part, by the standard potentials listed in Table 3. The behavior of these halogens in solution is complicated by the fact that they form a wide variety of oxy anions; hence we are forced to rely upon thermodynamic data of the type listed in Table 3.

The data in Table 3 are thermodynamic data, and they tell us *nothing* about rates of reaction. Thus, from the data in Table 3, we can only compare the relative tendency of two half-cell reactions to take place given an infinite amount of time. Suppose we couple the two half-cell reactions x and y and write $\mathcal{E}_r^\circ = \mathcal{E}_x^\circ - \mathcal{E}_y^\circ$. If the $\mathcal{E}^\circ$ value for x is greater than the $\mathcal{E}^\circ$ for y, then the reaction x will go as written and the half-cell reaction for y will go in the reverse direction, provided that all the species are at unit concentrations. Often, reactions that occur take place very slowly. Whether this will be the case for a particular reaction *cannot* be determined by data shown in Table 3. On the other hand, if the $\mathcal{E}_r^\circ$ is negative, we can be sure that this reaction will not occur appreciably.

(There will always, of course, be an equilibrium wherein we will expect to find minute traces of product. In those cases where $\mathcal{E}_r^\circ$ is negative and large in magnitude, the amount of product will be undetectable; in those cases where the $\mathcal{E}_r^\circ$ is small and negative, the amount of product may be detectable.) In summary, therefore, this table tells us what reactions *may* occur if the kinetics are favorable, and it also tells us what reaction *cannot* occur.

In the previous section, we noted that the $\mathcal{E}_r^\circ$ value for the formation of chlorine from chloride ion is such that we would expect chlorine gas to react with water to form oxygen. We can show by the following calculation that this reaction is thermodynamically favorable:

$$4OH^-(10^{-7}M) = O_2 + 2H_2O + 4e^- \qquad \mathcal{E}^\circ = -.82$$
$$4Cl^- = 2Cl_2 + 4e^- \qquad \mathcal{E}^\circ = -1.36$$
$$4OH^-(10^{-7}M) + 2Cl_2 = O_2 + 2H_2O + 4Cl^- \qquad \mathcal{E}_r^\circ = +0.54$$

In spite of the fact that this reaction is thermodynamically favorable, it does not take place readily. To see why, we must look at the kinetics of the reaction. A one-step process in which oxygen is produced directly from a hydroxide ion does not occur. The mechanism of the oxidation of water is not clear even today, but it is thought to involve the direct oxidation of OH^- groups to OH radicals that combine to form peroxide; subsequently, the peroxide decomposes to yield oxygen. The standard potential of the half-cell reaction for the formation of hydrogen peroxide is about -1.8. Thus, to accomplish this first step readily, we need a couple with an $\mathcal{E}^\circ$ of about -1.8 or less. This is found among the halogens only for fluorine.

Table 3 Oxidation-Reduction Chemistry of Halogens

Reaction	$\mathcal{E}^\circ$
(*a*) $6OH^- + I^- = IO_3^- + 3H_2O + 6e^-$	-0.26
(*b*) $8OH^- + I^- = IO_4^- + 4H_2O + 8e^-$	-0.37
(*c*) $2OH^- + I^- = IO^- + H_2O + 2e^-$	-0.49
(*d*) $8OH^- + Cl^- = ClO_4^- + 4H_2O + 8e^-$	-0.51
(*e*) $2I^- = I_2 + 2e^-$	-0.53
(*f*) $6OH^- + Br^- = BrO_3^- + 3H_2O + 6e^-$	-0.61
(*g*) $6OH^- + Cl^- = ClO_3^- + 3H_2O + 6e^-$	-0.61
(*h*) $4OH^- + Cl^- = ClO_2^- + 2H_2O + 4e^-$	-0.76
(*i*) $2OH^- + Br^- = BrO^- + H_2O + 2e^-$	-0.77
(*j*) $2OH^- + Cl^- = ClO^- + H_2O + 2e^-$	-0.94
(*k*) $2Br^- = Br_2 + 2e^-$	-1.06
(*l*) $2Cl^- = Cl_2 + 2e^-$	-1.36
(*m*) $2F^- = F_2 + 2e^-$	-2.85

In view of the fact that the oxidation of water by chlorine is kinetically unfavorable, we might expect that other reactions would take place when chlorine gas is bubbled through a basic solution. If we look at reactions *j* and *l* in Table 3, we find that the following should occur:

$$(j\text{-}l) \qquad 2OH^- + Cl_2 = Cl^- + ClO^- + H_2O \qquad \mathcal{E}_r^\circ = 0.42$$

The same data suggest that there are several reactions that might occur once the hypochlorite ion, ClO^-, is formed. Thus we can write

$$(g\text{-}j) \qquad 3ClO^- = 2Cl^- + ClO_3^- \qquad \mathcal{E}_r^\circ = +0.33$$

and

$$(d\text{-}j) \qquad 4ClO^- = 3Cl^- + ClO_4^- \qquad \mathcal{E}_r^\circ = +0.43$$

Once again, in practice, the kinetics of these reactions must be considered. The formation of the hypochlorite ion takes place in cold, basic solutions. When we warm the solution and increase the rates of reaction, the hypochlorite ion further decomposes to form chloride ion and the chlorate ion according to the equation (*g-j*).

The reaction (*d-j*), although favorable thermodynamically, does not readily occur in aqueous solution. It will occur, however, when wet crystalline $KClO_3$, which is formed from ClO^-, is heated to high temperatures.

There are a wide variety of auto-oxidation-reduction reactions that occur with the oxygen compounds of the halogens. In general, it is best either to know the experimental facts or to make liberal use of a table of oxidation-reduction potentials. Attempts to deduce what will occur purely on the basis of horizontal and vertical relations in the periodic chart often lead to wrong predictions. Let us consider several examples to show the inadequacies of an intuitive approach for predictions in a complex system of reactants.

The electronegativity of the halogens decreases in the order: fluorine, chlorine, bromine, and iodine. It is tempting, therefore, to argue that, *in solution*, the fluorine molecule has a stronger tendency to form an anion than the chlorine molecule. If this were so, fluorine gas in contact with chloride ion would remove the electron from the chloride ion and, thereby, form fluoride ion. Similar reactions would be expected between chlorine and bromide ions and bromine and iodide ions. The data in Table 3 bear this out, but actually we have arrived at the right answer for the wrong reason.

Processes that occur in solution are generally the result of a combination of terms involving addition and subtraction of large quantities. For sound predictions of what will occur, it is necessary to consider each of these terms. Below we have listed in parallel columns the elementary processes which must be considered in the calculation of the $\mathcal{E}^\circ$ values for chlorine and fluorine:

	$\Delta H°$ $x = \text{F}$	$\Delta H°$ $x = \text{Cl}$
$\frac{1}{2}x_{2g} \rightarrow x_g$	0.7 ev	1.3 ev
$x_g + e^- \rightarrow x_g^-$	−3.7 ev	−4.0 ev
$x_g^- \rightarrow x_{\text{aq.}}^-$	−5.1 ev	−3.7 ev
$\frac{1}{2}x_{2g} + e^- \rightarrow x_{\text{aq.}}^-$	−8.1 ev	−6.4 ev

In these equations, we have not indicated the free energy of reactions, but rather the $\Delta H°$. (It seems reasonable to suppose that the difference in the $\Delta F°$ values for fluorine and chlorine are due to differences in $\Delta H°$ rather than $\Delta S°$.) The difference in these two values, i.e., 1.7 volts, represents the difference in $\mathcal{E}°$ potentials. In the comparison represented by the $\mathcal{E}°$ values, the source of electrons is not gaseous electrons, as indicated above; it is hydrogen molecules. If the halogens obtain the electrons from hydrogen molecules and, by so doing, produce dissolved hydrogen ions, the $\Delta F°$ values for both the above processes would be raised four to five electron volts. If we examine each of the steps above, we find that the more negative $\mathcal{E}°$ value for fluorine is *not* a result of the greater electronegativity; it is the result of two other factors. First, it is easier to form fluorine atoms than chlorine atoms; second, the smaller fluoride ion has a greater heat of hydration than the larger chloride ion.

Of all the oxygen anions of chlorine, perchlorate is the most stable in aqueous solution, and the stability of the oxy anions progressively decreases with decreasing oxidation number. A similar order prevails for the oxy anions of bromine; but, for anions of iodine, we find that the IO_3^- ion is the most stable. The periodate is next in stability, and the hypoiodous is the least stable. Such a reversal is not easily predicted.

Finally, let us compare the relative stability of the halate anions. Since iodine is the most metallic of the halogens, we expect it to be the most stable in the higher oxidation state. By similar reasoning, we expect bromate ion to be more stable than chlorate ion. Let us look at the data.

$$I_2 + 6H_2O = 12H^+ + 2IO_3^- + 10e^- \qquad \mathcal{E}° = -1.20$$
$$Cl_2 + 6H_2O = 12H^+ + 2ClO_3^- + 10e^- \qquad \mathcal{E}° = -1.47$$
$$Br_2 + 6H_2O = 12H^+ + 2BrO_3^- + 10e^- \qquad \mathcal{E}° = -1.52$$

It is clear from these data that, although the iodate is the most stable ion, chlorate ion is more stable than bromate ion. Thus, even though chlorate ion readily oxidizes iodine to form iodate ion and elementary chlorine, the following reaction will occur only to a limited extent:

$$Br_2 + 2ClO_3^- = 2BrO_3^- + Cl_2 \qquad \mathcal{E}_r^\circ = -0.05$$

This reaction constitutes a rather interesting case. Although it is true that there is a driving force for the reaction of chlorine with bromate when all the species are of unit concentrations, this tendency is so slight that the equilibrium can shift to the right or to the left by changes in concentration; hence high concentrations of chlorate ion favor the reaction going to the right, whereas high concentrations of chlorine gas favor the reaction going to the left.

Elementary halogens. The halides occur in nature as the X^- anion. Alkali metal and alkaline earth halides are found in sea water and in salt beds deposited by the evaporation of the inland seas. Separation of the elements from these salts involves oxidation of the anion. This can be done electrolytically in concentrated aqueous solution for chloride, bromide, and iodide ions, but there is some loss of the halogen due to hydrolysis. All the halogens, including fluorine, can be prepared by electrolysis of a fused salt.

Chemical means can be used for preparation of halogens other than fluorine. For example, HCl and MnO_2 (a basic anhydride) react according to the equation

$$4HCl + MnO_2 \rightarrow MnCl_4 + 2H_2O$$

The manganese tetrachloride undergoes an auto-oxidation-reduction, and the following reactions occur:

$$2MnCl_4 \rightarrow 2MnCl_3 + Cl_2$$
$$2MnCl_3 \rightarrow 2MnCl_2 + Cl_2$$

This method of preparation can be used not only for chlorine, but also for bromine and iodine. Other methods for the preparation of chlorine include the oxidation of HCl to water and Cl_2 with elementary oxygen or the oxidation in aqueous solution with $KMnO_4$; that is,

$$16H^+ + 2MnO_4^- + 10Cl^- \rightarrow 2Mn^{+2} + 5Cl_2 + 8H_2O$$

The heavier halogens can be oxidized with relatively weak oxidizing agents.

Hydrides. All hydrogen halides can be prepared by direct reaction of the elements. Such a preparation of hydrogen iodide, however, would be contaminated with both hydrogen and iodine; at the temperature required for the reaction to proceed at a reasonable rate, some dissociation of hydrogen iodide occurs. The bond strength of the lighter hydrogen halides is sufficiently great that contamination due to decomposition is not a serious problem.

Preparation of the lighter hydrogen halides, HF and HCl, is usually carried out by treatment of the salts with sulfuric acid; for example,

$$CaX_2 + H_2SO_4 \rightarrow CaSO_4 + 2HX$$

The heavier hydrogen halides cannot be prepared by this procedure because both bromide and iodide ions are oxidized to the elements by concentrated sulfuric acid:

$$2KI + 3H_2SO_4 \rightarrow SO_2 + I_2 + 2KHSO_4 + 2H_2O$$

The preparation of hydrogen bromide and hydrogen iodide is generally carried out by the hydrolysis of covalent phosphorus halides:

$$PBr_3 + 3HOH \rightarrow H_3PO_3 + 3HBr$$

or

$$PI_5 + 4HOH \rightarrow H_3PO_4 + 5HI$$

These halides of phosphorus may be prepared by direct combination of the elements.

A possible way of preparing hydrogen iodide is

$$H_2S + I_2 \underset{\text{gas}}{\overset{\text{aqueous}}{\rightleftarrows}} 2HI + S$$

Iodine can be prepared by the reaction of gaseous sulfur and gaseous hydrogen iodide. This reaction can be reversed by carrying it out in aqueous solution, because the heat of solution of hydrogen iodide is far greater than the heat of solution of hydrogen sulfide.

Oxy anions. The hypochlorite oxy anion can be prepared by treating water with chlorine gas; under these conditions the following equilibrium is set up:

$$Cl_2 + H_2O \rightleftarrows Cl^- + ClO^- + 2H^+$$

If the solution is made basic, this equilibrium is shifted to the right. Commercially, hypochlorites are prepared by treating limewater, i.e., calcium hydroxide, with chlorine gas. This material, called bleaching powder, is a mixture of $CaCl_2$ and $Ca(OCl)_2$. Hypochlorous acid, HClO, is a weak acid that decomposes to form chlorine and water; hence elementary chlorine can be prepared by treatment of bleaching powder with a strong acid.

If the solution containing hypochlorite ion is treated with base and heated, the following reaction occurs:

$$3ClO^- \rightarrow ClO_3^- + 2Cl^-$$

If potassium ion is present, the sparingly soluble potassium chlorate precipitates when the solution is cooled; in this way, $KClO_3$ free of chloride ion can be obtained. Potassium chlorate is unstable at high temperatures and decomposes into KCl and oxygen at about 400°. It is an excellent oxidizing agent and is used as a solid source of oxygen in explosives and safety matches.

Chloric acid, $HClO_3$, can be formed by treatment of $KClO_3$ with a strong acid. $HClO_3$ readily decomposes, especially when concentrated, and cannot be prepared in concentrations in excess of 50%. The concentrated acid decomposes as follows:

$$3HClO_3 \rightarrow HClO_4 + 2ClO_2 + H_2O$$

ClO_2 is a free radical. It is reactive and explosive. In water it undergoes this auto-oxidation-reduction reaction:

$$2ClO_2 + H_2O \rightarrow HClO_3 + HClO_2$$

This reaction proceeds rapidly in light; in the dark, the ClO_2 dissolves in water but the reaction is so slow that the ClO_2 can be recovered unreacted. The acid, $HClO_2$, is unstable and reactive. Its transition metal salts detonate when heated.

If $KClO_3$ is heated carefully to the melting point, the decomposition occurs as follows:

$$4KClO_3 \rightarrow KCl + 3KClO_4$$

If it is heated more strongly or in the presence of MnO_2, which acts as a catalyst, decomposition occurs according to the equation

$$2KClO_3 \rightarrow 2KCl + 3O_2$$

The last reaction is often used for the preparation of oxygen in the laboratory.

The salts of ClO_4^- ion are the most stable of those formed by the oxychlorides. For this reason, they are often used as oxidizing agents in the place of the less stable potassium chlorate. When potassium perchlorate is acidified, $HClO_4$ is formed. This is one of the strongest acids. It is far more stable than the other acids formed by oxychlorides and can be distilled under reduced pressures. When pure perchloric acid is brought into contact with organic matter, it explodes. At lower concentrations, however, perchloric acid finds many uses in analytical and preparative chemistry. In view of the fact that perchlorate anion has very little tendency to associate with the very small hydrogen cation (i.e., $HClO_4$ is a very strong acid), it is not surprising to find that perchlorate ion forms few, if any, complex ions.

The halates and hypohalites of bromine or iodine are formed by the reaction of elementary bromine or iodine in a basic solution. The hypobromite and bromate ion are quite similar in oxidizing power to the hypochlorite and chlorate ion, as may be seen on reference to Table 3; hence we expect their reactions to be similar. On the other hand, periodate and iodate ions are much more stable in aqueous solution than the corresponding oxy anions of chlorine. In view of the very great stability of the iodate ion (Table 3), it is not surprising to find that the hypoiodite ion decomposes rapidly into iodate and iodide ion even at room temperature.

The chemistry of the periodate ion is complex. Although we have indicated the IO_4^- ion in Table 3, this species usually exists combined with water of hydration. The form usually found in solution is H_5IO_6 and the five associated anions. These periodates can be formed by the oxidation of iodates with chlorine as indicated:

$$IO_3^- + 6OH^- + Cl_2 \rightarrow (IO_6)^{-5} + 2Cl^- + 3H_2O$$

A large number of oxy anions of iodine in the $+7$ oxidation state are not stable in solution but form salts of univalent cations which have the formulas

$$M_5(IO_6), \quad M_3(IO_5), \quad M(IO_4), \quad M_4(I_2O_9)$$

Table 4 Common Oxides of the Halogens

F_2O	Cl_2O	Br_2O	
	ClO_2	BrO_2	I_2O_4
	$(ClO_3)Cl_2O_6$	Br_3O_9	I_4O_9
	Cl_2O_7		I_2O_5

For purposes of identification, the first is referred to as an orthoperiodate; the second as a mesoperiodate; the third as a metaperiodate, and the fourth as the diperiodate. Various acid salts of these anions are also found. These salts have strikingly different properties. For example, the orthoperiodates decompose slowly to form iodine when heated; the metaperiodates, on the other hand, decompose violently with the evolution of oxygen. Some metaperiodate salts are so unstable that they explode when touched.

Oxides of halogens. The oxides of the halogens (Table 4) are unstable and often decompose violently. Iodine, the most metallic halogen, forms the most stable oxides. The relative stability of the oxides of bromine and chlorine is especially noteworthy. On the basis of electronegativity, we might expect that the oxides of chlorine would be less stable than the oxides of bromine; actually, the reverse is observed. The oxides of bromine are so unstable that not many of the compounds have been well characterized; in fact, these oxides have only been discovered in the last thirty years.

Chlorine monoxide, Cl_2O, can be prepared by the passage of chlorine gas over mercuric oxide:

$$Cl_2 + HgO \rightarrow Cl_2O + Hg$$

If it is heated or even poured from one vessel to another, it explodes. It is the anhydride of hypochlorous acid, and it reacts readily with water.

We have already discussed the preparation of gaseous ClO_2 by the decomposition of chloric acid. It is unstable and shows magnetic properties consistent with the fact that the molecule contains an odd number of electrons.†

If chlorine dioxide is treated with the mixture of O_2 and O_3 (ozone), an oxide is formed in which chlorine is in the +6 oxidation state. Since the liquid is diamagnetic, it appears that the formula for this oxide is Cl_2O_6. This picture is supported by molecular-weight measurements which yield the values expected for the molecule Cl_2O_6. Thus we find that the odd-electron molecule ClO_2 exists largely unassociated, but the odd-electron molecule ClO_3 exists largely as Cl_2O_6. When liquid Cl_2O_6 is vaporized, there is probably some decomposition

† All substances are diamagnetic, paramagnetic, or ferromagnetic. Substances that are diamagnetic are repelled when placed between the poles of a horseshoe magnet; substances that are paramagnetic are attracted. Broadly speaking, paramagnetism is found in compounds with unpaired spins; diamagnetism is found in compounds with paired spins. We will not discuss ferromagnetism.

of Cl_2O_6 into ClO_3, but this gas rapidly decomposes; hence the existence of ClO_3 is questionable. Cl_2O_6 reacts with water violently and undergoes an auto-oxidation-reduction to form $HClO_3$ and $HClO_4$.

If $HClO_4$, perchloric acid, is treated with phosphorus pentoxide, P_2O_5, to remove the water, a colorless oil, Cl_2O_7, is formed. Like all the oxides of chlorine, it is explosive, but the $+7$ oxidation state appears to form the most stable of the oxides. For example, when the pure liquid is poured onto paper or other organic materials at ordinary temperatures, no reaction occurs. It slowly dissolves in cold water to form $HClO_4$.

The oxides of bromine are not so well established as those of chlorine. Bromine monoxide can be prepared by the procedure used for the preparation of chlorine monoxide. It is the anhydride of hypobromous acid. The other oxides, i.e., BrO_2 and Br_3O_8, require exotic preparative techniques. Although the oxides of bromine are even less stable than the oxides of chlorine, they decompose without explosion.

Oxides of iodine, I_2O_4, I_4O_9, and I_2O_5, are stable solids. I_2O_4 is prepared by the decomposition of iodic acid in hot concentrated sulfuric acid; I_2O_5 is made by the dehydration of iodic acid; I_4O_9 is formed by the reaction of ozone and iodine. All three of these oxides are stable below 100°C.

Not only are the oxides of iodine surprisingly unreactive, but also there is evidence that these are ionic compounds; for example, it has been suggested that I_4O_9 should be written $I(IO_3)_3$. The fact that iodine will function as a cation in an ionic compound should not come as a surprise; iodine is the most metallic of the halogens. Since these are ionic compounds, the oxides are more stable than the other oxides of the halogens. There are other compounds in which iodine behaves like a cation. Both the $+1$ and the $+3$ oxidation states are found, but the $+3$ oxidation state is the more stable. Among the salts that have been isolated are: $I(ClO_4)_3 \cdot 2H_2O$, $I_2(SO_4)_3$, $I(C_2H_3O_2)_3$, $I(CNO)$, $I(NO_3)$, and $I(ClO_4)$. The last two salts named are usually stabilized by the presence of an organic base such as pyridine.

Interhalogen compounds. A whole series of compounds is formed by reactions between the elementary halogens; those that are well established are listed in Table 5. The coordination number of a given halogen toward, say, fluorine, appears to be controlled largely by the atomic size; for example, we find that iodine has the highest coordination number toward fluorine. It is somewhat surprising that iodine is not found with a lower coordination number than 5 toward fluorine. Presumably, iodine, when it is coordinated with 5 or 7 fluorine atoms, is so stable relative to compounds with the lower coordination number that the latter, if formed, immediately undergo an auto-oxidation-reduction to form the more highly coordinated compounds.

Table 5 Principal Interhalogen Compounds

Iodine	Bromine	Chlorine
IF_7, IF_5	BrF_5, BrF_3, BrF	ClF, ClF_3
ICl_3, ICl	BrCl	
IBr		

The interhalogen compounds occur not only as neutral molecules, but also as anions. For example, polyhalides analogous to the tri-iodide ion often occur as salts of the heavier alkali metals. In some of these anions, the iodine functions as a centrally located metallic element; e.g., ICl_2^-, IF_4^-, IF_6^-, and ICl_4^-. Bromine can also function as the centrally located element as in BrF_4^-.

Complex ions. All the halides form complex ions. If the central ion is small, like aluminum ion, the most stable complex will be that with the fluoride ion. On the other hand, if the central ion is relatively large, like mercuric ion or silver ion, the iodide complex will often be the strongest. The first case can be rationalized on the basis of the relative acid strengths of the halogen hydrides. We have already noted that the weakest acid among these is hydrofluoric acid. On this basis, it appears that, in solution, the fluoride ion is the most strongly attached to a small positively charged cation for the same reasons that it forms the strongest bonds to the proton. The reversal of this behavior for heavy metal cations is not easily rationalized, but it is well established that, with an ion like silver, which does form complexes with the halides, the strength of the complex increases with the atomic weight of the halide. Presumably, this type of behavior is due to the formation of a covalent rather than an ionic bond.

Complex ions formed between small cations and the fluoride anion rival the oxy anions in stability. Two of these are

$$AlF_6^{-3} \quad \text{and} \quad BF_4^-$$

From purely coulombic considerations, we expect that the acids corresponding to these anions to be quite a bit stronger than their oxygen analogs, i.e., HBF_4 and H_3AlF_6 versus H_3BO_3 and H_3AlO_3. Not many data are available on this point, but fluoroboric acid is roughly as strong as the hydrogen halides, whereas boric acid is a weak acid.

The stability of the chloroaurate ion, $AuCl_4^-$, is responsible for the efficiency of *aqua regia* in dissolving gold. A similar situation holds for platinum. To see this, consider the equilibrium that is established when metallic gold is in contact with nitric acid:

$$6H^+ + 3NO_3^- + Au \rightleftarrows Au^{+3} + 3NO_2 + 3H_2O$$

Thermodynamically, this equilibrium lies far to the left. Even in very strong nitric acid, the amount of gold oxidized to the +3 state is undetectable. If, however, we add chloride ion to the nitric acid, the following reaction occurs:

$$Au^{+3} + 4Cl^- \rightleftarrows AuCl_4^-$$

With the formation of the complex, the concentration of Au^{+3} ion is reduced to a very low value. In accord with Le Chatelier's principle, this shifts the first equilibrium to the right, and, since the complex of chloride ion with gold is very strong, the oxidation proceeds to completion.

CHEMISTRY OF THE GROUP VI ELEMENTS

In this section, we shall focus our attention on the chemistry of sulfur, selenium, and tellurium. The chemistry of the various compounds of oxygen is discussed elsewhere.

Solution chemistry. The more common oxidation states of sulfur in solution are −2, +4, and +6, as found in the anions, S^{-2}, SO_3^{-2}, and SO_4^{-2}. Like many of the tetrahedral anions, sulfate can condense to form chains of the type

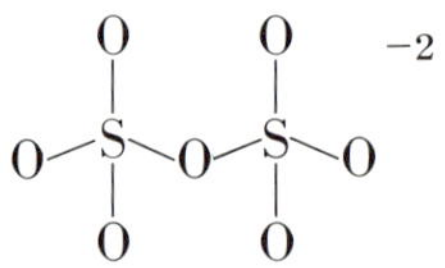

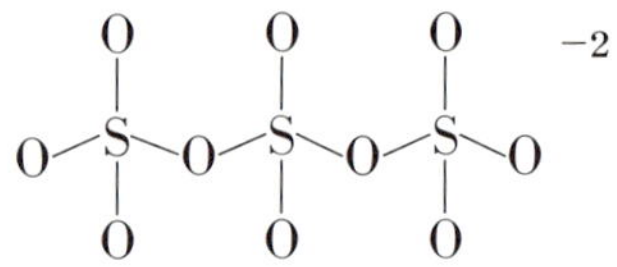

In both these condensed anions, however, the oxidation state of the sulfur remains +6; only the various linkages change. Such condensed structures are also found for the elements to the left and to the right of sulfur in the periodic chart. In general, the stability decreases with increasing electronegativity; those formed with the more electronegative elements tend to add water and reform the parent ion.

In addition to the above-mentioned anions, sulfur forms a number of thio anions in which sulfur is substituted for one of the oxygen atoms. Because these anions are more or less unstable in aqueous solution, we defer until later the discussion of their behavior in solution.

Of the more common species found in solution, i.e., SO_4^{-2}, SO_3^{-2}, and S^{-2}, only the sulfate is stable in contact with air. In acid solution, the sulfite ion decomposes to form the gaseous anhydride, SO_2; in basic solution, the sulfite ion is slowly oxidized to form sulfate ion. In acid solutions, the sulfide anion decomposes to form gaseous H_2S; in basic or nearly neutral solution, the sulfide anion is slowly oxidized to sulfur by atmospheric air.

Selenium and tellurium form the analogs of sulfide, sulfite, and sulfate ions. As might be suspected, the selenide and telluride ions are even more susceptible to atmospheric oxidation. Selenates and tellurates are less stable than sulfates, and they function as powerful oxidizing agents. (This trend in the Group VI elements is just the opposite of that for the Group VII elements. The periodates are more stable than the perchlorates.) The formula for telluric acid, H_6TeO_6, is not analogous to that for sulfuric acid, H_2SO_4; hence, although selenic acid, H_2SeO_4, is about as strong as sulfuric acid, telluric acid is a weak acid. Selenous and tellurous acids and the corresponding selenites and tellurites are the most stable compounds of these elements.

Oxides and oxy anions. A variety of compounds are formed containing only oxygen and sulfur. The more common of these are

$$SO_4, \quad S_2O_7, \quad SO_3, \quad SO_2, \quad S_2O_3, \quad SO$$

Since some of these oxides are unstable or are formed by rather peculiar reactions, we shall discuss only the two stable ones, SO_3 and SO_2.

Sulfur dioxide, SO_2, can be formed by the reaction of elementary sulfur with oxygen, by acidifying a solution containing sulfite ion, or by the air oxidation of pyrites, FeS_2; that is,

$$4FeS_2 + 11O_2 \rightarrow 2Fe_2O_3 + 8SO_2$$

Although, thermodynamically, sulfur dioxide should combine with oxygen to form the trioxide, this reaction does not occur readily in the absence of a catalyst. Commercially, the trioxide is made by oxidation of sulfur dioxide in the presence of a V_2O_5 or platinum catalyst:

$$2SO_2 + O_2 \xrightarrow[\text{Pt}]{V_2O_5} 2SO_3$$

Sulfur trioxide can also be made by the decomposition of various sulfates such as $Fe_2(SO_4)_3$ and $K_2S_2O_7$.

Sulfur dioxide and sulfur trioxide are the anhydrides of sulfurous and sulfuric acid. SO_2 readily reacts with water; the reaction of SO_3 with water is violent. Sulfur trioxide is so avid for water that it will extract water from wood, cellulose, and other carbohydrates to form carbon.

Sulfuric acid is prepared commercially from the SO_3 produced by catalytic oxidation of SO_2. The SO_3 is absorbed by a solution of sulfuric acid, and water is continuously added to maintain the concentration constant. Sulfuric acid forms a number of definite hydrates. Like SO_3, it has such a strong tendency to combine with water that it will char carbohydrates.

Sulfuric acid, when hot and concentrated, is a powerful oxidizing agent. This is somewhat surprising in view of the low value of the standard potential:

$$H_2O + H_2SO_3 = SO_4^{-2} + 4H^+ + 2e^- \qquad \mathcal{E}^\circ = -0.20$$

At unit concentration, sulfuric acid *is* a poor oxidizing agent; but, when concentrated (18 molar), the solution departs strikingly from ideal thermodynamic behavior; the sulfurous acid more readily decomposes to form SO_2; and the increase in hydrogen-ion concentration shifts the above equilibrium to the left. These three factors combine to yield an oxidizing agent that will oxidize charcoal to CO_2 and sulfur to SO_2. This is why concentrated sulfuric acid dissolves many metals that stronger acids, like HBr, will not dissolve. The oxidation of copper is indicated below:

$$Cu + 2H_2SO_4 \rightarrow CuSO_4 + SO_2 + 2H_2O$$

Sulfuric acid can condense to form dimers and trimers as follows:

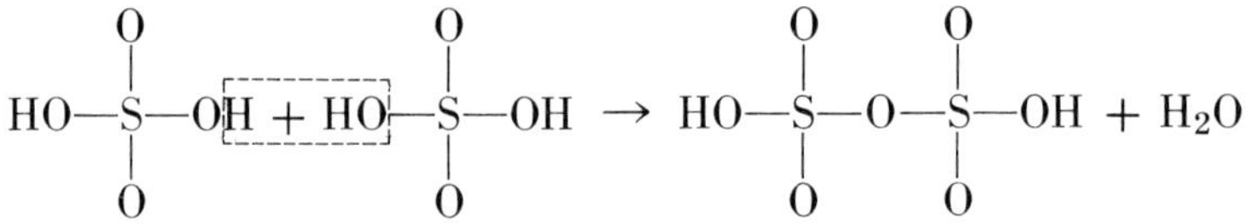

But pyrosulfuric acid, $H_2S_2O_7$, is usually formed by the addition of sulfur trioxide to sulfuric acid. In addition to pyro- or disulfuric acid, we can also form trisulfuric acid, $H_2S_3O_{10}$, by this procedure.

If sulfuric acid is electrolyzed below room temperature, peroxydisulfuric acid is formed. This compound is somewhat similar to pyrosulfuric acid, but the two sulfurs are linked by *two* oxygen atoms, rather than one; that is,

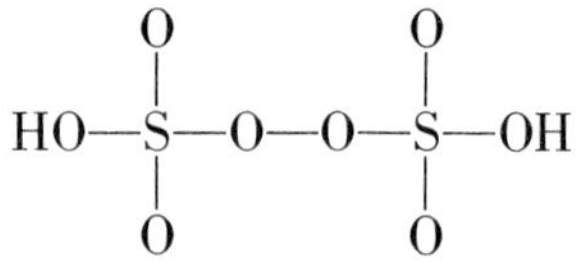

Peroxydisulfuric is a powerful oxidizing agent. Its salts are stable, but the acid slowly decomposes in water to form sulfuric acid and peroxysulfuric acid, which has the formula H_2SO_5. This acid, called "Caro's acid," will undergo further reaction to form H_2O_2:

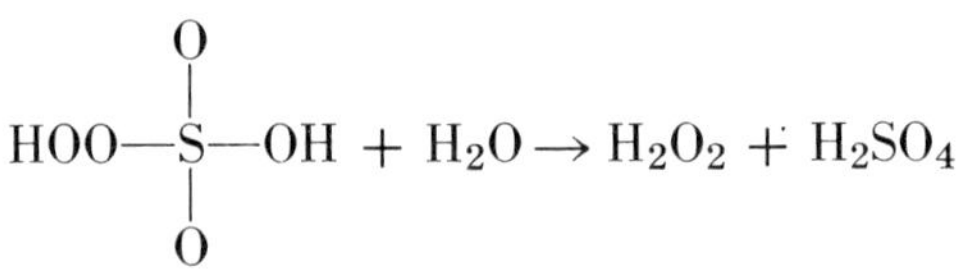

Concentrated hydrogen peroxide is prepared by this procedure.

If sodium sulfite is treated with powdered sulfur, the thiosulfate ion is formed. The structure is

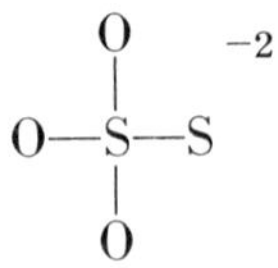

If a solution containing thiosulfate ion is treated with a strong acid, it decomposes according to the equation

$$S_2O_3^{-2} + 2H^+ \rightarrow S + SO_2 + H_2O$$

Many other thio salts are known. If we treat sodium thiosulfate with hydrogen peroxide, the following reaction occurs:

$$2Na_2S_2O_3 + 4H_2O_2 \rightarrow Na_2SO_4 + Na_2S_3O_6 + 4H_2O$$

The anion $S_3O_6^{-2}$, called the trithionate ion, is presumed to be the sulfur analog of pyrosulfuric acid; that is,

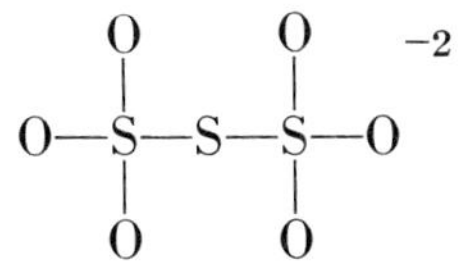

The thiosulfate salts find many uses in analytical chemistry. Sodium thiosulfate readily reduces iodine and forms the tetrathionate ion, as shown by the following equation:

$$2S_2O_3^{-2} + I_2 \rightarrow S_4O_6^{-2} + 2I^-$$

The tetrathionate ion is analogous to the peroxydisulfuric acid anion. Since iodine, which is colored itself, forms an intense blue color with starch, the completion of the reaction can be detected in the presence of starch by a disappearance of blue color.

Selenium and tellurium dioxides are prepared by the reaction of elementary selenium or tellurium with oxygen or by burning the hydrides in oxygen. Neither of the trioxides can be prepared by direct combination with oxygen. The +6 oxidation states of these two elements are obtained in solution by the reaction of the selenite or tellurite with a strong oxidizing agent, such as perchlorate ion or elementary chlorine. The equation for the latter reaction is

$$SeO_3^{-2} + Cl_2 + H_2O \rightarrow SeO_4^{-2} + 2H^+ + 2Cl^-$$

The trioxide of tellurium can be prepared by the careful dehydration of telluric acid. The corresponding reaction does *not* occur with selenic acid. The preparation of SeO_3 is achieved by heating sulfur trioxide with potassium selenate, so that an exchange takes place, as indicated below:

$$K_2SeO_4 + SO_3 \rightleftarrows K_2SeSO_7 \rightleftarrows K_2SO_4 + SeO_3$$

Care must be taken not to overheat selenium trioxide; it is so unstable that even at moderate temperatures the following occurs:

$$SeO_3 \rightarrow SeO_2 + \tfrac{1}{2}O_2$$

Except for the instability of SeO_3, its reactions are similar to those of SO_3; for example, it combines vigorously with water to form H_2SeO_4. Reactions of selenic acid are similar to those of sulfuric acid, except that H_2SeO_4 is a better oxidizing agent. As might be expected, selenic acid condenses to form compounds corresponding to the pyrosulfates.

Oxygen compounds of tellurium, in contrast to those of selenium, differ sharply from those of sulfur. The dioxide of selenium, like that of sulfur, dissolves readily in water to form H_2SeO_3; the dioxide of tellurium reacts very slowly with water, and the acid formed is much weaker than the corresponding sulfurous and selenous acids.

Tellurium trioxide is also different from the other trioxides. Not only is it almost insoluble in water, but it is also unreactive toward other chemicals. As might be expected, the oxygen acid of tellurium in the $+6$ state, H_6TeO_6, bears little resemblance to sulfuric acid.

Hydrides of Group VI elements. The hydrides of sulfur and selenium can be prepared by combination of the elements at moderate temperatures. Since both hydrides are unstable thermally, these products are contaminated with reactants. Hydrogen telluride is so unstable that preparation from the elements cannot be used; it is prepared by the electrolysis of water with a tellurium cathode. None of these hydrides is completely stable in aqueous solution. The sulfide, selenide, and telluride ions are oxidized to the elements by reaction with atmospheric oxygen. This reaction, as expected, proceeds most readily with tellurium and least readily with sulfur. When the hydrides are burned in an excess of oxygen, the product is water and the dioxide. If, however, they are burned in a limited supply of oxygen, the products are water and the element.

All of these hydrides are weak acids, but H_2S is the weakest.

Two salts are formed by the hydrides, the acid salt NaHS, and Na_2S. Bisulfides and sulfides are similar to the hydroxides and oxides. Thus a solution of sodium sulfide in water is strongly basic owing to the reaction

$$Na_2S + H_2O \rightarrow NaOH + NaHS$$

(This reaction represents the hydrolysis of the sulfide ion in aqueous solution.) The reaction is about 90 per cent complete. The similarity of this reaction to the reaction of sodium oxide with water, i.e.,

$$Na_2O + H_2O \rightarrow 2NaOH$$

is obvious. Many other similarities are found in the behavior of the oxides and sulfides. Just as we have the combination of the oxide of a metal with the oxide of a non-metal, i.e., the reaction of basic and acidic anhydrides, we also have combination of the sulfide of a metal with the sulfide of a non-metal. Such type reactions are indicated below:

$$Na_2O + MO_3 \rightarrow Na_2MO_4$$
$$Na_2S + MS_3 \rightarrow Na_2MS_4$$

The second reaction is the reaction of a basic sulfide with an acidic sulfide, just as the first reaction is the reaction of a basic oxide with an acid oxide. Unlike many oxy anions, the thio anions are unstable in acid solutions; for example, when we acidify the compound Na_2MS_4, we do not obtain the acid H_2MS_4; the MS_4^{-2} breaks down to form the sulfide we started with and H_2S.

In strongly basic solution, the bisulfide ion is oxidized to S_2^{-2}, a species that in acid solutions forms the sulfur analog of hydrogen peroxide. This reaction can go even further to form compounds of the type H_2S_3 and H_2S_5 consisting of chains of sulfur with hydrogens located at the ends. Selenium and tellurium form similar species.

Halogen compounds of Group VI elements. The Group VI elements form a rich variety of compounds with halogens by direct combination of the elements. The common halides are listed in Table 6, together with some of the oxyhalides. The halogen compounds of the Group VI elements are similar to the interhalogen compound in so far as there seems to be a size effect. By this we mean that, whereas sulfur forms fluorides and chlorides readily, not many bromides or iodides are formed. On the other hand, the larger tellurium atom forms compounds with all of the halogens, including iodine. In line with this, there is a trend in stability such that, when iodine combines with a smaller atom, the compound is unstable. Thus compounds formed between selenium and iodine are stable only in solution, but compounds formed between tellurium and iodine are stable in pure form.

The fluorides of sulfur, SF_6 and S_2F_{10}, are so stable that they do not react in strong alkali and strong acids. They are not soluble in water but do dissolve readily in organic solvents. Even at elevated temperatures, these fluorides are

Table 6 Halogen Compounds of Sulfur

S_2F_2	S_2Cl_2	S_2Br_2	(S_2I_2)
SF_2	SCl_2		
SF_4	SCl_4		
SF_6	S_xCl_2		
S_2F_{10}			
SOF_2	$SOCl_2$	SO_2Br_2	
SO_2F_2	SO_2Cl_2		
$SO_2(OH)F$	$SO_2(OH)Cl$		
SeF_4	Se_2Cl_2	$SeBr_2$	(Se_2I_2)
SeF_6	$SeCl_4$	$SeBr_4$	(SeI_4)
TeF_4	$TeCl_2$	$TeBr_2$	TeI_2
TeF_8	$TeCl_4$	$TeBr_4$	TeI_4
Te_2F_{10}			

unreactive and do not combine with hydrogen or oxygen. At these high temperatures, they do combine slowly with H_2S as shown in the equation

$$SF_6 + 3H_2S \rightarrow 6HF + 4S$$

The other fluorides of sulfur and all the fluorides of selenium and tellurium slowly hydrolyze in water to yield HF and the oxy anions of sulfur.

The chlorides of sulfur are prepared by the direct union of the elements:

$$2S + Cl_2 \rightarrow S_2Cl_2$$

Further addition of chlorine leads to the formation of SCl_2 and SCl_4. S_2Cl_2 dissolves in sulfur and reacts to form S_3Cl_2 and S_4Cl_2. The structures of these are not known, but it is probable that S_3Cl_2 and S_4Cl_2 are sulfur chains with chlorine atoms attached to the ends.

Hydrolysis of these chlorides of sulfur yield the expected products. The equations for the decomposition of sulfur tetrachloride are

$$SCl_4 + 4HOH \rightarrow 4HCl + S(OH)_4$$
$$S(OH)_4 \rightarrow H_2O + H_2SO_3$$

In contact with the hydrogen and hydroxide ions in water, the more electronegative chlorine atoms associate with the hydrogen ions, whereas the more electropositive sulfur combines with hydroxide ions. The sulfur-containing species formed in this particular instance splits off water to form H_2SO_3; this can further decompose in acid solution to form SO_2. The hydrolysis products of other chlorides, e.g., S_2Cl_2, are not so obvious. We might assume that the first step in the hydrolysis of S_2Cl_2 is

$$S_2Cl_2 + 2HOH \rightarrow S_2(OH)_2 + 2HCl$$

But $S_2(OH)_2$ is not a stable compound of sulfur; hence we expect it to decompose by auto-oxidation-reduction into two stable compounds. The oxidation states closest to $+1$ are $+4$ and -2; consequently, we *guess* that further reaction takes place according to the equation

$$S_2(OH)_2 \rightarrow H_2S + SO_2$$

It does.

The oxychlorides of sulfur are important in organic synthetic chemistry because they are used to incorporate a chlorine atom into an organic molecule. Thionyl chloride, $SOCl_2$, is prepared by the reaction of sulfur dioxide with PCl_5:

$$PCl_5 + SO_2 \rightarrow SOCl_2 + POCl_3$$

In this reaction, there is no change in the oxidation number of any of the atoms involved. Sulfuryl chloride, SO_2Cl_2, is prepared by the direct reaction of sulfur dioxide and chlorine:

$$SO_2 + Cl_2 \rightarrow SO_2Cl_2$$

In this reaction, a change in the oxidation states occurs. Both oxychlorides react with water to yield hydrochloric acid and sulfurous or sulfuric acid.

Tellurium, selenium, and sulfur do not form compounds with each other, but some are known in which oxygen is also present; for example, tellurium and selenium form the sulfites $TeSO_3$ and $SeSO_3$. The existence of these compounds is consistent with the fact that selenium and tellurium are more metallic than sulfur.

CHEMISTRY OF THE GROUP V ELEMENTS

Solution chemistry. The X_3^- anions of nitrogen, phosphorus, and arsenic are unstable in aqueous solution. Their salts, formed with the more active metals, hydrolyze as follows:

$$Mg_3N_2 + 6HOH \rightarrow 2NH_3 + 3Mg(OH)_2$$

Because of this, the solution chemistry of this group is the chemistry of their oxy anions and +3 cations.

Arsenic, antimony, and bismuth are found in the +3 state in solids. In solution, these ions are strongly hydrated; in fact, arsenic generally hydrolyzes to form the oxy anion. Bismuth and antimony do exist in acid solutions as +3 cation, but they have a strong tendency to form the oxy cation in neutral or basic solutions. Thus antimony trichloride in solution establishes the equilibrium

$$Sb^{+3} + 3Cl^- + H_2O \rightleftharpoons SbO^+ + 2H^+ + 3Cl^-$$

If the hydrogen ion concentration is reduced, SbO(OH) precipitates.

All these elements form hydrides, but only the hydride of nitrogen is stable in aqueous solution; the others decompose.

The elements. Nitrogen constitutes 80% of the atmosphere. It can be prepared pure in small quantities by heating a solution of ammonium nitrite. Decomposition takes place according to the equation

$$NH_4NO_2 \rightarrow N_2 + 2H_2O$$

Phosphorus occurs in nature as $Ca_3(PO_4)_2$. The element itself can be prepared in two steps. First, the phosphate is fused with SiO_2 and the following reaction occurs:

$$2Ca_3(PO_4)_2 + 6SiO_2 \rightarrow 6CaSiO_3 + P_4O_{10}$$

Since the oxide P_4O_{10} is volatile at these temperatures, the equilibrium is shifted to the right in accord with Le Chatelier's principle. Elementary phosphorus is then prepared from the oxide by treatment with carbon:

$$P_4O_{10} + 10C \rightarrow P_4 + 10CO$$

Phosphorus vapor consists of P_4 molecules, in contrast to nitrogen which occurs as a diatomic gas.

Let us compare elementary nitrogen and phosphorus. In the atomic form, nitrogen is definitely more reactive than phosphorus. Even in the combined form, nitrogen is as reactive, thermodynamically, as phosphorus, but the strength of the nitrogen-nitrogen triple bond, which must be split prior to reaction, is so very great that nitrogen reacts very slowly at ordinary temperatures. If it were not for this fact, we would have very little nitrogen in the atmosphere; all of it would combine with the ocean or with substances present in the crust of the earth. The phosphorus-phosphorus bonds are not so strong as the nitrogen-nitrogen triple bond; for this reason, phosphorus appears to be more reactive. It is readily oxidized by the air, even at room temperature. (This reaction gives rise to the glow or phosphorescence which surrounds phosphorus in contact with air.) Elementary phosphorus also reacts in basic solutions:

$$P_4 + 3KOH + 3H_2O \rightarrow PH_3 + 3KH_2PO_2$$

Reaction with water occurs rapidly with white phosphorus, one of the common forms of phosphorus, but, if red phosphorus is used, the reaction proceeds slowly. The two modifications of phosphorus differ strikingly in properties. Red phosphorus is harmless; white phosphorus is poisonous. Red phosphorus is stable in air at temperatures below 250°; white phosphorus ignites spontaneously in air at about 40°. More will be said about the *allotropy* (the existence of several structures) of elementary phosphorus in the next chapter.

Arsenic, antimony, and bismuth are generally found combined with oxygen or as anions. Elementary arsenic is prepared from the arsenide by air oxidation of the anion to the oxide; this oxide is then reduced to the metal with carbon. Similar preparations are used for antimony and bismuth.

Oxides and oxy anions. The oxide corresponding to the +1 oxidation state of nitrogen is formed by the decomposition of ammonium nitrate:

$$NH_4NO_3 \rightarrow N_2O + 2H_2O$$

This gas, which supports the combustion, is used as an anesthetic. Because it causes convulsive laughter, it is known as "laughing gas."

The oxide of nitrogen in the +2 oxidation state is formed by the direct combination of the elements at high temperatures, i.e., about 3000°C. In order to use the nitric oxide at lower temperatures, it is necessary to cool the products. Since the combination of nitrogen and oxygen is an endothermic reaction, the equilibrium,

$$N_2 + O_2 \rightleftarrows 2NO$$

shifts to the left at lower temperatures and reforms nitrogen and oxygen. This process for the production of NO is feasible only if the products are suddenly chilled so that the amount of reversal of the equilibrium is small. Nitric oxide, a colorless gas, can also be prepared by the oxidation of ammonia or by the reaction of copper with dilute nitric acid.

Since NO has an odd number of electrons, it is paramagnetic. It combines rapidly with oxygen at room temperature, forms complex ions with a number of the transition metal cations, and replaces one of the hydrogen atoms in sulfuric acid. The presence of NO can be detected by exposure to a solution containing ferrous sulfate (the brown ring test). The complex formed with Fe^{+2} has a characteristic dark-brown color. This complex is also used to detect NO_3^- ion, but it is first necessary to reduce the nitrate to NO; this is done by ferrous ion in concentrated sulfuric acid.

Nitrogen dioxide like nitric oxide is a gas, a free radical, and paramagnetic. It is rapidly formed by air oxidation of nitric oxide. Once formed, this odd-electron molecule can dimerize to form the other oxide of nitrogen in +4 state. Thus, if we treat NO with oxygen, the following equilibria are set up:

$$2NO(g) + O_2 \rightleftarrows 2NO_2(g) \rightleftarrows N_2O_4(g)$$

Both NO and N_2O_4 are colorless; NO_2 is brown. If NO_2 is exposed to water, the following reaction occurs:

$$2NO_2 + H_2O \rightarrow HNO_3 + HNO_2$$

The HNO_2 formed by this reaction further decomposes to form more NO and nitric acid. Thus nitrogen dioxide can be used for the production of nitric acid. Commercially, nitric acid is prepared by the oxidation of ammonia to nitric oxide; this nitric oxide is further oxidized with air to nitrogen dioxide, which upon treatment with water yields nitric acid.

When nitric acid is treated with a dehydrating agent such as phosphorus pentoxide, the following reaction occurs:

$$2HNO_3 \xrightarrow{P_2O_5} N_2O_5 + H_2O$$

N_2O_5 is a solid that melts slightly above room temperature and readily decomposes into N_2O_4 and oxygen.

Several other unstable oxides of nitrogen are known; for example, when NO_2 and NO are condensed together, the following reaction occurs:

$$NO_2 + NO \rightarrow N_2O_3$$

This oxide, N_2O_3, the anhydride of nitrous acid, generally exists in equilibrium with NO_2 and NO. The oxide NO_3 can be prepared by the action of ozone on NO_2 as indicated by the equation

$$NO_2 + O_3 \rightarrow NO_3 + O_2$$

We expect nitrogen trioxide to be paramagnetic, but it is so unstable that pure NO_3 is not available for studies of this type.

Nitric acid is not only a strong acid, it is also a strong oxidizing agent. If it is pure, it is a colorless liquid, but ordinary nitric acid is slightly colored owing to the presence of traces of NO_2 produced (in the presence of light) by the following reaction:

$$4HNO_3 \rightarrow 4NO_2 + 2H_2O + O_2$$

Concentrated nitric acid is such a powerful oxidizing agent that it will oxidize elementary sulfur to sulfuric acid, and elementary phosphorus to phosphoric acid. Many metals that are not dissolved by acids like hydrochloric acid with the evolution of hydrogen will dissolve in nitric acid with the evolution of oxides of nitrogen. The only common metals not dissolved by nitric acid are gold and platinum.

Reactions of nitric acid with metals are complex. For oxidations carried out with dilute nitric acid, the principal product is nitric oxide; for oxidations carried out with concentrated acid, the principal product is nitrogen dioxide. Thus, in the oxidation of metallic copper by nitric acid, we find the following reactions:

$$8H^+ + 2NO_3^- + 3Cu \xrightarrow{\text{dilute}} 3Cu^{+2} + 2NO + 4H_2O$$

and

$$4H^+ + 2NO_3^- + Cu \xrightarrow{\text{concentrated}} Cu^{+2} + 2NO_2 + 2H_2O$$

Since the NO evolved in the first reaction reacts with air, brown fumes due to NO_2 are seen in both reactions.

Iron, chromium, aluminum, and calcium readily dissolve in dilute nitric acid, but, when these metals are treated with concentrated nitric acid, no apparent reaction occurs because the metals are made "passive" by an oxide skin formed by the concentrated acid. If the pacified metals are removed from solution and struck with a hammer or scratched with a pin, the reaction proceeds readily; the rupture of the brittle oxide coating allows contact between the unoxidized metal and the nitric acid.

Nitrate ion can yield reduction products other than nitrogen dioxide and nitric oxide. If the nitrate ion is treated directly with hydrogen produced by the action of alkali on zinc, the nitrate ion is reduced to ammonia.

The salts of the nitrates are unstable when heated. Those of the alkaline metals decompose in two stages. If they are heated mildly, the following reaction occurs:

$$2KNO_3 \xrightarrow{350^\circ C} 2KNO_2 + O_2$$

This reaction occurs cleanly if metallic lead is added to react with the oxygen produced. When the salts are heated to higher temperatures, more complete decomposition occurs:

$$4KNO_3 \rightarrow 2K_2O + 2N_2 + 5O_2$$

Thus nitrate salts are a convenient solid source of oxygen for a chemical reaction. For this reason, nitrates are often used in the manufacture of explosives. (Black powder is a mixture of sulfur, charcoal, and potassium nitrate. The function of the potassium nitrate is to provide oxygen for the combustion of the charcoal and the sulfur.)

If nitrite salts are treated with a strong acid, nitrous acid, a weak acid, is formed. This acid is unstable; aqueous solutions are blue when first formed but slowly turn brown because of the reactions

$$3HNO_2 \rightarrow HNO_3 + 2NO + H_2O$$
$$2NO + O_2 \rightarrow 2NO_2$$

Since the nitrogen in HNO_2 is in an intermediate oxidation state, it can function either as an oxidizing agent or as a reducing agent. When nitrite ion functions as a reducing agent, it is oxidized to nitrate ion. When it functions as an oxidizing agent, the products depend on the particular reaction. Thus reaction of nitrite and iodide ion produces iodine and nitric oxide. In the presence of stronger reducing agents, nitrite ion may be reduced to ammonia.

Hyponitrous acid, $H_2N_2O_2$, is a weak acid with the probable structure

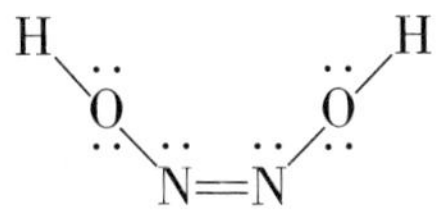

It is produced by the reaction of hydroxyl amine (NH_2OH) and nitrous acid. Although it decomposes to form nitrous oxide and water, it is not produced by hydration of N_2O.

Oxides of phosphorus, arsenic, and bismuth are prepared by the direct reaction of the element with oxygen. P_4O_{10} is prepared by the oxidation of phosphorus in an excess of oxygen; P_4O_6 is prepared by the oxidation of phosphorus in a limited amount of oxygen. P_4O_{10} and P_4O_6 are acid anhydrides and, hence, react with water. In practice, a mixture of polymeric phosphorous and phosphoric acids are formed on hydration, but, if the conditions are such that the hydration is complete, the following reactions occur:

$$P_4O_{10} + 6H_2O \rightarrow 4H_3PO_4$$
$$P_4O_6 + 6H_2O \rightarrow 4H_3PO_3$$

The oxides of arsenic dissolve in water to form arsenic and arsenous acids. Antimony oxides, like those of its neighbor, tellurium, are sparingly soluble. The oxides of bismuth are quite insoluble.

There are three common acids of phosphorus, H_3PO_4, H_3PO_3, and H_3PO_2. Phosphorus appears to have a very strong preference for a coordination number of 4. With H_3PO_4, this arrangement is achieved by the phosphorus forming bonds with each of the oxygen atoms; with H_3PO_3, this is achieved by bonds to one oxygen atom, two hydroxide groups, and a hydrogen atom; with H_3PO_2, this is achieved by bonds to one oxygen atom, one hydroxide group, and two hydrogen atoms. Similar behavior is *not* observed for H_3AsO_4 and H_3AsO_3. In the +5 oxidation state, arsenic is four-coordinated like H_3PO_4; in the +3 oxidation state, arsenic is bonded directly to three OH groups.

For both nitrogen and phosphorus, the +5 oxidation state is the more stable and the +3 oxidation state can function as an oxidizing or reducing agent. (The +1 oxidation state of phosphorus is a particularly powerful reducing agent and reduces many metal ions to the element.) For arsenic, on the other hand, the +5 oxidation state is the least stable and is a very strong oxidizing agent. Compared to arsenic, the +5 oxidation state of antimony is stable, whereas that of bismuth is unstable.

Arsenic shows evidence of a transition from non-metallic behavior to metallic behavior. The +5 oxidation state is non-metallic in so far as the oxide is acidic. By way of contrast, arsenious oxide is amphoteric.

The hydrous oxide of antimony in the +5 oxidation state shows a higher coordination with water than the other members of this group; in fact, antimonic acid can best be represented by the formula $HSb(OH)_6$. The sodium salt, $NaSb(OH)_6$, is one of the few insoluble sodium salts. Antimonous acid, on the other hand, has the formula H_3SbO_3.

Bismuth is a metal and its oxides are basic.

Hydrides. Ammonia is, by far, the most important hydride of this group. The chemistry of reactions in liquid ammonia is as varied as the chemistry of reactions in aqueous solution. Although differences exist, there are many features common to water and ammonia. This should not surprise us; ammonia should be similar to water. If we carry out an experiment (imaginary, of course) in which we pull the proton out of the oxygen nucleus of a water molecule, we form an ammonia molecule. Thus, in a crude sense, ammonia is a distorted water molecule with NH playing the role of an oxygen atom.

Liquid ammonia, like water, is a hydrogen-bonded liquid and, also like water, it dissociates to a limited extent. The analog of this dissociation to that in water is

$$2NH_3 \rightleftharpoons NH_4^+ + NH_2^-$$

and

$$2H_2O \rightleftharpoons H_3O^+ + OH^-$$

Thus we see that, in liquid ammonia, NH_2^- takes over the role of hydroxide ion and NH_4^+ takes over the role of proton. The validity of this analogy may be clearly demonstrated by comparison of the electrolysis of liquid ammonia with a mercury cathode to that of water with a palladium cathode. When the water is electrolyzed, hydrogen atoms are formed and dissolve in the palladium cathode. When ammonia is electrolyzed, the ammonium ion gains an electron to form neutral ammonium "atoms" that dissolve in the mercury cathode. Both the hydrogen atoms in palladium and the ammonium "atoms" in the mercury are unstable species. We will consider the analogy between liquid ammonia solutions and aqueous solutions in more detail in the next chapter.

Just as ammonia is analogous to water, hydrazine, $NH_2 \cdot NH_2$ is analogous to hydrogen peroxide. It can be prepared by the reactions

$$NaOCl + HNH_2 \rightarrow NaOH + NH_2Cl$$
$$NH_2Cl + HNH_2 \rightarrow HCl + NH_2 - NH_2$$

In this preparation, the hydrochloric acid produced by the second step is neutralized by the sodium hydroxide produced in the first step. According to Le Chatelier's principle, this shifts the equilibrium for both reactions to the right. NH_2—NH_2, like hydrogen peroxide, is a hydrogen-bonded liquid that dissolves many salts because of its high dielectric constant. Also like hydrogen peroxide, it is an extremely reactive compound and finds use as a rocket fuel. Hydrazine is a weak base that is capable of neutralizing two equivalents of acid.

In addition to the amino analog of hydrogen peroxide, a mixed analog occurs in the form of hydroxylamine, NH_2OH. This compound, produced by the reduction of nitric acid, is even more unstable than hydrazine or hydrogen peroxide and decomposes explosively; moreover, it is unstable in aqueous solution and decomposes to form either nitrogen and ammonia, or hyponitrite salts. The reaction observed depends on the conditions.

Hydrazoic acid, HN_3, is almost pure nitrogen by weight. Although it is isoelectronic with N_2O, it is an extremely unstable compound. Its sodium salt can be prepared by the treatment of sodium amide with N_2O, as indicated by the equations

$$NaNH_2 + N_2O \rightarrow NaN_3 + H_2O$$

and

$$NaNH_2 + H_2O \rightarrow NaOH + NH_3$$

Water produced in the first equation is removed by reaction with sodium amide. The acid itself, HN_3, is very weak and stable in dilute solution. It reacts with basic hydroxides to form azide salts. Those of the alkaline metals and alkaline earth metals are stable, and, if heated to high temperatures, they decompose smoothly to yield nitrogen. The azides of transition metals wherein the bonding may be covalent are sensitive to shock, and they detonate.

If hydrazoic acid is passed through a quartz tube at 1000°C, the following reaction occurs:

$$HN_3 \xrightarrow{1000^\circ C} NH + N_2$$

The imine radical, NH, is an unstable reactive species; although its mean life is only 10^{-3} second, it can be condensed at $-196^\circ C$ to form the species $(NH)_n$.

The hydrides of the other Group V elements are prepared by the decomposition of the arsenides, antimonides, bismuthides, and phosphides in water or dilute acid. As we remarked earlier, phosphine can also be prepared by the reaction of white phosphorus and potassium hydroxide in aqueous solutions. All of these hydrides slowly decompose in water. Their stability decreases with increasing atomic weight.

If phosphine is produced by the reaction of white phosphorus with aqueous alkali, the product is contaminated with another hydride of phosphorus with the formula P_2H_4. This hydride decomposes with the elimination of hydrogen to form a chain of indefinite structure with the general formula $(P_2H)_n$. It is insoluble in water and most organic solvents but dissolves in liquid phosphorus.

Both PH_3 and P_2H_4 should be analogous to the corresponding compounds of nitrogen. The hydrazine analog is too unstable in water for a detailed study of its properties, but phosphine in water forms a nearly neutral rather than a basic solution. It is like ammonia in so far as salts of the type PH_4Cl and PH_4Br are known.

Covalent compounds. Nitric acid reacts with organic compounds in two ways:

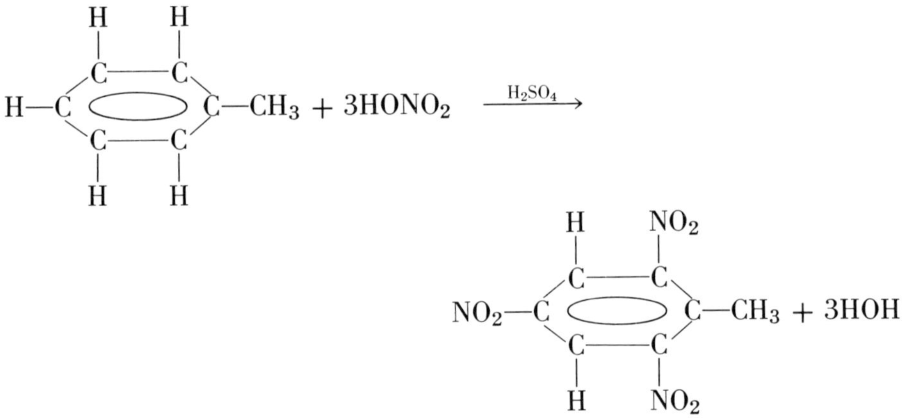

and

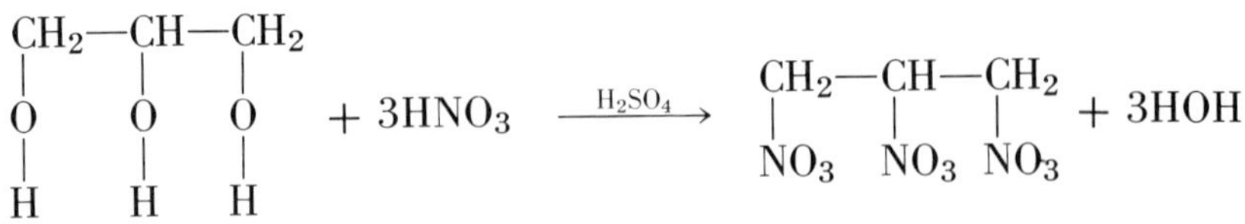

In the first reaction, we split out water between the hydrogen of the toluene and the hydroxide of the nitric acid. The product, called trinitrotoluene, or TNT, contains nitrogen as the NO_2 group. If nitric acid reacts with organic molecules that already contain OH groups, the nitrogen appears in the molecule as the NO_3

group. This compound is called glycerol trinitrate, or nitroglycerin. Both compounds are important as explosives.

As important as the nitro and nitrate compounds are in organic chemistry, the amines are an even more important group of nitrogen-containing molecules. We have already remarked that the NH_2 group is analogous to the OH group; hence, in many organic compounds like acids and alcohol, we form analog compounds like those indicated below:

The NH_2 group functions as a stronger base than the OH group. Thus molecules containing NH_2 groups are the bases of organic chemistry.

An acetic acid is known in which one of the hydrogens on the CH_3 group is replaced by a hydroxide. This compound is referred to as glycolic acid. We also find a nitrogen analog to this compound, an *amino acid* called glycine. The structures of these are indicated below:

Glycolic acid — Glycine (amino acid)

Amino acids contain on one end an acidic group, which can give up protons, and on the other end a basic group, which can react with the proton. Thus glycine can form a "zwitter ion" that is charged negative on the acid end and positive on the amino end; that is,

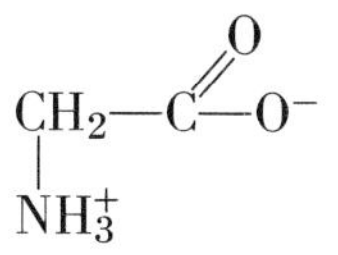

Glycine is found in this form in the crystal. In addition to the internal neutralization that occurs within the amino acid molecules, we also find external neutralization, wherein the acid tail of one molecule reacts with the basic head of the other with the elimination of water. When this occurs, we form long chains of the type

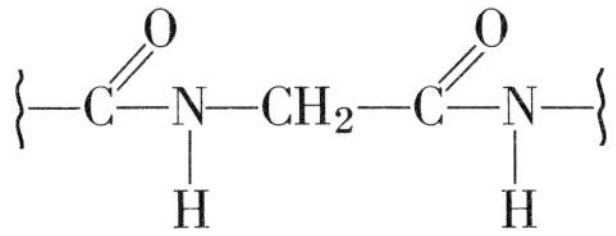

These polymers are *proteins* and are the stuff of which we are made. Since there are a number of amino acids, and proteins are made up of thousands of amino acid links, fantastic variety of protein chains is possible.

Halides of nitrogen. There are two principal types of nitrogen halides: the first has the formula XN_3; the second has the formula NX_3. The first of these can be viewed as the azide salts of electropositive halogens and can be prepared by the reaction

$$AgN_3 + I_2 \rightarrow AgI + IN_3$$

To understand this reaction, imagine that the iodine molecule splits into a positively charged atom and a negatively charged atom. The negatively charged atom pairs with the silver ion, and the positively charged iodine pairs with the azide ion. These haloazides are unstable compounds and explode easily.

The NX_3 type of halogen is a derivative of ammonia, and all of these (but NF_3) are formed by the direct action of the halogen on ammonia. (NF_3 is prepared by the electrolysis of ammonium fluoride.) The hydrolysis of these halides provides an indication of their structure. Here are the equations for the hydrolysis of NF_3 and NCl_3:

$$2NF_3 + 3HOH \rightarrow 6HF + N_2O_3$$

and

$$NCl_3 + 3HOH \rightarrow 3HOCl + NH_3$$

In NF_3, the nitrogen functions as if it were in the $+3$ oxidation state, because, compared to fluorine, nitrogen is the less electronegative element. On the other hand, in NCl_3, nitrogen functions as the more electronegative compound; and it reacts as if it were in the -3 oxidation state. Only the fluorides of nitrogen are moderately stable; the others detonate when subjected to shock. (NI_3 and NBr_3 exist only in the presence of excess ammonia.)

Dinitrogen difluoride, N_2F_2, which is highly explosive, occurs in *cis* and *trans* forms. Both structures are indicated in the accompanying formulas.

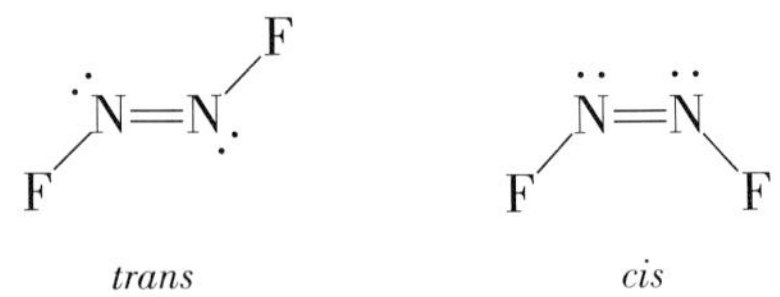

The two forms are possible because the sp^2 hybrid with the double bond locks the molecules in the planar configuration; that is, free rotation is impossible.

In addition to nitrogen-halogen compounds, nitrogen also forms oxyhalides. These compounds are generally more stable than the halides.

Phosphorus forms a variety of halides by direct combination with the elements. The stable iodides are

$$P_2I_4, \quad PI_3, \quad PI_5$$

It is noteworthy that, although we do not find a halide for nitrogen in the +5 state, we do find one for phosphorus in the +5 state. The reason is that, in the nitrogen atom, only *s* and *p* orbits are available for bonding, whereas, in the phosphorus atom, *s*, *p*, and *d* orbits are available for bonding. The octet rule is strictly true only for the family heads of the periodic chart.

The phosphorus compounds listed for iodine are also found with the other halogens. In addition, there are compounds with the formulas PBr_7, PCl_2Br_4, PCl_3Br_8, etc. These halogen compounds are analogous in structure to the polyhalides discussed on p. 521.

The best-characterized halides of phosphorus are those in the +3 and the +5 oxidation states. (In the lower oxidation states, only P_2I_4 is definitely established.) As expected, these halides hydrolyze in water to produce the hydrogen halide and the hydrated oxides of phosphorus. They can be looked upon as the mixed anhydride of hydrohalic and phosphorus oxyacids. The +5 halides hydrolyze to phosphoric acid; the +3 halides hydrolyze to phosphorous acid. If sufficient water is not present for complete hydrolysis, the oxyhalides form; for example, PCl_5 treated with a limited amount of water forms $POCl_3$ and hydrochloric acid.

Phosphorus halides provide a convenient source of chlorine for a number of reactions. They are used in organic chemistry to substitute chlorine groups for hydroxide groups, but, when they are used for this purpose, the pentachloride is first converted to $POCl_3$ to limit the extent of chlorination. Phosphorus halides are also used as chlorinating agents in inorganic chemistry. For example, thionyl chloride is prepared by the reaction of sulfur dioxide with PCl_5:

$$PCl_5 + SO_2 \rightarrow SOCl_2 + POCl_3$$

The halides of arsenic, antimony, and bismuth show a peculiar variation in the stability of the +5 oxidation state. Thus arsenic, antimony, and bismuth all form halides in the +3 oxidation state. In the +5 oxidation state, arsenic forms only the fluoride, whereas antimony forms all the pentahalides except iodine, and bismuth forms no pentahalides.

Other covalent compounds. When sulfur is treated with liquid ammonia, the following reaction occurs:

$$10S + 16NH_3(l) \rightarrow S_4N_4 + 6(NH_4)_2S$$

This reaction is complete if some means are used to remove the ammonium sulfide. The compound S_4N_4 is insoluble in water but is quite soluble in a number of organic solvents. The molecule probably consists of an eight-membered ring with alternating nitrogen and sulfur atoms. Although it is insoluble in water, it hydrolyzes slowly in hot water; the nitrogen forms ammonia, whereas the sulfur forms a mixture of the oxygen acids.

Phosphorus forms a variety of sulfides; a few of these are

$$P_4S_3, \quad P_4S_5, \quad P_4S_7, \quad P_2S_5$$

Molecular-weight measurements in solution indicate that P_2S_5, like "P_2O_5," actually occurs as the P_4S_{10} molecule. Thus, in each of these compounds, the molecule preserves a group of four phosphorus atoms together. Some of these compounds are unstable; P_4S_3 is used in the manufacture of "strike-anywhere" matches.

The hydrolysis products of the sulfides of phosphorus are expected to be different from the hydrolysis products of the sulfides of nitrogen. The relative electronegativities of phosphorus and sulfur are such that phosphorus is the more electropositive element; hence, on hydrolysis of phosphorus sulfide, we form H_2S and the oxides of phosphorus.

If PCl_5 is treated with ammonium chloride, the following reaction occurs:

$$PCl_5 + NH_4Cl \rightarrow (NPCl_2) + 4HCl$$

The $NPCl_2$ does not occur as a single unit but is linked with others to form three- or four-membered rings in which the building unit is

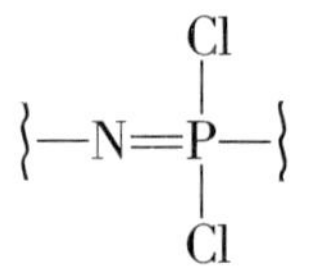

If the compound is heated, these rings open and long chains with rubber-like elasticity form. All of these compounds are insoluble in water and are particularly stable toward acid and alkali, but they are soluble in organic solvents. In these solvents, they hydrolyze to form HCl. The ring is *not* broken, but the chlorine atoms are replaced by OH groups to form a variety of acids with the generic name *phosphimic* acid. If these acids are further hydrolyzed by extreme conditions, the ultimate products are ammonia and phosphoric acid.

Arsenic, antimony, and bismuth also form sulfides. On the basis of the oxygen analogs to the sulfur compounds, the sulfides of arsenic and antimony should be amphoteric. In the presence of excess sulfide ion, they form the thio anions, as indicated by the equation

$$As_2S_3 + 3S^{-2} \rightarrow 2AsS_3^{-3}$$

PROBLEMS

1. Indicate the electronic structures and estimated bond angles for the following:

(*a*) P_4O_{10} (*b*) NO_2
(*c*) N_2O_3 (*d*) H_3PO_3
(*e*) Cl_2O_7 (*f*) H_2S_3
(*g*) ClO_2

2. Indicate the diamagnetic species in Problem 1. *Ans.* (*a*), (*c*), (*d*), (*e*), (*f*).

3. If a compound is ductile, it can be drawn into wires and beaten into thin sheets. Would you expect metallic crystals or molecular crystals to be more ductile? Explain your answer.

4. The bond energies for triple bonds in N≡N and P≡P are 224 kcal and 120 kcal, respectively. The single-bond energies are 38 kcal for N—N and 51 kcal for P—P. On the basis of these data, calculate whether the diatomic or tetratomic form

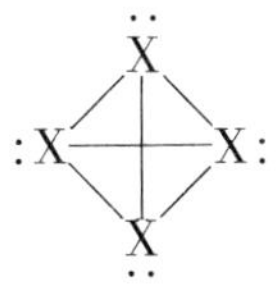 or :X≡X:

is the more stable. Make these calculations for both nitrogen and phosphorus. Phosphorus exists in both forms. The stable form depends on the temperature. Is: P≡P: favored by high temperatures or low temperatures? Explain your answer.

Ans. $X_4 \rightarrow 2X_2$; $\Delta E = -220$ kcal for N_2; $\Delta E = +66$ kcal for P_2. Stable forms are N_2 and P_4; P_2 is favored by high temperature.

5. Confirm by calculation that

$$H_2 = 2H^+\ (10^{-7}M) + 2e^- \qquad \mathcal{E} = 0.414$$

given

$$H_2 = 2H^+ + 2e^- \qquad \mathcal{E}^\circ = 0.000$$

6. Compare the thermal stability and ionization constant of HAt with those for the other hydrogen halides.

7. A dissociated salt dissolved in liquid hydrogen peroxide shows ideal behavior at higher concentration than when it is dissolved in water. Why?

8. There is restricted rotation about the O—O bond in hydrogen peroxide. Assume that the nature of this restricted rotation is similar to that found in ethane and sketch a three-dimensional view of the hydrogen peroxide molecule.

9. On the basis of structure, is H_3PO_3 or H_3AsO_3 the stronger acid? *Ans.* H_3PO_3.
Which is stronger, telluric or sulfuric acid? *Ans.* H_2SO_4.

★ **10.** All anions of the halides, including the oxy anions, are colorless. On the basis of the discussion of the solution chemistry of the halogens, explain the following observations.

Test tube A contains a solution of potassium iodide. After the addition of 1 drop of water saturated with chlorine, the solution turns brown, the characteristic color of iodine in the presence of iodide ion. After the addition of 10 drops of chlorine water, the solution becomes colorless.

Test tube B contains a solution of potassium bromide. After the addition of 1 drop of chlorine water, the solution turns a reddish brown, characteristic of free bromine. This color persists when more chlorine water is added.

★ **11.** Iodine is only sparingly soluble in water. In the presence of iodide ion, it forms an I_3^- complex which is quite soluble. Gaseous iodine is purple. A solution of iodine in aqueous potassium iodide is brown. If this solution is shaken with CCl_4, an inert non-polar solvent immiscible with water, the aqueous solution turns clear and the carbon tetrachloride turns purple. Explain these observations.

12. Given the data

$$Fe^{+2} \rightarrow Fe^{+3} + e^- \qquad \mathcal{E}° = -0.771$$

devise a test to see if a solution of KBr contains some KI. You may want to refine the test on the basis of information given in Problems 10 and 11.

13. How much chlorine is produced when 5.0 gm of MnO_2 is treated with excess HCl at 0°C? Under these conditions, the end product is $MnCl_3$.

★ **14.** Excess pure bromine is added to a closed vessel containing 10 moles of $KClO_3$ dissolved in enough water to form 100 liters of solution. How many moles of chlorine will form if the partial pressure of chlorine is maintained at 1 atm? *Ans.* 3.0×10^{-4}.

15. Devise a chemical process for isolating bromine from sea water. Assume that it is present as $MgBr_2$.

16. Outline a method for preparing pure $KClO_3$ starting with elementary chlorine. $KClO_4$ is even less soluble at 0°C than $KClO_3$. Outline a method for preparing pure $KClO_4$.

17. List the potassium salts of the periodates.

18. Give balanced equations for the following:

(*a*) The auto-oxidation-reduction of KIO.
(*b*) The preparation of Br_2O.
(*c*) The reaction of Br_2O with water.
(*d*) The reaction of Cl_2O_6 with water.
(*e*) The reaction of NO_2^- with I^- in acid solution.

19. Which of the halides would form the strongest complex with Hg^{+2}? *Ans.* I^-.
Which of the halides would form the strongest complex with Al^{+3}? *Ans.* F^-.

20. Indicate the structure and bond angles of Cl_2O_7 and $P_2O_7^{-4}$. Which is the more stable in aqueous solution?

★ **21.** Write balanced equations for the following reactions in aqueous solution:

(*a*) Ag + conc. HNO_3 (*b*) C + conc. H_2SO_4
(*c*) Ag + dil. HNO_3 (*d*) $Se^{-2} + O_2$
(*e*) $SeO_3^{-2} + HClO_4$ (*f*) $H_2S_2O_8$ + heat

22. In moderately strong acid solutions, NO_3^- reacts with S^{-2} to form free sulfur. Write a balanced equation for this reaction.

Ans. $4H^+ + 2NO_3^- + S^{-2} \rightarrow 2NO_2 + S + 2H_2O$.

23. The $\Delta F°$ for the reaction

$$2SO_2 + O_2 \rightarrow 2SO_3$$

is -33.5 kcal at 25°C. The $\Delta H°$ is -47.0 kcal. If, as usual, $\Delta H°$ and $\Delta S°$ are independent of temperature, what is the equilibrium constant at 25°C and at 500°C?

24. Indicate the likely structure and bond angles of the following:

(*a*) $H_2S_3O_{10}$ (*b*) AsO_3^{-3}
(*c*) S_3Cl_2 (*d*) P_4S_6

25. List all the potassium salts of H_3PO_4, H_3PO_3 and H_3PO_2.

26. Write equations for the following reactions.

(*a*) Hydrolysis of SO_2Cl_2. *Ans.* $SO_2Cl_2 + 2H_2O \rightarrow 2HCl + H_2SO_4$.
(*b*) Hydrolysis of $SOCl_2$.
(*c*) Hydrolysis of NI_3.
(*d*) Complete hydrolysis of 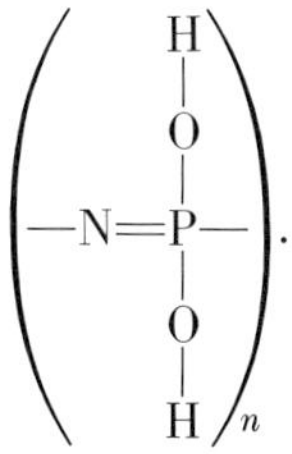.

27. Explain the following observation: A solution of Sb_2O_3 is dissolved completely in concentrated HCl. When the solution is diluted, a precipitate forms.

28. If 10 gm of P_4 are prepared by heating $Ca_3(PO_4)_2$ with SiO_2 and reducing the resulting P_4O_{10} with carbon, how much $CaSiO_3$ is produced as a by-product?

29. Phosphorus occurs in two structural forms. At room temperature, white phosphorus is unstable compared to red phosphorus, but both can be kept for long periods of time at room temperature. When they are vaporized, they form identical P_4 molecules. Which form has the higher vapor pressure?

30. Write the structure of the product formed when NO reacts with H_2SO_4.

31. Nitric acid is prepared by the air oxidation of ammonia followed by treatment of the oxidation products with water. Write a series of equations for the reactions that occur.

32. If 5 gm of HNO_3 acid is formed by the hydrolysis of NO_2 followed by subsequent decomposition of the nitrous acid formed, how many grams of NO_2 hydrolyzed?

33. Write the equation for the reaction that occurs when a solution of ferrous sulfate and KNO_3 is acidified with concentrated H_2SO_4. (It turns brown at the interface of the concentrated H_2SO_4 and $FeSO_4$-KNO_3 solution.)

34. A student tests for NO_3^- with the procedure in Problem 33. He finds a brownish

red ring at the interface and assumes that NO_3^- ion is present. He later finds that NO_3^- was not present but I^- was. Explain his result.
Hint. Ferrous salts in solution are readily oxidized to ferric ion by dissolved air.

35. If you could obtain a pure sample of NO_3 dissolved in CCl_4, what tests could you perform to see if the structure is NO_3 rather than N_2O_6?

36. Write equations for:

(*a*) The reaction of Zn with NO_3^- in basic solution.
Ans. $6H_2O + 4Zn + NO_3^- \rightarrow 4Zn^{+2} + NH_3 + 9OH^-$.
(*b*) The preparation of $H_2N_2O_2$ from HNO_2 and NH_2OH.
(*c*) The reaction of As_2O_3 with NaOH.
(*d*) The reaction of As_2O_3 with H_2SO_4.
(*e*) The decomposition of NH_2OH.
(*f*) The reaction of AsS_3^{-3} with H_2SO_4. *Ans.* $2AsS_3^{-3} + 6H^+ \rightarrow 3H_2S + As_2S_3$.

37. How much $NaNH_2$ is required to prepare 2 gm of NaN_3?

38. A sample of NaN_3 is heated and decomposed completely. The resulting gas is collected over water at 100°C in a 1000-cc bulb at a total pressure of 1.20 atm. How much NaN_3 decomposed?

39. A protein molecule consists of ten links. If eight of these are glycine and two are alanine,

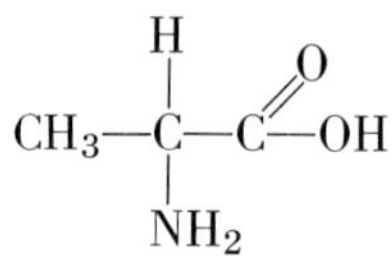

how many different proteins can exist with these constituents? (Proteins contain as many as 10^6 links. There are nine essential amino acids. The number of different combinations of these in a large protein of definite length is comparable to the number of Ångstroms in the distance across the Milky Way.)

21

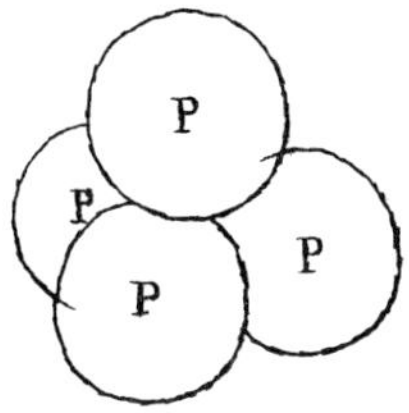

Structure and reactions of non-metals

The reactions of an element or a compound are governed by its structure. For example, even though atomic nitrogen is a reactive species, molecular nitrogen is relatively inert because of the strong nitrogen-nitrogen triple bond. Many such observations are made reasonable when we know the structures of the species involved. In this chapter, we discuss the structural chemistry of non-metals in order to obtain a better understanding of their chemical reactions.

ALLOTROPY

The uncombined elements themselves constitute a fascinating area of structural chemistry. As we have seen earlier, most metallic elements crystallize in the body-centered, face-centered, or hexagonal close-packed lattice. A wider variety of structures is encountered among the non-metals and metalloids. Of these, only the halogens, oxygen, and nitrogen crystallize to form molecular crystals in which diatomic molecules are arranged in a regular array. The other non-metals have structures that differ radically even for neighbors in the periodic chart. For example, sulfur forms a molecular lattice consisting of S_8 rings; on the other hand, phosphorus occurs with a linked zig-zag chain of atoms. Thus, whereas the metals are systematic in so far as just a few structures are observed, the non-metals exhibit a variety of structures. As a result, the chemistry of the metals is relatively simple, while that of the non-metals is much more difficult to systematize but, at the same time, much more interesting.

Many of the non-metallic elements occur with more than one structure; this phenomenon is referred to as *allotropy*. The common form of oxygen is the diatomic molecule, O_2, which is found in the atmosphere. Another form of oxygen is the triatomic molecule, O_3, called *ozone*, which is produced in an electrical discharge. As might be expected, such changes in structure are accompanied by spectacular changes in reactivity: the diatomic form of oxygen is essential for life processes; the triatomic form is a poison.

Except for nitrogen, all Group V elements show allotropy. Phosphorus occurs in at least three allotropes: white, red, and black. White phosphorus is reactive and poisonous; red phosphorus is only moderately reactive and not poisonous; black phosphorus is practically inert. The details of the structure and the interrelation of these various forms of phosphorus are not well established. (There appear to be two forms of white phosphorus and six forms of red phosphorus.) At low temperature, phosphorus vapor consists of tetratomic molecules; at higher temperatures, this changes to a diatomic form, similar to nitrogen, but with a much lower bond strength; at still higher temperatures, the diatomic form dissociates into atoms. Arsenic, antimony, and bismuth also exist in several allotropic modifications, some of which are similar to those of phosphorus, but at the present time their structures are even more doubtful than those of phosphorus.

Sulfur, selenium, and tellurium also exist in several allotropic modifications. We shall discuss the allotropy of sulfur in order to provide you with an appreciation for the complexity of the chemistry of the elements.

At ordinary temperatures, sulfur occurs as a molecular unit containing eight sulfur atoms linked together in a ring. This structure is

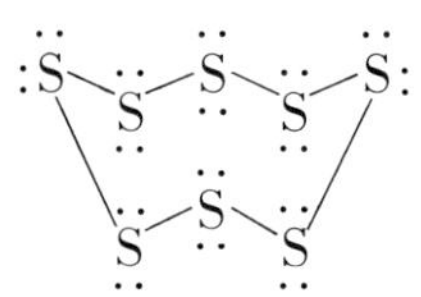

In the stable solid form at room temperature, *rhombic* sulfur, these crown-shaped molecules are packed together in a molecular crystal. Rhombic sulfur melts at 110° to 120°. After cooling the melt, we find a different variety of crystalline sulfur, *monoclinic* sulfur. Although the structure is not known with certainty, monoclinic sulfur is also made up of S_8 molecules, but the lattice arrangement is different from that for rhombic sulfur. At room temperature, monoclinic sulfur slowly changes back to rhombic sulfur.

The behavior of sulfur as a function of pressure and temperature is represented by the *phase diagram* shown in Fig. 1. This diagram summarizes the changes that occur when sulfur is confined in the cavity under an airtight piston at a pressure P and a temperature t. Thus, at the pressure and temperature indicated by arrow 1, the sulfur is present as the rhombic modification. If we keep the temperature constant and reduce the pressure to the point indicated by X, the rhombic sulfur changes to monoclinic sulfur. Monoclinic sulfur persists as the pressure is

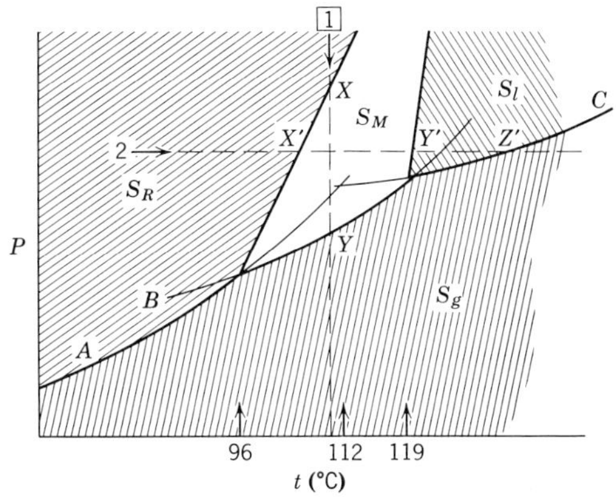

Fig. 1. Sulfur phase diagram.

lowered until we reach the point labeled *Y;* then it vaporizes to form gaseous S_8 molecules. Further reduction in the pressure merely results in the expansion of the gaseous sulfur. At the point *Y,* the pressure of sulfur gas is about 0.01 mm.

Now consider a process in which sulfur is at a temperature of about 90°. Let us maintain constant pressure but slowly raise the temperature as indicated by the arrow labeled 2. The rhombic sulfur persists until the temperature reaches the point labeled *X′;* then it changes into monoclinic sulfur. As we raise the temperature further, we reach the point labeled *Y′*, and the monoclinic sulfur melts to form liquid sulfur. Liquid sulfur persists until the temperature indicated as *Z′* is reached; then it vaporizes and a further increase in temperature simply heats the gaseous sulfur.

What occurs when we heat an evacuated flask containing rhombic sulfur? If we keep it at 96° a long time, rhombic sulfur is converted to monoclinic sulfur. The monoclinic sulfur persists up until 119°, at which point liquid sulfur appears. Under the usual conditions, however, the transition from rhombic to monoclinic sulfur takes place so slowly that equilibrium is not attained. Thus, when we heat rhombic sulfur rapidly, the vapor pressure is that indicated by the line *A* and the *extension of the line A* up to 112°. At this point, there is a transition from superheated rhombic sulfur to supercooled liquid sulfur. The lines representing this metastable behavior are the extrapolated vapor pressure curve for rhombic sulfur (line *A*) and the extrapolated vapor pressure curve for liquid sulfur (line *C*). If we rapidly heat rhombic sulfur to 112°, it melts; if we maintain it at this temperature, the liquid formed slowly solidifies to form monoclinic sulfur. If we chill the monoclinic sulfur thus formed to room temperature, the monoclinic sulfur, which is metastable at room temperature, persists for days before it transforms into rhombic sulfur.

According to the phase diagram, if we gradually increase the pressure at 110°C, there is a transition from monoclinic sulfur to rhombic sulfur. This observation is in line with Le Chatelier's principle; rhombic sulfur is denser than monoclinic sulfur.

Not only is the structure of solid sulfur complex, the liquid also consists of several different types of molecules. These are labeled S_λ, S_μ, and S_π. S_λ and S_π are both soluble in CS_2, but the third variety, S_μ, is insoluble. Immediately above the melting point, sulfur is thought to consist of chains of sulfur formed by opening the S_8 ring together with some unopened S_8 ring molecules. The S_μ variety of sulfur appears to be a polymeric material formed by linking together the eight-atom chains of sulfur atoms.

The melting point of monoclinic sulfur is about 119°. If we allow sulfur to stand at this temperature, we find that the melting point slowly drifts down to about 114°. Apparently, when monoclinic sulfur first melts, the liquid sulfur consists almost entirely of S_λ. With time, an equilibrium is established between the S_λ and the S_μ. The fraction of S_λ and S_μ in the liquid changes with temperature, and, accordingly, a variety of changes occur in the molten sulfur. Directly above the melting point, liquid sulfur is a yellow mobile liquid; as it is heated, the liquid gradually turns red and becomes *more* viscous. When pure sulfur is heated from 160° to 166°, the viscosity of the liquid increases by a factor of 2000. This increase continues up to about 180°; then, between 180° and the boiling point, 445°, the viscosity again decreases.

One interpretation of the behavior of molten sulfur is as follows: directly above the melting point, liquid sulfur consists primarily of S_8 ring molecules; at slightly higher temperatures, the S_8 ring opens to form

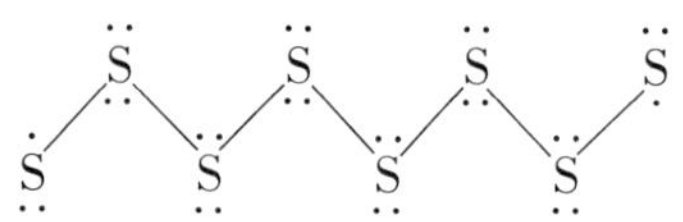

This chain is reactive because of the unpaired electron on the end sulfur atom; it can combine with similar chains to form a giant molecule (presumably, S_μ). These giant chains have a very high viscosity since they tangle in each other and, thereby, decrease mobility. The increase in viscosity continues until most of the molten sulfur is in long-chain form. As the temperature is further increased, there is no longer an increase in the fraction of melt that is polymeric and the average chain length of the polymeric sulfur decreases owing to the rupture of sulfur-sulfur bonds; hence there is a decrease in viscosity.

A simple observation lends considerable weight to the picture above. If we add to the molten sulfur a molecule that can react with chain ends, this should prevent the formation of long chains and thereby reduce the viscosity. The viscosity of the purest sulfur at around 200° corresponds to a chain length of 100,000 atoms. If we add 0.02% iodine to liquid sulfur and then heat it to 200°, the viscosity drops to a value consistent with a chain length of only 3000 atoms.

If the molten polymeric sulfur is chilled to room temperature, a different form of solid sulfur, termed *plastic* sulfur, is obtained. This sulfur exhibits rubber-like elasticity and is soft and pliable. Plastic sulfur slowly hardens with time as it transforms into rhombic sulfur.

Sulfur vapor is complex. Depending on the temperature, sulfur exists in one of the following forms: S_8, S_6, S_4, S_2, S. The magnetic properties of S_2 reveal that diatomic sulfur, like diatomic oxygen, contains two electrons that are unpaired.

NON-AQUEOUS SOLVENTS

Understandably, we think of solution chemistry in terms of reactions in which water is the solvent. If, however, we examine other solvents, such as liquid ammonia, we find that such reactions as neutralization and amphoterism are typical of solution chemistry in general, rather than only of aqueous solutions. In this section, we shall re-examine water in order to see what properties lead to its solvent action. We shall find that the properties of water are not unique; consequently, other solvents can be found which undergo reactions similar to those of water in general detail, but different from water in specifics. Such studies yield a broader view and a better understanding of reactions in solution.

Solvents. The principal property of water that makes it effective as a solvent for electrolytes is its high dielectric constant. Water is also a good solvent for hydrogen-bonded organic liquids such as alcohols. With alcohols, the solvent ability of water stems from the fact that it forms a hydrogen bond to an alcohol molecule almost as well as other alcohol molecules; hence hydrogen-bonded liquids separate and mingle with water molecules because the hydrogen bond lost on their separation is regained on solution. There are a number of liquids that have a high dielectric constant and show strong hydrogen bonding; among these are hydrazine, hydrogen peroxide, ammonia, hydrogen fluoride, and hydroxylamine. All of these can be used for water-like solvents, but, for reasons of stability, ammonia and hydrogen fluoride are the ones commonly used. Sulfur dioxide is also used as a solvent. This solvent differs from those above in that it does not contain a proton. It is similar to those we have discussed above in so far as it has a moderately high dielectric constant (about 14).

In discussions of aqueous solutions, we found it useful to classify reactions as acidic or basic, depending on their effect on the equilibria:

$$2HOH \rightleftarrows H_3O^+ + OH^- \qquad K = 10^{-14}$$

If a reactant was such that it increased the concentration of hydronium ion, we called it an acid reactant. If, on the other hand, it increased the concentration of hydroxide ion, we called it a basic reactant. Thus it was essential to this description that water undergo this auto-dissociation. In other words, one of the characteristic properties of water as a solvent is the dissociation into hydrogen ions and hydroxide ions. These other solvents also undergo auto-dissociation reactions of the type

$$2NH_3 \rightleftarrows NH_4^+ + NH_2^- \qquad K = 10^{-33}$$
$$2HF \rightleftarrows H_2F^+ + F^- \qquad K > 10^{-14}$$
$$2SO_2 \rightleftarrows SO^{+2} + SO_3^{-2} \qquad K \sim 10^{-13}$$

Thus the similarity between water and these three solvents stems from the high dielectric constant *and* this auto-dissociation process. When we say an aqueous solution is acidic, we imply that there is an increase in the concentration of positive cations formed by the dissociation of the solvent. On such a basis, NH_4^+ is acidic in liquid ammonia, H_2F^+ is acidic in liquid hydrogen fluoride, and SO^{+2} ion is acidic in liquid sulfur dioxide.

LIQUID AMMONIA CHEMISTRY

Before we start the study of chemistry in liquid ammonia, examine Table 1. There are marked differences between water and ammonia. First of all, ammonia boils well below water; consequently, reactions in liquid ammonia will be carried out at much lower temperatures than those in aqueous solution. Secondly, the dielectric constant of ammonia at the boiling point is about half that of water at the boiling point; consequently, interionic attraction would be expected to be much stronger in ammonia. Finally, the dissociation constant of ammonia is far less than that of water; consequently, reactions that depend on the concentration of solvent anion and cation, i.e., reactions analogous to hydrolysis, are not so common in liquid ammonia as in water.

In liquid ammonia, the NH_2^- ion takes on the role of hydroxide ions; the NH_4^+ ion takes on the role of the hydrogen ion. In Table 2, we have listed some of the water-ammonia analogs; lithium hydroxide corresponds to *lithium amide;* lithium oxide corresponds to *lithium imide.* There is no direct analogy to lithium nitride and aluminum nitride; they correspond to the completely deammoniated amides and imides.

The ammonia analog of hydrochloric acid is ammonium chloride. There are two analogs of nitric acid. For one, we substitute an ammonium ion for the hydrogen ion to obtain ammonium nitrate. If we substitute an NH group for each oxygen atom, we obtain an unstable compound, which would immediately lose ammonia to form hydrazoic acid, HN_3. In addition to the analogs of inorganic acids, we also have analogs of organic acids.

Salts often crystallize from aqueous solution with a definite number of water molecules as part of their structure. Likewise, salts crystallize from liquid

Table 1 Properties of Pure Water and Ammonia

	Ammonia	Water
Melting point	195°K	273°K
Boiling point	240°K	373°K
Dielectric constant at boiling point	22	48
Density, gm/cc, at boiling point	0.68	0.96
$K_{diss.}$	10^{-33} (−50°C)	10^{-14} (25°C)

Table 2 Ammonia-Water Analog

H_2O	NH_3
$LiOH$	$LiNH_2$
Li_2O	$Li_2(NH)$, Li_3N
$Al_2O_3 \cdot xH_2O$	$Al_2(NH)_3 \cdot xNH_3$, AlN
$HCl \cdot H_2O$	NH_4Cl
HNO_3	$HN(NH)_3$, HN_3, NH_4NO_3
$CH_3C(=O)$—OH (acetic acid)	CH_3—$C(=NH)$—NH_2
$CaCl_2 \cdot 6H_2O$	$CaCl_2 \cdot 6NH_3$

ammonia solution with a definite number of ammonia molecules as part of their structure; analogs for calcium chloride are listed in Table 2.

The solubilities of salts in liquid ammonia do not necessarily parallel those found in aqueous solution. Thus the chlorates, nitrates, and nitrites are quite soluble, but the sulfates are all insoluble; sodium hydroxide is readily soluble in aqueous solution, but sodium amide is only moderately soluble in ammonia. Not all of the halides are soluble. In particular, those of the alkaline earth elements, which are quite soluble in water, are virtually insoluble in liquid ammonia. On the other hand, the solubility order of silver halides in ammonia is the reverse of that in water. In water, the most soluble is silver fluoride; the least soluble is silver iodide. In liquid ammonia, the most soluble is silver iodide; the least soluble is silver fluoride.

Once we take into account the different solubilities in liquid ammonia, we find that reactions in this medium are similar to those in water. For example, if we treat an aqueous solution of calcium chloride with silver nitrate, a white curdy precipitate of silver chloride forms and the remaining solution contains dissolved calcium nitrate. This reaction goes to completion because the formation of an insoluble precipitate drives the reaction to the right. Similar reactions occur in liquid ammonia, and two of these are

$$Ca(NO_3)_2 + 2AgI \xrightarrow{NH_3} \underline{CaI_2}_s + 2AgNO_3$$

$$Ag(NO_3) + KNH_2 \xrightarrow{NH_3} \underline{AgNH_2}_s + KNO_3$$

These reactions occur because an insoluble precipitate is formed. The first reaction is surprising; in water, calcium iodide is soluble and silver iodide is insoluble; consequently, in an aqueous medium, we get the reverse of the first of the reactions above. The second reaction is more directly related to the reaction we observe in water. If we treat an aqueous solution of silver nitrate with KOH, we form an insoluble hydrated silver oxide. A similar reaction occurs with many of the heavy metals because the insolubility of their hydroxides. In the ammonia, the analog of this reaction is quite common with the amide ion group taking the role of the hydroxide ion.

Acid-base reactions are common in liquid ammonia. These are pictured below:

$$KNH_2 + NH_4Cl \xrightarrow{NH_3} 2NH_3 + KCl$$

$$\underline{AgNH_2} + NH_4NO_3 \xrightarrow{NH_3} AgNO_3 + 2NH_3$$

In the first equation, the soluble base and acid react to form solvent and a soluble salt; in the second reaction, an insoluble base reacts with a soluble acid and, thereby, dissolves.

When aqueous solutions of zinc salts are treated with base, an insoluble precipitate is formed that dissolves in an excess of base or acid because of complex-ion formation and neutralization, respectively. This phenomenon is termed amphoterism. Similar behavior is found in liquid ammonia. If we treat insoluble zinc amide with an excess of base, the following reaction occurs:

$$\underline{Zn(NH_2)_2} + 2KNH_2 \xrightarrow{NH_3} K_2Zn(NH_2)_4$$

On the other hand, if we treat the insoluble base with an excess of acid, it dissolves according to the equation

$$\underline{Zn(NH_2)_2} + 2NH_4NO_3 \xrightarrow{NH_3} Zn(NO_3)_2 + 4NH_3$$

If aluminum chloride is added to a beaker of water, the solution turns cloudy because the aluminum chloride hydrolyzes to form a suspension of insoluble aluminum hydroxide. As a consequence of this hydrolysis, the solution becomes acidic. A parallel reaction occurs when we add aluminum chloride to liquid ammonia; this *ammonolysis* reaction can be represented by the equation

$$AlCl_3 + 6NH_3 \xrightleftharpoons{NH_3} Al(NH_2)_3 + 3NH_4Cl$$

Reactions in liquid ammonia are usually analogous to those in water, but these analogies should not be overextended. The fact that a specific reaction takes place in water does not *guarantee* that the analogous reaction takes place in liquid ammonia. In addition, analogous compounds can differ enormously in properties. For example, the oxides and hydroxides formed in aqueous solution are stable thermally or, at worst, decompose smoothly at elevated temperature. By way of contrast, the amides and imides formed in liquid ammonia solutions are generally unstable thermally and often decompose with violence when heated.

ACIDS AND BASES

In the initial discussion of acids and bases, we defined an acid as any substance that yields hydrogen ion in water, and a base as any substance that yields hydroxide ions in water. This definition suffers because it is too specific. In the preceding section, we saw that, even in liquid ammonia, acid-base reactions occur. Similar reactions take place in organic solvents; for example, in perfectly dry benzene where there is no dissociation, dissolved HCl reacts with dissolved ammonia to form ammonium chloride. It is our purpose in this section to reexamine critically the concept of an acid-base reaction and to extend it to similar reactions in non-aqueous media.

Brönsted acids. Consider the following typical acid-base reaction:

$$\text{H}_3\text{N}: + \text{H—}\ddot{\underset{..}{\text{Cl}}}: \rightarrow \text{H}_3\text{N}:\text{H}^+ + :\ddot{\underset{..}{\text{Cl}}}:^-$$

$$\text{base}_1 + \text{acid}_2 \rightarrow \text{acid}_1 + \text{base}_2$$

According to this, the acid, HCl, has given up a hydrogen ion to the base, NH_3. In any such reaction, there is a finite equilibrium constant, and, in the reverse reaction, the ammonium ion gives up a proton and the chloride ion gains the proton. On the basis of this reaction, let us define an *acid as a substance that gives up hydrogen ions, and a base as a substance that takes up hydrogen ions.* This is Brönsted's definition of an acid and a base.

We did not specify in our discussion whether the reaction took place in water, in benzene, or in liquid ammonia. In water or benzene, the reactants are solutes. In liquid ammonia, however, only HCl is a solute; the formation of the ammono acid NH_4Cl involves reaction of the acid, HCl, with the solvent. This provides us with a broader view for the role of the solvent in an acid-base reaction.

In aqueous solution, the fundamental reaction between an acid and a base is always

$$H_3O^+ + OH^- \rightarrow 2H_2O$$

Thus, for a base to react with an acid in aqueous solution, the base must first react with water to produce hydroxide ion, and the acid must react with water to produce hydronium ion. Therefore the precursor of neutralization is

$$\underset{B_1}{NH_3} + \underset{HB_2}{H_2O} \rightleftarrows \underset{HB_1}{NH_4^+} + \underset{B_2}{OH^-}$$

and

$$\underset{HB'_1}{HCl} + \underset{B'_2}{\text{H—}\ddot{\underset{|}{\text{O}}}\text{:}\atop\text{H}} \rightleftarrows \underset{HB'_2}{H_3O^+} + \underset{B'_1}{Cl^-}$$

In the first of these reactions, water acts as an acid in so far as it gives up a hydrogen ion to the base, ammonia. In the second reaction, water acts as a base in so far as it takes up a hydrogen ion from the HCl. (B is used as a general symbol for a base.) In this sense, water is amphoteric: it can function either as an acid, as it did with respect to ammonia, or it can function as a base, as it did with respect to hydrochloric acid.

The fact that ammonia is a weak base, i.e., that the reaction, $NH_3 + H_2O \rightleftarrows NH_4^+ + OH^-$ does not take place to a great extent, tells us that water is a weaker acid than ammonium ion or, equivalently, hydroxide is a stronger base than ammonia. In a similar fashion, the fact that hydrochloric acid is a strong acid tells us that HCl is a stronger acid than the hydronium ion or, equivalently, water is a stronger base than chloride ion.

According to the Brönsted definition of an acid and a base, we always find acids and bases in pairs. Thus ammonia is a base, B_1, and ammonium ion is its *conjugate acid,* HB_1; HCl is an acid, and chloride ion is its *conjugate base.*

The acidity of HCl in water depends not only on the intrinsic ability of HCl to give up its hydrogen ion, but also on the intrinsic ability of water to take up the hydrogen ion; thus we have working in each acid-base reaction a push from the acid and a pull from the base. This situation is similar to that in the oxidation-reduction reaction:

$$Ox_2 + Red_1 \rightarrow Red_2 + Ox_1$$

Coupled with each oxidizing agent we have, if you like, a "conjugate" reducing agent. The ability of a substance to act as an oxidizing agent depends not only on the intrinsic ability of an oxidizing agent to receive electrons, but also on the intrinsic ability of the other reactant to give up the electrons. In dealing with oxidation-reduction reactants, we recognized this push-pull relation and set up an electromotive force series wherein the strongest reducing agents were at the top and the weakest were at the bottom. With such a table available, we were able to say that a reaction would take place as written whenever combined with products of reactants listed below it. We can do a similar thing for acids; such a partial series is shown below:

$$\begin{aligned} HClO_4 &\rightarrow H^+ + ClO_4^- \\ H_2Ac^+ &\rightarrow H^+ + HAc \\ HCl &\rightarrow H^+ + Cl^- \\ H_3O^+ &\rightarrow H^+ + H_2O \\ HAc &\rightarrow H^+ + Ac^- \\ NH_4^+ &\rightarrow H^+ + NH_3 \\ H_2O &\rightarrow H^+ + OH^- \end{aligned}$$

This series does not refer to an aqueous solution or a solution in liquid ammonia, but rather to the pure components themselves. Accordingly, we would say that, if pure acetic acid is mixed with ammonia, ammonium ion, and acetate ion, the acetic acid produces protons and acetate ions, and the ammonia combines with the protons to produce ammonium ions; on the other hand, in a solution that contains perchloric acid, acetic acid, acetate ion, and perchlorate ion, the perchloric acid ionizes to produce hydrogen ions that combine with acetate ion to form acetic acid and also combines further to form H_2Ac^+. In other words, this acid-base scale is similar to the table of oxidation-reduction potentials.

Our partial series of acid-base reactions has some interesting implications. Both $HClO_4$ or HCl react almost completely with water to form hydronium ion; on the other hand, relatively little acetic acid combines with water to form hydronium ion. If, instead of combining these three acids individually with water, we combine them individually with ammonia, we find that all three of the acids are equally strong in so far as all three react almost completely with ammonia to form ammonium ion, which plays the role of hydronium ion in ammonia solution. Thus, in ammonia solution, we have these three acids reduced to the same level. In other words, in spite of the fact that, intrinsically, these acids differ in their ability to give up a proton, the ammonia is such a good base that there is substantially 100 per cent conversion to the solvent cation, NH_4^+. A similar thing happens in water for perchloric acid and hydrochloric acid. Since both of these lie above H_3O^+, they are rated as strong acids. On the other hand, if we were to add perchloric acid to pure acetic acid and compare its acid strength to that of a similar solution of hydrochloric acid, we would find that this solvent would differentiate between these two acids. The hydronium equivalent in pure acetic acid is H_2Ac^+. In this solution, perchloric acid would be a strong acid, whereas HCl would be a weak acid.

Lewis acids and bases. It is possible to generalize further the definition of acid and base. Consider again the neutralization of NH_3 with HCl. Fundamentally, we are exchanging the covalent bond of the hydrogen with chlorine for a covalent bond of hydrogen with nitrogen. If we consider both ammonia and chloride ion to be bases, the neutralization process evidently involves the formation of a covalent bond between the electron-poor hydrogen and the electron-rich ammonia. G. N. Lewis expanded the definition of acids and bases so that it included Brönsted acid reactions, together with a large number of other reactions. In effect, Lewis said that an acid can be viewed as an electron-poor atom; a base can be viewed as an electron-rich atom; neutralization occurs when the two combine with the formation of a covalent bond. Such a reaction, involving *no hydrogen atom transfer*, is

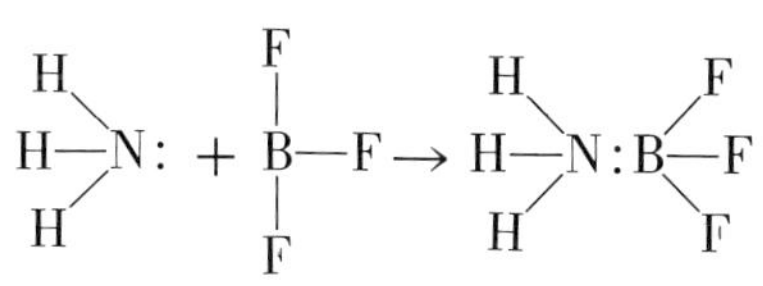

Although this definition has proved valuable in the description of many types of reaction in organic chemistry, many object to it because it is too broad. In effect, it classifies the formation of a covalent bond as an acid-base reaction and ignores the fact that covalent bonds with protons must be fundamentally different from those of other atoms, wherein the nucleus is shielded by electrons. All of this, however, does not detract from its value in systematizing reactions in non-aqueous media.

STRUCTURAL GEOMETRY

Thus far, our discussion of the structure of covalent molecules has been largely restricted to those formed by the second row elements. The structures of compounds of these elements are relatively simple because the octet rule is strictly obeyed. For elements in the third period and heavier elements, the situation is more complex; the octet rule is frequently violated because the *s*, *p*, and *d* orbitals of the third shell are available and as many as nine bonds can be formed with the given atom.

The variety of complex structures encountered with non-metals in the later periods of the periodic chart can be systematized in a rational manner. Assume that only pairs of electrons are involved and that these pairs of electrons, whether they bond or not, arrange themselves as far apart as possible in order to reduce coulombic repulsions. If we consider only bonds involving two to seven pairs of electrons, these are arranged as indicated in Fig. 2. The arrangements are: (*a*) the two pairs collinear with the central atom; (*b*) three pairs at the corners of an equilateral triangle with the atom in the center; (*c*) four pairs at the corners of the tetrahedron with the atom in the center; (*d*) five pairs at the corners of a trigonal bipyramid; (*e*) six pairs at the corners of an octahedron; and (*f*) seven pairs at the corners of a pentagonal bipyramid. With such arrangements, the symmetry is perfect only if all the pairs are bonded to the same kind of atom. Distortions of the indicated symmetries occur whenever some of the pairs of electrons are lone pairs or are bound to different kinds of atoms.

In Table 3, we have indicated some typical structures encountered with Group V, VI, and VII elements. In the paragraphs that follow, we shall discuss the listed structures in order to illustrate the principles involved.

The triangular structures are typified by SO_3. An electron-dot structure for this is written

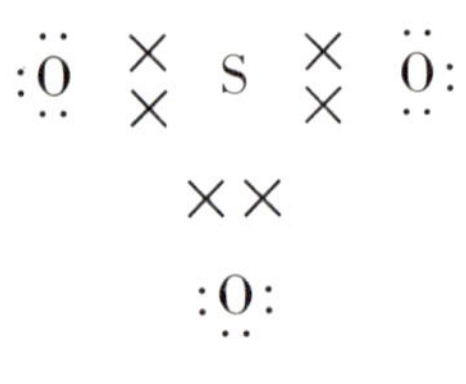

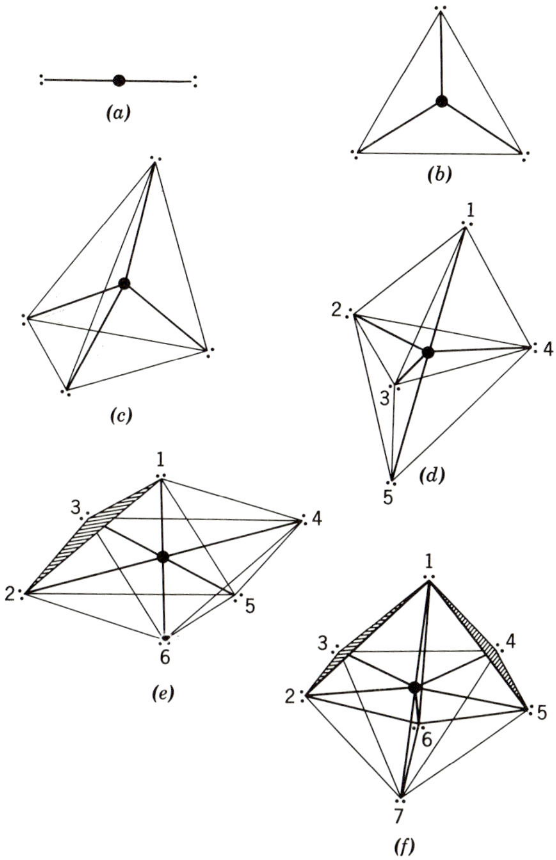

Fig. 2. Common geometric arrangements.

We could, of course, have indicated this and also achieved the octet structure for sulfur by moving two of the lone-pair electrons from the oxygen to a position between the sulfur and the oxygen, but this is not necessary for our purposes. From what we have said earlier, we expect a triangular arrangement with an OSO bond angle of 120°, and that is what we have. SO_2 can be visualized as an SO_3 from which an oxygen atom is removed. This leaves three pairs of electrons around the sulfur; hence we find a kinked configuration for the SO_2 molecule with the OSO angle 120°.

The tetrahedral configuration is found for a number of ions; some of these are listed in Table 3. The $P_2O_7^{-4}$ and $S_2O_7^{-2}$ and Cl_2O_7 structures consist of tetrahedra sharing a corner; all the others listed as having four bonds are also tetrahedral. In some cases, e.g., $POCl_3$, we find that the tetrahedron is distorted; the observed ClPCl bond angle is 103°. We can rationalize this by assuming

Table 3 Structures of Compounds Formed with Group V, VI, and VII Elements

Structure	Number of Bonds	Group V	Group VI	Group VII
Triangular	3		SO_3	
	2		SO_2	
Tetrahedral	4	PO_4^{-3} $P_2O_7^{-4}$ NH_4^+ $POCl_3$ $PO_2F_2^-$	SO_4^{-2} $S_2O_7^{-2}$	Cl_2O_7
	3	NH_3	SO_3^{-2}	
	2		H_2O F_2O H_2O_2 S_8	
	1			HCl
Trigonal bipyramid	5	PCl_5 PF_3Cl_2		
	4	SbF_4^-		
	3			ClF_3
Octahedral	6	PCl_6^-	SF_6 S_2F_{10} $Te(OH)_6$	
	5	$SbCl_5^{-2}$	—	IF_5
	4			
Pentagonal bipyramid	7			IF_7
	6	$SbBr_6^{-3}$		

that, since the PO bond has double-bond character, the O-P-Cl bond angle increases and the Cl-P-Cl bond angle decreases, owing to the added repulsions.

The sulfite ion has the structure

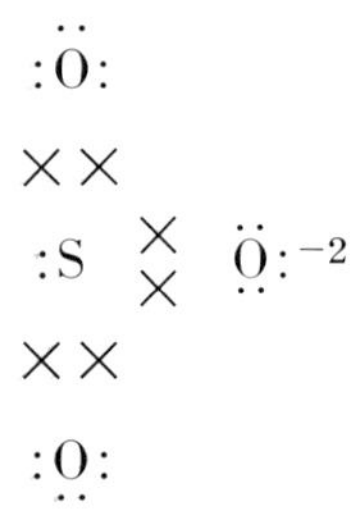

This provides an example of a molecule that has a nearly tetrahedral arrangement of pairs of electrons, even though one of the electron pairs is not bonded. The O-S-O bond angle is 114°.

Molecules like PCl_5 have a symmetrical trigonal bipyramid arrangement; molecules like PF_3Cl_2 have a distorted arrangement of this type. The structure of ClF_3 is

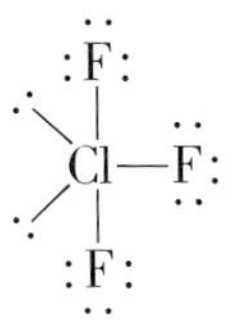

It can be seen that this molecule has two lone pairs and three bonded pairs of electrons. The arrangement which minimizes the repulsion is one in which the fluorine atoms are located at the 1, 3, and 5 positions in Fig. 2*d;* hence this molecule has an odd "T" shape.

The octahedral structure is found for such complex ions as PCl_6^- and the pyro-antimonate ion. Rather interesting examples are sulfur hexafluoride and sulfur decafluoride. The first of these is simply octahedral; in the second, two octahedra share corners. Another example of this type of configuration is IF_5. The electron-dot structure is

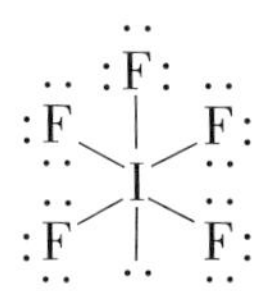

Six pairs of electrons on the iodine atom are involved; five of these are shared with fluorine atoms and one is unshared. Such an arrangement leads to a structure in which the unshared pair is located at the 6 position in Fig. 2*e*. Another molecule found in this configuration is $SbCl_5^{-2}$.

IF_7 is an example of the pentagonal bipyramid structure. A more interesting structure is that of $SbBr_6^{-3}$. This ion has seven pairs of electrons surrounding it, six of which are shared with the bromine. This leads to a structure in which one of the positions indicated in Fig. 2*f* is occupied by the lone pair. If this be the case, the arrangement of the six chloride ions around the centrally located antimony is that of a distorted octahedron. Investigations on this compound suggest such a configuration.

The simple rules we set forth at the beginning of this section are capable of correlating a lot of structural data: however, since our procedures were founded on qualitative principles, we should look upon these predictions as educated guesses, rather than necessary consequences of the principles of structural chemistry.

PROBLEMS

★ **1.** If sulfur vapor consists of S_8 rings, and pure S_λ and S_μ have ring and chain structures suggested, which of these two liquid forms of sulfur would have the greater heat of vaporization? Is this answer consistent with the change in melting point with time? Explain your answer.

2. The density of ice at ordinary pressures is less than that of water. Sketch a phase diagram for water between 0 and 10 atm and $-10°$ and $+10°C$. Assume that the relative density of ice and water do not change with temperature. Sketch the phase diagram between $-20°$ and $400°C$ and 0 and 250 atm. The critical pressure is 220 atm at $375°C$.

3. The following thermodynamic data are available for sulfur:

$$\text{S (rhombic)} \rightarrow \text{S (monoclinic)}$$
$$\Delta H° = +66 \text{ cal} \qquad \Delta F° = 18 \text{ cal}$$

These values were determined at $25°C$. Estimate from these data the temperature at which the transition above occurs. *Ans.* 413°K.

4. Given the data

$$S_8 \text{ (rhombic)} \rightarrow S_8(g)$$
$$\Delta S° = 48.24 \text{ cal/°mole at } 25°C$$
$$\Delta F° = 5700 \text{ cal/mole at } 25°C$$

Calculate the vapor pressure of rhombic sulfur at $25°C$. With the data in Problem 3, calculate the vapor pressure of monoclinic sulfur at $25°C$. Estimate the vapor pressure of rhombic sulfur at $90°C$.

★ **5.** Carbon exists in two allotropic forms, graphite and diamond. The density of graphite is 2.26 gm/cm^3, that of diamond is 3.51 gm/cm^3. Available thermodynamic data at $25°C$ and 1 atm for the reaction

$$\text{C (graphite)} \rightarrow \text{C (diamond)}$$

indicate $\Delta H° \doteq 450$ cal, $\Delta S° \doteq -0.81$ cal/° mole. Which form is stable at room temperature? On the basis of the usual assumptions, would there be a transition from one form to another at any temperature at a pressure of 1 atm? If you wanted to induce such a transition, what pressure would you need? Assume for simplicity that $\Delta E°$, $\Delta S°$, and the densities do not vary with pressure.

Ans. A pressure of 1.43×10^5 psi is needed to produce this transition.

6. Write equations for reactions analogous to neutralization and amphoterism when liquid HF is the solvent.

7. Prepare a table analogous to Table 2 for liquid SO_2. Assume that all analogous compounds exist.

8. The dielectric constants of pure H_2O, pure NH_3, and pure HF are 48, 22, and ~60 at the boiling point. If you wanted to use one of these solvents to determine molecular weights of salts by assuming that ideal solution laws apply to the boiling point elevation of moderately concentrated solutions, which would be the best choice on the basis of dielectric constants? *Ans.* HF.

9. In the following set of equations identify the Lewis acids and bases:

$$SO_2 + O \rightarrow SO_3$$
$$NH_3 + HCl \rightarrow NH_4Cl$$
$$Ag^+ + 2NH_3 \rightarrow Ag(NH_3)_2^+$$
$$Ag^+ + Cl^- \rightarrow AgCl_s$$
$$H^+ + OH^- \rightarrow H_2O$$
$$BF_3 + F^- \rightarrow BF_4^-$$
$$NO_2^- + O \rightarrow NO_3^-$$

10. In water, H_2SO_4 is a strong acid and $H(ClCH_2C(=O)—O)$ is a weak acid. In pure acetic acid both are weak. In liquid ammonia both are strong. Explain.

11. Acetic acid is sometimes called an amphoteric solvent. What is meant by this?

12. Indicate the likely structures and bond angles for the following compounds.

(*a*) ClO_4^- (*b*) $TeCl_4$
(*c*) ClO_3^- (*d*) $IO_2F_2^-$
(*e*) ClO_2^- (*f*) IO_5^{-5}
(*g*) PCl_3 (*h*) $Sb(OH)_6^-$
(*i*) N_2H_4 (*j*) ICl_4^-
(*k*) $SOCl_2$

Ans. (*a*) Tetrahedral; (*g*) pyramidal (tetrahedral with one corner occupied by a lone pair of electrons).

22 Complex equilibria in solution

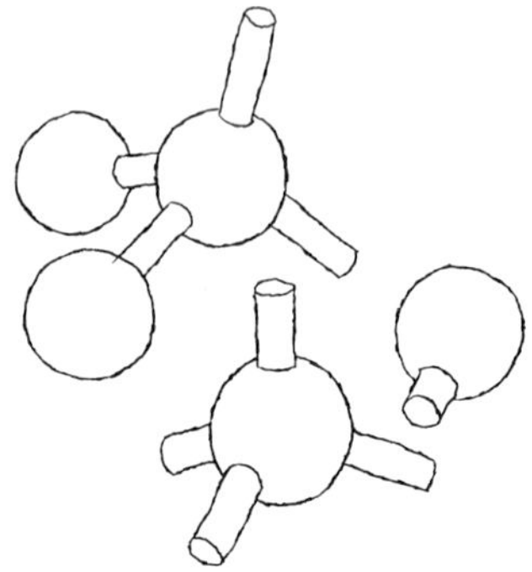

Description of complex equilibria involves the formulation of the equilibrium expressions in terms of appropriate unknowns and the solving of these equations. The first step, which provides an algebraic summary of the problem, utilizes the balanced chemical equations that define the equilibria; the second step involves the application of simple but often involved algebra. The trick in solving such problems is to make appropriate approximations that will simplify the equations and reduce the labor required for the solution. For this, an awareness of the chemistry of the problem is indispensable. By the proper formulation of the summary we often reduce the algebraic part of the problem from the solution of an involved equation to the solution of a simple one. Thus, in dealing with complex equilibria, our goal is to use chemistry to make the solution of equilibrium problems as easy as possible.

We do not discuss complex equilibria solely to give you an opportunity to develop skill in solving these problems. The solution of these problems gives you an insight into the nature of such chemical phenomena as amphoterism and complex-ion formation that would be unattainable from purely verbal descriptions; moreover, a quantitative understanding of the solubility relations for the various sulfides provides you with the basis of the theory of qualitative analysis.

ALGEBRAIC COMPLICATIONS

When we deal with simple equilibrium problems, we often find it convenient to make approximations rather than solve quadratic equations. For example, in calculations involving the dissociation of acetic acid, we assume that essentially all of the acetic acid remains undissociated and, thereby, simplify the resulting equilibrium expression. Such a simplification is possible only if there is very little dissociation, i.e., if the value of the dissociation constant is small. It should be emphasized, however, that the equilibrium constant is not always small enough that this approximation is valid; then we must solve the resulting equations without approximations. This necessity arises whenever the acid is only moderately weak and, correspondingly, the value of the ionization constant is only one or two orders below unity.

Example Problem 1: What is the hydrogen ion concentration of a $0.10M$ $NaHSO_4$ solution?

Summary:

$$\begin{array}{lcccccl} HSO_4^- & \rightleftarrows & H^+ & + & SO_4^{-2} & & K = 1.2 \times 10^{-2} \\ 0.10 & & 0 & & 0 & & \\ 0.10 - x & & x & & x & & \end{array} \qquad (a)$$

This leads to the equilibrium expression:

$$\frac{x^2}{0.10 - x} = 1.2 \times 10^{-2} \qquad (b)$$

If we solve this with the approximation that x is negligible compared with 0.10, we obtain

$$x^2 = 12 \times 10^{-4}$$

or $\qquad (c)$

$$x = 0.035$$

This value of x, however, shows that x is *not* negligible compared with the 0.10. (We should have realized this from the fact that the dissociation constant of the bisulfate ion is fairly large.) To solve this problem correctly, we must solve Eq. (b) exactly. Equation (b) can be rewritten in the form

$$x^2 + 1.2 \times 10^{-2}x - 12 \times 10^{-4} = 0 \qquad (d)$$

With the help of the quadratic formula, we write

$$x = \frac{-1.2 \times 10^{-2} \pm \sqrt{49.44 \times 10^{-4}}}{2} \qquad (e)$$

which yields for x

$$x = -4.1 \times 10^{-2}; \qquad +2.9 \times 10^{-2} \qquad (f)$$

Since x represents the concentration of hydrogen ions, a positive quantity, we can rule out the first solution shown in Eq. (f). Thus the correct value of the hydrogen-ion concentration is 2.9×10^{-2}, not 3.5×10^{-2}.

COMPETING EQUILIBRIA

Often there are several equilibria in a given solution. If the equations for these equilibria have no species in common, the presence of one has no effect on the equilibrium reached by the other except for those effects due to departure from ideal behavior. In the event that two equations contain a species in common, reactants compete for the common species. This section contains some examples of competing equilibria. In such problems, the *exact* algebraic solution often requires a graphical method. Thus, if we are to handle these problems easily, we must make approximations; to make approximations that are valid, we must carefully appraise the chemistry of the problem.

Solubility of mixtures. First consider a competitive equilibrium involving the solubility of two different solids containing a common ion. The treatment here is straightforward, provided we keep in mind that all the ions are in the same beaker, and, consequently, for the ion common to both pure crystalline solids, the concentration is the same in both expressions of the K_{sp}.

Example Problem 2: If we add 1 mole of $PbCl_2$ and 1 mole of AgCl to 1 liter of water, what will be the amount of dissolved $PbCl_2$ and AgCl? The value of the K_{sp} for $PbCl_2$ is 1.0×10^{-4}, and that for AgCl is 1.56×10^{-10}.

Let us assume that x moles of $PbCl_2$ and y moles of AgCl dissolve.

Summary:	$PbCl_2$	$\rightleftarrows$	Pb^{+2}	+	$2Cl^-$
	S		0		0
	$S - x$		x		$y + 2x$
	AgCl		Ag^+	+	Cl^-
	S		0		0
	$S - y$		y		$2x + y$

If we substitute the indicated values into the two expressions for the K_{sp}, we obtain

$$(x)(2x + y)^2 = 1.0 \times 10^{-4}$$
$$y(2x + y) = 1.56 \times 10^{-10}$$

We can, of course, solve these two equations in two unknowns exactly, but it is much easier if we make use of approximations. It is apparent from the values of the K_{sp} that $PbCl_2$ is more soluble than AgCl. Therefore the amount of chloride ion from $PbCl_2$ will be much larger than that from AgCl. In terms of the quantities in these equations, this means that y is much less than $2x$; consequently, to a first approximation, we can neglect y with respect to $2x$. If we do so, the first equation becomes

$$4x^3 = 1.0 \times 10^{-4}$$

or

$$x = 2.9 \times 10^{-2} = (Pb^{+2})$$

and the second equation yields

$$2xy = 1.56 \times 10^{-10}$$

or

$$y = 2.7 \times 10^{-9} = (Ag^+)$$

a value consistent with our assumption.

Very weak acids. In the previous discussions of acid equilibrium, we usually made the assumption that the hydrogen-ion concentration is determined solely by the dissociation of the acid. This approximation is valid only when the contribution of the dissociated acid to the hydrogen ion concentration is so great that the hydrogen ion contributed by the dissociation of water is negligible. We have already seen (p. 448) that, when we are dealing with a very dilute solution of a strong acid, we can no longer ignore the contribution of the water. When we are dealing with a dilute solution of a weak acid, the contribution of the water may again be important. In this case, however, we have to consider *two* simultaneous equilibria in order to obtain an accurate answer.

Example Problem 3: Glycine, $HC_2H_4O_2N$, is one of the amino acids that make up body proteins. The ionization constant is 1.7×10^{-10}. (For the sake of brevity in what follows, we write HG for glycine.) What is the value of (H^+) and (G^-) in a solution 10^{-4} molar in glycine?

If we ignore the dissociation of water, the summarizing statement of this problem is

$$\begin{array}{ccccc} HG & \rightleftarrows & H^+ & + & G^- \\ 10^{-4} & & 0 & & 0 \\ 10^{-4} - x & & x & & x \end{array}$$

This leads to the equilibrium expression,

$$\frac{x^2}{10^{-4} - x} = 1.7 \times 10^{-10}$$

With the usual approximation, this yields

$$x = 1.3 \times 10^{-7}$$

This concentration of hydrogen ion, however, is about the same as that in pure water; consequently, we cannot ignore the hydrogen ion from the dissociation of water. For a correct solution of this problem, we *must* consider the simultaneous dissociation of glycine and water. Then the summary is

$$\begin{array}{ccccc} HG & \rightleftarrows & H^+ & + & G^- \\ 10^{-4} & & y & & 0 \\ 10^{-4} - x & & x + y & & x \\ HOH & \rightleftarrows & H^+ & + & OH^- \\ - & & y & & y \\ & & x + y & & y \end{array}$$

In these expressions, x is the amount of HG that dissociates, and y is the amount of water that dissociates. For the equilibrium expressions we have

$$\frac{(x + y)(x)}{(10^{-4} - x)} = 1.7 \times 10^{-10} \qquad (a)$$

and

$$(x + y)(y) = 1.0 \times 10^{-14} \qquad (b)$$

Let us, in this instance, give the details of the exact algebraic solution. First, we solve (*b*) for x in terms of y; this yields

$$x = \frac{1.0 \times 10^{-14}}{y} - y \qquad (c)$$

If we substitute (*c*) into (*a*), we obtain

$$\frac{\left(\frac{1.0 \times 10^{-14}}{y}\right)\left(\frac{1.0 \times 10^{-14}}{y} - y\right)}{\left(10^{-4} + y - \frac{1.0 \times 10^{-14}}{y}\right)} = 1.7 \times 10^{-10} \qquad (d)$$

This leads to a complicated algebraic expression involving a cubic equation; fortunately, we can simplify this by assuming that the dissociation of glycine is so small that the denominator is approximately 10^{-4}. If we do so, (*d*) reduces to

$$\left(\frac{1.0 \times 10^{-14}}{y}\right)\left(\frac{1.0 \times 10^{-14}}{y} - y\right) = 1.7 \times 10^{-14} \qquad (e)$$

This can be simplified to

$$(1.0 \times 10^{-14})(1.0 \times 10^{-14} - y^2) = 1.7 \times 10^{-14}\, y^2$$

If we solve this for y, we find

$$1.0 \times 10^{-28} = 2.7 \times 10^{-14}\, y^2$$

and

$$y = 0.61 \times 10^{-7} \qquad (f)$$

Once we know y, we need only make a simple substitution. If we substitute (*f*) into (*b*), we obtain

$$x = 1.03 \times 10^{-7}$$

a value consistent with the approximation. From the values of x and y we can calculate the concentrations of (H^+) and (G^-), viz.:

$$(H^+) = 1.6 \times 10^{-7}$$
$$(G^-) = 1.0 \times 10^{-7}$$

Complex hydrolysis. Hydrolysis problems already treated have involved salts in which either the anion alone or the cation alone hydrolyzed. We have deferred until now those cases in which both the anion and the cation hydrolyze. Consider, for example, what happens in a solution of NH_4CN. Obviously, there is a tendency for the cation to hydrolyze and produce hydrogen ions; but, at the same time, there is a tendency for the anions to hydrolyze and form hydroxide ions. What occurs depends on the relative affinity of the cation for hydroxide ions and that of the anions for hydrogen ions. Differences in these affinities provide a wide variety of results that are often unexpected.

Example Problem 4: What is the hydrogen-ion concentration in $1M$ NH_4CN?
The ionization constants for NH_4OH and HCN are

$$NH_4OH \rightleftarrows NH_4^+ + OH^- \qquad K_i = 1.8 \times 10^{-5}$$

and

$$HCN \rightleftarrows H^+ + CN^- \qquad K_i = 4 \times 10^{-10}$$

Thus, in a solution of HCN, there is far less dissociation than in a solution of NH_4OH. This means that the cyanide ion has a stronger affinity for hydrogen ion than the ammonium ion has for hydroxide ion; hence, hydrolysis of NH_4CN results in a basic solution. This is qualitatively correct; but any analysis based solely on the separate equilibria is misleading because the hydroxide ion produced by the hydrolysis of the cyanide ion gives the ammonium ion a new source of hydroxide ions with which to form ammonium hydroxide. To do this problem in a precise manner, we must consider both equilibria. Let us write down a complete summary first and then explain each part of the summary.

	CN^-	+	HOH	$\rightleftarrows$	HCN	+	OH^-
1.	$1 - x$				x		x
2.	$1 - x$				x		α

	NH_4OH	$\rightleftarrows$	NH_4^+	+	OH^-
A.	0		1		x
B.	x		$1 - x$		0
C.	$x - \alpha$		$1 - x + \alpha$		α

Step 1 depicts the hydrolysis of the CN^- ion when no other reaction occurs. Since there are both ammonium and hydroxide ions present, it is in principle possible for all the hydroxide to react with the ammonium ions as depicted in step B. If this occurred, a small amount of the ammonium hydroxide, α, would dissociate to form ammonium ion and hydroxide ion as shown in step C. We now have to modify the summary involving the hydrolysis of the cyanide ion in order for the (OH^-) to be the same in both equilibria. Therefore we write the *final* equilibrium condition for the cyanide ion as shown in part 2. Thus the two equilibrium expressions are

$$\frac{(HCN)(OH^-)}{(CN^-)} = \frac{(x)(\alpha)}{(1 - x)} = 2.5 \times 10^{-5} \qquad (a)$$

and

$$\frac{(NH_4^+)(OH^-)}{(NH_4OH)} = \frac{(1 - x + \alpha)(\alpha)}{(x - \alpha)} = 1.8 \times 10^{-5} \qquad (b)$$

We now have two equations in two unknowns and can solve for both unknowns exactly. As usual, we can achieve simplification if we examine the chemistry of the problem. Equation (*a*) tells us that, as long as $1 - x$ never gets small (this is unlikely), either x or α must be very small. In (*b*), the first factor is probably close to 1. Therefore α must be less than $x - \alpha$, and, if this is true, α will have to be small compared to x. Thus it would be a reasonable guess to neglect α with respect to x.

Before we proceed further, let us note that the fact that we can probably neglect α with respect to x is not accidental. The steps involving NH_4OH dissociation could just as easily have been formulated as

NH_4OH	$\rightleftarrows$	NH_4^+	+	OH^-
0		1		x
β		$1 - \beta$		$x - \beta$

(This would also lead to changes in the CN^- summary.) We know, however, from *chemical* principles, that most of the available NH_4^+ and OH^- ion will combine to form NH_4OH; hence β would not be small compared with x and simplifying approximations could not be made. To make valid approximations, it is generally necessary to have one of the variables much smaller than the others. Clearly, in the equilibrium above, the amount of dissociated NH_4OH or, equivalently, the (OH^-) will be small; hence we formulate our problem so that (OH^-) represents one of the variables, i.e., α.

When we neglect α with respect to x, Eq. (*b*) becomes

$$\frac{\alpha(1 - x)}{x} = 1.8 \times 10^{-5} \qquad (c)$$

Now we multiply (*a*) by (*c*) to obtain

$$\frac{(x)(\alpha)}{(1 - x)} \times \frac{\alpha(1 - x)}{x} = 2.5 \times 10^{-5} \times 1.8 \times 10^{-5}$$

or

$$\alpha^2 = 4.5 \times 10^{-10}$$

and

$$\alpha = 2.1 \times 10^{-5} = (OH^-) \qquad (d)$$

From this, it follows that

$$(H^+) = 4.8 \times 10^{-10}$$

and, if we substitute (*d*) into (*c*), we obtain $x = 0.54$, a result that shows that our neglect of α with respect to x was valid.

One feature of this problem is rather surprising. If there were only hydrolysis of the cyanide ion, as in the case of a 1 molar solution of sodium cyanide, we would find that approximately 0.5% of the cyanide ion combined with a proton from the water to form molecular HCN. On the other hand, when we have a solution of ammonium cyanide, we find that nearly 54% of the cyanide is present as molecular hydrogen cyanide, and, similarly, about 54% of the ammonium ion is present as ammonium hydroxide. Therefore a solution of ammonium cyanide is actually more nearly a solution of ammonium hydroxide and hydrogen cyanide than ammonium ions and cyanide ions.

We can describe complex hydrolysis qualitatively by two successive equations of the type

$$CN^- + HOH \rightleftarrows HCN + OH^- + NH_4^+ \rightleftarrows NH_4OH$$

The results obtained in the preceding paragraphs really stem from Le Chatelier's principle. The fact that cyanide ion increases the hydroxide-ion concentration shifts the second equation to the right; the fact that the hydroxide ion produced by cyanide hydrolysis is constantly removed by the ammonium ions shifts the first equation to the right. Since both of these equilibria are shifted to the right, we can expect a rather large amount of undissociated acid and undissociated base.

SUCCESSIVE DISSOCIATION

Polyprotic acids. Some acids, such as sulfuric acid, can dissociate to form two or more hydrogen ions per molecule. Dissociation to form one hydrogen ion may be essentially complete for a strong acid, but further dissociation is usually far from complete. This is readily understandable since, when a proton is removed from the bisulfate ion, a positively charged ion is being separated from the doubly charged SO_4^{-2} ion. As might be expected, this is a more difficult step than the separation of a positively charged proton from the singly charged HSO_4^- ion.

Polyprotic acids can be weak as well as strong. For example, H_2S or H_3PO_4 can dissociate to form two and three hydrogen ions, respectively, but, in both cases, even the first dissociation occurs only to a limited degree. There are a number of such weak polyprotic acids. In discussing their equilibria in solution, we must consider the equilibria of each of the dissociation steps.

Example Problem 5: What is the concentration of hydrogen ion, bisulfide ion (HS^-), and sulfide ion in a 0.01 molar solution of H_2S? Let us write down a summary of the steps that occur. These are

	H_2S	$\rightleftarrows$	H^+	$+$	HS^-
1.	0.01		0		0
2.	$0.01 - x$		x		x
3.	$0.01 - x$		$x + y$		$x - y$
	HS^-	$\rightleftarrows$	H^+	$+$	S^{-2}
A.	x		x		0
B.	$x - y$		$x + y$		y

The first step of the dissociation of hydrogen sulfide forms a hydrogen ion and a bisulfide ion, to give the situation depicted in step 2. The bisulfide ion can also dissociate for form sulfide ion as indicated in step B, and this further dissociation will, of course, necessitate a change in the equilibrium summary for H_2S. From the final summary for each equilibrium, i.e., steps 3 and B, we can write the two equations

$$\frac{(x - y)(x + y)}{(0.01 - x)} = 1.1 \times 10^{-7} \qquad (a)$$

$$\frac{(x + y)(y)}{(x - y)} = 1.0 \times 10^{-14} \qquad (b)$$

From the value of the ionization constant, we see that the amount of dissociation of bisulfide ion is small. This means that we can simplify our equations by neglecting y wherever it appears with x in a sum or a difference. Thus (*b*) reduces to

$$y = 1.0 \times 10^{-14}$$

Similarly, the amount of dissociation of hydrogen sulfide is probably small enough that we can neglect x with respect to 0.01. With these approximations, solution of (*a*) yields

$$x = 3.3 \times 10^{-5}$$

In view of the values of x and y, it is apparent that both approximations are valid. By combining x and y as indicated in the summary, we can calculate the concentration of the three ions present. If we do, we find

$$
\begin{aligned}
(H^+) &= 3.3 \times 10^{-5} \\
(HS^-) &= 3.3 \times 10^{-5} \\
(S^{-2}) &= 1.0 \times 10^{-14}
\end{aligned}
$$

Example Problem 6: What is the concentration of hydrogen ion in a 1 molar solution of H_3PO_4? What are the concentrations of the other ions present?

Let us first write the summary.

	H_3PO_4	$\rightleftarrows$	H^+	$+$	$H_2PO_4^-$
1.	1		0		0
2.	$1 - x$		x		x
3.	$1 - x$		$x + y$		$x - y$
4.	$1 - x$		$x + y + z$		$x - y$
	$H_2PO_4^-$	$\rightleftarrows$	H^+	$+$	HPO_4^{-2}
A.	x		x		
B.	$x - y$		$y + x$		y
C.	$x - y$		$y + x + z$		$y - z$
	HPO_4^{-2}	$\rightleftarrows$	PO_4^{-3}	$+$	H^+
i.	y		0		$x + y$
ii.	$y - z$		z		$x + y + z$

The initial conditions are shown in step 1. Dissociation of H_3PO_4 (step 2) forms $H_2PO_4^-$, and this dissociates to form HPO_4^{-2} (steps B and 3); likewise, HPO^{-2} ionizes to form PO_4^{-3} (steps ii, C, and 4). Thus the combined result of these three dissociation steps are the following equilibrium expressions:

$$\frac{(x - y)(x + y + z)}{(1 - x)} = 7.5 \times 10^{-3} \tag{a}$$

$$\frac{(y - z)(x + y + z)}{(x - y)} = 6.2 \times 10^{-8} \tag{b}$$

$$\frac{(z)(x + y + z)}{(y - z)} = 4.8 \times 10^{-13} \tag{c}$$

A little consideration shows that it is likely that

$$
\begin{aligned}
1 - x &\approx 1 \\
x - y &\approx x \\
y - z &\approx y \\
x + y + z &\approx x
\end{aligned}
$$

On this basis, (a), (b), and (c) simplify to

$$x^2 = 7.5 \times 10^{-3}$$

$$\frac{yx}{x} = 6.2 \times 10^{-8}$$

$$\frac{zx}{y} = 4.8 \times 10^{-13}$$

And if we solve for x, y, and z, we find

$$x = 8.7 \times 10^{-2}$$
$$y = 6.2 \times 10^{-8}$$
$$z = 3.4 \times 10^{-19}$$

These values show that our approximations are valid. From x, y, and z, we can calculate the concentration of each of the ions in solution. This yields

$$(H^+) = 8.7 \times 10^{-2}, \quad (H_3PO_4) = 0.91, \quad (H_2PO_4^-) = 8.7 \times 10^{-2},$$
$$(HPO_4^{-2}) = 6.2 \times 10^{-8}, \quad (PO_4^{3-}) = 3.4 \times 10^{-19}$$

Note that in Examples 5 and 6 we could have computed the concentration of hydrogen ion by considering only the first step in the ionization of these acids. This is generally true provided that the second dissociation constant is far less than the first.

Problems dealing with solutions of the acid salts of polyprotic acids are often involved. Consider, for example, a solution containing sodium bicarbonate, $NaHCO_3$. In such a solution, there are sodium ions and bicarbonate ions. The latter not only dissociate to form hydrogen ions and carbonate ions, but also hydrolyze to form carbonic acid and hydroxide ions. Whether the resulting solution is acidic or basic depends on the values of the successive dissociation constants.

Example Problem 7: What is the hydrogen-ion concentration of $1M$ $NaHCO_3$?

The dissociation steps of carbonic acid are

$$H_2CO_3 \rightleftarrows H^+ + HCO_3^- \qquad K = 4.3 \times 10^{-7}$$
$$HCO_3^- \rightleftarrows H^+ + CO_3^{-2} \qquad K = 5.6 \times 10^{-11}$$

Thus the bicarbonate ion is a weak acid; moreover, carbonic acid itself is so weak that we expect some hydrolysis of the bicarbonate ion. If we compute the hydrolysis constant of the bicarbonate ion, we find that its value is 2.3×10^{-8}. Since the hydrolysis constant of the bicarbonate ion is greater than the ionization constant of the bicarbonate ion, there should be more hydrolysis than dissociation, and the resulting solution should be basic.

In general, if we know that a solution is acidic (basic), it is desirable to arrange the equilibrium expression so that it involves *only* the hydrogen (hydroxide) ion concentration. In the present case, therefore, it is best to describe this basic solution in terms of the hydrolysis equation for bicarbonate ion and carbonate ion, because then both of these will involve only the concentration of hydroxide ion. On this basis, we write the following summary:

	HCO_3^-	+	H_2O	$\rightleftarrows$	H_2CO_3	+	OH^-	$K = 2.3 \times 10^{-8}$
1.	$1 - x$				x		x	
2.	$1 - 2x + y$				x		y	
	CO_3^{-2}	+	H_2O	$\rightleftarrows$	HCO_3^-	+	OH^-	$K = 1.8 \times 10^{-4}$
A.					$1 - x$		x	
B.	x				$1 - 2x$		0	
C.	$x - y$				$1 - 2x + y$		y	

Step 1 corresponds to the hydrolysis of the bicarbonate ion. Then hydroxide ion produced in step 1 reacts with the bicarbonate ion to form carbonate ion (step B). Reaction B, however, is incomplete; some of the carbonate ion produced would hydrolyze to yield the situation depicted in step C. This, of course, requires a change (step 2) in the summary for HCO_3^- hydrolysis, and the resulting equilibrium expressions are

$$\frac{(x)(y)}{(1 - 2x + y)} = 2.3 \times 10^{-8} \qquad (a)$$

and

$$\frac{(1 - 2x + y)(y)}{(x - y)} = 1.8 \times 10^{-4} \qquad (b)$$

From what we have said about the chemistry of this problem, it seems reasonable to neglect y with respect to x. Then (*b*) becomes

$$\frac{(1 - 2x + y)(y)}{(x)} = 1.8 \times 10^{-4} \qquad (c)$$

If we multiply (*a*) by (*c*), we obtain

$$\frac{(x)(y)}{(1 - 2x + y)} \times \frac{(1 - 2x + y)(y)}{(x)} = 2.3 \times 10^{-8} \times 1.8 \times 10^{-4}$$

$$y^2 = 4.1 \times 10^{-12}$$

$$y = 2.0 \times 10^{-6} = (OH^-)$$

This yields

$$(H^+) = 5 \times 10^{-9}$$

We solve for x by noting that in line with our approximation $1 - 2x + y \approx 1 - 2x$. Then, if we substitute for y in (*c*), we obtain

$$\frac{(1 - 2x)}{(x)} 2.0 \times 10^{-6} = 1.8 \times 10^{-4}$$

or

$$x = 1.1 \times 10^{-2}$$

From these values of x and y, we see that the approximations are valid ones. When we substitute these values into the equations indicated by the summary, we obtain

$$(CO_3^{-2}) = (H_2CO_3) = 1.1 \times 10^{-2}$$

and

$$(HCO_3^-) = 0.98$$

Complex ions. The dissociation of a complex ion such as $CuCl_4^{-2}$ is often represented by

$$CuCl_4^{-2} \rightleftarrows Cu^{+2} + 4Cl^- \qquad K_d = 2.4 \times 10^{-6}$$

The simplest interpretation of this equation is that the complex ion, $CuCl_4^{-2}$, dissociates in solution to form cupric ions and chloride ions and *no* other species.

This is no more correct than to say that the dissociation of phosphoric acid can be represented by the equation

$$H_3PO_4 \rightleftarrows 3H^+ + PO_4^{-3}$$

which implies that the only ions present are hydrogen ions and phosphate ions. It *is* true that we can write an equilibrium expression for the equation and that this is equal to a constant, but it is *not* true that all of the dissociating phosphoric acid forms a phosphate anion, nor is it true that all the dissociating $CuCl_4^{-2}$ forms Cu^{+2} ion.

Like the dissociation of phosphoric acid, the dissociation of $CuCl_4$ takes place in steps. These steps can be represented by the equations

$$CuCl_4^{-2} \rightleftarrows CuCl_3^- + Cl^-$$
$$CuCl_3^- \rightleftarrows CuCl_2 + Cl^-$$
$$CuCl_2 \rightleftarrows CuCl^+ + Cl^-$$
$$CuCl^+ \rightleftarrows Cu^{+2} + Cl^-$$

These equilibria have been studied in some detail, and the individual equilibrium constants are known to be

$$\frac{(CuCl_3^-)(Cl^-)}{(CuCl_4^{-2})} = 1.9 \times 10^{-1}$$

$$\frac{(CuCl_2)(Cl^-)}{(CuCl_3^-)} = 3.3 \times 10^{-1}$$

$$\frac{(CuCl^+)(Cl^-)}{(CuCl_2)} = 2.5 \times 10^{-2}$$

$$\frac{(Cu^{+2})(Cl^-)}{(CuCl^+)} = 1.6 \times 10^{-3}$$

The similarity of these equilibria to those of a polyprotic acid is clear. In the dissociation of this complex ion, the dissociation products are a chloride anion and a copper complex, whereas, in the dissociation of a polyprotic acid, the dissociation products are protons and one of the acid anions.

Note the last two steps in this dissociation. They represent steps in the ionization of cupric chloride. Thus, when we dissolve cupric chloride, we have present not only the complex ions but also the undissociated cupric chloride and $CuCl^+$. To a greater or lesser extent, similar species are present in solutions of any salt. Thus the assumption that the only species in the solution are ions, made in the simple treatment of solubility equilibria, is an approximation; the validity of this approximation varies from one system to another. In recent years, it has become evident that the concentration of undissociated salt in solution is often comparable to the concentration of ions. This is certainly true for AgCl and $PbCl_2$; an appreciable fraction of these salts in solution is present as undissociated or partially dissociated molecules.

Example Problem 8: Calculate the concentration of the various species present in 1 liter of solution which contains 0.1 mole of cupric sulfate and 1.4 moles of sodium chloride.

This problem involves four simultaneous equilibria; therefore we must be especially careful in the formulation of the problem. We assume that the cupric and chloride ions combine completely and approach the final equilibrium by dissociation. On that basis, the summary is

	$CuCl_4^{-2}$	$\rightleftarrows$	$CuCl_3^-$	+	Cl^-
(*a*)	0.1		0		1
(*b*)	$0.1 - w$		w		$1 + w$
(*c*)	$0.1 - w$		$w - x$		$1 + w + x$
(*d*)	$0.1 - w$		$w - x$		$1 + w + x + y$
(*e*)	$0.1 - w$		$w - x$		$1 + w + x + y + z$

	$CuCl_3^-$	$\rightleftarrows$	$CuCl_2$	+	Cl^-
(*a*)					
(*b*)	w				$1 + w$
(*c*)	$w - x$		x		$1 + w + x$
(*d*)	$w - x$		$x - y$		$1 + w + x + y$
(*e*)	$w - x$		$x - y$		$1 + w + x + y + z$

	$CuCl_2$	$\rightleftarrows$	$CuCl^+$	+	Cl^-
(*a*)					
(*b*)					
(*c*)	x				$1 + w + x$
(*d*)	$x - y$		y		$1 + w + x + y$
(*e*)	$x - y$		$y - z$		$1 + w + x + y + z$

	$CuCl^+$	$\rightleftarrows$	Cu^{+2}	+	Cl^-
(*a*)					
(*b*)					
(*c*)					
(*d*)	y				$1 + w + x + y$
(*e*)	$y - z$		z		$1 + w + x + y + z$

Step *a* corresponds to the initial conditions as defined. In this artificial condition, we have present only the chloride ions and the tetrachlorocuprate ions. We assume that a small amount of the tetrachlorocuprate dissociates; this is shown in step *b*. When this occurs, we form *w* moles of the $CuCl_3^-$, and we must then consider the dissociation of the $CuCl_3^-$; this is shown in step *c*. Once this occurs, we form $CuCl_2$; this, too, will dissociate (step *d*). The dissociation of $CuCl_2$ yields $CuCl^+$, and we also must consider the dissociation of this species as indicated in part *e*. The final equilibrium conditions for each of these equations are shown in step *e;* this leads to the following expressions:

$$\frac{(w - x)(1 + w + x + y + z)}{(0.1 - w)} = 1.9 \times 10^{-1} \qquad (A)$$

$$\frac{(x - y)(1 + w + x + y + z)}{(w - x)} = 3.3 \times 10^{-1} \qquad (B)$$

$$\frac{(y - z)(1 + w + x + y + z)}{(x - y)} = 2.5 \times 10^{-2} \qquad (C)$$

$$\frac{(z)(1 + w + x + y + z)}{(y - z)} = 1.6 \times 10^{-3} \qquad (D)$$

We have four equations and four unknowns, and from the principles of algebra we *know* that we can solve for each one of these unknowns. As is usual, however, we wish to simplify the problem as much as possible by making any reasonable approximation. Let us refer to the summary. Ordinarily, we might guess that w is negligible with respect to 0.1; the dissociation this describes, however, has a rather large equilibrium constant, which, of course, means that a fair amount of the $CuCl_4^{-2}$ ion dissociates; hence, in this instance, this is not a good approximation. On the other hand, the concentration of chloride ion is initially ten times that of the $CuCl_4^{-2}$ ion; hence w might be negligible in the expression for the chloride-ion concentration. Thus, as a guess, we might make the following approximation:

$$1 + w + x + y + z \approx 1$$

As is usually the case in such an approximation, we have to check its validity after we have completed the problem. On such a basis, equations *A, B, C* and *D* reduce to

$$\frac{w - x}{0.1 - w} = 19 \times 10^{-2}$$

$$\begin{aligned} (x - y) &= 3.3 \times 10^{-1}(w - x) \\ (y - z) &= 2.5 \times 10^{-2}(x - y) \\ z &= 1.6 \times 10^{-3}(y - z) \end{aligned}$$

This leads to the following set of simultaneous equations:

$$\begin{aligned} 1.19w - x &= 0.019 \\ -0.33 \cdot w + 1.33x - y &= 0 \\ -0.025x + 1.025y - z &= 0 \\ -0.0016y + 1.0016z &= 0 \end{aligned}$$

If we apply the usual methods in order to obtain a solution to these equations, we find

$$\begin{aligned} w &= 0.020 \\ x &= 5.1 \times 10^{-3} \\ y &= 1.25 \times 10^{-4} \\ z &= 2.0 \times 10^{-7} \end{aligned}$$

These values can be translated into concentrations of the species present through the use of our summary. This yields the following set of values:

$$\begin{aligned} (Cu^{+2}) &= 2.0 \times 10^{-7} \\ (CuCl^{+}) &= 1.2 \times 10^{-4} \\ (CuCl_2) &= 5.0 \times 10^{-3} \\ (CuCl_3^{-}) &= 1.5 \times 10^{-2} \\ (CuCl_4^{-2}) &= 8.0 \times 10^{-2} \\ (Cl^{-}) &= 1.02 \end{aligned}$$

It is clear from these values that our single approximation is a valid one.

The results show that the assumption that the $CuCl_4^{-2}$ ion dissociates completely in a single step is not valid. The concentration of each of the partially dissociated complex ions is far greater than that of the cupric ion; moreover, approximately one-fifth of the complex copper exists in complexes other than $CuCl_4^{-2}$.

AMPHOTERISM

A large number of acids and bases are only sparingly soluble in water. The equilibrium depicting a saturated solution can be written

$$\text{Acid} \qquad HA(s) \rightleftarrows HA(aq.) \rightleftarrows H^+ + A^-$$
$$\text{Base} \qquad MOH(s) \rightleftarrows MOH(aq.) \rightleftarrows M^+ + OH^-$$

HA and MOH are symbols representing acids and bases in general. We can increase the amount of dissociation of HA by decreasing the hydrogen-ion concentration. This can be done by the addition of a base to the solution. Then, in accord with Le Chatelier's principle, the equilibrium shifts to the right and more of the solid acid dissolves. Similar considerations hold for a base, i.e., sparingly soluble bases dissolve more readily in acids than in pure water.

The difference between an acid and a base is a difference in degree rather than kind. Thus, in any given period, we find a gradual transition from the base-forming elements (on the left) to the acid-forming elements (on the right). In the transition region, it is difficult to predict whether an oxide forms an acid or a base in water. Elements in this area of the periodic chart often show *amphoteric* behavior; i.e., their oxides dissolve more readily in an acid *or* a base than in pure water.

A hydrated oxide that shows amphoteric behavior undergoes the reaction

$$MOH_s + OH^- \rightleftarrows MO^- + HOH$$
$$MOH_s + H^+ \rightleftarrows M^+ + HOH$$

The first equation depicts the typical reaction of an acid with a strong base; the second depicts the typical reaction of a base with a strong acid. Actually, in water, the compound MOH is either very slightly acidic or very slightly basic. Since acidity and basicity represent two extremes of a continuous transition, MOH, compared to a strong base, appears acidic and, compared to a strong base, appears basic. Accordingly, a very weak acid always "looks" like a base to an extremely strong acid and vice versa.

Amphoteric behavior can be viewed as the result of complex-ion formation with hydroxide ions. Thus $Zn(OH)_2$ is a sparingly soluble amphoteric solid that dissolves via a series of steps the first of which is

$$Zn(OH)_2(s) \rightleftarrows Zn(OH)_2(aq.) \qquad K_1 = 10^{-6} \qquad (A)$$

Molecular zinc hydroxide can further dissociate according to the equations

$$Zn(OH)_2(aq.) \rightleftarrows Zn(OH)^+ + OH^- \qquad K_2 = 10^{-7} \qquad (B)$$
$$Zn(OH)^+ \rightleftarrows Zn^{+2} + OH^- \qquad K_3 = 10^{-4} \qquad (C)$$

In addition to this, however, the zinc ion can form the complex $Zn(OH)_3^-$ and $Zn(OH)_4^{-2}$. The formation of these species can be represented by the equations

$$Zn(OH)_2(aq.) + OH^- \rightleftarrows Zn(OH)_3^- \qquad K_4 = 10^{+3} \qquad (D)$$
$$Zn(OH)_3^- + (OH^-) \rightleftarrows Zn(OH)_4^{-2} \qquad K_5 = 10 \qquad (E)$$

These equations show how amphoterism occurs. The total amount of zinc hydroxide that dissolves in a liter of water is

$$\text{Dissolved } Zn(OH)_2(s) = (Zn^{+2}) + (Zn(OH)^+) + Zn(OH)_2(aq.) + (Zn(OH)_3^-) + (Zn(OH)_4^{-2}) \qquad (F)$$

If the hydroxide concentration is very small, then, on applying Le Chatelier's principle to *B* and *C*, we find that the species Zn^{+2} and $Zn(OH)^+$ are present in high concentrations. On the other hand, the species $Zn(OH)_3^-$ and $Zn(OH)_4^{-2}$ (*D* and *E*) are present at high hydroxide concentrations. The over-all solubility of solid $Zn(OH)_2$ is the concentration of ionic zinc species plus $Zn(OH_2)$(aq.), which is fixed at 10^{-6} mole/liter. For a nearly neutral solution, the concentration of ionic species is low and the solubility is about 10^{-6} mole/liter. In an acid solution, the concentration of cationic species increases and the over-all solubility increases above 10^{-6} mole/liter; in a basic solution, the concentration of anionic species increases and the solubility again increases above 10^{-6} mole/liter. Thus the solubility of $Zn(OH)_2$ is greater in an acidic or basic solution than it is in neutral water, and $Zn(OH)_2$ is amphoteric.

Example Problem 9: Calculate the solubility of solid zinc hydroxide at a pH of 5, 9, and 13.

We first solve the equilibrium expression for each of the five species in solution. According to reaction *A*, the concentration of molecular zinc hydroxide is always the same provided it is in contact with solid zinc hydroxide; thus we write

$$(Zn(OH)_2(aq.)) = 10^{-6} \qquad (a)$$

In terms of reaction *B*, the concentration of singly dissociated zinc hydroxide is

$$(Zn(OH)^+) = \frac{(Zn(OH)_2(aq.))}{(OH^-)} \times 10^{-7}$$

With the help of Eq. (*a*), we write

$$(Zn(OH)^+) = \frac{10^{-13}}{(OH^-)} \qquad (b)$$

Similarly, according to reaction *C*, the concentration of zinc ion is

$$(Zn^{+2}) = \frac{(Zn(OH)^+)}{(OH^-)}10^{-4}$$

This simplifies with the help of (*b*) to

$$(Zn^{+2}) = \frac{10^{-17}}{(OH^-)^2} \qquad (c)$$

The other anion concentrations can be calculated in a similar fashion. Thus we find, from reaction *D*, that the concentration of $Zn(OH)_3^-$ is

$$(Zn(OH)_3^-) = 10^{+3}(Zn(OH)_2(aq.))(OH^-)$$
$$(Zn(OH)_3^-) = 10^{-3}(OH^-) \qquad (d)$$

The concentration of $Zn(OH)_4^{-2}$ can be calculated from reaction *E*:

$$(Zn(OH)_4^{-2}) = 10(Zn(OH)_3^-(OH^-)) = 10^{-2}(OH^-)^2 \qquad (e)$$

We can combine Eqs. *a*, *b*, *c*, *d*, and *e* in order to obtain an expression for the total amount of dissolved zinc hydroxide. This yields

$$\text{Dissolved } Zn(OH)_2 = (Zn^{+2}) + (Zn(OH)^+) + (Zn(OH)_2(aq.)) + (Zn(OH)_3^-)$$
$$+ (Zn(OH)_4^{-2}) = \frac{10^{-17}}{(OH^-)^2} + \frac{10^{-13}}{(OH^-)} + 10^{-6} + 10^{-3}(OH^-) + 10^{-2}(OH^-)^2$$

From this general expression, we can calculate the solubility at various pH's:

$pH = 13,\ (OH^-) = 0.10$:
$\text{Dissolved } Zn(OH)_2 = 10^{-15} + 10^{-12} + 10^{-6} + 10^{-4} + 10^{-4} \approx 2 \times 10^{-4}$ mole/liter

$pH = 9,\ (OH^-) = 10^{-5}$:
$\text{Dissolved } Zn(OH)_2 = 10^{-7} + 10^{-8} + 10^{-6} + 10^{-8} + 10^{-12} \approx 10^{-6}$ mole/liter

$pH = 5,\ (OH^-) = 10^{-9}$:
$\text{Dissolved } Zn(OH)_2 = 10 + 10^{-4} + 10^{-6} + 10^{-12} + 10^{-20} \approx 10$ moles/liter

Thus, at low pH corresponding to an acidic solution, the solubility is high. With increasing pH this solubility decreases to a rather low value, but, when the solution is made strongly basic (pH = 13), the solubility again increases. The behavior of the zinc hydroxide as a function of pH affords a typical example of amphoteric behavior.

THE EFFECT OF pH ON SOLUBILITY

General considerations. The solubility of sparingly soluble salts in which the anion is that of a weak acid represents one more important application of the principles of complex equilibrium. The solubility equilibrium for the sulfide MS involves the equation

$$MS \rightleftarrows M^{+2} + S^{-2}$$

However, since H_2S is a weak acid, we also have to consider the equations that govern the equilibrium in aqueous solutions between S^{-2}, HS^-, and H_2S; that is,

$$H_2S \rightleftarrows H^+ + HS^-, \qquad \frac{(HS^-)(H^+)}{(H_2S)} = 1.1 \times 10^{-7}$$

$$HS^- \rightleftarrows H^+ + S^{-2}, \qquad \frac{(S^{-2})(H^+)}{(HS^-)} = 1.0 \times 10^{-14}$$

There are two types of problems that we often encounter. The first deals with the calculation of the pH necessary to dissolve completely the salt MS in a solution saturated with gaseous H_2S. For the purposes of such calculations, we first multiply together the two equilibrium expressions in order to get

$$\frac{(H^+)^2(S^{-2})}{(H_2S)} = 1.1 \times 10^{-21} \approx 1.0 \times 10^{-21}$$

From this, it follows that, if we know the (H^+) and (H_2S), we can calculate (S^{-2}), and, if we know (S^{-2}), we can determine whether the salt MS will dissolve. The second type of problem involves the calculation of the solubility of MS in a buffered solution that contains no H_2S.

Example Problem 10: A solution containing both zinc ion and manganous ion at a concentration of 0.010 molar is saturated with H_2S. (A saturated solution of H_2S is 0.10 molar.) If the K_{sp} of zinc sulfide is $\sim 1.0 \times 10^{-22}$ and the K_{sp} of manganous sulfide is $\sim 1.0 \times 10^{-16}$, what is the pH at which the manganous sulfide will just form a precipitate? Under these conditions, what will be the concentration of zinc ion remaining?

Assume that at the start the $(S^{-2}) = 0$ so that there is no solid manganous sulfide present. As we gradually increase the sulfide-ion concentration, we eventually reach the point at which manganous sulfide precipitates. When we *just* precipitate manganous sulfide, we have an equilibrium of the type

$$MnS_s \rightleftarrows Mn^{+2} + S^{-2}$$

and

$$(Mn^{+2})(S^{-2}) = 1.0 \times 10^{-16}$$

Thus the minimum S^{-2} ion concentration for the *start* of the precipitation is that which satisfies the K_{sp} with $(Mn^{+2}) = 0.010$. Therefore we write

$$(Mn^{+2})(S^{-2}) = 1.0 \times 10^{-16} = (0.010)(S^{-2}).$$

From this, we find that this minimum of concentration of sulfide ion must be

$$(S^{-2}) = \frac{1.0 \times 10^{-16}}{0.010} = 1.0 \times 10^{-14}$$

If we substitute the required values of the (S^{-2}) and the H_2S concentration into the H_2S equilibrium expression, we find

$$\frac{(H^+)^2(S^{-2})}{(H_2S)} = \frac{(H^+)^2(1.0 \times 10^{-14})}{(0.10)} = 1.1 \times 10^{-21}$$

or

$$(H^+)^2 10^{-13} = 1.1 \times 10^{-21}$$

and

$$(H^+) = 1.1 \times 10^{-4} \quad \text{or} \quad pH = 4.0$$

We must also calculate that the concentration of Zn^{+2} remaining in the solution; this is 10^{-8} when the $(S^{-2}) = 10^{-14}$. Thus, by properly adjusting the hydrogen-ion concentration in the solution, we can precipitate effectively all of the zinc from this solution without precipitating any manganous ion.

Example Problem 11: What is the solubility of manganous sulfide in a solution buffered to a pH of 7? The buffer contains no sulfide ions.

In this case, since the solution is *buffered* to a pH of 7, the hydrogen-ion concentration will always be 10^{-7}. Moreover, since there are no sulfide ions in this solution initially, we can write

	MnS	$\rightleftarrows$	Mn^{+2}	+	S^{-2}
(*a*)	s		0		0
(*b*)	$s - x$		x		x
(*c*)	$s - x$		x		y

In this summary, (*a*) corresponds to the initial conditions; (*b*) corresponds to what the situation would be if we had no reaction of sulfide ion with hydrogen ion; and (*c*) summarizes what will actually happen if there are hydrogen ions present and if some of the sulfide produced from MnS forms HS^- and S^{-2}. In this particular case, for each molecule of H_2S or HS^- ion or S^{-2} that appears, we must have formed one manganous ion. Because of this, we can write

$$x = (S^{-2}) + (HS^-) + (H_2S)$$

For these ions the following equilibrium is established:

$$\frac{(H^+)(HS^-)}{(H_2S)} = 1.1 \times 10^{-7}$$

and

$$\frac{(H^+)\,(S^{-2})}{(HS^-)} = 1.0 \times 10^{-14}$$

Since the hydrogen-ion concentration is 10^{-7}, we find, after substitution,

$$(HS^-) = 1.1(H_2S)$$
$$10^{+7}(S^{-2}) = (HS^-)$$

Now the relation between x and the concentration of the sulfide, bisulfide, and H_2S in solution becomes

$$x = y + 10^7y + 0.9 \times 10^7y$$

or

$$x = 1.9 \times 10^7y$$

In spite of the complexity, the K_{sp} expression still holds; that is,

$$(Mn^{+2})(S^{-2}) = 5.6 \times 10^{-16} = (x)(y)$$

If we substitute the values for the concentration of sulfide ion, y, in terms of x, we obtain

$$\frac{x^2}{1.9 \times 10^7} = 5.6 \times 10^{-16} \quad \text{or} \quad x = 10.3 \times 10^{-5}$$

Note that the amount that dissolves in this buffered neutral solution is about 5000 times as great as the amount which would dissolve if there were no reaction of sulfide ion with hydrogen ion.

Applications in qualitative analysis. Qualitative analysis is that area of analytical chemistry which involves the identification of the ions present in an aqueous solution or a solid. In such studies, we are faced with the problem of proving the presence or absence of a number (usually about thirty) of cations. Since many of these cations are similar, it is difficult to obtain a test that is specific to only one; therefore, the unknown mixture is first separated into five or six groups, each of which may contain a half-dozen cations. Further separations are effected within each group until it is possible to perform a unique test for each of the ions that might be present. Thus one of the problems of qualitative analysis is to provide the basis for this group separation.

In some cases, it is possible to perform the group separation on the basis of solubility class. For example, the Group I cations, which include silver, mercurous, and lead ions, are separated from the others by the formation of an insoluble chloride. But the cations found in Groups II and III fall into the same solubility class in so far as they all form soluble chlorides and insoluble sulfides. These two groups are separated by making use of the effect of pH on the solubility.

In Table 1, we have listed the solubility products of the sulfides of the ions that make up Groups I, II, and III. In order to illustrate concretely the implication of this wide range of K_{sp} values, let us calculate the minimum concentration

Table 1 Solubility Products of Sulfides

	Compound	K_{sp}
Group I	Ag_2S	1.0×10^{-51}
	Hg_2S	1.0×10^{-45}
	PbS	1.0×10^{-29}
Group II	CuS	4×10^{-38}
	HgS	3×10^{-53}
	SnS	8×10^{-29}
	SnS_2	—
	PbS	1.0×10^{-29}
	Sb_2S_3	10^{-80}
	Bi_2S_3	1.6×10^{-72}
Group III	ZnS	$10^{-20}(4.5 \times 10^{-24})$
	Cr_2S_3	—
	MnS	5.6×10^{-16}
	FeS	1.0×10^{-19}
	Fe_2S_3	10^{-44}
	CoS	$7 \times 10^{-23}(1.9 \times 10^{-27})$
	NiS	$3 \times 10^{-21}(1.9 \times 10^{-27})$

of sulfide ion required to precipitate the ions of Group I, II, and III if they are present in solution at a concentration of 0.01 mole per liter.

Consider the precipitation of cupric sulfide. In a solution in contact with solid cupric sulfide, we have the equilibrium

$$CuS \rightleftarrows Cu^{+2} + S^{-2}$$

and

$$(Cu^{+2})(S^{-2}) = 4 \times 10^{-38}$$

If the cupric ion in solution is at a concentration of 0.01 molar and we gradually increase the sulfide concentration, the K_{sp} relation will not be fulfilled until the solid precipitates. Before precipitation, the product of the cupric-ion and the sulfide-ion concentration will be less than the K_{sp}; once we have formed the solid, the amount of cupric ion *added* times the amount of sulfide ion *added* must be *greater* than the K_{sp}. (The process of forming a precipitate will reduce both of these concentrations so that their products satisfy the K_{sp} expression.) This, then, is the *criterion for precipitation: the solubility product must be exceeded.* Therefore the minimum concentration of sulfide ion necessary to start the precipitation of cupric sulfide is

$$(0.01)(S^{-2}) = 4 \times 10^{-38}$$
$$(S^{-2}) = 4 \times 10^{-36}$$

In an analogous fashion, we have calculated the minimum concentration of sulfide ion required to precipitate the other cations, and listed the results in Table 2.

In Table 2, there is also a column of figures for the maximum hydrogen-ion concentration at which the solid forms in a solution saturated with H_2S. These values were computed on this basis: In a solution containing dissolved H_2S, the following expression must hold:

$$\frac{(H^+)^2(S^{-2})}{(H_2S)} = 1.1 \times 10^{-21}$$

In an aqueous solution that is in equilibrium with H_2S gas at 1 atm, $(H_2S) = 0.1$; hence we can say that, for a saturated solution of H_2S, the following relations must hold:

$$(H^+)^2(S^{-2}) = 1.1 \times 10^{-22}$$
$$(H^+) = \sqrt{\frac{1.1 \times 10^{-22}}{S^{-2}}}$$

Thus, for a saturated solution of H_2S, the sulfide-ion concentration is determined by the hydrogen-ion concentration, and, by adjusting the hydrogen-ion concentration, we can adjust the sulfide-ion concentration to the desired value. If we use the values for the minimum sulfide concentration indicated in Table 2, we can

Table 2 Solubility of Sulfides

Solid	Min. (S^{-2})	Max. (H^+)	pH
HgS	3×10^{-51}	2×10^{14}	-14.3
CuS	4×10^{-36}	5×10^{6}	-6.7
PbS	1.0×10^{-27}	3×10^{2}	-2.5
SnS	8×10^{-27}	1×10^{2}	-2.0
Sb_2S_3	$\sim 10^{-25}$	3×10	-1.5
Bi_2S_3	2.5×10^{-23}	2	-0.3
CoS	7×10^{-21}	0.13	0.9
NiS	3×10^{-19}	2×10^{-2}	1.7
ZnS	10^{-18}	1×10^{-2}	2.0
FeS	10^{-17}	3×10^{-3}	2.5
MnS	5.6×10^{-14}	4×10^{-5}	4.4

calculate the maximum hydrogen-ion concentration that yields sufficient sulfide ion to cause precipitation for each case. The values thus calculated are listed in the third column of Table 2.

The data in Table 2 show how we effect a separation of Groups II and III. If we saturate a solution with H_2S *after* the hydrogen-ion concentration has been buffered to between 0.13 and 2, only cations in Group II precipitate as the sulfides. The solid precipitate can be separated from the rest of the solution by filtration or centrifugation. Then individual separation within Group II sulfides can be made and specific analyses for the individual ions can be carried out. The Group III cations in solution can be separated from those cations that do not form insoluble sulfides by making the solution less acidic. Under these conditions, all of the sulfides listed in Table 2 under Group III precipitate. (In addition, the insoluble hydroxides of chromium and aluminum ions, also included in Group III, precipitate.) Subsequently, this precipitate can be redissolved and further separated for analysis of individual cations.

It is clear from Table 2 that ions such as Ag^+ and Hg^{+2} will precipitate with the Group II ions if they are present. For this reason, the unknown solution is first treated with HCl to precipitate and remove the Group I cations. The subsequent analyses are designed on the assumption that these cations are removed; if they are present, they may interfere. Therefore, in testing an unknown, it is essential that the procedure be carried out in the order indicated.

PROBLEMS

The data on acid and base ionization constants refer to Table 4, p. 445, and Table 5, p. 447.

1. Calculate the pH of $0.2M$ H_2SO_4.

2. The solubility product for $AgNO_2$ is 6×10^{-4}. Calculate the solubility in moles per liter of $AgNO_2$ in distilled water and $0.01M$ $NaNO_2$.

3. Calculate the pH of $0.01M$ HClAc.

4. Calculate the (Ag^+) and (Hg_2^{+2}) when 1 mole of $AgNO_3$ and 1 mole of $Hg_2(NO_3)_2$ are added to a liter of solution containing 3 moles of NaCl. K_{sp} AgCl $= 1.6 \times 10^{-10}$; K_{sp} $Hg_2Cl_2 = 2 \times 10^{-18}$.

Ans. $(Ag^+) = 1.3 \times 10^{-5}$ mole/liter; $(Hg_2^{+2}) = 1.2 \times 10^{-8}$ mole/liter.

5. One mole of $AgNO_3$ is added to a liter of solution containing 1 mole of sodium acetate and 1 mole of sodium nitrite. Given

$$K_{sp}\ AgNO_2 = 6 \times 10^{-4}$$
$$K_{sp}\ AgAc = 2 \times 10^{-3}$$

Calculate the composition of the precipitate.

6. Calculate the pH of $10^{-3}\ M$ HCN.

7. Calculate the pH of $10^{-3}\ M$ NH_2OH. (K_{base} $NH_2OH = 1.0 \times 10^{-8}$.)

★ **8.** Calculate the pH of the following solutions:

(*a*) $0.1M$ NH_2CN *Ans.* pH = 7.7.
(*b*) $0.2M$ NH_4Ac
(*c*) $0.2M$ N_2H_5Ac
(*d*) $0.1M$ $NH_4(ClAc)$

9. Calculate the concentration of NH_2OH and HCN in Problem 8*a*.

Ans. $(NH_2OH) = (HCN) = 0.098$ mole/liter.

10. Indicate whether the following solutions are acid, basic, or neutral:

(*a*) NaAc
(*b*) NH_4Ac
(*c*) NH_2Ac
(*d*) $NaHSO_4$
(*e*) $NH_4(HSO_4)$
(*f*) NH_4ClAc
(*g*) $(N_2H_5)_2CO_3$
(*h*) $(NH_4)(HS)$
(*i*) $CaCO_3$
(*j*) $Li(NO_3)$

11. Calculate the pH of the following solutions:

(*a*) $1M$ H_2CO_3
(*b*) $2M$ H_2SO_3 $K_I = 1.7 \times 10^{-2}$, $K_{II} = 6.0 \times 10^{-8}$

★ **12.** Fumaric acid is a diprotic acid with the formula $H_2(C_4H_2O_4)$. The ionization constants are $K_I = 0.9 \times 10^{-4}$ and $K_{II} = 4 \times 10^{-5}$. Calculate the pH of a 1 molar solution of fumaric acid.

13. Calculate the pH of $1M$ NaH_2PO_4. *Ans.* pH = 4.67.

★ **14.** Calculate the pH of $1M$ NH_4HCO_3.

15. Calculate all the species present in 1 liter of solution containing 0.2 mole of cupric nitrate and 1.8 moles of potassium chloride.

16. Calculate the amount of $Zn(OH)_2(s)$ that dissolves in solutions buffered to a pH of 4, 6, 8, 10, 12, 14, and 15.

17. Calculate the concentration of Zn^{+2} and $Zn(OH)_4^{-2}$ for each pH in Problem 16.

18. Calculate the solubility of AgAc ($K_{sp} = 2 \times 10^{-3}$) in a solution buffered to a pH of (*a*) 3, (*b*) 5, (*c*) 9. *Ans.* (*a*) 0.34 mole/liter.

19. The K_{sp} of $CaCO_3$ is 5×10^{-9}. Calculate the solubility of $CaCO_3$ in a solution buffered to a pH of (*a*) 11, (*b*) 6, (*c*) 3. *Ans.* (*a*) 7.7×10^{-5} mole/liter.

20. A solution is 0.01 molar in $AgNO_3$. HCl is added until the chloride-ion concentration is 0.1 molar. Then the solution is filtered and the supernatant solution is buffered to a pH of 0 and saturated with H_2S. How many milligrams of Ag_2S precipitate from 100 ml of the solution?

$$K_{sp}\ AgCl = 1.6 \times 10^{-10}$$
$$K_{sp}\ Ag_2S = 1.0 \times 10^{-51}$$

21. The solubility product of $Fe(OH)_2$ is 1.64×10^{-14}; the solubility product of FeS is 1.0×10^{-19}. A solution containing $0.01M$ $Fe(NO_3)_2$ is buffered to a pH of 5 and then saturated with H_2S. What is the composition of the solid formed?

★ **22.** Calculate the solubility of MnS in: (*a*) a solution saturated with H_2S buffered to a pH of 5; (*b*) a solution buffered to a pH of 8; (*c*) unbuffered pure water.

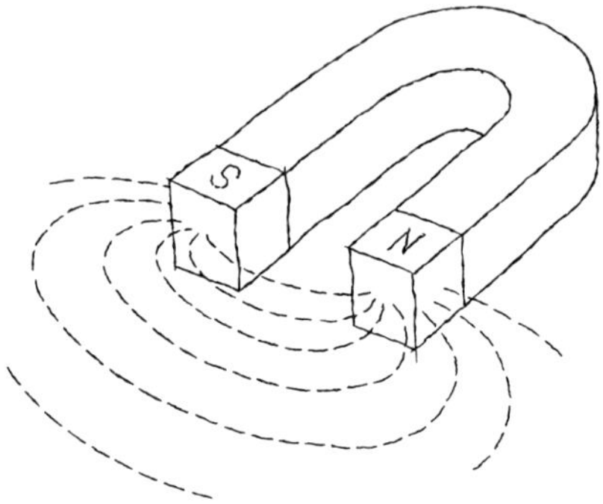

23

The transition elements

GENERAL NATURE

The *transition elements* occupy the places in the periodic table (Table 1) lying between group II on the left and group III on the right side. Chemically these transition elements have much in common; differences are determined by the nature of the *underlying* shell of the electrons rather than by the *outer* valence electrons. Proceeding from left to right across the periodic table in each row, the transition elements occupy the places where the underlying shell is being filled stepwise while the number of valence electrons remains at the constant value of 2, roughly speaking. In rows 4, 5, 6, the ten places in the *d* orbitals are being filled. In the elements lying between La and Lu it is the fourteen places

Table 1 The Transition Elements (Enclosed in Box)

Group →		I	II									IB[a]	IIB[a]	III	IV	V	VI	VII	VIII
Row	1	H																	He
↓	2	Li	Be											B	C	N	O	F	Ne
	3	Na	Mg											Al	Si	P	S	Cl	Ar
	4	K	Ca	Sc	Ti	V	Cr	Mn	Fe	Co	Ni	Cu	Zn	Ga	Ge	As	Se	Br	Kr
	5	Rb	Sr	Y	Zr	Nb	Mo	Tc	Ru	Rh	Pd	Ag	Cd	In	Sn	Sb	Te	I	Xe
	6	Cs	Ba	*	Hf	Ta	W	Re	Os	Ir	Pt	Au	Hg	Tl	Pb	Bi	Po	At	Rn
	7	Fr	Ra	**															
L		*		La	Ce	Pr	Nd	Pm	Sm	Eu	Gd	Tb	Dy	Ho	Er	Tm	Yb	Lu	
A		**		Ac	Th	Pa	U	Np	Pu	Am	Cm	Bk	Cf	Es	Fm	Md	No		

[a] Groups IB and IIB are discussed in Chapter 19.

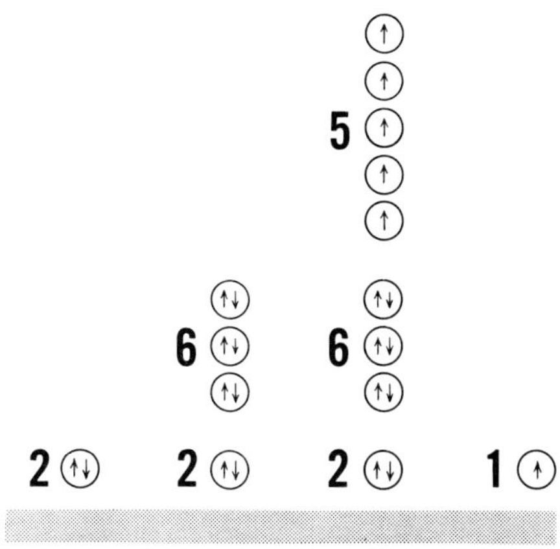

Fig. 1. Chromium atomic orbitals showing unpaired electrons.

in the *f* orbitals that fill up, and these elements are generally classified in a special subdivision called the *rare earths* or the *lanthanides* after the element lanthanum that initiates them. The elements in the last row of the table have overlapping energy levels, so that the filling takes place in a less regular manner oscillating between the *d* orbitals and the *f* orbitals. This series is called the *actinides* after the element actinium at their head. The nature of the transition filling may be seen from the electronic structure diagrams in Table 2 and in the Appendix. The order of the filling depends on the energy levels of the various orbitals as explained in Chapter 4. This will be reviewed in more detail as the elements are discussed individually.

All these elements are metals, without exception; and this property is due to the *s* electrons which move about and conduct electricity. However, there is also the possibility of covalent binding between the *d* electrons in neighboring atoms; and most of these elements are *hard* metals because of covalence.

Nearly all of these elements form compounds which are paramagnetic, indicating that not all of the electrons are paired. As an example, see the diagram for chromium in Fig. 1. When combined with other elements, the transition elements display many oxidation states because there are so many different possible electron arrangements that have about the same energy. This results in many colored compounds; because a photon of light in the visible region, which is relatively low in energy, can shift an electron to another energy level, the light is absorbed and this leaves the transmitted or reflected light colored. There is a strong tendency to form complex ions that accounts for much of the solution chemistry.

In Fig. 2 are plotted the melting points and boiling points for the transition metals. As the number of *d* electrons increases, we expect the number of

Table 2 Electronic Structure of the Principal Transition Elements

1 6 6 2 2 2 2 Sc	2 6 6 2 2 2 2 Ti	3 6 6 2 2 2 2 V	5 6 6 2 2 2 1 Cr	5 6 6 2 2 2 2 Mn	6 6 6 2 2 2 2 Fe	7 6 6 2 2 2 2 Co	8 6 6 2 2 2 2 Ni
10 1 6 6 6 2 2 2 2 2 Y	10 2 6 6 6 2 2 2 2 2 Zr	10 4 6 6 6 2 2 2 2 1 Nb	10 5 6 6 6 2 2 2 2 1 Mo	10 6 6 6 6 2 2 2 2 1 Tc	10 7 6 6 6 2 2 2 2 1 Ru	10 8 6 6 6 2 2 2 2 1 Rh	10 10 6 6 6 2 2 2 2 0 Pd
10 10 1 6 6 6 6 2 2 2 2 2 2 La[a]	14 10 10 2 6 6 6 6 2 2 2 2 2 2 Hf	14 10 10 3 6 6 6 6 2 2 2 2 2 2 Ta	14 10 10 4 6 6 6 6 2 2 2 2 2 2 W	14 10 10 5 6 6 6 6 2 2 2 2 2 2 Re	14 10 10 6 6 6 6 6 2 2 2 2 2 2 Os	14 10 10 9 6 6 6 6 2 2 2 2 2 0 Ir	14 10 10 9 6 6 6 6 2 2 2 2 2 1 Pt

[a] Between La and Hf the fourteen electrons fill in the f orbitals in the lanthanide elements.

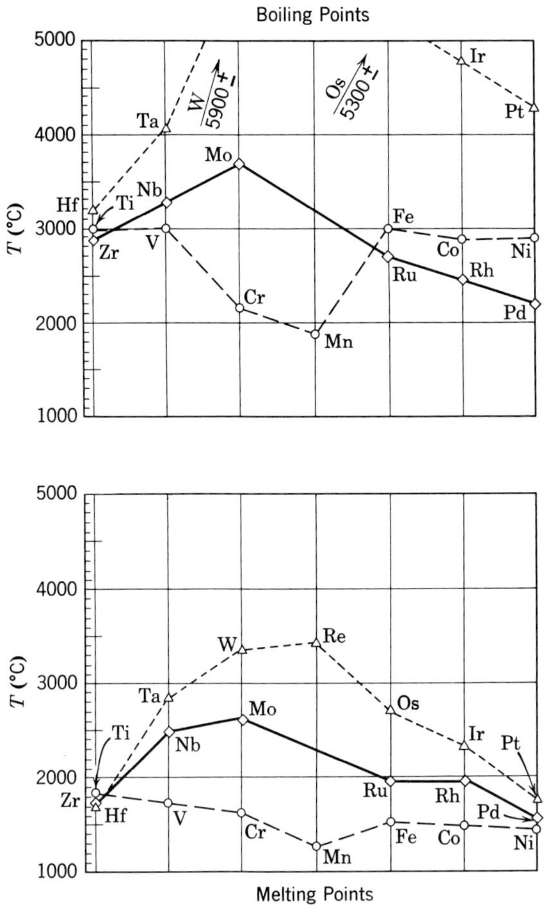

Fig. 2. Boiling points and melting points of transition elements.

covalent bonds between atoms to increase up to the Cr-Mo-W family where each of the *d* orbitals has only one electron and the opportunity for covalent sharing is greatest. Beyond this family, some of the *d* orbitals have two electrons and are prevented from covalent sharing. So we might expect melting points and boiling points to rise until we reach Cr, Mo, and W, and then fall again. This is observed for the second and third transition series.

Increasing nuclear charge reduces the covalent radius; the radii fall sharply as we go from the Sc family to the Cr family. Beyond this point, the radii remain constant, as shown in Fig. 3.

The 4*s* electrons are not affected substantially by these changes in the number of electrons in the *d* levels. As shown in Fig. 4, the ionization potentials for the removal of the first (I) and the second (II) of the 4*s* electrons changes 10–30%

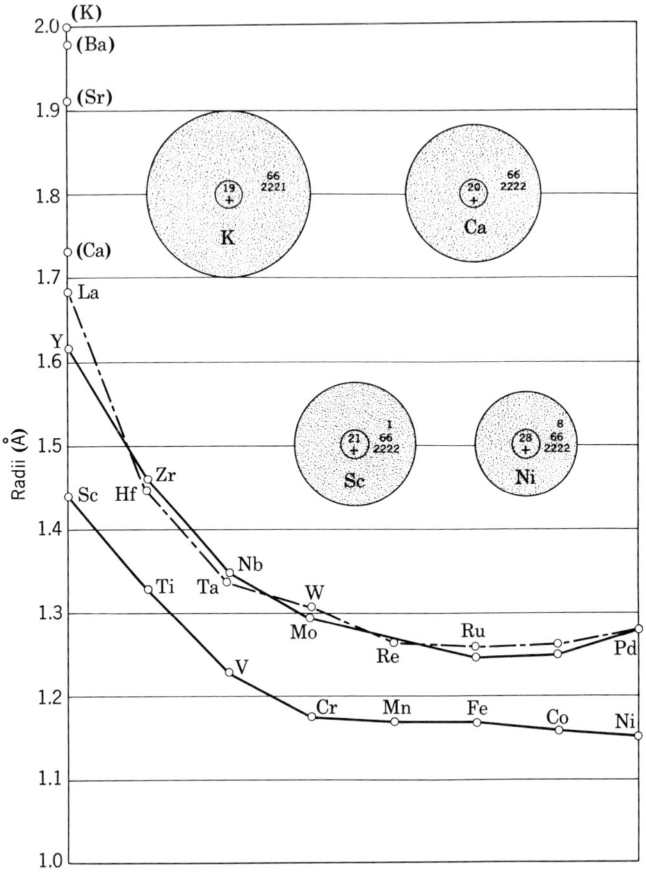

Fig. 3. Covalent radii of transition elements.

from Sc-Y-La to Ni-Pd-Pt. On the other hand, the ionization potentials of the third period elements change threefold from Na to Cl (p. 111).

We now consider the chemistry of some of these different families of transition elements in more detail. These families will be referred to as subgroups and will be designated by the name of the element that occurs at the top of the appropriate column in the fourth row of Table 1. The more important compounds are listed in subsequent tables and arranged by oxidation states.

THE SCANDIUM SUBGROUP

The electron structure diagrams for the three primary members of the scandium subgroup are shown in the left-hand column of Table 2. Like the alkali elements and the alkaline earth elements, these atoms donate all electrons outside the rare gas core and, therefore, occur only in the +3 oxidation state in their compounds. If we include not only scandium, yttrium, lanthanum, and actinium, but also all the lanthanides and actinides as well, we have quite a large family for this subgroup. All these 30 elements are metallic, showing good electrical conductivity and exhibiting a high reflectivity to light. Chemically, they are all reactive and have oxidation potentials in the neighborhood of 2 volts. Their compounds show low solubility in general; this includes phosphates, hydroxides, and carbonates. The +3 oxidation state is strongly favored. In other words, both the two *s* electrons and the one *d* electron can be donated readily. The ions have a tendency to hydrolyze, and the solutions are slightly acid. In most respects, their chemistry is like that of aluminum. The more important compounds are shown in Table 3.

Scandium is found in the minerals *gadolinite* (a silicate) and *monazite* (a phosphate). For both this element and its neighbors, yttrium and lanthanum, only

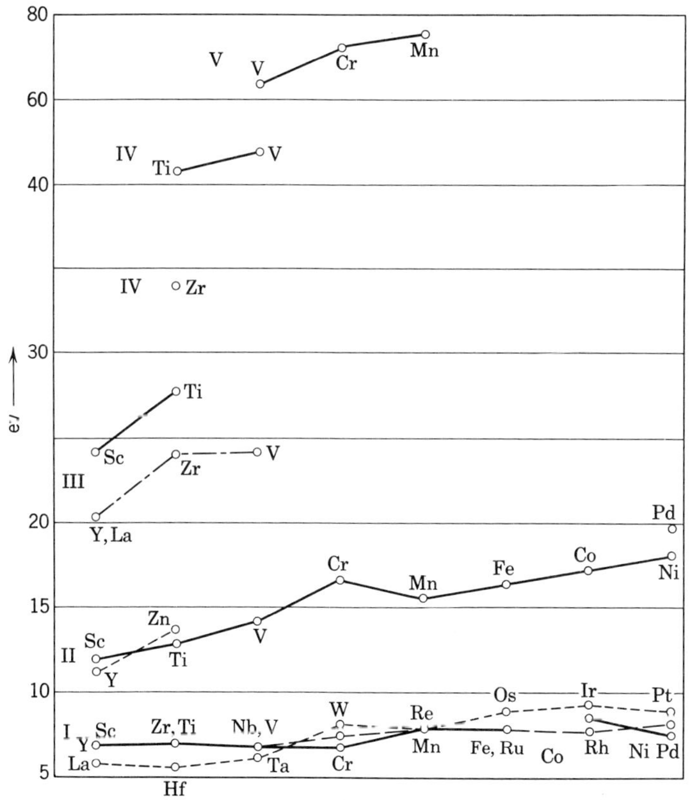

Fig. 4. Ionization potentials of transition elements.

Table 3 Oxidation States and Compounds

	Sc	Ti	V	Cr
+6				Cr^{+6}(0) CrO_4^{-2} $Cr_2O_7^{-2}$ CrO_3 CrO_2Cl_2 K_2CrO_4 $PbCrO_4$ $K_2Cr_2O_7$
+5			V^{+5}(0) V_2O_5 V_2S_5 VF_5 VOF_3 NH_4VO_3 HVO_3	
+4		Ti^{+4}(0) TiO^{+2} TiO_2 TiS_2 $TiO(OH)_2$ $TiO(SO_4)$ $TiCl_4$ TiC	V^{+4}(1) blue VCl_4 $VOCl_2$ VC $VOSO_4$	Cr^{+4}(2) CrO_2
+3	Sc^{+3}(0) Sc_2O_3 Sc_2S_3 ScH_3 $Sc(OH)_3$ $ScCl_3$, etc. ScN $Sc_2(CO_2)_3$	Ti^{+3}(1) purple Ti_2O_3 Ti_2S_3 $TiCl_3$ $TiBr_3$, etc. TiN TiP	V^{+3}(2) green V_2O_3 V_2S_3 VCl_3 VOBr VN $KV(SO_4)_2$	Cr^{+3}(3) violet Cr_2O_3 $Cr(OH)_3$ $CrCl_3$ CrN CrP CrAs $Cr(NO_3)_3$ $KCr(SO_4)_2$
+2		Ti^{+2}(2) black TiO TiS $TiCl_2$ $TiBr_2$, etc.	V^{+2}(3) violet VO VS VCl_2 VSO_4	Cr^{+2}(4) blue CrO CrS $Cr(OH)_2$ $CrCl_2$ $Cr(H_2O)_6^{+2}$

Note: The number in parentheses, e.g., Ti^{+3}(1), indicates the number of unpaired electrons. Where the color of an ion is given, it is the color in aqueous solution. Not all the ions listed in the table are stable in aqueous solution.

the +3 oxidation state exists and the compounds are not paramagnetic and are colorless. Yttrium is also found in gadolinite.

THE LANTHANIDES

The fourteen elements that follow lanthanum were formerly called the rare earths; today they are called the *lanthanides.* The richest source of these elements is the mineral *monazite.* These fourteen elements resemble each other

Table 3 *(Continued)*

	Y	La	Zr	Hf	Nb	Ta	Mo	W
+8							MoS_4 $MoOF_6$	
+7								
+6							$Mo^{+6}(0)$ MoO_3 MoS_3 MoF_6 MoO_2Br_2 $MoOCl_4$	$W^{+6}(0)$ WO_3 WS_3 WF_6 WCl_6 WAs_2 H_2WO_4
+5					$Nb^{+5}(0)$ Nb_2O_5 NbF_5 $NbCl_5$ $NbOCl_3$	$Ta^{+5}(0)$ Ta_2O_5 TaF_5 $TaCl_5$	$Mo^{+5}(1)$ Mo_2O_5 $MoO(OH)_3$ $MoCl_5$ $MoOCl_3$	$W^{+5}(1)$ WCl_5
+4			$Zr^{+4}(0)$ ZrO_2 $ZrOS$ ZrS_2 $ZrOCl_2$, etc. $ZrCl_4$ $Zr(SO_4)_2$	$Hf^{+4}(0)$ HfO_2 $HfOCl_2$ HfC	$Nb^{+4}(1)$ NbO_2	$Ta^{+4}(1)$ TaO_2 TaS_2 TaC	$Mo^{+4}(2)$ MoS_2 $MoCl_4$	$W^{+4}(2)$ WO_2 WS_2 WCl_4 WC
+3	$Y^{+3}(0)$ Y_2O_3 Y_2S_3 YH_3 $Y(OH)_3$ YCl_3, etc. YN	$La^{+3}(0)$ La_2O_3 LaH_3 $La(OH)_3$, etc. $LaCl_3$ LaN	$Zr^{+3}(1)$ $ZrCl_3$, etc. ZrN		$Nb^{+3}(2)$ Nb_2O_3 NbN	$Ta^{+3}(2)$ TaN	$Mo^{+3}(3)$ Mo_2O_3 $MoCl_3$	$W^{+3}(3)$
+2			$Zr^{+2}(2)$ $ZrCl_2$, etc.				$Mo^{+2}(4)$ $MoCl_2$	$W^{+2}(4)$ WCl_2 W_2C

closely because they differ only in the degree to which the 4*f* sub-shell is filled. They are usually separated from each other in solution by a lengthy ion-exchange process. As the charge on the nucleus increases in this series, there is a correspondingly stronger pull on the electrons and the ionic radius decreases, a

Table 3 ***(Continued)***

	Tc	Re	Ru	Os	Rh	Ir	Pd	Pt
+8			Ru^{+8}(0) RuO_4	Os^{+8}(0) OsO_4 OsF_8				
+7	Tc^{+7}	Re^{+7}(0) Re_2O_7 ReO_3Br Re_2S_7						
+6		Re^{+6}(1) ReO_3 ReO_2F_2 $ReCl_6$ ReF_6 $ReOCl_4$	Ru^{+6}(2)	Os^{+6}(2) OsF_6		IrF_6		PtO_3
+5		Re^{+5}(2) $ReCl_5$	Ru^{+5}(3) RuF_5					
+4		Re^{+4}(3) ReO_2 ReS_2 $ReCl_4$	Ru^{+4}(4) RuO_2 RuS_2 $RuCl_4$ RuSi	Os^{+4}(4) OsO_2 OsS_2 $OsCl_4$	Rh^{+4}(5) RhO_2	Ir^{+4}(5) IrO_2 IrS_2 $IrCl_4$	Pd^{+4}(4) PdO_2 PdS_2 PdSi	Pt^{+4}(4) PtO_2 $PtCl_4$ PtS_2
+3		Re^{+3}(4) Re_2O_3 $ReCl_3$	Ru^{+3}(5) $Ru(OH)_3$ $RuCl_3$	Os^{+3}(5) Os_2O_3 $OsCl_3$	Rh^{+3}(4) Rh_2O_3 Rh_2S_3 $RhCl_3$	Ir^{+3} Ir_2O_3 Ir_2S_3 $IrCl_3$	Pd^{+3}(3) PdF_3	Pt^{+3}(3) $PtCl_3$
+2			Ru^{+2}(4)	Os^{+2}(4) $OsCl_2$	Rh^{+2}(3) RhO RhS	Ir^{+2}(3) $IrCl_2$	Pd^{+2}(2) PdO PdS $PdCl_2$ $PdBr_2$ $Pd(CN)_2$	Pt^{+2}(2) PtO PtS $PtCl_2$

change often called the *lanthanide* contraction. The elements can be prepared in elementary form by electrolysis of the fused salts. They oxidize in moist air. Because the fourth shell contains seven *f* orbitals, many of the electrons are unpaired, and the compounds are, therefore, largely paramagnetic and colored. They all show the +3 oxidation state; other states have been found: Ce^{+4}, Pr^{+4}, Tb^{+4}, Sm^{+2}, Eu^{+2}, and Yb^{+2}.

The oxides are formed by the reaction of the metals (M) with oxygen:

$$4M + 3O_2 \rightarrow 2M_2O_3$$

Halides are formed similarly:

$$2M + 3X_2 \rightarrow 2MX_3$$

where X is a halogen. Cerium is anomalous, forming CeF_4 and CeO_2. The +3 oxidation state is shown in the sulfides (M_2S_3), the nitrides, (MN), and hydrides (MH_3); but the carbides are reported to be MC_2. The metals react with water to give $M(OH)_3$ and with H^+ to give M^{+3} as the usual product.

THE ACTINIDES

Passing to the right from actinium to the series of elements called the *actinides,* we find considerably greater chemical differences than between the lanthanides. Because these elements are all radioactive, it has been much more difficult to study their chemistry, and the details of electron assignments and oxidation states are consequently much less certain. In Table 4 are listed the more important compounds and oxidation states. From Ra to U, the evidence for oxidation states indicates that chemical behavior is about what we would expect for the two 7*s* electrons in Ra, with *d* electrons participating as they are added, increasing the maximum oxidation states in unit steps. From U onward, there is a great variety of states owing to the different degrees of participation by the *d* electrons; there is some evidence that *f* orbitals from the fifth shell may also play a part. In a compound like UF_6, the binding is almost certainly covalent to a considerable degree as this compound boils at about 56°C, a surprisingly low temperature in view of the high atomic weight of uranium.

THE TITANIUM SUBGROUP

The three members of this family are titanium (Ti), zirconium (Zr), and hafnium (Hf). All three are metals with boiling points above 3000°, indicating a considerable degree of covalency between the atoms in the elementary state. In forming compounds, there is a much larger variety of oxidation states than in the scandium subgroup; and, with titanium, the +2, +3, and +4 states are all observed as shown in Table 3. Presumably, this means donating the two *s* electrons in the lowest of these states with the additional possibility of donating or sharing either one or two of the *d* electrons. Titanium is distributed widely; it is found in the mineral *rutile,* TiO_2, and in a number of different iron ores. It is the tenth among the elements in abundance in the earth's crust, more common than sulfur, carbon, or chlorine. It forms surface coatings of oxides and nitrides and, for this reason, resists corrosion. This fact, coupled with its high melting point, makes it a valuable material for the construction of heat-resistant surfaces such as are needed in the orifices for jet engines.

Since the remaining *d* electrons occupy separate orbitals, they are unpaired.

Table 4

	Ra	Ac	Th	Pa	U	Np	Pu	Am
	14 10 10 10 6 6 6 6 6 2 2 2 2 2 2 2	14 10 10 10 1 6 6 6 6 6 2 2 2 2 2 2 2	14 10 10 10 2 6 6 6 6 6 2 2 2 2 2 2 2	14 2 10 10 10 1 6 6 6 6 6 2 2 2 2 2 2 2	14 3 10 10 10 1 6 6 6 6 6 2 2 2 2 2 2 2			
+6					U^{+6} UF_6 UO_3 UO_2Cl_2	Np^{+6}	Pu^{+6}	Am^{+6}
+5				Pa^{+5}	U^{+5}	Np^{+5}	Pu^{+5}	Am^{+5}
+4			ThO_2 ThOS ThS_2 $ThBr_4$ $Th(NO_3)_4$ $Th(CO_3)_2$		U^{+4} UO_2 US_2 UCl_4 U_3N_4	Np^{+4}	Pu^{+4}	Am^{+4}
+3		Ac_2O_3 Ac_2S_3 AcF_3 AcOF $AcCl_3$ AcI_3			U_2S_3 UH_3 UD_3 UCl_3	Np^{+3}	Pu^{+3}	Am^{+3}
+2	RaO $RaBr_2$ $RaCl_2$ $RaSO_4$ $RaCO_3$							

Whenever this occurs, it usually leads to closely spaced energy levels with the result that an electron can be excited by a low-energy photon (~2 ev) with a wavelength corresponding to that of visible light. Thus compounds formed with titanium in the +2 or +3 oxidation state absorb light of a specific wavelength and are colored. The titanous ion (Ti^{+3}) gives a violet color to solutions and readily donates the remaining *d* electron so that it is a good reducing agent. In aqueous solution, the +4 oxidation state is usually in a hydrolyzed form called the titanyl ion, TiO^{+2}. Since it has lost its *d* electrons, it is colorless. This ion is also found in ionic salts such as titanyl sulfate, $TiOSO_4$. Another example of the +4 oxidation state is titanium tetrachloride, $TiCl_4$, which is a liquid at room temperature and vaporizes, readily combining with water to form a fog for which the formula $TiCl_4(4H_2O)$ has been suggested. Titanium tetrachloride can be produced by heating TiO_2 with carbon and chlorine.

Titanium dioxide, TiO_2, is a white powder that makes a good pigment for paint. In crystalline form, it has a higher refractive index than diamond but without the hardness of diamond. It can be formed spontaneously from the black titanium hydroxide by the reaction

$$2Ti(OH)_3 \rightleftarrows H_2(g) + 2H_2O + 2TiO_2(s)$$

Zirconium (Zr) and hafnium (Hf) are practically twin elements. Although the electron core of hafnium contains far more electrons than that of zirconium (see second column in Table 3), nevertheless the lanthanide contraction that takes place in the period between zirconium and hafnium results in a size for the hafnium atom that is almost identical with that of zirconium. For this reason, these two elements have similar chemistry. Hafnium can replace zirconium in crystals and is found in nearly all zirconium minerals. The commonest of these are *baddeleyite*, ZrO_2, and *zircon*, $ZrSiO_4$, found in the Florida sand dunes and in Brazil and Australia. With these elements, there is a great tendency to form carbides, nitrides, and silicides at high temperatures and, consequently, the elements as metals are difficult to prepare. The chemical reactions are about what one would expect. It is possible to make the tetrahalides with fluorine and chlorine. It is also possible to make ZrH_2 and HfH_2. As sulfides, the metals are found as ZrS_2 and HfS_2. Nitrides are found both as Zr_3N_4 and ZrN, with similar compounds for hafnium. In a half-cell, the metals react with water according to the equation

$$M + 2H_2O \rightleftarrows MO_2 + 4H^+ + 4e^-$$

For this cell reaction, the potentials in volts for the three members of this group are Ti, 0.95; Zr, 1.43; Hf, 1.57.

Hafnium is noteworthy because it becomes superconducting at 0.35°K. Superconductivity was discovered by a Dutch physicist, H. Kammerlingh Onnes, and is a special state in which the conduction electrons can pass through the metal without encountering any electrical resistance. About twenty of the metallic chemical elements exhibit superconductivity; and a large number of alloys also show this phenomenon. As the temperature is lowered, these substances suddenly change from ordinary electrical conductivity to superconductivity. Among the elements, this change takes place at the highest temperature for niobium (about 9°K) and at the lowest for hafnium.

Recently, superconducting amplifier elements have been developed to replace vacuum tube amplifiers when great sensitivity is required. The advantage of having amplifying elements at these extremely low temperatures lies in the reduction of electrical noise that originates from the thermal motions of the atoms and electrons. The lower the temperature, the less noise there is and the more sensitive the element becomes for the detection of minute energy changes. Even with superconductors operating at 14°K, it is possible to measure the energy imparted when an individual α-particle strikes the metal. By going to a temperature proportionally almost 50 times lower, as in the case of hafnium, it may be possible in the future to measure correspondingly smaller energy changes resulting from the impacts of individual atoms or from chemical action involving single molecules. Thus a special property possessed by an unusual element like hafnium may be the key to whole new areas of physical and chemical research.

There has been considerable interest in titanium and zirconium and their compounds because of the high melting and boiling points, which suggest possibilities for use where resistance to high temperatures is required, as in rocket engines. Both experimental observations and theoretical calculations have been made to determine thermodynamic properties. In Fig. 5, some of these results are tabulated. As might be expected, there is a parallelism between similar compounds of Ti and Zr, the latter lying below the former and thus possessing greater stability. It is also interesting to see the parallelism between pairs like TiO and $TiCl_2$. The oxides have the greater stability; but the change in oxidation state produces a far greater change in ΔF_f° than the shift from Cl to O as the partner to these metals.

THE VANADIUM SUBGROUP

The three elements of this group are vanadium (V), niobium (Nb), and tantalum (Ta), all commercially of considerable importance and all interesting from the point of view of their physical and chemical properties. Their electron structure diagrams are shown in the third column of Table 2. It appears that the *d* electrons produce covalent binding in the metallic state, as these metals again all have extremely high melting points. They are less abundant in nature than the members of the titanium family but are produced commercially in considerable quantity nevertheless. Vanadium is an important component of many kinds of steel, giving

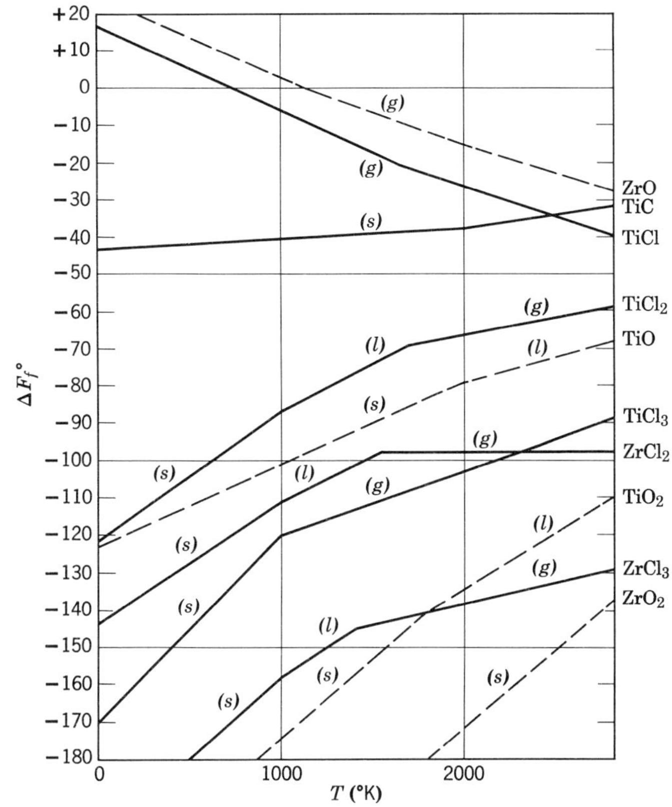

Fig. 5. Free energies of formation for some Ti and Zr compounds (Ref. 1).

to the alloy most desirable properties. Tantalum has high ductility and can be drawn into filaments such as are used in electric light bulbs and electron tubes; it was employed before tungsten for this purpose. It is also resistant to corrosion and is used to make vessels for handling acids. Niobium resembles tantalum but, in addition, has the remarkable property of having the *highest* superconducting temperature of all the elements, just as hafnium has the lowest. Niobium becomes superconducting at 9.22°K, and by the alloying with nitrogen this transition temperature can be raised 5° to 10° higher. This brings the superconductivity in the range that can be attained with liquid hydrogen. Use of this property has been made in a number of devices for detecting small amounts of energy from photons or particles.

Vanadium is found in the mineral *vanadinite*, $Pb_5(VO_4)_3Cl$. Niobium (formerly called columbium) is found in *columbite*, $FeNb_2O_6$, and tantalum is found in *tantalite*, $FeTa_2O_6$. As shown in Table 3, these metals occur in the +5 oxidation state as V_2O_5, Nb_2O_5, and Ta_2O_5. Vanadium pentoxide is important as a catalyst where O_2 is the oxidizing agent, as in the process for making SO_3 from SO_2. It is made by heating ammonium vanadate, NH_4VO_3. Some of the characteristic ions are: +2, V^{+2} (vanadous, violet); +3, V^{+3} (vanadic, green); +4, $V(OH)_2^{+2}$, blue; +5, $V(OH)_4^+$. Vanadium in the +5 oxidation state can be progressively reduced with corresponding changes in color as it passes down to the +2 state. Vanadium forms a hydroxide, $V(OH)_3$ with a green color, and an oxychloride, VOCl, in the same oxidation state. It forms sulfates, VSO_4 and $V_2(SO_4)_3$, and a nitride, VN, in the +3 state, as do also Nb and Ta in the compounds NbN and TaN.

Niobium forms compounds in the +5 state with F, Cl, and Br, as does Ta also; Nb also forms +3 compounds that in solution are easily oxidized. Thus $NbCl_3$ is oxidized by nitric acid and, when heated with carbon dioxide, reacts according to the equation

$$NbCl_3 + CO_2 \rightleftarrows NbOCl_3 + CO$$

Niobium also readily unites with hydrogen to form NbH, which burns in air. It also forms oxysulfides such as $Nb_2O_2S_3$ and $NbOS_3$. Tantalum behaves very much like niobium in aqueous solution; the +5 oxidation state is the only stable one.

THE CHROMIUM FAMILY

The members of the chromium subgroup are chromium (Cr), molybdenum (Mo), and tungsten (W); the latter symbol is derived from the earlier continental name, wolfram. In the elemental state, these are all metals of great industrial importance. They are hard and resistant to corrosion and in small amounts impart desirable properties to steels. Chromium plating is well known. Molybdenum steel is common, and tungsten is in wide use as the metal in the filaments of electric light bulbs. As shown by the diagrams in Table 2, these elements have six electrons which can be donated to or shared with other elements in compound formation. In chromium, the +2, +3, and +6 oxidation states are quite common, as shown in Table 3. In molybdenum, the common states are +3, +4, and +6. Tungsten is found almost exclusively as +6. Chromium is obtained from *chromite* ($FeCr_2O_4$). Steels with high chromium content are frequently referred to as stainless steels and are highly resistant to corrosion. Chromium gets its name from the Greek root, *chroma,* meaning color, because so many chromium compounds are highly colored. Looking at the orbital diagram in Fig. 1, one sees that there are five orbitals for *d* electrons in the third shell. Since there are five electrons, this provides one unpaired electron in each orbital with the possibility for a large number of energy levels and combinations with other atoms. In mildly basic solution, we find the chromite ion CrO_2^-; in strongly basic solution, this changes to CrO_3^{-3}. The chromate ion, CrO_4^{-2}, is also stable in basic solution. In acid solution, we have the chromous ion, Cr^{+2}; the chromic ion, Cr^{+3}; and the dichromate ion, $Cr_2O_7^{-2}$. The chromous ion is blue and is obtained by reducing the chromic ion or the dichromate ion; it oxidizes readily in aqueous solution even from the dissolved air. It can be precipitated as chromous hydroxide, which again oxidizes readily with air to chromic hydroxide, $Cr(OH)_3$.

Chromic salts are predominantly violet in color; examples are chromic nitrate, $Cr(NO_3)_3$, and chromic perchlorate, $Cr(ClO_4)_3$. The color is believed to be due to the hydrated chromic ion, $Cr(H_2O)_6^{+3}$. This is an example of the way in which the chromic ion forms many complexes with the chromium atom surrounded by six atoms or molecules at the corners of an octahedron. An example of this is CrF_6^{-3}. One can think also of these fluorine atoms as located on the six surfaces of a cube with a chromium atom at the center. Other examples are $Cr(NH_3)_4Cl_2^{+}$ and $Cr(H_2O)_5Cl^{+2}$. Two important ions containing chromium are the chromate, CrO_4^{-2}, and the dichromate, $Cr_2O_7^{-2}$. The former are generally yellow and the latter orange in color. In these ions the chromium is in the +6 oxidation state and the ions are so small that the coordination number is 4 rather than 6. The equation for the conversion is

$$2CrO_4^{-2} + 2H^+ \rightleftarrows Cr_2O_7^{-2} + H_2O$$

In line with Le Chatelier's principle, the principal species in acid solution is the $Cr_2O_7^{-2}$ ion.

As shown in Table 3, molybdenum is generally found in the +6 oxidation state, the most important compound being molybdenum trioxide, MoO_3. In basic solution, this forms an oxy anion, MoO_4^{-2}. This atom forms oxygen bridges much as do silicon and phosphorus; giant poly anions are formed that are linked together by Mo—O—Mo bonds. Molybdenum is also found in the +4 state as molybdenite, MoS_2, which resembles graphite in appearance.

Tungsten is found in the mineral wolframite, $(FeMn)WO_4$, a mixture of iron and manganese tungstates.

THE MANGANESE FAMILY

As contrasted with the previous groups that we have been discussing, only the head of this family, *manganese* (Mn), is at all familiar. The other two members are extremely rare; technetium (Tc) is radioactive and does not occur naturally; rhenium (Re) is so rare that its chemistry has not been explored to any great extent.

As shown in Table 5, manganese is found in the form of the oxides, MnO_2 and Mn_2O_3, as well as in the form of Mn_3O_4. Manganese is valuable in the production of steel because it imparts a special hardness and resistance to wear, and also because it combines with oxygen and sulfur to form compounds that separate out from the main body of the molten metal and thus can be removed easily. As

Table 5 Oxidation States and Compounds

	Mn	Fe	Co	Ni
+7	Mn^{+7}(0) $HMnO_4$ MnO_4^-			
+6	Mn^{+6}(1) MnO_3 MnO_4^{-2}	Fe^{+6}(2) FeO_4^{-2} K_2FeO_4		
+5				
+4	Mn^{+4}(3) MnO_2 MnSi MnS_2 $MnO(OH)_2$		Co^{+4}(5) CoS_2 CoSi	Ni^{+4}(4) NiO_2
+3	Mn^{+3}(4) violet Mn_2O_3 $Mn(CN)_6^{-3}$ $MnCl_3$ MnO(OH) MnP MnAs	Fe^{+3}(5) yellow Fe_2O_3 Fe_2S_3 $Fe(OH)_3$ $FeCl_3$ FeP FeAs	Co^{+3}(4) Co_2O_3 Co_2S_3 $CoCl_3$	Ni^{+3}(3) Ni_2O_3 NiAs NiSb
+2	Mn^{+2}(5) pink MnO MnS $Mn(OH)_2$ $MnCl_2$ $Mn(CN)_6^{-4}$ $MnCO_3$ Mn_3P_2 Mn_3As_2	Fe^{+2}(4) green FeO FeS $Fe(OH)_2$ $FeCl_2$	Co^{+2}(3) pink CoO CoS $CoCl_2$ $CoCO_3$	Ni^{+2}(2) green NiO NiS $Ni(OH)_2$ $NiCl_2$ $NiCO_3$

shown in the diagram in Table 2, there are the two *s* electrons and the five *d* electrons that can be donated to or shared with other atoms. Of course, the two *s* electrons are easily given up; but we find, in addition to this +2 state, the +3, +4, +6, and +7 states with only the +5 unrepresented. The +7 state is one of the strongest oxidizing agents, MnO_4^-. This permanganate ion has a deep violet color that makes it readily recognizable; and, because of the dramatic change in color when it is reduced to the almost colorless Mn^{+2} ion, it makes an excellent indicator for titration. The half-cell reaction is

$$Mn^{+2} + 4H_2O \rightleftarrows MnO_4^- + 8H^+ + 5e^- \quad (-1.52 \text{ volts})$$

The manganate ion, MnO_4^{-2}, representing the +6 state, has a deep green color. It is unstable in acid solution and disproportionates to give MnO_2 and MnO_4^-. The MnO_2 is called manganese dioxide and represents the +4 state. This is one of the components of the familiar dry battery. In the +3 state, we find the manganic ion, Mn^{+3}, in solids and complex ions. This ion liberates oxygen from water and is consequently a very useful oxidizing agent. Like the manganate ion, it is unstable and tends to go to Mn^{+2} and MnO_2, but this can be prevented by forming complexes with the oxalate ion, $C_2O_4^{-2}$, such as $Mn(C_2O_4)_3^{-3}$, or with CN^-, forming $Mn(CN)_6^{-3}$. The +2 state is represented by the manganous ion, Mn^{+2}, pink in color. It is found in manganous sulfate, $MnSO_4$, and manganous chloride, $MnCl_2$. Manganous hydroxide, $Mn(OH)_2$, is a white substance that can be oxidized by air to $Mn(OH)_3$.

Technetium and rhenium resemble each other in a manner similar to the way zirconium and hafnium resemble each other. We find potassium perrhenidate, $KReO_4$, similar to potassium permanganate, $KMnO_4$. Rhenium forms a stable oxide in the +7 state called rhenium septoxide, with the formula Re_2O_7. In contrast, the compound Mn_2O_7 is violently explosive.

THE TRIADS

The element iron plays a special part in the chemical family. First of all, it is about halfway across the first long row of the periodic table; and at this point there is more resemblance between elements horizontally rather than vertically; the groups of elements are, therefore, called *triads* instead of families. Thus it is usual to discuss the chemistry of iron and its two right-hand neighbors, cobalt and nickel, as a related group. The same kind of relationship exists for the elements that lie in the other two long rows of the periodic table just below these three; there we find the triad of ruthenium, rhodium, palladium and the triad of osmium, iridium, platinum where there are again strong resemblances. We shall, therefore, discuss these nine elements next and the resemblances found in each triad.

These elements are all hard metals and are unusually dense or heavy. Iron, cobalt, and nickel are moderately active chemically but the other six are almost inert. As may be seen from the diagrams in Table 6, the *d* level in the third shell is more than half-filled in the iron triad; this is true of the *d* level in the fourth shell for the ruthenium triad and of the *d* level in the fifth shell in the osmium triad. Hund's rule tells us that the electrons will be distributed as shown in Fig. 6 for Fe^{+3}, with the *d* orbitals singly occupied in so far as possible.

It is quite likely that in the *elemental* state all of these elements utilize some of their *d* electrons to form covalent bonds, and this leads to the hardness of the metals. In the ruthenium and osmium triad, the bonding makes the metals un-

Table 6 Electronic Structure of the Triad Elements

6 6 6 2 2 2 2 Fe	7 6 6 2 2 2 2 Co	8 6 6 2 2 2 2 Ni
6 6 6 2 2 2 + + Fe^{+2} Ferrous ion +2 state	7 6 6 2 2 2 + + Co^{+2} Cobaltous ion	8 6 6 2 2 2 + + Ni^{+2} Nickelous ion
5 + 6 6 2 2 2 + + Fe^{+3} Ferric ion +3 state	6+ 6 6 2 2 2 + + Co^{+3} Cobaltic ion	7 + 6 6 2 2 2 + + Ni^{+3} Nickelic ion
10 7 6 6 6 2 2 2 2 1 Ru Ruthenium	10 8 6 6 6 2 2 2 2 1 Rh Rhodium	10 10 6 6 6 2 2 2 2 0 Pd Palladium
14 10 10 6 6 6 6 6 2 2 2 2 2 2 Os Osmium	14 10 10 7 6 6 6 6 2 2 2 2 2 2 Ir Iridium	14 10 10 9 6 6 6 6 2 2 2 2 2 1 Pt Platinum

reactive. They have been referred to as the "noble" metals in that their lack of reactivity suggests the inactivity of the gases in the eighth row of the periodic table where the outer shell has the octet and there is no chemical reactivity. Why *inactivity* should be regarded as *noble* it is difficult to say.

Because of their unpaired *d* electrons, these elements form colored ions. The unpaired *d* electrons also lead to the phenomenon of ferromagnetism. At room temperature or lower, the unpaired electrons in the iron, cobalt, and nickel atoms tend to orient in such a way that their magnetic fields combine and produce a strong magnetic field outside the metal itself; but, at sufficiently high temperatures, the atoms are knocked about by the heat and the electrons become oriented at random so that most of the magnetic fields cancel each other out, and the magnetism disappears.

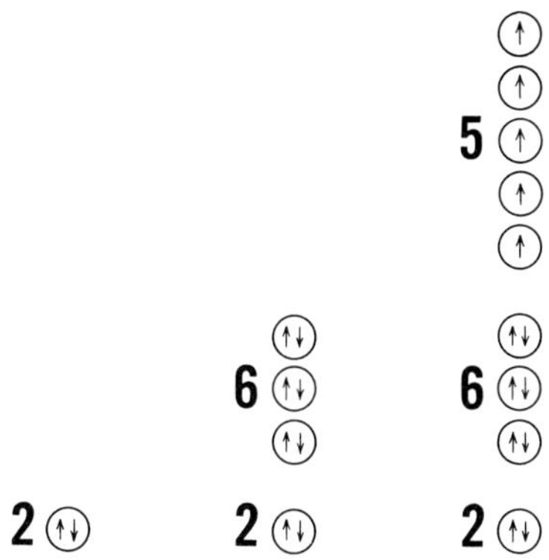

Fig. 6. Orbital diagram for Fe^{+3} showing unpaired electrons.

Iron. The name iron comes from the Anglo-Saxon; it was applied to this substance many centuries ago. The symbol for iron, Fe, comes from the Latin word for iron, *ferrum*. Iron has a melting point of 1535°C, a boiling point of 2700°C, and an ionization potential ($I_{\frac{1}{2}}$) of 7.9 ev. Its density is 7.9 gm/cm^3; it is the fourth most abundant element in the crust of the earth. Iron has about the lowest nuclear energy of all the elements. If one could wave a magic wand over the universe so that the nuclei of all elements would change by fission or fusion to the lowest possible energy state, a universe completely made of iron might result. The central core of the earth is probably composed of iron that is in the liquid state because of the conditions of temperature and pressure. Thus, in addition to being the *fourth* most abundant element in the crust of the earth, iron is the *most* abundant element if the total substance of the planet is taken into account.

Some of the most important sources of iron are the minerals *hematite*, Fe_2O_3; *magnetite*, Fe_3O_4; *limonite*, $Fe_2O_3 \cdot H_2O$; and *siderite*, $FeCO_3$. Iron is also found abundantly in the form of *pyrites*, FeS_2, but this is not a useful source of the metal because the sulfur is difficult to remove and traces of it impart undesirable properties to iron. There is an extensive chemistry of iron and steel production that is very complicated. The principal steps generally employed in the manufacture of iron from iron ore involve heating the ore in a stream of gas produced by the combination of carbon and oxygen. It seems reasonably certain that the active agent is carbon monoxide, CO, and that this reduces the iron oxide of the ore stepwise, the final product consisting of iron with considerable traces of different impurities. Calcium carbonate is generally added in this process because the ore contains a great deal of silica, and in the heat of the furnace the calcium carbonate decomposes to give carbon dioxide and calcium oxide, CaO, which combines with the silica to form calcium silicate, $CaSiO_3$. At the temperature of the furnace, this is still a liquid and it can be withdrawn easily as slag. The iron as directly produced is referred to as *pig iron,* and articles can be made of it in the form called *cast iron.* Usually, the iron is further purified by heating with carbon and other elements, and small amounts of other metals such as Mn, Cr, V, and Mo are added. Depending on the composition, the treatment, and the working, one gets varieties of *steel.*

The most stable form for the iron ion in solution is the $+3$ state called *ferric.* The ferric ion is shown in Fig. 6. We see that there are five electrons in the *d* level, exactly one for each orbital. This gives a stable configuration to the outer shell. This oxidation state of iron is also found in the oxide, Fe_2O_3, and in many compounds such as $FeCl_3$ and $Fe(OH)_3$; the latter is a slimy red-brown and gelatinous substance which can be dehydrated to give the yellowish or reddish oxide, Fe_2O_3. Iron is also relatively stable when the atom loses the two electrons in the fourth shell and is left with the outer shell of 2, 6, 6 in the third layer; this is called the *ferrous* ion and it represents the $+2$ oxidation state. Both types of ion in solution are definitely acid. When a base is added to a *ferrous* solution, a light green precipitate of ferrous hydroxide, $Fe(OH)_2$, is formed. This precipitate turns brown in air, forming hydrated ferric oxide, $Fe_2O_3 \cdot H_2O$.

These ions have a strong tendency to form complexes. Six CN^- ions will join with Fe^{+2} to make the ferrocyanide ion, $Fe(CN)_6^{-4}$; six of these cyanide ions will also join with Fe^{+3} to form the ferricyanide ion, $Fe(CN)_6^{-3}$. These six CN^- groups are arranged octahedrally around the central iron atom; that is, the iron can be thought of as surrounded by a cube with a CN^- group on each face. The ion SCN^- will combine with the ferric Fe^{+3} ion to form the red complex $FeSCN^{+2}$. Another interesting complex ion, $Fe(C_2O_4)_3^{-3}$, is formed when Fe^{+3} is combined with the oxalate ion, $C_2O_4^{-2}$. One can think of the oxalate ion as consisting of the group $^-OOCCOO^-$. This is a U-shaped ion, and it can form a bridge between two faces of the imaginary cube surrounding the Fe; so that, with each U-shaped ion bridging two faces, it takes only three of these bridges to cover the six faces of the cube. Oxalate ions have such a strong tendency to combine with iron in this complex that solutions of oxalic acid are useful for removing rust stains.

Many of the complex ions of iron are strongly colored. The potassium salts of ferrocyanide are yellow; those of ferricyanide are red. If ferric ion is added to a solution of potassium ferrocyanide, a blue pigment (Prussian blue) precipitates; if ferrous ion is added to potassium ferricyanide a different blue pigment (Turnbull's blue) precipitates. Both of these pigments, which are different, have the same approximate composition, $KFeFe(CN)_6 \cdot H_2O$. If instead of combining iron atoms with different oxidation states we combine those with the same oxidation states, the result is less colorful. Ferrous ion and ferrocyanide ion form a white precipitate; ferric ion and ferricyanide form a greenish brown solution.

Metallic iron has a tendency to rust. Rain water always contains a little carbon dioxide and, in industrial areas, can also contain traces of many other substances that make it acidic. Under these conditions, moist iron rusts badly, combining with oxygen in the air to form ferric oxide which has no structural strength. Once the rust process starts, it is autocatalytic in nature and proceeds rapidly. For this reason, it is important to use protective coatings to prevent the initial formation of rust.

Cobalt. Cobalt is much less abundant than iron, though not really a rare element, constituting about 0.002% of the earth's crust. The name may come from a German word, *kobold*, meaning *devil*, and probably is associated with the difficulty encountered in smelting ores containing Ni and Co. It is found combined with elements like sulfur and arsenic, and the poisonous properties of the fumes complicate purification. Cobalt has a melting point of 1490°C and a boiling point of 2900°C. Its oxidation potential for the metal to form the ion in the +2 oxidation state is +0.28 volt. It dissolves slowly in acids, liberating hydrogen; it will not dissolve in nitric acid because of the formation of an unreactive coating. As contrasted with ferrous ion, the cobaltous ion is quite stable. The cobaltous solutions are, in general, pink; and, indeed, this is the color of all crystals that contain the hydrated ion, $Co(H_2O)_6^{+2}$. In basic solutions, the dark-blue insoluble hydroxide is precipitated, and this can be dehydrated to give the greenish yellow cobaltous oxide, CoO. This will oxidize in air, when heated, to give Co_3O_4, which is similar in structure to Fe_3O_4. The cobaltic ion is shown also in Table 6.

The potential for the change from cobaltous to cobaltic ion is −1.84 volts. This is strong enough to oxidize water with the formation of oxygen. There are not many examples of cobaltic ions; CoF_3 and $Co_2(SO_4)_3 \cdot 18H_2O$ have been formed, but they decompose in aqueous solution to give cobaltous ions. When the cobaltic ion unites with other ions to form complexes, it can be stabilized. Examples of this are $Co(CN)_6^{-3}$ and $Co(NH_3)_6^{+3}$.

Nickel. Nickel, like cobalt, is difficult to extract in the pure state from its mineral source. The name comes from the Swedish word, *kopparnickel*, which means *false copper.* It is found in the form of a sulfide mixed with sulfides of iron, or as a silicate mixed with magnesium. It is normally purified electrolytically; but there is an unusual method available employing a volatile compound, called nickel carbonyl, $Ni(CO)_4$, that nickel forms with carbon monoxide at room temperature. When the vapor of this substance is heated to 200°C, it decomposes into nickel and carbon monoxide. Of the three members of this iron triad, nickel is the least subject to oxidation; for this reason, nickel-plating of iron produces resistance to corrosion. It also increases the strength of iron when present in small amount. A great deal of nickel is also used in coins. In practically all nickel compounds, the Ni atoms have lost only the two *s* electrons in the fourth level and are in the +2 oxidation state. The oxide in this state is NiO, nickelous oxide, which is black in color.

The nickelous complex ions are quite stable in air with respect to oxidation; they can be either square-planar or octahedral in shape: $Ni(NH_3)_4^{+2}$ and $Ni(NH_3)_6^{+2}$; and other ions are found with tetrahedral coordination: $Ni(CO)_4$. Solutions of nickelous ion are usually pale green, but in the presence of complexing agents the color changes; for example, in the presence of NH_3 nickelous solutions are blue. Nickelous oxide combines with oxygen to give a substance with the rock salt structure that has a larger percentage of oxygen than the formula, NiO, indicates; it is believed that this is the result of a complex solid solution in which some of the nickel ions are present as nickelic ions and some nickelous ions are missing from their lattice positions. Nickel has a boiling point almost the same as that of iron and a melting point slightly lower; its ionization potential is the lowest of the three members of the triad, 7.6 volts, and its density is the highest, 8.9 gm/cm. It also has the lowest oxidation potential for the change from the metal to the nickelous ion, 0.25 volt.

THE PALLADIUM TRIAD

The palladium triad is made up of the three elements ruthenium (Ru), rhodium (Rh), and palladium (Pd). In contrast with the first triad, which is named for its first member, iron, the other two triads are named for their last members, palladium and platinum, because these are by far the most important members of their respective groups. In fact, because the elements in the palladium triad resemble platinum so closely, this triad is sometimes referred to as the *light platinum triad.* The electron structure diagrams for the two triads are shown in Table 6. In the gaseous state, it appears that the outer shells of the palladium triad do not have the orbitals for the *s* electrons completely filled. In the metallic state, however, these elements behave so much like those of the platinum triad that there is every reason to believe the electronic structures must be essentially identical. In other words, as in the iron triad, the *s* electrons wander freely to produce metallic properties and part of the other electrons participate in covalent binding between neighboring atoms, making these metals hard. The covalency leads to the high melting points characteristic to this group, as shown in Table 7.

Table 7 Physical Properties of the Palladium Triad

	Ru	Rh	Pd
Melting point	2450°	1970°	1550°C
Boiling point above	2700°	2500°	2200°C
Ionization potential, ev	7.5	7.7	8.3
Oxidation potential to M^{+2}, volts	−0.5	−0.6	−1.2

Because of the similarity in properties, these elements are usually found combined in nature. Their most striking characteristics are the negative values for the oxidation potentials, in contrast with the positive values for the iron triad. This means that the tendency to oxidize is much less and, as a result, we find these elements occurring in the free metallic state in nature.

Ruthenium. Ruthenium is found in nature alloyed with osmium and iridium. It is separated from these elements by oxidation of the mixture under basic conditions, heating with KNO_3 and KOH, which forms K_2RuO_4, potassium ruthenate. On boiling, osmium peroxide (OsO_4) distills off; on making this solution basic again, it is possible to distill off ruthenium peroxide, RuO_4. This separation illustrates how the oxides of these elements differ from those of the preceding triad. Thus FeO has a melting point of 1420°C, and a boiling point so high that it almost certainly decomposes before it boils. This is due to the ionic linking in the oxide crystal, which is not made up of molecules but has a structure like NaCl. The same is true of the higher oxide of iron, Fe_2O_3, which melts at 1565°C. By way of contrast, ruthenium tetraoxide more closely resembles the oxides of elements at the far right of the periodic table, which are true molecules. Of course, it is much heavier than sulfur dioxide, SO_2, which is a gas at room temperature; hence ruthenium tetraoxide melts at 25°C; but it has a vapor pressure sufficiently high that it can be boiled off as a gas and thus separated from the solution. The volatile compound, RuO_4, is in the +8 oxidation state. The +5 oxidation state is illustrated by RuF_5; the +4 oxidation state is found in RuS_2 and RuSi; the +3 oxidation state occurs in $RuCl_3$.

Rhodium. Rhodium is generally found with platinum. The name comes from the Greek word meaning rose, appropriate because of the rose color of some of its salts, such as $KRh(SO_4)_2$. The precipitation of this salt is the basis for one of the methods for getting pure rhodium. In this salt, the oxidation state is +3, a state also found in Rh_2S_3, $RhCl_3$ and in the complex salt K_3RhCl_6. Rh is also found in the +2 oxidation state in compounds like RhS and RhO. Added in small amounts to platinum, it makes an alloy that is used in thermocouples and metallic vessels designed to withstand extreme heat.

Palladium. Palladium has the lowest melting point of any of the elements of these two triads and is also the most reactive. It dissolves in concentrated nitric acid. The name comes from the Greek and refers to a statue of the goddess Athena, who was also called Pallas. The name was given to one of the planetoids, and by this devious route it was assigned to this element. Palladium forms an insoluble cyanide, $Pd(CN)_2$, in the +2 oxidation state that can be precipitated out and, by vigorous heating, will decompose to give the metal.

Palladium has the remarkable property of absorbing large amounts of hydrogen. This is similar to the way in which some of the lighter transition metals take up gases like hydrogen and nitrogen, giving substances that properly might be called gaseous alloys; they are frequently assigned formulas like NbN and TaN, even though the alloy can have a relative composition far different from that indicated by the formula. In the case of palladium, the hydrogen absorbed will amount to as much as one thousand times the volume of the palladium crystal itself. The hydrogen enters into the lattice and is held there as atomic hydrogen by an electron arrangement involving a binding between the proton and mobile metallic electrons.

The majority of well-recognized palladium compounds are in the +2 oxidation state, such as $PdBr_2$, $PdCl_2$, PdF_2, PdO, and PdS. Other compounds occur in the +1 oxidation state, such as Pd_2S; and in the +3 oxidation state, such as PdF_3; and even in the +4 oxidation state, such as PdSi and PdO_2.

THE PLATINUM TRIAD

The members of the platinum triad are osmium (Os), iridium (Ir), and platinum (Pt). These elements are all metals with exceptionally high densities and high melting points. They are extremely unreactive. Compared with iron, cobalt, and nickel, these metals have a more stable structure, probably owing to more covalent bonding in a three-dimensional network.

Osmium. Osmium occurs in nature associated with ruthenium and iridium. In the +8 oxidation state, it forms the volatile compound OsO_4, and use of this fact is made in the usual process of purification. This compound can be reduced easily as shown by the reaction

$$Os + 4H_2O \rightleftarrows OsO_4 + 8H^+ + 8e^- \quad -0.85 \text{ volt}$$

The metal is so inert that it is not even attacked by a mixture of concentrated nitric and hydrochloric acids. Compounds typifying the oxidation states are: +8, OsO_4 and OsS_4; +6, OsF_6; +4, $OsCl_4$, OsF_4, OsO_2, OsS_2; +3, Os_2O_3; +2, OsO, $OsSO_3$.

Iridium. This element is obtained from the mixture with ruthenium and osmium; it is left behind after the latter elements are separated as volatile oxides. Just as palladium can be prepared by the decomposition of its cyanide, iridium in metallic form can be prepared by decomposing the compound $(NH_4)_2IrCl_6$.

Table 8 Physical Properties of the Platinum Triad

	Os	Ir	Pt
Melting point	2700°	2450°	1774°C
Boiling point above	5300°	4800°	4100°C
Ionization potential, ev	8.7	9.2	9.0
Oxidation potential, to M^{+2}, volts	−0.9	−1	−1.2

It hardens platinum when combined with the latter in small amounts. Some typical compounds are: +6, IrF_6; +4, $IrBr_4$, $IrCl_4$, IrI_4, IrO_2, IrS_2; +3, IrI_3, $IrCl_3$, Ir_2S_3.

Platinum. Of the six elements in these two triads, platinum is the most useful because it can be employed to make vessels that withstand high temperature and are quite unreactive to almost all chemicals. It is found in relatively concentrated deposits so that it is not prohibitively expensive from a commercial point of view. It can be dissolved in concentrated mixtures of hydrochloric acid and nitric acid and then it precipitates as $(NH_4)_2PtCl_6$; and, on heating, this compound decomposes to give platinum metal. Platinum also dissolves in melted alkali and alloys with some metallic substances like lead and antimony, and with metalloids like arsenic and silicon, and even with phosphorus. Because of its industrial and scientific importance, a somewhat larger number of compounds of platinum have been studied. Some examples are: +2, $PtCl_2$, PtI_2, PtO, PtS; +3, $PtCl_3$, Pt_2S_3; +4, PtO_2, PtS_2; +6, PtO_3.

One of the most important uses of platinum is as a catalyst. It is employed for this purpose in the oxidation of ammonia to nitric acid and in the *platforming* of petroleum to make gasoline.

Platinum has a great tendency to form complex compounds, and a number of these are important for many different chemical purposes. A few examples are: $Pt(NH_3)_4Cl_2 \cdot H_2O$; $Pt(NH_3)_4PtCl_4$; $Pt(NH_3)_2Cl_4$.

THERMODYNAMIC PROPERTIES

Values of the free energy of formation from the elements at 25°C (ΔF_f°) are plotted for a number of the compounds of the transition elements in Fig. 7. In the compounds like the carbides and nitrides, there is very little difference between the free energy of the compound and the free energy of the atoms combined separately in their elemental states; although the stability of the compound with respect to the elements is somewhat greater at the left side of the diagram, as shown by the more negative values of ΔF_f°. The same trend can be observed throughout the table, almost without exception. In the compounds shown, these metallic transition elements are combined with non-metallic elements lying at the right of the periodic table. Therefore one expects a combination of a transition element on the left of the table having lower electronegativity with a more electronegative element from the right of the table with higher electronegativity to yield the more negative values of free energy.

The change in moving from right to left for similar compounds follows the same trend, with the slope of the line getting greater toward the bottom of the diagram. The sulfides fall off more sharply than the carbides and nitrides, and the chlorides more sharply than the sulfides. The fall in ΔF_f° in passing from

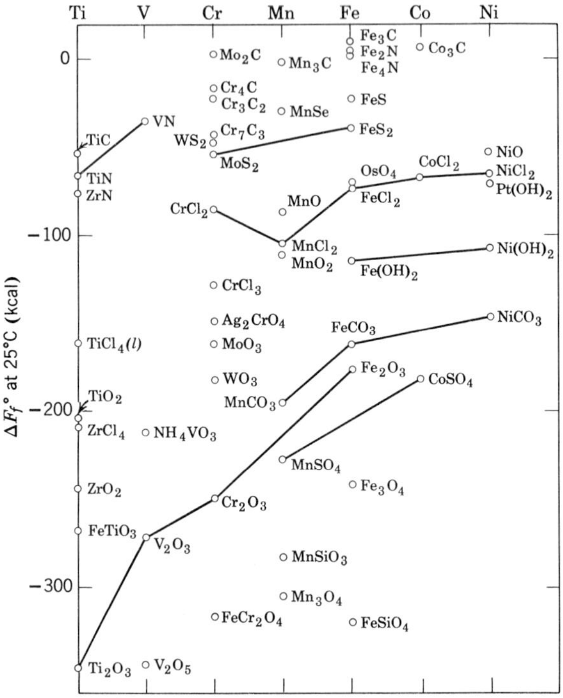

Fig. 7. ΔF_f^0 for compounds of transition elements (Ref. 2).

$CrCl_2$ to $MnCl_2$ is explicable. The diagram shows the resemblance between compounds of a triad like Fe, Co, and Ni with much greater differences appearing in passing to the neighboring elements Mn, Cr, V, and Ti. The series M_2O_3 shows the drop in ΔF_f° from M = Fe to M = Ti. In all these changes it seems clear that variations in ΔF_f° are due largely to variations in ΔH_f° and that the entropy factor plays a minor role. The variations in ΔH_f° are in turn due largely to the variations in the character of the binding and its dependence on electronegativity. Keeping in mind that at 25°C it takes only a 1.4-kcal change in ΔF_f° to produce a ten-fold change in the equilibrium constant, one can see the profound change in the chemistry brought about by these changes in electronegativity.

PROBLEMS

1. If pure Fe_2N and pure Fe are present in a closed flask, what will be the pressure of N_2 above the mixture at 25°C? *Ans.* 2.6×10^4 atm.

2. What will be the pressure of N_2 over a mixture of TiN and Ti at 25°C?

★ **3.** One mole of pure Fe and one mole of pure Ti are present in a closed flask and N_2 gas is equilibrated with both solids. Make a plot of the logarithm of the pressure against n, the number of moles of gas admitted, taking the volume of the flask as 100 liters, the temperature as 25°C, and regarding the volume occupied by the Fe and Ti as negligible. Indicate the points where TiN, Fe_2N, and Fe_4N are formed.

4. Calculate the equilibrium constant at 25°C for the reaction

$$3Fe_2O_3 \rightleftarrows 2Fe_3O_4 + \tfrac{1}{2}O_2$$

Ans. 3.4×10^{-35}.

5. State what happens when 1 mole of Fe and 1 mole of $CoCl_2$ interact and come to equilibrium at 25°C.

★ **6.** Make a plot of the logarithm of the pressure of Cl_2 over $TiCl_2$ for the range 500° to 1500°K, using $1/T$ as the second coordinate.

7. Estimate the value of ΔH_f° for the reaction in Problem 6.

8. Compare the values of ΔS_f° for TiCl, $TiCl_2$, $TiCl_3$, $ZrCl_2$, and $ZrCl_3$ when this series of compounds is at 1000°K, at 2000°K, and at 3000°K.

9. Estimate the heat of melting and the heat of vaporization for $ZrCl_3$ from the graphs in Fig. 5. Give the deviation from Trouton's rule for this compound.

10. Calculate the equilibrium constant for the following reaction at 2000°K:

$$ZrCl_2 + \tfrac{1}{2}Cl_2 \rightleftarrows ZrCl_3$$

11. Assume that an iron ore consists of a mixture of Fe_2O_3 and SiO_2. Write balanced chemical equations for the likely reactions that occur in a blast furnace.

12. Platinum does not dissolve in concentrated nitric acid but does dissolve in aqua regia. Explain this and write appropriate chemical equations.

13. The energies required for the following reactions at 0°K are as given:

$$Fe(g) \rightarrow Fe^{+2}(g) + 2e^-(g) \qquad \Delta E = -24 \text{ ev}$$
$$Fe(g) \rightarrow Fe^{+3}(g) + 3e(g)^- \qquad \Delta E = -54 \text{ ev}$$

Yet, it is well established that ferric ion in solution is even more stable than ferrous ion. Explain.

14. A solution formed by dissolving $Ni(NO_3)_2$ in distilled water is green. If ammonia is added to this solution, it turns blue. Indicate a mechanism by which such a color change could occur.

15. Imagine that the octahedrally coordinated ferricyanide ion is formed by the approach of six cyanide ions to a ferric ion. If complete covalent bond formation occurs, the $3d$, $4s$, and $4p$ subshells of the iron ion are completely occupied by six pairs of electrons from the cyanide ions. If perfect ionic bonding occurs, the electrons of the ferric ion are arranged in the same way as in the free ion. Magnetic measurements can show the number of unpaired electrons in a compound. (*a*) How many unpaired electrons would be found in "ionic" ferricyanide ion? (*b*) How many would be found in "covalent" ferricyanide ion? *Ans.* (*a*) 5; (*b*) 1.

16. The density of palladium is 12.0 gm/cm^3. If palladium forms the definite compound Pd_2H, calculate (*a*) the weight per cent of hydrogen in the compound; (*b*) the volume of hydrogen gas STP that will react with 1 cm^3 of Pd.

★**17.** Real "FeO" contains at most 48.6 at. % iron. Ideally, FeO has the rock salt structure. What must be the condition of the iron ions if the iron-deficient structure is to be physically realizable? Such an iron-deficient structure could come about if: (*a*) a perfect FeO lattice were formed with extra oxygen atoms squeezed between lattice positions; (*b*) the lattice were a rock salt type in which some of the cation sites are not filled. Design a simple test that you could perform in the laboratory that would enable you to distinguish between these two provided you knew the physical properties of ideal stoichiometric (and imaginary) FeO.

18. Both Turnbull's and Prussian blue have the composition

$$K(Fe)[Fe(CN)_6] \cdot H_2O$$

If the cyanide complex is covalent in the sense of Problem 15 and the other iron atom behaves as a free ion, would magnetic measurements allow you to distinguish between the two? Explain. *Ans.* No difference in number of unpaired electrons.

19. A neutral solution is yellow, and you suspect the presence of CrO_4^{-2} ion. How could you test your hypothesis?

20. A solution containing only one cation is a very pale pink. When base is added, a white precipitate forms that slowly turns brown. What could the cation be?

21. A basic solution has a deep green color. When acidified, a black precipitate forms and the solution turns purple. What could be in the solution? *Ans.* MnO_4^{-2}.

22. A sample contains a mixture of cobalt and nickel. How could you prepare a small sample of pure metallic nickel from this mixture?

23. Give a balanced equation for the preparation of V_2O_5 by heating NH_4VO_3.

24. What might happen when the following are mixed?

(*a*) $MnO_4^{-2}(aq.) + \text{acid}$
(*b*) $MnO_4^{-}(aq.) + \text{acid} + Fe^{+2}(aq.)$
(*c*) $I^{-}(aq.) + Fe^{+3}(aq.)$
(*d*) $CrO_4^{-2}(aq.) + \text{acid}$
(*e*) $Co^{+3}(aq.) + \text{water}$
(*f*) $Ni^{+2}(aq.) + Zn(s)$
(*g*) $Co(NO_2)_6^{-3}(aq.) + K^{+}(aq.)$
(*h*) $MnO_2(s) + H_2O_2(aq.) + \text{acid}$
(*i*) $Fe_3O_4(s) + CO(g)$
(*j*) $Ni(s) + CO(g)$
(*k*) $Ni(s) + HNO_3(\text{conc.})$

Ans. (*a*) $3MnO_4^{-2}(aq.) + 4H^{+}(aq.) \rightarrow 2MnO_4^{-}(aq.) + MnO_2 + 2H_2O$
(*d*) $2CrO_4^{-2}(aq.) + 2H^{+}(aq.) \rightarrow Cr_2O_7^{-2}(aq.) + H_2O$
(*i*) $Fe_3O_4(s) + 4CO(g) \rightarrow 4CO_2(g) + 3Fe(s)$

24 The nature of metals and alloys

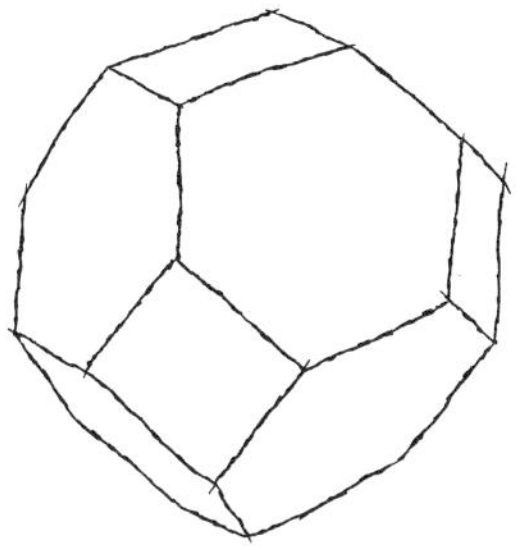

THE METALLIC STATE

Although eighty-one of the one hundred and two chemical elements are metals, the metallic state is in many ways the least understood of all the varieties of elemental matter. Even after the development of a coherent theory of chemical bonds, it was clear that the strong forces binding together metal atoms could not be explained in terms either of the ionic or the covalent bond. Different concepts, which also have their origins in quantum mechanics, must be used to explain the metallic bond.

The difficulty in accounting for the properties of metals by familiar chemical concepts can be seen by recalling the nature of some of the metals with simple electronic structures, like the lighter elements in Groups I, II, and III of the periodic table. Lithium has the structure 2-1, a pair of electrons in the 1*s* level and a lone electron in the 2*s* state outside of the stable 1*s* core. It is reasonable to conclude that this single 2*s* electron provides the link that binds lithium atoms together; but what is the nature of this link?

Now, if we were to reason by simple analogy, we might conclude that this link is a covalent bond, and that lithium should be a gas like hydrogen. The hydrogen atom has a single electron outside the nucleus with its single positive charge. Two hydrogen atoms can share their electrons, and the electron pair in this combination forms the covalent bond that holds the two atoms together in the dumbbell-shaped molecule. Once this covalent bond is formed, the residual force on the outside of the molecule is so small that the molecules have little tendency to cling together; hydrogen is a gas boiling at 20°K. Actually, lithium does resemble hydrogen closely in its electronic structure. The core of the 2*s* electrons shields the lithium nucleus with its positive charge of three units, so that the effective field outside of the core is very nearly that of a single charge; and the lone 2*s* electron moves in this unit field, a situation almost identical with that of hydrogen. Yet, whereas hydrogen boils at 20°K, lithium boils at 1336°C! Hydrogen is a gas; lithium is a metal.

From this it is clear that, in the solid state, lithium cannot consist of pairs of lithium atoms, each pair linked by a covalent bond. Alternatively, the high boiling point might suggest multiple covalent bonding, but, in simple chemical terms, the one electron on the lithium atom can only link in a single bond, and a single bond can only join a single pair of atoms; so simple covalent bonding is not the answer. Again, a high boiling point also suggests ionic bonding; but ionic bonding is found only when the crystal is made up of both positive and negative ions arranged in alternating fashion. It is the attraction between opposite charges that provides the force of ionic bonding; and there is no reasonable mechanism by which lithium atoms can give and receive electrons to form the Li^+ and Li^- necessary to make an ionic crystal.

To resolve this difficulty, the suggestion was made some years ago that the lone 2*s* electrons play the part of negative ions while the positively charged lithium cores, Li^+, provide the positive ions. This arrangement might lead to something like an ionic binding force. It was further suggested that the 2*s* electrons move freely like a fluid through this pseudo-lattice, thus accounting for the electrical conductivity of the metallic lithium. However, this concept of an electron fluid was vague; one could say that it acted like atomic glue and bound the lithium cores together; but such a statement was hardly a satisfactory substitute for a rigorous quantitative theory. The first clue to a better theory for metals came from considerations of the way atoms are packed together in the metallic state. So let us recall some of the aspects of packing.

Metallic packing. With the exception of boron, which must be classed as a semimetal because of its low electrical conductivity, the elements of Groups I, II, and III are all metallic and normally are found packed in one of the three types of crystal lattice shown in Fig. 1. Calcium, strontium, and aluminum have the first type, face-centered cubic packing; all the alkali metals, together with barium, are found with body-centered cubic packing; and beryllium and magnesium exhibit hexagonal close packing. These three types of packing have been discussed in Chapter 9 in connection with our study of solids, but we review them here to show their relation to the nature of the metallic state.

If we could look at the face perpendicular to the body diagonal of a crystal arranged with face-centered close packing, using the supermicroscope to see the individual particles, we would see alternate rows of staggered atoms. Figure 1*b* shows a block of fourteen atoms taken from such a lattice. The whole group is

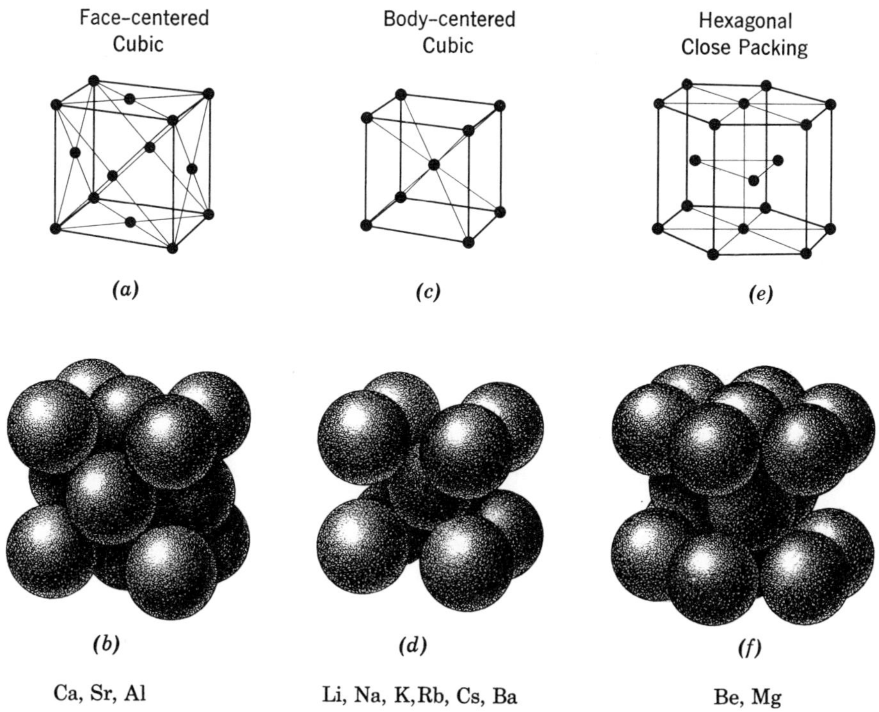

Fig. 1. Metallic packing.

in the form of a cube with an atom at each corner and an atom in the middle of each face, hence the name *face-centered cubic.* Tennis balls or any spheres can be packed together in this type of lattice with a minimum of empty space; it is one kind of spherical close packing. Of course, the atom does not have a sharply defined spherical surface like a tennis ball, but the potential field surrounding it is frequently nearly spherical, because of the spherical symmetry of the electron waves.

Another common arrangement is the body-centered cubic structure shown in Figs. 1*c* and 1*d*. Here, instead of the atoms at the centers of the *faces* of the cube, there is one atom in the center of the cube itself, in addition to the atoms at each of the corners of the cube. This central atom can be seen in the more open representation in Fig. 1*c;* and, in the more closed representation in Fig. 1*d*, a portion of it can be seen tucked away at the center between the four atoms at the top of the cube and the four atoms at the bottom. All the alkali metals and also barium exhibit this type of packing.

The third type of packing is the hexagonal close packing shown by beryllium and magnesium. Here, instead of a cube, we find a structure based on the polygon represented by the open drawing shown in Fig. 1*e*. Atoms are shown as dots at each of the points of the upper hexagon with one atom placed at the center of the hexagon. The same arrangement is found in the lower face. Halfway between there are the three atoms shown at the points of a triangle; the atoms lie in these positions because they fit naturally into the depressions left between the atoms in the upper and the lower faces when packed as shown; in Fig. 1*f*, the atoms are shown as balls.

Now consider the case of the atom at the center in Fig. 1*d*, tucked in the little central cavity between the four atoms above it and the four atoms below. Suppose that these are atoms of lithium. We know that each atom will have the two 1*s* electrons in its inner core and a single 2*s* electron outside in the form of a spherical wave with two loops. Now, when the atoms are packed close together as shown in Fig. 1*d*, these 2*s* spherical waves will overlap so that there will be a tendency to form a bond. With which of its neighbors will this central atom form a bond? There are eight atoms closest to it, the four above and the four below. In addition, there are other atoms almost as close, the atoms that are in the centers of the similar cubes, above, below, in front, in back, and at the left and at the right of the cube we are considering. These atoms are not included in the drawing, but would be represented if, instead of showing only one unit cell of the whole crystal, we had drawn in the whole crystal extending indefinitely to the right and left and above and below. Thus, in all, there are fourteen neighboring atoms surrounding the atom at the center, and there is no reason for it to select any one of these in preference to any of the others in the formation of a bond. In order to understand what happens in such a situation, let us review briefly a few of the facts considered before in connection with the formation of covalent bonds.

In Fig. 2, there are represented the waves for the hydrogen atom with electrons of the 1*s*, 2*s*, and 3*s* types, together with the plots that show the density of charge on a cross section through a typical spherical atom. When two hydrogen atoms are brought so close to each other that their wave patterns overlap, there is a redistribution of the electron density and the formation of a covalent bond with a wave pattern as shown in Fig. 2*c*. The density plot for these two atoms bound by a covalent bond is shown in Fig. 2*d*.

Covalent binding for lithium is slightly more complicated because of the presence of both the 1*s* and 2*s* electrons; the individual atom is shown both as a wave diagram and crudely in a density plot in Fig. 3. When two lithium atoms are brought together, the 2*s* waves merge just as in the previous example of the hydrogen molecule, and we get the formation of Li_2, shown both in the wave diagram and in the density plot (Figs. 3*c* and 3*d*).

If we were dealing only with *pairs* of lithium atoms in metallic lithium, the situation would be simple. We would regard the metallic state as made up of

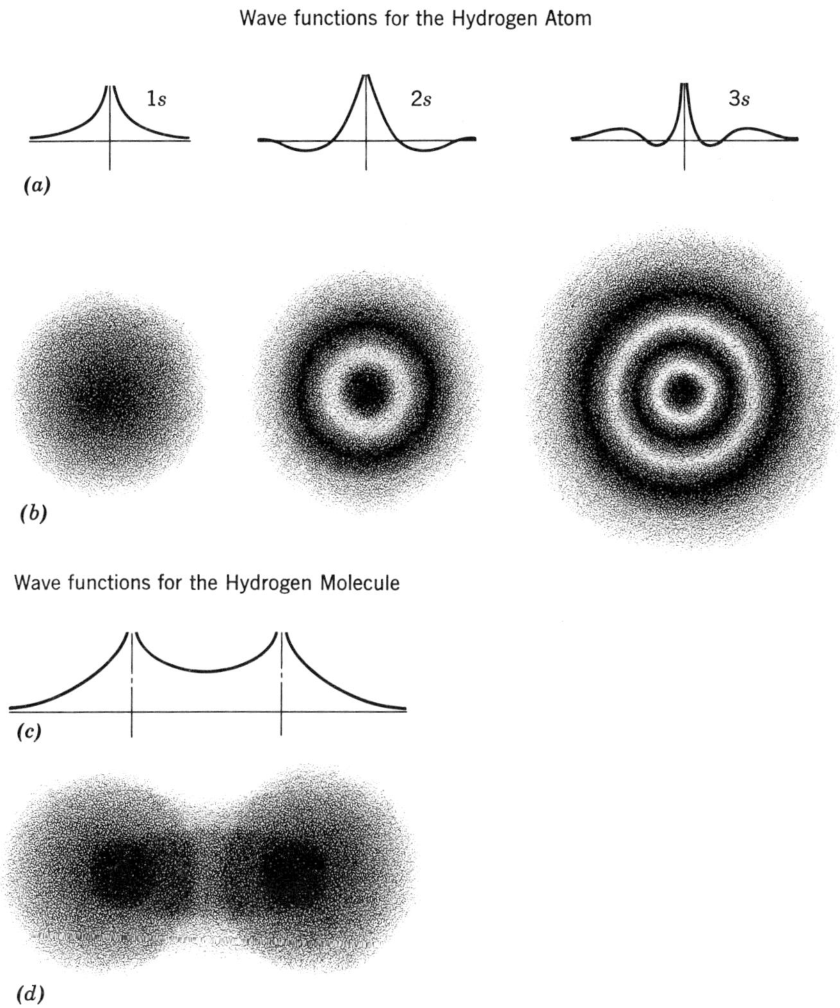

Fig. 2. Charge distribution in hydrogen.

pairs of atoms strongly bound to each other, but with only loose association between different molecules; but, because any one lithium atom in the crystal lattice has not *one* nearest neighbor, but fourteen neighbors, we find a very new and different bonding. Recall what happens in the case of benzene, where six carbon atoms are arranged in a ring. There, the electron waves spread themselves out around the six carbon atoms. Because a wave is, by nature, *periodic,* because it repeats itself with alternate waves and loops in space, a special situation arises when such a wave occurs in a periodic structure. Around

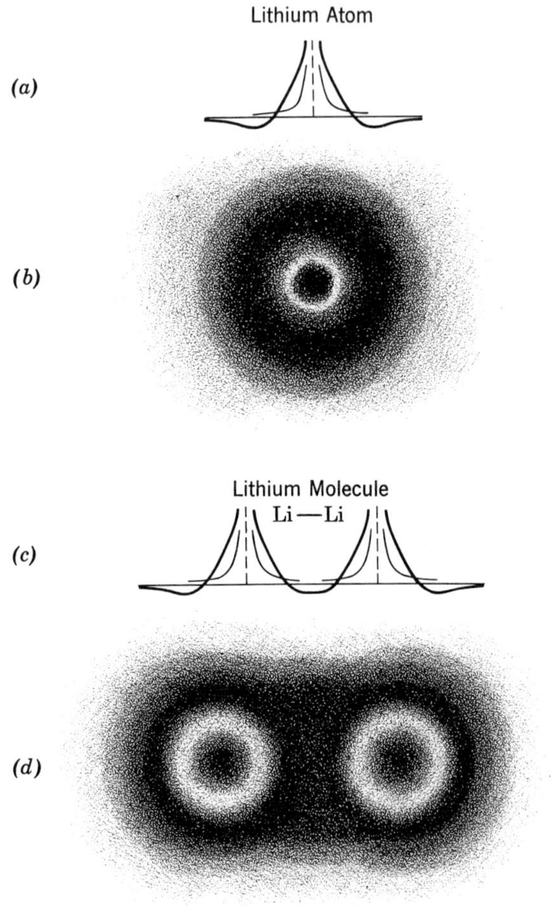

Fig. 3. Charge distribution in lithium.

the edge of the hexagonal benzene ring, we find a repetition at each of the six successive atoms. In other words, the benzene ring represents a periodic structure in terms of the edge of the hexagon. The electron wave in this periodic structure is, of course, determined by the structure; as a result the nodes and loops of the wave arrange themselves in a periodic relation with respect to the periodic nuclei around the edge of the ring. This is shown in Fig. 4; it was described in detail in Chapter 6. Linus Pauling pointed out the analogy between metallic binding and resonance binding as found in molecules like benzene. It now seems clear that it is this large-scale resonance binding that provides the force that ties atoms together in the metallic lattice.

In the metal, there is a similar situation. There are the waves that are periodic in nature; the atoms are arranged in the periodic lattice; and it is natural to expect something like the condition found in benzene but on a larger scale. Imagine a lithium crystal that is a perfect cubic crystal, with each of its edges 1 mm long. Then there will be a line of atoms from one face through the crystal to the other, more than six million in number. In contrast to benzene, where there are *six* atoms in a ring ready to share a single electron, in this metallic crystal there are six million atoms in a line ready to share a single electron; and in the x, y, and z directions taken together there are 10^{20} atoms ready to share the same single electron!

In Fig. 5, the dashed lines show the type of wave for a single 1*s* electron on hydrogen atoms marked 1 and 2. These waves overlap when the atoms are brought this close together in the lattice. Because of this overlap, the wave does not take the form shown by the dashed lines, but the form taken by the solid lines; in other words, there is periodic array of nodes and loops extending across the lattice for six million atoms. Thus the electron is spread out just as it was in benzene, but extended millions of times farther; and we must remember that this electron is spread out not only along the x direction as shown, but also in all the other directions because the lattice is periodic along all three axes.

Thus the wave nature of the electron produces a situation that is impossible to visualize precisely. This illustrates again the fact that we cannot make a visual model which corresponds exactly to nature; and this talk of waves must be regarded as only one incomplete perspective of a situation with a structure beyond the scope of any visual model. It is helpful and stimulating to our thinking, however, to take a look at the visual aspect of this underlying structure of the

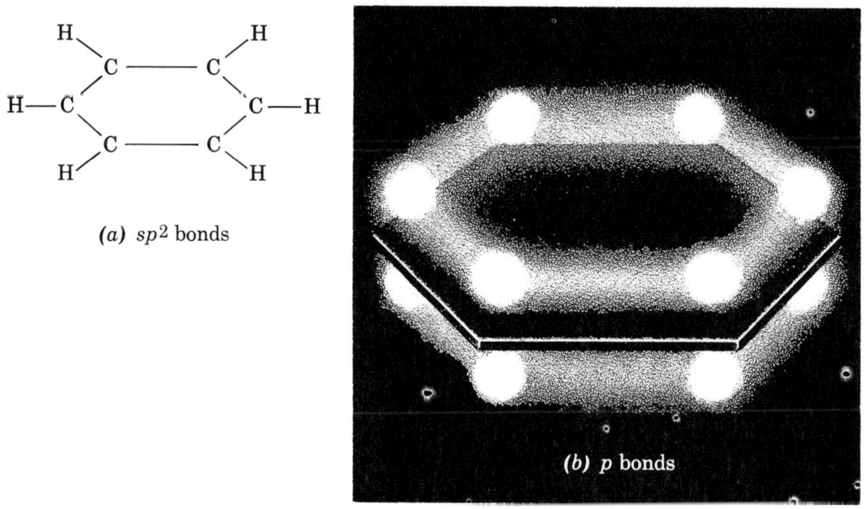

Fig. 4. Resonance binding in benzene (schematic).

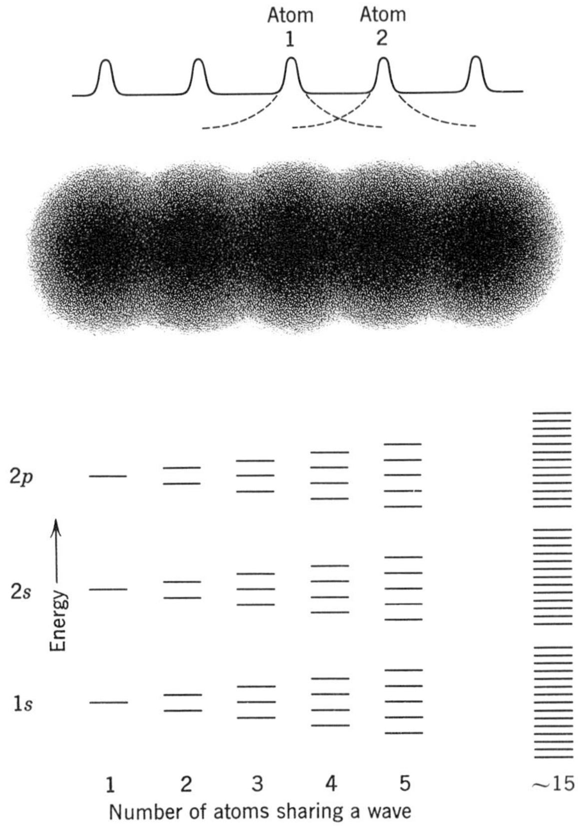

Fig. 5. Resonance binding in a crystal lattice made of imaginary metallic hydrogen. This drawing is meant to represent principles only. For hydrogen atoms, *s* and *p* levels have the same energy; and the number of levels arising from *p* levels is greater than that from *s* levels.

electron even though this view is seriously limited, for it helps us to coordinate our knowledge in a more coherent and logical way. And we get a better understanding by expressing mathematically the structure of this wave situation; but we must still remember that mathematics, itself, is still only a kind of model, somewhat more sophisticated and subtle than a model that can be visualized in three-dimensional space, but still essentially a model that can never approach the complete physical reality. Such a model does guide us in showing how other aspects, such as energy of these waves, can be logically related to their space pattern.

In the lower part of Fig. 5 there is a diagram which illustrates the way in which the sharing of waves between different atoms changes the energy levels. Consider first the electron wave focused around a single nucleus, one wave associated only with one atom. This is the situation shown at the left in column 1. The energy levels are drawn in schematically to illustrate the principle of sharing, so that no significance can be attached to the actual values of the energy shown. In a general way, we expect to find a single 1*s* level, a single 2*s* level, and a single 2*p* level related to each other roughly as shown on the diagram, with energy increasing vertically. For example, these might be the wave types associated with a hydrogen or a lithium atom. Now, as soon as two hydrogen atoms approach one another so that there is an overlap of their 1*s* waves, the resultant *shared* wave pattern has two energy levels, one a little higher and the other a little lower than the value of the energy for the 1*s* waves as they existed on each of the atoms separately as shown in column 1. In other words, when there are two *individual* atoms, the 1*s* electron on each atom will have the value of the energy as shown by the *single* line. When we bring those two atoms together, there are two joint wave patterns formed with different energy. The wave pattern having the lower energy corresponds to the sum of the individual waves and yields a density plot like that shown in Fig. 3. The wave pattern with the higher energy corresponds to the difference of the two waves and yields a density plot like that shown on p. 139 for the σ^* orbital. Both of these levels can accommodate two electrons. When the two electrons are paired in the lower level, we form an H_2 molecule in the ground state. Other distributions of electrons correspond to unstable excited states.

When the wave is shared between three hydrogen atoms, then there are three centers of positive charge (one on each of the three nuclei); and there are three levels in the wave now spread out over all three nuclei. In the same way, four atoms have four waves spread out over these atoms with energy levels as shown in the fourth column; and for five atoms there are five different types of waves.

Because the energy levels are spread out both above and below the initial level for a wave on a single atom, the levels approach one another as the number of sharing atoms increases. In the far right-hand column for waves spread out over 15 atoms, there are bands of waves where the separation between the band originating in the 1*s* level and the band originating in the 2*s* level becomes very small; and, in a crystal with millions of atoms sharing a single wave, the patterns of energy levels associated originally with a single level spread out to the point where the bands may overlap.

In Fig. 6*a*, this kind of spreading is shown as a function of the distance of separation of the lithium nuclei. In an imaginary crystal of lithium, where the atoms were separated by distance of more than 12 Å between neighbors, each atom would behave essentially like an isolated lithium atom; there would be a sharp 2*s* level and a sharp 2*p* level as shown at the right-hand side of the diagram. As the distance between neighbors becomes less, the overlapping of the waves between neighboring atoms is sufficient to split these sharp levels into bands; in other words, decreasing the distance between neighbors has the same effect as having more and more neighbors. Thus, when the atoms approach to

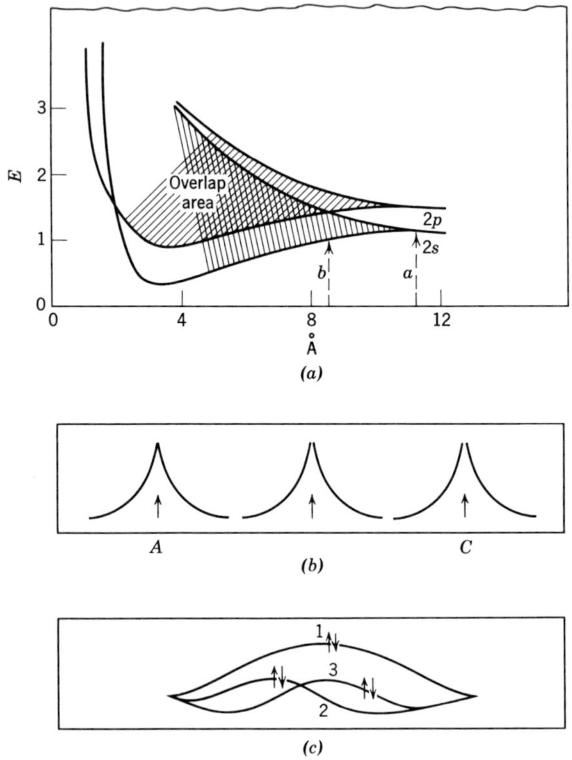

Fig. 6. (*a*) Overlap effect in lithium as a function of interatomic distance; (*b*) waves in metallic hydrogen—no overlap; (*c*) waves in metallic hydrogen—large overlap.

a distance of about 11 Å, at the point shown by the dotted line labeled *a*, the bands begin to split in exactly the same way as indicated in column 2 in the energy level diagram. When the distance between neighboring atoms drops to about 9 Å, the 2*s* band has widened and the 2*p* band has widened, and the two bands begin to overlap. This overlapping is at a distance shown by the dotted line that is labeled *b*.

Now, the question is, can we interpret these new wave patterns found in a periodic crystal in a way that will give us some kind of visual perspective of the nature of the wave? Again keeping in mind the limitations of any model, we can say that, with waves in a crystal like this, we can consider each wave as extending from one face of the crystal to the opposite face. In a metallic hydrogen crystal with the atoms separated by large distances and each wave extended indefinitely in each direction, there would be so little overlap that one wave would not appreciably affect another. In other words, there is a collection of independent waves, each associated with a separate atom, as in Fig. 6*b*. Here a one-loop symmetrical wave is focused on each of the individual atoms. After we have brought these atoms close together, the waves overlap and the effect of periodicity is seen, so that the wave pattern of three individual waves all looking alike changes to the wave patterns in Fig. 6*c*. Thus, instead of the three familiar one-loop spherical wave of the type 1*s*, there are waves of an entirely new type extending from one face of the crystal to the other. These are waves very much like the waves in a one-dimensional string.† They would resemble these waves closely if we were dealing only with a single line of atoms and not with an entire crystal. In other words, we would have the wave of lowest energy with one loop, as shown in the wave labeled one. We would have the wave with the energy next above this, a two-loop wave, as shown in the wave labeled 2, and the next wave would have three loops as shown in the wave labeled 3. Thus there would be a collection of waves extending from crystal face to crystal face, related to one another in a way closely like the standing wave patterns discussed in connection with the vibrations of a string.

How does the Pauli exclusion principle enter into this new situation found in a crystal? With individual atoms, the exclusion principle allows two electrons in each of the spherical one-loop waves (1*s*) provided that the two electrons have spins of opposite sign. In the same way, for the waves in a complete crystal, we can have again two electrons associated with each wave pattern. We have indicated this by drawing two arrows on top of each of the three kinds of waves shown.

With regard to energy, however, there is a very different situation in the compact crystal with overlapping waves as contrasted with the expanded imaginary crystal of isolated atoms. In the expanded crystal with isolated independent atoms, each electron wave has the same energy. If we had a crystal consisting of 10^{20} atoms, each with a 1*s* electron, each of these electrons would have the same energy. But, in the real crystal *where there is overlap*, the wave pattern is like that shown in Fig. 6*c;* we can only have two electrons with the lowest energy, namely that of the part of Fig. 6*c* labeled 1; the next two electrons which are in the crystal will have an energy slightly higher, that associated with the two-loop wave, labeled 2; and the next two electrons will have an energy still slightly higher associated with the three-loop wave, labeled 3, and so on.

Thus, the exclusion principle results in an energy distribution entirely different from that expected on the basis of classical physics applied to an electron gas. In an ideal gas, the molecules are completely independent and on the average each gas particle has $\frac{3}{2}kT$ energy. The increase in energy per degree is the heat capacity; for an ideal monatomic gas, this is $\frac{3}{2}k$ per molecule or $\frac{3}{2}R$ per mole. In the older picture of metals, the electrons *were assumed to move* in the crystal lattice much like gaseous molecules. By analogy, each mole of these electrons would

† See pp. 83–84 for a discussion of these waves.

have associated with it ($\frac{3}{2}$)RT of energy; and the heat capacity of the metal, therefore, should contain a term with a value $\frac{3}{2}R$ for the energy absorbed by the gaseous electrons as the temperature of the whole gram mole of metal is increased. According to experimental data, the electrons do *not* absorb energy as do free gaseous particles. This is expected on the basis of quantum mechanics. What, then, are the properties of metallic electrons? First, the electrons must be regarded not as gas-like particles but as waves distributed over the entire crystal. Just as we had to give up the idea of electrons on individual atoms as particles circulating in orbits, so we have to give up the idea of electrons in metallic crystals as gas-like particles. If, then, the electrons are waves, they must obey the Pauli exclusion principle, applied now not to an individual atom or to an individual molecule, but to the entire crystal. Every electron in the entire crystal must have a set of quantum numbers different from every other electron in the crystal. If there are 10^{20} of these electrons in the crystal, we must have 10^{20} different sets of quantum numbers.

Next we observe that, as the quantum number goes up, the number of loops in the wave goes up just as in atomic orbitals. The more loops we have, the shorter will be the wavelength; and the shorter the wavelength, the higher will be the energy. So, in order to accommodate all the 10^{20} electrons, we completely fill all the possible energy levels up to a point where the electrons with the highest energy have far more than they would if they were behaving as gas particles.

When temperature increases from 0°K, the only way in which the electrons can absorb energy is by a process in which the electrons with the highest energy move to still higher energy levels. The electrons in the low and middle levels cannot go to the next higher levels because all the levels above them are filled. There is room only at the top for electrons to go still higher on the energy level diagram. When we heat a mole of an ideal gas we (statistically) promote *all* the gas molecules to higher energy levels, and this results in a specific heat of $\frac{3}{2}R$ per mole of promotable gas atoms. When we heat a metal we still need about $\frac{3}{2}R$ cal per degree for one mole of promotable electrons. But only a fraction f of all the electrons present are promotable; hence the heat capacity per mole of electrons (promotable and unpromotable) is roughly $f \times \frac{3}{2}R$, where f is a small fraction that varies with temperatures. At very low temperatures, only the electrons near the top of the band contribute to the heat capacity. As the temperature increases, more and more electrons move to higher levels, leaving empty levels behind them so that f gradually increases with temperature. At high temperatures, all electrons can find nearby empty levels above them. Then they all contribute and f becomes 1. Calculations show that the temperature at which f becomes 1 (and the electrons become truly gas-like) is roughly 50,000°K. Such temperatures cannot be attained before the metal vaporizes.

ALLOYS

The nature of the chemical bond in metals gives rise to an important class of metallic substances called *alloys.* These are substances in which two or more metals† are combined in a mixture. Sometimes this is a true compound formation between the elements; and the alloy has a definite chemical formula like AgSr or Ag_2Sr_3. More frequently, the sharing of the spread-out electrons in metallic bonding makes it possible for the two elements to merge over a wide continuous range of solid composition exactly like substances mutually miscible in liquid solution. These alloys are examples of *solid solutions.* As an example, Au and Ag are miscible in the solid state in all proportions and form a continuous series of solid solutions ranging all the way from pure Au to pure Ag.

Just as the thermodynamic approach to familiar liquid solutions like alcohol in water helps us to understand the nature of the solution process and solution behavior, so the thermodynamic perspective enables us to classify the thousands of known alloys and to understand their various patterns of behavior. Since alloys are frequently formed by melting the two original pure components like Bi and Sn and then cooling to obtain the solid, we examine first the nature of the solidification or crystallization process where the two components form a thermodynamically ideal solution when in the liquid state, but do not interpenetrate or dissolve mutually in the crystalline state.

Ideal thermodynamic solution. Even though metallic atoms either in the crystalline or liquid state are bound by interaction between ion cores and quasi-free conduction electrons, they frequently behave as if they were molecules attracted to each other by simpler forces in so far as the same equations may be applied to interpret their behavior.

In discussing solutions, we have seen how two similar compounds have similar fields of force around each molecule. The same holds true for the atoms of two similar metallic elements like bismuth and cadmium. These two metals mix to form an ideal solution. When a bismuth atom leaves the solution to go into the vapor state, it gains about the same enthalpy as if it were evaporating from pure liquid bismuth. The vapor pressure of bismuth is lower in a solution of bismuth and cadmium because the *entropy* of the bismuth in solution is greater. Hence the probability of passing into the vapor state is less.

In the description of such solutions, the term *phase* refers to the different states of matter like the crystalline state or the liquid state, and also refers to liquid solutions or solid solutions. In general, one *phase* is separated from another by a sharply defined physical boundary. In liquid-vapor equilibrium, this phase

† Alloys are sometimes formed by mixing metals and non-metals.

boundary is the *surface* of the crystal. Occasionally there is a phase boundary between two different liquids having different compositions and not mutually miscible. Thus the term *physical state* generally refers to solid, liquid, or gas, whereas phase may refer to these and also to two different kinds of liquids or two different kinds of crystals that can be brought into equilibrium with each other.

The equilibrium between the crystal and the vapor phase of bismuth is expressed by the equation

$$\mathrm{Bi}(s) \rightleftarrows \mathrm{Bi}(g)$$

$$K = \frac{(\mathrm{Bi}\ g)}{(\mathrm{Bi}\ s)}$$

Here the factor for the crystalline state s is taken as unity, so that

$$K = (\mathrm{Bi}\ g) = P_{\mathrm{Bi}}$$

The variation of K with temperature is given by the expression

$$\ln K = -(\Delta H_s/R)(1/T - 1/T_s)$$

where T is the absolute temperature at which K has the value under consideration, T_s is the absolute temperature at which $K = 1$, ΔH_s is the heat of sublimation, and R is the gas constant with the same energy units in which ΔH_s is expressed. Since $K = P_{\mathrm{Bi}}$, $\ln P_{\mathrm{Bi}} = -(\Delta H_s/R)(1/T - 1/T_s)$, the familiar equation for the variation of vapor pressure with temperature that gives us a graph of the type shown in Fig. 7.

Now we can regard the equilibrium between liquid solution and crystal in exactly the same way:

$$\mathrm{Bi}_{\mathrm{crystal}} \rightleftarrows \mathrm{Bi}_{\mathrm{in\ solution}}$$

$$K = \frac{[\mathrm{Bi}_{\mathrm{in\ solution}}]}{[\mathrm{Bi}_{\mathrm{crystal}}]}$$

Just as a certain pressure of Bi vapor will be in equilibrium with pure liquid bismuth at a given temperature, a certain concentration of Bi in solution $[\mathrm{Bi}_{\mathrm{in\ solution}}]$ will be in equilibrium with pure crystalline bismuth at a given temperature. As before, the factor for the pure crystal in the equilibrium expression is taken as unity, so

$$K = [\mathrm{Bi}_{\mathrm{in\ solution}}] = x_{\mathrm{Bi}}$$

where x_{Bi} denotes the concentration of Bi in equilibrium with pure crystalline Bi at a given temperature; x is normally expressed as mole fraction.

$$\ln x_{\mathrm{Bi}} = -(\Delta H_f/R)\ (1/T - 1/T_m)$$

Just as the vapor pressure equation contained ΔH_s, the enthalpy change in going from one phase (crystal) to the other (vapor), here the expression for variation of

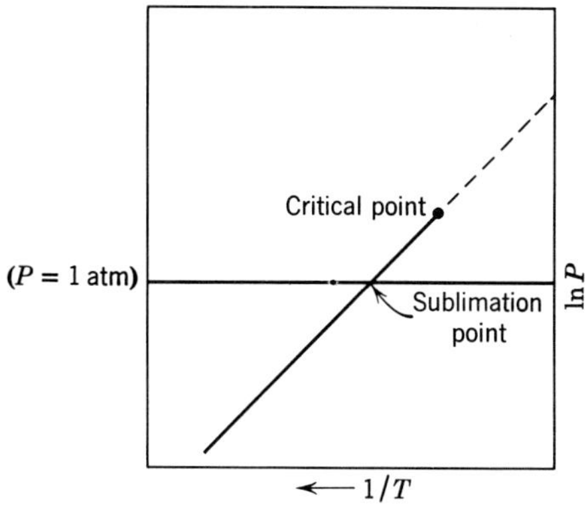

Fig. 7. Typical vapor pressure graph on ln P against $1/T$ plot.

concentration in the liquid (x) contains ΔH_f, the enthalpy change in going from crystal to the liquid solution phase.† In the vapor pressure equation, the constant temperature term $(1/T_s)$ contained the temperature at which P_{Bi} is unity, namely the sublimation temperature T_s; so, in the concentration equation, the constant-temperature term $(1/T_m)$ contains the temperature at which solution with $x_{Bi} = 1$ will be in equilibrium with the pure crystal; in other words, $x_{Bi} = 1$ means pure liquid, and the temperature at which pure liquid Bi is in equilibrium with pure crystalline Bi is the melting point (T_m).‡

If we take some pure liquid Bi at T_m and drop some crystalline Bi into it, the two phases will remain in equilibrium at the melting temperature. If we now drop some pure cadmium into the liquid phase, it will dissolve and the temperature at which the solution and crystalline bismuth are in equilibrium will fall; x_{Bi} is now less than unity and the equilibrium temperature is lower. The relation between x_{Bi} and T is given, of course, by the equation above.

Just as with vapor pressure, we can plot the values of x_{Bi} on a logarithmic scale against $1/T$, giving the *straight line* graph shown in Fig. 8. The logarithmic equation relating x_{Bi} and $1/T$ holds for all values of $x_{Bi} < 1$ that are described by the equilibrium

$$\mathrm{Bi}_s \rightleftarrows \mathrm{Bi}_{\mathrm{sol'n}}$$

It will break down when other phases are present or one of these is absent. For example, a solution prepared by mixing a given amount of bismuth and cadmium

† As contrasted with ideal solutions where ΔH_f is independent of x, there may be considerable variation of ΔH_f with x in non-ideal solutions.

‡ The equation above is clearly valid only for $x_{Bi} \leqslant 1$. Although the mathematical expression predicts values of x_{Bi} greater than unity, this is not physically realizable for a mole fraction. On the other hand, the vapor pressure expression applies even for values of $P_{Bi} > 1$ since such values are physically realizable.

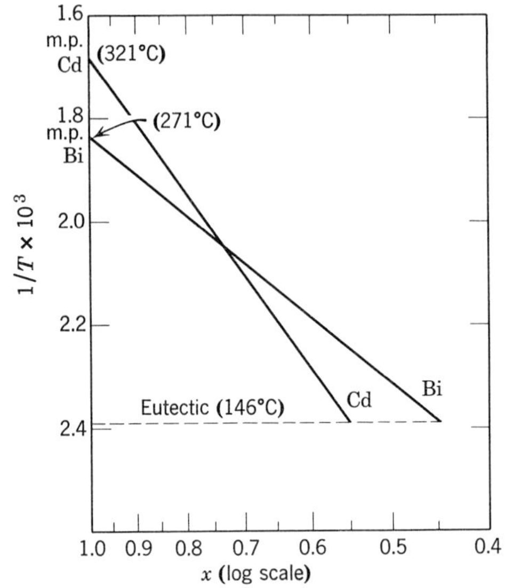

Fig. 8. The binary solution system—bismuth-cadmium.

is stable at a variety of temperatures provided solid bismuth and cadmium are absent, but if it is in equilibrium with crystalline bismuth the temperature is specified by the logarithmic equation. A similar situation holds for sodium chloride in water. At a given concentration a salt solution can exist for a variety of temperatures, but it is in equilibrium with solid sodium chloride at only one temperature. Thus, on the basis of the relations shown in Fig. 8, we can make a plot of the equilibrium temperature as a function of composition. The left half of Fig. 9 ($x_{Cd} < 0.55$) is a plot of the temperature at which a solution (melt) of composition x_{Cd} is in equilibrium with solid bismuth; the right half ($x_{Cd} > 0.55$) is a plot of the temperature at which a melt of composition x_{Cd} is in equilibrium with solid cadmium.

As the temperature is lowered below the melting points of both pure metals, we finally reach the point where a solution of 45 atomic per cent Bi and 55 atomic per cent Cd is in equilibrium with *both* crystalline Bi and crystalline Cd as shown by the intersection of the curves at this point. This point is called the *eutectic* composition and the mixture of solid Bi and Cd separating out there as more heat is abstracted is called the eutectic mixture.

Thus the formula relating $\log x$ to $1/T$ gives the complete quantitative relationship of solubility to temperature for this ideal system. As with the plot of $\log K$ or $\log P$, the slope of the straight line in the log plot is equal to a heat factor, here ΔH_f, divided by $2.303R$. It is frequently far easier to get a value for the heat of fusion from the slope of this graph than to measure it by direct experiment. Of course, the linearity of the graph depends on the fact that ΔH_f does not vary appreciably over the range of temperature considered.

Phase diagrams. The temperature-composition curve for the Bi-Cd system is a *phase* diagram. We used phase diagrams earlier to describe the allotropy of sulfur. In that case, the only variables were temperature and pressure. For the bismuth-cadmium system we also have the variable of concentration, so we need a three-dimensional (pressure, temperature, and composition) diagram to show all possibilities. Let us therefore content ourselves with what happens at a fixed pressure of 1 atm; then we can draw the phase diagram shown in Fig. 9.

There are five regions in the diagram: M, M + Bi, M + Cd, Eu + Bi, Eu + Cd. For any composition and temperature in the region M, we find that the liquid alloy (melt) is stable; in the region M + Bi, the melt plus crystalline bismuth is stable; in the region M + Cd, melt plus crystalline cadmium is stable; in the region Eu + Bi, we find bismuth crystals and the solid eutectic containing a mixture of bismuth and cadmium crystals; in the region Eu + Cd, we find cadmium crystals and the solid eutectic mixture.

Suppose we start with a mixture containing 0.8 gm atom of Bi and 0.20 gm atom of Cd at a temperature corresponding to point *a* and allow it to cool. The melt will persist until we reach the temperature T_1; then a small amount of pure crystalline bismuth forms. When this occurs, the melt becomes richer in cadmium and, as we cool to lower temperatures, the composition of the melt

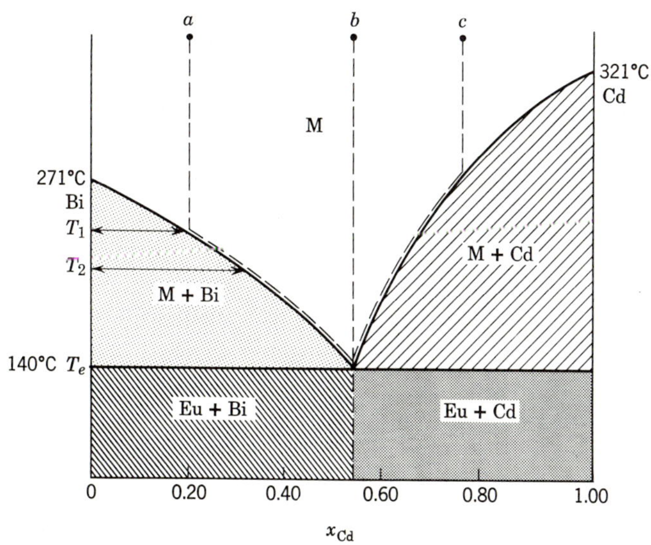

Fig. 9. Phase diagram, ideal system, for Bi and Cd (Ref. 6). (Based primarily on data from *Constitution of Binary Alloys* by Max Hansen, published by McGraw-Hill Book Co., Inc., New York, 1958. Printed with permission from the publisher.)

changes along the indicated dashed line as more and more bismuth freezes and the melt becomes richer in cadmium. At the point T_2, the composition of the melt is such that $x_{Cd} = 0.30$, and from our starting conditions we can compute, Y, the number of moles left in the melt. This is done as follows:

$$\text{Cd} + \text{Bi in melt} = Y \qquad\qquad \text{Solid Bi} = 1 - Y$$

$$0.70Y + 1 - Y = \text{total amount Bi} = 0.80 \text{ mole}$$

$$1 - 0.30Y = 0.80$$

$$Y = \frac{1 - 0.80}{0.30} = 0.67 \text{ mole}$$

Thus at T_2 we have a melt containing 0.20 mole of Cd and 0.47 mole of bismuth and 0.33 mole of crystalline bismuth. The composition of the melt is $x_{Cd} = 0.30$, but the composition of the melt plus solid is, as it must be, $x_{Cd} = 0.20$. Precipitation of bismuth continues as we cool the mixture and the melt becomes richer in cadmium until we reach the eutectic temperature T_e. At this point we have initially a melt of the eutectic composition $x_{Cd} = 0.55$ and solid bismuth. As soon as we cool below T_e, the whole melt solidifies to form a mixture of small crystals of bismuth and cadmium. Thus the net result of cooling from point *a* to below T_e is the eutectic mixture and crystalline bismuth.

If we start with the eutectic mixture at point *b* and cool it, no solid appears until we reach T_e. At T_e, the melt solidifies to yield the eutectic mixture.

If we start at point *c*, the course is that indicated by the dotted line. The behavior is much like that described for point *a* except now the roles of the cadmium and bismuth are interchanged.

Cooling curves. One of the simplest ways of determining the phase diagram of a substance is to cool the melt by the steady removal of heat while measuring the temperature. Since phase transitions are always accompanied by a gain or loss of heat, phase changes are indicated by a break in the cooling curve.

A series of idealized cooling curves are shown in Fig. 10. In each curve we assume that we are removing a specific amount of heat per minute. The curve in Fig. 10*a* represents a plot of temperature versus time (calories removed) for pure *solid* cadmium. Since the heat capacity of cadmium is roughly constant over a small range of temperature, the temperature drop per calorie removed (or per unit time) is constant and a straight line results.

Now consider Fig. 10*b*, which represents a cooling curve for bismuth initially present as a *liquid*. As long as only liquid bismuth is present, the plot is a straight line. As soon as bismuth starts to freeze, however, heat is given off by this phase transition. If equilibrium is maintained, the temperature cannot fall below the melting point until all the liquid has frozen. Hence the heat supplied by the phase transition from liquid to solid bismuth just compensates for the constant removal

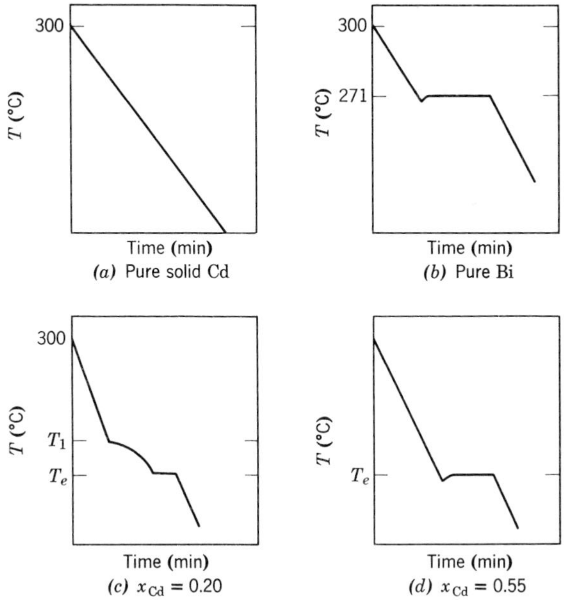

Fig. 10. Cooling curves.

of heat and the temperature stays constant until all the bismuth has frozen. Then the rate of temperature drop again becomes constant. Thus in Fig. 10*b*, the flat portion corresponds to the melting point of the bismuth. In practice, the flat portion of the curve is seldom absolutely flat and is usually preceded by the small dip shown in Fig. 10*b*. What is the significance of this small dip?

Figure 10*c* corresponds to cooling a melt with the composition marked *a* in Fig. 9. As the sample is cooled to T_1, the drop is uniform. At T_1, a small amount of bismuth crystallizes out and some heat is liberated so that the drop in temperature is less pronounced. No more solid bismuth is formed until the temperature drops slightly below T_1, so the temperature continues to drop; but between T_1 and T_e the rate is less than it would be for cooling a pure compound because in this region bismuth crystallizes out and continuously liberates heat to offset partially the constant heat loss during cooling. When T_e is reached, the temperature can fall no further until the remaining melt freezes to form the eutectic mixture. Therefore at this point the curve is again flat, just as in Fig. 10*b*, until all the melt solidifies.

Figure 10*d* corresponds to the cooling of the melt *b* in Fig. 9. Since the only phase transition occurs at T_e, the curve is flat at T_e.

Phase rule. The behavior of such two component systems is restricted by thermodynamics. If we call the number of species that must be specified to describe the composition the components C, and the number of phases P, then, for a *constant-pressure phase diagram*, we can say

$$V = C - P + 1$$

where V is the number of independent variables. In other words, the variables are temperature and composition. If the allowed composition is fixed by the temperature, we have $V = 1$ corresponding to one independent variable. If neither one can vary, $V = 0$; if both can vary, $V = 2$. In these diagrams $V = 2$ corresponds to a region in which both T and X can change; $V = 1$ corresponds to a line since one specifies the other; $V = 0$ corresponds to a point.

Let us apply this to the bismuth-cadmium phase diagram for which $C = 2$. If we are considering a situation where we have only the molten phase present, $P = 1$ and $V = 2$. In other words, this corresponds to the region M in which we can change T without changing x_{Cd}; hence T and x_{Cd} are independent and $V = 2$. If we have two phases present, $V = 1$. We can vary the composition x_{Cd} so that it has any value, but the value of T is fixed for a given value of x_{Cd}. In other words, for two phases we get a line on the phase diagram. Suppose now we have three phases present: solid bismuth, solid cadmium, and melt. Then $V = 0$ and this corresponds to the eutectic point on the phase diagram.

The phase rule is a useful guide in constructing a phase diagram of any complexity. If multidimensional diagrams are made in which pressure is not fixed, the general rule becomes

$$V = C - P + 2$$

This applies to any equilibrium involving different phases. For example, it tells us that the point where liquid, solid, and gaseous water are in equilibrium ($C = 1$, $P = 3$) is fixed at a single temperature and pressure. But it also tells us that the melting point of a salt solution ($C = 2$, $P = 3$) is not fixed, unless the composition is specified (for this case $V = 1$); hence there is one independent variable. If we fix this one, the others (temperature and pressure) are fixed. The number of variables V is often called the number of *degrees of freedom* of the system.

Alloy compounds. In the ideal solution just discussed, the conditions existed: (*a*) that no fixed combination of components formed a stable crystalline compound; and (*b*) that neither component was able to penetrate into the lattice of the other to form solid solutions. Let us consider next the case where intermediate compounds are formed. If magnesium and zinc are melted together, it is found that a crystal lattice can be formed where the proportion of Mg to Zn is given by the specific formula $MgZn_2$. A plot of x_{Mg} and x_{Zn} is shown in Fig. 11. Because of favorable size and energy relations, Mg and Zn atoms can arrange themselves in this stable crystal lattice which has a precise ratio of the two components. This lattice behaves exactly like a stable compound with respect to solubility. Imagine that the diagram is split down the middle at $x_{\text{Mg}} = 0.33$; then the left half and the right half of the original diagram each look like ideal solubility diagrams. In many cases of compound formation, these separate diagrams are given by the ideal equation relating x and $1/T$. The behavior of the compound alloy can thus be predicted quantitatively by the thermodynamic equation.

Consider now what happens when melts corresponding to a and b in Fig. 11

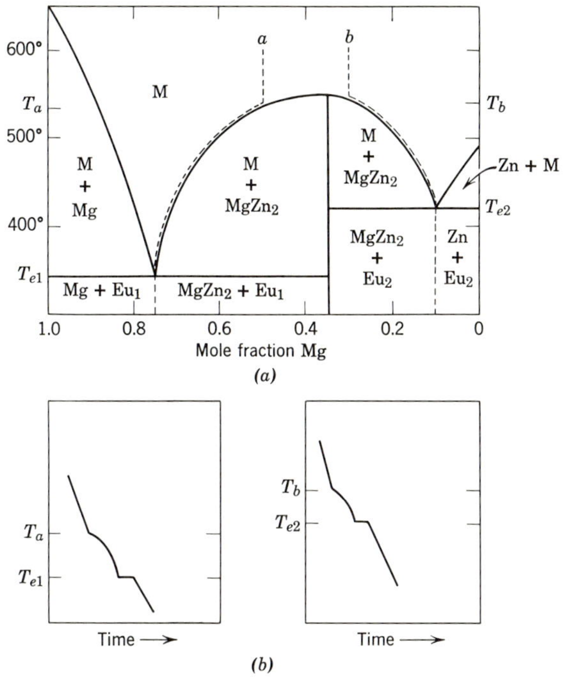

Fig. 11. (*a*) Simplified phase diagram for Mg-Zn (Ref. 6); (*b*) cooling curves. (Based primarily on data from *Constitution of Binary Alloys* by Max Hansen, published by McGraw-Hill Book Co., New York, 1958. Printed with permission from the publisher.)

are cooled. For the composition *a*, we have the melt, provided the temperature is above T_a. At T_a a solid compound with the composition $MgZn_2$ begins to crystallize, and the melt becomes richer in Mg. At any temperature between T_a and T_{e1} this alloy will contain melt and solid $MgZn_2$; the changing composition of the melt is indicated by the dotted line. At T_{e1} the melt has the composition of the eutectic Eu_1 and, if it is cooled below this temperature, mixed crystals of $MgZn_2$ and Mg in eutectic proportions form, together with the solid $MgZn_2$ already present. Two cooling curves for this phase diagram are shown in Fig. 11*b*.

Partial solid solution. When the two atomic species in solution are composed of atoms of about the same size and energy field, it is frequently found that there is penetration by one into the crystal lattice of the other. The system Bi-Sn is an example of this partial solid solution; that is, from a liquid where $x_{Sn} = 0.65$ and $x_{Bi} = 0.35$, bismuth atoms will penetrate into crystalline tin and replace about 10% of the tin atoms, so that the composition of the solid is only 90% Sn when it is in equilibrium with the solution at the temperature of 155°C. The solubility diagram for Bi and Sn is given in Fig. 12. This shows that, at the temperature of 155° C, it is possible to replace the Sn atoms in the crystal lattice of Sn by Bi atoms up to 10 atomic per cent of Bi. Beyond this point, additional Bi atoms melt some of the alloy and begin to form a liquid. This is consequently a case of *limited solid solution.* Such an alloy is a solution in the sense that the composition of the Bi can be varied continuously from 0 to a limiting percentage according to the graph; but it is still a true crystal lattice with certain Sn atoms displaced by Bi atoms that occupy the corresponding places in the crystal lattice.

Now consider what will happen as we cool the melts of composition a, b, c, indicated in Fig. 12. The cooling curves are shown in Fig. 13. Point a is simplest: for $T > a_1$ we have the melt; for $a_2 < T < a_1$, we find the melt in equilibrium with crystalline bismuth; at $T = a_2$, the composition is that of the eutectic mixture; and below a_2 we find the solid eutectic mixture and bismuth crystals. The new aspect of this case is that for $T < a_2$ the solid does not consist of crystals of bismuth and crystals of tin, but of crystals of pure bismuth and crystals of tin with some dissolved bismuth. Moreover, the composition of the tin-bismuth solid solution varies as the temperature is lowered below a_2. This feature is (in principle) reflected in the cooling curve. If we were cooling a mixture of bismuth and tin-bismuth crystals and the composition of each remained fixed, the cooling curve below a_2 would be a straight line, just as it is in Fig. 11. Since bismuth is precipitating out of the tin-bismuth solid solution and a heat is connected with this change, the cooling curve below a_2 is not quite linear if equilibrium is attained. (In practice, equilibrium in solids is so sluggish that it would not be attained unless the cooling were very slow.)

As we cool the alloy b, a solid first appears when $T = b_1$. The solid in equi-

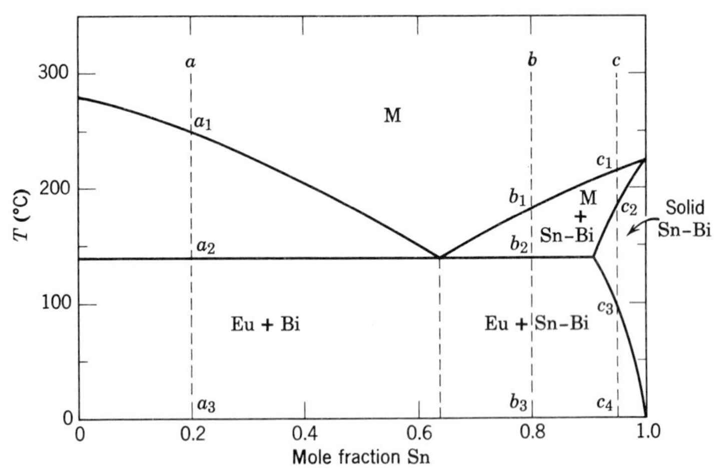

Fig. 12. Simplified phase diagram for Sn-Bi (Ref. 6). (Based primarily on data from *Constitution of Binary Alloys*, by Max Hansen, published by McGraw-Hill Book Co., New York, 1958. Printed with permission from the publisher.)

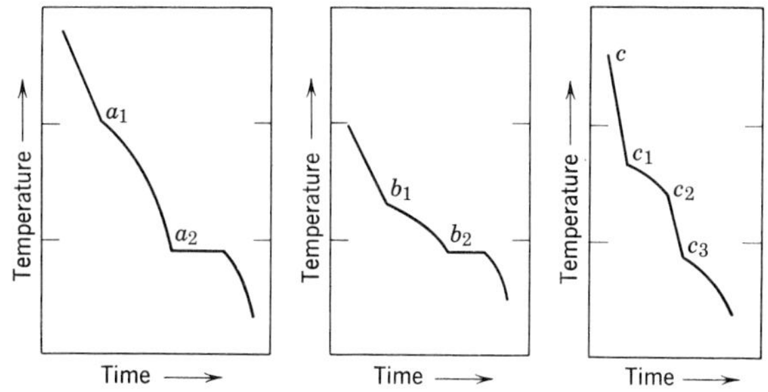

Fig. 13. Cooling curves, system Sn-Bi.

librium with the melt is not pure tin but a solid solution of bismuth in tin. The composition of the solid solution in equilibrium with the melt becomes richer in bismuth as the temperature is lowered to b_2; and at b_2 the whole melt solidifies to form the solid eutectic mixture together with the bismuth-tin solid solution. As the temperature is lowered below b_2, the bismuth-tin solid solution now becomes richer in tin; and, if the temperature is lowered to b_3, the equilibrium solid consists of crystals of pure tin and pure bismuth. This behavior is reflected in the cooling curve in Fig. 13*b*.

The most complicated case is found for alloy *c*. When we cool the melt to c_1, a tin-bismuth solid solution forms. With further cooling, the solid solution becomes richer in bismuth and finally at c_2 no melt is left; all the alloy is in the form of a solid solution. Further cooling to c_3 only cools the solid solution, but below c_3 pure bismuth crystals begin to form and the solid solution becomes richer in tin. When $T = c_4$, we find we have a mixture of pure tin crystals and pure bismuth crystals.

The cooling curve in Fig. 13*c* reflects these changes. Above c_1 we are just cooling the melt, so we find a straight line. Between c_2 and c_3, the solid solution slowly forms and we obtain a curved line. At c_2 all the melt is solid; this occurred without an isothermal freezing of the melt, so the curve is not horizontal here. From c_2 to c_3 we are simply cooling the solid solution, and since no phase change occurs a straight line is found. Below c_3, bismuth precipitates so that the cooling curve is again non-linear owing to heat absorbed as the result to this phase change.

Partial solid solution with compound formation. There are many instances where both types of deviation from ideal systems are found. One of the most important of these is the system Cu-Zn, shown in Fig. 14. As may be seen, there are possibilities for many alloys of different compositions to be formed from these two chemical elements. In general, these are known as brasses and are of great industrial importance.

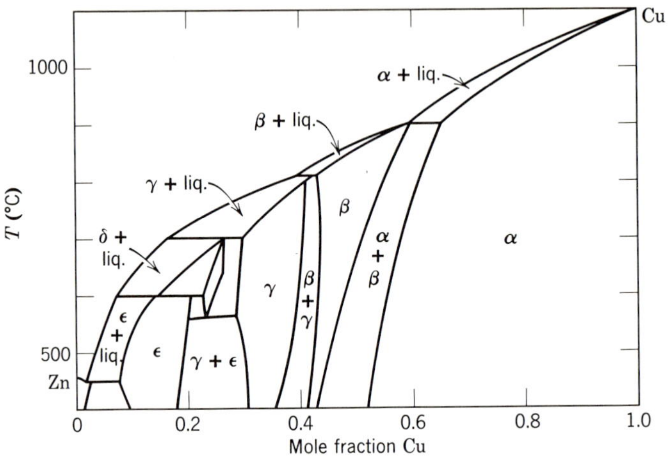

Fig. 14. Simplified phase diagram for Zn-Cu—partial solid solution with compound formation (Ref. 6). This figure and Fig. 15 are based primarily on data from *Constitution of Binary Alloys*, by Max Hansen, published by McGraw-Hill Book Co., New York, 1958. Printed with permission from the publisher.

Ideal solid solution. At one extreme we considered systems like Bi-Cd, where neither the Bi lattice or the Cd lattice would permit significant replacement of its atoms by the other partner. At the other extreme, there are systems like Au-Ag where continuous replacement is possible all the way from pure Au to pure Ag without a break. These two atoms are sufficiently similar in size and in character of charge cloud so that one can replace the other in the common type of crystal lattice and alloys can be formed with any composition all the way from pure gold to pure silver. The graph is shown in Fig. 15. The lattice has cubic close-packed structure. These alloys are widely used in jewelry, frequently with the addition of some copper.

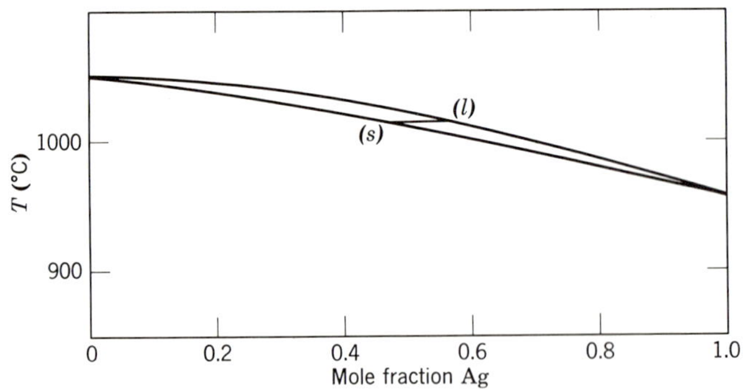

Fig. 15. System Au-Ag—complete ideal solid solution (Ref. 6). *Note:* At any temperature such as that shown by the horizontal line near the center of the figure, curve (*s*) gives the composition of the solid and curve (*l*) the composition of the liquid that are in equilibrium with each other at that temperature.

Composition and properties. There are many alloys where a small percentage of a second component has a far-reaching effect on properties like electrical conductivity and tensile strength. The hardness and strength of iron are greatly affected by components in the alloys commonly called *steels.* The addition of 12 to 14% of Mn produces an especially hard steel; about 0.15% of vanadium added with 5 to 10% of chromium produces a tough steel that has unusual elasticity; 18% Cr + 8% Ni gives a steel that resists corrosion and is sometimes called *stainless steel.* Nearly all steels contain 0.1 to 1.5% of carbon.

The effect of alloying constituents on such properties as hardness, resistance to corrosion, and ductility is explicable on the basis of modification of the electronic structure, even though today the theory is qualitative and *ad hoc* rather than predictive. On another scale, however, it is found that minute amounts of impurities profoundly modify metals. For example, truly pure iron, free of gases, is a silvery metal that will not corrode for years. It is far more ductile and has a higher tensile strength than ordinary "pure" iron that contains traces (parts per million) of various impurities and dissolved gases. It is interesting to ask why such small amounts of added components should produce such striking effects on alloy properties.

There is clear evidence today from many sources, such as X-ray studies and electron micrographs, that all crystals of macroscopic size have faults in the lattice, called dislocations. When a crystal grows, the deposit of the additional atoms, ions, or molecules is controlled by many factors. In general, the reason why atoms add as they do to a growing metal crystal is that these positions leave the system with minimum energy. Just as a chemical system tends to go to the position of minimum free energy, so the on-coming atoms adding to a growing crystal seek to minimize free energy. However, after a certain number of layers are added, irregularities occur in the additional layers, leaving these *dislocations* in the interior of the crystal, probably because of a combination of the influences that no longer favor regular growth. These dislocations can be regions of weakness in the lattice. If there were a tendency for the added components to promote the development of these regions, there could be a large effect produced by a small percentage. If a chain is only as strong as its weakest link, a crystal may be only as strong as its dislocations, and a component which forms these can have a big effect on the strength. This is a gross oversimplification of the problem, but it points toward the importance of understanding the microscopic atomic aspects of chemical systems as well as the macrothermodynamic aspects.

ELECTRON-DEFICIENT MOLECULES

The periodicity in atomic arrangement, coupled with overlapping of atomic orbitals, leads to the smearing out of electrons over tremendous volumes in space. For example, in metallic sodium the 3*s* electrons are delocalized in the crystal lattice, and spread over space more than 10^{10} times the volume of a single sodium atom. This produces a "sea" of electrons which, through their joint action, tie all the sodium atoms together in the metallic crystal. Periodicity in a more limited structure like the benzene ring spreads out the electron wave over six different atoms.

This idea of a delocalized electron has been applied recently to electron-deficient molecules. In order to see the way in which this idea has developed, recall first some of the properties of the normal covalent bond. Figure 16 shows first the electron-dot formulas for the three familiar compounds, water, hydrogen peroxide, and ammonia. The pair of electrons forming the covalent bond is indicated by a pair of dots in the customary way. Just below these structural formulas, there is listed the number of atoms in each compound (N_a). In the second line below, there are listed the number of bonds (N_b). The number of bonds is always one less than the number of atoms. When a number of atoms are linked together by bonds with each bond joining two atoms, the number of bonds required is always one less than the number of atoms; we are not linking together the atoms at the end of the chain. If we take six atoms and link them in the form of a circle, then, of course, the number of bonds is the same as the number of objects. For the moment omit consideration of cyclic compounds and think in terms either of straight-chain compounds or compounds like ammonia. Because there are two electrons in each bond, the number of electrons required to form such linking bonds will be double the number of bonds. Thus the number of electrons required is related to the number of atoms by the formula in Fig. 16*b*.

Boron, the first member of Group III, has three electrons available in the external shell for forming bonds. If we form a compound of boron with hydrogen, we expect three hydrogen atoms to link with one boron atom to form the compound BH_3. Experimental evidence shows that this compound of boron has a more complex formula, B_2H_6. There are not enough electrons to form the number of bonds expected on the basis of the equation just given, $N_e = 2(N_a - 1)$. As may be seen from the left-hand diagram in Fig. 16*d*, there are no electrons left to form a bond between the two boron atoms if we assign the normal number of two electrons to each of the covalent bonds between the boron atoms and hydrogen. Then how are these boron atoms tied together?

If this question is to be answered in terms of electron waves, we have to develop a broader concept of the covalent bond, a picture that includes not only an electron-pair bond between two atoms, but an electron bond that can embrace *more* than two atoms. As we have seen from the previous discussion, this broader concept has been forced upon us by the considerations of the nature of the forces in metallic crystal lattices, such as that found for the elements in the first three groups of the periodic table. Also we have seen that the behavior of

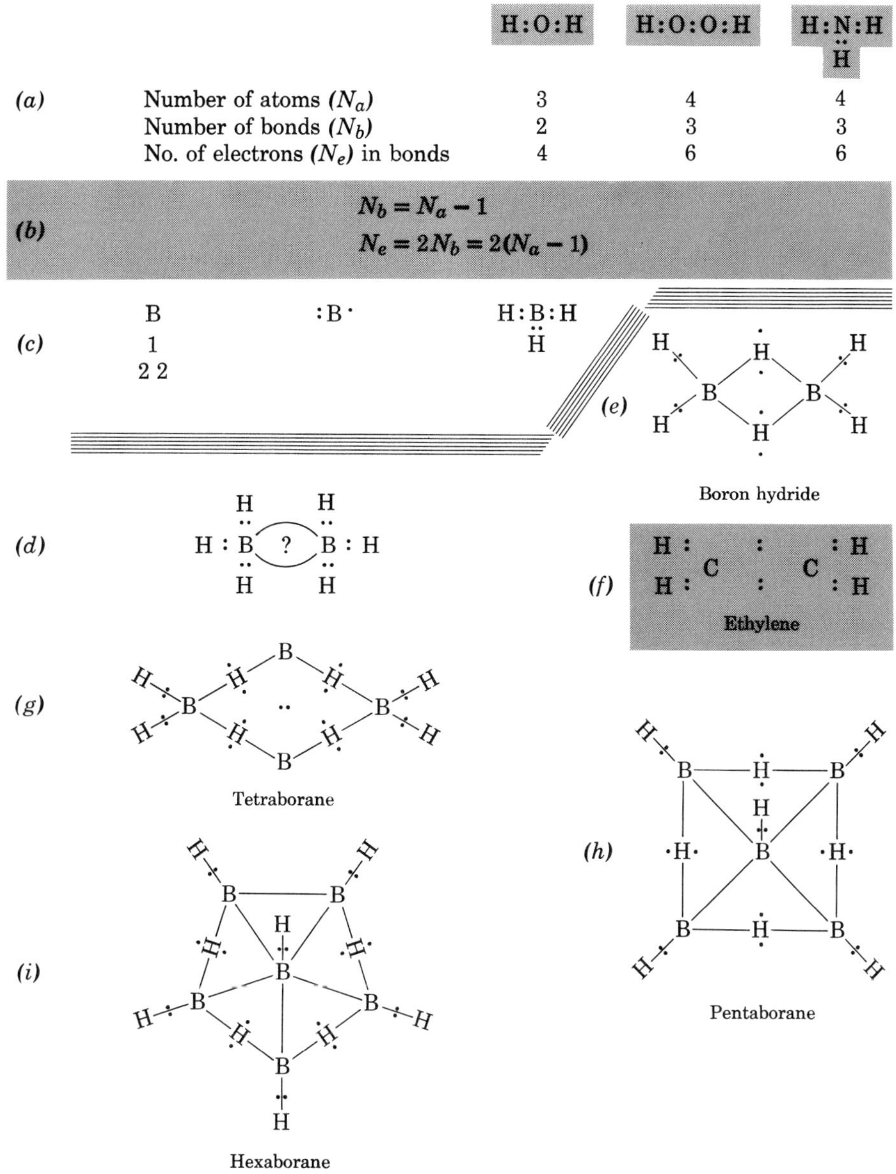

Fig. 16. Electron-deficient molecules, after W. N. Lipscomb. *Note:* In Figs. *g*, *h* and *i* the central rings are puckered. In Figs. *h* and *i* the central B lies above the ring.

the bonds in the hexagonal ring of carbon atoms found in benzene also points in the direction of an electron bond that can embrace more than two atoms. Can B_2H_6 be explained in terms of extended electron wave bonds? If we admit the possibility of such bonds, then we can write a formula for this compound as shown in Fig. 16*e*. Here, the cloud of the electron pair spreads out and enfolds a boron atom at each end; embedded in each electron pair is a positively charged hydrogen atom. So this spread-out electron-pair wave with a proton in the middle forms a hydrogen bridge between the two boron atoms. Since there are two hydrogen nuclei, we get two bridges between the boron atoms as shown in the figure. Because of repulsion between the positive charges on the two hydrogen nuclei, the bridges are bent away from each other in the middle so that we get the two electron waves localized roughly, as suggested in Fig. 16.

This bridge structure is confirmed by a variety of evidence. The infrared absorption bands which have been observed for this compound correspond closely to those found in the spectrum of ethylene; and, as may be seen from the structural formula for ethylene (Fig. 16*f*), there is a striking resemblance to this proposed formula for boron hydride, the only difference being that the two bonds formed by the pair of electrons in ethylene do not have hydrogen atoms in the middle. However, the geometry for the two compounds is essentially the same and, therefore, one would expect the infrared bands to have similar structures. Evidence from other sources supports this idea. Pitzer has proposed the name "protonated double bond" for this type of bonding where there are two electron-pair waves, each containing a proton in its center and embracing another atom at each end. Thus the essential difference between this type of electron wave and the normal double-bond wave lies in the fact that each of these waves embraces three atoms instead of two as in the normal covalent bond.

Originally it was thought that the protons would not be completely shielded by this type of three-center wave and, therefore, they should have an acidic character; actually, this is not indicated by the chemistry of boron hydride. A more detailed treatment of this structure has been proposed that results in molecular orbitals indicated roughly in Fig. 17. Here are presented a typical one-center atomic orbital (Fig. 17*a*), a typical two-center orbital as found in the covalent bond between two hydrogen atoms (Fig. 17*b*), and a typical three-center orbital which can explain the nature of the binding between the boron atoms B_2H_6 (Fig. 17*c*).

The most convincing evidence for the correctness of this interpretation comes from its extension to the high boron hydrides, B_4H_{10}, B_5H_9, and B_5H_{11}. Here the structure can be explained also in terms of multiple center bonds as illustrated in Fig. 17. The same idea has been applied to the boron monochloride, B_4Cl_4. Here the boron atoms are at the corners of a regular tetrahedron with an outlying chlorine atom attached to each corner. Again, there are insufficient electrons to explain this molecule in terms of normal two-center covalent bonds.

There are a number of other molecules containing atoms from the first three groups of the periodic table which are also electron-deficient. Al_2Me_6 is an example and also $(BeMe_2)_n$, where the symbol Me stands for the radical group

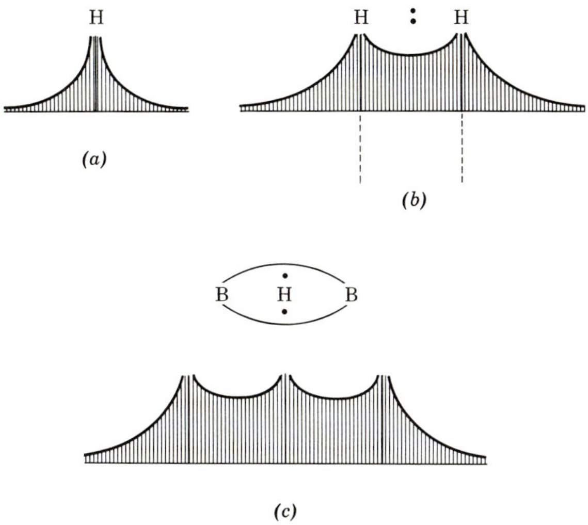

Fig. 17. Wave patterns in boron hydrides.

methyl, with the formula CH_3. There is a possibility that certain species of molecules containing carbon also may be members of this general class of compounds. For example, the ion CH_5^+ has been identified and may have a bipyramidal configuration. The propyl cation, usually considered as $CH_3 \cdot CH_2 \cdot CH_2^+$, in actuality may not be a straight chain, but may be a cyclopropane molecule, where two carbon atoms are joined by a three-center protonated bond. It may even be that a sort of bipyramidal structure can be formed from an isobutyl radical, C_4H_9.

Apart from their interest as unusual chemical compounds, these molecules illustrate the fact that the localized electron-pair bond, like the "rule of eight," serves as a good working rule but cannot apply to all compounds. An electron cloud can spread to cover one, two, six, or even 10^{20} atoms; hence the picture of point electrons in orbits circling atoms is misleading. If we accept the concept of an electron as a wave and think of the electron as a deformable cloud of electricity, single-atom, two-atom, six-atom, and million-atom orbitals differ only in the degree of diffusness of the electron cloud. Such an outlook provides the most satisfactory concept presently available for coordinating all aspects of chemical bonding.

PROBLEMS

1. A melt containing 20 mole % Bi and 80 mole % Cd is cooled to a little above the eutectic temperature. What is the composition of the melt?

2. A 10-gm sample of a bismuth-cadmium alloy is melted and cooled to slightly above the eutectic temperature. The weight of solid present is 3 gm. What can you say about the composition of the original sample? *Ans.* 72 mole % Cd or 42 mole % Cd.

★ **3.** A chemist prepares two samples of supposedly pure bismuth. Cooling curves for samples A and B are given below:

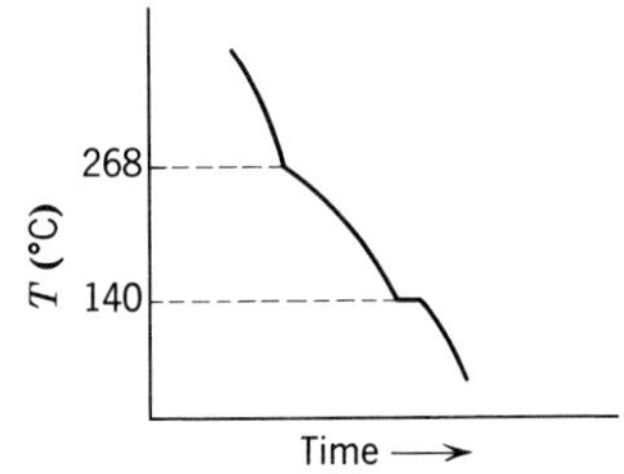

Sample A

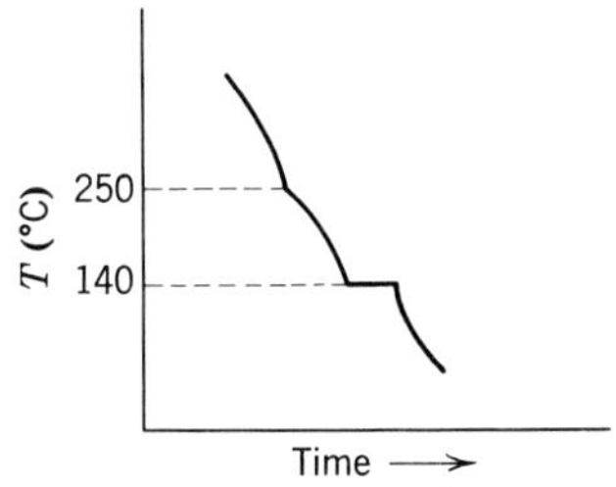

Sample B

(*a*) If the only possible impurity is Cd, which is purer? Why?

(*b*) The heat of fusion of bismuth is 2550 cal/atom. If the impurity forms an ideal solution like cadmium, what is the composition of A and B? *Ans.* B is 91 mole % Bi.

(*c*) According to these curves is it likely that cadmium is the impurity? Why?

★ **4.** Cooling curves are obtained for an alloy of A and B. The data in the accompanying table are obtained. By a gradual break in the curve we mean one of the type seen at T_1 in Fig. 10*c*. Construct a likely temperature composition phase diagram for this system.

Mole %	Sharp Break in Cooling Curve at	Gradual Break in Cooling Curve at
0	200°	none
10	100°	170°
20	100°	130°
30	100°	none
40	100°	175°
50	250°	none
60	50°	150°
70	50°	none
80	50°	125°
90	50°	250°
100	300°	

5. If you had a bismuth-cadmium alloy containing 10 mole % cadmium and wanted to obtain from this a pure sample of pure crystalline bismuth, how would you proceed?

6. Zone melting is now used to purify metals. In this procedure a heater is placed at one end of the bar and the metal is just melted. The heater is then moved slowly down the bar, so that in the heater the metal is just molten but in front of and in back of the heater the metal is solid. In this procedure the impurities accumulate in the molten zone, and the metal in back of the heater is purified. Explain how this would work for bismuth containing 1 atom % cadmium. For purifying this metal, what would be the ideal temperature to use? (See Problem 3*b*.)

Ans. Start at 269°C and let temperature fall as impurities get more concentrated.

7. Given the phase diagram shown in the figure for the system A-B.

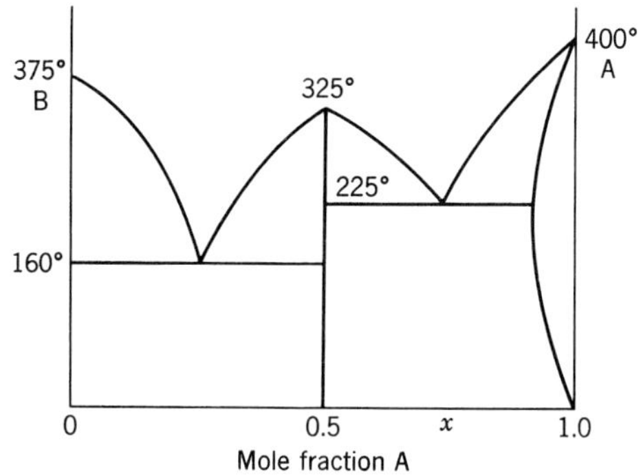

(*a*) Label each region of the diagram to show what phases are present.

(*b*) Sketch cooling curves for a melt containing 0% A, 10% A, 25% A, 40% A, 50% A, 60% A, 80% A. Indicate the origin of any irregularities in the cooling curves.

(*c*) Which would be easier to purify by zone melting, A or B?

(*d*) Would zone melting be at all helpful in the less favorable case?

25

Complex ions

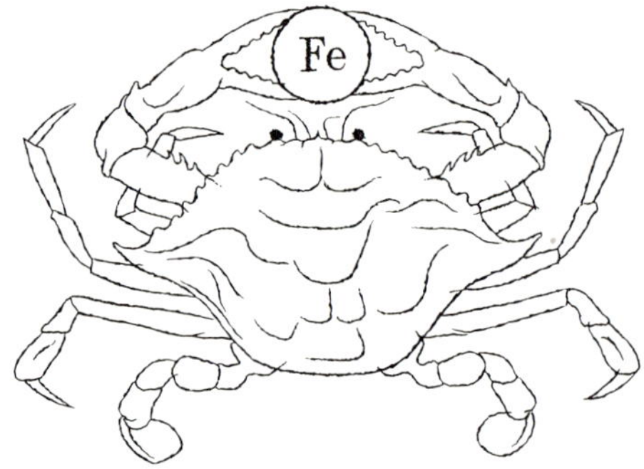

Reactions in aqueous solutions are reactions of complex ions. For example, a solution of ferric chloride is really a solution containing hydrated ferric and chloride ions and a number of complexes formed from the ferric and the chloride ions, viz.:

$$FeCl^{+2},\quad FeCl_2^+,\quad FeCl_3,\quad FeCl_4^-,\quad FeCl_5^{-2},\quad \text{and}\quad FeCl_6^{-3}$$

In addition, since the chloride and ferric ion have a hydration shell attached, even these are complex ions. Thus, to understand the reactions of ferric chloride, we must understand the factors that determine the structure and reactivity of complex ions.

AQUO COMPLEXES

Aquo ions. The equation

$$NaCl(s) \rightleftarrows Na^+(aq.) + Cl^-(aq.)$$

which is used to describe the equilibrium between solid and dissolved sodium chloride is an abbreviated one in which the (aq.) indicates the hydration shell surrounding each ion. A more detailed notation is

$$Na^+(aq.) = Na^+(H_2O)_n \quad \text{and} \quad Cl^-(aq.) = Cl^-(H_2O)_n$$

which indicates that the water molecules in the hydration shell of an ion differ from the other water molecules. The average number of water molecules in a hydration shell, n, is largely determined by the size of the cation. The lithium ion holds four molecules in the hydration shell, but each of these is firmly held; on the other hand, the cesium ion holds eight water molecules, but each of these is loosely held.

Water of hydration is in dynamic equilibrium with solvent water, and in this respect aquo ions are like other complex ions. If the hydration shell of a dissolved sodium ion contains six water molecules, we express the equilibria as follows:

$$\mathrm{Na(H_2O)_6^+ \rightleftharpoons Na(H_2O)_5^+ + H_2O\ (solvent)}$$
$$\mathrm{Na(H_2O)_5^+ \rightleftharpoons Na(H_2O)_4^+ + H_2O\ (solvent)}$$

etc.

This equilibrium is not established instantaneously; it involves a rate process that depends on the electronic structure of the cation. Some aquo complexes are as short-lived as the clusters in non-ideal gases; others are so long-lived that there is little exchange between the water molecules of the hydration shell and the water molecules of the solvent. Studies with tracer and other techniques show that the time required for exchange between solvent water molecules and those in the hydration shell varies from microseconds to weeks.

Since aquo ions are a chemical species, it is possible to calculate the bond energies between water molecules and the ions. A gaseous sodium ion dissolves in water with release of energy; that is,

$$\mathrm{Na^+}(g) + \mathrm{6H_2O(aq.) \rightleftharpoons Na(H_2O)_6^+(aq.)} \qquad \Delta E = -100\ \mathrm{kcal}$$

Therefore, for each of the water-sodium bonds, we can write

$$\mathrm{Na:\ddot{\underset{..}{O}}:H \longrightarrow Na^+ + :\ddot{\underset{..}{O}}:H} \qquad E_B \approx \tfrac{100}{6}\ \mathrm{kcal}^\dagger$$
$$\mathrm{\quad\ H \qquad\qquad\qquad\quad H}$$

where E_B is the average bond energy. Compared to other bond energies, this value is small; for example, the OH bond energy is about 110 kcal.

With increasing charge and decreasing size of the cation, the hydration energy increases. For aluminum ions,

$$\mathrm{Al^{+3}}(g) + \mathrm{6H_2O(aq.) \rightleftharpoons Al(H_2O)_6^{+3}(aq.)} \qquad \Delta E = -1100\ \mathrm{kcal}$$

corresponding to a bond energy of nearly 200 kcal. With such a high bond energy, the electrons concentrate between the aluminum and oxygen atoms. This shift of electrons makes the hydrogen atom more positive than in normal water and leads to repulsion of the proton by the positively charged aluminum ion coupled with weakning of the OH bond; hence the following hydrolysis occurs:

† These are not true bond energies, since not all the species are in the gaseous state. For qualitative comparisons this makes little difference.

$$(H_2O)_5\overset{+3}{Al}:\underset{\cdot\cdot}{\overset{\overset{\textstyle H}{\cdot\cdot}}{O}}:H \rightleftharpoons (H_2O)_5Al:\underset{\cdot\cdot}{\overset{\overset{\textstyle H}{\cdot\cdot}}{O}}:^{+2} + H^+$$

This is one way of explaining why solutions containing aluminum salts hydrolyze to yield acid solutions, whereas solutions of sodium salts (for which the cation hydration energy is low) do not hydrolyze but yield neutral solutions.

Hydrates. During the evaporation of a solution containing dissolved sodium chloride, the hydration sheath around each sodium ion disappears before the formation of solid sodium chloride. For doubly and triply charged cations, however, the hydration shell is more firmly bound, and crystallization from aqueous solution results in a solid containing water of constitution; such solids are called *hydrates*. Examples of crystalline hydrates are listed in Table 1. For small ions, such as beryllium, the hydration shell contains four water molecules; hence the corresponding solid salt contains four water molecules attached each cation. With increasing size and charge, the coordination number increases; cations like aluminum form salts containing six molecules of water per cation; still larger cations like neodymium form salts containing nine molecules of water per cation.

Although the water in crystalline hydrates is usually associated with the cations, many sulfates form hydrates in which one or more molecules are associated with the anion. Typical of this type of compound is magnesium sulfate, which contains seven water molecules of hydration: six of these are associated with the cation; the seventh is associated with the anion.

There is a class of compounds called *alums* with the general formula $M^IM^{III}(SO_4)_2 \cdot 12H_2O$ wherein M^I stands for a unipositive cation (e.g., K^+, NH_4^+) and M^{III} stands for a tripositive ion (e.g., Al^{+3}, Fe^{+3}, Cr^{+3}). Each cation possesses its own hydration shell and the resulting hydrate contains twelve water molecules.†

COMPLEX IONS

Formation of complex ions. Formation of a complex ion involves competition between the water molecules in the hydration shell and the complexing agent (e.g., NH_3). Thus, if the water of hydration is tightly held by the centrally located ion, the aquo complex persists and no complex ion is formed; on the other hand, if the complexing agent (also called the *ligand*) forms the stronger bond with the central ion, a complex ion results. Such competition is typified by the equation

$$Zn(H_2O)_4^{+2} + 4NH_3 \rightleftharpoons Zn(NH_3)_4^{+2} + 4H_2O$$

† All alums have the same crystal structure. Different compounds with the same crystal structures are referred to as an *isomorphous* series. The existence of such a series suggests that all the possible M^I ions are similar chemically and that all possible M^{III} ions are similar chemically.

Table 1 Simple Hydrates

$BeCl_2 \cdot 4H_2O$	$MgSO_4 \cdot 7H_2O$
$AlCl_3 \cdot 6H_2O$	$KAl(SO_4)_2 \cdot 12H_2O$
$Nd(BrO_3)_3 \cdot 9H_2O$	

Since water molecules are attached to the zinc ion by donation of the unshared electron pair on the oxygen atom, and, in addition, the ammonia molecules are attached to the zinc ion by donation of the unshared electron pair on the nitrogen atom, we can categorize complex ion formation as a Lewis type of acid-base reaction. In this language, the reaction between the aquo zinc complex and ammonia involves the displacement of the weak base, water, by the stronger base, ammonia. Although this nomenclature is useful for purposes of classification, it does not permit us to predict results with confidence. The difficulty is that the Lewis basicity of a compound depends on the acid. For example, we might expect that ammonia would always displace water from the hydration shell since ammonia is the stronger base for reactions involving protons. In point of fact, however, this is not always true; for example, the alkali metal and alkaline earth cations do not form ammonia complexes in aqueous solution. The Lewis approach to complex formation is useful only when cations (acids) are similar in structure.

Complex ions of the fourth period elements. The tendency of cations to form complexes with various ligands is so dependent on the details of the atomic structure of the individual cations that it is difficult to make generalizations. Nevertheless, if we survey the complexes formed by ions in a given period, a pattern does emerge. Table 2 summarizes the stability of common complexes of the

Table 2 Complex Ions of Fourth Period Elements

	Oxygen	Fluoride	NH_3	Halide	CN	CO
K	—	—		—	—	—
Ca	w	—	v.w.	—	—	—
Sc	m	m	v.w.	—	—	—
Ti	s	s	—	m	—	—
V	s	s	—	m	v.w.	—
Cr	s	s	—	m	—	s
Mn	s	m	s	m	s	s
Fe	m	s	w	m	s	s
Co	m	m	s	m	s	s
Ni	m	m	s	m	s	s
Cu	m	—	s	m	s	—
Zn	s	w	s	s	s	—
Ga	s	s	—	w	—	—
Ge	s	s	—	w	—	—
As	s	—	—	m	—	—
Se	s	s	—	m	—	—

s = strong; w = weak; m = medium-strong; v.w. = very weak.

fourth period elements. Those designated as strong or medium-strong form readily in aqueous solutions; those designated as weak or very weak form only to a slight extent in aqueous solution.

In general, the stability of the complexes indicated in Table 2 parallels that found for members of the fifth and sixth period elements in so far as cations of similar charge in the same group tend to complex with the same ligands. Thus complex ions of platinum in the +2 oxidation state resemble those formed by nickel. On the other hand, those formed by platinum in the +4 oxidation state are more closely related to the complexes of Fe^{+2}, since these two ions have similar electronic structures.

The stability of the halide complex ions offers the most clear-cut trend encountered for complexes with transition metal cations. For the lighter elements, the complexes with the lighter halides are the most stable; for the heavier elements, this order is reversed, viz.:

Stability order for lighter elements: $F^- > Cl^- > Br^- > I^-$
Stability order for heavier elements: $I^- > Br^- > Cl^-$

COORDINATION COMPOUNDS

In coordination compounds, the cation or anion is a complex ion. The chemistry of these compounds was placed on a firm basis by the studies of Werner at the turn of the century. On the basis of conductivity measurements he found that the compound $Co(NH_3)_6Cl_3$ dissociated in aqueous solution to yield four ions. When this compound was heated slightly above 100°C, it lost one molecule of ammonia to form the new compound, $Co(NH_3)_5Cl_3$. This also dissociated in aqueous solution but yielded only three ions. He postulated that the cobalt ion forms a complex with six groups so firmly attached that they do not dissociate in solution. Thus, according to Werner, these two compounds establish the following equilibria in aqueous solution:

$$Co(NH_3)_6Cl_3 \rightleftarrows Co(NH_3)_6^{+3} + 3Cl^-$$
$$Co(NH_3)_5Cl_3 \rightleftarrows Co(NH_3)_5(Cl)^{+2} + 2Cl^-$$

The first compound forms four ions, and the second forms only three ions. This picture is supported by the fact that, if silver nitrate is added to a solution containing the first compound, all of the chloride precipitates, whereas, in a solution containing the second compound, only two-thirds of the chloride precipitates immediately.

The whole series of coordination compounds has been prepared for the complexes of platinum and chlorine. These are listed in Table 3. The brackets indicate the hexa-coordinated species, which does not dissociate in solution.

Table 3 Coordination Compounds

Compounds	Number of Ions in Solution
$[Pt(NH_3)_6]Cl_4$	5
$[Pt(NH_3)_5Cl]Cl_3$	4
$[Pt(NH_3)_4Cl_2]Cl_2$	3
$[Pt(NH_3)_3Cl_3]Cl$	2
$[Pt(NH_3)_2Cl_4]$	0
$K[Pt(NH_3)Cl_5]$	2
$K_2[PtCl_6]$	3

STEREOCHEMISTRY

Bivalent platinum forms tetra-coordinated compounds in which the platinum is at the center of the square and the four ligands are at the corners. Compounds of the type PtX_4 and PtX_3Y occur in only one form, but the type PtX_2Y_2 occurs as two isomers. The isomers of $Pt(NH_3)_2Cl_2$ are pictured here:

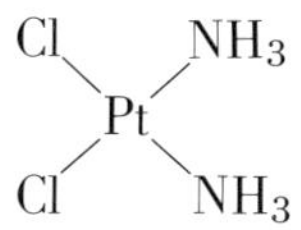

cis

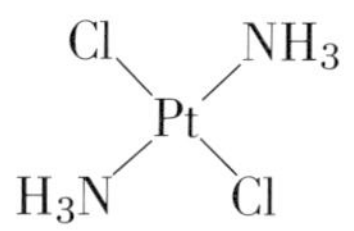

trans

Hexa-coordinated species have the geometry indicated in Fig. 1. Consider the compounds listed in Table 3. For which of these are stereoisomers possible? The various types of complexes are

$$[Pt(NH_3)_6]^{+4} - PtX_6 - (PtCl_6)^{-2}$$
$$[Pt(NH_3)_5Cl]^{+3} - PtXY_5 - [Pt(NH_3)Cl_5]^-$$
$$[Pt(NH_3)_4Cl_2]^{+2} - PtX_2Y_4 - [Pt(NH_3)_2Cl_4]$$
$$[Pt(NH_3)_3Cl_3]^+ - PtX_3Y_3 - [Pt(NH_3)_3Cl_3]^+$$

For the first and second types each species forms only one molecule. The third species forms the two isomers indicated in Fig. 2; in one of these the X groups are as far apart as possible, and, in the other, they are adjacent. The fourth species also forms two possible isomers; in one of these the X's grouped together and the Y's grouped together; the other has an alternative arrangement. Both are shown in Fig. 3.

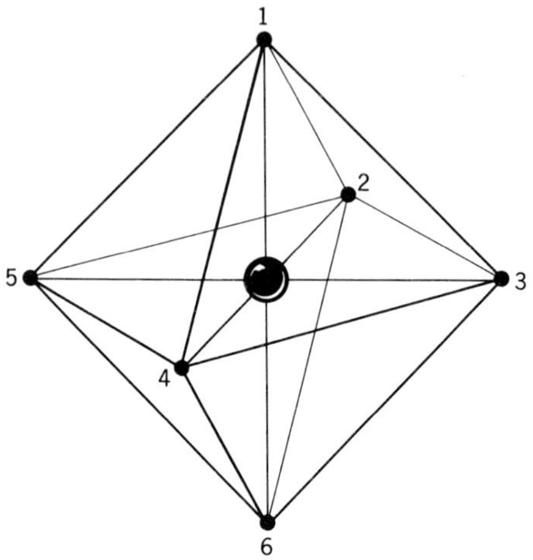

Fig. 1. Octahedral configuration.

Chelation. In the complexes we have talked about so far, each ligand contributes one pair of electrons. There are, however, ligands that contribute more than a single pair of electrons; examples of these are the oxalate ion and the ethylenediamine group:

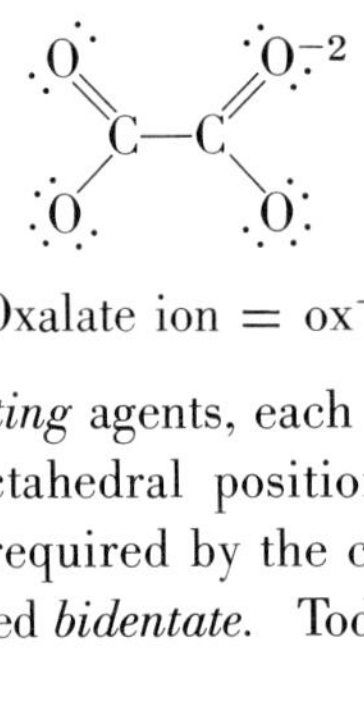

Oxalate ion = ox^{-2}

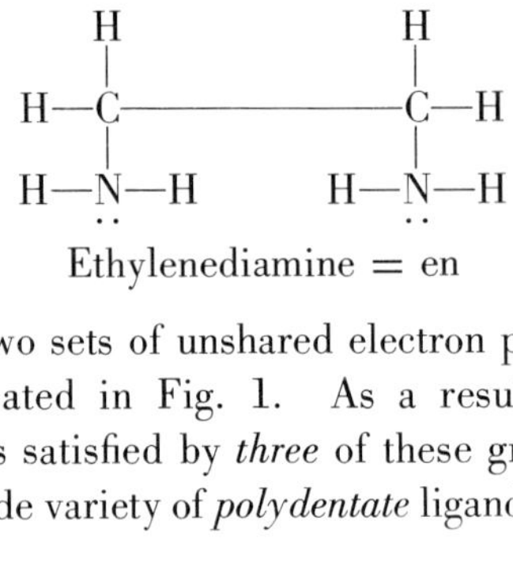

Ethylenediamine = en

In these *chelating* agents, each of the two sets of unshared electron pairs assumes one of the octahedral positions indicated in Fig. 1. As a result, the hexacoordination required by the cations is satisfied by *three* of these groups. Such groups are called *bidentate*. Today, a wide variety of *polydentate* ligands are known,

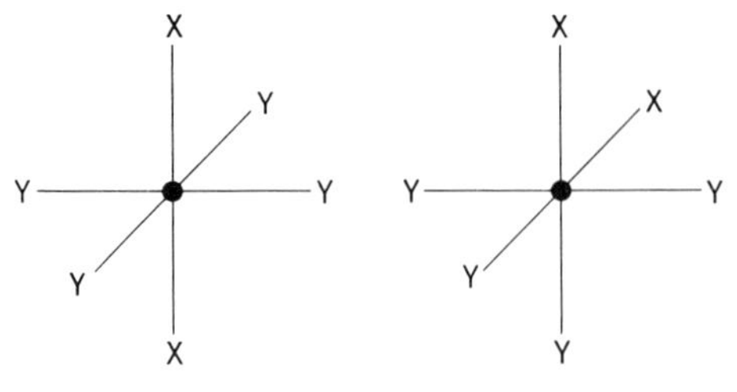

(a) *(b)*

Fig. 2. MX_2Y_4 isomers.

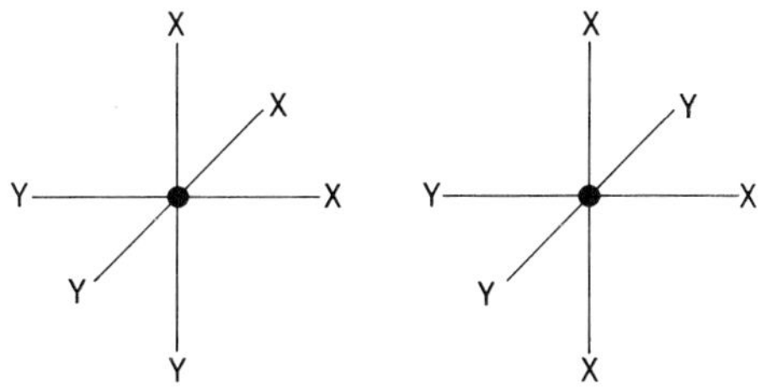

(a) *(b)*

Fig. 3. MX_3Y_3 isomers.

and complexes exist wherein the ligand is a tridentate, tetradentate, and even hexadentate group.

Bidentate groups occur in two varieties. For either the oxalate ion or the ethylenediamine molecule, the electron-pair donors are equivalent in the sense that one donor has the same environment as the other donor. We indicate this by the symbol AA. Bidentate groups are also known in which one donor pair has an environment different from the other donor pair; an example of this is propylenediamine,

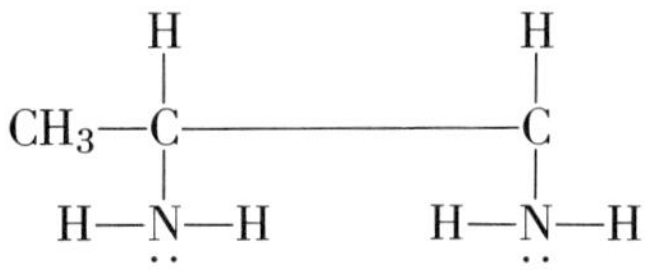

We indicate this by the symbol AB.

The possible isomers formed by bidentate groups (also called *chelating* agents) depends on whether it is of the AA or AB type. Thus, for a square coplanar complex of the type $Pt(AA)_2$, there is only one type of molecule formed. For the complex with unsymmetrical bidentate groups, $Pt(AB)_2$, there are two possible isomers. In one of these, the AB groups are attached with the A groups, *cis;* in the other the AB groups are attached with the A groups, *trans.*

OPTICAL ISOMERS

Some isomers differ in that they are mirror-images. These were first encountered in organic chemistry, but, early in the study of inorganic complex compounds, it was found that these also can form optical isomers.

Organic optical isomers exist whenever all four groups attached to the carbon atom are different. Consider the compound

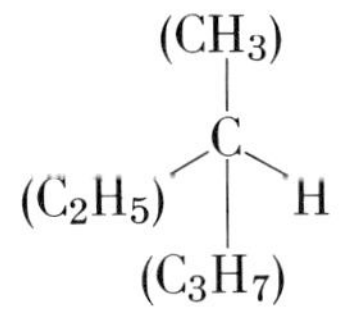

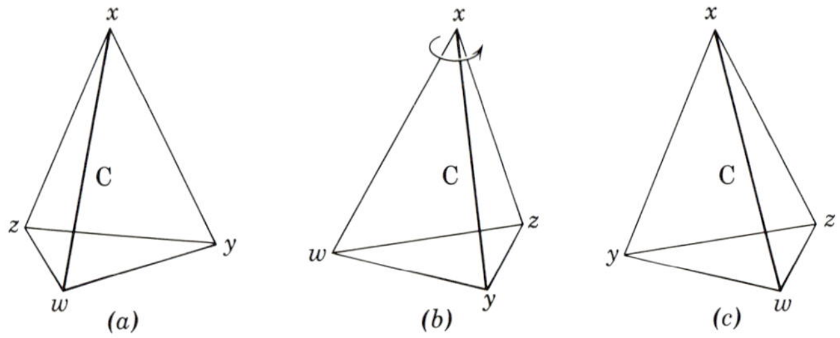

Fig. 4. Optical isomers.

a substituted methane. This and other compounds of this type can be represented by the non-specific formula $C(x)(y)(z)(w)$. The spatial arrangement is shown in Fig. 4. Figures 4*a* and 4*b* show different views of the same molecule. Thus, if we rotate Fig. 4*b* as indicated, we find that the x, y, z, and w positions coincide with those in Fig. 4*a*. The arrangement shown in Fig. 4*c*, however, is the mirror-image of that in Fig. 4*a*. Just as no rotation can make your right hand look like your left hand (these, of course, are mirror-images), no rotation can make Fig. 4*a* look like Fig. 4*c*. Thus these molecules are truly different; the (*a*) molecule is the *enantiomorph* of the (*c*) molecule.

Optical isomers differ in chemical reactivity only if the other reactant has a right-handedness or left-handedness. The isomers have the same melting point and boiling point and, in general, the physical properties are identical. The principal difference between these two forms is in their optical properties. (Polarized light is rotated to the left by one form and to the right by the other.)

Some inorganic hexa-coordinated compounds occur as optical isomers, but the planar tetra-coordinated compounds do not.† We can appreciate this from the drawings in Fig. 5. These molecules are mirror-images of each other, but they are the same molecule; if we rotate the right-hand molecule 180 degrees around the z-x axis, the positions of the four groups coincide. On the other hand, this is not true with hexa-coordinated molecules. For example, suppose we had a compound of the type PtABCDEF. This forms the two mirror-image compounds indicated in Fig. 6. There are no rotations that bring these six groups into register; hence the mirror-images are optical isomers.

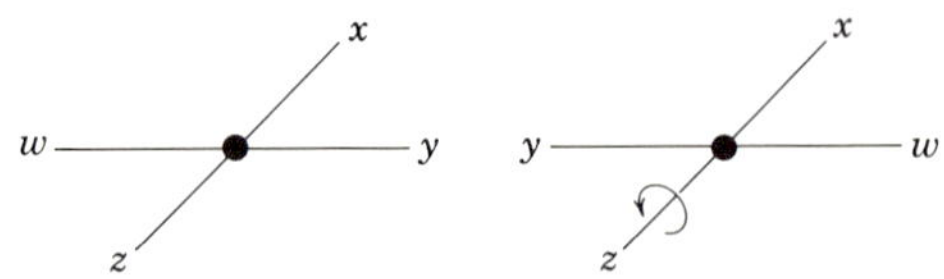

Fig. 5. Mirror images.

† Tetrahedral tetra-coordinated species do form optical isomers.

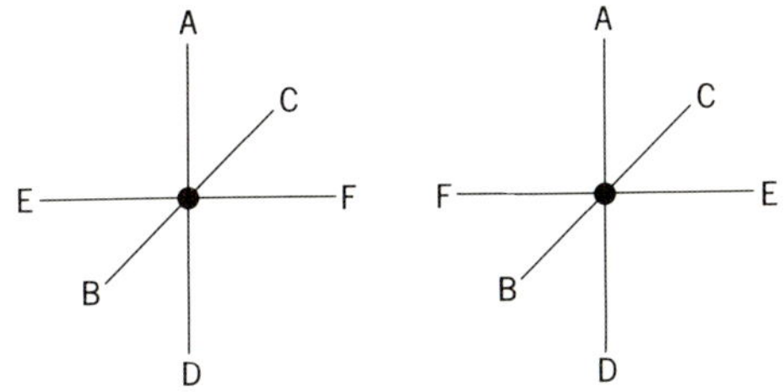

Fig. 6. Optical isomers.

Optical isomers are found for organic compounds only when *all* of the attached groups are different. This need not be the case with hexa-coordinated compounds. For example, even if the groups F and D (Fig. 6) are the same, the molecule still forms optical isomers.

Optical isomers are most common in complexes with bidentate groups. For example, $Pt(C_2O_4)_2Cl_2$ has the two optical isomers shown in Fig. 7. Optical isomers even occur for compounds in which all the bidentate ligands are the same. Thus, for a complex ion of the type $Fe(C_2O_4)_3^{-3}$, we find the two optical isomers indicated in Fig. 8.

In a solution containing $Fe(C_2O_4)_3^{-3}$, we find both optical isomers. Since the two forms have essentially the same properties, separation of the two forms is difficult, but changes from one form to another can be followed by changes in optical properties. For the sake of identification, one form is labeled *d* (for *dextro,* right-handed); the other is labeled *l* (for *levo,* left-handed).

A solution containing pure *d* molecules rearranges in time to form an equal number of *l* molecules so that, eventually, the solution contains both the *d* and *l* isomers. Such a solution is called *racemic;* the formation of this solution is called *racemization;* the separation of the isomers is termed *resolution.*

Studies of the racemization rate have contributed significantly to our knowledge of the stability of the various complex ions. For example, it has been found that

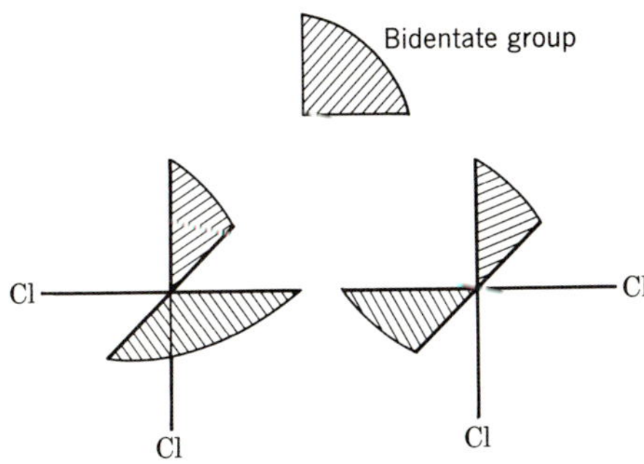

Fig. 7. Optical isomers.

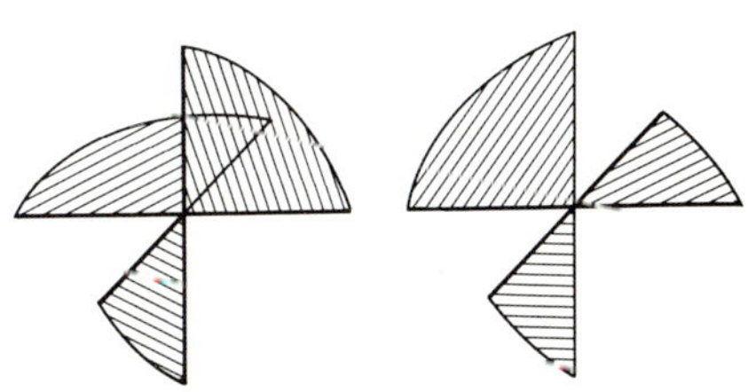

Fig. 8. Optical isomers.

the pure *d* form of $Fe(C_2O_4)_3^{-3}$ racemizes fast, whereas the corresponding complex of chromium does not. From this, we reason that the opposing rates in the dynamic equilibrium $Fe(C_2O_4)_3^{-3} \rightleftarrows Fe(C_2O_4)_2^{-} + (C_2O_4)^{-2}$ are large for the iron compound but small for the chromium compound. Thus the iron complex is labile, but the chromium complex is not. (This conclusion includes the assumption, which is not always true, that the complex ion has to dissociate partially for conversion from the *d* to the *l* form to occur.)

ELECTRONIC STRUCTURE

The complex ions of considerable current interest are those formed by elements in which *d* orbitals are available either in the outermost shell or the next to outermost shell. A sketch of the 3*d* orbitals is given in Fig. 9. They are not used directly for bonding, but they hybridize with the *s* and *p* orbitals to form d^2sp^3 and dsp^2 hybrids. The d^2sp^3 hybrids result in an octahedral structure; dsp^2 hy-

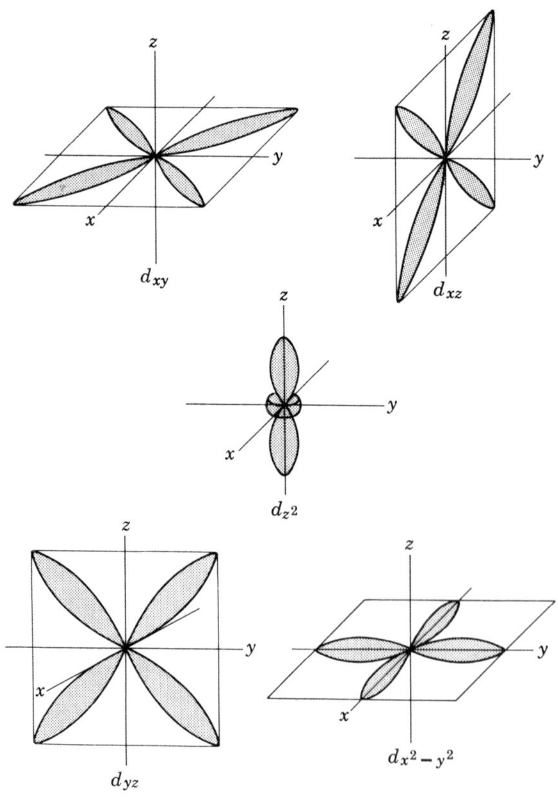

Fig. 9. Types of 3*d* orbitals.

brids result in a square coplanar structure. Occasionally, a tetahedral tetra-coordinated species is formed, e.g., $Ni(CO)_4$, $Cu(CN)_4^{-3}$. It is presumed that these structures result from sp^3 hybridization.

Hybridization. Consider the hexa-coordinated complexes of Fe^{+3}. A gaseous iron atom has the inert-core structure of argon, plus eight additional electrons. Six of these outer electrons are in the $3d$ subshell and two are in the $4s$ subshell. When three electrons are removed, an ion is formed that has five electrons in addition to the argon core. These five electrons are unpaired in the $3d$ subshell. What occurs when this ion is complexed by six water molecules or six chloride ions. Each of the ligands contributes two electrons to the ion; hence we can obtain the d^2sp^3 hybridization required for the hexa-coordinated species by either of the following arrangements:

$3d$	$4s$	$4p$	$4d$
⇅ ⇅ ↑ 2 2	2	2 2 2	_ _ _ _ _
↑ ↑ ↑ ↑ ↑	2	2 2 2	2 2 _ _ _

(The arrows correspond to the non-bonding electrons in the iron ion; the numeral 2 corresponds to the bonding electrons from the ligand.) From this, it is clear that the necessary hybridization can be achieved by use of either the $3d$ subshells or the $4d$ subshells. Both types of bonding are encountered. Compounds with bonding involving the $3d$ subshells are referred to as *inner-orbital* complexes; those involving the $4d$ subshells for binding are called the *outer-orbital* complexes.

Table 4 lists various ions in the fourth period that form inner-orbital, hexa-coordinated complexes. The small d's refer to the electrons not used in bonding; the capital letters refer to those used in hybridization; the exponents give the number of electrons in each of the subshells. Thus, in Ti^{+3}, one electron in

Table 4 Inner Complexes

Ions	Configuration	Number of Unpaired Electrons
Sc^{+3}, Ti^{+4}	$d^0d^0d^0D^2SP^3$	0
Ti^{+3}, V^{+4}	$d^1d^0d^0D^2SP^3$	1
Ti^{+2}, V^{+3}	$d^1d^1d^0D^2SP^3$	2
V^{+2}, Cr^{+3}, Mn^{+4}	$d^1d^1d^1D^2SP^3$	3
Cr^{+2}, Mn^{+3}	$d^2d^1d^1D^2SP^3$	2
Fe^{+3}, Mn^{+2}	$d^2d^2d^1D^2SP^3$	1
Fe^{+2}, Co^{+3}	$d^2d^2d^2D^2SP^3$	0
Co^{+2}, Ni^{+3}	$d^2d^2d^2D^2SP^3d^1$	1
Ni^{+2}	$d^2d^2d^2D^2SP^3d^1d^1$	2

the *d* shell is not used for bonding. In Mn^{+3}, four *d* electrons are not used in bonding; two of the four non-bonding *d* electrons are paired and two are unpaired.

Magnetism and bond type. In the discussion of atomic structure (Chapter 4), we found that the spin associated with an electron causes it to act like a small magnet. If an atom has an unpaired electron, it is attracted to a magnet (paramagnetism). The strength of this attraction depends on the number of unpaired electrons; the greater the number of unpaired electrons, the greater is the attraction to the magnet. If all the electrons in an atom are paired, their spins cancel, and the atom is repelled by a magnet (diamagnetism). (If the atoms are gaseous, there is also a contribution to the magnetism from the orbital motion of the electrons. In solutions of transition elements this contribution is reduced to a small value and will be ignored.)

Unpaired electrons in molecules and in ions also function like magnets. Thus ClO_2, a molecule with one unpaired electron, is attracted by a magnet as strongly as Li, an *atom* with one unpaired electron. Studies of the magnetic properties of molecules show us the number of unpaired electrons, and this information provides a firm basis for the discussion of electronic structures.

Now let us reconsider the possible electronic structures of Fe^{+3} in a complex. An outer-orbital complex has five unpaired electrons; an inner-orbital complex has only one unpaired electron. The complex formed with cyanide ion must be an inner-orbital complex since the measured paramagnetism corresponds to one unpaired electron. On the other hand, the complex with fluoride or chloride ion must be an outer-orbital complex since the paramagnetism corresponds to five unpaired electrons. In this manner, magnetic measurements provide us with the means of differentiating between inner- and outer-orbital complexes.

Cyanide ions have the strongest tendency to form inner-orbital complexes, and fluoride ions have the weakest tendency to form inner-orbital complexes. We can order the common ligands according to their tendency to form inner-orbital complexes; this yields the following scale:

$$CN^- > NH_3 > H_2O > F^-$$

We interpret this order as follows: The formation of an inner-orbital complex requires the donation of an electron pair to the central ion coupled with the pairing of non-bonding electrons. The first step yields a more stable structure; the second yields a less stable structure. Thus the formation of an inner-orbital complex occurs only if the donor electron pair is loosely held by the donor atom. If the ligand is firmly attached to its electron pair as in the fluoride ion, it will be a poor donor and will form an outer-orbital complex. Accordingly, this order represents the order of basicity of these ligands (in the Lewis sense) for transition metal ions of the fourth period.

Crystal field theory. An alternative approach, called the *crystal field theory*, is now being used to explain the electronic structure of complex ions. Let us use this approach to describe the binding in the $Fe(CN)_6^{-3}$ complex.

In the gaseous Fe^{+3} ion, the five *d* subshells are occupied by five electrons with parallel spins. For the free ion all five electrons have the same energy, but, after the donor cyanide ions form the octahedral complex, this is no longer true. The cyanide ions in effect inject negative charge into the Fe^{+3} ion. Thus all *d* electrons are repelled by the cyanide groups, but the *d* electrons whose charge cloud lies along the *x*, *y*, and *z* axes, i.e., along the Fe^{+3}—CN^- bonds, are repelled the most. Accordingly, electrons in the d_z and $d_{x^2-y^2}$ orbitals (cf Fig. 9) become less stable than electrons in the d_{xy}, d_{xz}, and d_{yz} subshells. The original five *d* subshells, which had equal energies, then separate into two groups with different energies. The d_{xy}, d_{xz}, and d_{yz} levels have the lower energy, and the d_{z^2} and $d_{x^2-y^2}$ levels have the higher energy. The process is summarized in Fig. 10.

There are two possibilities: the first is that the electrons remain unpaired as in the free Fe^{+3}; this is indicated in Fig. 10*a*. (If it were not for the fact that the pairing of electrons requires energy, the electrons in the higher *d* level would pair in order to assume positions in the lower level.) As the energy separation of the two sets of levels gets greater and greater, we eventually reach the point at which the energy gained when all the electrons are fitted into the lowest level, *with pairing*, is more than enough to offset the loss in energy associated with the pairing of electrons. This situation is indicated in Fig. 10*b*.

If the ligands have only a slight tendency to donate their unshared pairs, the

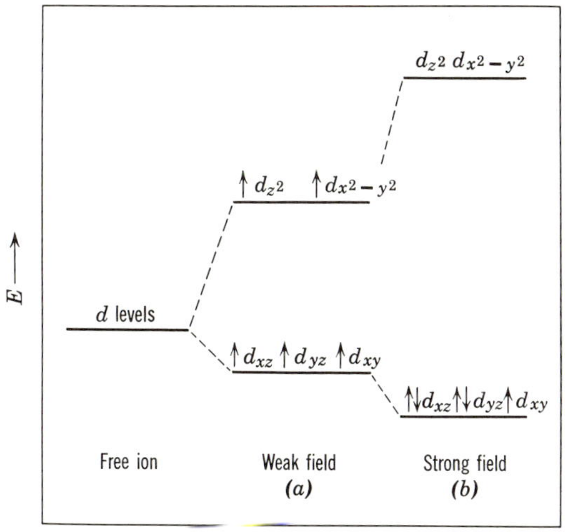

Fig. 10. Crystal field splitting.

separation of the two sets of levels will be small, and the electrons will have the unpaired arrangement as in the free ion. If, on the other hand, the donor ligands readily yield their electron pair, the separation becomes large enough that the electrons fill the lower levels in preference to the upper levels. This leads to pairing and the situation depicted in Fig. 10*b*. As before, a ligand like cyanide favors the spin-paired condition, whereas a ligand like fluoride favors the spin-unpaired condition. The spin-paired complex (sometimes referred to as the strong field case) is equivalent to the inner-orbital complex. The spin-unpaired complex (sometimes referred to as the weak field case) is equivalent to the outer-orbital complex.

REACTIONS OF COMPLEX IONS

Much has been learned about the relative stability of complex ions through tracer studies of the following type of reaction:

$$Cr(H_2O^{16})_6^{+3} + H_2O^{18} \rightarrow Cr(H_2O^{16})_5(H_2O^{18})^{+3} + H_2O^{16}$$

Experimentally, a salt with the chromium ion hexa-coordinated with water molecules is added to water labeled with O^{18}. Samples of this solution are withdrawn periodically in order to determine how much of the labeled water has entered the hydration shell of the chromium ion. These experiments show that the water in the hydration shell of the chromic ion exchanges very slowly with that of the solvent water; in fact, it takes 30 to 40 hours for half of the water in the hydration shell to be replaced by water from the solvent. The situation is not the same for neighboring tripositive transition metal ions. Available evidence indicates that the water in the hydration shell surrounding Fe^{+3} ions exchanges so rapidly with solvent water that complete replacement is effected every 2 minutes. Thus, in the kinetic sense, the hydration shell of chromic ion is relatively inert, and that of ferric ion is labile.

Tracer studies have established that the extremes of kinetic lability found for aquo ions are also found for complex ions. For example, the reaction

$$FeCl_6^{-3} + Cl^{-*} \rightarrow FeCl_5Cl^{*-3} + Cl^-$$

(wherein the asterisk indicates a radioactive tracer) takes place rapidly, but the corresponding reaction of $CrCl_6^{-3}$ takes place very slowly. Such exchange reactions always occur more slowly with Cr^{+3} than with Fe^{+3}.

It is tempting to ascribe the inertness of the chromic complexes and the lability of the ferric complexes to differences in thermodynamic stability. We must remember, however, that these exchanges are rate processes, and a slow rate of exchange does not require thermodynamic stability. In fact, the aquo complexes for Fe^{+3} and Cr^{+3} are approximately equal in thermodynamic stability; the hydration energies are 1105 kcal for Cr^{+3} and 1072 kcal for Fe^{+3}. The contrast in kinetic lability and thermodynamic stability is even more pronounced in the chlorides. From the kinetic standpoint, the complex chlorides of chromic ion are more inert than the complex chlorides of ferric ion. Thermodynamically, on the other hand, the ferric-chloride complexes are more stable than the corresponding complexes of chromium. Thus the thermodynamic and kinetic stability are diametrically opposed.

In line with the recently established usage, we shall denote kinetic instability by "lability" and kinetic stability by "inertness." These two labels need not parallel thermodynamic stability and instability.

Lability of complex ions. The coordination compound $[Co(NH_3)_5Cl]Cl_2$ dissolves in water to form three ions. The conductivity, however, slowly increases as the solution stands. Such changes are evidence that the following reaction occurs:

$$[Co(NH_3)_5Cl]^{+2} + H_2O \rightarrow [Co(NH_3)_5(H_2O)]^{+3} + Cl^-$$

From the rate of this increase in conductivity, or from the rate of this *substitution* reaction, one can estimate the lability of the complex. Such reactions as these, together with studies of isotopic exchange reactions and rates of racemization, have been used to measure the lability of many complex ions. Table 5 summarizes the results obtained for a variety of hexa-coordinated fourth period elements.

Table 5 Lability of Complex Ions

Ions	Labile	Inert
Sc^{+3}, Ti^{+4}	all	—
Ti^{+3}, V^{+4}	all	—
Ti^{+2}, V^{+3}	all	—
V^{+2}, Cr^{+3}, Mn^{+4}	—	all
Mn^{+3}	F^-Cl^-	CN^-
Cr^{+2}	H_2O	Cl^-
Fe^{+3}	F^-, Cl^-, $C_2O_4^{-2}$	CN^-
Mn^{+2}	—	CN^-
Fe^{+2}	$C_2O_4^{-2}$	CN^-
Co^{+3}	—	CN^-, $C_2O_4^{-2}$, NH_3
Co^{+2}	Cl^-, NH_3, CN^-?	CN^-?
Ni^{+2}	NH_3, $C_2O_4^{-2}$, CN^-	—
Cu^{++}	all	—
Zn^{++}	all	—
As^{+5}	—	F^-
Se^{+6}	—	F^-

The lability of complex ions as measured by an exchange of the type

$$MA_6 + B \rightleftarrows MA_5B + A$$

must be determined by the mechanism of the exchange. One of the simplest plausible mechanisms is one in which the hexa-coordinated species becomes a hepta-coordinated unstable species. In other words, the reaction occurs via $MA_6 + B \rightarrow (MA_6B) \rightarrow MA_5B + A$. Hepta-coordinated species are known; the bonding is described as D^3SP^3 hybridization.

For such hybridization to occur readily, the complex MA_6 should have empty d orbitals available. If we compare Table 4 to Table 5, we find that the first three sets of ions listed, which form only labile compounds, do have empty $3d$ orbitals. The ions between the V^{+2} ion and the Fe^{+2} ion do not have empty d orbitals if they form inner-orbital complexes. The inner-orbital complexes of these ions are all inert; on the other hand, the outer-orbital complexes, like FeF_6^{-3}, are labile. On this basis, we conclude that all inner complexes with unoccupied $3d$ levels form labile complex ions; inner complexes in which all the $3d$ levels are occupied are inert; and, finally, outer-orbital complexes are labile. The last statement can be rationalized if we assume that, when $4d$ levels are already utilized for bonding, they can also be utilized for bonding to form the hepta-coordinated intermediate. For the sake of completeness, we must also consider those cases (cobaltous ion and nickelous ion) in which an inner-orbital complex forms with $4d$ orbitals occupied by non-bonding electrons. These are also labile, for the same reasons that the outer-orbital complexes are labile. As the apparent charge on the central ion increases, the outer-orbital complexes become increasingly inert; when the charge is five or greater, they are completely inert.

Oxidation-reduction reactions. Oxidation-reduction reactions such as $Fe(X)_6^{+3} + Fe^*(X)_6^{+2} \rightarrow Fe^*(X)_6^{+3} + Fe(X)_6^{+2}$, have been extensively studied with radioactive tracers. It seems reasonable to assume that reaction occurs by an electron jump during a collision. If this is so, why doesn't the electron jump occur after every collision? This may be the answer: The tripositive ions generally hold the donor ligand much closer than the larger dipositive ions does. For example, the hydration energy of Fe^{+3} is 1072 kcal, compared to 500 kcal for Fe^{+2}. If we rip the electrons from the ferrous ion and place it on the ferric ion, *without changing the bond length to the water molecules,* the species formed will be unstable; hence such a reaction is unlikely. Now, we must keep in mind that, although we speak about Fe-H_2O distance, there are vibrations due to thermal energy that make the bond distance shorter than, and longer than, the equilibrium distance. Thus, for the ferric ion, there will be an expansion of the bond distance during part of the vibration; for the ferrous ion, there will be a compression of this bond distance during part of the vibration. If these two molecules collide just when the Fe^{+2}-X distance is a minimum, and the Fe^{+3}-X distance is a maximum, the instability effected by electron transfer will be a minimum and reaction can occur. This can happen for only a very small fraction of the collisions between aquo ions with appropriate vibrational energies. Because of this, the reaction does not occur as fast as it would if an exchange took place on every collision.

Electron-transfer reactions are most likely when the complex is inert; if it is labile, a more likely path for oxidation-reduction reactions involves transfer of an atom or group of atoms.

The following reaction† has been studied in detail:

$$Co^{+3} + {}^{\circ}Cr^{+2} \rightarrow Cr^{+3} + {}^{\circ}Co^{+2}$$

If we refer to Table 5, we find that the cobaltic and the chromic ion are inert, whereas cobaltous ion forms labile complexes. In addition, note that the exchange for the chromous aquo ion is rapid. One of the reactions studied was was

$${}^{\circ}Cr^{+2} + (NH_3)_5Co\!:\!\ddot{\underset{\cdot\cdot}{Cl}}\!:^{+2} \rightarrow [CrCl(H_2O)_5]^{+2} + {}^{\circ}Co^{+2}$$

(In writing this equation, we have left off the water of hydration for the chromous and cobaltous ion; from isotopic substitution studies, we know these aquo ions are labile.) When Cr^{+2} and $(NH_3)_5CoCl^{+2}$ reacted, the product Cr^{+3} appeared as the complex shown. Since the chloro and aquo complexes of the chromic ion are inert, the formation of chloride complex, $[CrCl(H_2O)_5]^{+2}$, after oxidation of Cr^{+2} would be extremely slow. Therefore the formation of $[CrCl(H_2O)_5]^{+2}$ takes place in the same step that brings about oxidation, and oxidation occurs with the formation of the intermediate $(NH_3)_5Co^{\circ}\,\vdots\!:\!\ddot{\underset{\cdot\cdot}{Cl}}\!:\!Cr^{+4}$. Reaction occurs when the break-up of this intermediate occurs along the dotted line. The electron transfer is accomplished by transfer of the chlorine atom.

If this is the correct interpretation, then, for the reaction to be rapid, we must have a bridging group that is capable of donating electrons not only to the cobalt atom but also to the chromium atom. Thus, if we compare the reactivity of the two species $(NH_3)_5Co(NH_3)^{+3}$ and $(NH_3)_5Co(H_2O)^{+3}$, the reaction of the aquo species should proceed more rapidly than the reaction of the ammono species because nitrogen, having a single lone pair of electrons, cannot function as a bridge, whereas the oxygen, with its two lone pairs of electrons, can function as a bridge. Measurements shown that the aquo complex reacts at least a hundred times faster than the ammono complex.

† For clarity, the extra electron on the divalent ion is suggested by a small circle.

PROBLEMS

1. Assume that the effective radius of a water molecule is the same as that of O = (1.40 Å) and estimate the most likely coordination number of aquo Li^+ and K^+ ions. *Hint:* See Table 2 in Chapter 9.

2. Calcium chloride forms three hydrates: $CaCl_2 \cdot 6H_2O$, $CaCl_2 \cdot 4H_2O$, and $CaCl_2 \cdot 2H_2O$. The transition from $CaCl_2 \cdot 6H_2O$ to $CaCl_2 \cdot 4H_2O$ occurs at 30°C (in the presence of pure water). The transition from $CaCl_2 \cdot 4H_2O$ to $CaCl \cdot 2H_2O$ occurs at 50°C. Assume for simplicity that the solubility of all three increase linearly with temperature, and sketch a plot of the equilibrium solubility versus temperature. Which salt is most soluble at 25°C? Why? *Ans.* $CaCl_2 \cdot 2H_2O$.

3. On the basis of the stability of the ammonia complexes of Ca^{+2} and Ag^+ in aqueous solutions, can you predict the relative solubility of CaI_2 and AgI in liquid NH_3 and liquid H_2O? (In aqueous solution, Ca^{+2} forms weak NH_3 complexes and Ag^+ forms strong NH_3 complexes.)

4. In aqueous solution, Ag^+ has more of a tendency to form complexes than does Ca^{+2}. Would you expect the same to hold in liquid NH_3 solutions? Why?

5. By a suitable use of solvents, design an experiment to separate $Fe(NO_3)_3$ and $Co(NO_3)_2$. Explain. *Ans.* Use liquid NH_3 and precipitate $Fe(NO_3)_3$. See Table 2.

6. A pure compound is known to be $Co(NH_3)_6Cl_3$ or $Pt(NH_3)_6Cl_4$. When it is heated, it loses 2 moles of NH_3 to form a new compound. A 1.0-gm sample of the salt is dissolved in 1 kg of water and the freezing point is found to be −0.0158°C. Which salt is it? Which of these would be the more paramagnetic?

7. If $Pt(NH_3)_2Cl_2$ were tetrahedral instead of square coplanar, how many isomers would exist?

★ **8.** At least part of a series of phosphorus halides with the formula PCl_nF_{5-n} can be formed. In all of them the phosphorus atom is at the center of a trigonal bipyramid with the halogen atoms at the corners. Which of these compounds would have stereo-isomers? Sketch the isomers. *Ans.* $PClF_4$; $PFCl_4$; PF_2Cl_3; PCl_3F_2.

9. How many stereoisomers would the compound $Pt(AB)_3$ have? (AB stands for a non-symmetric bidentate group.)

10. How many isomers would $Pt(AAA)_2$ form? (AAA is a tridentate group.)

11. Would any of the halides discussed in Problem 8 form optical isomers? Which ones?

★ **12.** Draw the two isomers formed when

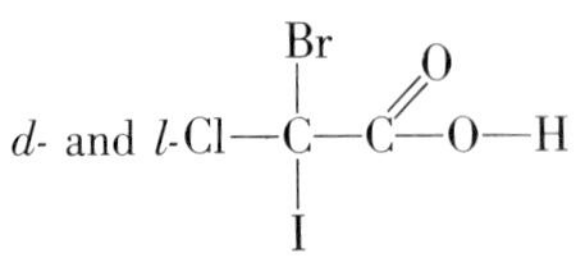

reacts with

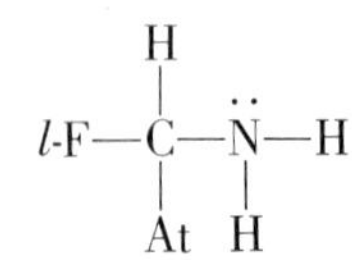

to form the amide. Assume for simplicity that the CNCC bond is linear and that there is no free rotation about the bond. If the end groups are opposed so that the Cl-F distance is a maximum, show a view looking along the CNCC axis. Choose your own convention for d and l. Would you expect the two product isomers to have different chemical and physical properties?

13. Indicate by means of a diagram like that on p. 715 the dsp^2 configuration of the Pt^{+2} ion.

Ans.	$3d$	$4s$	$4p$	$4d$
	⇅ ⇅ ⇅ ⇅ $\underline{2}$	$\underline{2}$	$\underline{2}$ $\underline{2}$ _	_ _ _ _ _

14. If Fe^{+3} formed a dsp^2 complex, what would be the number of unpaired electrons?

15. Indicate by means of a diagram like that on p. 715 the inner- and outer-orbital complexes for Pt^{+4}.

16. What do you think the evidence is that suggests Co^{+2} forms an inner-orbital complex with CN^-? Explain. (Inertness alone is insufficient evidence.)

17. Since $Fe(C_2O_4)_3^{-3}$ forms a labile ion, how many unpaired electrons are probably found in the ion? *Ans.* 5.

18. Predict the magnetic properties of $Cr(H_2O)_6^{+2}$ and $Cr(Cl)_6^{-4}$. Explain these properties with a diagram similar to the one in Fig. 10.

19. If Cr^{+3} were to form a stable heptavalent species, what do you think its magnetic properties would be?

20. The ion CrX_6^{-4} (X = a univalent anion) has a paramagnetism consistent with four unpaired electrons. Should it be labile?

21. Discuss the magnetic properties of Cr^{+3} and Mn^{+2} complexes in terms of the crystal field theory.

22. The potential developed by the battery

$$Zn(metal)/Zn^{+2}(1M)//Cu^{+2}(1M)/Cu(metal)$$

is 1.10 volts. If the right-hand side is made 4 molar in NH_3 initially, the voltage drops to 1.02 volts owing to the formation of the complex $Cu(NH_3)_4^{+2}$. If no complexes of the forms $Cu(NH_3)^{+2}$, $Cu(NH_3)_2^{+2}$, and $Cu(NH_3)_3^{+2}$ are formed, what is the dissociation constant of the $Cu(NH_3)_4^{+2}$ complex?

23. The $Ni(CN)_4^{-2}$ complex has a square planar structure. $Ni(CO)_4$ is tetrahedral. Explain this on the basis of their atomic structure. Would they be paramagnetic?

24. Tetra-coordinated cupric ions are usually planar. Tetra-coordinated cuprous ions are tetrahedral. From the relative energies of the atomic levels in the ions rationalize this behavior. Would they be paramagnetic?

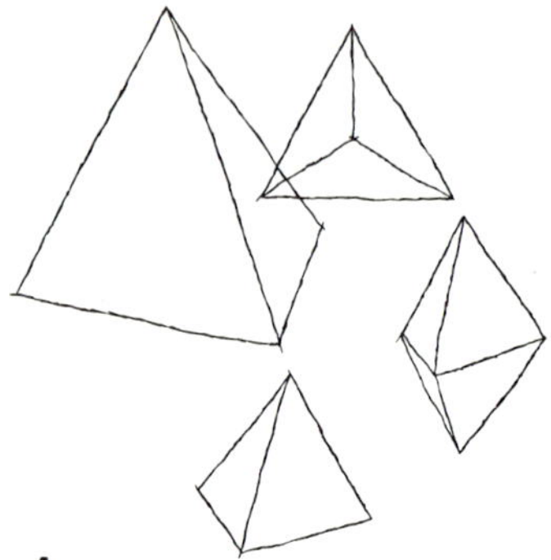

26
Organic chemistry and biochemistry

THE NATURE OF ORGANIC CHEMISTRY

Of all the hundred-odd elements, carbon has the most complex chemistry. Under the conditions prevalent on earth the carbon atom forms branched and cyclic chain structures far more numerous than the compounds of any other chemical element. These carbon structures can contain other elements like oxygen, nitrogen, phosphorus, and sulfur either as links in the skeletal chains or as *functional groups* attached as side chains that produce a wide variety of chemical properties. Among these properties there is the ability to propagate themselves, to reproduce their own kind, through the dynamics of the life process. This link with living matter suggested the name *organic* for the chemistry of carbon.

Could any other element take the place of carbon?

In order for an element to form compounds of the variety and complexity required for life processes, the element must be capable of forming thousands of complex branching chain structures that are more or less stable in the moist, oxidizing atmosphere characteristic of our planet. To build stable branching structures two requirements must be met. *First,* an atom must form covalent bonds with its own kind, directed in space and energetically stable at ordinary temperatures. *Next,* it must be possible to form at least *three* such covalent bonds on an atom in order to have branching. In the light of these requirements, many elements can be eliminated from our consideration. The metallic elements are eliminated because no metallic element ever forms several strong covalent bonds with its own kind. Many of the non-metallic elements are also eliminated. For example, hydrogen and the halides cannot form chains because it takes at least two bonds on each atom to string them together; and, when multiply bonded, these atoms do not form stable covalent bonds with their own kind. Similarly, the second requirement rules out the elements of Group VI, the members of the oxygen family that are normally divalent. The Group V elements, nitrogen and phosphorus, can have the required three covalent bonds; and moderately complicated branching structures are formed by phosphorus. Nitrogen is also an important link in thousands of organic compounds where the predominant atom is carbon. But no structures are known of even moderate complexity where the skeleton consists of nitrogen atoms linked together. The energy of the N—N bond is far too low to impart the necessary stability. The P—P bond, however, is stronger than the N—N bond; hence complex compounds involving P—P bonds are known. For the most part, however, these P—P bonds hydrolyze readily and therefore such compounds are not stable enough in a moist atmosphere to form sufficiently complex structures.

On the left of carbon in the periodic table the only non-metallic element besides hydrogen is boron. This is a metalloid and does have the ability to form fairly stable covalent bonds. Some branched-chain structures are known containing several boron atoms; but, again, the requisite energy conditions for stability are lacking.

This leaves only the two elements, carbon and silicon, with the necessary properties for complex structure formation. Both of these elements have slightly metallic properties. Carbon in the form of graphite exhibits a small conductivity for electricity, and the same is true of silicon. But the predominant behavior pattern is the formation of four covalent bonds ideally suited both in spatial arrangement and in bonding-energy pattern for making stable branched chains. In Table 1, the values of the bond energy are given for the links that we have just discussed. These values show the stability of the C—C bond as contrasted with other bonds that might be found in branched-chain structures. When a C—C chain is formed and the unused carbon bonds are linked to hydrogen, the chain is insulated by this hydrogen skin against chemical attack. Yet, if re-

Table 1 Bond Energies (kcal)

Covalence:	4		3		2		1	
	C—C	83	N—N	38	O—O	33	H—H	104
	Si—Si[a]	42	P—P	51	S—S	51	F—F	37
	C—N	70						
	C—O	84						
	C—H	98	N—H	93	O—H	110	F—H	135
	C—F	105						
	C—Br	66	N—Cl	48	S—Cl	86		

[a] Silicon can form as many as six bonds.

active groups like —OH, —CHO, or —COOH are substituted at a few points for these relatively inert hydrogen atoms, then loci of chemical activity are introduced into the structure, and the threads are provided for weaving a complex chemical tapestry.

In the brief survey of organic chemistry presented in this chapter, we begin with a short discussion of the more important types of these reactive groups; and we examine their effects on the energy of compounds and on the entropy or probability relationships. When two of these chemical compounds come in mutual contact, the energy considerations may be favorable for several different reactions; but the reaction that actually takes place is frequently determined by the kinetic factors. Though one reaction may have a larger $-\Delta F$ than another and possess a greater thermodynamic tendency to go, the second less-favored reaction may occur because its potential barrier is lower. This may be due to the presence of an inorganic catalyst or to a type of organic catalyst called an *enzyme*. After reviewing some of the active organic groups, we conclude with a brief consideration of some of the chemical aspects of the life process.

FUNCTIONAL GROUPS

Hydrocarbon radicals.† Among the frequently encountered *functional groups* or *radicals* attached to organic molecules to alter their properties are the groups made up solely of carbon and hydrogen, like the methyl group or radical, $—CH_3$. Groups like the methyl and ethyl radicals are themselves fragments of saturated hydrocarbons. In general, when found as side groups, they are not active chemically, resembling the parent saturated hydrocarbon, but their presence as side groups may alter the chemical properties of the resulting compound.

From the thermodynamic point of view, these radicals can alter both chemical and physical properties to a considerable extent, because of the influence on the symmetry and, thus, on the entropy of a compound. Consider benzene and toluene. *Benzene* is a hexagonal ring structure with a symmetry of 12, six-fold in front and six-fold in back. If a methyl group is attached to the side of a benzene ring, the compound *toluene* is formed with a symmetry of 2. As may be seen from the figures in Table 2, the boiling point is affected only slightly by the addition of the methyl radical but the melting point is dropped by about 100 degrees.

The reason for this may be seen by examining the nature of the condensation and crystallization processes. In condensation from the vapor to the liquid state, there is little change in the orderliness of the states; in both vapor and liquid states, the molecules are orientated at random, spinning in the vapor and either spinning or rocking back and forth in a random orientation in the liquid. So there is no change in entropy *due to orientation change* in passing from vapor to liquid. Since the addition of the relatively inactive methyl group has only a small effect on the energy field around the molecule, there is little change in the boiling point.

† The term radical may refer to a functional group *or* an odd-electron molecule. Today the former meaning has fallen into disfavor, but it is still current in many texts.

Table 2 Effect of Hydrocarbon Radical

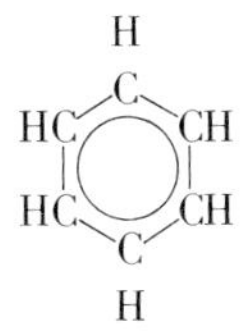

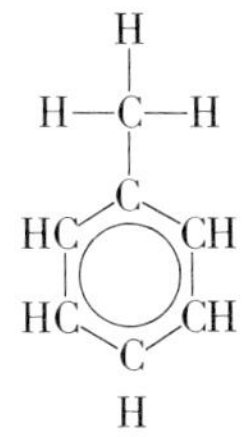

	Benzene		Toluene	
m.p.	5.5°C	(278°K)	−95.0°	(168°K)
b.p.	80.0°C	(353°K)	110.6°	(383°K)

With crystallization, the situation is different. In crystallizing, the molecule passes from a state of random orientation in the liquid to a state of ordered orientation in the crystal. As a benzene molecule in the liquid approaches the crystal surface, there are twelve ways in which it can present the proper aspect to fit into the crystal because of its twelve-fold symmetry. On the other hand, toluene has only two orientations in which it can approach the crystal and fit into place because of its reduced two-fold symmetry. Thus with other factors equal, the *probability* of the crystalline state for benzene is $\frac{12}{2}$ or 6 times greater for benzene than for toluene. Thus between −95° and 5°C we find benzene as a crystal and toluene as a liquid because the symmetry of the former is so much higher and the crystalline state consequently more probable. Since entropy and probability are directly related, the different symmetries alone yield

$$\Delta_1^2 S = R \ln (\sigma_2/\sigma_1)$$

where $\Delta_1^2 S$ is the entropy difference between benzene and toluene and σ_1 and σ_2 are numbers characteristic of the symmetry of the molecules; the ratio would be $\frac{12}{2}$ for benzene and toluene. This formula gives the difference in the entropy factor between two compounds that differ only in symmetry; and, with the help of thermodynamics, the entropy factor tells how both physical and chemical properties will be affected. Especially in the complicated ring structures encountered in biochemistry, the presence or absence of hydrocarbon side chains can have a profound effect on the chemical properties because of the way in which side chains affect the probability of reaction with other compounds.

Hydrocarbon radicals may be added to aromatic hydrocarbons by the Friedel-Crafts reaction in which an organic halide compound yields a hydrocarbon radical attached to a benzene ring when the two are heated together with aluminum chloride.†

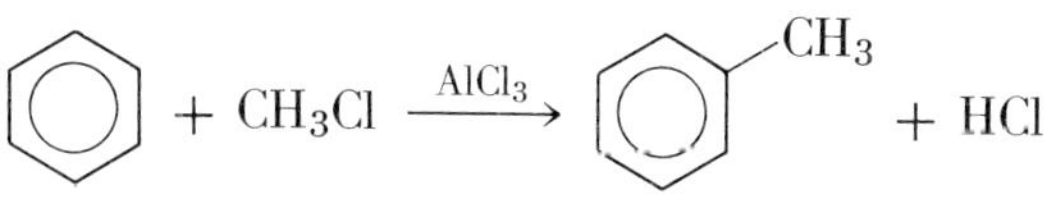

† In writing equations for organic reactions, it is customary to indicate special conditions or catalysts over the arrow. We follow this procedure in this chapter.

If the methyl halide is present in too high a concentration, a second addition frequently occurs:

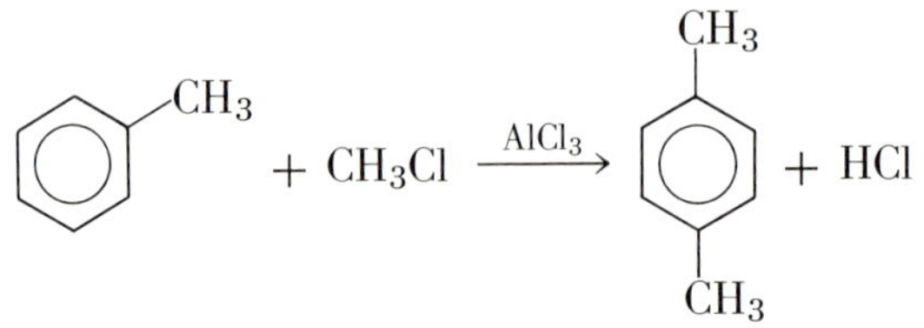

But, if the benzene is present in sufficient excess, toluene will be the preponderant product.

A hydrocarbon radical can also be added by passing a mixture of benzene and ethylene vapors through a liquid mixture of aluminum chloride and hydrogen chloride dissolved in a hydrocarbon solvent:

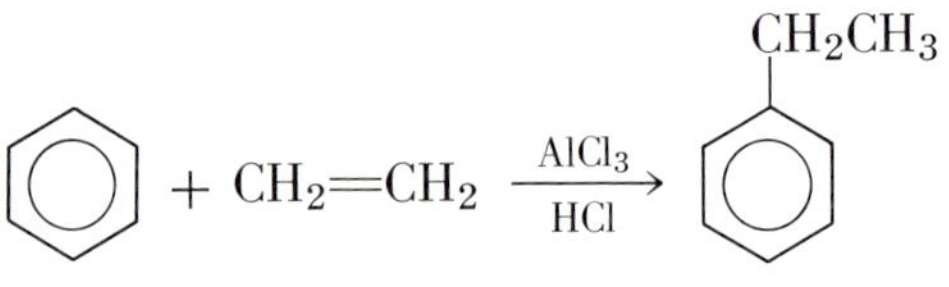

ethylbenzene

The Wurtz reaction provides another way of coupling a hydrocarbon radical to another organic compound. In this reaction, each of the molecules to be coupled must contain a chlorine or other halogen atom. The two are then mixed with metallic sodium and the hydrocarbon radicals couple, with the elimination of sodium chloride. In such a reaction, there is a tendency for each of the compounds to couple with themselves as well as with each other, so that the product is a mixture and the separation of the desired compound from it may be difficult. Thus the Wurtz reaction is of historic rather than practical interest.

Alcohols. Among the simplest of the active groups encountered in organic chemistry is the group —OH. Compounds characterized by this group are called *alcohols.* The simplest alcohols consist of the saturated hydrocarbon compounds in which a single hydrogen has been replaced by this OH group. Other compounds are found that contain 2, 3, or more OH groups. These OH groups are also found in compounds that contain other functional groups.

A few alcohols are listed in Table 3. They may be regarded as derivatives of water prepared by the replacement of one of the hydrogen atoms by the organic group. The boiling points all lie much higher than those of the related hydrocarbons because hydrogen bonding between the OH groups produces a force that must be overcome to get individual molecules into the vapor state. Since it takes energy to break the hydrogen bond, this increment of energy is added to the heat of vaporization; thus the compound must be brought to a higher

Table 3 Alcohols

		Boiling Point (°C)
(Water)	(HOH)	100
Methyl alcohol	$CH_3—OH$	66
Ethyl alcohol	$CH_3—CH_2—OH$	78
Propyl alcohol	$CH_3—CH_2—CH_2—OH$	87
Butyl alcohol	$CH_3—CH_2—CH_2—CH_2—OH$	116
Amyl alcohol	$CH_3—CH_2—CH_2—CH_2—CH_2—OH$	138
Hexyl alcohol	$CH_3—CH_2—CH_2—CH_2—CH_2—CH_2—OH$	157
Cyclohexyl alcohol	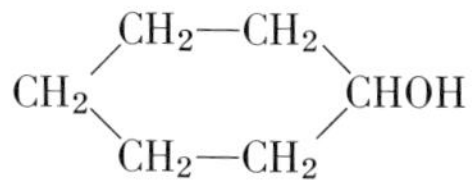	160
Ethylene glycol	$CH_2OH—CH_2OH$	197
Glycerol (glycerin)	$CH_2OH—CHOH—CH_2OH$	290

temperature to get it into the vapor state at 1 atm. If, on the other hand, the compound goes into the vapor state as pairs of molecules linked together by hydrogen bonds between OH groups, the molecular weight of the compound is twice that to be expected on the basis of its normal formula; and this in turn means that a higher temperature must be reached in order to get this heavy double molecule into the vapor state at 1 atm.

One of the common ways of making an alcohol is the addition of water to a double bond in an unsaturated compound:

$$CH_2{=}CH_2 + H_2O \xrightarrow{H_2SO_4} CH_3CH_2OH \qquad \text{(ethyl alcohol)}$$

$$(CH_3)_2C{=}CH_2 + H_2O \xrightarrow{H_2SO_4} (CH_3)_3COH \qquad \text{(tertiary butyl alcohol)}$$

A mixture of hydrogen and carbon monoxide forms the simplest of the alcohols, *methyl alcohol* or *methanol:*

$$2H_2 + CO \xrightarrow[200\text{ atm, } ZnO\text{-}ZnCrO_4]{300°} CH_3OH$$

Alcohol can also be produced by the fermentation of sugar in the presence of a very complex organic compound called an *enzyme;* the sugar molecules break up to give ethanol and carbon dioxide gas:

$$C_6H_{12}O_6 \xrightarrow[20\text{–}25°C]{\text{Zymase}} 2CH_3CH_2OH + 2CO_2(g)$$

As is to be expected, the chemistry of alcohol is closely related to the chemistry of water; the simplest alcohol, methanol, can substitute for the water molecule in forming coordination compounds:

$$CaCl_2 + 4CH_3OH \rightarrow CaCl_2 \cdot 4CH_3OH$$

Alcohols also react with metallic sodium and form a compound, called an *alkoxide*, that is analogous to sodium hydroxide:

$$2CH_3OH + 2Na \rightarrow 2NaOCH_3 + H_2$$

When an aromatic group like that of benzene is substituted for the aliphatic group like *ethyl*, the resultant compound with OH takes on quite a different character. The compound formed by adding an OH group in place of one of the hydrogen atoms of benzene is called *phenol;* its more common name is *carbolic acid;* the hydrogen atom of an OH group attached to an aromatic ring shows typical acid behavior. This is of significance when various kinds of alcohol are mixed. Two alcohols like methanol and ethanol form almost ideal solutions thermodynamically. The hydrogen bonding of a methanol molecule to an ethanol molecule has just about the same strength as that between two methanol molecules or between two ethanol molecules. On the other hand, the binding of a methanol molecule to phenol is far stronger than the binding either of methanol to methanol or phenol to phenol. The result is a large negative deviation from Raoult's law that determines the behavior of ideal solutions.

The effect of hydrogen bonding on physical properties is particularly striking when more than one OH group is present in the same molecule. Among the more familiar and widely used compounds of this variety are *glycol,* $HOH_2C \cdot CH_2OH$ and *glycerol,* $CH_2OH \cdot CHOH \cdot CH_2OH$. The opportunity for hydrogen bonding between different molecules is greatly enhanced by the presence of more than one OH group on each molecule. While the simple alcohols like CH_3OH have a tendency to form pairs or dimers, molecules like glycerin tend to link up with each other in complicated networks. The problem is similar to that we discussed at the beginning of this chapter. If an atom can form only two covalent bonds, it can link up in a long chain but not in a complicated network; when it can form three or more covalent bonds, then the possibilities for a network are high. In the same way, the simpler alcohols do not form complicated networks of hydrogen bonding, but molecules like glycerin can link up with several other molecules by means of hydrogen bonds. The result is that in the liquid state the molecules of glycerin are strongly tied together; the displacement of a molecule from one position to another in the liquid is correspondingly more difficult; and we find that glycerin has a high *viscosity* or resistance to flow.

Glycol can be made by the direct addition of oxygen and water to ethylene:

$$2CH_2{=}CH_2 + O_2 \xrightarrow{Ag} 2\underset{\diagdown O \diagup}{CH_2{-}CH_2}$$

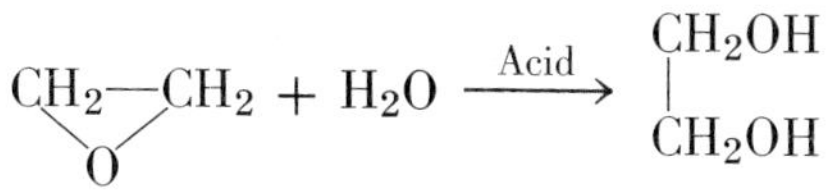

Glycerin is one of the principal constituents in animal fat. When soap is prepared from fat, glycerin remains behind as an important by-product.

In general, alcohols react with acids; for example, methanol reacts with hydrochloric acid to form methyl chloride:

$$CH_3OH + HCl \rightarrow CH_3Cl + H_2O$$

glycerin reacts with nitric acid to form trinitroglycerin:

$$\begin{array}{l} CH_2OH \\ | \\ CHOH \\ | \\ CH_2OH \end{array} + 3HNO_3 \rightarrow \begin{array}{l} CH_2ONO_2 \\ | \\ CHONO_2 \\ | \\ CH_2ONO_2 \end{array} + 3H_2O$$

This product is an important commercial explosive. An explosion is, by definition, a process in which there is a sudden release of energy. As a rule, chemical explosions take place because of an extremely rapid oxidation reaction that releases large amounts of energy. For example, a mixture of hydrogen gas and oxygen gas in the proportions 2 to 1 is stable at room temperature; but, when a small portion of the gas is heated by a spark or flame, the oxidation of the hydrogen with the formation of water releases energy; and this energy in turn is passed on to neighboring molecules. A branching chain reaction is set up, and the energy of oxidation in the whole mixture is released so suddenly that an explosion takes place and a shock wave is generated. When oxygen, hydrogen, and carbon are in even more intimate proximity as in nitroglycerin, the oxidation reaction can take place by the rearrangement of the atoms and an even quicker explosive process takes place. In particular, the oxygen atoms which are attached to the nitrogen remain more or less indefinitely in position unless disturbed; but, when there is a sudden input of energy that activates them, they shift over and combine with the carbon and hydrogen to form products like carbon monoxide, carbon dioxide, and water with this sudden release of energy.

Another important class of compounds, called the carbohydrate class, has several OH groups in the same molecule. The members of this group have about twice as much hydrogen present as oxygen, the same proportions found in water; and this suggested the name. These compounds play a key role in the life cycle. The leaves of plants contain chlorophyll, the coloring matter that makes the leaves green (see Fig. 4 for structural formula). This compound acts as a catalyst for combining carbon dioxide from the surrounding air and water, either from the air or the plant's sap, into a compound of the carbohydrate class.

Light is needed since the photons of light provide the energy that is needed to make the carbon dioxide and the water combine.

$$6CO_2 + 6H_2O \xrightarrow[\text{Chlorophyll}]{\text{Light}} C_6H_{12}O_6 + 6O_2$$

Certain of the carbohydrates such as the common sugar sucrose have a sweet taste when placed on the tongue. Melted sugar is viscous like glycerin because of multiple hydrogen bonding between its many OH groups. Many sugars frequently contain different molecules that are members of the same class but differ slightly in details of their chemical structure. Some forms of glucose occur as a straight chain resembling glycerin. In such a chain the OH groups may bear different relations to each other, lying on different sides of the chain as roughly indicated by the two-dimensional formula in the center of the diagram. Because of the tetrahedral angle between the bonds of carbon, a chain like that in the center formula sometimes bends around, bringing the oxygen atom in the CHO group (aldehyde) at the top close to the carbon atom next to the bottom. The aldehyde functional group will be discussed in the next section. The H in the OH group in the number 5 position can shift over and attach itself to the oxygen of the CHO group when the two are brought into proximity by the bending of the chain. Meanwhile, the change of the CHO group to CHOH leaves a free bond on the number 1 carbon atom; the oxygen in the number 5 position attaches itself at this position and a six-membered ring is formed consisting of five carbon atoms and one oxygen atom. In such a ring, the OH groups are held either above or below the ring and do not shift from one position to another under the influence of thermal energy. Thus there can be several *space-geometric or stereoisomers* of these carbohydrate rings as shown in the first and the last formulas. Suppose that the ring shown at the top of the diagram breaks at the bond between C and O, shown as a dotted line. The molecule

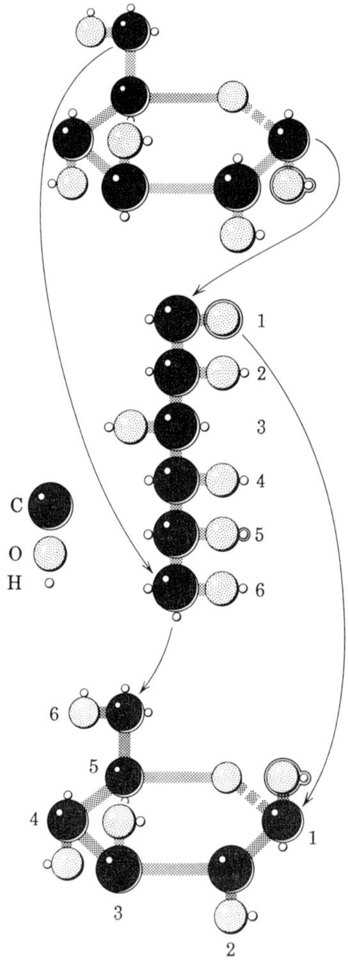

straightens out to the form shown in the center of the diagram and then reforms the ring shown at the bottom, with the OH group on the number 1 carbon atom now pointing up instead of down. This process illustrates the way in which passages from ring to chain to ring can shift an OH group (shown in the double circles) and thus make an isomer.

These carbohydrates oxidize to produce CO_2 and H_2O, the materials from which carbohydrates are formed in the plant leaves. Carbohydrates dissolve in water and circulate readily in the bloodstream in the human body. They provide one of the important sources of energy for biochemical reactions in the body such as the contraction of muscle tissue to perform work. Such rings also appear in the complex structures that transmit hereditary influences. We shall discuss them further in a section on biochemistry at the end of this chapter.

Aldehydes. Alcohols can be advanced by one stage of oxidation to give the group CHO:

$$CH_3CH_2OH + \tfrac{1}{2}O_2 \rightarrow CH_3\cdot C\begin{matrix} \diagup H \\ \diagdown\!\!\diagdown O \end{matrix} + H_2O$$

CHO is the functional group characteristic of the class of compounds called *aldehydes.*

The CHO group is fairly active but cannot form hydrogen bonds like the OH group, as shown by the boiling points. Methanol boils at 66°C, while the comparable aldehyde, methanal, boils at −21°C. Similarly, ethanol boils at 78°C while ethanal boils at 21°C.

Ethanal can be produced both by the careful oxidation of ethanol and the direct addition of water to acetylene:

$$HC{\equiv}CH + H_2O \xrightarrow{Hg_2SO_4} CH_3CHO$$

Aldehydes undergo a number of important characteristic reactions, only a few

Table 4 Aldehydes

Systematic Name	Common Name	Structural Formula	Boiling Point (°C)
Methanal	Formaldehyde	$HCHO$	44
Ethanal	Acetaldehyde	CH_3CHO	21
Propanal	Propionaldehyde	CH_3CH_2CHO	49
Butanal	*n*-Butyraldehyde	$CH_3CH_2CH_2CHO$	76
Benzaldehyde	Benzaldehyde	$C_6H_5—CHO$	180
Trichloroethanal	Chloral	Cl_3CCHO	98

of which can be mentioned here. First of all, there is a strong tendency for the molecules to couple with each other to form polymers:

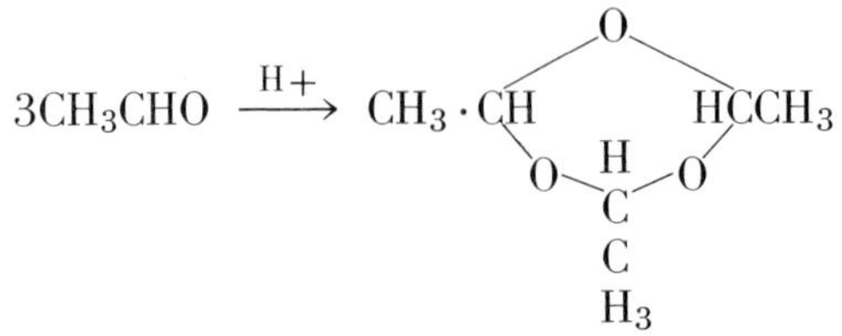

Acetaldehyde Paraldehyde

Aldehydes also combine with other molecules by a simple addition process:

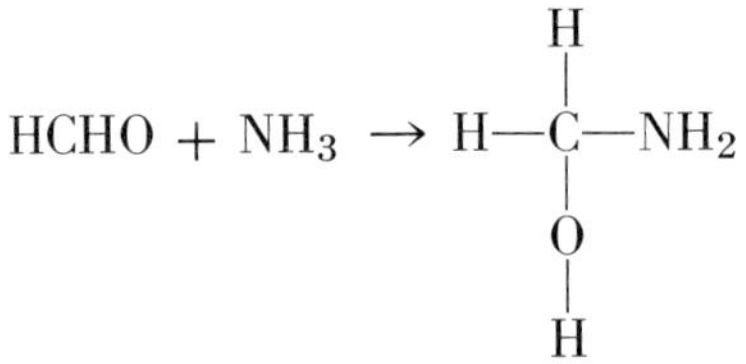

Aldehyde ammonia

Aldehydes can be oxidized to give a class of compounds called *carboxylic acids* (characterized by the functional group COOH); for example,

$$2CH_3CHO + O_2 \rightarrow 2CH_3 \cdot COOH$$

Acetic acid

or they can be reduced to alcohols; for example,

$$CH_3CHO + H_2 \rightarrow CH_3CH_2OH$$

Formaldehyde is a preservative, and solutions of it are used to store medical specimens. It has an unpleasant odor and is poisonous, but the higher aldehydes are used in perfume. The polymerizing ability of the CHO group makes it an important constituent in the formation of plastics.

Ketones. When the hydrogen atom in the CHO group is replaced with an organic radical like CH_3 it is possible to form another distinctive class of compounds called *ketones*—characterized by the functional group

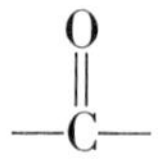

The names and formulas of some typical ketones are listed in Table 5, together with their boiling points.

Ketones may be prepared by the oxidation of the CHOH group in a secondary alcohol:

Table 5 Ketones

Formula	Name	Boiling Point (°C)
$CH_3 \cdot CO \cdot CH_3$	Acetone	56.1
$CH_3 \cdot CO \cdot C_2H_5$	Methyl ethyl ketone	79.6
$C_2H_5 \cdot CO \cdot C_2H_5$	Diethyl ketone	101.7
$CH_3 \cdot CO \cdot C_3H_7$	Methylpropyl ketone	101.8
$C_3H_7 \cdot CO \cdot C_3H_7$	Di-*n*-propyl ketone	143.5
$C_3H_7 \cdot CO \cdot C_3H_7$	Di-isopropyl ketone	123.7

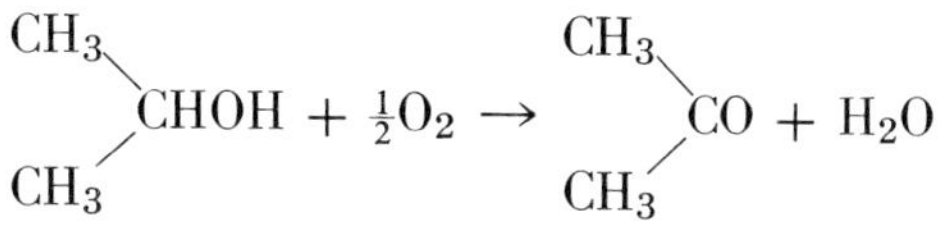

Isopropyl alcohol — Acetone

The same product may be obtained by dehydrogenation of alcohol at relatively high temperatures:

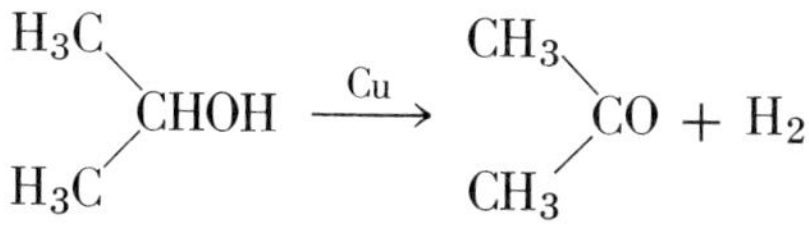

Like aldehydes, ketones undergo addition reactions, though they are less active in both this and other chemical aspects than aldehydes:

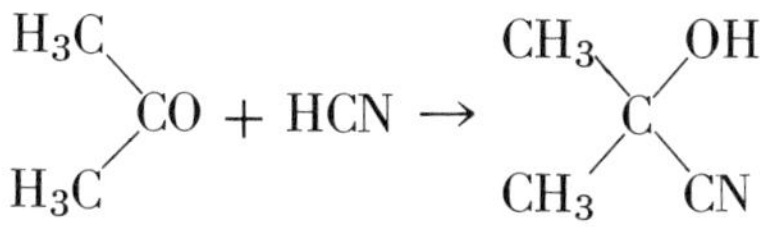

Acetone cyanohydrin

They can also be reduced by the addition of hydrogen to give secondary alcohols:

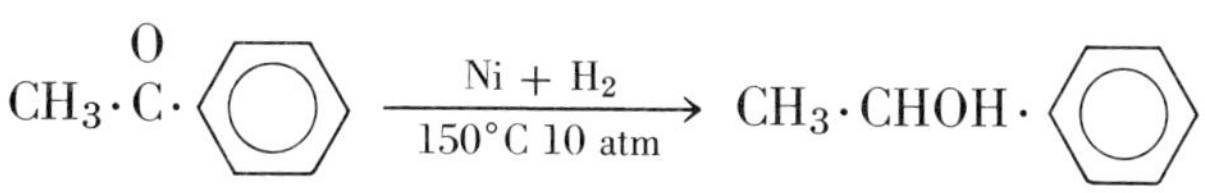

Acetophenone

Carboxylic acids. Aldehydes characterized by the functional group CHO can be oxidized to produce *carboxylic acids*—characterized by the group

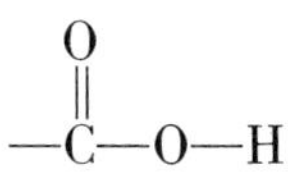

The names and formulas of a few of the commoner acids are given in Table 6.

Table 6 Carboxylic Acids

Systematic Name	Common Name	Structural Formula	Boiling Point (°C)	K_a
Methanoic	Formic	$HCOOH$	101	21×10^{-5}
Ethanoic	Acetic	CH_3COOH	118	1.8×10^{-5}
Propanoic	Propionic	CH_3CH_2COOH	141	1.5×10^{-5}
Butanoic	Butyric	$CH_3(CH_2)_2COOH$	163	1.3×10^{-5}
Pentanoic	Valeric	$CH_3(CH_2)_3COOH$	186	1.7×10^{-5}
Hexanoic	Caproic	$CH_3(CH_2)_4COOH$	206	1.4×10^{-5}

The boiling points for the acids run even higher than the boiling points of the corresponding alcohols, showing strong hydrogen bonding. Values for the dissociation constant (K_a) are also given in the table. Formic acid is the strongest acid; acetic acid is considerably weaker; and, as the length of the hydrocarbon chain is increased, the acid strength goes through a minimum and then starts climbing slightly.

These acids may also be prepared by the oxidation of alcohols:

$$CH_3CH_2OH + O_2 \rightarrow CH_3COOH + H_2O$$

but a more unusual preparative procedure is Grignard's reaction. Aliphatic halides like CH_3CH_2Cl can be coupled to metals like magnesium to form Grignard reagents (CH_3CH_2MgCl).† If such a compound is dissolved in ether and carbon dioxide is passed in at low temperatures, an acid is formed on hydrolysis.

$$CH_3CH_2MgCl + CO_2 \rightarrow CH_3 \cdot CH_2 \cdot C\begin{matrix} /\!\!/ O \\ \backslash O{-}MgCl \end{matrix} \xrightarrow{+\,H_2O} C_2H_5COOH + Cl \cdot Mg \cdot OH$$

These carboxylic acids react like familiar inorganic acids. They form salts like CH_3COONa. They react with alcohols to form the important class of substances known as *esters.*

$$\underset{\text{Acetic acid}}{CH_3COOH} + \underset{\text{Ethanol}}{CH_3CH_2OH} \rightleftarrows \underset{\text{Ethyl acetate}}{CH_3COOCH_2CH_3} + H_2O$$

The carboxylic acids are found widely distributed in natural products; and the common names for these acids are based on their origin. *Formic acid* is found in ants, the name being derived from the Latin word for this insect. *Butyric acid* is found in rancid butter. *Caproic acid* is found in goat-butter, the name being another Latin derivative.

† Today evidence is accumulating that suggests the Grignard reagent is really $Mg(C_2H_5)_2 \cdot MgCl_2$.

Table 7 Amines

		Boiling Point (°C)
Ammonia	NH_3	−33
Methyl amine	$CH_3—NH_2$	− 6
Ethyl amine	$CH_3—CH_2—NH_2$	17
Propyl amine	$CH_3—CH_2—CH_2—NH_2$	49
Isopropyl amine	$CH_3—CHNH_2—CH_3$	34
Dimethyl amine	$CH_3—NH—CH_3$	7.4
Trimethyl amine	$CH_3—N—CH_3$ | CH_3	3.5
Aniline		184.3
Methylaniline		196
Diethylaniline	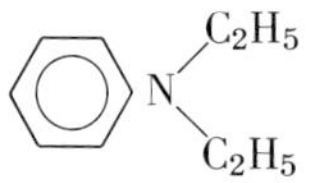	215.5
Diphenylamine	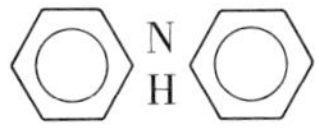	256
Triphenylamine	$(C_6H_5)_3N$	345

Amines. Just as a hydrogen atom on an organic compound may be replaced by an OH group, it may also be replaced by an NH_2 group. Compounds that contain the functional group NH_2 are called *amines.* These compounds resemble the alcohols, as we might expect since NH_2 and OH are isoelectronic; however, there are definite differences.

Table 7 lists a few of the amines. The amines are derived from ammonia by replacing, respectively, one, two, or three hydrogen atoms by organic radicals. Their characteristic behavior stems largely from the remaining hydrogen attached to the nitrogen after one or more organic radicals have been attached. If three radicals are present so that no hydrogen remains on the nitrogen, then the compound changes its behavior from that of the true amine. Just as we have *alcohols* with an OH group and *ethers* with a group ROR, we have the primary amines RNH_2, the secondary amines R_2NH, and the tertiary amines R_3N.

Amines may be prepared by the reaction of organic halides with ammonia:

$$CH_3Cl + NH_3 \rightarrow CH_3NH_3Cl \rightarrow CH_3NH_2 + HCl$$

Another method for replacing hydrogen by nitrogen starts with the nitration of a compound:

$$C_6H_6 + HNO_3 \rightarrow C_6H_5NO_2 + H_2O$$

Then the product (in this case nitrobenzene) is reduced. This may be carried out by treating the nitro compound with iron in the presence of steam:

$$8C_6H_5NO_2 + 18Fe + 8H_2O \xrightarrow[\text{HCl}]{\text{Trace}} 8C_6H_5NH_2 + 6Fe_3O_4$$

Amines react with water much as ammonia does:

$$NH_3 + H_2O \rightleftarrows NH_4^+ + OH^-$$
$$RNH_2 + H_2O \rightleftarrows RNH_3^+ + OH^-$$
$$R_2NH + H_2O \rightleftarrows R_2NH_2^+ + OH^-$$
$$R_3N + H_2O \rightleftarrows R_3NH^+ + OH^-$$

They are typical bases and form salts with acids:

$$CH_3NH_2 + H_2SO_4 \rightarrow (CH_3NH_3)^+(HSO_4)^-$$

They react with chloroform to form compounds called isocyanides:

$$3KOH + C_6H_5NH_2 + CHCl_3 \rightarrow \underset{\text{Isocyanide}}{C_6H_5NC} + 3KCl + 3H_2O$$

When heated, the carbon and nitrogen atoms swap places and a compound called a cyanide results:

$$\underset{\text{Isocyanide}}{C_6H_5NC} \xrightarrow{\text{Heat}} \underset{\text{Cyanide}}{C_6H_5CN}$$

Other groups. Before going on to consider the part that organic compounds play in the life process, we call attention to a few other types of organic compounds that are frequently encountered. These are listed in Table 8. The compounds in Table 8 are a small sample of the almost infinite variety of structures that may be obtained by combining carbon, hydrogen, nitrogen, oxygen, and sulfur in various ways. Most of the other elements in the periodic table may also be coupled into these structures in a stable way. There appears to be no limit to the complexity of the pattern that can be woven by these atomic fibers. The limit is set only by the skill and patience of the synthetic organic chemists. Large molecules con-

Table 8 Other Types of Compounds

Dibasic Acids		
COOH–COOH	COOH–CH_2–COOH	COOH–$(CH_2)_2$–COOH
Oxalic acid	Malonic acid	Succinic acid

Table 8 Other Types of Compounds (Continued)

	Unsaturated Dibasic Acids	
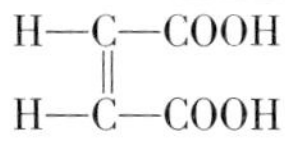 Maleic acid (*cis*)	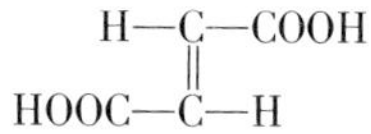 Fumaric acid (*trans*)	
	Hydroxy Acids	
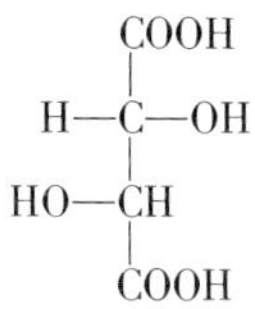 Dextrotartaric acid	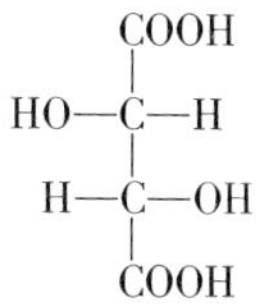 Levotartaric acid	
	Keto Acids	
H—CO—COOH Glyoxylic acid	CH_3—CO—COOH Pyruvic acid	
	Mercaptans	
CH_3SH Methyl mercaptan	C_2H_5SH Ethyl mercaptan	
	Ethers	
CH_3OCH_3 Dimethyl ether	C_2H_5—O—C_2H_5 Diethyl ether	
	Esters	
$CH_3 \cdot COO \cdot CH_3$ Methyl acetate	$CH_3 \cdot COO \cdot CH_2CH_3$ Ethyl acetate	
	Ring Compounds	
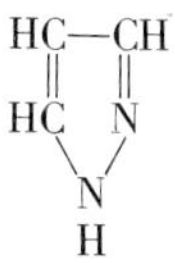 Pyrazole	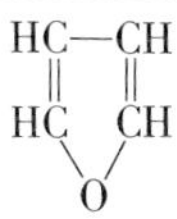 Furan	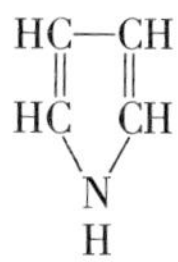 Pyrrole
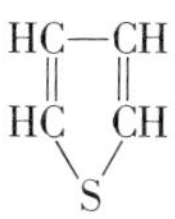 Thiophene	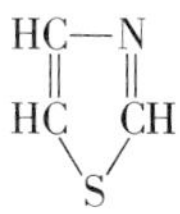 Thiazole	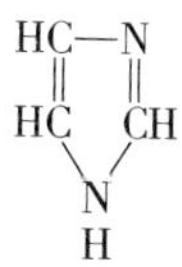 Imidazole

taining thousands of atoms "stitched" into place already have been synthesized in the laboratory. Yet, in spite of all our scientific knowledge, nature still remains the master weaver. Even in the simplest forms of life, molecular tapestries are put together through natural biochemical actions that are more intricate than the scientist can fabricate.

SOME ASPECTS OF BIOCHEMISTRY

Life goes on within us and all around us constantly, both on a microscopic and on a macroscopic scale; it is *commonplace.* But consider for a moment the complex chemical reactions that must occur for you to move your little finger, to blink your eye, to read this sentence; life is commonplace but it is miraculously complex. During each second of your lifetime, 10^{24} atoms in your body undergo thousands of different chemical reactions. Taken as a whole, the life processes are so intricate that the most complex constructions of man—a telephone continental network or a giant electronic computer—are toys by comparison. But, just as the telephone network and the computer are made up of simple components like radio tubes, transistors, capacitors, and resistors, the components of the life process are also simple. And, just as the telephone network is essentially a means of the precise and intricate channeling of energy and information, so the biochemical network in living matter is also a precise and intricate mechanism for channeling energy and entropy, the thermodynamic analog of information.

Recent developments in the theory of communication have shown that there is a direct relation between *entropy* and *information* when each is defined with mathematical consistency. In the simplest form of this relation, entropy content is the negative of information content when each is a logarithmic measure. Entropy content is a measure of the probability of a state; and greater probability is associated with greater disorder or lack of structure. Information content is a measure of the improbability of a state; and greater improbability is associated with greater order or presence of structure. Thus there is more than a superficial analogy between the telephone network in living society and the biochemical network in living matter. It is no exaggeration to say that molecules communicate with one another; and it is important to see how molecules store this communicated information or entropy. Before examining the communicative biochemical processes, let us first have a look at some of the components through which they function. Many of these are familiar molecules like water, sugar, alcohols and acids together with many of the other species that we have discussed previously. One especially important class is made up of molecules that are at the same time both acidic and basic in character, the *alpha-amino acids.* These molecules are characterized by an amino group (NH_2) and an acid group (COOH) attached to the same carbon atom; that is,

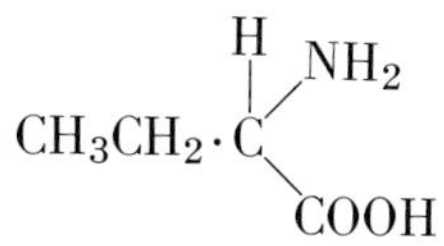

Since the amino group is basic, internal neutralization often occurs and the amino acid appears as a zwitter ion:

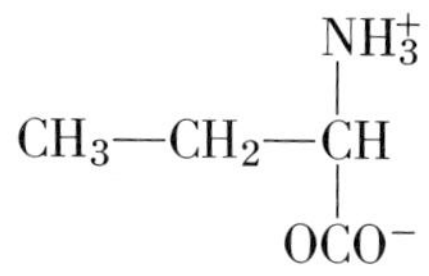

The more important of these amino acids are shown in Fig. 1. The designation *alpha* denotes that each is a molecule that has NH_2 attached to a carbon atom next to the COOH group. As may be seen from the figure, the molecules to which the COOH and NH_2 groups are attached may vary considerably in structure. Sometimes there also are present more than one of each of these active functional groups. In addition, there may be OH groups present together with atoms of other elements such as sulfur and iodine.

The COOH and the NH_2 provide means for coupling these molecules to other molecules. There can be actual salt or ester formation at the COOH group; or there may be a somewhat less firm binding by means of the formation of a hydrogen bond. But the fact that we have an acidic group at one point and a basic group at another means that such a molecule is a building block of great flexibility.

In living tissue we find these amino acids combined into larger molecules which are called *proteins.* These have molecular weights ranging roughly from 10^4 to 10^7. These amino acid chains are also called polypeptides. In these chains, the principal link between molecules is formed between an NH_2 group on one molecule and a COOH group on another; these two join with the elimination of water, and form the polypeptide bond.

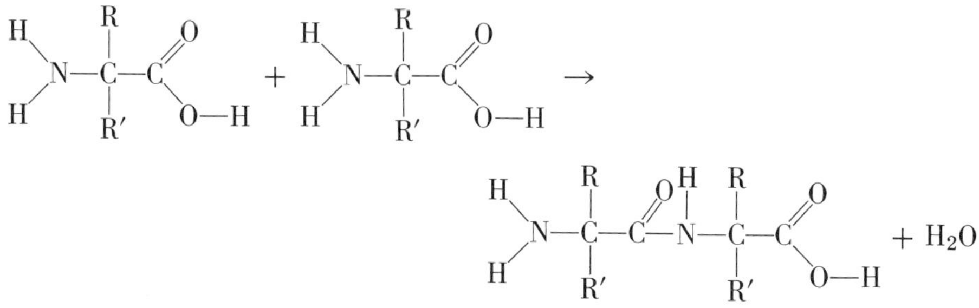

When two amino acids join together, we still have one free amino group and one free acid group on the resulting molecule; hence further reaction can occur with the formation of still longer chains. In the protein called *insulin,* there are

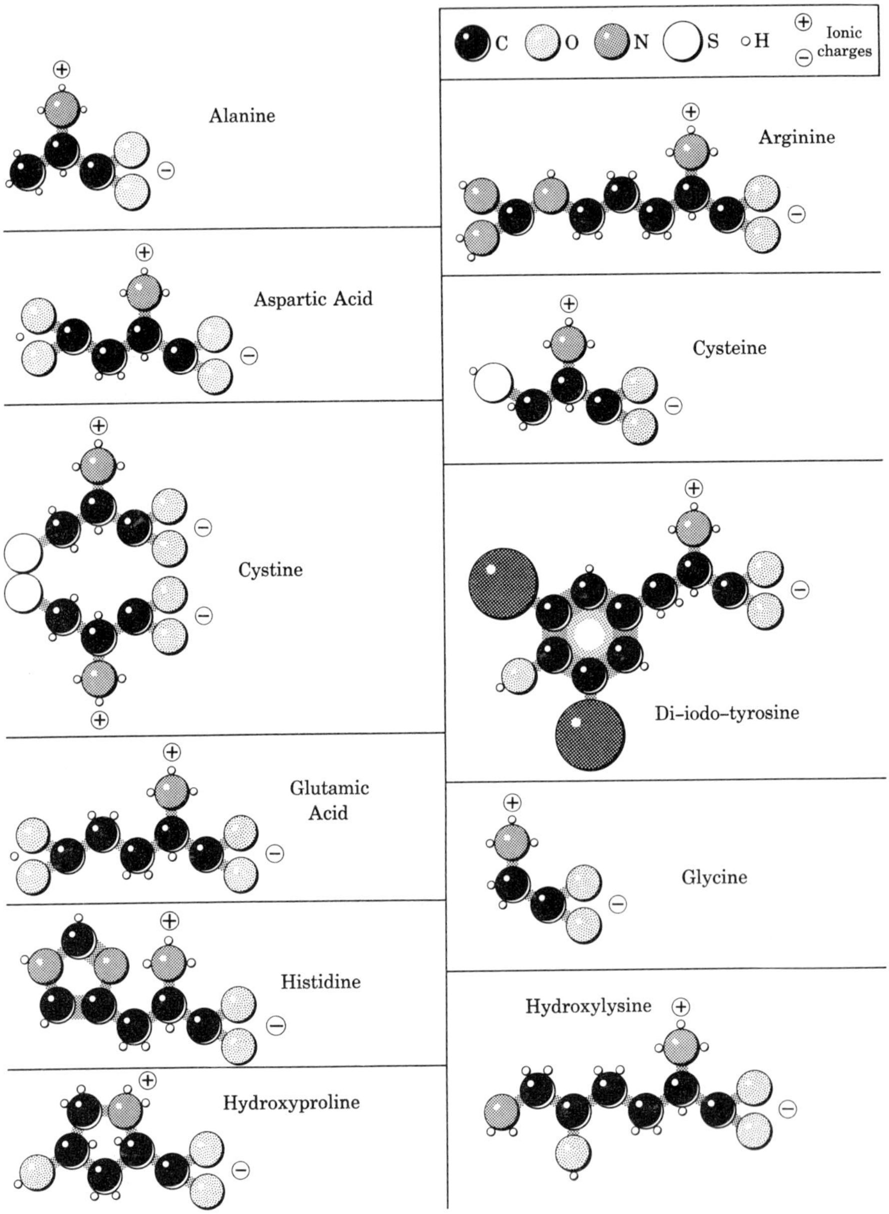

Fig. 1. Amino acids; + and − are ionic charges.

four polypeptide chains; two of these contain thirty amino acid molecules each and the other two twenty-one. The chains are attached to one another by S—S bonds coming from the sulfur in cystine (see Fig. 1). The sequence of the amino acids in these chains has been determined by the English biochemist Sanger and his co-workers. These polypeptide chains have a specific geometric form. Evidence on the nature of this form obtained by Linus Pauling and his collaborators shows that the chains twist in a helix now generally called the α-helix. When the protein chains are linked together to form material like hair or animal horn, a number of chains may be twisted together, much like the way that strands of fiber make up a rope (see Fig. 1*d*, p. 2).

The amino acids, together with sugars, alcohols, phosphates, and many other related molecules, comprise the components of the biochemical network that constitutes a single biological cell. The uniqueness of a cell stems more from the way that the component molecules are linked together than from the number and kinds of these components. To achieve individuality, it is necessary to have a variety of sequences of combination. It is this sequential variation that provides the unique pattern of entropy or information. Different sequences are characteristic of different cells. Different sequences determine different patterns of reaction for the small molecules that join together to form macromolecules. In this molecular society, the certain molecules with these sequences are the rulers that give the orders to the simpler subject molecules, so that an orderly pattern of activity results.

One of the most important types of biological activity is growth. The foci of the forces that direct growth are called the genes. We believe that a gene is a molecule of deoxyribose nucleic acid, usually abbreviated DNA. While a molecule of DNA always has the same constituent parts, these parts may be joined together in millions of different sequences. Thus, while all benzene molecules are exactly alike, there are untold millions of varieties of DNA, all made of similar parts but each varying in the sequence of these parts.

The central constituents of DNA are four kinds of molecules: (1) adenine, (2) thymine, (3) cytosine, (4) guanine. These are hooked like rungs on a ladder between rows of alternating pentose sugar molecules and phosphate molecules that form the upright parts of the ladder on the left and on the right. The outline structure of this ladder is shown in Fig. 2*a;* Fig. 2*b* shows the separated component parts. One of the rungs of the ladder with the atoms represented individually is shown in Fig. 3. There is considerable evidence that *only* these four constituents are used in the ladder rungs of DNA. The almost endless variety of biological cells differ in their hereditary components only in the *order* in which these occur. In Fig. 2*a*, the steps of the ladder consist of

A, adenine-thymine
T, thymine-adenine
C, cytosine-guanine
G, guanine-cytosine

Labeling each rung by the initial letter, we see that the particular sample of DNA shown in Fig. 2*a* spells out the message: ACGTGCAC. We have not yet learned

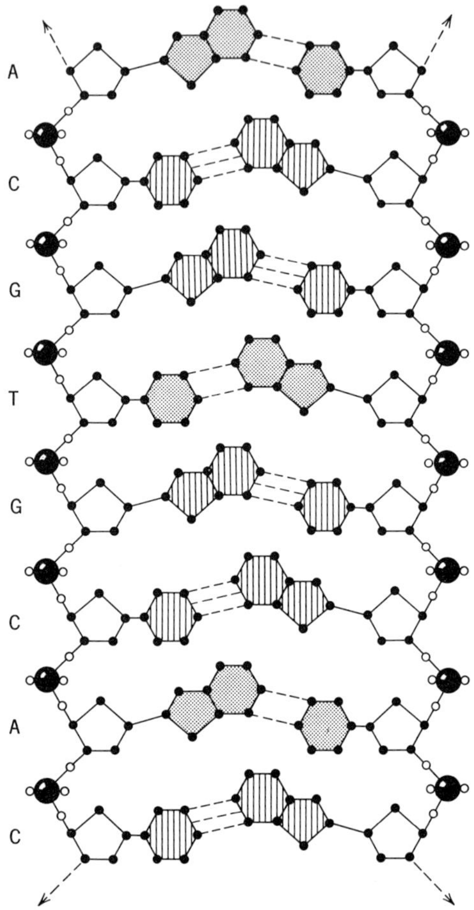

Fig. 2*a*. Section of DNA chain. T = thymine-adenine; C = cytosine-guanine; A = adenine-thymine; G = guanine-cytosine. (Simplified representation.)

the language of the genes and so cannot translate this message; it may be the order to build glycyl proline or threonyl glutamine. We do know that this code specifies ultimately the location of a kind of amino acid in the chain of a protein. Thus DNA serves as the governor of growth.

But, if a cell is to be reproduced, somehow in the process the DNA itself also must be reproduced. This is the core of the mechanism of heredity. It has been suggested that, when reproduction occurs, the DNA unwinds and breaks the bonds indicated by dotted lines in Fig. 2*a*. Since the phosphate-pentose filaments are preserved in each half, these serve as templates for building two new DNA molecules. In the fluid containing the simpler component molecules of DNA, the top part of the left-hand strand can only bond to thymine, the next rung can only

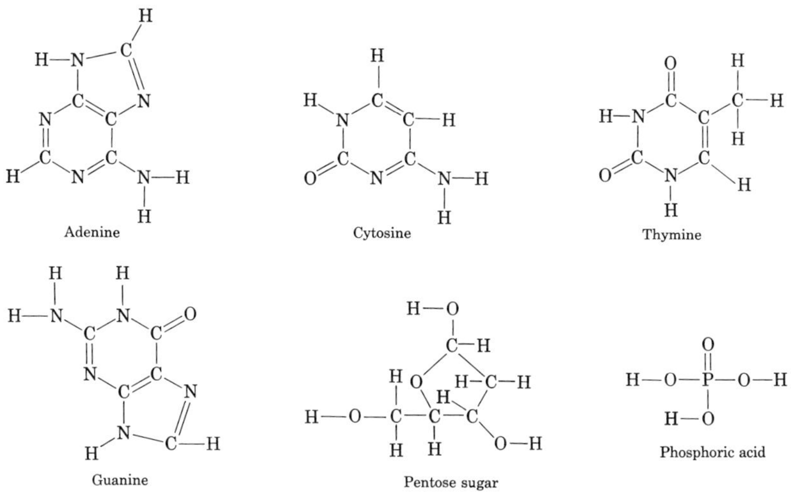

Fig. 2b. Constituents of DNA.

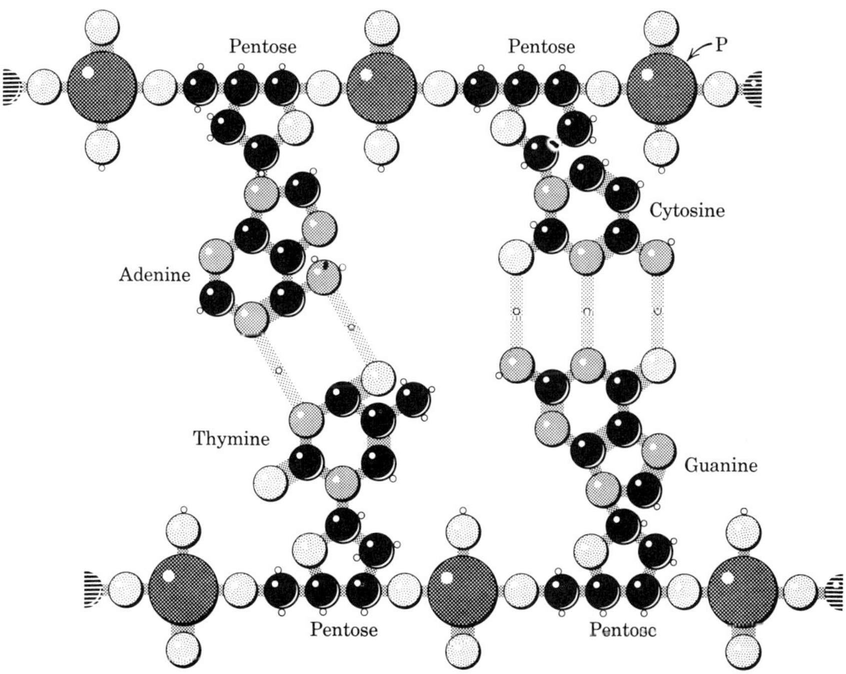

Fig. 3. Cross link in DNA. See Fig. 1, p. 742, for key.

bond to guanine, and so on down the ladder. Thus the left-hand strand reproduces the whole DNA molecule when it reacts. Similarly, the right-hand strand couples to the appropriate molecules available in the surrounding fluid and builds still another replica of the original DNA; so there are now two molecules of identical DNA where there was only one when the process started.

The information in the genes is transmitted to the proteins via a messenger molecule called *ribose nucleic acid,* RNA. The DNA serves as a template both

Vitamins

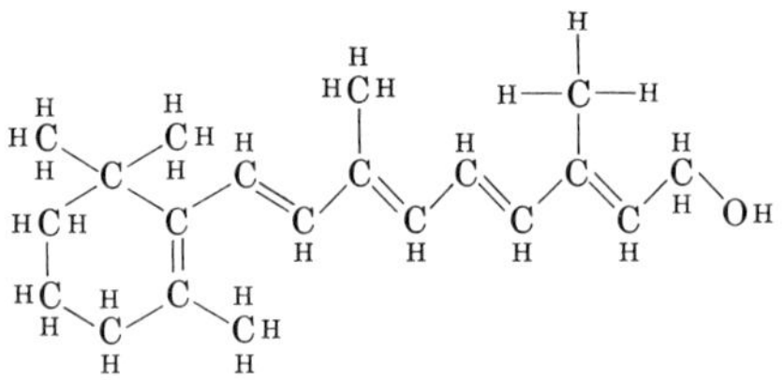

Vitamin A

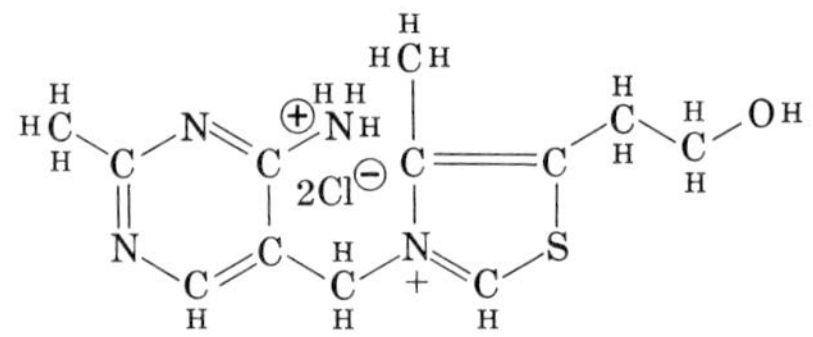

Vitamin B_1
Thiamine hydrochloride

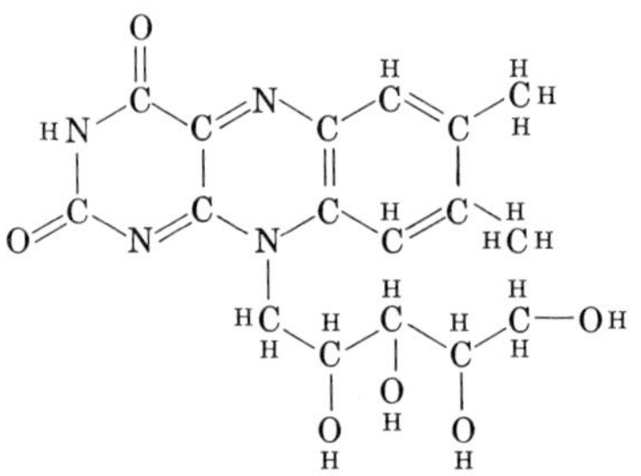

Vitamin B_2

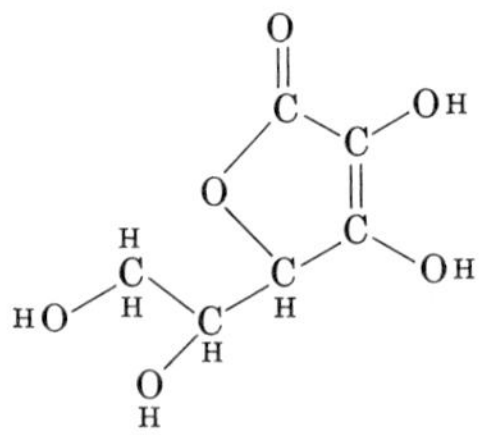

Vitamin C
Ascorbic acid

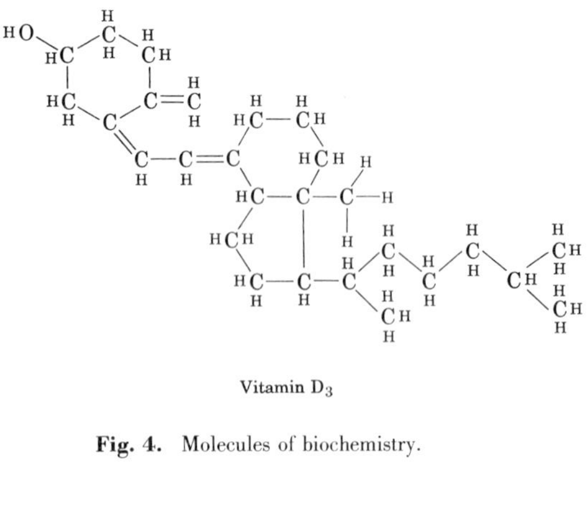

Vitamin D_3

Fig. 4. Molecules of biochemistry.

Sex Hormones

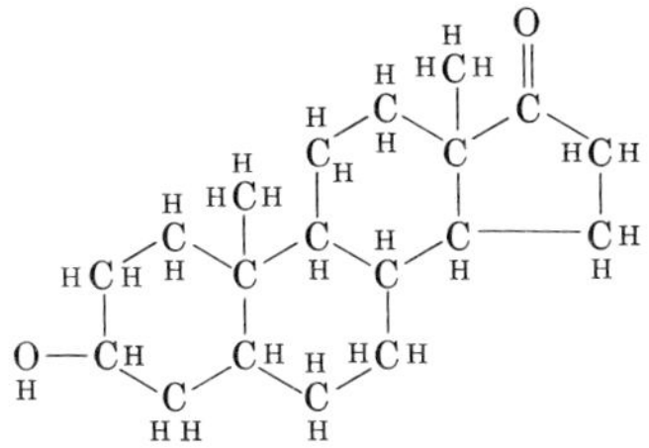

Androsterone

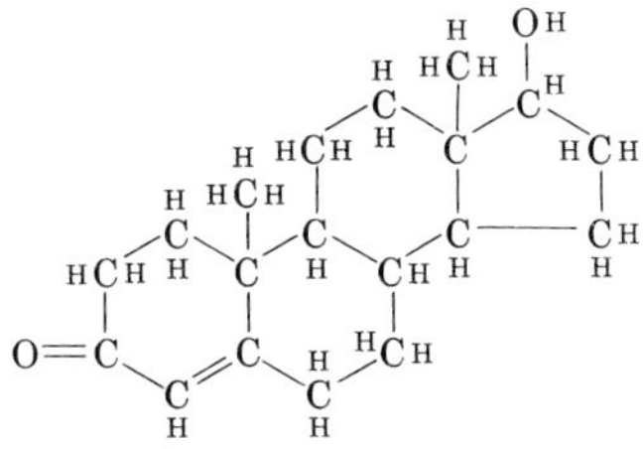

Testosterone

Stimulant Drugs

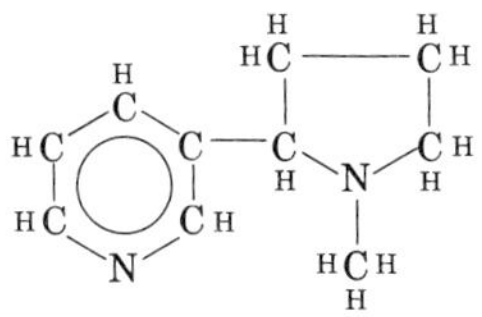

Nicotine

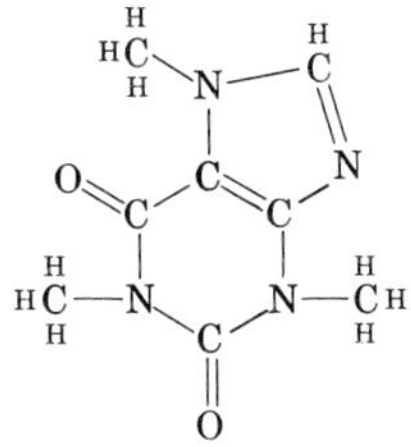

Caffeine

Local Anesthetics

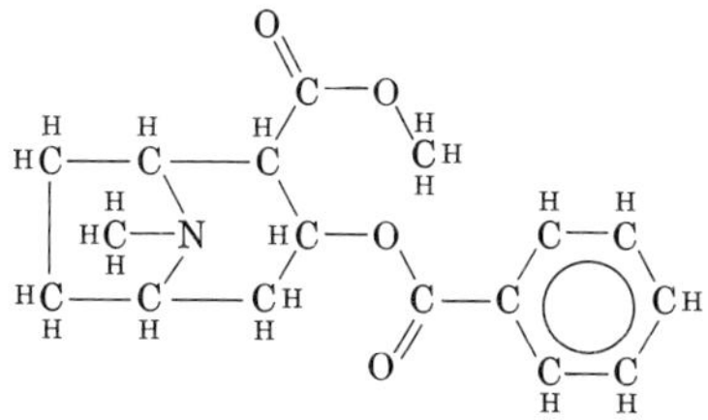

Cocaine

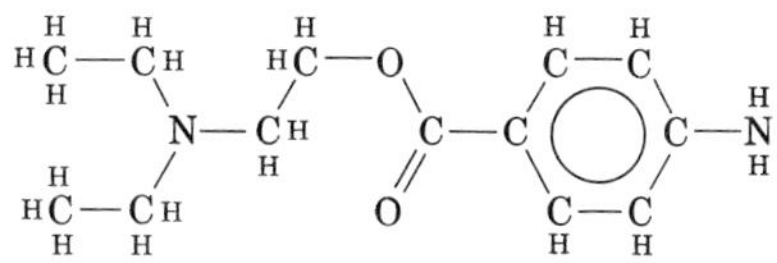

Procaine

Antibiotics

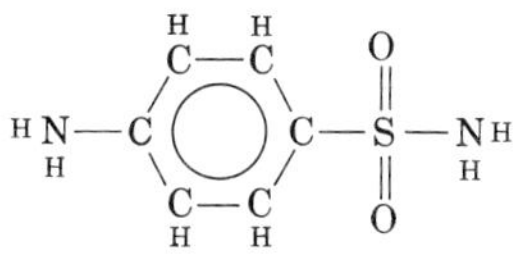

Sulfanilamide

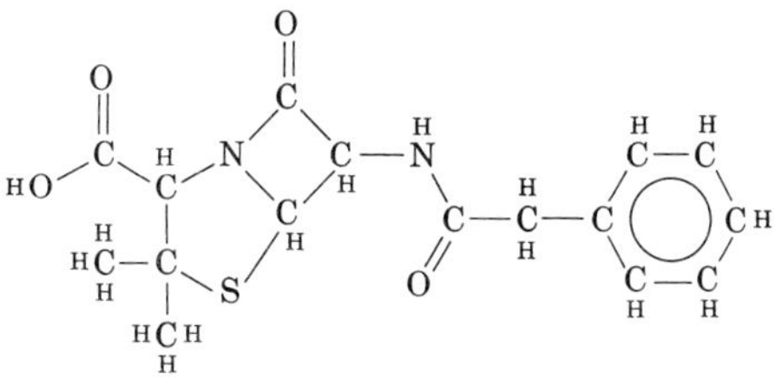

Benzylpenicillin G

Fig. 4 (Continued).

$O_2 \rightleftarrows CO_2$ Cycle Partners

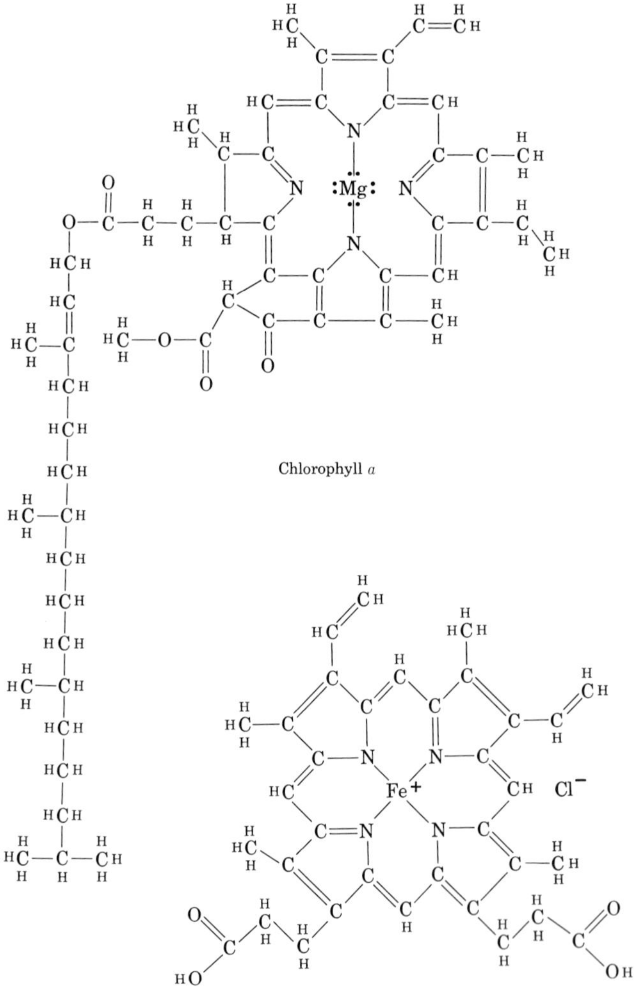

Chlorophyll *a*

Hemin

Fig. 4 (Continued).

for reproducing itself and for producing RNA. The RNA serves as a template for linking the proteins. This appears to be the basis of heredity.

Molecules of pharmacological interest. Chemistry has made great progress not only in throwing light on the nature of the substances from which living material is made, but also in elucidating the structure of those special substances like vitamins, hormones, and drugs that influence living material. In our brief survey there is not space to take up these topics in any detail, but we include the structural diagrams of some of the more familiar of these molecules in Fig. 4. These structural formulas illustrate the variety of space-forms found in molecules of biochemical interest.

27

Nuclear chemistry

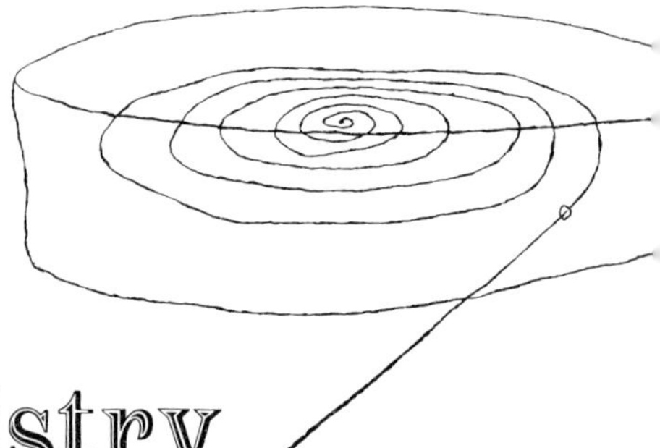

Just as the concept of the atom as a particle dominated scientific thinking from the time of the earliest Greeks until the present century, the particle concept also dominated the earlier theories of the atomic nucleus. Because α-particles and β-particles were emitted from the nucleus, scientists were inclined to interpret the structure of the nucleus as resembling a collection of particles, much as a molecule was regarded as a collection of atoms.

Even today, many discussions of nuclear structure are phrased in these particulate terms. We talk about the nucleus of a heavy atom as made up of a combination of protons and neutrons; we say that the neutron breaks up spontaneously into a proton and an electron; and we speak of the electron, proton, neutron, and their relatives as *fundamental particles.* Such terms are necessary to provide a scientific language for describing nuclear phenomena. But, as science has progressed, it has been increasingly apparent that we have to consider not only the particle aspect but also the complimentary wave aspect both outside and inside the nucleus. In this chapter, we shall use both these points of view to elucidate nuclear phenomena.

FUNDAMENTAL PARTICLES

What is a fundamental particle? Some forty years ago the answer seemed plain. It was a particle that never changed into anything else, that was not derived from anything else, that consequently had a kind of invariant behavior throughout the course of time, and therefore was entitled to be called *fundamental.* There were about a half-dozen such particles out of which the universe appeared to be made. Among these, the electron and the proton were the most important pair, and it was thought that combinations of these two accounted for the nucleus. One put together enough protons to correspond to the mass of the nucleus; then added enough electrons to neutralize part of the positive charge and yield the actual *net* positive charge observed; and one thus accounted for both mass and positive charge. But, after many observations of nuclear decay were made, and particularly after it was possible to produce artificial radioactivity and nuclear transmutations, it became clear that it was misleading to regard the electron as existing independently inside the nucleus.

Scientists next adopted the picture of a nucleus made up of protons and neutrons. In many ways, this view provided a logical explanation of nuclear chemistry; but the disquieting fact remained that the neutron was not a *stable* particle. Inside the atomic nucleus it might be stable for billions of years; but, when it passed outside the nucleus, it had a relatively short life, a *half-life* of about 13 minutes, before decomposing into a proton and an electron; so it seems somewhat inconsistent to call the neutron a *fundamental* particle. Yet in the universe there probably are far more stable neutrons inside the nucleus than there are unstable ones outside. Moreover, the more we look into the relations between mass and energy and the way they change in time, the more we realize the complexity of the pattern of the universe. In view of the interaction between particles and the spreading out of the particle into waves, the term *particle* must be used with reservations. And the same limitation applies to the term *fundamental.* This word expresses only one fleeting aspect of the dynamic flux of matter and energy in the universe. It is a useful concept to emphasize the invariant way in which matter and energy behave at certain times, as in the action of independent electrons. But, just as the particle concept is wholly inadequate to express the way electrons act when shared by atoms or spread out in metals, the concept of an eternally invariant fundamental particle is inadequate to express the many changes of form that matter and energy display both within the nucleus and throughout the universe.

In order to make this clearer, we take a look at a remarkable theory of matter and energy proposed by a British physicist, P. A. M. Dirac, shortly after de Broglie's theory of the wave nature of the electron was verified experimentally and the exclusion principle was discovered by Pauli. Dirac suggested that the entire universe is a single quantum system; in this system there is a vast set of energy levels, just as there is a simpler set of levels in an atom or a more complicated set in a metallic crystal. From this elaborate yet elegant treatment, which included corrections stemming from relativity theory, he concluded that there must be present in such a universe not only the familiar particles like the electron and the proton but also anti-particles which in their properties were symmetrical with and therefore opposite to our familiar particles. For example, he predicted the existence of the positron that should have the same mass as the electron but a positive instead of a negative charge. He hypothesized that the entire universe was packed with this anti-matter that normally did not influence ordinary matter. This anti-matter occupied all the possible energy levels of the universe up to a certain value of energy, just as the conduction electrons in a metal at 0°K occupy all the energy levels in accordance with the exclusion principle. And, just as the electrons in the uppermost energy levels can acquire a little extra energy and move through the metal, conducting electricity, Dirac postulated that some of the "particles" in the universe are at the top level where they can acquire energy and, therefore, move through the universe and carry out the types of motion that we observe as the familiar motion of matter around us.

This concept was criticized by other theoretical physicists at the time; but almost immediately the positron was discovered experimentally, confirming at least one prediction of Dirac's remarkable suggestion. Since that time, there have been a number of other discoveries which show that this theory does contain some truth, although there are many aspects of the universe that, for their explanation, will require a far more comprehensive and penetrating theory.

In Table 1 are summarized some of the facts regarding the particles that are usually called *fundamental* particles today; nearly all of them transform back and forth into other particles or other aspects of matter and energy. At each place in the table, there is listed the name of the particle, such as *photon,* followed by the symbol for the particle, such as γ. Directly below each name there is listed the rest mass relative to a rest mass of unity for the electron. Below this figure is listed the approximate half-life for the unstable particles. Thus the photon and proton are stable, but the neutron has a half-life of about 800 seconds when it is free outside the nucleus.

The lightest particle is the photon, found at the top of the table. It is the bundle or quantum of energy that travels from one atom to another, vibrating with a frequency lying somewhere in the electromagnetic spectrum. This may be a frequency that makes the photon visible to the eye; or the frequency may correspond to a radio wave with energy only one ten-thousandth that of a visible photon; or it may correspond to energy a million times greater than that of a visible photon and thus lie in the X-ray region or the gamma-ray region.

Below the photon, there is the particle called the neutrino. The existence of the neutrino was first postulated by the theoretical physicist Enrico Fermi to explain how the principle of conservation of energy could hold in certain transformations like that of the neutron to an electron and proton. Careful observation showed that the total amount of energy at the end of this process was not the same as at the beginning, even when one took into account the relativistic relation between mass and energy. Apparently, the neutrino interacts only slightly with any other

Table 1 Sub-atomic Particles

	Negative Charge	Neutral			Positive Charge
			photon $m = 0$ stable		
		neutrino ν $m = 0$ stable		anti-neutrino $\bar{\nu}$ $m = 0$ stable	
	electron e^- $m = 1$ stable				positron e^+ $m = 1$ stable
Mesons	neg. muon $m = 206$ 2×10^{-6} sec				muon $m = 206$ 2×10^{-6} sec
Mesons	neg. pion $m = 273$ 2.6×10^{-8} sec		pion $m = 264$ 2×10^{-16} sec		pion $m = 273$ 2.6×10^{-8} sec
Mesons	neg. kaon K^- $m = 967$ 10^{-8} sec		kaon $K_1{}^0$ $m = 973$ 10^{-10} sec		kaon K^+ $m = 967$ 10^{-8} sec
			kaon $K_2{}^0$ $m = 973$ 10^{-8} sec		
	anti-proton p^- $m = 1836$ stable				proton p^+ $m = 1836$ stable
		neutron n $m = 1839$ 8×10^2 sec		anti-neutron $\bar{n}$ $m = 1839$ 10^3 sec	
Baryons		lambda Λ $m = 2182$ 2×10^{-10} sec		anti-lambda $\bar{\Lambda}$ $m = 2182$ 2×10^{-10} sec	
Baryons	neg. sigma Σ^- $m = 2342$ 10^{-10} sec	sigma Σ^0 $m = 2326$?		anti-sigma $\bar{\Sigma}$ $m = 2326$?	plus-sigma Σ^+ $m = 2324$ 10^{-10} sec
Baryons	neg. xi Ξ^- $m = 2585$ 10^{-10} sec	xi Ξ^0 $m = ?$ 10^{-10} sec		anti-xi Ξ^0 $m = ?$ 10^{-10} sec	†

† In March 1962 the discovery of the predicted *anti-xi minus* was reported as the result of a cooperative effort of scientists working in Switzerland, the United States, and France.

form of matter, so it is very difficult to prove its existence directly by experiment. As a partner to the neutrino (symbol ν) there is also the anti-neutrino (symbol $\bar{\nu}$). Where there is no difference in the sign of the charge to indicate anti-particles, the less familiar anti-particle is denoted by a bar over the symbol.

Directly below these particles with essentially zero mass are listed the electron and positron, the particle and anti-particle of electricity. These are given a single unit of mass in the scale of mass that is commonly used to measure the intrinsic mass of the various basic particles. Both are stable. Below the electron and positron, there is a series of particles known as *muons, pions* and *kaons.* Some of these are charged and some are neutral. The evidence for their existence is derived from recent experiments in high-energy physics. They are nearly all highly unstable, decaying in periods that range from a minute to 10^{-15} second or less.

We come next to the familiar proton (p) and the less familiar anti-proton (p^-). These are stable particles. Below them, with mass only slightly greater, there are the neutron and anti-neutron. Until quite recently, it was believed that there were no particles both entitled to be called *fundamental* and also possessing mass greater than that of the neutron; but recently evidence has been found for the existence of a number of these that are often called *baryons,* from *barys* meaning *heavy.* Studies of these particles are giving us new insight into other subtle aspects of the interaction of mass and energy.

Because of the rapid advances that are currently being made in the study of sub-atomic particles, any table of properties is partially out of date even a few weeks after publication; so that especially with respect to mesons and baryons, Table 1 must be regarded not as a definitive list but as a glimpse of a changing scene.

NUCLEAR STRUCTURE

The liquid drop model. The particle model of the nucleus, known as the "liquid drop" model, was first proposed by Niels Bohr in 1936. The nucleus is thought of as an aggregate of particles held together by mutual attraction such as a drop of water consists of an aggregate of water molecules held together by the surface tension that is also caused by mutual attraction. As the drop gets larger, the surface tension decreases and ultimately "fission" of the drop may result when it is agitated. Because nuclear density appears to be relatively constant all the way from the lightest nuclei like those of hydrogen and helium to the heaviest nuclei that are neighbors of uranium in the periodic table, the picture of all nuclei as clusters of common fundamental particles is convincing; but it leaves unsolved the problem of the origin of force that holds the particles in this "drop" form.

The shell model. Why are these high-energy phenomena associated with the nucleus? High-energy states of nuclei seem reasonable when we recall some of the wave equations of the quantum theory and realize that in a consistent universe there should be wave aspects of these nuclear particles. The two basic equations of the quantum theory are $E = h\nu$ and $\lambda = h/p$. Since λ is the wavelength and is related to the momentum (p), by the second of these equations, we see that, as the wavelength gets shorter, the momentum gets higher. If one thinks of a particle circulating in an orbit, a higher momentum implies a higher energy. When the wavelengths are of the order of the diameter of the *atom*, we find that the energy corresponding to these wavelengths has the order of magnitude observed in ordinary chemical changes where electrons are exchanged between atoms and visible light is frequently emitted. When the wavelength drops to a figure one ten-thousandth of the atomic diameter or less, a length comparable with the diameter of the *nucleus*, or simple fractions thereof, then the energy goes up by a corresponding factor. Thus it is reasonable to find energy of millions of electron volts associated with nuclear changes together with the emission of gamma-rays. It is also reasonable that the forces within the nucleus should be much stronger than the forces between atoms and should operate over much shorter ranges.

We still know almost nothing about the nature of these nuclear forces. There is some evidence that they are due to interaction between the waves in the nucleus just as covalent bond forces are due to the interaction between electron waves. Just as we conclude that there is a single type of wave in a resonance bond and not a flipping back and forth between different waves, we conclude that the binding forces in the nucleus are due to a combined wave that represents simultaneously states of matter and energy sharing the aspects of several of the more familiar particles.

There is still another similarity between the nucleus and familiar atomic orbitals; for there are numbers associated with the energy levels in the nucleus in many ways analogous to the numbers associated with the electronic energy levels in the atom. In studying the electronic structure of the atom, we found that the numbers 2-6-10-14 are derived from the mathematically defined types of vibration that are possible in the four-dimensional continuum that includes both the three dimensions of ordinary space and a spin space for the electron. These numbers are the basis of the periodic properties of the chemical elements and of the repeated appearance of the inert rare gas electronic structure. The total numbers of electrons in these peculiarly stable rare gas atoms are 2, 10, 18, 36, 56, and 84.

In studying the properties of the *nuclei* of the chemical elements, similar quasi-periodic properties can be seen, as first made clear by M. Goeppert-Mayer. This relationship is described by Goeppert-Mayer and Jensen in the following way:

"Except for the odd-even difference (and spins and moments of nuclei of odd A) the general properties of nuclei vary rather smoothly with Z (the number of protons) and N (the number of neutrons). On closer inspection, however, some obvious sharp irregularities are found. These are associated with neutron numbers $N = 2, 8, 20, 28, 50, 82$, and 126, and with the same proton numbers. There are many empirical data that suggest that at the 'magic numbers' there occurs something analogous to the closure of shells in the electronic structure of the noble-gas atoms." (Ref. 5.)

One of the more direct bits of evidence for the magic numbers is found in the table of the numbers of naturally occurring isotopes. In Fig. 1, the chemical elements are listed in order of increasing nuclear charge, and a line is drawn above each one that shows by its length the number of natural stable isotopes associated with each atomic number. Each natural radioactive isotope is denoted by an ×. The latter should be given some weight in estimating the stability of a nuclear arrangement, because the fact that they can exist at all, even for a brief moment, is an indication of some stability as contrasted with an arrangement that is never observed naturally.

The simplest and most obvious regularity in this figure is the greater number of isotopes observed for *even* atomic numbers as contrasted with *odd.* Except for the anomaly of beryllium, this relation holds throughout the figure and is particularly striking in the middle of the figure. This suggests that odd protons in the nucleus produce nuclear instability, just as odd electrons in the outer part of the atom produce chemical instability. Similar instability is produced by an odd number of neutrons in the nucleus.

Among the even atomic numbers corresponding to an even number of protons in the nucleus, there are certain places that are clear maxima in the series of natural isotopes; the most striking are at Ca(20), Sn(50), and Pb(82). Other evidence, to be discussed in a moment, points to similar unusual stability for He(2), O(8), and Ni(28). There is comparable evidence for stability in nuclei containing 126 neutrons. Taken together, these values constitute the *magic numbers;* the corresponding lines in the figure are given extra width to emphasize their location.

Binding energy. Just as we talk of the *bonding energy* of molecules that is released when we combine the constituents atoms, we can also talk of the *binding energy* of the nucleus that is released when we combine its constituent particles. Just as a high bonding energy indicates unusual molecular stability, a high binding energy indicates unusual nuclear stability. The series of binding energies for increasing *numbers of neutrons* (denoted by N) in the nucleus also show irregularities that provide evidence for the significance of the magic numbers. On the whole, the binding energies vary with N in a smooth curve, but as N passes the magic numbers the value of the binding energy falls markedly, in-

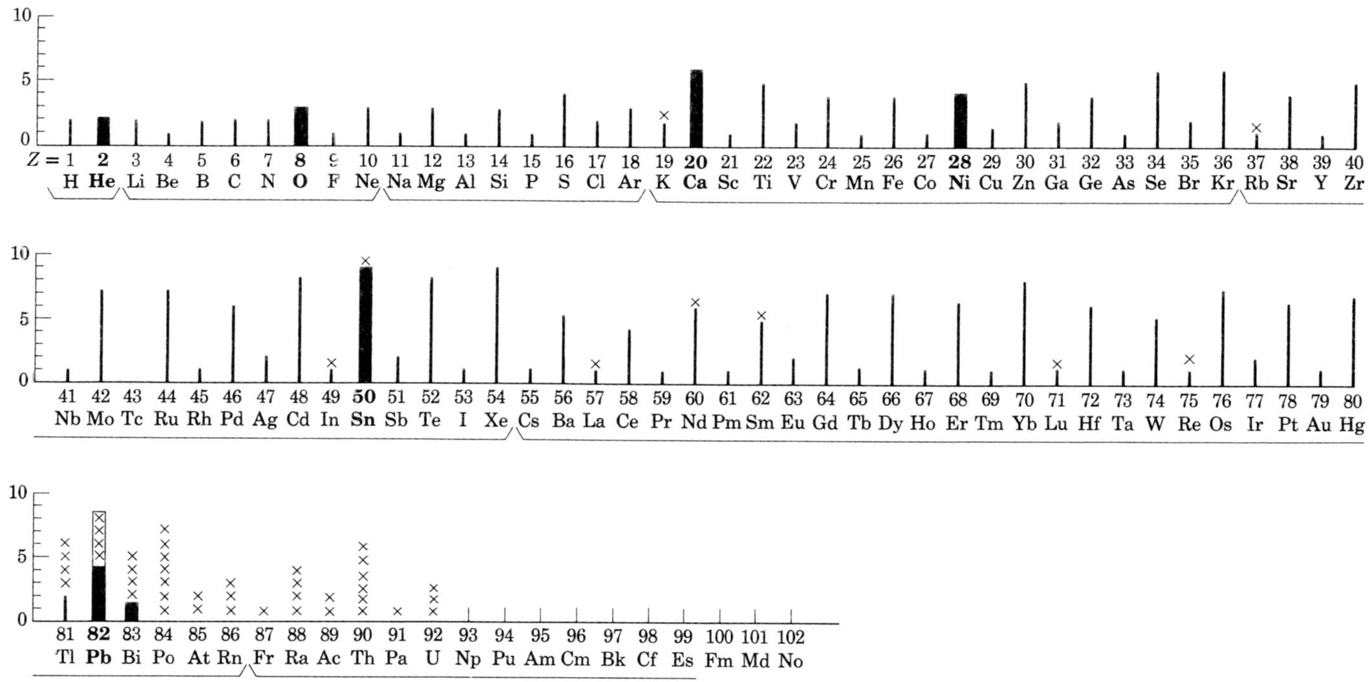

Fig. 1. Number of natural stable isotopes occurring in significant amounts. Magic numbers are indicated by heavy lines: 2, 8, 20, 28, 50, 82, (126); natural radioactive isotopes by ×.

dicating that there is a passage from a stable shell to one distinctly less stable. The same type of change is noted in the binding energy of the proton as number of protons (Z) in the nucleus rises through the magic numbers. Similar discontinuities are found for alpha- and beta-particle decay energies, for neutron capture cross section, and for the threshold of gamma-ray energy that stimulates neutron emission. Taken altogether, the evidence is convincing that the magic numbers do denote specially stable shells, and considerable progress has been made in explaining this in terms of quantum mechanical relations roughly similar to those that explain the stable shells for electrons in the outer part of atoms.

NUCLEAR TRANSFORMATIONS

Because of the equivalence of energy and mass expressed by Einstein's equation, $E = mc^2$, whenever energy leaves an atom it takes with it a little mass; and, therefore, the mass of the atom left behind is less. In *ordinary* chemical reactions, the mass equivalent of the energy lost or gained is so small as to be almost undetectable. When a liter of water is frozen into ice, it possesses less energy and, consequently, less mass; it has lost the heat of crystallization in the process of freezing; but no chemical balances have ever been constructed sufficiently sensitive to detect this loss of mass. On the other hand, when hydrogen atoms join together to form helium by the nuclear process operating in the sun, the energy loss is so great that there is an appreciable change in mass. We can see this from the atomic weights. The mass of the proton in atomic mass units is 1.00728. The mass of the neutron is 1.00867. We know that the nucleus of the helium atom has four mass units and two units of positive charge, indicating that it is made of two protons and two neutrons. If we take twice the mass of the proton and twice the mass of the neutron, this gives us a total of 4.0319. However, the mass of the helium nucleus is observed to be 4.0015, and this indicates that there is less energy in this nucleus than there was when the particles were separate in the form of two protons and two neutrons.

As we go across the periodic table from hydrogen toward heavier nuclei, these nuclei have less energy than the separate protons and neutrons of which they are made; this remains true until we reach nuclei with mass about that of iron. Beyond iron, a comparison of the observed atomic weight with the sum of the masses of the protons and neutrons indicates that the energy of the nucleus climbs up again toward the sum of the energies of the separate particles. The term binding energy is applied to this difference between the observed mass of the nucleus and the sum of the masses of its separate particles regarded as an equivalent energy difference.

In order to see these relations more precisely, we set them down in the form of equations. Let us now use a more complete convention for designating isotopes:

$$\text{chemical symbol} \rightarrow {}_{\text{atomic number} \rightarrow Z}X^{A \leftarrow \text{mass number } (Z+N)}_{N \leftarrow \text{number of neutrons}}$$

The actual mass of the common isotope of carbon (${}_6C^{12}_6$) is arbitrarily assigned the value of exactly 12. We then designate the atomic mass of ${}_1H^1_0$ by the symbol ${}_1m^1_0$ and find that ${}_1m^1_0 = 1.0073$; the mass of the neutron similarly turns out to be ${}_0m^1_1 = 1.0087$. The difference between the mass of the nucleus and the mass number (A) is called the *mass defect* and is often denoted by the symbol ${}_Z\Delta^A_N$. It is usually more informative to express this difference in *energy* units rather than *mass* units; so we will use the symbol ${}_Z\Delta^A_N$ for the mass defect in *energy* units; it is proportional to the binding energy of the nucleus. The constant of proportionality is the velocity of light (c) squared and divided by Avogadro's number ($\mathfrak{N}$). Thus one atomic mass unit equals 931.5 Mev in customary atomic energy units.

In this way we find:

$$\begin{aligned} &\text{Hydrogen: } {}_1\Delta^1_0 = 7.3 \text{ Mev} \\ &\text{Neutron: } \ {}_0\Delta^0_1 = 8.1 \text{ Mev} \end{aligned}$$

For any isotope, the binding energy released in forming the nucleus (A, Z, N) from its constituent protons and neutrons is

$$W = \frac{c^2}{\mathfrak{N}}\left[Z\left({}_1m^1_0\right) + N\left({}_0m^1_1\right) - {}_Zm^A_N\right]$$

Thus

$$W = Z\left({}_1\Delta^1_0\right) + N\left({}_0\Delta^1_1\right) - {}_Z\Delta^A_N$$

Frequently it is more informative to examine the binding energy *per nucleon* in the nucleus. This is sometimes called the binding fraction (f) and is equal to W/A. Numerically we find

$$f = \frac{{}_1\Delta^1_0 + {}_1\Delta^1_1}{2} + \frac{N-Z}{A}\,\frac{{}_0\Delta^1_1 - {}_1\Delta^1_0}{2} - \frac{1}{A}\,{}_Z\Delta^A_N$$

In Fig. 2 the value of f is plotted against the value of A for elements from H to U. Thus the curve shows the value of the average energy released per nucleon (proton or neutron) when the correct number of each of these particles combines to form the isotope in question. Just as in ordinary chemical reactions, we expect that in nuclear reactions the most stable product will be formed when the energy released is greatest. From the curve in Fig. 2, we see that the greatest release of energy is found when nuclei in the neighborhood of iron are formed. Consequently these nuclei have the greatest stability. There is a tendency for lighter nuclei like H to join together by *fusion* to form heavier nuclei lying nearer iron; there is a similar tendency for the heaviest nuclei like U to split apart by *fission*, likewise yielding nuclei to the left in the figure nearer iron. Both processes take place because they produce energy and yield nuclear products in

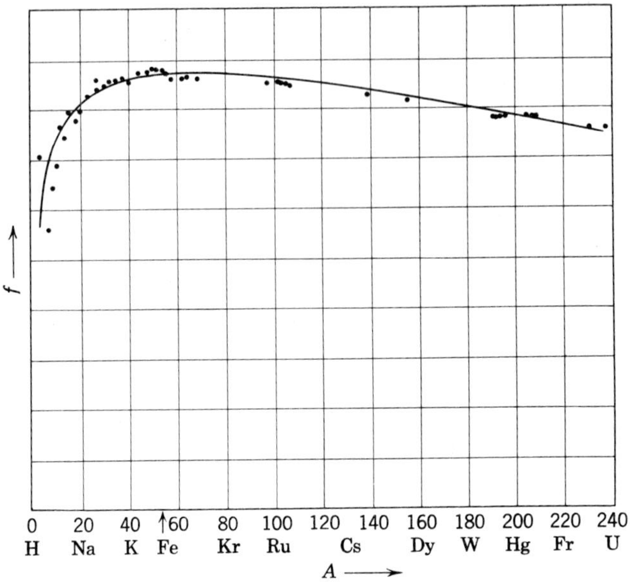

Fig. 2. Average energy released per nucleon in the formation of isotopes with mass numbers A. (The approximate location of a few of the chemical elements is shown.) (From Mayer and Jensen, *The Shell Structure of the Nucleons*, p. 5, John Wiley and Sons Inc., 1955.)

lower energy levels than originally found in the reactants. We now examine these two processes of fusion and fission in greater detail.

Fusion. For many centuries, man has speculated about the source of the enormous energy streaming from the stars. Our sun gives out 10^{41} ergs per year, and does this century after century without significantly lowering its temperature. Where does this energy originate? One of the most promising clues came from some observations of nuclear transmutation. In 1932, two physicists, Cockroft and Walton, found that a hydrogen nucleus (${}_1H^1$) and a lithium nucleus (${}_3Li^7$) can combine to form two atoms of helium, with resultant loss in mass and release of energy:

$${}_1H^1 + {}_3Li^7 \rightarrow {}_2He^4 + {}_2He^4 + \text{energy}$$

Since there are both H and He in the sun, and since the temperature in the interior is almost certainly high enough to produce transmutation, this suggests that the process producing the sun's energy may include two cycles as shown here:

Hydrogen Cycle

$${}_1H^1 + {}_1H^1 \rightarrow {}_1D^2 + e^+$$

$${}_1H^1 + {}_1D^2 \rightarrow {}_2He^3$$

$${}_2He^3 + {}_2He^3 \rightarrow {}_2He^4 + {}_1H^1 + {}_1H^1$$

Carbon Cycle

$$_{6}C^{12} + {_1}H^1 \rightarrow {_7}N^{13} \rightarrow {_6}C^{13} + e^+$$
$$_{6}C^{13} + {_1}H^1 \rightarrow {_7}N^{14}$$
$$_{7}N^{14} + {_1}H^1 \rightarrow {_8}O^{15} \rightarrow {_7}N^{15} + e^+$$
$$_{7}N^{15} + {_1}H^1 \rightarrow {_6}C^{12} + {_2}He^4$$

Fission. Just as there is a tendency for the lighter elements to go downhill energy-wise by the fusion of their nuclei, there is also a tendency for heavy elements to go down toward the minimum of the energy curve by breaking up their nuclei. This is shown in the series of changes by which elements like uranium break up in small steps with the emission of one or more particles at a time to move in the direction of lighter nuclei. However, it is possible for a much more spectacular change to take place in which a heavy element like uranium actually splits practically in two with the formation of two *much* lighter nuclei. This large-scale splitting is called "fission." The first evidence for it was found by two German scientists, Hahn and Strassman, in January, 1939; they discovered that it was induced by the impact of a neutron on U^{235}:

$$_{92}U^{235} + {_0}n^1 \rightarrow {_{56}}Ba^{140} + {_{36}}Kr^{93} + 3\,{_0}n^1$$

This is a type of equation where the significance of the process is lost if we cancel out the neutrons from both sides, as we might be tempted to do, following the pattern of simplifying ordinary chemical equations. Here it is necessary to show on the left the entry of the neutron into the uranium nucleus which sets in motion the process of splitting; as a result of the splitting, a barium nucleus and a krypton nucleus are produced and *three* neutrons are given out. If each of these neutrons happens to hit another nucleus of uranium 235, then, in turn, each one of these will generate three more neutrons so that the initial neutron in the first step has now produced nine neutrons. Such a process is known as a *chain reaction* of the branching type; it was immediately apparent that this offered the possibility of releasing atomic energy on a large scale. It is exactly like knocking over bowling pins on a floor. Suppose the pins are four or five feet apart; if one pin falls over, it may possibly knock over one other pin; but there is little chance of more than that happening. However, if the bowling pins are close together, one pin falling down may knock down three, and each of those may knock three more for a total of nine, and soon all the pins are lying on the floor.

The uranium fission is a process much like this. The isotope of uranium having the mass 235 makes up a little less than 1% of natural uranium ore. When a U^{235} nucleus is hit by a neutron, it may decompose and give out three neutrons; but the chances are very small that these will hit other U^{235} nuclei. They are very apt to collide with the more abundant U^{238} nuclei and terminate the incipient chain reaction. However, it is possible to concentrate the rarer isotope and make a ball of uranium metal that is practically pure U^{235}; if these nuclei are brought sufficiently close together, the density of neutron emission becomes so high that the chain reaction sets in. This was first shown experimentally by a test explosion at Alamogordo, New Mexico, in July 1945; atomic bombs were used shortly thereafter. From the reports that have been made public, it appears that a little ball of uranium was encased in a shell of a chemical explosive made in the usual way from carbon, nitrogen, and oxygen atoms. When the chemical explosive was detonated, the uranium ball was compressed to the point where the density of neutrons was sufficient to set off the chain reaction. The relatively few pounds of uranium in this bomb then released the equivalent explosive force of 20,000 tons of TNT. The isotope, U^{235} is the only naturally occurring nucleus that undergoes chain fission of this type. However, it has been possible to make use of artificially induced transmutation to produce other nuclei that will decompose similarly. We now wish to turn to a more detailed examination of such nuclear transmutations by which one element is changed into another.

The natural radioactive series. Two of the most important series of naturally occurring radioactive transformations were discovered shortly after the initial discovery of radioactivity. One of these starts from uranium itself and is known as the uranium-actinium series. The other starts from the element thorium and is known as the thorium series. In Fig. 3 are shown the changes that yield the uranium series. First we notice that uranium (U^{238}) emits an α-particle and changes to Th^{234}. This, in turn, emits a beta-particle yielding an element with one more nuclear charge, Pa^{234}. From that point on, the various steps in the transfor-

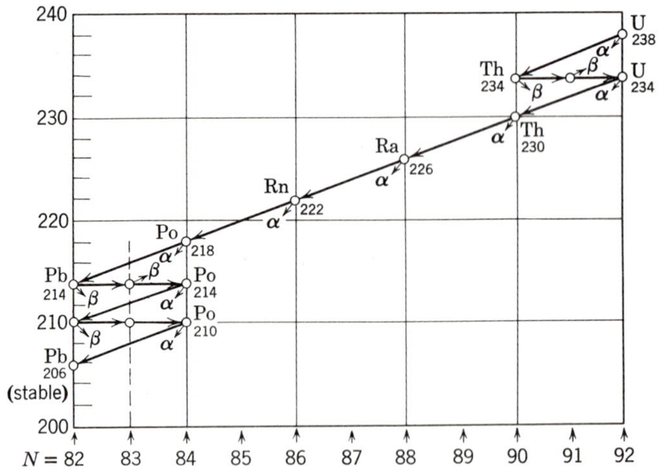

Fig. 3. Uranium radioactive series.

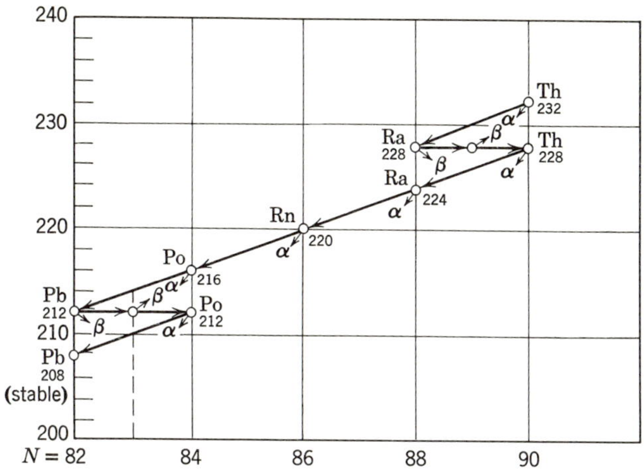

Fig. 4. Thorium radioactive series.

mation can be followed that yield Pb^{206} as the final product. In Fig. 4, there is shown a similar series starting from thorium.

Artificial transmutation. Ever since the days of the early alchemists, man dreamed about the possibility of changing one element into another. Because gold was so valuable, many attempts were made by ordinary chemical means to transform some of the more common metals like iron and lead into gold. During the centuries when most of these attempts were made, there was no real understanding of the nature of an atom or of a chemical process, and today we can see why all the alchemists' efforts were doomed to failure from the start. Centuries later, when the discovery of radioactivity provided a clearer view of the nature of the atom and its nucleus, scientists turned their attention to this old alchemist's problem in a more rational way. Success was achieved in 1919 by the English physicist, Lord Rutherford, who used the alpha-particles from a radium source to bombard nitrogen and found that occasionally an alpha-particle combined with a nitrogen nucleus to produce a nucleus of oxygen and one of hydrogen. This was the first artificially induced transmutation:

$$_2He^4 + {_7N^{14}} \rightarrow {_1H^1} + {_8O^{17}}$$

With alpha-particles from natural sources, it would be difficult to carry out transmutation on anything like a large scale; in order to get a gram atom transmuted, a process 10^{21} times more effective than Rutherford's is required. About the year 1930, two independent lines of attack on the problem brought success nearer. Cockroft designed an apparatus called a *linear accelerator* in which charged atoms could be accelerated through a large electrostatic field. At about the same time, Ernest O. Lawrence was experimenting at the University of California with an instrument called the *cyclotron.* In this instrument, charged particles are accelerated with a pulsing electrostatic field while constrained to move in a circular path by a magnetic field. By timing the pulsation of the field with the motions of the particles, they are accelerated each time they go around the circle and, in this way, whipped up to speeds again comparable with those of alpha-particles from radium, where the energy is of the order of a million electron volts (1 Mev). It was at this time that Urey, Brickwedde, and Murphy were able to isolate and show the existence of the heavy isotope of hydrogen known as deuterium, ${}_1D^2$. With ${}_1D^2$ in the cyclotron, the study of transmutations increased rapidly and a large body of data was acquired, giving scientists for the first time a comprehensive view of the problems of nuclear chemistry. The improvement of these machines also brought about the possibility of using particles with far larger amounts of energy than are normally obtained from the common radioactive disintegrations. The most powerful new instrument of this sort is called a *strong-focusing synchrotron.* It appears that with this apparatus it is possible to get particles with energies 25 to 50 thousand times greater than those possessed by the normal alpha-particles, i.e., in the range of billions of electron volts (25 to 50 Bev). This is probably only a small beginning in opening the door to an investigation of the basic transmutations of matter and energy.

While these studies were being carried on, other investigators found that nature also provides particles and waves with energy much higher than is available from the natural radioactive sources. The earth is continually bombarded with high-energy particles from outer space. These are known collectively as cosmic rays. At first, the nature of the cosmic rays was somewhat uncertain, and we still are far from knowing the complete story, but we know these high-energy particles come in from outer space, collide with atoms in the upper regions of the atmosphere, and set in motion series of reactions which produce many of the fundamental particles listed in Table 1. There is evidence that cosmic particles have been detected with energy as high as 10^{18} ev. It is noteworthy that our two natural sources of high-energy particles are the cosmos, the very large, and the atomic nucleus, the very small.

In order to study these high-energy particles, it is necessary to have many kinds of instruments that detect *individual* particles. One of these is the *cloud chamber* where a charged particle passing through leaves fog track that marks its path; the nature of the paths through magnetic and electrostatic fields yields valuable information about the particle. Another device is called the *bubble chamber.* This is a chamber filled with liquid hydrogen under conditions such that the passage of a particle leaves behind a string of bubbles much like the track of fog in the cloud chamber. *Photographic emulsion* also provides, in a more limited way, a means of getting particle tracks; a moving particle can produce a string of energized atoms in the emulsion that, on development, results in a black line. In addition to all these methods, there are the well-known counters like the *Geiger counter* where the passage of a single particle can set in motion a chain reaction of ionization resulting ultimately in an electrical pulse. With these various means of study, it has been possible to observe thousands of nuclear transformations and, in this way, to make a start in the direction of understanding the laws which govern the interaction of matter and energy in the nucleus.

In order to make clear the general classes of nuclear transformation, we summarize a number of them below, grouped in accordance with the nature of the reaction. Radioactive products are indicated by an asterisk.

Reactions with Alpha-Particles

$${}_{7}\mathrm{N}^{14} + {}_{2}\mathrm{He}^{4} \rightarrow {}_{8}\mathrm{O}^{17} + {}_{1}\mathrm{H}^{1}$$
$${}_{7}\mathrm{N}^{14} + {}_{2}\mathrm{He}^{4} \rightarrow {}_{9}^{*}\mathrm{F}^{17} + {}_{0}n^{1}$$
$${}_{18}\mathrm{Ar}^{40} + {}_{2}\mathrm{He}^{4} \rightarrow {}_{20}\mathrm{Ca}^{43} + {}_{0}n^{1}$$
$${}_{21}\mathrm{Sc}^{45} + {}_{2}\mathrm{He}^{4} \rightarrow {}_{22}\mathrm{Ti}^{48} + {}_{1}\mathrm{H}^{1}$$
$${}_{46}\mathrm{Pd}^{108} + {}_{2}\mathrm{He}^{4} \rightarrow {}_{47}^{*}\mathrm{Ag}^{111} + {}_{1}\mathrm{H}^{1}$$
$${}_{92}\mathrm{U}^{238} + {}_{2}\mathrm{He}^{4} \rightarrow {}_{94}^{*}\mathrm{Pu}^{241} + {}_{0}n^{1}$$
$${}_{94}\mathrm{Pu}^{239} + {}_{2}\mathrm{He}^{4} \rightarrow {}_{96}^{*}\mathrm{Cm}^{242} + {}_{0}n^{1}$$
$${}_{94}\mathrm{Pu}^{239} + {}_{2}\mathrm{He}^{4} \rightarrow {}_{96}^{*}\mathrm{Cm}^{240} + 3\,{}_{0}n^{1}$$
$${}_{95}\mathrm{Am}^{241} + {}_{2}\mathrm{He}^{4} \rightarrow {}_{97}^{*}\mathrm{Bk}^{243} + 2\,{}_{0}n^{1}$$
$${}_{96}\mathrm{Cm}^{242} + {}_{2}\mathrm{He}^{4} \rightarrow {}_{98}^{*}\mathrm{Cf}^{244} + 2\,{}_{0}n^{1}$$

Special Source Reactions for Neutrons

$${}_{4}\mathrm{Be}^{9} + {}_{2}\mathrm{He}^{4} \rightarrow {}_{6}\mathrm{C}^{12} + {}_{0}n^{1}$$
$${}_{4}\mathrm{Be}^{9} + {}_{2}\mathrm{He}^{4} \rightarrow 3\,{}_{2}\mathrm{He}^{4} + {}_{0}n^{1}$$
$${}_{1}\mathrm{D}^{2} + {}_{1}\mathrm{D}^{2} \rightarrow {}_{2}\mathrm{He}^{3} + {}_{0}n^{1}$$

Reactions with Neutrons

$${}_{7}\mathrm{N}^{14} + {}_{0}n^{1} \rightarrow {}_{6}^{*}\mathrm{C}^{14} + {}_{1}\mathrm{H}^{1}$$
$${}_{8}\mathrm{O}^{16} + {}_{0}n^{1} \rightarrow {}_{6}\mathrm{C}^{13} + {}_{2}\mathrm{He}^{4}$$
$${}_{1}\mathrm{H}^{1} + {}_{0}n^{1} \rightarrow {}_{1}\mathrm{H}^{2} + \gamma\text{-rays}$$
$${}_{19}\mathrm{K}^{39} + {}_{0}n^{1} \rightarrow {}_{19}^{*}\mathrm{K}^{38} + 2\,{}_{0}n^{1}$$

Tritium Production

$${}_{5}\mathrm{B}^{10} + {}_{0}n^{1} \rightarrow {}_{1}^{*}\mathrm{H}^{3} + 2\,{}_{2}\mathrm{He}^{4}$$
$${}_{3}\mathrm{Li}^{6} + {}_{0}n^{1} \rightarrow {}_{1}^{*}\mathrm{H}^{3} + {}_{2}\mathrm{He}^{4}$$

Other Reactions

$${}_{29}\mathrm{Cu}^{65} + {}_{0}n^{1} \rightarrow {}_{29}^{*}\mathrm{Cu}^{66} + \gamma\text{-rays}$$
$${}_{12}\mathrm{Mg}^{24} + {}_{0}n^{1} \rightarrow {}_{11}^{*}\mathrm{Na}^{24} + {}_{1}\mathrm{H}^{1}$$
$${}_{27}\mathrm{Co}^{59} + {}_{0}n^{1} \rightarrow {}_{27}^{*}\mathrm{Co}^{60}\ (\text{radioactive, } t_{1/2} = 5.3\text{ yr})$$

Reactions with High-Speed Protons

$${}_{1}\mathrm{H}^{1} + {}_{4}\mathrm{Be}^{9} \rightarrow {}_{3}\mathrm{Li}^{6} + {}_{2}\mathrm{He}^{4}$$
$${}_{1}\mathrm{H}^{1} + {}_{20}\mathrm{Ca}^{44} \rightarrow {}_{21}^{*}\mathrm{Sc}^{44} + {}_{0}n^{1}$$

Reactions with High-Speed Deuterons

$$ {}_{1}H^{2} + {}_{83}Bi^{209} \rightarrow {}^{*}_{83}Bi^{210} + {}_{1}H^{1} $$
$$ {}_{1}H^{2} + {}_{1}H^{2} \rightarrow {}^{*}_{1}H^{3} + {}_{1}H^{1} $$
$$ {}_{1}H^{2} + {}_{6}C^{12} \rightarrow {}^{*}_{7}N^{13} + {}_{0}n^{1} $$
$$ {}_{1}H^{2} + {}_{78}Pt^{196} \rightarrow {}^{*}_{78}Pt^{197} + {}_{1}H^{1} $$
$$ {}_{1}H^{2} + {}_{15}P^{31} \rightarrow {}^{*}_{15}P^{32} + {}_{1}H^{1} $$

Artificially induced radioactivity. The possibility of causing nuclear transformations by laboratory reactions opened the door for the production of new types of nuclei that either did not occur in nature or were present in virtually undetectable amounts. Just as the knowledge of the basic principles of organic chemistry made it possible for scientists to synthesize in the laboratory thousands of new organic molecules never before available for study or use, so the first knowledge of the basic principles of nuclear physics and chemistry made possible the production or *synthesis* of new types of nuclei. Among these, some of the most interesting are the new nuclei that are radioactive. The first of these was produced in 1934 by Irene Curie (daughter of Marie) and Frederic Joliot. Thus radioactivity was artificially induced for the first time. Today there are hundreds of processes known by which new artificially radioactive isotopes can be produced.

The following equations illustrate the process:

(*a*)
$$ {}_{2}He^{4} + {}_{12}Mg^{24} \rightarrow {}^{*}_{14}Si^{27} + {}_{0}n^{1} $$
$$ {}^{*}_{14}Si^{27} \rightarrow {}_{13}Al^{27} + {}_{+1}e^{0} $$

(*b*)
$$ {}_{92}U^{238} + {}_{2}He^{4} \rightarrow {}^{*}_{94}Pu^{241} + {}_{0}n^{1} $$

Among the other nuclear equations cited in this chapter there are many more examples of induced radioactivity.

Transuranic elements. The processes for laboratory transmutation of nuclei opened the door to another important possibility, the synthesis of new elements heavier than uranium, the *transuranic* elements. Until the second quarter of our century, uranium was the heaviest known chemical element, i.e., the element with the highest atomic weight. After the discovery of radioactivity and the emission of alpha-particles by uranium, scientists believed that any nucleus heavier than that of uranium or with a larger positive charge would be inherently so unstable as to be non-existent. No such heavier elements were found in nature.

When the high-speed particles produced by the cyclotron were applied to uranium, it was found that the nucleus of uranium would absorb a neutron, giving an isotope with a half-life of 23 minutes, decomposing with the emission of an electron to give an element with atomic number 93; this in turn decomposed to give an element with atomic number 94. These two elements were named after two of the outermost planets, neptunium (Np) and plutonium (Pu). The nuclear reactions are

$$_{92}U^{238} + {_0n^1} \rightarrow {}^{*}_{92}U^{239}\ (t_{1/2} = 23\text{ min}) + \gamma\text{-rays}$$
$${}^{*}_{92}U^{239} \rightarrow {}^{*}_{93}Np^{239} + {_{-1}e^0}\ (\text{electron})$$
$${}^{*}_{93}Np^{239} \rightarrow {}^{*}_{94}Pu^{239} + {_{-1}e^0} + \gamma\text{-rays}$$

This synthesis was accomplished by E. M. McMillan and P. H. Abelson in 1940 at the Radiation Laboratory of the University of California in Berkeley. There followed a remarkable series of researches at this laboratory through which Glenn Seaborg and his associates produced four more transuranic elements, americium (95, Am), curium (96, Cm), berkelium (97, Bk), and californium (98, Cf). Since then four more have been added to the list: einsteinium (99, Es), fermium (100, Fm), mendelevium (101, Md) and nobelium (102, No). Some of the reactions associated with the production of transuranic elements are

$${}^{*}_{94}Pu^{241} \rightarrow {}^{*}_{95}Am^{241} + {_{-1}e^0}$$
$$_2He^4 + {}^{*}_{94}Pu^{239} \rightarrow {}^{*}_{96}Cm^{242} + {_0n^1}$$
$$_2He^4 + {}^{*}_{95}Am^{241} \rightarrow {}^{*}_{97}Bk^{243} + 2\,{_0n^1}$$
$$_2He^4 + {}^{*}_{96}Cm^{242} \rightarrow {}^{*}_{98}Cf^{244} + 2\,{_0n^1}$$

RADIOACTIVITY AS A SCIENTIFIC TOOL

Tracer studies. Isotopes can be used to study the nature of many familiar or unusual chemical reactions. In our earlier studies of the phenomena of vapor pressure, we pointed out that, when a gas is in equilibrium with its own kinds of atoms in the liquid form, there is a dynamic factor involved. Although we can look at a tube with liquid water on the bottom and water vapor at the top, and see only a static system with an unmoving meniscus separating the two phases, we are confident that there is a continuous interchange of water molecules between the liquid and the vapor at the rate of 10^{18} per second. We also viewed chemical equilibrium in the same light; we can have H_2O in equilibrium with hydrogen and oxygen, while, on the other hand, the gaseous molecules are continually recombining to form water. Isotopes make it possible to trace the course of individual molecules in such processes. Not only can we use the isotopes that are artificially radioactive, but also we can use isotopes of abnormal weight. When these isotopes pass from one state to another, the sample can be taken from the second state and placed in a mass spectrograph and we can detect there the presence of the isotopes. Thus, if we have a sample of liquid water in equilibrium with its vapor, we can inject a small amount of water labeled with O^{18} into the liquid, and then later periodically take samples of the vapor; and we find that the concentration of the H_2O^{18} will gradually increase in the vapor until the concentration in the vapor and in the liquid are equal. As another example, there is a radioactive iron, $_{26}Fe^{59}$, with a half-life of 36 days for a process in which beta-particles are emitted. One of the principal constituents of blood is hemoglobin, in which there is a small amount of iron. By putting radioactive iron into hemoglobin, and injecting this hemoglobin into the bloodstream, measurements can be made of the rate at which blood circulates to different parts of the body.

Another important radioactive tracer is carbon 14, which has a half-life of 5580 years, again with respect to a process in which beta-rays are emitted. This can be produced by the reaction

$$_7N^{14} + _0n^1 \rightarrow {}^*_6C^{14} + _1H^1$$

Through the bombardment of nitrogen by cosmic rays in the upper atmosphere, a certain amount of C^{14} is constantly generated and there is a constant percentage of this radioactive isotope of carbon in the air in which we breathe. During life, animals and plants ingest this radioactive carbon. Thus the concentration of C^{14} reaches a standard value in all living things. On the other hand, after the animal or plant dies and the dead matter becomes part of the earth and out of equilibrium with the atmosphere, the C^{14} continues to decay; so that, after a given number of years, only a fraction of it is left. Through calculations similar to those that we made in Chapter 3, it is possible to tell from the beta-ray emission from dead matter how many years back death took place. Careful measurements even make it possible to estimate a time of death about 5000 years ago with an accuracy of approximately 1 per cent or better. The basic studies along these lines have been carried out by the chemist W. S. Libby and his collaborators. They made an initial check on the process by taking wood from a Sequoia tree where it was possible to fix the date at which the wood had been formed from the rings marking the years of growth in the tree. The beta-particle count confirmed the age of the tree shown by the rings. Similar studies have shown that the Dead Sea scrolls which have recently been discovered in the Near East were actually prepared and buried at about the time of the birth of Christ.

Atomic energy. During the four decades that followed the discovery of radioactivity in 1896, the accumulation of experimental evidence showed conclusively that many atomic nuclei contained almost unbelievably large amounts of energy. The heavier nuclei yielded some of this energy in natural radioactive processes when transformed to slightly lighter nuclei. The lightest of the nuclei, found in hydrogen, appeared capable of giving out even larger amounts of energy by coupling or *fusing* together to form the helium nucleus. At least such a process was the most probable source of the large amounts of energy streaming from the stars, as discussed earlier in this chapter.

Clearly, if the energy of the nucleus could be released on a large scale and put to work, it could be a vast new source of wealth for the human race; if it were released explosively, it could be a weapon to dwarf all previous military might. In view of this, scientists began to speculate on the possibility of the large-scale release of atomic energy, long before any clues had been discovered to suggest a practical way of harnessing the atom.

At first the outlook was not promising. The disintegration of a single nucleus might give out thousands of times more energy than that ever released by an atom by combining with another atom in a process like combustion. But, although it was easy to get 10^{30} atoms per second to react in combustion, there was no known way of inducing enough nuclei to transmute fast enough (say 10^{20} per second) to produce appreciable power. Even as recently as the 1930's, the greatest authorities in the field of nuclear physics believed that the large-scale release of atomic energy never would be accomplished by human effort. Lord Rutherford had produced the first artificial transmutation of a nucleus; he stated flatly in 1937 that the outlook for obtaining a large-scale release of nuclear energy was not promising. Professor Ernest O. Lawrence had invented the cyclotron, the most effective way of producing laboratory transmutation; he went on record in 1938 with an assertion that the large-scale production of atomic energy was impossible.

But less than a year after these discouraging pronouncements, a discovery by Hahn and Strassman in Germany completely changed the outlook. They found that the nucleus of one of the isotopes of uranium underwent a completely new type of transmutation, not merely expelling a single alpha- or beta-particle, but splitting in two to form the far lighter nuclei of barium and krypton, and at the same time releasing a large amount of energy. If the spreading of a nuclear fission chain reaction is controlled so that it rises to a constant level where useful power is produced at a steady rate, then one has a nuclear power source or *reactor.* Many such reactors are today actually at work producing electrical power for commercial use.

When the decision was made to try to make a uranium bomb, the first step required was the production of pure ${}_{92}U^{235}$ by separating this rare isotope from the far more abundant ${}_{92}U^{238}$. Several processes were tried, such as the deflection of a beam of uranium ions in a combined electrostatic and magnetic field by centrifugal separation and by thermal diffusion in the liquid state. The successful process was based on diffusion in the gaseous state, using the compound UF_6.

Meanwhile, a way had been found to change ${}_{92}U^{238}$ into plutonium on a large scale, and this new element, now available in quantity for the first time, was another suitable fuel for a fission chain reaction. In the last two decades an extensive technology has been developed for employing fission to produce useful power as well as weapons.

Before the impact of these discoveries on our civilization had hardly begun to be evaluated, an even more startling advance was made: the man-made large-scale release of energy from nuclear *fusion.* Using fission as a detonator to produce the required high temperature (about 10^8 degrees) to start fusion, it was possible to carry out a spreading or explosive chain reaction based on the equation

$$ {}_{1}H^{3} + {}_{1}H^{2} \rightarrow {}_{2}He^{4} + {}_{0}n^{1} \quad (18\text{ Mev}) $$

Once the reaction was started, the amount of energy released was so great that the chain was sustained, energy from the fusion of one pair of atoms being sufficient to trigger the fusion of several other pairs. Such an explosion is called *thermonuclear.*

From the information made public, it appears that the explosive force of uranium or plutonium bombs (atomic bombs) has been about the equivalent of 2×10^4 tons of a conventional explosive like TNT. On the other hand, figures like 10^8 tons equivalent or higher have been obtained for thermonuclear or "hydrogen" bombs. The possible use of weapons of this sort raises critical problems for the survival of human life on earth.

If the energy from fusion can also be harnessed, this will be an even greater contribution to future world power sources because hydrogen is so plentiful. Extensive research on this problem is now in progress. One of the most promising approaches is the study of the reactions involving deuterium, similar to that given above; other reactions of interest are

$$_1D^2 + {}_1D^2 \rightarrow {}_2He^3 + {}_0n^1 \quad (3\ \text{Mev})$$
$$_1D^2 + {}_1D^2 \rightarrow {}_1^*T^3 + {}_1H^1 \quad (4\ \text{Mev})$$

The difficulties of the problem may be summarized in a somewhat oversimplified but informative way. First, the two particles must approach each other with a velocity sufficient to penetrate the potential barrier of 0.1 Mev. At the same time, the particles must be confined so that they will continue to bounce against one another, but will not lose energy in bouncing against walls. Calculations indicate that a density of 10^{14} to 10^{18} particles per cubic centimeter may be most favorable. Under the temperature required, the nuclei will be completely stripped of electrons; such a gas is generally called a *plasma.* There is some hope that the plasma can be confined by a properly shaped magnetic field providing a small enough energy loss that the temperature can be maintained at a level necessary to sustain the reaction. For the $_1D^2 + {}_1D^2$ reaction this appears to be about 4×10^8 °K, and for the $_1D^2 + {}_1T^3$ reaction about 4.5×10^7 °K.

Paradoxically, one of the most promising means for producing the necessary intense magnetic field are superconductors held at temperatures below 1°K. There is evidence that an alloy of NbSn at sufficiently low temperature will conduct a high electrical current without generation of heat and will produce magnetic fields far more intense than can be obtained from any conventional magnets. If it is possible to couple these two extremes of temperature effectively, there is a chance that the energy of fission can be converted directly into electrical power.

Thus research at the most advanced frontiers of science has a continually increasing impact on civilization. The current rate of energy consumption on earth is about 10^{10} kw. If there are no global wars, it may be expected to increase by an order of magnitude every fifty years, leveling off when a world population balance is established. At this rate, conventional sources of fuel will almost certainly be exhausted by A.D. 2100, perhaps much sooner. On the other hand, if we can "burn" $_1D^2$, there should be enough power available to last until A.D. 1,000,000. One ton of $_1D^2$ has as much power as 10^7 tons of conventional fuel.

There is every expectation that in less than a century nuclear fuels will be available for space ships and that man will be exploring far beyond the boundaries of our planetary system. In considering the future of the human race, we may not be restricted to the supply of fuel on earth. Such consideration may appear to be not so much science as science-fiction. But these problems are being studied intensively today in dozens of laboratories all over the world.

PROBLEMS

1. Using the data in Table 1, calculate the difference in energy in calories between the neutron and the pair of particles into which it decomposes when the latter are at rest with respect to the former.

2. Calculate the energy in calories equivalent to the change in mass when two $_1H^1$ nuclei and two neutrons unite to form $_2He^4$.

3. If the energy released in the reaction in Problem 2 were emitted as a single photon, what would be the wavelength? What is the ratio of this wavelength to the approximate radius of the helium nucleus?

4. What is the energy released when 1 mole of $_2He^4$ nuclei is formed from protons and neutrons? Express the answer in joules. *Ans.* 2.7×10^{12} joules.

5. Calculate the net mass change and the net change in charge when 1 mole of U^{238} changes into 1 mole of Pb^{206}. What is the average energy change in calories per step for this change?

6. Make nuclear reaction charts for the hydrogen cycle and the carbon cycle similar in form to the chart for the decay of U and Th.

7. Make a nuclear reaction chart for the production of $_{98}Cf^{244}$ from $_{92}U^{238}$.

8. If a sample of cloth from the wrappings of an Egyptian mummy has about two-thirds the rate of beta-ray emission as a sample of cloth woven today, what is the approximate date of the burial of the mummy? *Ans.* 1300 B.C.

9. If a rocket propelled by the oxidation of liquid hydrogen requires 5×10^4 kg of fuel to leave the earth's gravitational field, calculate the weight of hydrogen required to produce the same amount of energy by fusing $4\,_1H^1$ to $_2He^4$ and positrons.

★**10.** The isotope $_{11}Na^{24}$ decays to $_{12}Mg^{24}$ with the emission of a beta-ray and has a half-life of about 15 hr. An animal is fed meat sprinkled with salt containing 10 μgm of NaCl formed from this isotope. A 0.001-gm sample of blood is tested 24 hr later and gives a count of 90 beta-particles per second. The total weight of blood in the animal is 10 kg. What fraction of undecayed salt remains in the blood 24 hr after its ingestion? *Ans.* 2×10^{-3}.

Appendix A

Fundamental constants

In 1961, the International Union of Pure and Applied Chemistry adopted a new table of atomic weights based on the value 12.0000 for the isotope of carbon designated C^{12}. This table replaces the former table based on the value 16.0000 for oxygen of isotopic composition found in the earth's atmosphere. This new basis for the table not only shifts the values for the atomic weights but also changes the values of a number of the fundamental constants that depend on the number of atoms in one gram atom. Although this may prove to be a small inconvenience during the period of readjustment, there is the advantage that it eliminates the difference between values used by chemists and by physicists. The basic values quoted below are based on a list by J. A. Bearden, J. S. Thomsen, and W. C. Sauder, reported in the *Bulletin of the American Physical Society*, **6**, 297 (1961), and on preliminary values from the Cohen-Dumond–McNish Tables soon to be published. For most purposes the shift in values is not significant for chemical calculations, but care should be taken to use the correct values when accuracy requires them.

Symbol	Name	Value
c	Speed of light	2.997930×10^{10} cm sec^{-1}
h	Planck's constant	6.6252×10^{-27} erg sec
k	Boltzmann's constant	1.38046×10^{-16} erg deg^{-1} molecule^{-1}
e	Charge on the electron*	4.80286×10^{-10} esu
	Charge on the electron	1.60206×10^{-19} coulomb
m_e	Rest mass of electron	9.1083×10^{-28} gm
m_p	Rest mass of proton	1.6724×10^{-24} gm
	Rest mass of proton	1.007276 a.m.u.
m_n	Rest mass of neutron	1.008665 a.m.u.
N_0	Avogadro's number	6.0229×10^{23} molecules mole^{-1}
R	Gas constant	82.055 atm cm^3 deg^{-1} mole^{-1}
	Gas constant	0.082054 liter atm deg^{-1} mole^{-1}
	Gas constant	8.3143 joule deg^{-1} mole^{-1}
	Gas constant	1.9872 def'd cal deg^{-1} mole^{-1}
V_0	Standard gas volume	2.24134×10^4 cm^3 mole^{-1}
T_0	Ice point (°C) (0°C)	273.15°K
	Triple point of H_2O (Base of International Temperature Scale)	273.16°K
g	Standard gravity	980.665 cm sec^{-2}
$\mathcal{F}$	Faraday constant	9.6489×10^4 abs coulomb equivalent^{-1}
	Faraday constant	2.3061×10^4 cal volt^{-1} equivalent^{-1}
ev	Electron volt	1.60206×10^{-12} erg molecule^{-1}
	Electron volt	2.3061×10^4 cal mole^{-1}
V_L	Volume of one liter	1.000028×10^3 cm^3
π	Ratio of radius to circumference	3.1415926536
π^2		9.869604
	$\ln x/\log x$	2.302585
	$R \ln x/\log x$	4.576
Wave number of a photon with energy of 1 ev		8.0658×10^3 cm^{-1}
Wavelength of a photon with energy of 1 ev		1.2398×10^{-4} cm

* The charge on the electron must be expressed in esu units when using Coulomb's Law, $F = Q_1Q_2/r^2$, where F is expressed in dynes and r in cm and Q_1 and Q_2 are the charges in esu.

Appendi B

Conve s

1 kilometer (

1 meter (m) (mm)

1 cm = 10⁴

1 kilogram lb)

1 gm = 1 gm)

1 atomic mass

Volume

1 liter (l) = 10^3 milliliters (ml) = 1.000028×10^3 cm^3

Pressure

1 bar = 10^6 dyne cm^{-2} = 0.98692 atm

1 atmosphere (atm) = 1.013250×10^6 dyne cm^{-2} = 1.0332275 kg (wt) cm^{-2} = 760.0 mm Hg

Energy

1 absolute joule = 0.23901 calorie (cal) = 10^7 ergs = 0.9478 British thermal unit (Btu)

1 liter atmosphere (liter atm) = 24.2179 cal = 2.815 kilowatthours (kwhr)

Appendix C

Units

Numbers by themselves mean very little. If you were told that a year ago John had 3250 and that now he has only 1700, your immediate reaction would be to ask "1700 what?" Obviously, you have not been given useful information until you are told what is being talked about. The same holds true for the answer that you may give to a problem in chemistry. If the units are not given, the answer is incomplete and has almost no meaning.

In science, it is customary to use the metric system of units to denote length, time, energy, and the many other kinds of properties that are discussed. The definitions of a number of these units were given in Chapter 2. In the metric system, the ratios of different units for the same quantity are equal to integral powers of 10, as contrasted with other systems where many other ratios are found. As pointed out in Chapter 2, the meter is the standard of length and is defined in terms of the wavelength of light emitted by an isotope of Kr. This is an example of the way in which it is frequently necessary to relate quantities of widely different magnitudes. In this discussion, we wish to review a few of the basic unit relations and point out ways in which units can help in solving problems.

In science we deal both with the very large and the very small. We consider aspects of the universe where distances may be as great as 10^{30} kilometers; and we work with the atomic nucleus where the diameter may be as small as 10^{-12} cm. Large ranges of magnitude are encountered in dealing with time, force, energy, and many other quantities. We frequently have several different units for the same quantity to fit either the big or the small aspects conveniently.

To relate large and small units systematically, the following usage has been adopted:

mega means *million*	(1 megabar $=$ 10^6 bars)
micro means *one millionth*	(1 microgram $=$ 10^{-6} gm)
kilo means *thousand*	(1 kilometer $=$ 10^3 meters)
milli means *one thousandth*	(1 millimeter $=$ 10^{-3} meter)

These terms are sometimes loosely used, for example, as *megamolecule* for a very large molecule and *micrometer* for an instrument to measure very small distances; but, when found in the name of a unit, they have the precise meaning given above.

Example Problem 1: Calculate the number of miles in a kilometer.

Suppose that the only conversion factor we have between the English and the metric system of length is: 0.0254 meter $=$ 1 inch. Under these conditions we will have to translate all the English units to inches, translate from inches to meters, and finally translate from meters into whatever metric units we desire. The answer called for should have the units of miles per kilometer. Therefore the factors we start with should have mi in the numerator or km in the denominator. Let us start with 1 mile/5280 ft $\times$? $=$? mi/km. From this starting point it is obvious that we are going to use the units of the various conversion factors as a guide to solve the problem. We already have miles in the numerator, as we will in the answer; therefore we should perform all conversions on the denominator. First, let us convert feet to inches so we can use our conversion factor to change to metric units.

$$\frac{1\text{ mi}}{5280\not{\text{ft}}} \times \frac{1\not{\text{ft}}}{12\text{ in.}} \times ? = \frac{\text{mi}}{\text{km}}$$

And, with the help of our conversion factor, this becomes

$$\frac{1\text{ mi}}{5280\not{\text{ft}}} \times \frac{1\not{\text{ft}}}{12\not{\text{in}}.} \times \frac{1\not{\text{in}}.}{0.0254\text{ m}} \times ? = \frac{\text{mi}}{\text{km}}$$

The units of the product are now miles per meter. To convert to the desired units, we must convert the m in the denominator to km. Thus,

$$\frac{1\text{ mi}}{5280\not{\text{ft}}} \times \frac{1\not{\text{ft}}}{12\not{\text{in}}.} \times \frac{1\not{\text{in}}.}{0.0254\not{\text{m}}} \times \frac{1000\not{\text{m}}}{1\text{ km}} = 0.62\text{ mi/km}$$

Notice that, when we are using the unit notation, we perform the same operation on the units as we perform on the numbers, and, because of this, conversion problems like this one follow a logical sequence which is self-checking.

Proper use of units allows you to dispense with memorization of many factors. For example, do you know the number of cubic inches in a gallon? It is easy to calculate.

Example Problem 2: What is the number of cubic inches in a gallon of liquid?

We do not have any relation between the English units for length and the English units for volume, i.e., between quarts and cubic inches. Therefore first we are going to convert from the English system of length to the metric system of length; from the metric system of length to the metric system of volume; from the metric system of volume to the English system of volume. We note immediately, however, that we do not have any factors involving cubic inches. This presents no great difficulties. We do have the factor 1 in./0.0254 m or, equivalently, 39.4 in./m. If we cube this factor, we get

$$\frac{6.1 \times 10^4 \text{ in.}^3}{\text{m}^3}$$

Note that when we cube the number we also cube the units. Now let us proceed stepwise along the path we have already outlined above. This yields

$$\frac{6.1 \times 10^4 \text{ in.}^3}{\text{m}^3} \times \frac{1 \text{ m}^3}{10^6 \text{ cm}^3} \times \frac{1 \text{ cm}^3}{1 \text{ ml}} \times \frac{1000 \text{ ml}}{1 \text{ l}} \times \frac{.946 \text{ l}}{\text{qt.}} \times \frac{4 \text{ qt.}}{1 \text{ gal.}} = 231 \text{ in.}^3/\text{gal.}$$

The second factor in this product was obtained in the manner similar to that used to obtain the first factor.

Units are used not only for problems involving conversions from one system of units to another; they also serve as a useful guide for dealing with all sorts of problems. To see this, let us consider one more example.

Example Problem 3: There are about 3×10^{22} molecules in 1 ml of water. These molecules have a diameter of about 2 Å (1 Å $= 10^{-8}$ cm). If you piled enough of these water molecules end on end in a column one molecule wide, how many milliliters of water would you need to reach the nearest star? The nearest star is 4.3 light years from the earth. A light year is the distance light travels in one year. (Velocity of light $= 3 \times 10^{10}$ cm/sec.)

This problem consists of three more or less separate steps. First, we must compute how far the star is from the earth, in Angstroms. Secondly, we must calculate how many molecules will be needed to cover this distance. And, lastly, we must compute how many milliliters of water correspond to this number of molecules. These three steps are indicated below.

$$\frac{3 \times 10^{10} \cancel{\text{cm}}}{\cancel{\text{sec}}} \times \frac{60 \cancel{\text{sec}}}{1 \cancel{\text{min}}} \times \frac{60 \cancel{\text{min}}}{1 \cancel{\text{hr}}} \times \frac{24 \cancel{\text{hr}}}{1 \cancel{\text{day}}} \times \frac{365 \cancel{\text{days}}}{\cancel{\text{yr}}} \times \frac{10^8 \text{ Å}}{1 \cancel{\text{cm}}} \times 4.3 \cancel{\text{yr}} = 4.06 \times 10^{26} \text{ Å}$$

$$4.06 \times 10^{26} \text{ Å} \times \frac{1 \text{ molec}}{2 \text{ Å}} = 2.03 \times 10^{26} \text{ molec}$$

$$2.03 \times 10^{26} \text{ molec} \times \frac{1 \text{ ml}}{3 \times 10^{22}} = 6600 \text{ ml}$$

or about

$$1.5 \text{ gal } H_2O$$

Appendix D

Mathematics

EXPONENTS AND LOGARITHMS

Science deals with very large and very small numbers. Numbers like 0.0000000667 and 5,980,000,000,000,000,000,000,000 are more the rule than the exception. Because of this it becomes important to devise a shorthand for writing these numbers that will help us keep track of the zeros. We can achieve this by the use of exponential expressions. As an example, the symbol x^n stands for the number obtained when x is multiplied by itself n times. The quantity $x^{1/n}$ is a factor of x; and n such factors multiplied together equal x. By definition, x raised to the 0 power, i.e., x^0, is unity for any value of x except 0. The rules for multiplying and dividing exponentials are summarized below.

1. To multiply exponentials that have the same base, i.e., equal values of x, we add exponents, $x^a \cdot x^b = x^{a+b}$; $yx^a \cdot zx^b = yzx^{a+b}$.
2. To divide exponentials that have the same base, we subtract the exponent of the divisor from the exponent of the dividend. For example, $x^a/x^b = x^{a-b}$; or

$$\frac{yx^a}{zx^b} = \frac{y}{z} \cdot x^{a-b}$$

From these two rules and the definitions, we can derive three more rules:

(*a*) When we raise an exponential to a power, we multiply the exponent by the power, e.g., $(x^a)^n = x^{an}$, or $(yx^a)^n = y^n x^{an}$.

(*b*) To obtain the *n*th root of any exponential, we divide the exponent by *n*, e.g., $(x^a)^{1/n} = x^{a/n}$, or $(yx^a)^{1/n} = y^{1/n}x^{a/n}$.

(*c*) A base raised to a negative exponent is the same as 1/base raised to the positive exponent. We can see this by the following identity.

$$x^{-a} = x^{0-a} = \frac{x^0}{x^a} = \frac{1}{x^a}$$

Let us apply these rules to a typical example problem.

Example Problem 1: What is the value of the following quantity?

$$\frac{10^{-8} \times 10^5 \times 10^{-2} \times (10^7)^{1/2}}{10^2 \times 10^{-7} \times (10^5)^{1/2}}$$

As our first step, let us get rid of all the powers and all the roots. This yields

$$\frac{10^{-8} \times 10^5 \times 10^{-2} \times 10^{3.5}}{10^2 \times 10^{-7} \times 10^{2.5}}$$

We will now add exponents in the numerator and the denominator and get

$$\frac{10^{-1.5}}{10^{-2.5}} = 10$$

Logarithms. Operations like multiplying, dividing, extracting roots, and raising to a power are much simpler with exponentials than with ordinary numbers. For example, let us perform some of these operations on the numbers $2613 = 10^{3.41714}$ and $2.738 = 10^{0.43743}$. The product of these two numbers is

$$10^{3.41714} \times 10^{0.43743} = 10^{3.85457}$$

If, instead, we divide the first by the second, we get

$$10^{3.41714} \div 10^{0.43743} = 10^{2.97971}$$

If we take the square root of the first and the square of the second, we get

$$(10^{3.41714})^{1/2} = 10^{1.70857} \text{ and } (10^{0.43743})^2 = 10^{0.87486}$$

Unfortunately, although these computations are simple, they yield the answers in a form that is generally not directly usable. It is helpful, therefore, to have a table that enables us first to translate ordinary numbers into exponentials and then to translate the answer back to ordinary numbers. This is precisely what a table of logarithms does. *When we say log of a number we mean the power to which 10 must be raised to give us that number. When we speak of the antilog of a number we mean the quantity that corresponds to 10 raised to that number.* In other words, in the two equations above, 3.41714 is the logarithm of 2613, and 2613 is the antilog of 3.41714.

Let us consider some examples that illustrate the use of logarithms. Tables of logarithms list the exponent to which 10 must be raised to yield numbers between 1 and 10. To get the log of any number, including those greater than 10, we must apply our knowledge of the properties of exponentials. For example, let us carry out a typical multiplication with the use of logarithms.

Example Problem 2: What is the product of the two following numbers?

$$873.2 \times 0.0002123 \quad (1)$$

First, express both of these quantities as numbers from 1 to 10 multiplied by 10 raised to an integral power; that is,

$$\begin{aligned} 873.2 &= 8.732 \times 10^2 \\ 0.0002123 &= 2.123 \times 10^{-4} \end{aligned} \quad (2)$$

We can determine the exponential of 10 corresponding to the numbers between 1 and 10 by referring to a four-place log table. For example, let us look up the log of the number 8.732. First look on the left-hand margin to locate 8.7. In this row, the number under the column marked 3 corresponds to the log of 8.73, i.e., 0.9410, provided we place the number to the right of the decimal point. Similarly, 0.9415 is the log of 8.74. Log 8.732 is one-fifth of the way between 0.9410 and 0.9415; hence log 8.732 = 0.9411. From five-place tables, we would get the value 0.94111. Thus

$$8.732 = 10^{0.94111} \quad \text{and} \quad 2.123 = 10^{0.32695} \quad (3)$$

This, however, is not exactly what we want. We want

$$8.732 \times 10^2 \quad \text{or} \quad 10^{0.94111} \times 10^2 \quad (4)$$

which is

$$10^{2.94111}$$

and

$$2.123 \times 10^{-4} \quad \text{or} \quad 10^{0.32695}\, 10^{-4}$$

which is

$$10^{-3.67305} \quad (5)$$

Thus $873.2 \times 0.0002123 = 10^{2.94111} \times 10^{-3.67305}$, which on application of the rules for multiplication of exponents yields $10^{-0.73194}$. To convert this to a number between 1 and 10 times a power of 10, we can write

$$10^{-0.73194} = 10^{0.26806} \times 10^{-1}$$

We now use our tables to work backwards and get $10^{0.26806} = 1.8538$; hence $873.2 \times 0.0002123 = 0.18538$.

Normally in using logarithms we do not write $8.732 = 10^{0.94111}$, but we work directly with the exponents and write log 8.732 = 0.94111. Let us solve this problem again and use the new notation. Step 3 and the subsequent steps become

$$\begin{aligned} &\log 8.732 = 0.94111 \\ &\log 873.2 = 2.9411 \\ &\log 2.123 = 0.32695 \\ &\log 0.0002123 = 0.32695 - 4 = -3.67305 \\ &\log (873.2 \times 0.0002123) = \log (873.2) + \log (0.0002123) \\ &\qquad = -0.73194 \\ &-0.73194 = 0.26806 - 1 \\ &\log^{-1} (0.26806 - 1) = 1.8538 \times 10^{-1} = 0.18538 \end{aligned}$$

In working with logarithms, we usually denote the number to the right of the decimal place in the logarithm as the *mantissa* and the whole number to the left of the decimal place as the *characteristic.* Much of the trouble in finding a logarithm of the number is getting the right characteristic. As a working rule, let us say the characteristic (if positive) is the number of places we must move the decimal point to the left to get a number between 1 and 10. If it is negative, it is the number of places we must move the decimal point to the right to get a number between 1 and 10. For example, if we have 7,000,000, we must move the decimal point to the left six places to get a number between 1 and 10. In this case, therefore, the characteristic will be 6. On the other hand, if we have a number like 0.0007, we must move the decimal point to the right four places to get a number between 1 and 10. This means the characteristic for the latter number is -4. By convention, in the last case, we would write 6 (mantissa) -10, which, of course, is the same as writing (mantissa) $+(-4)$.

We can summarize the useful operations which can be performed with logarithms by

(a) $$\log (a \times b \times c) = \log a + \log b + \log c$$

(b) $$\log \left(\frac{a \times b}{c \times d}\right) = \log a + \log b - \log c - \log d$$

(c) $$\log (a^n) = n \log a$$

QUADRATIC EQUATIONS

A quadratic equation has the form

$$ax^2 + bx + c = 0$$

The solution is

$$x = \frac{-b \pm (b^2 - 4ac)^{1/2}}{2a}$$

There are two *roots* or values of x that make the sum of the three terms equal to 0. It is informative to make a graph with x as the abscissa and plot two lines, one being ax^2 and the other bx. The first is a curve, and the second a straight line with the slope equal to the value of b. If the sum of the two ordinates is added to c and called y so that $ax^2 + bx + c = y$, then a plot of y shows that it crosses the zero axis at two points, one corresponding to the choice of $+$ and the other to the choice of $-$ before the term in parentheses. Usually, only one of these values or *roots* will have physical significance in a problem and the choice is obvious from the context.

EXACT DIFFERENTIALS

In many of the thermodynamic equations we find two classes of the small quantities called *differentials*, such as dE, dS, DW, and DQ. The quantities designated by the small d are called exact differentials. They are formed from quantities like T, P, E, S, H, F, V that are *state functions*. Whenever a given system is brought into a given state, each of these properties called state functions always has the real value that is characteristic of that state. For example, a gram mole of water vapor that is in the state that is in equilibrium both with ice and with liquid water will always have $T = 273.16°\text{K}$, $P = 4.615$ mm Hg, $V = 3691.2$ liters, etc. If it is heated to a higher temperature, expanded, cooled, and then contracted again to $V = 3691.2$ liters at $T = 273.16°\text{K}$, it will always have the same value of P, of E, of S, etc., that it had when it was initially in this state. Because there is no net change in a state variable when the system goes through a cyclic change and is restored to its initial state, we write mathematically

$$\oint dP = 0, \quad \oint dE = 0, \quad \oint dS = 0, \quad \text{etc.}$$

These formulas express mathematically the fact expressed in words above. The sign $\oint$ is called a circuit integral; if $\oint dx = 0$, then dx is said to be an *exact* differential.

By way of contrast, there are other quantities encountered in thermodynamics that are not exact differentials. We have seen that one can take a system consisting of a gas around the Carnot cycle circuit, and, when the system returns to its initial state, the work done is found to be not zero but equal numerically to the area enclosed by the path of the circuit on the P-V diagram. If we were to associate a certain value of W with a gas in the initial state, and then change W as energy in the form of work leaves or enters the system, just as we change E when energy leaves or enters, then we would find at the end of the circuit when the system was back in its initial state that $E_{\text{final}} = E_{\text{initial}}$ but $W_{\text{final}} < W_{\text{initial}}$ because more energy has left the system in the form of work than has entered it as work. So, mathematically, $\oint DW \neq 0$ and DW is not an exact differential; neither is DQ, because in the same manner $Q_{\text{initial}} < Q_{\text{final}}$. Therefore we designate these quantities by a capital D to distinguish them from the exact differentials designated by a small d.

THE LOG $x - 1/T$ EQUATION

One of the most important equations of chemistry is the expression that relates the logarithm of a quantity x to $1/T$. It has the form

$$\log x = \frac{-\Delta E}{2.303R}\left(\frac{1}{T} - \frac{1}{T_0}\right)$$

where T and T_0 are in degrees Kelvin. If x is vapor pressure (P), then $\Delta E = \Delta H_v$, the heat of vaporization. But the equation also applies to a number of other intensive properties more or less associated with vapor pressure. If x stands for the mole fraction of a substance in solution, then $\Delta E = \Delta H_f$, the heat of fusion. If x is the equilibrium constant (K), then ΔE is the heat of reaction and T_0 is the temperature at which $K = 1$. If x is the rate constant of reaction (k), then ΔE is the energy of activation, and T_0 is the temperature at which $k = 1$. If x is the viscosity (η), then ΔE is the energy of activation for flow and T_0 is the temperature at which $\eta = 1$.

Because of the far-reaching importance of this equation, we give here a brief derivation of the form applicable to vapor pressure. From the combined first and second laws of thermodynamics,

$$dE = T\,dS - P\,dV$$

For an isothermal process,

$$P = T\left(\frac{\partial S}{\partial V}\right)_T - \left(\frac{\partial E}{\partial V}\right)_T$$

where the partial derivative sign (∂) is used because the derivative is restricted to the value at constant temperature, as indicated by the subscript T outside the parentheses. Differentiating with respect to temperature, we get

$$\left(\frac{\partial P}{\partial T}\right)_V = T\frac{\partial^2 S}{\partial T\,\partial V} + \left(\frac{\partial S}{\partial V}\right)_T - \frac{\partial^2 E}{\partial T\,\partial V}$$

But, from the second law,

$$dS = \left(\frac{\partial E}{\partial T}\right)_V \frac{dT}{T} \quad \text{and} \quad \frac{\partial^2 S}{\partial V\,\partial T} = \frac{1}{T}\,\frac{\partial^2 E}{\partial V\,\partial T}$$

So the double differentials cancel each other, giving

$$\left(\frac{\partial P}{\partial T}\right)_V = \left(\frac{\partial S}{\partial V}\right)_T$$

For the evaporation of a mole of liquid at constant temperature into the vapor state, the pressure is independent of the volume and we can write

$$\frac{dP}{dT} = \frac{\Delta S}{\Delta V}$$

If we neglect the volume in the liquid state, then $\Delta V = V_{\text{gas}} = nRT/P$, assuming that the vapor is an ideal gas. Also, $\Delta S = \Delta H_v/T$. Substituting, we get

$$\frac{dP/P}{dT} = \frac{\Delta H_v}{RT^2}$$

Changing to $d \ln P$ and $d(1/T)$, we get

$$\frac{d \ln P}{d(1/T)} = \frac{-\Delta H_v}{R}$$

If ΔH_v is assumed to be constant over the temperature range under consideration,

$$\ln P = \frac{-\Delta H_v}{R}\left(\frac{1}{T} - \frac{1}{T_0}\right) \quad \text{or} \quad \log P = \frac{-\Delta H_v}{2.303R}\left(\frac{1}{T} - \frac{1}{T_0}\right)$$

Appendix E

Oxidation-reduction potentials

Data taken from *Oxidation Potentials* (Second Edition) by Wendell M. Latimer, published by Prentice-Hall, Inc., Englewood Cliffs, N. J., 1952, with permission from the publisher.

		$\mathcal{E}^0$(volts)
$Li(s)$	$= Li^+ + e^-$	+3.05
$K(s)$	$= K^+ + e^-$	2.93
$Rb(s)$	$= Rb^+ + e^-$	2.93
$Cs(s)$	$= Cs^+ + e^-$	2.92
$Ra(s)$	$= Ra^{+2} + 2e^-$	2.92
$Ba(s)$	$= Ba^{+2} + 2e^-$	2.90
$Sr(s)$	$= Sr^{+2} + 2e^-$	2.89
$Ca(s)$	$= Ca^{+2} + 2e^-$	2.87
$Na(s)$	$= Na^+ + e^-$	2.71
$Mg(s)$	$= Mg^{+2} + 2e^-$	2.37
H^-	$= \frac{1}{2}H_2 + e^-$	2.25
$Al(s) + 6F^-$	$= AlF_6^{-3} + 3e^-$	2.07
$Be(s)$	$= Be^{+2} + 2e^-$	1.85
$U(s)$	$= U^{+3} + 3e^-$	1.80
$Hf(s)$	$= Hf^{+4} + 4e^-$	1.70
$Al(s)$	$= Al^{+3} + 3e^-$	1.66
$Ti(s)$	$= Ti^{+3} + 3e^-$	1.63
$Zr(s)$	$= Zr^{+4} + 4e^-$	1.53
$Si(s) + 6F^-$	$= SiF_6^{-2} + 4e^-$	1.2
$Ti(s) + 6F^-$	$= TiF_6^{-2} + 4e^-$	1.19
$Mn(s)$	$= Mn^{+2} + 2e^-$	1.18

		$\mathcal{E}^0$(volts)
$V(s)$	$= V^{+2} + 2e^-$	ca. 1.2
$Nb(s)$	$= Nb^{+3} + 3e^-$	ca. 1.1
$Zn(s)$	$= Zn^{+2} + 2e^-$	0.76
$Cr(s)$	$= Cr^{+3} + 3e^-$	0.74
U^{+3}	$= U^{+4} + e^-$	0.61
$Tl + Cl^-$	$= TlCl + e^-$	0.56
$H_3PO_2 + H_2O$	$= H_3PO_3 + 2H^+ + 2e^-$	0.50
$Fe(s)$	$= Fe^{+2} + 2e^-$	0.44
Cr^{+2}	$= Cr^{+3} + e^-$	0.41
$Cd(s)$	$= Cd^{+2} + 2e^-$	0.40
Ti^{+2}	$= Ti^{+3} + e^-$	0.4
$Pb + SO_4^{-2}$	$= PbSO_4 + 2e^-$	0.356
$In(s)$	$= In^{+3} + 3e^-$	0.342
$Tl(s)$	$= Tl^+ + e^-$	0.336
$Co(s)$	$= Co^{+2} + 2e^-$	0.277
$H_2O + H_3PO_3$	$= H_3PO_4 + 2H^+ + 2e^-$	0.276
V^{+2}	$= V^{+3} + e^-$	0.255
$Sn + 6F^-$	$= SnF_6^{-2} + 4e^-$	0.25
$Ni(s)$	$= Ni^{+2} + 2e^-$	0.250
$Cu + I^-$	$= CuI + e^-$	0.185
$Ag + I^-$	$= AgI + e^-$	0.151
$Sn(s)$	$= Sn^{+2} + 2e^-$	0.136
$Pb(s)$	$= Pb^{+2} + 2e^-$	0.126
$Hg + 4I^-$	$= HgI_4^{-2} + 2e^-$	+0.04
H_2 (1 atm)	$= 2H^+ + 2e^-$	0.00
$Ag + Br^-$	$= AgBr + e^-$	−0.095
H_2S (*g*, 1 atm)	$= S + 2H^+ + 2e^-$	−0.141
Np^{+3}	$= Np^{+4} + e^-$	−0.147
Sn^{+2}	$= Sn^{+4} + 2e^-$	−0.15
Cu^+	$= Cu^{+2} + e^-$	−0.153
$Ag + Cl^-$	$= AgCl + e^-$	−0.222
$Cu(s)$	$= Cu^{+2} + 2e^-$	−0.337
$Cu(s)$	$= Cu^+ + e^-$	−0.521
$2I^-$	$= I_2 + 2e^-$	−0.536
$3I^-$	$= I_3^- + 2e^-$	−0.54
$2H_2O + HAsO_2$	$= H_3AsO_4 + 2H^+ + 2e^-$	−0.56
MnO_4^{-2}	$= MnO_4^- + e^-$	−0.56
H_2O_2	$= O_2 + 2H^+ + 2e^-$	−0.68
Fe^{+2}	$= Fe^{+3} + e^-$	−0.77
$2Hg$ (*l*)	$= Hg_2^{+2} + 2e^-$	−0.79
$Ag(s)$	$= Ag^+ + e^-$	−0.80
$Rh(s)$	$= Rh^{+3} + 3e^-$	ca. −0.8
Hg_2^{+2}	$= 2Hg^{+2} + 2e^-$	−0.92
$NO + 2H_2O$	$= NO_3^- + 4H^+ + 3e^-$	−0.96
$Pd(s)$	$= Pd^{+2} + 2e^-$	−0.99
$Au(s) + 4Cl^-$	$= AuCl_4^- + 3e^-$	−1.00
$2H_2O$	$= O_2 + 4H^+ + 4e^-$	−1.23

		$\mathcal{E}^0$(volts)
$Mn^{+2} + 2H_2O$	$= MnO_2 + 4H^+ + 2e^-$	−1.23
$7H_2O + 2Cr^{+3}$	$= Cr_2O_7^{-2} + 14H^+ + 6e^-$	−1.33
$2Cl^-$	$= Cl_2 + 2e^-$	−1.36
$Au(s)$	$= Au^{+3} + 3e^-$	−1.50
Mn^{+2}	$= Mn^{+3} + e^-$	−1.51
Ce^{+3}	$= Ce^{+4} + e^-$	−1.61
$2H_2O$	$= H_2O_2 + 2H^+ + 2e^-$	−1.77
Co^{+2}	$= Co^{+3} + e^-$	−1.84
$H_2O + O_2$	$= O_3 + 2H^+ + 2e^-$	−2.07
$2F^-$	$= F_2 + 2e^-$	−2.87
$2HF(aq.)$	$= F_2 + 2H^+ + 2e^-$	−3.06

For additional values of $\mathcal{E}^0$ for the oxyhalide ions see p. 571.

Appendix F

Thermodynamic data

(Data from Circular 500, National Bureau of Standards, for compounds at 25°C.)

	$\Delta Hf°$ kcal/mole	$\Delta Ff°$ kcal/mole		$\Delta Hf°$ kcal/mole	$\Delta Ff°$ kcal/mole
$O_3(g)$	+ 34.0	+ 39.06	$(NH_4)_2SO_4(c)$	-281.86	-215.19
$H_2O(l)$	- 68.32	- 56.69	$PH_3(g)$	+ 2.21	+ 4.36
$H_2O_2(l)$	- 44.84	—	$PCl_3(g)$	- 73.22	- 68.42
$HF(g)$	- 64.2	- 64.7	$PCl_5(g)$	- 95.35	- 77.59
$ClO_2(g)$	+ 24.7	+ 29.5	$POCl_3(g)$	-141.5	-130.3
$Cl_2O(g)$	+ 18.20	+ 22.40	$PBr_3(g)$	- 35.9	- 41.2
$HCl(g)$	- 22.063	- 22.769	$PN(g)$	- 20.2	- 25.3
$HBr(g)$	- 8.66	- 12.72	$CO(g)$	- 26.4157	- 32.8079
$HI(g)$	+ 6.20	+ 0.31	$CO_2(g)$	- 94.0518	- 94.2598
$SO_2(g)$	- 70.96	- 71.79	$CH_4(g)$	- 17.889	- 12.140
$SO_3(g)$	- 94.45	- 88.52	$CH_3OH(g)$	- 48.08	- 38.69
$H_2S(g)$	- 4.815	- 7.892	$CH_3OH(l)$	- 57.02	- 39.73
$SF_6(g)$	-262	-237	$CCl_4(g)$	- 25.5	- 15.3
$H_2Se(g)$	+ 20.5	+ 17.0	$CCl_4(l)$	- 33.3	- 16.4
$SeF_6(g)$	-246	-222	$CHCl_3(g)$	- 24	- 16
$H_2Te(g)$	+ 36.9	+ 33.1	$CHCl_3(l)$	- 31.5	- 17.1
$TeF_6(g)$	-315	-292	$CS_2(g)$	+ 27.55	+ 15.55
$NO(g)$	+ 21.6	+ 20.719	$CS_2(l)$	+ 21.0	+ 15.2
$NO_2(g)$	+ 8.091	+ 12.390	$HCN(g)$	+ 31.2	+ 28.7
$N_2O(g)$	+ 19.49	+ 24.76	$HCN(l)$	+ 25.2	+ 29.0
$N_2O_4(g)$	+ 2.309	+ 23.491	$C_2H_6(g)$	- 20.236	- 7.860
$NH_3(g)$	- 11.04	- 3.976	$C_2H_5OH(g)$	- 56.24	- 40.30
$NH_4Cl(c)$	- 75.38	- 48.73	$C_2H_5OH(l)$	- 66.356	- 41.77
$NOCl(g)$	+ 12.57	+ 15.86	$SiF_4(g)$	-370	-360
$NOBr(g)$	+ 19.56	+ 19.70	$SiCl_4(g)$	-145.7	-136.9

Appendix G

Chart of variation of free energy with temperature

(Corresponding to values of the equilibrium constant equal to integral powers of ten.)

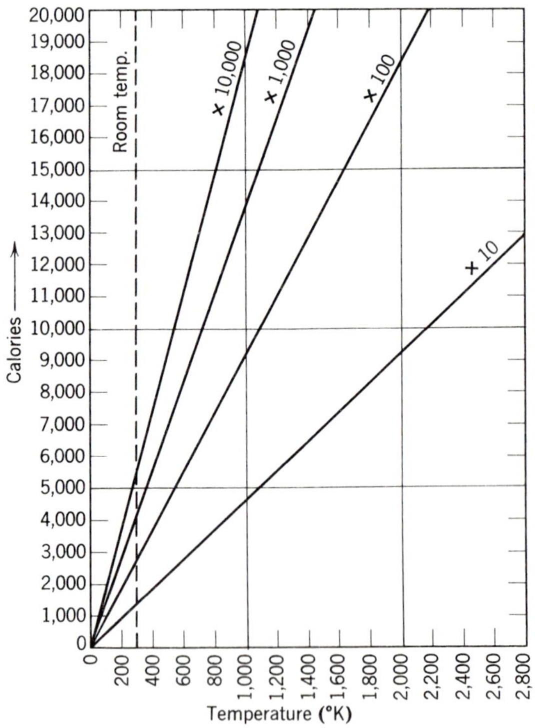

The graphs express the relation between $-\Delta F^\circ$ and T corresponding to $K = 10$, 100, 1000, and 10,000, based on the equation: $-\Delta F^\circ = RT \ln K$. *Examples:* At 1000°K we find that $K = 10$ when $\Delta F^\circ = -4576$ cal; $K = 10{,}000$ when $\Delta F^\circ = -18.5$ kcal. Thus one can see quickly that at a given temperature (say 1000°K) a value of about -5 kcal for ΔF° means that $K \cong 10$.

Appendix H

Vapor pressure of water at various temperatures

Pressure (mm Hg)	Temperature (°C)	Pressure (mm Hg)	Temperature (°C)
0.00001 (ice)	−100	44.6	36.0
0.0004 (ice)	−80	55.3	40.0
0.008 (ice)	−60	68.3	44.0
0.097 (ice)	−40	83.7	48.0
0.776 (ice)	−20	102.1	52.0
1.24 (ice)	−15	123.8	56.0
1.95 (ice)	−10	149.2	60.0
3.01 (ice)	−5	179.2	64.0
4.58 (ice)	0.0	214.2	68.0
6.1	4.0	254.5	72.0
8.0	8.0	301.2	76.0
10.5	12.0	355.0	80.0
13.6	16.0	416.7	84.0
17.5	20.0	487.0	88.0
22.4	24.0	567.0	92.0
28.3	28.0	657.5	96.0
35.7	32.0	760.0	100.0
		1520.0	120.6

References

1. *JANAF Interim Thermochemical Tables*, The Dow Chemical Company, Midland, Mich., December 31, 1960.
2. *Circular of the National Bureau of Standards* No. 500, by F. D. Rossini, D. D. Wagman, W. H. Evans, Samuel Levine, and Irving Jaffe, Washington, D. C., February 1, 1952.
3. *Oxidation Potentials* Second Edition, by W. M. Latimer, Prentice-Hall, Inc., Englewood Cliffs, N. J., 1952.
4. "Values of the Fundamental Constants of Physics," by J. A. Bearden, J. S. Thomsen and W. C. Sauder, *Bulletin of the American Physical Society*, **6**, 297 (1961).
5. *Elementary Theory of Nuclear Shell Structure*, by Maria Goeppert Mayer and J. Hans D. Jensen, John Wiley and Sons, Inc., New York, 1955.
6. *Constitution of Binary Alloys*, Second Edition, by Max Hansen with the cooperation of Kurt Anderko, McGraw-Hill Book Company, Inc., New York, 1958.
7. *The Nature of the Chemical Bond*, Third Edition, by Linus Pauling, Cornell University Press, Ithaca, N. Y., 1960.
8. *Inorganic Chemistry*, by Therald Moeller, John Wiley and Sons, Inc., New York, 1952.
9. *An Introduction to Transition Metal Chemistry: Ligand Field Theory*, by Leslie E. Orgel, Methuen and Co., Ltd., London, 1960.
10. *Steric Effects in Organic Chemistry*, Edited by Melvin S. Newman, John Wiley and Sons, Inc., New York, 1956.
11. *Resonance in Organic Chemistry*, by G. W. Wheland, John Wiley and Sons, Inc., New York, 1955.
12. *Physical Chemistry*, by Farrington Daniels and Robert A. Alberti, John Wiley and Sons, Inc., New York, 1955.
13. *Chemical Thermodynamics*, by Frederick D. Rossini, John Wiley and Sons, Inc., New York, 1950.
14. *Thermodynamics*, by H. B. Callen, John Wiley and Sons, Inc., New York, 1960.
15. *Structural Inorganic Chemistry*, Second Edition, by A. F. Wells, Oxford University Press, London, 1950.
16. *Inorganic Reactions and Structure*, by Edwin S. Gould, Henry Holt and Co., New York, 1955.
17. *Treatise on Inorganic Chemistry*, by H. Remy, translated by J. S. Anderson and edited by J. Kleinberg, Elsevier Publishing Co., New York, 1956.
18. *Reference Book of Inorganic Chemistry*, by Wendell M. Latimer and Joel H. Hildebrand, The Macmillan Co., New York, 1940.
19. *Systematic Inorganic Chemistry*, by Don M. Yost and Horace Russell, Jr., Prentice-Hall, Inc., Englewood Cliffs, N. J., 1944.

Index

ELECTRONIC STRUCTURE DIAGRAMS

f				×			
d			×	×	×		
p		×	×	×	×	×	
s	×	×	×	×	×	×	×
Shell No. →	1	2	3	4	5	6	7

H 1 1								
Li 3 2 1	Be 4 2 2							
Na 11 6 2 2 1	Mg 12 6 2 2 2							
K 19 6 6 2 2 2 1	Ca 20 6 6 2 2 2 2	Sc 21 1 6 6 2 2 2 2	Ti 22 2 6 6 2 2 2 2	V 23 3 6 6 2 2 2 2	Cr 24 5 6 6 2 2 2 1	Mn 25 5 6 6 2 2 2 2	Fe 26 6 6 6 2 2 2 2	Co 27 7 6 6 2 2 2 2
Rb 37 10 6 6 6 2 2 2 2 1	Sr 38 10 6 6 6 2 2 2 2 2	Y 39 10 1 6 6 6 2 2 2 2 2	Zr 40 10 2 6 6 6 2 2 2 2 2	Nb 41 10 4 6 6 6 2 2 2 2 1	Mo 42 10 5 6 6 6 2 2 2 2 1	Tc 43 10 6 6 6 6 2 2 2 2 1	Ru 44 10 7 6 6 6 2 2 2 2 1	Rh 45 10 8 6 6 6 2 2 2 2 1
Cs 55 10 10 6 6 6 6 2 2 2 2 2 1	Ba 56 10 10 6 6 6 6 2 2 2 2 2 2	La 57 10 10 1 6 6 6 6 2 2 2 2 2 2	Hf 72 14 10 10 2 6 6 6 6 2 2 2 2 2 2	Ta 73 14 10 10 3 6 6 6 6 2 2 2 2 2 2	W 74 14 10 10 4 6 6 6 6 2 2 2 2 2 2	Re 75 14 10 10 5 6 6 6 6 2 2 2 2 2 2	Os 76 14 10 10 6 6 6 6 6 2 2 2 2 2 2	Ir 77 14 10 10 9 6 6 6 6 2 2 2 2 2
Fr 87 14 10 10 10 6 6 6 6 6 2 2 2 2 2 2 1	Ra 88 14 10 10 10 6 6 6 6 6 2 2 2 2 2 2 2							

LANTHANIDES

Ce 58 2 10 10 6 6 6 6 2 2 2 2 2 2	Pr 59 3 10 10 6 6 6 6 2 2 2 2 2 2	Nd 60 4 10 10 6 6 6 6 2 2 2 2 2 2	Pm 61 5 10 10 6 6 6 6 2 2 2 2 2 2	Sm 62 6 10 10 6 6 6 6 2 2 2 2 2 2	Eu 63 7 10 10 6 6 6 6 2 2 2 2 2 2	Gd 64 7 10 10 1 6 6 6 6 2 2 2 2 2 2
Tb 65 9 10 10 6 6 6 6 2 2 2 2 2 2	Dy 66 10 10 10 6 6 6 6 2 2 2 2 2 2	Ho 67 11 10 10 6 6 6 6 2 2 2 2 2 2	Er 68 12 10 10 6 6 6 6 2 2 2 2 2 2	Tm 69 13 10 10 6 6 6 6 2 2 2 2 2 2	Yb 70 14 10 10 6 6 6 6 2 2 2 2 2 2	Lu 71 14 10 10 1 6 6 6 6 2 2 2 2 2 2